FROM ALCHEMY TO CHEMISTRY IN PICTURE AND STORY

THE WILEY BICENTENNIAL–KNOWLEDGE FOR GENERATIONS

Each generation has its unique needs and aspirations. When Charles Wiley first opened his small printing shop in lower Manhattan in 1807, it was a generation of boundless potential searching for an identity. And we were there, helping to define a new American literary tradition. Over half a century later, in the midst of the Second Industrial Revolution, it was a generation focused on building the future. Once again, we were there, supplying the critical scientific, technical, and engineering knowledge that helped frame the world. Throughout the 20th Century, and into the new millennium, nations began to reach out beyond their own borders and a new international community was born. Wiley was there, expanding its operations around the world to enable a global exchange of ideas, opinions, and know-how.

For 200 years, Wiley has been an integral part of each generation's journey, enabling the flow of information and understanding necessary to meet their needs and fulfill their aspirations. Today, bold new technologies are changing the way we live and learn. Wiley will be there, providing you the must-have knowledge you need to imagine new worlds, new possibilities, and new opportunities.

Generations come and go, but you can always count on Wiley to provide you the knowledge you need, when and where you need it!

William J. Pesce
PRESIDENT AND CHIEF EXECUTIVE OFFICER

Peter Booth Wiley
CHAIRMAN OF THE BOARD

FROM ALCHEMY TO CHEMISTRY IN PICTURE AND STORY

ARTHUR GREENBERG
University of New Hampshire
Durham, NH

WILEY-INTERSCIENCE
A JOHN WILEY & SONS, INC., PUBLICATION

The artwork on the cover of the present book, which depicts an impoverished and ragged alchemist, is from an engraved plate attributed to Augsburg printmaker Martin Engelbrecht (1684–1756) in the early eighteenth century. It is one print in a series on the theme *Die Ursachen der Verarmung* (the causes of impoverishment). We are grateful to William Schupbach, Wellcome Library (London), for providing this information. The full plate, included as the first of 24 color plates in this book, has two brief poems below the figure.

German (left-hand side; courtesy Heinz D. Roth)

One Who Was Impoverished Making Gold

From now on let laboratory work be cursed by me,
Ah, if only I had never tried it,
I have searched for the Philosopher's Stone in the fire,
And now I have found the Stone of Fools in my head,
Nobody ever got rich from making gold,
But many have ended up on a beggar's staff.

French (right-hand side; Arthur Greenberg)

A Pauper for the Sake Of Alchemy

I have searched in the fire to find a treasure,
And for that I have finally lost all my gold,
I am poor now and have reclaimed my life,
Easing the pain. Alas! What folly!
Take an example from this great misfortune,
Ah! I thus counsel you with all my heart.

Published by John Wiley & Sons, Inc., Hoboken, New Jersey.
Published simultaneously in Canada.

Library of Congress Cataloging-in-Publication Data is available.

ISBN-10: 0-471-75154-5
ISBN-13: 978-0-471-75154-0

Printed in the United States of America.

10 9 8 7 6 5 4 3 2 1

This book is dedicated to my wife Susan
and our children, David and Rachel.

CONTENTS

SECTION V. THE CHEMICAL REVOLUTION 269

SECTION VI. A YOUNG DEMOCRACY AND A NEW CHEMISTRY 381

PREFACE

Amiable reader, the purpose of *From Alchemy to Chemistry in Picture and Story* is to treat you to a light-hearted tour through selected highlights of chemical history. The physician and writer Oliver Sacks has written that "Chemistry has perhaps the most intricate, most fascinating, and certainly most romantic history of all of the sciences." His autobiographical book, *Uncle Tungsten*, speaks to the joys of learning chemistry as an adolescent. It is my hope to provide an entertaining, attractive, and informative tour through this history for high school and college chemistry teachers and students, practicing professionals in science and medicine, as well as the lay public interested in science and appreciative of artwork and illustration. We are increasingly an image-oriented culture and I have provided a picture book with sufficient text to explain details and context. Like any tour, the book is idiosyncratic in the highlights that it chooses to show the tourist. *From Alchemy to Chemistry in Picture and Story* is the result of consolidating its two well-received progenitor books: *A Chemical History Tour,* published in 2000, and *The Art of Chemistry,* published in 2003. Not coincidentally, the two books were complementary in the topics they covered. The current book has merged some essays, eliminated a few, added new essays and artwork, and updated the original essays.

From Alchemy to Chemistry in Picture and Story is meant to be skimmed as well as read. It includes almost 200 brief essays, over 350 figures, and 24 color plates. The ten sections begin with the practical, medical, and mystical roots of chemistry and trace, in pictures and words, its evolution into a modern science. Our tour starts with the metaphorical frontispiece of the 1738 edition of *Physica Subterranea,* describing the "birth of metals" in the bowels of the Earth. Practical metallurgical chemistry is accompanied by symbolism introduced centuries ago in cultures trying to understand the true nature and character of matter. Iron, the metal of choice for making sharp weapons, was equated with Mars, the god of war and the red planet. Many centuries later, scientists would discover that iron-containing hemoglobin is responsible for the red color of blood and decades later that the Martian surface is covered with oxides of iron.

The spiritual and allegorical representations of alchemy in the second section include a menagerie of fantastic creatures: lions and winged dragons; wolves; the feared basilisk that kills at great distance with a single glance; the ouroboros, continuously devouring and regenerating itself; passionate birds of prey; and the fabulous phoenix, the very symbol of the Philosopher's Stone.

The third section introduces Renaissance medicinal chemistry. Distillations, in warm boar dung, of plant and animal matter produced medications of widely varied efficacies. The bombastic sixteenth-century physician and alchemist Paracelsus developed his own coherent theories of medication. He believed in a vital force called the *Archaeus*, a kind of Alchemist of Nature, having a head and hands only and inhabiting the stomach. The *Archaeus* separates the nutritive from the poisonous. Illness occurs when the *Archaeus* is poisoned. The cure for poison is poison. Paracelsus pioneered the chemical syntheses of effective medicines, such as calomel, derived from toxic heavy metals.

The fourth section begins in the seventeenth century, a period in which chemistry started to become a science. Johann Baptist Von Helmont is, in many respects, a missing link between alchemy and superstition on the one hand and science on the other. Although he coined the term "gas," and can be said to have discovered carbon dioxide, his famous "tree experiment" completely missed the point that a considerable percentage of a tree's mass is contributed by carbon dioxide. Van Helmont was a believer in the concept of "sympathy," whereby a wound is treated by sprinkling the sword that caused it with powder of sympathy. Although Isaac Newton founded physics and codiscovered calculus, and Robert Boyle forever vanquished the four ancient Greek elements and is considered to be the father of chemistry, both were fully credulous about and practiced alchemy. During the early seventeenth century, the German scientist Daniel Sennert formulated a chemical concept of atoms based upon experimentation. Pierre Gassendi, a French clergyman, described air pressure in terms of collision of atoms. While Boyle's corpuscles suggest atoms, his belief in alchemy suggests that such corpuscles could transmute from one substance to another. Thus, it has little relation to our modern concept. During this period, chemistry's first true unifying concept, phlogiston theory, was introduced by Johann Joachim Becher. It was later extended by Georg Ernst Stahl. We commonly think of Becher as the *ur*-father of chemical theory. However, he was also the foremost mercantilist of his era and the economic advisor to Leopold I, Emperor of the Holy Roman Empire.

The fifth section of this book is the largest. It covers the chemical revolution that began quietly in 1727 when Stephan Hales learned to collect gases produced by chemical reactions, accelerated when Joseph Black isolated and fully characterized carbon dioxide, and literally exploded when Henry Cavendish isolated hydrogen. The brilliant Cavendish thought he had actually isolated the elusive phlogiston itself. Separate and independent discoveries of "fire-air" by Carl Wilhelm Scheele and "dephlogisticated air" by Joseph Priestley, both firmly anchored in phlogiston theory, would set the stage for Antoine Laurent Lavoisier to formulate the modern synthesis: combustion (and respiration) involves combination with oxygen from the air, not loss of phlogiston to the air. Lavoisier was a wealthy partner in the *Ferme Générale*, which collected taxes and helped manage the treasury for Louis XVI. On May 8, 1794, Lavoisier, his father-in-law, and 26 other members of the *Ferme* were guillotined in the space of 35 minutes. Some two decades later, John Dalton would formulate atomic theory and the modern science of chemistry was fully born.

The book's next section explores the role of chemistry in early pre- and post-colonial America. The roots of early American chemistry lie in Edinburgh, Scotland where Joseph Black influenced the first generation of American professors of chemistry. Benjamin Franklin was very knowledgeable about chemistry and also a friend of the Lavoisiers (Madame Lavoisier painted a beautiful portrait of him). John Adams and Thomas Jefferson publicly commented on the uses and limitations of chemistry, and James Madison taught the subject in Virginia.

Section VII traces the specialization of chemistry that occurred during the nineteenth century as organic, inorganic, physical, and analytical chemistries emerged as distinct disciplines. The systematization of the vast jungle known as organic chemistry led to the discovery of valence and the importance of the third dimension in molecular structure and chemical behavior.

Section VIII ("Teaching Chemistry to the Masses") recognizes the development of chemical pedagogy that began during the nineteenth century. Madame Jane Marcet's *Conversations on Chemistry,* first published anonymously in London in 1806, employed Socratic dialogue with young female pupils to teach science. The book went through many printings and modifications and is reputed to have sold some 160,000 copies in the United States. Michael Faraday proudly proclaimed Madame Marcet as his teacher, since her book drew him into the field of chemistry. I have also included an essay about a book of chemical psalms, titled *Chemistianity,* the goal of which was to teach chemistry to adolescents and octogenarians, both groups presumed to have short attention spans. The rhymes in this book are as enjoyable as the sound of a fingernail scraping across a blackboard. Another Victorian-era book, *Fairyland of Chemistry,* describes the comings and goings of hydrogen fairies and oxygen fairies, for example, as they flit about and link hands to form water molecules.

The light coverage of the twentieth century will certainly draw the attention of some not-so-amiable reviewers. I would defend this admitted weakness by noting that the exponential explosion of information during modern times would overwhelm the contents in this book. For example, in its first year of publication (1907), *Chemical Abstracts* presented summaries of 7,994 papers and 3,853 patents. In the year 2000, it abstracted 573,469 papers and 146,590 patents (see www.cas.org). Moreover, the significant modern findings that continue to matter are included in current chemistry texts. *From Alchemy to Chemistry in Picture and Story* is meant to supplement and enliven the coverage in a modern course. It makes no pretense of completeness. Nevertheless, we include the discoveries of subatomic structure, X-ray crystallography, the Kossel–Lewis–Langmuir picture of bonding based on the octet rule, the development of the quantum mechanics (the underlying basis of the periodic table), as well as resonance theory. The DNA double helix is included because it is a triumph of structural chemistry and its structure immediately explained its function. Indeed, DNA's function—duplication—implied that its structure would likely have "two-ness." The twentieth century "concludes" with brief visits to chemistry at its smallest (nanotechnology) and its fastest (femtochemistry). The use of the scanning tunneling microscope (STM and its modifications) to view individual atoms and move them one by one is certainly a crowning achievement of twentieth century science.

One leitmotif in our tour is the resistance from many distinguished scientists to the reality of atoms that continued for over one hundred years after Dalton's theory was postulated in 1803. Indeed, in the "minutes" before its universal acceptance in the first decade of the twentieth century, Ludwig Boltzmann committed suicide due in part, it is believed, to his failure to convince all physicists and chemists of the reality of atoms. Eighty years later, scientists "lassoed" together a circle of 48 iron atoms, one by one, to form a "quantum corral."

The final section ("Some Brief Chemical Amusements") includes clairvoyant images of atoms, a faux James Thurber short story shamelessly derived from "The Secret Life of Walter Mitty," a comparison between Babe Ruth and Antoine Lavoisier with musings on the low monetary value of collectors cards of famous chemists compared to baseball cards, and the long- (and well-) forgotten 92-chapter novel titled *White Lightning*. Yes, Virginia, it has a brief but dramatic chapter for each known or anticipated element up to and including uranium.

The book concludes with an Epilogue consisting of two brief, more personal, essays. One of these is about a friend from my adolescent years, Robert Silberglied, a quirky and ingenious butterfly collector and admired mischief-maker, who became a world-renowned entomology professor at Harvard and conservationist before he died at a young age in an airplane crash. The second is a brief essay whimsically visiting my own chemical genealogy. Although these two essays may appear to be exercises in self-indulgence and self-aggrandizement, they are not meant to be. The purpose is to give the reader a taste for our scientific culture—the early signs of "a natural scientist," and the interest in our personal scientific roots and the desire to connect with them.

In composing this work, I came to realize that one important theme is our very human need to pictorialize matter: four elements, three principles, platonic solids such as the cube, corpuscles or atoms with and without hooks, two-dimensional "clumps" of atoms, two-dimensional molecules, three-dimensional molecules, fairies linking arms, "ball-and-stick" and "space-filling" models, solar-system atoms, cubic atoms with electrons at the corners, resonating structures, atoms hooked together by springs, atomic and molecular orbitals, and electron-density contours on computer screens. Such images will recur throughout the book.

My first university chemistry teaching assignment included a "Chemistry for Non-Science Majors" course that sparked a lifelong interest in communicating chemistry to the public. In this type of endeavor, the question of "how did we come to believe or know this?" arises almost naturally and we take tentative steps to explore the historical development and context. It immediately becomes clear how little we practicing scientists understand about the histories of our own fields and, in any case, why *should* we understand more? In chemistry, the early beliefs and theories are now known to be incorrect, the symbols are outdated, and the language arcane, often deliberately so. It is so challenging to learn the modern canons of chemical knowledge as a student and then battle obsolescence as a practicing chemist, that it does not seem wise or practical to learn "this superfluous, outdated stuff."

I anticipate justified criticism of this idiosyncratic tour due to the numerous sites not visited and admit that there are countless other paths through chemical history and apologize in advance for numerous discoveries omitted or given short shrift. However, I want this book to be useful, and to fulfill this mission it must be read and enjoyed by nonspecialists as well as experts. A more thorough or encyclopedic approach will not help to achieve this goal. Although I have attempted to recognize contributions beyond those of Western culture, I am aware of the weak coverage given to early science in Chinese, Indian, African, Moslem, and other cultures. This is really more an artifact of the availability of printed books rather than intent.

Although this tour is meant to be both lighthearted and light reading, it tackles some of the important topics that are often too lightly or confusingly broached in introductory courses and are difficult to teach. We do, however, try our hand at humor and some of the earthiness so evident in the Renaissance works of Chaucer and Rabelais. Why not include Van Helmont's recipe for punishment of anonymous "slovens" who leave excrement at one's doorstep? By providing such vignettes, I hope to reengage chemists, other scientists, and the public in the history of our field, its manner of expressing and illustrating itself, and its engagement with the wider culture. I hope to provide teachers of introductory chemistry courses with some assistance through difficult teaching areas and a few anecdotes to lighten the occasional slow lecture. And if a few students *are* caught snickering over a page of Rabelaisian chemical lore or some bad puns, would that be such a bad thing?

SUGGESTIONS FOR FURTHER READING AND TOURING

I am not formally trained as a chemical historian. Fortunately, there are a number of truly wonderful books treating chemical history. The most authoritative is the inspirational four-volume reference work, *A History of Chemistry*, by James R. Partington. It is rigorous, amply referenced, engagingly written, and nicely illustrated. It extensively covers the period through the end of the nineteenth century and the decades up to the mid-twentieth century. Partington's reference work has been a major source of information and insight for me. I have also relied heavily on the book by Aaron J. Ihde, *The Development of Modern Chemistry*, published in 1964, and the book by William H. Brock, *The Norton History of Chemistry*, published in 1992. John Hudson's *The History of Chemistry*, published in 1992, also provides detailed and accessible coverage of chemical history. Two books that briefly outline chemical history from its earliest roots to the end of the twentieth century are *The Last Sorcerers: The Path from Alchemy to the Periodic Table*, by Richard Morris (2003), and *Creations of Fire*, by Cathy Cobb and Harold Goldwhite (1995). Although there are numerous excellent scholarly books referenced in specific essays in the present work, I wish to mention some that "cross cut" the field and its history. *Ideas in Chemistry*, by David Knight (1992), *The Atom in the History of Human Thought*, by Bernard Pullman (1998), *The Enlightenment of Matter*, by Marco Beretta (1995), *Instruments and Experimentation in the History of Chemistry*, edited by Frederic L. Holmes and Trevore H. Levore (2000), and *From Classical to Modern Chemistry: The Instrumental Revolution*, edited by Peter J.T. Morris (2002) are five such books. Levore has authored a more recent (2006) book titled *Transforming Matter: A History of Chemistry from Alchemy to Buckyball*. The book *Women in Chemistry*, by Marelene and Geoffrey Rayner-Canham, published in 1998, provides authoritative and well-balanced coverage to a long-neglected topic. Mary Ellen Bowden, at the Chemical Heritage Foundation, has produced a series of highly accessible works, including *Chemical Achievers: The Human Face of the Chemical Sciences* (1997) and *Joseph Priestley, Radical Thinker* (2005, edited with Lisa Rosner). I have also recently completed a book titled *Twentieth-Century Chemistry: A History of Notable Research and Discovery*. There are also a number of extraordinary books about the seventeenth century including the alchemy of Boyle and Newton authored by William R. Newman [*Gehennical Fire: The Lives of George Starkey* (1994); *Promethean Ambitions: Alchemy and the Refashioning of Nature* (2004); *Atoms and Alchemy* (2006) and Lawrence M. Principe [*The Aspiring Adept: Robert Boyle and his Alchemical Quest* (1998)] and co-authored by Newman and Principe [*Alchemy Tried in the Fire: Starkey, Boyle and the Fate of Helmontian Chymistry* (2002)].

The first Nobel Prizes were awarded in 1901 and the Nobel Foundation site (www.nobelprize.org) is a wonderful source for complete coverage, including full Nobel Prize lectures, often full of insights and humor that do not usually appear in the primary literature. The 1975 Smithsonian Institution pamphlet by Jon Eklund, titled *The Incompleat Chymist*, is a wonderful source for deciphering the names of chemicals and equipment during the eighteenth century, the period corresponding to the chemical revolution. Hopefully, this pamphlet will some day be either reissued or made available on line.

Although John Emsley's book *The 13th Element* (2000), first published in England under the title *The Shocking History of Phosphorus*, is devoted to a single chemical element, it beautifully evokes the atmosphere of late-seventeenth-century chemistry in its early chapters. The play *Oxygen*, by Carl Djerassi and Roald Hoffmann (2001), recreates the late eighteenth century and an imagined meeting of Joseph Priestley, Carl Wilhelm Scheele, and Antoine Laurent Lavoisier.

I am particularly fond of the 1927 book, *Old Chemistries*, by Edgar Fahs Smith. I imagine that I am in Professor Smith's den on a cold winter's night as he shows me his antiquarian book collection and gently reads selected passages as we are warmed by the fireplace. And how I wish that I could have met the erudite and ebullient John Read. His trilogy, *A Prelude to Chemistry*, *Humour And Humanism in Chemistry*, and *The Alchemist In Life, Literature and Art*, provides the reader with healthy doses of laughter and learning. In *Humour and Humanism*, Read gives us the "box score" of an Alchemical Rugby Match of All-Stars from the Bible (Noah, Moses), Greek and Roman mythology (Jupiter, Neptune, Aphrodite), ancient cultures (Cleopatra, Aristotle), the Renaissance (Paracelsus, Maier), and the early history of our science (Boyle, Lavoisier). The puns are deliciously low. He also writes a one-act play, "The Nobel Prize" ("A Chemic Drama In One Act"), and happily treats us to the bawdier moments in Ben Jonson's 1610 play, *The Alchemist*. Professor Read also arranged the first performance of Michael Maier's seventeenth-century alchemical music composed for his book, *Atalanta Fugiens* (performed by the "Chymic Choir" at St. Andrews College in 1935). I discovered John Read's books after I began this project and, thus, cannot blame any of my own excesses of ebullience on him.

The Chemical Heritage Foundation (CHF) published in 2002 an attractive pamphlet titled *Transmutations: Alchemy in Art, Selected Works from the Eddleman and Fisher Collections at CHF.* For decades, the beautiful catalogues of the then Aldrich Chemical Company featured artwork, particularly paintings of chemists by Dutch masters, collected by its founder, Alfred Bader. Bader's very noteworthy and dramatic autobiography is, fittingly enough, titled *Adventures of a Chemist Collector* (1995).

In the grand historical context of chemical history the United States is, of course, a latecomer, notwithstanding medicines and crafts developed by aboriginal cultures in the Americas and practical chemistries developed in Jamestown and in New England during the early seventeenth century. Visiting the worldwide websites of chemical societies in England, France, Germany, Canada, and other countries is a highly recommended activity. I will mention here two wonderful American resources for the potential chemical history tourist. The first is the Chemical Heritage Foundation located in Philadelphia. It holds a vast collection of artwork, equipment, artifacts, interviews with famous chemists, and books. The CHF sponsors scholars and conferences and is open to the public. It is now the home of the Roy G. Neville Historical Chemical Library, a collection in the Othmer Library. The CHF website (www.chemheritage.org) provides information for visitors and links to a great store of resources in chemical history. The Chemical Heritage Foundation has just published (2006) the magnificent two-volume work, *The Roy G. Neville Historical Chemical Library: The Annotated Catalogue of Printed Books on Alchemy, Chemistry, Chemical Technology, and Related Subjects*, written primarily by Neville. It compares favorably with the two large classics in the field: Denis I. Duveen's *Bibliotheca Alchemica Et Chemica,* and John Ferguson's *Bibiotheca Chemica*. The Chemical Heritage Foundation publishes a beautiful and inexpensive quarterly magazine titled *Chemical Heritage*.

The Edgar Fahs Smith Chemistry Collection at the University of Pennsylvania, the Duveen Collection at the University of Wisconsin, and the Lavoisier collection at Cornell University are three other sites very much worth visiting. Harding University, in Searcy, Arkansas has a comprehensive collection of eighteenth- and nineteenth-century American books on chemistry from the combined collections of William D. Williams and Wyndham D. Myles.

The American Chemical Society has recognized nearly 60 historical chemical landmarks accessible at its website, www.chemistry.org/landmarks. Each landmark has its own descriptive brochure. I hope readers will enjoy actual tours of these landmarks as well as virtual tours. Those members of the American Chemical Society who pay the small membership fee to join its Division of History of Chemistry receive a gratis subscription to the very useful and enjoyable *Bulletin for the History of Chemistry*. It is my profound hope that chemical history will once again find its way into both introductory and advanced courses in our field.

ACKNOWLEDGMENTS

I believe that my concept for *A Chemical History Tour*, the first of the two progenitors of the present book, was stimulated by *Chemistry Imagined*, written by Roald Hoffmann in collaboration with artist Vivian Torrence. It is my hope that the new book, *From Alchemy to Chemistry in Picture and Story*, includes some of the spirit of *Chemistry Imagined* along with essences of Edgar Fahs Smith's *Old Chemistries* and John Read's trilogy. I owe a special gratitude to Roald Hoffmann for his encouraging response to my partial manuscript and his generous support in discussions with potential publishers. Jeffrey Sturchio also provided early encouragement on this project. Barbara Goldman, then at John Wiley & Sons, accepted and recommended the project, providing moral support while encouraging creativity.

My daughter, Rachel, was employed to meticulously scan most of the images in the first book during the summer preceding her junior year at college. Happily, our friendship survived this one-time employer/employee experience and I confess that her healthy skepticism added to my own motivation. The artistic interests of my son David were another stimulus and I thank my wife Susan for tolerating early morning readings of the essays in both earlier books and the new ones in the present book. I am grateful for the comments and suggestions of my long-time friend Joel F. Liebman throughout these projects. My father, Murray Greenberg, was a proofreader for the two earlier books. Pierre Laszlo, my Ph.D. advisor, provided many useful comments concerning *The Art of Chemistry*, the second progenitor of the present volume. Artist Rita Shumaker provided three original works of interpretive artwork for this project. Dudley Herschbach provided some very stimulating suggestions concerning my coverage of Benjamin Franklin. Other dear and valued colleagues and friends are acknowledged throughout the present book. The John Wiley & Sons staff have been a joy to work with and I particularly acknowledge the efforts of Darla Henderson, Amy Byers, Christine Punzo, and intern Anna Pierrehumbert.

Unless otherwise noted, the figures are from books or artwork in my own collection. Roy G. Neville, chemist and renowned book collector, was most gracious in providing rare images from his extraordinary book collection. The Roy G. Neville Historical Chemical Library is now a collection within the Othmer Library of the Chemical Heritage Foundation in Philadelphia. The Chemical Heritage Foundation was also very helpful in supplying some images from the Othmer Library and I wish to express my thanks to Arnold Thackray and Elizabeth Swan.

SECTION I
PRACTICAL CHEMISTRY: MINING, METALLURGY, AND WAR

THE BIRTH OF METALS

What does this allegorical figure (Figure 1) represent? This bald, muscular figure has the symbols of seven original elements arrayed around (and likely including) the head. The all-too-perfect roundness of the head appears to correspond to the perfect circle that represents gold. The unique positions of male (sun) and female (moon) suggest the birth of metals.[1]

The elements, also including antimony and sulfur, are also buried in the intestines of the figure—literally its bowels—and now we have a hint of its nature. Any attempts at further interpretation are in the realm of psychology rather than science, and indeed the famous psychologist C.G. Jung owned a valuable collection of alchemical books and manuscripts and wrote extensively on the subject.[2]

At its heart, alchemy postulated a fundamental matter or state, the *Prima Materia*, the basis for formation of all substances. The definitions[2] of the *Prima Materia* are broad, partly chemical, partly mythological: quicksilver, iron, gold, lead, salt, sulfur, water, air, fire, earth, mother, moon, dragon, dew. At a more philosophical level, it has been defined as Hades as well as Earth.[2] Another figure from a seventeenth-century book on alchemy was identified by Jung as the *Prima Materia*—a similar muscular Earth shown suckling the "son of the philosophers."[2] This figure also has the breasts of a woman; the hermaphroditic being is reminiscent of the derivation of Eve from Adam and the subsequent seeding of the human species. The hermaphrodite is greater than the sum of its male and female natures.

Let us cling to the Earth analogy because it seems to help in understanding the presence of the elements in its bowels. The small figure in the upper abdomen, the homunculus, may be considered to be a type of Earth Spirit nurturing the growth of living things (see vegetation below it) and "multiplication" of the metals. The unique positions of gold (the head as well as the highest level in the intestines) implies *transmutation*—the conversion of base metals into noble metals. The figure holds a harp, representing harmony, and an isosceles triangle, representing symmetry. It is a metaphor for the unity that the true alchemists perceived between their art and nature.

This plate is the frontispiece from the book *Physica Subterranea* published by the German chemist and physician Georg Ernst Stahl in 1738.[3] It is the last edition of the famous book published by Johann Joachim Becher in 1669. Becher evolved chemistry's first unifying theory, the Phlogiston Theory, from alchemical concepts and it was subsequently made useful by Stahl. So in this plate are themes of alchemical transmutation, spiritual beliefs, and early chemical science that will begin our tour of alchemy and chemistry over two thousand years.

FIGURE 1. ■ Frontispiece from the final edition of *Physica Subterranea* by Johann Joachim Becher (Leipzig, 1738). The hermaphroditic figure may represent the Primary Matter (*Prima Materia*). The dwarf-like figure inside the body is the homunculus, the offspring of the "chymical wedding."

1. A. Roob, *The Hermetic Museum: Alchemy & Mysticism*, Benedikt Taschen Verlag GmbH, 1997, p. 183.

2. N. Schwartz-Salant, *Jung on Alchemy*, Princeton University Press, Princeton, NJ, 1995, pp. 25–30; 44–49.
3. A different interpretation of this figure, namely as Saturn, is to be found in C.A. Reichen, *A History of Chemistry*, Hawthorne Books, New York, 1963, p. 8.

THE ESSENCE OF MATTER: FOUR ELEMENTS (OR FIVE): THREE PRINCIPLES (OR TWO) OR THREE SUBATOMIC PARTICLES (OR MORE)

The ancient Greek philosophers were not scientists. They were, however, original thinkers who attempted to explain nature on a logical basis rather than by the whims of gods and goddesses. The father of this movement is considered to be Thales of Miletus, and during the sixth century B.C., he conceived of water as the essence of all matter. (We note later in this book that, in the mid-seventeeth century, Van Helmont had a somewhat similar view.) Thales is reputed to have predicted the total solar eclipse of 585 B.C., said to have occurred during a naval battle—although there is no basis for him having the knowledge to make such a prediction.[1] One of his successors in the Milesian School was Empedocles of Agrigentum (ca. 490–430 B.C.).[1] Empedocles is said to be the first to propose that all matter is composed of four primordial elements of equal importance,[2,3] although similar ideas appear to have formed in Egypt, India, and China (five elements) around 1500 B.C.[2] Figure 2 depicts the four earthly elements. It appears in *De Responsione Mundi et Astrorum Ordinatione* (Augsburg, 1472), a book derived from the writings of Saint Isidorus, Bishop of Seville, during the seventh century A.D.[4]

Although Empedocles wrote about the actual physical structure of matter, it was only during the fifth century B.C. that two philosophers of the Milesian School enunciated a coherent atomic cosmology. None of the writings of Leuccipus remain, but he is widely accepted as real and some of the writings of Democritus (ca. 460–ca. 370 B.C.),[1] his student, are known. For these scholars there were two realities in nature: Atoms (*atomos*, meaning not cuttable) and Void (derived from *vacuus*, meaning empty).[3] Void was considered to be as real as Atoms. Atoms of water were thought to be smooth and slippery; those of iron were jagged with hooks.

Aristotle (384–322 B.C.) is considered to be one of the two greatest thinkers of ancient times, the other being Plato.[1] Aristotle proposed a kind of primordial, heavenly element, "ether," and to each of the four earthly elements attributed two pairs of opposite or contrary "qualities" (wet versus dry; hot versus cold). The relationships between the elements and their qualities are depicted in a square that nicely places contrary qualities on opposite edges. The square is one of the fundamental symbols that often appear in alchemical manuscripts and books even as late as the eighteenth century. Thus, a liquid (rich in water) is cold and wet while its vapor (rich in air) is hot and wet. To vaporize a liquid, simply add heat—move from the cold edge to the hot edge of the square. To dissolve a solid (rich in earth), add wet; to burn the solid, add hot. Fire was not solid, liquid, or gas but a form of internal energy—perhaps related to the eighteenth-century concept of "caloric" propounded by Lavoisier.[2]

FIGURE 2. ■ The four elements of the ancients: Fire, Air, Earth, and Water from St. Isidore, *De Responsione Mundi Et Astrorum Ordinatione* (Augsburg, 1472) (courtesy of The Beinecke Rare Book and Manuscript Library, Yale University).

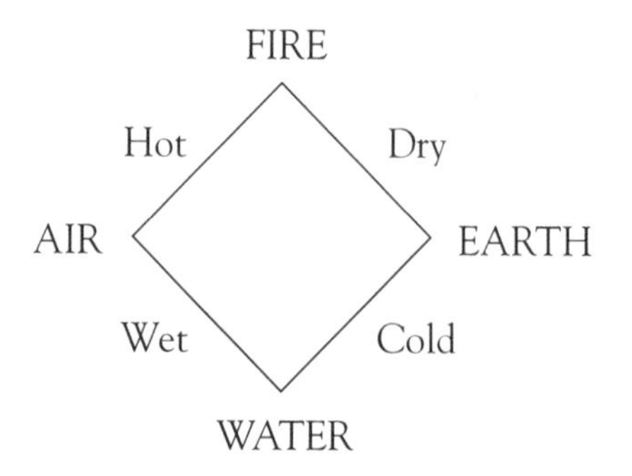

Aristotle was an anti-Atomist, in part, because he did not believe that space could be empty. This view was adopted by the great mathematician and philosopher Rene Descartes (1596–1650) who envisioned only two principles in matter (extent and movement) and rejected the four Aristotlean qualities. The idea of extent led him to reject the idea of finite atoms and the concept of void he considered ridiculous ("Nature abhors a vacuum"[5]). Thus, in the seventeenth and eighteenth centuries we have intellectual conflict between the Cartesians (school of Descartes) and the Corpuscular school (corpuscles were similar, yet fundamentally different, in concept to atoms), which included Robert Boyle and Isaac Newton.[6]

A 1747 oil-on-wood painting signed by a Johann Winckler[7] (Figure 3) joyously employs alchemical, spiritual, and religious symbolism characteristic of Rosicrucian beliefs. Most prominent are the four abbots whose activities symbolize earth, fire, air, and water. They are arrayed in the appropriate order of contrary properties—cold versus hot; wet versus dry.

The Cupid (or Mercurious) figure was said by the psychologist Carl Jung to represent "the archer who, chemically, dissolves the gold, and morally, pierces the soul with the dart of passion."[8] "Christian Rosencreutz in *The Chymical Wedding* is pricked with a dart by Cupid after stumbling upon the naked Venus."[8] The four abbots and the Venus figure each possess a vessel containing the Red Tincture, which represents the transmuting agent or Philosopher's Stone[9] or a preliminary stage of the Stone.[10] The castles may represent . . . well . . . castles. Or . . . they may symbolize the athanor or philosopher's furnace, which holds the hermetically sealed philosopher's egg.[11] The pair of doves represent the *albedo*, the white color that follows the *nigredo*, or the initial black color of The Great Work. Initially, metals and other substances are heated to form a black mass. Subsequent heating may calcine this mass to produce a white calx. Now, if that long-tailed bird attached to an abbot by a string is a peacock, we see represented the third color change of The Great Work, the rainbow hues. The fourth and final color is the ruby of the Red Tincture—four cucurbits—full and one goblets-worth in this painting. The phoenix also represents this final ruby red color but no phoenix is seen rising (or expiring) in the painting. No crows are in evidence either, so let's assume that the coals or the ashes in the athanor represent the *nigredo*.

Rosicrucians combine religious, occult, and alchemical beliefs.[12] Although the earliest writings date to the beginning of the seventeenth century, the origins of Rosicrucianism are commonly attributed to a Christian Rosenkreutz ("rosy cross"), allegedly born in 1378. Some consider the early sixteenth-century physician and alchemist Paracelsus to be the true founder. The alchemist Michael Maier appears to have been a Rosicrucian.[13]

The sign in the lower right of this painting may be translated as follows:

1. I search in the water here.
2. The air should give me
3. I search in the earth
4. The fires should become for me
5. Something here, you fools, here in the water, air and earths.
   In the fire, shall you busily search.
6. All here suddenly becomes.

During the Renaissance, the classical Greek views of nature were finally challenged by the likes of Paracelsus.[14] Paracelsus extended an earlier view of matter that held that it was a union between an exalted sulfur of the philosophers ("Sophic Sulfur"—characterized often as male) and an exalted mercury of

FIGURE 3. ■ Eighteenth-century painting by a Johann Winckler, exhibiting Rosicrucian influences, in which the four friars represent water, air, fire, and earth in a correctly ordered square. The red tincture (see color plates) in the possession of each friar, is an embodiment of the Philosopher's Stone—the mysterious agent of projection and transmutation.

the philosophers ("Sophic Mercury"—characterized often as female). These are not related to the chemical elements we now recognize as sulfur and mercury. To these Paracelsus added Salt as the third Principle. Now, Mercury is Spirit, Sulfur is Soul, and Salt is Material Body. The relationship is depicted as a triangle, the other great metaphor found in alchemical manuscripts and books

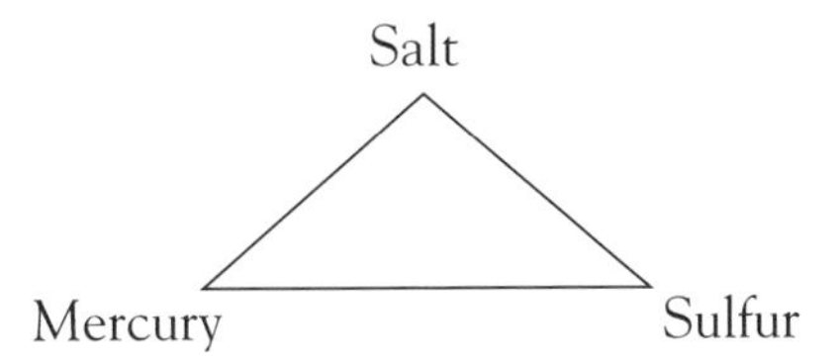

through the eighteenth century. All matter is composed of these three principles in various proportions. Later in this book (Figure 96) we see two such symbolic triangles in Oswald Croll's *Basilica Chymica*. Croll presented Paracelsan alchemy—the bottom triangle presents Life, Spirit, Body (or Fire, Air, Water or Animal, Vegetable, Mineral). Symbols of triangles and squares abound in alchemy. The Sioux view the circle as their high ideal: "circle of Life," the tipi, the campfire.[15] In his nineteenth-century satire *Flatland*, Edwin Abbot portrays increasing perfection through each successive generation as a triangle begets a square, which begets a pentagon, and so on. A *megagon* is close to the perfection of a circle—a kind of generational transmutation.

The modern view of the atom is that it *is* divisible and that the fundamental particles making up all atoms of all elements are protons (positive charge), neutrons (zero charge), in an unimaginably dense nucleus occupying a miniscule fraction of the atom's volume, and electrons (negative charge).[16] The positive nucleus and the negative electrons are our modern "contraries." (Incidentally, it was Benjamin Franklin who introduced the negative–positive nomenclature in the context of electricity.[17]) The electrons are considered to be fundamental particles of infinite lifetime and are actually one of six subatomic particles called leptons. Protons and neutrons are not considered fundamental and are two of a very complex class of subatomic particles called hadrons. Outside of the nucleus, a free neutron has a half-life of only 17 minutes and decays into a proton, an electron (β particle), and an antineutrino—another lepton.[16] So, based upon this modern view, we can draw a Paracelsan-style triangle, but not equilateral in the sense that the neutron can give rise to the other two. The modern *Prima Materia* could be a dense neutron star.

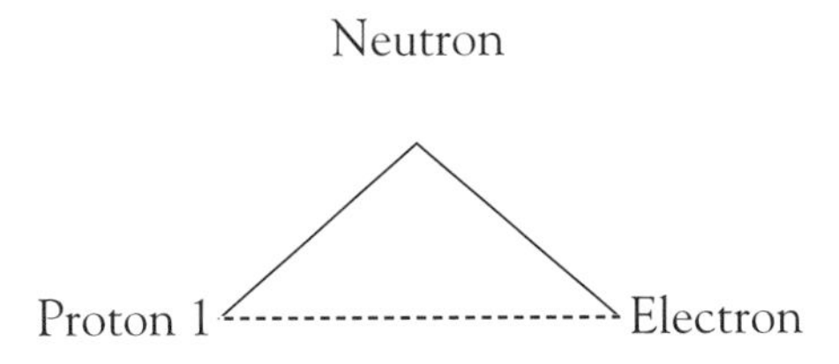

1. *Encyclopedia Brittanica*, 15th ed. Vol. 11, Chicago, 1986, p. 670.
2. J. Read, *Prelude to Chemistry*, MacMillan, New York, 1937, pp. 8–11.

3. B. Pullman, *The Atom in the History of Human Thought*, Oxford University Press, New York, 1998, pp. 2–47.
4. I. MacPhail, *Alchemy and The Occult*, Yale University Library, New Haven, 1968, Vol. 1, pp. 3–4.
5. Two sources for quotations simply refer this phrase (*Natura abhorret vacuum*) to a Latin proverb [B. Evans, *Dictionary of Quotations*, Delacorte Press, New York, 1968, p. 720, and *Dictionary of Foreign Quotations*, R. Collison and M. Collison (eds.), Facts on File, New York, 1980, p. 241]. One source attributes it to Gargantua in 1534 but from an ancient Latin source [A. Partington (ed.), *The Oxford Dictionary of Quotations*, 4th ed., Oxford University Press, New York, 1992, p. 534; *Bartlett's Familiar Quotations*, 16th ed., J. Kaplan (ed.), Little, Brown, Boston, 1992, p. 277] attributes the phrase to Spinoza in 1677. Just thought you'd want to know this one for the next Happy Hour.
6. B. Pullman, op. cit., pp. 140–142, 157–163.
7. I am not certain about the identity of the artist. One possibility is Johann Heinrich Winckler (1703–1770).
8. L. Abraham, *A Dictionary of Alchemical Imagery*, Cambridge University Press, Cambridge, UK, 1998, p. 51.
9. J. Read, op. cit., p. 12; p. 148.
10. Abraham, op. cit., p. 169.
11. Abraham, op. cit., pp. 31–32.
12. *The New Encyclopedia Britannica*, Encyclopedia Britannica, Inc., Chicago, 1986, Vol. 10, p. 188.
13. Read, op. cit., pp. 230–232.
14. J. Read, op. cit., pp. 21–30.
15. J. Lame Deer and R. Erdoes, *Lame Deer Seeker of Visions*, Simon and Schuster, New York, 1972, pp. 108–118.
16. B. Pullman, op. cit., pp. 343–353.
17. J.R. Partington, *A History of Chemistry*, MacMillan, London, 1962, Vol. 3, p. 66.

UNIFYING THE INFINITE AND THE INFINITESIMAL

It is human nature to try to harmonize our universe—to attempt to unify the infinite with the infinitesimal. Pythagoras and his followers developed a purely mathematical conception of the universe. As Pullman notes:[1] "Indeed, the Pythagoreans held that numbers are the essence of all things. Numbers are the source of what is real; they themselves constitute the things of the world."

Mendeleev developed the periodic table roughly 2400 years after Pythagoras died. He could not possibly have understood the origin of its order. But in 1926, the new quantum mechanics of Schrödinger explained the periodic table on the simple basis of four quantum numbers (n, l, m, and s) that students now learn in high school. Pythagoras would have been pleased but not surprised.

Figure 4 is from Johannes Kepler's *Harmonices Mundi* (1619). The fanciful drawings on the middle right depict the five platonic solids—polyhedra whose faces are uniformly composed of triangles, squares, or pentagons. The Pythagorean Philolaus of Tarentum (480 B.C.–?) is generally credited with equating the four earthly elements to these polyhedra.[1] Starting from the top center and moving counterclockwise, we have the tetrahedron (fire), octahedron (air), cube (earth), and icosahedron (water). Plato added the fifth solid, the dodecahedron,

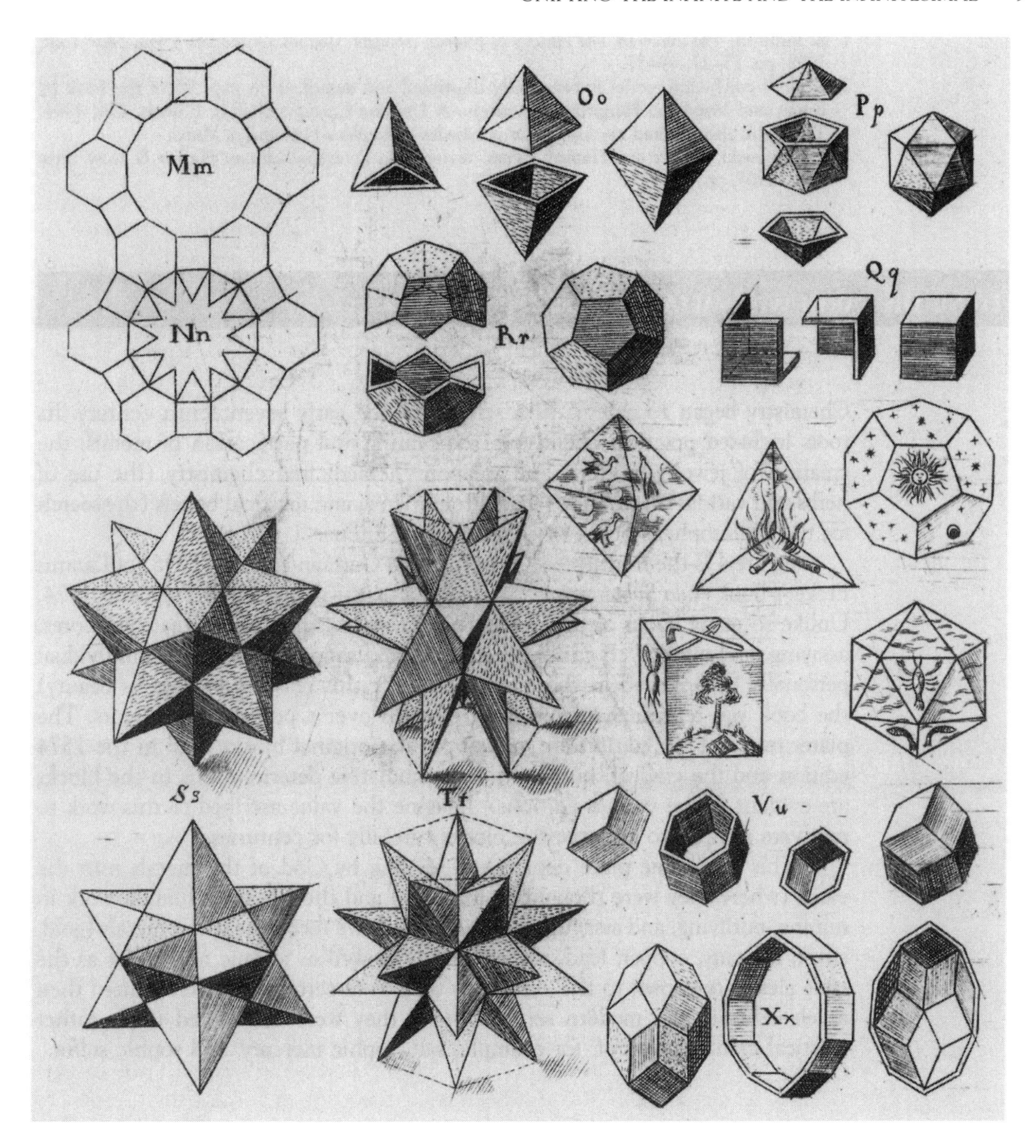

FIGURE 4. ■ Polyhedra in Johannes Kepler's *Harmonices Mundi* (Linz, 1619). Note the five Platonic solids on the middle right of this figure representing the four earthly elements Air, Fire, Water, and Earth as well as the fifth (heavenly) element Ether (courtesy of Division of Rare and Manuscript Collection, Carl A. Kroch Library, Cornell University).

to represent the universe (similar to Aristotle's ether). The tetrahedron is the sharpest of these polyhedra, and fire is, thus, the "most penetrating" element. The dodecahedron is most sphere-like, most perfect. Its pentagons are also unique—you cannot tile a floor with pentagons as you can with triangles, squares, and hexagons. Plato further imagined that the four earthly elements were themselves composed of fundamental triangles—an isosceles right triangle

A (derived from halving the square face of the cube) and a right-triangle B (derived from halving the equilateral triangular face of the tetrahedron, octahedron, or icosahedron). Earth was composed of triangle A. Air, fire, and water were composed of triangle B and could therefore be interconverted.[1]

In his 1596 book *Mysterium Cosmographicum*, Kepler proposed a solar system that placed the orbits of the six known planets on concentric spheres inscribed within and circumscribed on these five polyhedra arranged concentrically.[2] In the words of Jacob Bronowski:[3] "All science is the search for unity in hidden likenesses." He states further: "To us, the analogies by which Kepler listened for the movement of the planets in the music of the spheres are farfetched. Yet are they more so than the wild leap by which Rutherford and Bohr in our own century found a model for the atom in, of all places, the planetary system?"

1. B. Pullman, *The Atom In The History of Human Thought*, Oxford University Press, New York, 1998, pp. 25–27, 49–57.
2. Kepler's polyhedral model is beautifully illustrated and described on page 95 of the book by Istvan and Magdolna Hargittai, *Symmetry—A Unifying Concept*, Shelter, Bolinas, CA, 1994. This book also inspired my use of the polyhedra in Kepler's *Harmonices Mundi*.
3. J. Bronowski, *Science and Human Values*, revised ed., Perennial Library Harper & Row, New York, 1965, pp.12–13.

SEEDING THE EARTH WITH METALS

Chemistry began to emerge as a science in the early seventeenth century. Its roots included practical chemistry (the mining and purification of metals, the creation of jewelry, pottery, and weaponry), medicinal chemistry (the use of herbs and various preparations made from them), and mystical beliefs (the search for the Philosopher's Stone or the Universal Elixir).

Figure 5 is the frontispiece from the final German edition (1736) of Lazarus Ercker's book *Aula Subterranea* . . . , which was first published in Prague in 1574. Unlike so many books of the sixteenth century, this important treatise on ores, assaying, and mineral chemistry was clearly and simply written by an individual personally experienced in the mining arts. For this reason (and for its beauty) the book was reprinted in numerous editions over a period of 160 years. The plates in this 1736 edition are made from the original blocks used in the 1574 edition and the gradual, but slight and cumulative deteriorations in the blocks are evident in the various editions.[1] Imagine the value ascribed to this work to motivate printers to preserve the blocks carefully for centuries.

This handsome plate depicts the seeding by God of the metals inside the earth (only there can they multiply naturally) and the laborious human work in mining, purifying, and assaying them. The heat inside the Earth is singular in its nature with no counterpart on the surface. Although we recognize seven metals (gold, silver, mercury, copper, lead, tin, and iron) as well as arsenic and sulfur as the nine elements known to the Ancients, they were certainly not recognized then as elements in the modern sense. Instead they were considered to

FIGURE 5. ■ Frontispiece from the final edition of *Aula Subterranea* by Lazarus Ercker (Frankfurt, 1736) depicting God seeding the earth with metals and their harvesting and refining by people. (The first edition of this book was published in 1574; the original blocks were employed to strike the plates in all subsequent editions.)

be rather mystical combinations of, for example, salt, sophic mercury, and sophic sulfur.

---

1. A.G. Sisco and C.S. Smith, *Lazarus Ercker's Treatise on Ores and Assaying* (translated from the German Edition of 1580), The University of Chicago Press, Chicago, 1951.

CHYMICALL CHARACTERS

This table of chemical symbols (see Figure 6) is found in the book titled *The Royal Pharmacopoea, Galenical and Chymical, According to the Practice of the Most Eminent and Learned Physitians of France, and Published with their Several Approbations*, the English edition published in 1678. The author, Moses Charas, fled religious persecution in France to join the enlightened intellectual environment in the England of Charles II, who chartered the Royal Society. Its membership included Robert Boyle, Robert Hooke, and Isaac Newton.

The elements listed in the table include the nine ancient elements described previously and a few others readily separable. Gold, of course, being "inert," is commonly found in an uncombined state and its high density (about 9 times denser than sand) allows it to be panned. Actually, we now also know that inert gases such as helium, neon, argon, krypton, and xenon are also found uncombined in nature, but they are colorless and odorless. In any case, we are suddenly over 200 years ahead of ourselves and apologize to the reader for getting carried away by our enthusiasm.

The association of elements with planets and their symbols, evident in Figure 6, appears to have been adopted from the ideas of Arab cultures during the Middle Ages. Association of gold with the sun is too obvious. The others are more subtle. For example, of the planets, mercury appeared to the Ancients to move most rapidly in the sky and was most suited as a messenger. Mercury's wings nicely represent the metal's volatility. In contrast, Saturn was the most distant of planets observed by the Ancients (Uranus, Neptune, and Pluto were discovered in the eighteenth, nineteenth, and twentieth centuries, respectively). The apparent slow movement of this planet through the skies was likened to Saturn, the god of seed or agriculture, who is sometimes depicted with a wooden leg. Lead was dense, slow . . . leaden. A person who is *saturnine* is sluggish or gloomy (not to be confused with a person who is *saturnalian*—riotously merry or orgiastic after the Roman holiday Saturnalia).

But let's return to a modern use of metaphor, based upon the toxic element lead, and visit the book *The Periodic Table*, by Primo Levi,[1] who used 21 elements as metaphors in 21 stories. For example:

> My father and all of us Rodmunds in the paternal line have always plied this trade, which consists in knowing a certain heavy rock, finding it in distant countries, heating it in a certain way that we know, and extracting black lead from it. Near my village there was a large bed; it is said that it

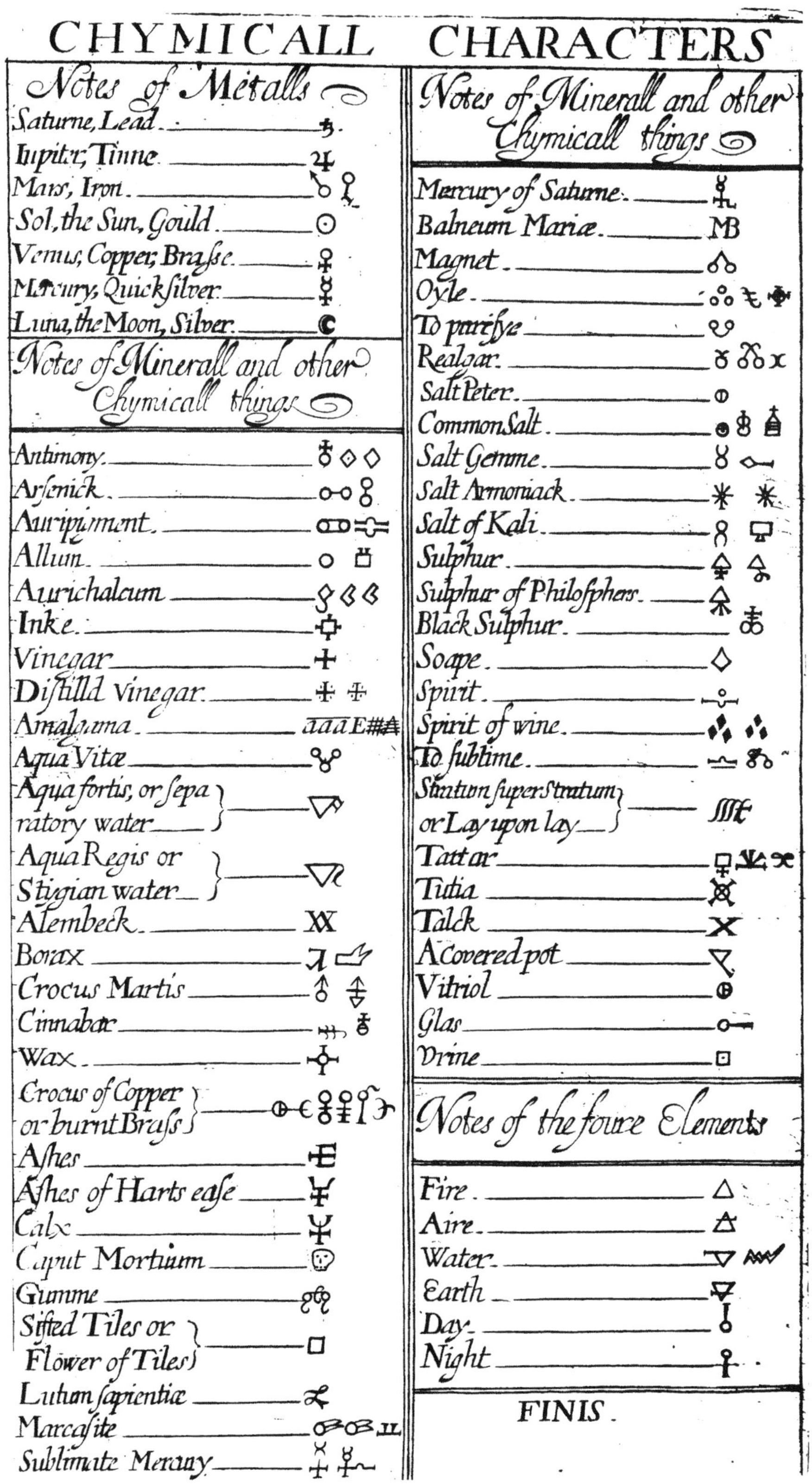

CHYMICALL CHARACTERS

Notes of Metalls

- Saturne, Lead.
- Iupiter, Tinne.
- Mars, Iron.
- Sol, the Sun, Gould.
- Venus, Copper, Brasse.
- Mercury, Quicksilver.
- Luna, the Moon, Silver.

Notes of Minerall and other Chymicall things

- Antimony.
- Arsenick.
- Auripigment.
- Allum.
- Aurichalcum
- Inke.
- Vinegar
- Distilld Vinegar.
- Amalgama.
- Aqua Vitæ
- Aqua fortis, or separatory water
- Aqua Regis or Stigian water
- Alembeck.
- Borax
- Crocus Martis
- Cinnabar
- Wax.
- Crocus of Copper or burnt Brass
- Ashes
- Ashes of Harts ease
- Calx
- Caput Mortuum
- Gumme
- Sifted Tiles or Flower of Tiles
- Lutum sapientiæ
- Marcasite
- Sublimate Mercury

Notes of Minerall and other Chymicall things

- Mercury of Saturne.
- Balneum Mariæ.
- Magnet.
- Oyle.
- To purifye
- Realgar.
- Salt Peter.
- Common Salt.
- Salt Gemme.
- Salt Armoniack.
- Salt of Kali.
- Sulphur.
- Sulphur of Philosphers.
- Black Sulphur.
- Soape.
- Spirit.
- Spirit of wine.
- To sublime.
- Stratum super Stratum or Lay upon lay
- Tartar
- Tutia
- Talck
- A Covered pot
- Vitriol
- Glas
- Vrine

Notes of the foure Elements

- Fire.
- Aire.
- Water.
- Earth
- Day.
- Night

FINIS.

FIGURE 6. ■ Chemical symbols from *The Royal Pharmacopoea* by Moses Charas (London, 1678).

> had been discovered by one of my ancestors whom they called Rodmund Blue Teeth. It is a village of lead-smiths; everyone there knows how to smelt and work it, but only we Rodmunds know how to find the rock and make sure it is the real lead rock, and not one of the many heavy rocks that the gods have strewn over the mountain so as to deceive man. It is the gods who make the veins of metals grow under the ground, but they keep them secret, hidden; he who finds them is almost their equal, and so the gods do not love him and try to bewilder him. They do not love us Rodmunds: but we don't care.
>
> All the men have resumed their former trades, but not I: just as the lead, without us, does not see the light, so we cannot live without lead. Ours is an art that makes us rich, but it also makes us die young. Some say that this happens because the metal enters our blood and slowly impoverishes it; others think instead that it is a revenge of the gods, but in any case it matters little to us Rodmunds that our lives are short, because we are rich, respected and see the world.
>
> So, after six generations in one place, I began traveling again, in search of rock to smelt or to be smelted by other people, teaching them the art in exchange for gold. We Rodmunds are wizards, that's what we are: we change lead into gold.

With the naked eye, ancient people could discern that the planet Mars is red, just as is the calx of iron ("rust"). Associating Mars—the god of war—with iron—the stuff of weapons, as well as with blood—is intuitively reasonable. Late twentieth-century business executives wore red "power ties" to meetings. But in an almost too wonderful confirmation of ancient intuition, the findings of the NASA Viking Mission, which landed two spacecraft on Mars in 1976, indicated a red surface composed of oxides of iron: eyeball chemical analysis by the Ancients at over 30 million miles—not bad!

But let us take irony one or two steps further. As of this writing, it appears that Mars sent its own messenger to Antartica 13,000 years ago in the form of Meteorite ALH84001.[2] Comparison of the carbon isotope content in the carbonate globules of the meteorite with Viking data indicated its Martian origin. Among the fragments of chemical evidence, the finding of iron (II) sulfide coexisting with iron oxides suggested to the investigators a biogenic origin since these two are essentially incompatible under abiotic conditions. The electrifying, although not widely accepted today, conclusion of the scientists[2]:

> Although there are alternative explanations for each of these phenomena taken individually, when they are considered collectively, particularly in view of their spatial association, we conclude that they are evidence for primitive life on early Mars.

---

1. P. Levi, *The Periodic Table* (English translation of the Italian text), Schocken Books, New York, 1984 (see pp. 80–81 for the three quotations employed here).
2. D.S. McKay, E.K. Gibson, Jr., K.L. Thomas-Keprta, H. Vali, C.S. Romanek, S.J. Clemett, X.D., F. Chillier, C.R. Maechling, and R.N. Zare, *Science*, 273(5277):924–930, 1996.

PRACTICAL METALLICK CHEMISTRY

Figure 7 depicts the inside view of an assay laboratory of the late sixteenth century. Figures 7 to 17, like Figure 5, are from the 1736 edition of Ercker's *Aula Subterranae* . . . and were printed using plates from the 1574 edition.[1] Figure 8 depicts a machine washing alluvial gold ores. The great density of gold, 19.3 g/cm^3 (the density of water, 1.0 g/cm^3; mercury "only" 13.6 g/cm^3), allows its ready separation from sand and other minerals. Figure 9 depicts the operations in making cupels. Cupellation was a technique for purifying gold or silver in ores. Cupels were cuplike objects made of ground bones in which ground ores were placed. The ores were principally sulfides and heating in air roasted the sulfides and formed oxides of the less noble (more reactive) metals while melting gold or silver. The oxides were absorbed into the cupel while a droplet of gold or silver remained on its surface.

To make cupels, calf or sheep bones are calcined (heated in open air), crushed, and ground to the texture of flour and the "ash" is moistened with strong beer. The ash is then placed in cupel molds (see A and C, Figure 9) and coated with facing ashes, best obtained according to Ercker, from the foreheads of calves' skulls. The molded ash is then pounded and shaped (see H, man pounding cupels), removed from the molds (see B and D and the stack of cupels E), and allowed to dry. In Figure 8, G depicts a man washing ashes and F is a ball of washed ashes.

Figure 10 depicts an assayer's balance including: (A) forged balance beam, (B) shackle, (C) half of shackle, (D) filed assay beam with half of shackle, (E)

FIGURE 7. ■ A sixteenth-century assay laboratory (Ercker, see Figure 5).

FIGURE 8. ■ A sixteenth-century machine washing alluvial gold ores (Ercker, see Figure 5). Gold's great density (19.3 g/cm^3 permits its ready separation from other, lighter minerals.

two little beads—upper end of shackle and pointer, (F) ends, (G) how the beam is suspended, (H) sleeves of shackle, (K) knots by which strings are hung, (L) pans of the balance, and (M) assay head forceps.

Figure 11 depicts the amalgamation of gold concentrates and recovery of mercury by distillation of the amalgam. One of the earliest precepts of chemistry is *like dissolves like*, which explains why oil floats on water while alcohol freely mixes with water. Mercury, being a liquid metal, dissolves other pure metals and forms alloys called amalgams. Relatively mild heating of the amalgam frees the volatile mercury from the metal of interest. However, mercury does not dissolve salts (calxes or oxides, sulfides) of metals. Thus, crushed ore was treated by Ercker with vinegar for 2 or 3 days and then washed and rubbed into mercury by hand and then with a wooden pestle by the amalgamator depicted in Figure 11(F). (*Note:* Elemental mercury is very toxic. It caused nerve damage in workers who made hats in England during the 19th century—this was "Mad Hatters' Disease"—the source of the madness of the tea party in Alice in Wonderland. There has been some concern late in the twentieth century that amalgams used to

FIGURE 9. ■ Making cupels from calcined, crushed bones ground into a paste with beer and molded. The oxides of baser metals such as iron are absorbed into the cupel while molten gold or silver remain on its surface (Ercker, see Figure 5).

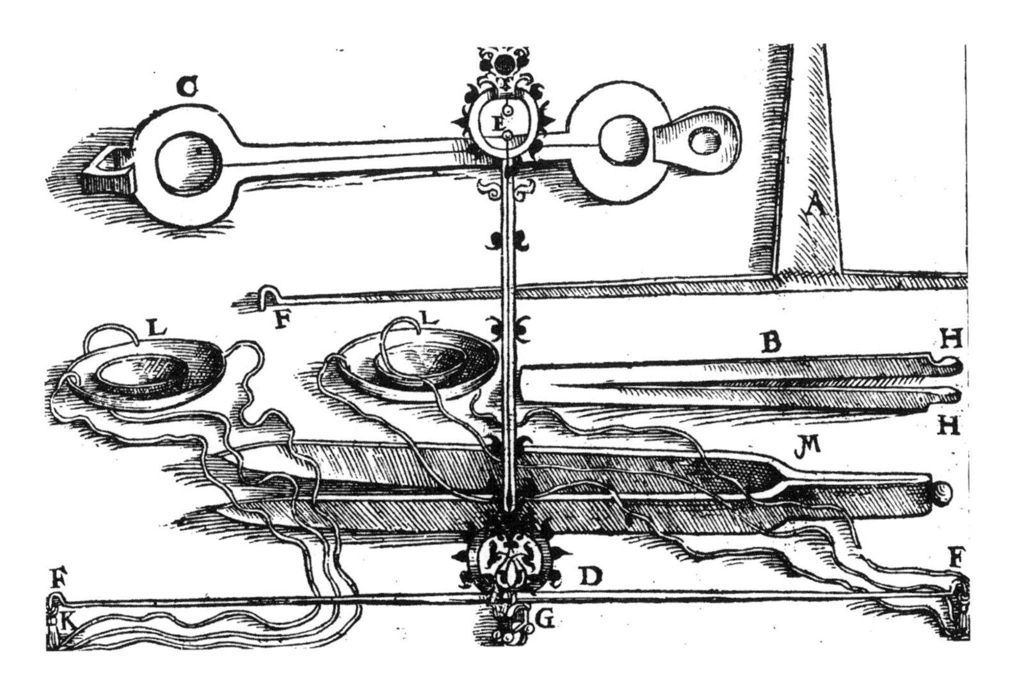

FIGURE 10. ■ A sixteenth-century assayer's balance (see text; Ercker, see Figure 5).

FIGURE 11. ■ Use of mercury to dissolve gold in ore concentrates. The gold amalgam is then heated and mercury distills (Ercker, see Figure 5).

make tooth fillings give off a steady stream of mercury vapor.) The mercury itself was purified by squeezing through a leather bag [see (L) and (G) in Figure 11]. Distillation of mercury from the amalgam employed a large furnace called an athanor (A), which supplied uniform and constant heat, side chambers (B), an earthenware receiver (C) and a still head (D), a blind head through which water can be poured for cooling purposes (E), and an iron pot [lower part (H); upper part (K)] to contain the amalgam to be heated. Also depicted (M) is a man who remelts gold using bellows.

Aqua regia (three parts hydrochloric acid to one part nitric acid) had the valuable property of dissolving gold and allowing its ready recovery (see our later discussion of this subtle chemistry). Figure 12 shows the distillation of *aqua regia* involving the athanor (A) and a chamber (B) for the flask, situated as in (C). (D) is the glass distillation head and (E) the receiver.

FIGURE 12. ▪ Distillation of *aqua regia* (3:1 HCl/HNO_3) (Ercker, see Figure 5). This "kingly water" is capable of dissolving gold. (See essay in Section IX "The Chemistry of Gold is Noble But Not Simple.")

Figure 13 depicts the use of parting acid to separate gold and silver. Parting acid (essentially nitric acid) "dissolves" silver but not gold and is obtained by melting pure saltpetre (potassium nitrate, KNO_3) with vitriol, $FeSO_4$, adding a small amount of water and distilling.

Figure 14 shows a self-stoking furnace for cementation—a process having some similarities to cupellation for purifying gold. The "cement" is made by taking four parts of brick dust, two parts of salt, and one part of white vitriol (zinc sulfate, $ZnSO_4$), grinding the mixed solid, and moistening the powder with urine or sharp wine vinegar. One-finger thickness of the cement is used to cover the bottom of the pot and upon this layer are placed thinly hammered strips of less pure gold, moistened with urine, for further purification. Then follows alternating layers of cement and gold strips finishing with a top layer, one-half-finger thick, of cement. The furnace is applied for 24 hours at a temperature lower than gold's melting point. At the conclusion, the powder is cleaned off and the resulting gold is said to be 23 carat. Pure gold is 24 carat.

Figure 15 depicts the smelting of bismuth in open air with the aid of a very stylized wind. Walnut-sized pieces of ore are placed in pans such that wind-blown fire will smelt the ore and cause liquid bismuth to flow in the pans.

Although saltpetre was used to make nitric acid (for research?) on a small scale, its largest demand was for its use in manufacturing gunpowder. Figure 16 depicts steps in the leaching and concentration, by boiling, of saltpetre. First, the best "earth" for obtaining saltpetre was said by Ercker to come from old sheep pens (which contain the remains of excrement and rotted building matter). Part (A) depicts the "earth" to be leached and (B) shows pipes containing water to

FIGURE 13. ■ The use of "parting acid" (mostly HNO_3) to separate silver from gold. Silver is soluble and gold is not soluble in this acid (Ercker, see Figure 5).

run into the vats. The vats are continuously drained into gutters (C) that run the leachate into a sump (D). Part (E) depicts a little vat from which the leachate runs into a boiler, and (F) to (L) depict parts of the furnace. The boilers distill off considerable water to make a concentrated "liquor."

Figure 17 shows pans (F) and tubs (G) for crystallizing concentrated leachate. One hundred pounds of this concentrate yield about 70 pounds of crystalline saltpetre upon standing.

The development of chemistry rests upon an ancient tripod. One leg of the tripod is formed out of the spiritual, mystical, and conceptual roots of chemistry, which began with the two contraries and four elements, evolved into the *tria prima* (mercury, sulfur, salt) out of which arose Becher's *terra pinguis* or "fatty earth," which, in turn, became Stahl's phlogiston. A second leg is comprised of the practical iatrochemical experience, techniques and apparatus derived from extractions and distillations using animal parts and plants that provided medications. This animal and plant chemistry eventually became our modern organic chemistry and biochemistry. The third leg is the metallurgical chemistry derived from mining and the ancient metallic arts. Aside from techniques learned and appara-

FIGURE 14. ■ A sixteenth-century self-stoking cementation furnace (see text; Ercker, see Figure 5).

tus developed, it was this chemistry—the chemistry of metals and minerals that first truly connected experiment and chemical theory. It ultimately evolved into inorganic chemistry.

In his 1671 book *Metallographia*,[2] John Webster uses biblical quotation to trace metallurgical chemistry back to Moses who, in turn, references Tubal-Cain (*Genesis* 4:22), "the eighth of Mankinde from Adam,"[3] a biblical worker of iron and brass. Here is how Tubal-Cain *might* have discovered metallurgical chemistry:[4]

> While through a Forest Tubal with his yew
> And ready Quiver did a Bore pursue,
> A burning Mountain from his fiery vain,
> An Iron River rolls along the Plain.
> The witty Huntsman musing, thither hies,
> And of the wonder deeply can devise.

FIGURE 15. ■ Smelting of bismuth ore in open wind; freshly formed molten bismuth flows into the pans (Ercker, see Figure 5).

> And first perceiving that this scalding mettle
> Becoming cold, in any shape would settle,
> And grow so hard, that with his sharpened side,
> The firmest substance it would soon divide.

Georg Bauer (1494–1555), Latinized as Georgius Agricola (the German *bauer* means "farmer"), studied medicine, probably obtained the M.D. degree in Italy and in 1526 returned to Germany, where he settled in a mining district in Bohemia.[5] Agricola served as a physician to the miners and developed an interest in mining and metallurgical chemistry. Although he wrote a Latin grammar, religious works and a medical work on the plague, his truly lasting works concern metallurgy. Figure 18 shows the title page of Agricola's first book on metallurgy, *Georgii Agricolae Medici Bermannvs, Sive De Re Metallica*, published in Basel in

FIGURE 16. ■ Steps in the leaching and concentration (by boiling) of saltpetre obtained from old sheep dung (Ercker, see Figure 5).

1530 by Froben.[6,7] The "Bermannus" was the first book on the science of mineralogy published in Europe and is of great rarity.[8] The first truly comprehensive book on metallurgical chemistry was the *De La Pirotechnia* of Vannuccio Biringuccio (Venice, 1540), and we will return to it soon.

Agricola died a quarter of a century after publication of the 1530 "Bermannus," and during the next year his most famous book, *De Re Metallica Libris XII*,[9] was published in Basel. Although significant sections were adapted from Biringuccio's work as well as other contemporary treatises, Agricola's work summarized a lifetime of experience, observation and learning. His work began with methods of surveying mountains and veins of ore and planning mine shafts. Figure 19, from *De Re Metallica*, illustrates the use of a carefully constructed hemicircle (protractor) and its use for surveying and planning a mine.[9,10] One can only assume that the mining company had progressive managers who encouraged their surveyors to be unencumbered in both their thinking and manner of dress. Figure 20 depicts a horse-driven apparatus for pumping water out of mines. The subterranean chamber was carefully reinforced by timbers to prevent its collapse and the death of the

FIGURE 17. ■ Pans and tubs from crystallizing concentrated leachate for saltpetre (see Figure 16). One hundred pounds of the concentrate yields about 70 pounds of saltpetre (Ercker; see Figure 5).

miners. The hollow plunger had a tightly sealed leather bag at the bottom that pushed air out in the downstroke and drew in drainage water in the upstroke.

Figure 21 shows a laboratory containing stills for the synthesis and purification of "*aqua valens*" or "powerful water."[9,10] Aqua valens was a term used by Agricola for powerful acidic agents, including both *aqua vita* (nitric acid) and *aqua regia* (hydrochloric acid–nitric acid, 3 : 1). (The etymology of "valens" is related to the modern term "valence," which refers to "combining power," for instance, of an atom with one hydrogen atom, two hydrogen atoms, etc.). In Figure 21, a typical distillation of *aqua valens* is represented. It includes an *ampulla* (or cucurbit, *K*) containing a mixture of niter or saltpeter, vitriol, and water along with some alum (aluminum sulfate–potassium sulfate), joining it to an *operculum* (or alembic, *H*). The *operculum* is heated by charcoals (stored in earthenware, *F*) in furnace A, red fumes are observed and liquid nitric acid is collected dropwise. Tiny quantities of silver are typically added to the distilled acid in order to precipitate small quantities of chloride that have co-distilled as a result of sea-salt impurities in the starting materials.

The purified nitric acid is used to "part" gold from silver and other base metals because gold is unreactive under these conditions. First, lead is added and

GEORGII
AGRICOLAE MEDICI
BERMANNVS, SIVE
DE RE METALLICA

Basileæ, in ædibus Frobenianis
Anno M. D. XXX.

FIGURE 18. ■ Title page from Agricola's first book on metallurgy—the 1530 *Bermannus*. The extensive mining and mineralogy book collection of President Herbert Hoover and Lou Henry Hoover lacked this exceedingly rare book. (From The Roy G. Neville Historical Chemical Library, a collection in the Othmer Library, CHF.)

the impure alloy heated in a cupel until the least reactive metals, gold and silver, form a melt while the "baser" metals have oxidized and merged with the bone cupel. The silver-gold alloy is then mixed with the nitric acid—silver dissolves, while gold sinks to the bottom, is filtered, and is then washed.

To this useful scientific knowledge, we must add Agricola's belief in mine "goblins" whose exhalations were deadly to miners.[5] A book concerning subterranean animals published by Agricola in 1549 includes a description of salamanders that survive fire[5] (perhaps taking the salamander allegory, see Figure 49 later in this book, a bit too seriously). Nevertheless, the value of Agricola's famous book was well stated by Webster—never a shy critic, in 1671, over a century after *De Re Metallica* was published:[11]

> As for the beating, grinding, sifting, and washing of Ores in general, from their earthy filthiness and superfluitities, Georgius Agricola hath written very largely and learnedly, more than any other Author that I know of. And I could with that some person that hath ability and leisure, would translate it into English; for it might be very serviceable to our common Miners, that in that particular have little to direct them, but what they learn from one another.

Webster's wish was finally answered some 241 years later by two people of very considerable ability and very little leisure: Herbert Hoover, the future president

A—STANDING PLUMMET LEVEL. B—TONGUE. C—LEVEL AND TONGUE.

FIGURE 19. ■ Surveying the coordinates of a mine (on a sweltering summer day?) from the 1912 Hoover translation of Agricola's most famous book, the 1556 *De Re Metallica Libris XII*.

FIGURE 20. ■ A horse-driven apparatus for pumping water out of mines (from the 1912 Hoover translation of Agricola's 1556 *De Re Metallica*). Agricola believed in mine goblins whose emanations (carbon monoxide?) were deadly to miners.

A—FURNACE. B—ITS ROUND HOLE. C—AIR-HOLES. D—MOUTH OF THE FURNACE. E—DRAUGHT OPENING UNDER IT. F—EARTHENWARE CRUCIBLE. G—AMPULLA. H—OPERCULUM. I—ITS SPOUT. K—OTHER AMPULLA. L—BASKET IN WHICH THIS IS USUALLY PLACED LEST IT SHOULD BE BROKEN.

FIGURE 21. ■ Distillation of *aqua valens* ("powerful water"), a general term used by Agricola to describe powerful acids such as *aqua fortis* (nitric acid) and *aqua regia* (hydrochloric acid/nitric acid, 3 : 1), depicted in the 1912 Hoover translation of Agricola's 1556 *De Re Metallica*.

of the United States of America, and his wife Lou Henry Hoover, the first female geology graduate of Stanford University (see the next essay).

Vannucio Biringuccio (1480–1539) is much less well known than Agricola.[5] However, his *Pirotechnia* (Venice, 1540)[12] was the first comprehensive book on mining and metallurgy and was sumptuously illustrated.[5] Amazingly, its first English translation appeared over four centuries later in 1942![13] Biringuccio was very much involved in the political affairs of his day, had military knowledge and skill and was Director of the Pope's (!) Arsenal[5] (much as Lavoisier over two centuries later would direct the Arsenal of Louis XVI). He did not believe in transmutation and was one of the earliest to note the increase in weight of lead upon its calcination:[14]

> The calcination of lead in a reverberatory furnace seems to me to be such a fine and important thing that I cannot pass by it in silence. For it is found in effect that the body of the metal increases in weight to 8 or perhaps 10 per

hundred more than it was before it was calcined. This is a remarkable thing when we consider that the nature of fire is to consume everything with a diminution of substance, and for this reason the quantity of weight ought to decrease, yet actually it is found to increase.

We know now that oxidation of lead to form lead oxide (PbO) should involve a weight increase relative to the metal of 7.7%.

Figures 22 and 23 are from the 1540 *Pirotechnia*. They depict five different types of large cupeling furnaces. Typically, we think of cupels as small molded cups made, for example, from calcined crushed bone ground into a paste with beer, molded, dried, and baked.[14] Crude silver ore may be heated to high temperature in these cupels. More reactive metals oxidize and their calxes are physically absorbed into the cupels leaving molten silver to be cooled and form the purified solid. The huge cupels in Figures 22 and 23 were made from wood ashes, crushed brick, limestone, and egg white and were employed to purify large quantities of silver.[15] The figure at the left (verso page) of Figure 22 shows a worker forming the hearth of a large cupeling furnace.[15] The upper and lower figures to the right (recto page) of Figure 22 are large cupeling furnaces with a brick dome and an iron hood, respectively. The upper figure on the verso page in Figure 23 is a cupeling hearth covered with clay plates, and the lower figure depicts a cover of wooden logs over a cupeling hearth.[15]

Agricola's *De Re Metallica* and Biringuccio's *Pirotechnia* are both recognized as "Heralds of Science—two hundred epochal books and pamphlets in the Dibner Library, Smithsonian Institution."[16] One additional mining and metallurgy book has also been included on this rarified list—*Beschreibung: Allerfürnemisten Mineralischen Ertzt Unnd Berckwercks Arten* ("Treatise Describing the Foremost Kinds of Metallic Ores and Minerals," 1574, Prague) by Lazarus Ercker described earlier in this section (Figures 7–17). Figure 24 is from the title page of the second (1580) edition of this beautiful folio book.[1,17] It depicts a full array of operations in a sixteenth-century mineral assayer's laboratory. The sumptuous book was published in eight Frankfort editions beginning in 1574 with the final one appearing in 1736.[1] The wonderful wood blocks used to print the illustrations in 1574 were preserved and used through 162 years in all eight editions.[1] A Dutch edition was published in 1745.

It is interesting that the first two of the three great mining, assaying, and metallurgy books of the sixteenth century were finally translated into English roughly four centuries after their original publication: Biringuccio's *Pirotechnia* (Venice, 1540; Chicago, 1942); Agricola's De Re Metallica (Basel, 1556; London, 1912). However, Ercker's *Beschreibung* (Frankfurt, 1574) was translated by Sir John Pettus[18] about one century (London, 1683) after the Frankfurt original. Anneliese Grünhaldt Sisco and Cyril Stanley Smith conjecture that the reason for the earlier English translation of Ercker's book in the seventeenth century might have been that it was the most recent, and thus current, of the three great texts. Pettus (1613–1690) played a significant military role in England's Civil War and, at one point, was held captive by Oliver Cromwell for 14 months. Following the restoration, Pettus served as Restoration Deputy to the Vice Admiral and was seriously wounded in the leg during a naval battle with the Dutch.[18]

An interesting aspect of Pettus' translation of Ercker's book (*Fleta Minor, or, the Laws of Art and Nature, In Knowing, Judging, Assaying, Fining, Refining and In-*

LIBRO TERZO

di sopra v'ho detto gia nella Alemagna viddi affinare a vn fornello che haueua in scambio di capello vna volta murata. & atorno vi haueuano ghettando a lauorare a [illegible] maestri. & questo talcenaraccio haueua tre gran mantici con canne & doppie canne lunghe & grosse, & alla bocchetta dell'uscita del vento ognuna haueua di ferro vna uentula, quale s'apriua quando ueniua il vento, & quando non cascando si [illegible]. & queste ventole [illegible] che potes copprende [illegible] il corpo di dentro de mantici che nel turare alle non v'entrassero carboni [illegible] che li bruciassero. & ancho perche tali impedimenti alle bocche [illegible] batter il lor vento piu nel mezzo del bagno. & di piu erano anchora di modo adattati, che mandar si poteuano in qua & in la, & far che'l vento arriuasse, doue piu li pareua a proposito.

ERA fatto di muro sotto doue posauano li mantici, & doue entrauano le canne era uno aperto a modo d'una finestra alto un braccio a circha, larga uno & mezzo, & a ognuna [illegible] era congegnato in due anelli di ferro un tuzzolo grande, sopra alquale si mettrua la'ponta d'un mezzo traue dabete o daltro legno grosso, longo un quatro o cinque braccia, & spingendolo quanto era largo il diametro del ceneraccio, facilmente il mandauano dentro, & queste erano le legna che adoperauano, che ueramente mi parse cosa bella, & considerando anchora conobbi che tal uia non potrua seruire bene, se non all'opere grandi & continuate come in que lochi si faceuano, la doue ogni settimana due uolte o almeno una non era che non se adoperasse, & che non ri ducessero a fino. 150. & 200. marche d'argento per uolta, & cosi si lauaua in affinare a gli edificii dell'Imperatore in Spruch.

DEL FAR LI CENERA

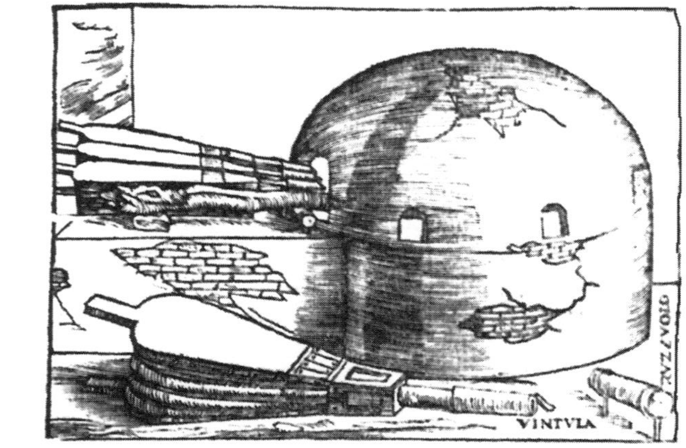

QVELL'ALTRO modo che s'adopera per coprire il ceneraccio, il cappel di ferro mi piace assai piu. Perche molto piu si puo ristregnere il fuoco & tenere il bagno caldo, & con esso si puo affinare il poco, & l'assai come al maestro piace.

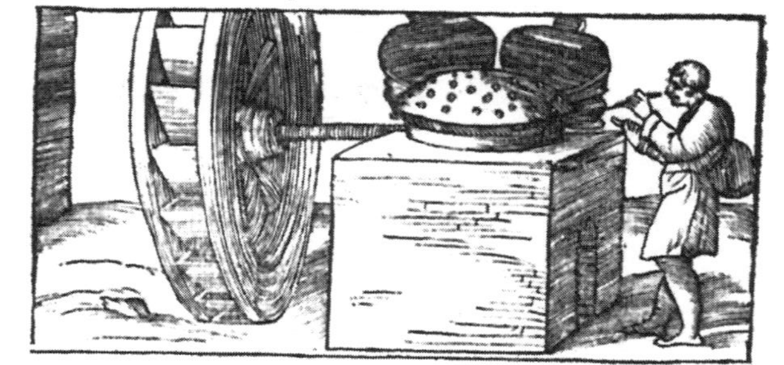

ET COME u'ho detto auanti si copreno, anchora quando s'affinano li ceneracci con certe piastre di terra cotta grosse tre dita, & larghe mezzo braccio, & longhe quanto il ceneraccio: & queste mi piacciono molto piu che alcuni de gli altri modi ch'io habbi ueduto adoperare, perche s'accostano meglio per tenerla calda secondo che la ua mancando.

H

FIGURE 22. ■ Large cupeling furnaces depicted in Vannucio Biringuccio's 1540 *Pirotechnia.* Cupels are cups made typically from crushed bone ground into a paste with beer, then molded, dried, and baked. Calxes of base metals (e.g., iron oxides) are absorbed into the cupel while molten gold is not and is thus readily separable. (From The Roy G. Neville Historical Chemical Library, a collection in the Othmer Library, CHF.)

IL SIMILE ſi fa anchora con li ceppi di quercia, ma non coſi bene, ne con tanta facilita.

ET PERCHE molte ſon le conſiderationi & l'auertentie che a condur perfetta l'opera biſogna hauere, & chi non ha uedute p eſperiẽtia, o che prima molto bene nõ ne ſia ſtato auertito, difficilmẽte ſi guarda dalli inconuenienti. PERO ſappiate ſe in q̃llo argento o piombo che affinate, ſara ſtagno, durarete gran fatiga a condurlo, & la via (quando queſto interueniſſe a purgarlo) e q̃ſta, che ſe gli ſtrenga il fuoco adoſſo, & ſcaldi bene il bagno, & come ſi uede che ſia ben caldo, vi ſi gitta ſopra della carbonige trita, & coſi ſoffiando con li mantici ſi fa il bagno ben gonfiare, & dipoi con vn caſtagniolo gentilmente ſcoprendolo ſe gli va leuando da doſſo la carbonige, cõ laquale tirãdola fuore ne vien con ſeco anchor lo ſtagno, ilquale prima tutto creſpo ſi ſta nel bagno, & non ſi diſtende in quella ſottigliezza che fa il piombo. ET ANCHO ſe aueniſſe che'l ceneraccio p troppa caldezza faceſſe li bollori

habbiate a mente di far allargare li ceppi, ouer fermare li mantici tanto, che ſi temperi. ET ANCHO ſe aueniſſe che'l bagno fuſſe molto ramigno come ſon le ritratte delle minere, o di ghette, o di loppe, auertite nel principio a ſopraſedere il gettare per fino à tanto che'l ceneraccio pigli certo neruo di ghetta, pche le materie ramigne gli fa teneri, per ilche ſono al ghettar pericoloſi, & pero auertirete di far ch'el taglio nel ceneraccio ſia ſottile & un poco appendino, & battete ſpeſſo la ponta del voſtro ferro accio non s'ingroſſi. APPRESSO di voi habbiate ſempre vn caſtagniuolo o due, & coſi ancho di quelli che nella ponta habbino legata con vn poco di fil di ferro vna pezzeta di panno bagnato per poter dare in ſul taglio & fermare quando vedeſſe che del bagno s'auiaſſe per volere vſcire fuore piu ghetta che quella che vorreſte, ouero per bagnare alle volte qualche luoco per li ceneracci fatti teneri dal piombo, ouer per inhumidire doue voleſte tagliare che fuſſe duro per farlo piu facile, Ricordateui anchora di fare il ceneraccio ſimile alle materie, cioe ſe le ſon dolci, dolce, & ſe le ſon dure, duro, & a ogni ceneraccio che farete, ricordateui di fregare ſpeſſo la verga alli ceppi, & di far caſcare di quella carbonigia acceſa ſopra il bagno, & maſſime quando non fuſſe alle ſponde ghetta che ſubito ve la vedrete apparire, & coſi ſe va ſeguitando tanto che l'arriuate al termine di fino quanto il ceneraccio per il ſuo ordinario puo.

MA VOLENDOLO anchora vn poco piu ſforzare, apparecchiate quando ſete a l'ultimo vn ceppo o due che nõ ſien ſtati in fuoco, & ſien ben ſecchi, & li mettete ſopra al ceneraccio aponto che coprino bene l'argento, & di nuouo li ridate una quantita di piombo ſecondo che uolete, & fate riuenire l'argento, liquali come gli uedrete inſieme uniti, & uoi con un caſtagnolo ſottile deſtramente gli rimenate & gli unite inſieme, & di poi pian piano menando li mantici sfumando il piombo, laſſarete l'argento ben chiarire, & dipoi fatto queſto, & che uedete che glie finito, leuate li ceppi & cauatene il uoſtro argento & lo fondete & nettate dal ceneraccio come auanti u'ho detto. MI VI RESTA a dire come nel leuar del ceneraccio adoperato, auertiate che non ſi meſcoli di quella cenere di ceppi che ſpeſſo reſta ſopra al ceneraccio con quella che ui meteſte per ſotto ricotta & ben diſpetta a rifare la compoſition del ceneraccio, perche la guaſtarebbe, & ſieui a mente per un de ricordi generale che mai con ferro freddo con carboni che non ſien prima acceſi o con legna, o coſe molli, non tocchare il uoſtro bagño, perche ui creſciarebbe fatiga a condurlo al ſuo fine, & in luoco d'utile ui darebbe forſe danno, et pero in ogni parte uſarete la diligentia et prudentia uoſtra.

H iii

FIGURE 23. ■ Large cupeling hearths from Biringuccio's 1540 *Pirotechnia* (from The Roy G. Neville Historical Chemical Library, a collection in the Othmer Library, CHF).

FIGURE 24. ■ A sixteenth-century mineral assayer's laboratory from the second (1580) edition of Lazarus Ercker's treatise on mining and metallurgy. The woodblock used to print the first edition (1574), and this (second) edition was preserved and employed for over 160 years through the final 1736 edition. (From The Roy G. Neville Historical Chemical Library, a collection in the Othmer Library, CHF.)

larging the Bodies of confin'd Metals), is the replacement of Ercker's sixteenth-century woodcuts with late-seventeenth-century costumed English figures engraved in copper plates. Figure 49 depicts a contemporary English assayer. Figures 25–29 include partial explanations with the plates. Figure 26 describes the making and molding of cupels made from crushed bone paste. Figure 27 is a scene in a gold-assaying laboratory. The cone-shaped vessel on the right is a parting flask, for assaying gold, seated on its stand. The wooden piece hanging to the right of the assayer in the rear of this figure has a slit through which to view the furnace while protecting the eyes. The person in the foreground is testing the density of "auriferous" silver in water. The *aqua fort* referred to in Figure 28 is nitric acid. Figure 29 depicts the smelting of bismuth in open air. It is fun to compare Figures 26, 28 and 29 with their sixteenth-century German counterparts (Figures 9, 12, and 15).

The curious title of Pettus' book, *Fleta Minor*, derives from the final years of his life that were spent in the "Fleta" or Fleet Prison wherein he wrote this work. Pettus informs his readers that "it seems a strange disposition of Providence that a man who had done so much for his King and Country should be suffered through the accusations of an unscrupulous woman, and that woman his own wife, to spend the closing years of an active and useful life a Prisoner in the Fleet."[18]

Lazarus Erskerus

aliàs

Erckern.

BOOK I.

CHAP. I.

Of Silver Oars.

Sculpture I.

Deciphered.

The *Aſſayer* 1. the *Scales* 2. the Caſes for *Weights* 3. *Glaſſes* for Aqua Regis, Aqua Fortis, Aqua Vitrioli, Aqua Argentea or Quickſilver, *&c.* 4.

FIGURE 25. ■ Depiction of an assayer in John Pettus' 1683 translation and extension of Ercker's treatise on metallurgy. Pettus' book title begins "*Fleta Minor*," referring to the Fleet Prison, in which he was an inmate while writing this book. Pettus was imprisoned "through the accusations of an un-scrupulous woman" who was, incidentally, his wife.

Now, how the *Copel-Caſe* and the Copel is to be ordered and performed the following *Sculpture* will ſhew:

Sculpture V.

FIGURE 26. ■ Manufacture of cupels (using a paste made from crushed bone and beer) from Pettus' 1683 *Fleta Minor.* Compare this figure with the corresponding one (Figure 9) in the 1736 edition (i.e., the original 1574 edition), and you will note that costumes have been updated by a century while the apparatus remains unchanged.

FIGURE 27. ■ Assaying gold ore in Pettus' 1683 *Fleta Minor*.

Of Gold Oars. 173

Sculpture XXII.

CHAP. XXIX

CHAP. XXIX.

To diſtil Aqua fort. *in* Retorts *with other* Advantages.

DISTILLING *Aqua fort.* in *Retorts* is no old *Invention*, and no long Labour, but a ſhort way; if *Retorts* may be had which are made of one piece, and will hold *Aqua fort.* and *Oyl*; then lute ſuch over with good and ſound *Clay*, let it be well dry, put in it the *Ingredients* or *ſtuff*, which ſhall be *calcin'd* and mingled with *Calx viva*, and lay the *Retort* in an *Oven* made on purpoſe(whoſe Deſcription ſhall follow hereafter) and fill a *Receiver* with water before it, then make a fire in the *Oven* (and ſpeedi-

Section 1.

Y y ly

FIGURE 28. ■ Distillation of aqua fortis (nitric acid) in Pettus' 1683 *Fleta Minor*. Compare this figure with the corresponding one (Figure 12) in the 1736 edition (i.e., the original 1574 edition), and you will note that the costumes have been updated by a century while the apparatus remains unchanged.

Of Lead Oars. 307

CHAP. X.

Sculpture XXXVII.

Deciphered.

1. *The little* Iron Pans *for Spelter or Wismet* Oar.
2. *The* fire *of* wood *for them.*
3. Melted Spelter *that is to be made clean in the* iron Pan, *and the workman that tends it.*
4. He *that draws the* Oar *out of the* Mine.

CHAP.

FIGURE 29. ■ Smelting of bismuth ore in the open wind from Pettus' 1683 *Fleta Minor*. Compare this figure with the corresponding one (Figure 15) in the 1736 edition (i.e., the original 1574 edition), and you will note that costumes have been updated by a century while the apparatus remains unchanged.

---

1. A.G. Sisco and C.S. Smith (transl.), *Lazarus Ercker's Treatise on Ores and Assaying* (translated by Anneliese Grünhaldt Sisco and Cyril Stanley Smith from the German edition of 1580), The University of Chicago Press, Chicago, 1951. I employed this source for interpretations.
2. J. Webster, *Micrographia: Or, An History of Metals*, Walter Kettilby, London, 1671.
3. J. Read, *Humour and Humanism in Chemistry*, G. Bell and Sons Ltd., London, 1947, pp. 3–4.
4. Webster, op. cit., p. 3.
5. J.R. Partington, *A History of Chemistry*, MacMillan and Co. Ltd., London, 1961, Vol. 2, pp. 32–66.
6. G. Agricola, *Georgii Agricolae Medici Bermannus, sive De Re Metallica*, Frobenianus, Basel, 1530. I thank The Roy G. Neville Historical Chemical Library (California) for supplying the image of the title page for this book.

7. Johann Froben (Johannes Frobenius, ca. 1460–1527) was a famous Basel printer-publisher whose techniques revolutionized printing. Among his gifted illustrators were Hans Holbein, and after 1513 he was the sole publisher of the great Dutch humanist-philosopher Desidarius Erasmus (*The New Encyclopedia Britannica*, Encyclopedia Britannica, Inc., Chicago, 1986, Vol. 5, p. 16.).
8. I thank The Roy G. Neville Historical Chemical Library (California) for supplying this image, and I am grateful to Dr. Neville for helpful discussions.
9. G. Agricola, *De Re Metallica Libri XII, Quibis Officia, Instrumenta, Machinae, Ac Omnia Denique Ad Metallicam Spectantia*, Basel, 1556. I am grateful to Ms. Elizabeth Swan, Chemical Heritage Foundation, for providing these images.
10. H.C. Hoover and L.H. Hoover (transl.), *Georgius Agricola De Re Metallica* (translated from the first Latin edition of 1556), *The Mining Magazine*, London, 1912 (reprinted by Dover Publications, Inc., New York, 1950), see pp. 439–447.
11. Webster, op. cit., p. 155.
12. V. Biringuccio, *De La Pirotechnia. Libri X.*, Venice, 1540. I am grateful to Ms. Elizabeth Swan, Chemical Heritage Foundation, for supplying images from this book.
13. C.S. Smith and M.T. Gnudi, *The Pirotechnia of Vannoccio Biringuccio Translated from the Italian with an Introduction and Notes by Cyril Stanley Smith and Martha Teach Gnudi*, The American Institute of Mining and Metallurgical Engineers, New York, 1942 (see also the 1959 reprint published by Basic Books, New York).
14. Smith, op. cit., p. 58.
15. Smith, op. cit., pp. 161–169.
16. *Heralds of Science*, revised edition, Burndy Library and Smithsonian Institution, Norwalk and Washington, DC, 1980. It has been duly noted that, of the *Great Books of the Western World*, published by Encyclopedia Britannica in 1952, only one work (of a collection of 130 authors and 517 works) is a treatise on chemistry (Lavoisier's *Traité élémentaire de Chimie*, Paris, 1789, first English translation, 1790) [R. Wedin, Chemistry (published by the American Chemical Society), Spring 2001, pp. 17–20]. Wedin surveyed a small, selected list of chemists and librarians to obtain his list of "The Great Books of Chemistry." The six books on "The Gold Shelf" included Lavoisier's Traité, Boyle's *The Sceptical Chymist*, Jane Marcet's *Conversations on Chemistry* (a useful and influential textbook that drew the young Michael Faraday into chemistry), Dalton's *A New System of Chemical Philosophy*, Mendeleev's *Osnovy Khimii*, and Pauling's *The Nature of the Chemical Bond*. "The Silver Shelf" comprised six additional books including Agricola's *De Re Metallica*. "The Bronze Shelf" included 12 books. Of the total of 24 books, thirteen were American publications, and two of these were published by the American Chemical Society itself. Hmmm.
17. L. Ercker, *Beschreibung Allefürnemisten Mineralischen Ertzt vnnd Bergwercks arten . . .* , Johannem Schmidt in verlegung Sigmundt Feyrabends, Frankfurt, 1580. I am grateful to The Roy G. Neville Historical Chemical Library (California) for supplying the image of the title page.
18. Sisco and Smith, op. cit., pp. 340–342.

A PROMISING PRESIDENT

The first English translation of Agricola's 1556 *De Re Metallica*[1] (Figure 30) was published in 1912 by Herbert C. Hoover (1874–1964),[2] the future president of the United States, and his wife, Lou Henry Hoover. It is hard to imagine a more promising future president. Born in rural Iowa to Quaker parents of extremely modest means, Herbert Hoover was orphaned by the age of nine. Although shy, he developed a very early sense of independence, rejected the choice of Quaker colleges suggested to him by his relatives, and chose to attend a brand new college, Stanford University. Hoover majored in geology and met Lou Henry, the

GEORGIUS AGRICOLA

DE RE METALLICA

TRANSLATED FROM THE FIRST LATIN EDITION OF 1556

with

Biographical Introduction, Annotations and Appendices upon the Development of Mining Methods, Metallurgical Processes, Geology, Mineralogy & Mining Law from the earliest times to the 16th Century

BY

HERBERT CLARK HOOVER

A. B. Stanford University, Member American Institute of Mining Engineers, Mining and Metallurgical Society of America, Société des Ingéniéurs Civils de France, American Institute of Civil Engineers, Fellow Royal Geographical Society, etc., etc.

AND

LOU HENRY HOOVER

A. B. Stanford University, Member American Association for the Advancement of Science, The National Geographical Society, Royal Scottish Geographical Society, etc., etc.

Published for the Translators by

THE MINING MAGAZINE

SALISBURY HOUSE, LONDON, E.C.

1912

FIGURE 30. ■ Title page from the first English translation of Agricola's 1556 *De Re Metallica*, written and tested for scientific accuracy by engineer Herbert Hoover, the future President of the United States, and his wife Lou Henry Hoover, the first female geology graduate of Stanford University.

only female geology major at Stanford. They married in 1899 and remained happily united until her death in 1944. She was a smart, independent, and forceful woman, raised as a tomboy and adept at riding horses, and later became a powerful champion of women's suffrage.[2] Not long after graduating from Stanford in 1895, Herbert Hoover began a career in mining engineering and management that soon would make him wealthy and possibly the world's most famous engineer. He spent most of his time overseas during the following decades and was in China during the Boxer Rebellion (1900), where he directed relief for foreigners.[2]

The Hoovers amassed a huge and famous mining book collection and, during the course of writing their translation, performed occasional experiments to test Agricola's veracity.[3] The challenge of the Hoovers' translation is not "merely" mastery of Latin but also a profound understanding of the engineering and chemistry, which allowed them to incorporate hundreds of now-defunct terms and concepts and make sense out of them for the modern reader. Not many years after this intellectual triumph, with the outbreak of World War I Herbert Hoover was appointed head of the Allied relief operation. After the American entry into the war in 1917, he was appointed national food administrator. Hoover's efforts at increasing food production, conserving foodstocks and relieving famine in Europe were so successful that the term "hooverize"[4] entered the vocabulary as an expression symbolizing the acts of being productive, economical, and generous with foodstocks. Even more generally, it became a term for efficiency, effectiveness, and compassion.

How ironic, then, that Herbert Clark Hoover, thirty-first president of the United States (1928–1932), is now principally remembered for his failure to ease the hardships of the Great Depression. Very strict ethical values inculcated in early childhood and reinforced by his own very early independence (and subsequent success) made widespread federal aid, particularly to the urban unemployed, anathema to him.[2] He was widely regarded as distant from the suffering populace.[2] And so, sadly enough, "hooverville,"[4] a shanty town populated by the unemployed poor, is a word both more recent and more widely remembered than "hooverize."

1. H.C. Hoover and L.H. Hoover, *Georgius Agricola De Re Metallic* (translated from the first Latin edition of 1556), *The Mining Magazine*, London, 1912.
2. J.H. Wilson, *Herbert Hoover—Forgotten Progressive*, Little, Brown and Co., Boston, 1975.
3. Wilson, op. cit., pp. 22–23.
4. *Oxford English Dictionary*, second ed., Vol. VII, Clarendon Press, Oxford, 1989, p. 374.

THESE ARE A FEW OF OUR NASTIEST THINGS

It is generally agreed that gunpowder (black powder) was invented in China over a thousand years ago.[1] It is a mixture consisting of about 75% saltpetre (potassium nitrate) with the remaining 25% containing comparable quantities of charcoal and sulfur. Saltpetre was readily obtained from old dung heaps; charcoal readily made by heating vegetables or wood under oxygen-poor conditions; sulfur was found in crystalline deposits and could also be obtained by heating many metal ores. William Brock has speculated, rather ironically, I think, that the Chinese accidentally discovered gunpowder through seeking an elixir of life by a combination of the "Yin-rich saltpetre and Yang-rich sulphur."[1] There is further irony in that gunpowder held secret keys to understanding the origins of fire and the very respiration that supports life. However, these would remain hidden for almost a millennium. Early hints would be provided by Boyle, Hooke, and Mayow in the mid-seventeenth century and the riddle solved by Lavoisier over a century later.

Gunpowder was introduced quite early into Western warfare. Figure 31 is from the first Stainer edition of the ancient book on the technology of warfare authored by Flavius Vegetius Renatus.[2] This excessively rare edition, published in 1529, contains the first printed text on making gunpowder, along with directions for purifying its components.[3] Figure 32 is from a 1598 work on artillery and fireworks by Alessandro Capo Bianco,[4] Captain of Bombadiers at Crema in the

FIGURE 31. ■ Figure from the first Stainer edition (Augsburg, 1529) of the ancient work by Flavius Vegetius Renatus on the technology of warfare (from The Roy G. Neville Historical Chemical Library, a collection in the Othmer Library, CHF).

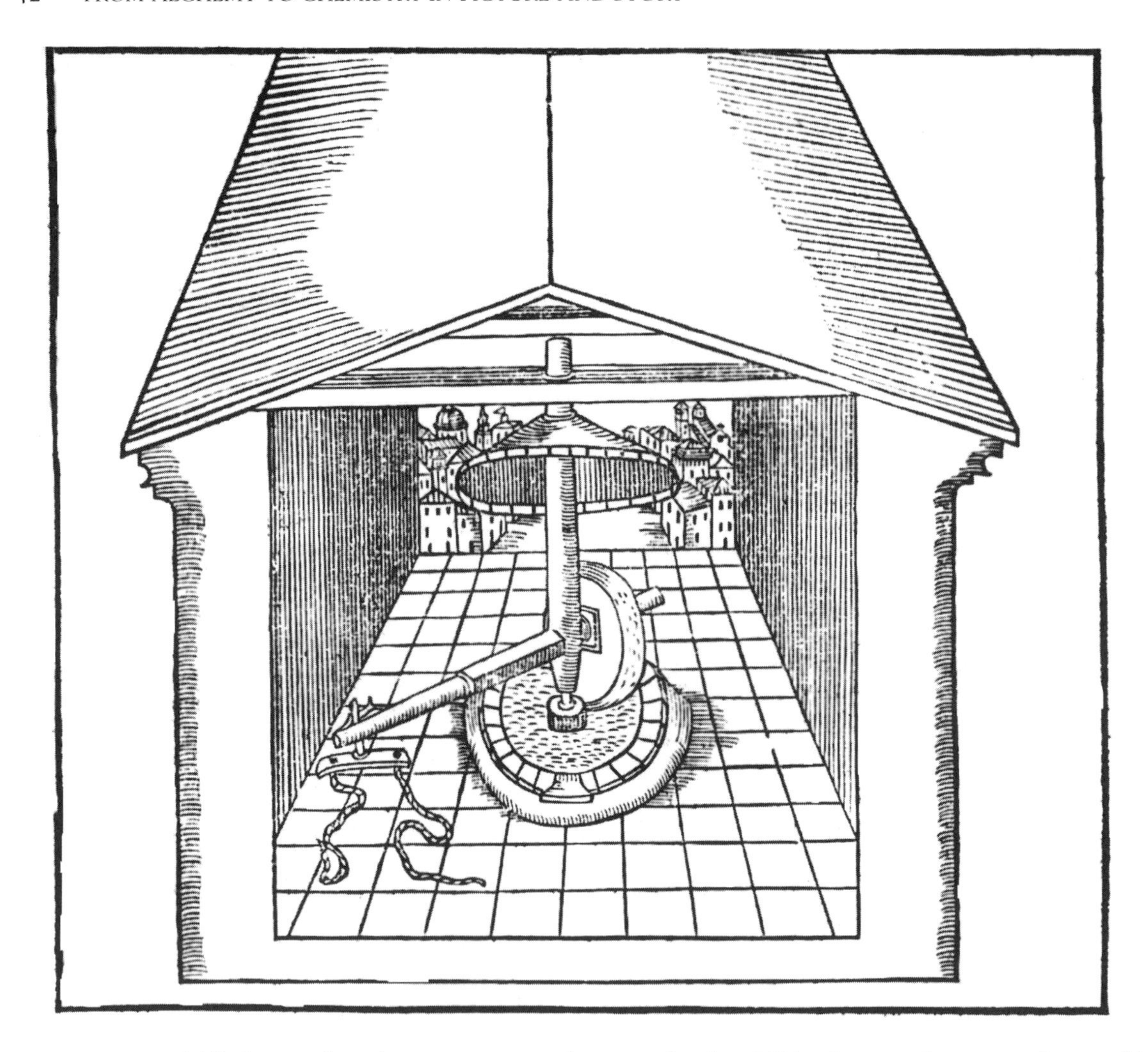

FIGURE 32. ■ Mill for grinding the components of gunpowder (ca. 75% saltpetre; the remainder roughly equal parts of charcoal and sulfur) depicted in Bianco's 1598 work on artillery and fireworks (from The Roy G. Neville Historical Chemical Library, a collection in the Othmer Library, CHF).

Veneto. The figure depicts a sixteenth-century mill for grinding components of gunpowder.[3] In Book Ten of his *Pirotechnia* (1540),[4] Biringuccio provides detailed directions for making gunpowder. Saltpetre is derived from the "manurous" soils of barns and the floors and walls of caves (rich in bat guano), which contain calcium nitrate as a decomposition product. If the "manurous" soil, once dried, is tasted and found to be "sufficiently biting," it is suitable for use.[5] The soil is added to boiling water, and wood ashes (rich in "pearl ash" or potassium carbonate) are stirred in. The hot solution is then filtered and allowed to cool; the resulting crystalline potassium nitrate is filtered and recrystallized once more with water and a bit of nitric acid.[5] Charcoal is made preferably from willow twigs by heating over fire in a large sealed earthen pot. The components of gunpowder must be moistened before being ground together, to avoid ignition, and Biringuccio recommends slow addition of finely ground sulfur to a paste of moistened charcoal and saltpetre.[5]

Biringuccio begins his chapter on gunpowder thus:[5]

> A great and incomparable speculation is whether the discovery of compounding the powder used for guns came to its first inventor from the demons or by chance.

At many points in Book Ten, Biringuccio laments the irony that learned and decent men discover and invent explosives that maim and kill. He then dutifully describes their fabrication in full detail. For example, Book Ten, Chapter Eight is titled "The Method of Preparing Fire Pots and of Making Balls of Incendiary Composition to Be Thrown by Hand." Biringuccio begins this chapter (in the 1559 edition):[6]

> There have always been in this world men of such keen intelligence that with their discourse they have been capable of infinite and various inventions that are as beneficial as they are simultaneously harmful to the human body.

He then describes pots made of dried clay filled with course gunpowder, pitch, and sulfur, and sealed with congealed pig fat mixed with powder (see Figure 33).[7] Prior to use, a small hole is bored into the fatty seal and either a fuse or black powder placed inside. The fuse or powder is lit, the pot tossed or launched with a sling and this penetrating, sticky mass will adhere to and burn its target.

Other early explosives and incendiary weapons included "Greek Fire" dating from the Hellenistic period. Chemical historian John Hudson describes "Greek Fire" as a liquid that caught fire on contact with water and speculates that calcium phosphide (from heating bones, lime and urine together), added to crude petroleum might have constituted the active ingredients.[8] Leonardo Da Vinci (1452–1519) described "Greek Fire" as consisting of charcoal, sulfur, pitch, saltpeter, spirit of wine, frankincense, and camphor boiled together and applied over Ethiopian wool.[9]

FIGURE 33. ■ Nasty munitions made from clay pots filled with coarse gunpowder, pitch, and sulfur, and sealed with congealed pig fat mixed with powder depicted in Biringuccio's *Pirotechnia*. These are manufactured to be lighted and launched with malice using a sling. (From The Roy G. Neville Historical Chemical Library, a collection in the Othmer Library, CHF.)

Fulminating gold (*aurum fulminans*) was first described at the beginning of the seventeenth century.[10] Gold was dissolved in an *aqua regia* derived from ammonium chloride and nitric acid. Addition of potassium carbonate led to a precipitate that, when dry, exploded readily with only the mildest application of heat. Johann Rudolph Glauber first described fulminating powder (*pulvis fulminans*) in 1648.[10] It is a mixture of potassium nitrate, potassium carbonate, and sulfur that violently explodes upon mild heating. Tenney Davis has described various similar mixtures discovered over the course of two centuries.[10] In the late seventeenth century Johann Kunckel made mercury fulminate by dissolving mercury in *aqua fortis* (nitric acid), adding spirit of wine and gently warming the mixture in horse dung.[11] The next day, the concoction exploded violently.

The nineteenth century would witness the development of nitrostarch, nitrocotton, nitroglycerin, trinitrotoluene (TNT), and pentaerythritol tetranitrate (PETN) and end with the discovery of RDX (cyclotrimethylenetrinitramine).[12] Contemporary studies of synthetic azides and picrates would also add to the armamentarium of war technology. From a modern perspective these developments appear to have grimly foreshadowed World War I. In 1867, Alfred Nobel (1833–1896), a Swedish chemist and industrialist, immobilized nitroglycerin onto diatomaceous earth, making it much safer to use, and thus made the first of many successful formulations of dynamite.[12] Just as hope often accompanies tragedy, Nobel willed most of his vast fortune to establish a series of Nobel Prizes—one of which is a prize to further the cause of world peace.

---

1. W.H. Brock, *The Norton History of Chemistry*, W.W. Norton & Co., New York, 1993, p. 6. Brock notes that in Taoism, "Yang" is the male, hot principle, "Yin" is the female, cool principle. In Western alchemical beliefs, sulfur is the male principle (Sol) and mercury the female principle (Luna).
2. F. Vegetius Renatus, *Vier Bũcher der Ritterschaft . . . Mit einem zũsatz von Bũchsen geschoss, Pulver, Fewrwerck, Auff ain newes gemeeret unnd gebessert, Gedruckt durch Heinrich Stainer*, Augsburg, 1529.The author is grateful to The Roy G. Neville Historical Chemical Library for supplying a copy of the woodcut in Figure 31.
3. The Roy G. Neville Historical Chemical Library; catalogue in preparation. I am grateful to Dr. Neville for helpful discussions.
4. Alessandro Capo Bianco, *Corona e Palma Militare di Arteglieria. Nella quale si tratta dell' Inventione di essa, e dell' operare nella fattioni da Terra, e Mare, fuochi artificiati da Giucco, e Guerra; & d'un Nuovo Instrumento per misurare di stanze. Con una giunta della fortificatione Moderna, e delli errori scoperti nelle fortezze antiche, tutto a proposito per detto essercitio dell' Artiglieria, con dissegni apparenti, & assai intendenti. Nova composta, e data in luce. Dallo strenuo Capitano Alessandro Capo Bianco . . .* Appresso Gio. Antonio Rampazetto, Venice, 1598.
5. C.S. Smith and M.T. Gnudi, *The Pirotechnia of Vannoccio Biringuccio* (English transl.), The American Institute of Mining and Metallurgical Engineers, New York, 1942, pp. 409–416. This is the first English translation of Biringuccio's *De La Pirotechnia* published in Venice in 1540.
6. Smith and Gnudi, op. cit., pp. 434–435.
7. The author thanks The Roy G. Neville Historical Chemical Library (California) for supplying this image from the 1540 edition of *De La Pirotechnia*.
8. J. Hudson, *The History of Chemistry*, The MacMillan Press Ltd, Hampshire and London, 1992, p. 22.
9. J.R. Partington, *A History of Chemistry*, MacMillan and Co. Ltd., London, 1961, Vol. 2, p. 6.
10. T.L. Davis, in *Chymia*, T.L. Davis (ed.), Vol. 2, University of Pennsylvania Press, Philadelphia, 1949, pp. 99–110.

11. Partington, op. cit., p. 377.
12. D.M. Considine (ed.), *Van Nostrand's Scientific Encyclopedia*, seventh edition, Van Nostrand Reinhold, New York, 1989, pp. 1104–1105.

"THE SUN RAINS GOLD; THE MOON RAINS SILVER"

In 1532, Spanish Commander Francisco Pizarro and a force of 168 soldiers compelled the surrender of the Nation of Incas that ruled a population of 12,000,000 people stretching from present-day Quito, Ecuador, to the south of Santiago, Chile. The conventional explanation for this has rested on the technical superiority of the Spanish and other European peoples, although it is admitted that the Europeans imported diseases such as smallpox, for which the aboriginal peoples had no resistance, which decimated their populations. In his compelling 1994 book, *Guns, Germs and Steel,*[1] the biologist and geographer Jared Diamond provides insights into the advantages enjoyed by Europeans: native horses rather than llamas too small to ride into combat, steel swords against bronze technology, and guns versus arrows.

The Incas, one of a number of ancient Andean native cultures, established their capitol in Cuzco in what is now southern Peru in the twelfth century. In the early fifteenth century, they began the series of conquests that would gain them the empire that Pizarro encountered and ruthlessly destroyed. It appears, however, that the technological advantages of the Europeans over the Incas may well be overstated.[2] Rifles of the sixteenth century were barely effective: they required about 2 to 3 minutes per shot and were notoriously inaccurate. The original name for these weapons, *donderbus* (German for "thunder box"), was soon changed to the more realistic "blunderbuss." In contrast, arrows could be launched at 10 per minute and were accurate at 200 yards.[2]

The rivers in the Andes were rich sources of gold and the mines rich sources of silver. It is fascinating that, much like ancient Asian and Arabic cultures, Andean cultures equated gold with the Sun and maleness, and silver with the Moon and femaleness. "The Sun rains gold; the Moon rains silver" is an ancient Andean invocation. Aboriginal cultures in the New World, such as those in the Andes, relied upon oral rather than written historical records and there are no manuscripts describing formulas and procedures. Only detailed investigations of the artifacts themselves provide clues about the level of technology. Recent evidence indicates that Andean metallurgy dates back at least 3,000 years. In 1998, researchers at Yale University discovered samples of gold foil that had been both hammered into thin sheets and heat treated (annealed) so as to gild metal objects, including those made of copper.[3,4] However, despite an abundant supply of iron in the mountains, Andean cultures did not fabricate steel. Heather Lechtman, at MIT, observes that Europeans optimized "hardness, strength, toughness, and sharpness" in metals.[5–7] In contrast, Andean cultures, including the Incas, valued "plasticity, malleability, and toughness." Rather than employing metals as weapons and machine parts, they fashioned objects that were employed to communicate social standing, wealth, and religious authority.[2,5–7] Most of the Andean metallurgical operations focused on gold, silver, and copper, although bronze and lead slingshot weapons were known. Early Andean metallurgists plated very thin (0.5–2 micron) and uniform layers of gold on copper metal and alloys despite their lack of aqua regia as well as electroplating technology[7] (see page 575).

Professor Lechtman demonstrated some of the sophistication of Andean metallurgy and engineering in her analysis of architectural cramps (metal bars bent at the ends to hold together stone blocks) from the Middle Horizon (ca. AD 600–1000).[8] The I-shaped cramps held together large stone blocks that formed the sides of canals for conducting collected rain water, serving the ancient city of Tiwanaku, presently a village in Bolivia, near the southern shore of Lake Titicaca. The design of the canal was quite sophisticated. Although the canal sloped 12°, the spaces for the cramps in adjoining blocks were cut horizontally so as to tightly lock the blocks into place. The canal is water-tight despite the fact that the blocks are not cemented together. In some cases, spaces have been cut into adjoining blocks that act as cramp molds to receive freshly molten bronze. Modern chemical analyses (neutron activation and ion-coupled emission analyses) have determined that these cramps are ternary bronze alloys of copper, arsenic,

FIGURE 34. ■ Thomas de Bry's 1596 engraving depicting Incan gold workers (from the first German edition of 1597).[8]

and nickel. The alloy was developed to provide a residual tensile stress when solidified, which held the blocks tightly together.

Although the Incas did not add much to the existing Andean metallurgical technology,[5,6] they were skilled at the craft. Figure 34 is from the 1597 first German edition of the *Americae pars sexta . . .*,[9] first published in Latin in 1596 by the renowned engraver Thomas de Bry (1528-1598). It depicts sixteenth century Incan gold workers heat-treating metals and hammering them into sheets. In contrast to the metal sculptures of Europeans, which were obtained using molds, the Incans fashioned art objects such as the statue in the lower right of Figure 34 from thin plates. Dr. Lechtman has furnished a beautiful example in which a miniature head only 1.3 cm high and 0.012–0.02 cm thick is made from 19 individual gold plates that have been hammered and then skillfully soldered or welded together.[5]

---

1. J. Diamond, *Guns, Germs, and Steel*, W.W. Norton & Co., New York, 1997.
2. C.C. Mann, "Native Ingenuity," *Boston Sunday Globe*, September 4, 2005, pp. E2–E3.
3. J. Quilter, *Science*, Vol. 282, pp. 1058–1059, 1998.
4. R.L. Burger and R.B. Gordon, *Science*, Vol. 282, pp. 1108–1111, 1998.
5. H. Lechtman, "Traditions and Styles in Central Andean Metalworking," in *The Beginning of the Use of Metals and Alloys, Papers from the Second International Conference on the Beginning of the Use of Metals and Alloys*, Zhengchou, China, 21–26 October, 1986, R. Madden, ed., The MIT Press, Cambridge, MA, 1988, pp. 344–378. (I am grateful to Professor Lechtman for correspondence on this topic and for supplying reprints of her work.)
6. H. Lechtman, "The Andean World," in *Andean Art at Dumbarton Oaks*, E.H. Boone, ed., 1996, pp. 11–43.
7. H. Lechtman, "Pre-Columbian Surface Metallurgy," *Scientific American*, Vol. 250, No. 6 (June 1984), pp. 56–63. The Author thanks Professor Roald Hoffmann for awareness of this work.
8. H. Lechtman, "Architectural Cramps at Tiwanaku: Copper–Arsenic–Nickel Bronze," in *Metallurgica Antiqual in Honour of Hans-Gert Bachmann and Robert Maddin*, T. Rehren, A. Hauptmann, and J.D. Muhly (eds.), Deutches Bergbau-Museum Bochum, 1998, pp. 77–92.
9. T. de Bry, *Americae pars sexta. Sive, Historiae ab Hieronymo Be[n]zono scriptae, section tertia . . . In hac . . . reperies qua ratione Hisponi . . . Peruäni regni provincias occuparint, capto rege Atabaliba . . . Additus est . . . de Fortunatis insulis co[m]mentariolis . . . Accessit Pervani regni chorographica tabula. Omnia figures in aes incises exprassa a Theodoro de Bry*, Frankfort, 1596. The figure shown is from the first German edition of 1597.

CATAWBA INDIAN POTTERY: FOUR COLORS AND A MIRACLE OF SURVIVAL

The *Ninth Key* of Basil Valentine [Figure 40(b)] describes four colors of trans-mutation in the Great Work: black, white, citrine (a yellow) and ultimately red, symbolized by the crow, swan, peacock, and phoenix, respectively. It is interesting that these are the four characteristic colors of earthenware fabricated for thousands of years by aboriginal peoples in diverse lands.

The Catawba Indians located in South Carolina spoke a Siouan language.[1] They were a once powerful nation that alternately coexisted and fought with the Cherokees in the Carolinas. However, as of June 1908, only nineteen houses and ninety-eight Catawbas were counted on the reservation and in its surroundings in York County.[2] Although pre-Columbian Catawba pottery was largely utilitarian (cooking pots, water jugs), starting in the eighteenth century it became a source of hard currency for the Indians. They began to fashion objets d'art in addition to traditional pieces. These were often taken to the port city of Charleston, South Carolina, traded, and sold. The very survival of Catawba culture came to depend to a significant extent on the sale and trade of pottery, largely fabricated by women.[3–5] This is elegantly stated by former Catawba Tribal Historian Tom Blumer:[3]

> [T]he Catawba pottery tradition has survived for over 4,500 years. That it has done so is a tribute to the tenacity of the people who make up the Catawba Nation and the power of pottery, as an art form, to define that Nation and help it endure. It is a miracle of survival that will take the Catawba to the Third Millenium and beyond.

Figure 35 (left) shows a mostly reddish-brown headed bowl with three running legs made by Master Potter Sara Ayers (1919–2002).[3,5] The legs are off center, and the broken symmetry provides a wonderful dynamic to the piece. Also shown in Figure 35 (right) is a two-headed, fluted bowl made by young master Monty ("Hawk") Branham (b. 1961). The heads were ultimately derived from a mold made over 100 years ago by the great Martha Jane Harris. These pieces are made almost the same way they were in prehistoric times. Clay is dug

FIGURE 35. ■ Two pieces of Catawba Indian pottery: left, two-headed fluted bowl with three "running" legs by master Potter Sara Ayers; right, two-headed, fluted bowl by young master Monty ("Hawk") Branham. Catawba pottery is still made essentially as it was 4500 years ago. [Photograph by Thomas W. ("Wade") Bruton.]

from holy and secret sites along the Catawba River which contain rich deposits of kaolinite, sifted, mixed, and dried in the sun and rolled and pounded to remove air pockets. Clean kaolinite is fluffy and white. Pipe clay has organic matter and is heavier. The two are usually mixed to make a pot. Larger pots are built using layers of coils of clay that are shaped and smoothed, then allowed to dry. A pot may be incised with symbols just before being totally dry and, when dry, it is laboriously burnished with smooth river stones that have usually been passed between generations of women. The pots are then wood-fired in pits in the ground, removed, and allowed to slowly cool. Open-pit firing is considered to be low temperature (1200°C or 2200°F) or soft firing as opposed to hard firing (1450°C or 2650°F).[6] Air pockets in the clay and even slight wind gusts often cause a high degree of breakage. The high shine in the finished product is due to hours of burnishing rather than to glaze, which is never used. Clay pots, which are unglazed, are not considered suitable for holding water since they "sweat" and will stain furniture. However, one can imagine taking a water jug into the field—its sweating and vaporization from the surface will cool the bulk of the water inside. Furthermore, the frequent heating and decomposition of fat as well as protein from sinew and meat will coat the inside of a cooking pot and seal it.

The colors in this pottery are largely due to the iron so abundant in all clays.[6] Iron is the fourth most abundant element in the earth's crust. It is largely found in the iron(II) (ferrous) or iron(III) (ferric) oxidation states. Iron(II)oxide (FeO), iron(III)oxide (Fe_2O_3, hematite), and ferroferric oxide (Fe_3O_4), which contains both Fe(II) and Fe(III), are the three oxides of iron commonly encountered. The mottled coloring of the pot depends upon the degree of oxidation and also reflects the smoke and soot of the wood employed in firing since different woods burn at different temperatures and oxygen levels.[7] One of my former professors at Princeton University, Tom Spiro, called the color changes associated with "tweaking" the environments of transition metals, such as iron, "tickling electrons." Under oxygen-rich conditions, the dominant colors are "white" (really buff), and yellow and red and are due to a greater abundance of Fe(III). Oxygen-poor conditions can be achieved by "smother-burning" pots by surrounding and covering them with wood. The presence of carbon monoxide (CO) causes more reducing conditions conducive to enrichment in Fe(II). This is the way to deliberately produce a shiny, black pot; otherwise, coloring is left largely to the fates. Traces of manganese also help to blacken pots as will soot.[7] When removed from the fire, the pieces are usually dark and then lighten as they cool. Dynamic chemistry is occurring, for example, disproportionation of FeO to Fe_3O_4 and Fe although Fe will further oxidize.[8] Sometimes a greasy-looking area can be seen on the surfaces of the pots. This is probably due to local vitrification perhaps by a local concentration of feldspar or mica.[6]

1. J.H. Merrell, *The Indians' New World,* The University of North Carolina Press, Chapel Hill, 1989.
2. M.R. Harrington, *American Anthropologist,* 10:399–407, 1989.
3. T. Blumer in Pamphlet *Catawba Pottery: Legacy of Survival, 7 Master Potters,* South Carolina Arts Commission and Catawba Cultural Preservation Project, Columbia, 1995.

4. T. Blumer, *The Catawba Indian Nation of the Carolinas*, Arcadia Publishing, Charleston, 2004.
5. T. J. Blumer, *Catawba Indian Pottery*, University of Alabama Press, Tuscaloosa, 2004.
6. *Encyclopedia Brittanica*, 15th ed., Chicago, 1986, Vol. 17, pp. 101–103.
7. I am grateful for discussions with Professor Victor A. Greenhut.
8. F.A. Cotton and G. Wilkinson, *Advanced Inorganic Chemistry*, 5th ed., Wiley, New York, 1988, pp. 711–713.

SECTION II
SPIRITUAL AND ALLEGORICAL ALCHEMY AND CHEMISTRY

EASTERN AND WESTERN SPIRITUAL ALCHEMY

How do we make sense of our very brief lives, our earthly domain, and the universe surrounding us? We humans have an instinct for symbolism that urges us to represent the tangible and intangible with thoughts, words, pictures, and music. Cave paintings predate our modern era by tens of thousands of years. My son David, barely two years old, called ice cream "um-num"—and it was an apt word-symbol from a toddler who did not yet know that it was frozen cream but certainly knew what tasted and felt good. We are also born with sexual instincts vital to our survival as a species, and we inherently recognize the duality of opposites and respond to sexual symbolism.[1,2] These instincts led to allegories and metaphors for understanding the nature and transformations of matter at least 2500 years ago in central and eastern Asia, biblical lands of the Middle East, and ancient Greece.[3]

Figure 36 shows a mandala painted in central Tibet during the fifteenth century.[4] Mandalas have their origins in Tantrist Hindhuism and Buddhism and are representations for contemplation of the universe. Mandalas are often made by Buddhist monks from painted sand over the course of many days, contemplated, and then returned to the sea—a symbolic act of enriching one's earthly life with thought rather than accumulation of worldly wealth, which is illusory and transient. Central to the Eastern mandala is the circle that represents unity and completeness. Circles also separate domains such as heaven and earth. Circle imagery can also be likened to very ancient ideas about the conservation of matter. Later in this book we encounter the ouroboros, a serpent that forms a circle by devouring its own tail even as it regenerates itself. The act of returning the sand of a mandala to the sea implies both conservation of matter and the cycle of life. Typically, a circle in a mandala will encompass a square surrounded by four gates representing the four cardinal directions (north—actually on the right here, south, east, and west). These four gates lead to an inner sphere inhabited by four gods. On the outside of this sphere are four goddesses in a complementary fourfold array (NE, SE, SW, NW). This male-female duality is also represented by female and male (mother-father) figures outside of the larger circles in positions of sexual embrace. The four elements, commonly coded for by a square, really represent pairs of opposing properties—hot versus cold, dry versus wet. Thus, fire is hot and dry and water is cold and wet. The center of this mandala depicts the boddhisatva ("enlightenment being") Vajrapani holding the thunderbolt and grasping a serpent.[4] The center of a mandala can be likened to the fifth ancient element—the ether.

Adam McLean has an interesting view concerning Western alchemical mandalas.[5] He analyzes 30 images of esoteric alchemy in European texts mostly

FIGURE 36. ■ Fifteenth-century Tibetan painted mandala in which gods and goddesses represent the dualities that are the origin of the four ancient elements (the square). The circle represents completeness, the cycle of life, and (*even*) the conservation of matter. See color plates. (Courtesy Rossi & Rossi, London and *Asianart.com*).

from the seventeenth century. Although many of these have the circular and square forms similar to Figure 36, others share only the essence but not the form of Eastern mandalas. For example, McLean considers the emblem from Libavius' 1606 *Alchymiae* (see Figure 50 in a later essay) to be a mandala.[6] In this light Figure 37, drawn by artist Rita L. Schumaker,[7] also contains the fundamental essence of a mandala. Earth, water, and air are clearly depicted by circular realms while fire penetrates these realms. Dualities are depicted by dark and light doves as well as dragons. Seeds of growth in the earth imply the multiplication of met-

FIGURE 37. ■ Pencil drawing, in the style of a mandala, by Ms. Rita L. Schumaker depicting dualities (entwined dragons, dark and light doves) as well as the four ancient elements (water, air, fire, and earth).

als. In the next essay we will encounter even more tangible metaphors for the four ancient elements.

---

1. N. Schwartz-Salant, *Encountering Jung on Alchemy*, Princeton University Press, Princeton, 1995.
2. I. MacPhail, *Alchemy and the Occult*, Yale University Library, New Haven, 1968, pp. xv–xxxii (essay by A. Jaffe).
3. B. Pullman, *The Atom in the History of Human Thought*, Oxford University Press, New York, 1998.

4. A.M. Rossi, F. Rossi, and J.C. Singer, *Selections 1994*, Rossi & Rossi, Ltd., London, 1994. I am grateful to Ms. Anna Maria Rossi, Rossi & Rossi, London, and Ian Alsop, Editor, *Asianart.com*, Santa Fe, New Mexico for permission to employ theVajrapani mandala image.
5. A. McLean, *The Alchemical Mandala*, Phanes Press, Grand Rapids, 1989.
6. McLean, op. cit., pp. 62–69.
7. The author thanks Ms. Rita L. Schumaker, Charlotte, North Carolina, for this original drawing and for discussion of its themes.

THE PHILOSOPHER'S STONE CAN NO LONGER BE PROTECTED BY PATENT

John Read's wonderful trilogy, *Prelude to Chemistry*,[1] *Humour and Humanism in Chemistry*,[2] and *The Alchemist in Life, Literature and Art*,[3] include many choice gems. For example, in *Prelude* we see publicly and plainly disclosed for the first time ever, and therefore no longer patentable, the recipe for The Philosopher's Stone[1] (also known as *Lapidus Philosophorum*, The Red Tincture, The Quintessence, The Panacea, The Elixir of Life, Virgins Milke, Spittle of Lune, Blood of the Salamander, The Metalline Menstruall, and hundreds of other straightforward names).[1] In *Humour* Read produces the box score for a cosmic cricket match between a timeless all-star team led by Hermes Trismegistos (223 runs) and another team captained by Noah (210 runs).[2] The game was umpired by Solomon and Ham and scored by the Bacon boys (Roger and Francis). For the winners, Aristotle contributed 4 runs (earthly elements) and Paracelsus 3 runs (the *tria prima* of sulfur, mercury, and salt)—it only gets worse!

In any case, and without further ado, here is the recipe for The Philosopher's Stone ("quicksilver" is the real element mercury):[1]

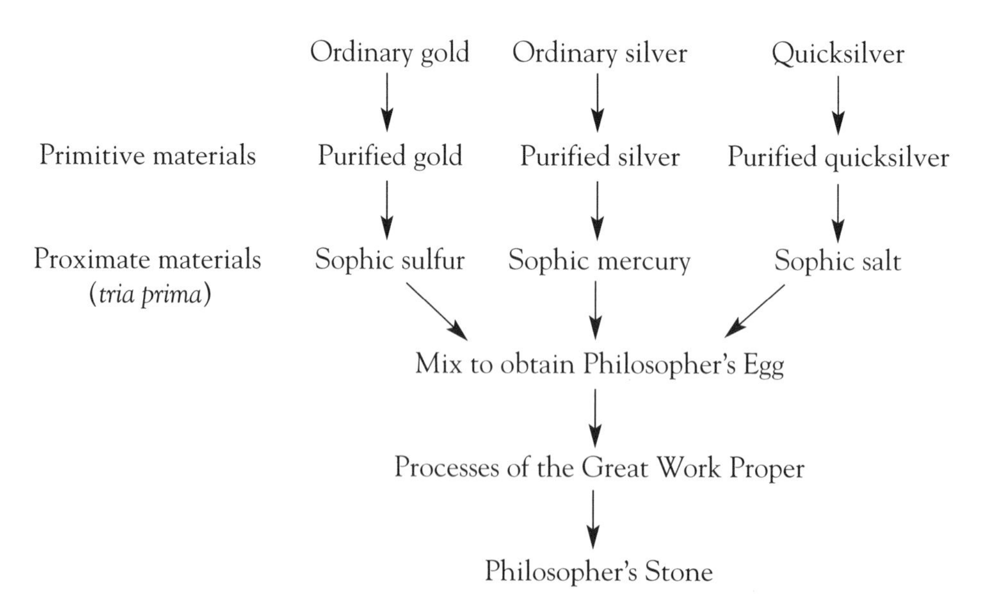

The Twelve Keys of Basil Valentine (see Figures 38–41) depict the Processes of The Great Work in the days before patent attorneys. The images are almost as obscure as legalese and clearly meant to protect his venture capital. Some of them refer to specific processes (twelve is not uncommon—one for each sign of the Zodiac).[1] Each process is best done under the appropriate sign (e.g., distillation under Virgo; digestion under Leo—what else?; the actual use of The Stone, projection, under Pisces). That means, optimally, one year for manufacturing each Quintessence and lots of "aging" space—inefficient use of time and commercial square-footage that will certainly vex the company's accountants.

1. J. Read, *Prelude To Chemistry*, MacMillan, New York, 1937, pp. 127–142.
2. J. Read, *Humour and Humanism in Chemistry*, Bell, London, 1947, pp. 12–14.
3. J. Read, *The Alchemist in Life, Literature and Art*, Nelson, Edinburgh, 1947.

MYSTICAL AND MAJESTIC NUMBERS

Certain simple numbers have for centuries been attributed great symbolic significance[1] in alchemy.[2] *One* connotes God or Allah in monotheistic religions as well as the *prima materia*—the origin of all matter. *Two* signifies the male and female principles (opposites—mercury and sulfur) that are present in all things. *Three*—the *tria prima* (mercury, sulfur, and salt—spirit, soul, and body) represents Paracelsus' (1493–1541) extension of the male and female principles; it also represents the Holy Trinity (Father, Son, and Holy Ghost). *Four*, as noted in the first two essays, is the number of ancient elements (earth, water, air, and fire), each one a composite of two opposing qualities or properties—hot/cold and dry/wet. There are also four seasons and four cardinal directions. Aristotle introduced a *fifth* element, the *quinta essencia*, also known as the *quintessence*, representing the heavens or the celestial ether. *Seven* metals known to the ancients (silver, gold, iron, copper, mercury or quicksilver, tin and lead) matched the number of visible "planets" (Moon, Sun, Mars, Venus, Mercury, Jupiter, and Saturn) as well as days of the week. *Twelve* signs of the Zodiac, equivalent to the twelve months in a year, were equated by the English alchemist George Ripley, Canon of Bridlington, to twelve "Gates" or operations en route to the Philosopher's Stone.[2] These were as follows:[2,3]

1. Calcination (action of fire on minerals in air)	Aries, the Ram
2. Congelation (a thickening by cooling)	Taurus, the Bull
3. Fixation (trapping a volatile as a solid or liquid)	Gemini, the Twins
4. Solution (dissolutions or reactions of substances)	Cancer, the Crab
5. Digestion (heat continuously applied; no boiling)	Leo, the Lion
6. Distillation (ascent and descent of a liquid)	Virgo, the Virgin
7. Sublimation (ascent and descent of a solid)	Libra, the Scales
8. Separation (isolation of insoluble liquids, solids)	Scorpio, the Scorpion
9. Ceration (bringing hard material into a soft state)	Sagittarius, the Archer
10. Fermentation (animation of a substance with air)	Capricornis, the Goat
11. Multiplication (increasing potency of the Stone)	Aquarius, the Water Carrier
12. Projection (mysterious action of the Stone)	Pisces, the Fishes

Basil Valentine ("Valiant King"), reputed to be born in 1394, is a perplexing literary forgery.[4] There appears to be fairly widespread agreement that he was in reality a certain sixteenth/seventeenth-century salt boiler and publisher named Johann Thõlde who "edited" Basil Valentine's works.[4] Whoever Basil Valentine was, he was quite knowledgeable about the chemistry of his times,[4] and he described Twelve Keys or operations defining The Great Work. The pictorial representations of the twelve keys are found in the next essay) see Figures 38–41) and have been repeated many times including in Manget's 1702 compendium.[5]

The *First Key* (*prima clavis*) signifies the chemical wedding and represents the creation of the primitive materials for the stone.[6,7] The wolf represents antimony sulfide, useful for separating gold from other metals. [A related figure was included in Michael Maier's 1618 book *Atalanta Fugiens* (see Figure 45) and is discussed later.] The old man may represent Saturn (lead) for separating sulfur. The *Second Key* represents watery separation, and the *Third Key* depicts the dragon as the *Prima Materia* and suggests a circular cycle of volatilization and fixation.[7] The *Fourth Key* symbolizes putrefaction, a heating, with fire that we now recognize as roasting of metal ores entailing a rather mixed and messy oxidation of sulfide ores to a dark mass. The *Fifth Key* is considered to represent a solution process (but this can, of course, signify chemical reactions). The *Sixth Key* represents conjunction—chemical marriage between sophic sulfur (the King) and sophic mercury (the Queen). The *Seventh Key* is a kind of alchemical mandala[8] representing the four earthly elements, the ether and the three Paracelsan principles. The *Eighth Key* is a resurrection scene symbolized by planting of seed. As John Read has noted,[2] if putrefaction (the *Fourth Key*) is associated with oxidation, the reverse process of regeneration or resurrection of metals corresponds to reduction (*reducere*—"to lead back" or "restore") and the return of their souls. The *Ninth Key* alludes to the three principles, the four ancient elements, and the consecutive colors of The Great Work in order of ascension—the crow (black, putrefaction), the swan (white, calcinations), the peacock (yellow or rainbow hues), and the red phoenix (symbolizing the Red Tincture or Philosopher's Stone). The *Tenth Key* represents the *tria prima* and has been discussed elsewhere.[6,7] The Hebrew may represent a verse in Psalms but with a kind of Kabbala letter-substitution code.[9] The *Eleventh Key* symbolizes the process of multiplication, and the *Twelfth* and final *Key* symbolizes calcinations, the mysterious fire in the wine barrel, and the fixation of the volatile is symbolized by the Lion (sulfur) eating the snake (mercury).[6,7]

---

1. Many faiths and philosophies include their own mystical numerological systems. For example, the Hebrew Kabbala, first appearing in the twelfth century, had its roots more than a millennium earlier. The Creation was a process involving 10 divine numbers and the 22 letters of the Hebrew alphabet. Together these two numbers provided the 32 paths to wisdom (see *The New Encyclopedia Britannica*, Encyclopedia Britannica, Inc., Chicago, 1986, Vol. 6, p. 671). A truly religious fan likewise intrinsically understands the Kabbala of Baseball: 3 (strikes for an out; outs for a half inning); 4 (balls for a walk); 7 (games in the World Series); 9 (innings; field positions); 13 (uniform number to avoid); 44 (the all-time home-run champ Hank Aaron's uniform number—a uniform number to seek); 60, 61, 70, 73 (home-run records established in turn by Babe Ruth, Roger Maris, Mark McGwire, and Barry Bonds); and 1955 (only year in the Modern Era in which Brooklyn won the World Series).

2. J. Read, *From Alchemy to Chemistry*, Dover Publications, Inc., New York, 1995, pp. 32–35.
3. J. Read, *Prelude To Chemistry*, The MacMillan Co., New York, 1937, pp. 139–142.
4. J.R. Partington, *A History of Chemistry*, MacMillan and Co. Ltd., London, 1961, pp. 183–203.
5. J.J. Manget, *Bibliotheca Chemica Curiosa, seu Rerum ad Alchemicum pertinentium Thesaurus Instructissimus . . .* , Sumpt. Chouet, G. de Tournes, Cramer, Perachon, Ritter, & S. de Tournes, Geneva, 1702.
6. Read (1937), op. cit., pp. 196–208.
7. S.K. De Rola, *The Golden Game—Alchemical Engravings of the Seventeenth Century*, Thames and Hudson Ltd., London, 1988, pp. 120–126.
8. A. McLean, *The Alchemical Mandala*, Phanes Press, Grand Rapids, MI, 1989.
9. Personal communication from Professor Laura Duhan Kaplan to the author.

THE TWELVE KEYS OF BASIL VALENTINE: THE IMPURE KING

The best evidence indicates that Basil Valentine (the Valiant King, a Benedictine cleric monk said to be born in 1394) never existed. Books attributed to him such as the ever-popular *Triumphal Chariot of Antimony*, first published in 1604, are generally attributed to the publisher Johann Tholde who, in turn, had perhaps "improved upon" earlier manuscripts that had come into his hands.[1] Nevertheless, they contain interesting images and even some useful information.

The Twelve Keys of Basil Valentine provide images pertaining to the Great Work and have been analyzed by many authors including John Read.[2] In Figure 38(a), we see the First Key. The wolf in this picture is generally considered to represent antimony (Sb or *stibium*), although, until recent centuries, that term really meant its ore stibnite (Sb_2S_3). Antimony had been referred to in the alchemical literature as *lupus metallorum* ("wolf of metals").[2] Actually, it is a *metalloid*. One of its forms (allotropes) is metallic—a brittle gray substance with relatively poor thermal and electrical conductivity, rather unlike typical metals. In his delightful book, Venetsky quips: "As if in revenge for the unwillingness of other metals to accept it in their family, molten antimony dissolves almost all of them."[3] In modern terms, we recognize a metal capable of dissolving other metals ("like dissolves like") but also something akin to a nonmetal, capable of oxidizing other metals.

We see the wolf near a figure of Saturn. [Remember the old man with the wooden leg in the essay on chemical symbols (Figure 6)? Antimony had also been called the "lead of the philosophers"[2]; the ancients described antimony as the progeny of lead through heating.] Now, if impure gold is heated in the fire (three times—the queen holds three flowers—pretty obvious, eh?), the king will emerge. The king is gold (perhaps seed of gold or sophic sulfur[2]). The queen represents purified silver from which is derived sophic mercury. The first purification gives the "primitive materials" of the Stone—derived from gold, silver, and mercury.[2] This picture actually represents the following chemistry occurring in the fire:

$$Au + Ag + Cu + Sb_2S_3 \rightarrow Au/Sb + Ag_2S + CuS$$

(Au + Ag + Cu) represents impure gold here; Au/Sb is gold alloy.

$$Au/Sb + O_2 \rightarrow Au + Sb_2O_3 \text{ (vapor)}$$

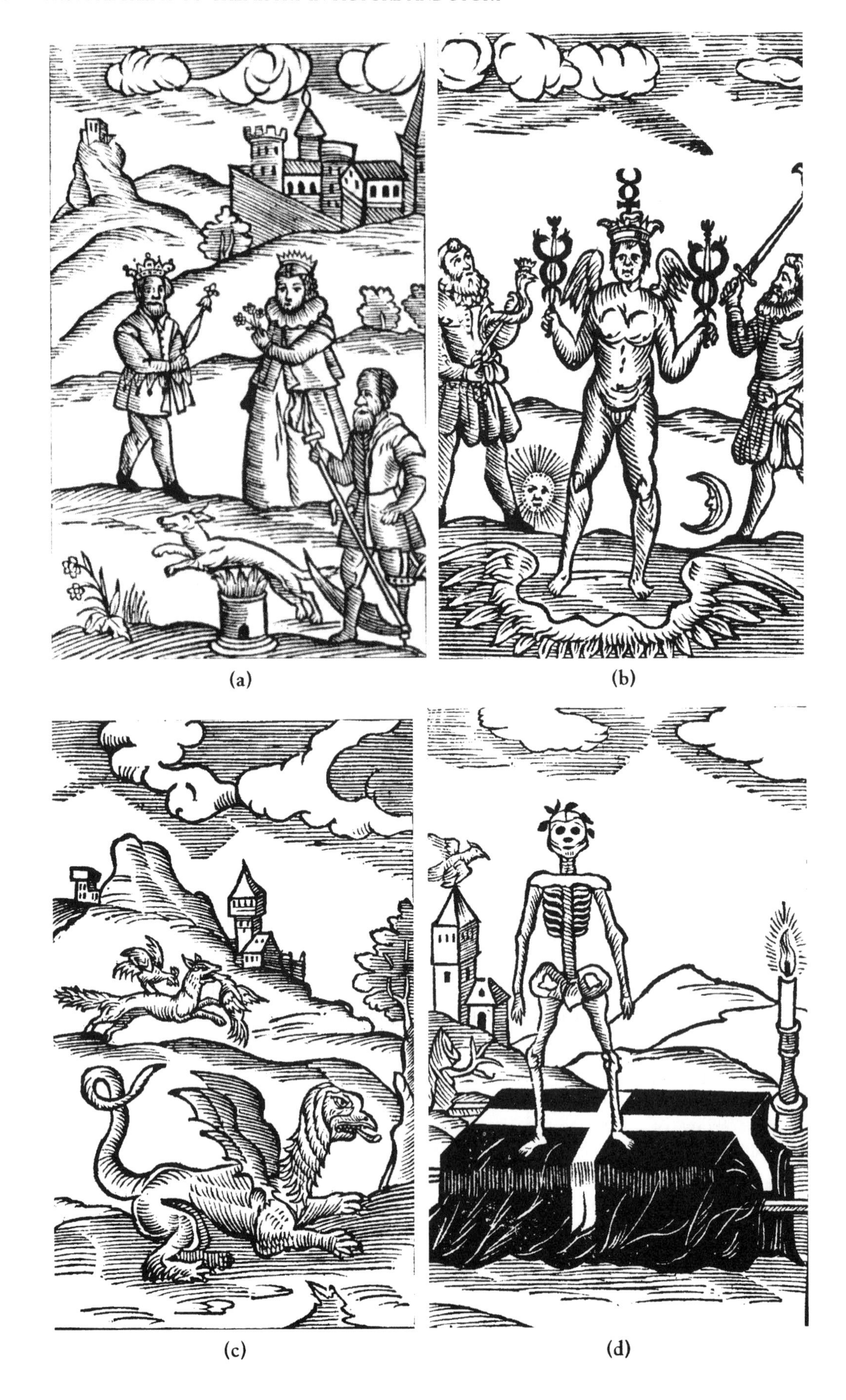

(a) (b)

(c) (d)

Similar chemistry occurs in the purification of antimony from stibnite through heating with metallic iron. It is also noteworthy that when metallic ("red") copper and antimony are alloyed (e.g., 6% Sb), the resulting metal looks very much like gold. The following is from a Syriac manuscript dating from the Crusades:[4]

> Throw in with red copper some antimony roasted in olive oil and it will become gold-like.

1. J.R. Partington, *A History of Chemistry*, MacMillan, London, 1961, pp. 183–203.
2. J. Read, *Prelude to Chemistry*, MacMillan, New York, 1937, pp. 196–211.
3. S.I. Venetsky, *On Rare and Scattered Metals—Tales About Metals*, (translated from the Russian by N.G. Kittell), Mir, Moscow, 1983, p. 83.
4. M.L. Dufresnoy and J. Dufrenoy, *Journal of Chemical Education*, 27:595–597, 1950.

RATZO RIZZO AND THE POET VIRGIL AS TRANSMUTING AGENTS?

Most of the remaining eleven keys of Basil Valentine have been analyzed by Read[1] and I will rely primarily upon his insights. The *Second Key* [Figure 38(b)] is said to represent the operation *separation*, a purifying of watery matter from its dregs. The matter appears to be volatile quicksilver which has undergone a kind of molting, under the influence of Sol (Sophic Sulfur) and Luna (Sophic Mercury), en route to Sophic Salt. The rooster (cock) held on the left is a male symbol implying the need for conjunction. The *Third Key* [Figure 38(c)] includes a winged dragon, fox, pelican, and cock. Dragons appear to have been used for many symbols and are sometimes used interchangeably with snakes. Winged dragons sometimes represent Sophic Mercury, sometimes Proximate Material.[2] Read did not discuss this key. My reading of the 1671 English edition indicates that this key pertains to purification of gold to Sophic Sulfur. The *Fourth Key* [Figure 38(d)] clearly refers to *putrefaction*, the necessary blackening that starts the operation. The symbols of the crow and skeleton are clear here. It was widely known from the work of the third-century alchemist Mary Prophetissa (or Maria the Jewess, for whom the hot-water bath or *Bain Marie* for controlled heating was named) that heating of a lead–copper alloy with sulfur produced a black mass.[3] So too did heating together of the four base metals lead, tin, copper, and iron.[3]

Putrefaction, as the first step toward transmutation to gold, is loaded with religious symbolism. The idea here is total abasement before salvation can begin. Impurities and imperfections must similarly be removed from metals in order for them to transmute to gold. Humans must remove their imperfections to achieve a state of grace. In *The Divine Comedy*, Dante Alighieri must first be guided by

FIGURE 38. ■ Depictions of the Twelve Keys of Basil Valentine (from Basil Valentine, *Letztes Testament* . . . , Strasburg, 1667). See text for interpretations, (a) *First Key*; (b) *Second Key*; (c) *Third Key*; (d) *Fourth Key*.

the classical Roman poet Virgil through Hell before he can enter Purgatory and thence to Paradise.[4] In the 1969 movie *Midnight Cowboy,* Joe Buck (Jon Voight) leaves the earthly sphere (rural Texas) and must first experience Hell (Times Square in the late 1960s New York City), in the company of Ratso Rizzo (Dustin Hoffman). He discovers his inner gentility and true nature and achieves salva-

FIGURE 39. ■ (a) *Fifth Key* of Basil Valentine; (b) *Sixth Key*; (c) *Seventh Key* (see Figure 38).

tion through a similar journey, by Greyhound bus, from colder to gentler climes (Florida as Paradise?). Virgil, who died before the birth of Jesus, was of course a nonbeliever in Christianity and could never enter Paradise. Ratso Rizzo dies of consumption on the bus and also never enters Paradise.

The *Fifth Key* [Figure 39(a)] is said by Read to represent the operation solution. The *Sixth Key* [Figure 39(b)] represents conjunction—the marriage of the King (Sophic Sulfur) and the Queen (Sophic Mercury), the conjunction of the Sun (Sol) and Moon (Luna), the fiery two-headed man and rain, condensation, and fertility. This operation is an excuse for lots of naughty pictures in alchemical manuscripts and texts—the golden seed follows coitus. The *Seventh Key* [Figure 39(c)] implies the four earthly elements, the Heavens (chaos is quite a complex concept), and the three Paracelsan principles. The double circle can symbolize the interaction between earthly and heavenly spheres. The stem at the top of the sphere appears again in the summary figure [Figure 41(c)] and Read interprets this figure [and implicitly the vessel in Figure 39(c)] as the Philosopher's Egg in which the Proximate Materials are joined: The stem is a kind of placenta. The Philosopher's Egg is hermetically sealed and may be placed for long periods in a special furnace termed an athanor—a kind of uterus. The *Eighth Key* [Figure 40(a)], another graveyard scene, is said by Read to represent *fermentation*.

The *Ninth Key* [Figure 40(b)] is a marvelous representation referring, in part, to the color changes that occur during the Great Work.[1,5] The falling figure is Saturn (base metals, notably lead); the rising figure is perhaps Luna (Sophic Mercury). The outer four-sided figure represents the four elements and the three snakes represent the *tria prima* (sulfur, mercury, salt). The four birds represent color changes. At the top, the crow represents blackness, then counterclockwise, the swan represents white, the peacock is multicolored, sometimes simply citrine, and the phoenix represents The Red Tincture (The Stone). It is interesting that the Sioux Nation recognizes four "true colors"—black, red, yellow, and white—the same as in the *Ninth Key*. Red is also the most important, representing earth, pipestone, and blood. The colors correspond to the four compass directions: west, north, east, and south.[6]

The *Tenth Key* [Figure 40(c)] represents the *tria prima*. The three symbols near the corners of the triangle (clockwise from top left) are: gold, silver and mercury. These are the three elements that are purified to make Sophic Sulfur, Sophic Mercury and Sophic Salt, respectively. The double borders of the circle (heavenly perfection) and the triangle represent the duality of the earthly and heavenly spheres. The German phrases translate thusly[7]: "From Hermogenes I was born" (top); "Hyperion has chosen me" (right); "Without Jamsuph I am lost" (left). In Gnostic mythology, Hermogenes developed the doctrine of the eternity of matter.[8] In Greek mythology, Hyperion was a Titan recognized as the Father of the Sun (Helios), the moon (Selene) and the dawn (Eos).[9] Jamsuph, from the Kabbalists, refers to the Red Sea—a sign of God's power—the parting of the Red Sea may refer to the splitting of matter.[10] Translation of the Hebrew has been more elusive and is possibly Kabbalistic in nature.[11] The *Eleventh Key* [Figure 41(a)] shows lion whelps and depicts the *multiplication* achievable by the Stone. The two vessels represent the Philosopher's Egg (Vase of Hermes) where *conjunction* takes place [also see the double pelican and its symbolism in Figure 27(c)]. Read has described the *Twelfth Key* [Figure 41(b)] as representing calcination (whitening, drying) with the lion and snake as fixed and volatile principles and the flowers as purified noble metals. The dragon here is said to represent the Proximate Material of the Stone; the circles around its wings and paws are the volatile and fixed principles. Figure 41(c) is the summary of the work.

1. J. Read, *Prelude To Chemistry*, MacMillan, New York, 1937, pp. 196–211, 260–267.
2. J. Read, op. cit., pp. 106–108, 208, 269–272.

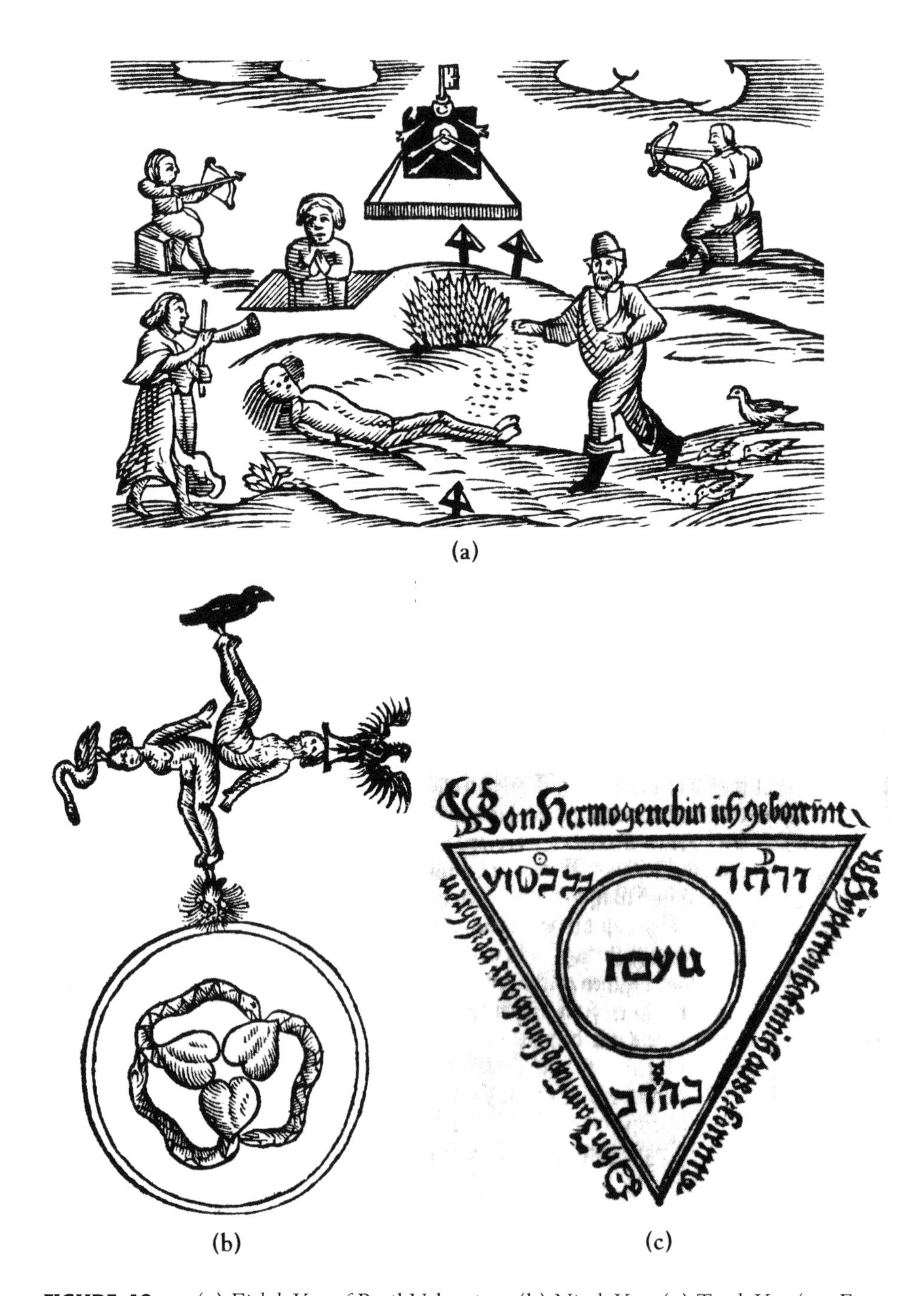

FIGURE 40. ■ (a) *Eighth Key* of Basil Valentine; (b) *Ninth Key*; (c) *Tenth Key* (see Figure 38).

3. J. Read, op. cit., pp. 13–17.
4. I am grateful to Professor Susan Gardner for this discussion.
5. J. Read, op. cit., pp. 145–148.
6. J. Lame Deer and R. Erdoes, *Lame Deer Seeker of Visions*, Simon & Schuster, New York, 1972, pp. 116–117.
7. I am grateful to Professor Ralf Thiede for this translation.

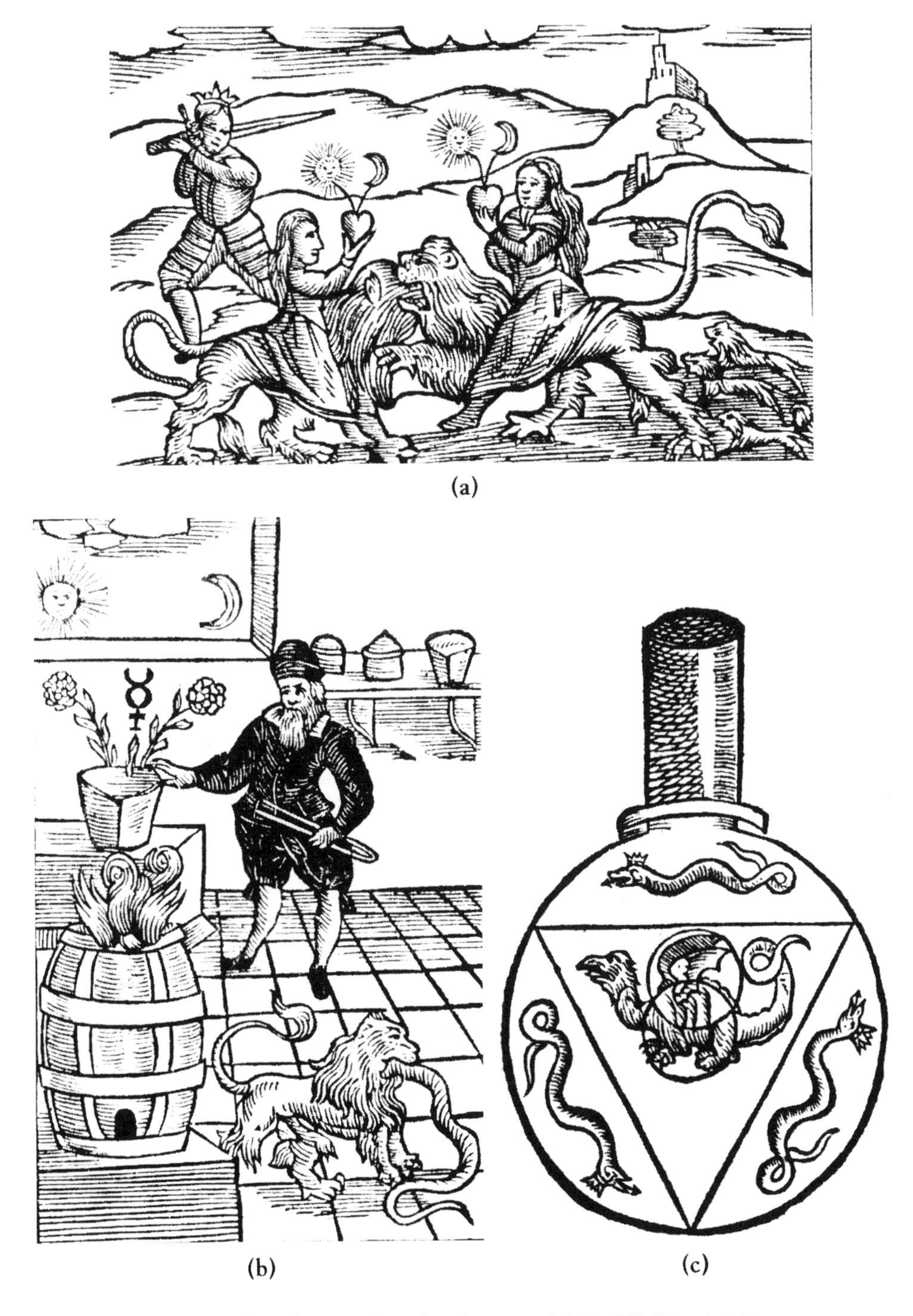

FIGURE 41. ■ (a) *Eleventh Key* of Basil Valentine; (b) *Twelfth Key*; (c) Summary image for the Twelve Keys of Basil Valentine (see Figure 38).

8. W. Doniger, Mythologies Compiled by Yves Bonnefoy, University of Chicago Press, Chicago, 1991, Vol. 2, p. 677.
9. W. Doniger, op. cit., Vol. 1, pp. 371, 375.
10. I am grateful to Professor Laura Duhan Kaplan for her help in interpretation.
11. R. Patai, *The Jewish Alchemists*, Princeton University Press, Princeton, 1994.

NATURAL MAGICK: METAMORPHOSES OF WEREWOLVES AND METALS

A twelfth-century ethnography of Ireland, written by a Gerald of Wales, describes an encounter by a Priest with a talking he-wolf in the wild, who entreaties him to give the Eucharist to his dying mate.[1] The Priest does so and the discussion of whether this action is sacrilegious depends on the nature of the creature—is it a true hybrid (like a griffin), a man in wolf's clothing, or a total change in identity? How did a man *change* to become a wolf? What humanity remained? Was the man "wolfish" before any change occurred?[2] Historian Caroline Walker Bynum treats the concepts of *identity* and *change* and posits that, in the final decades of the twelfth century, a new conceptual image took hold in European culture—the metamorphosis—gradual rather than sudden change.[1] For example, Bynum contrasts the ancient New Testament story of the sudden conversion of Saul from a persecutor of Christians to Paul, a disciple of Christ, with the twelfth-century version of his slow, evolutionary, and reasoned conversion on the road to Damascus.[1] Metamorphosis presents a more dynamic and complex story than sudden miraculous change or the mere appearance of a static hybrid. Ancient stories of change, including werewolf folklore, also took on new meanings toward the end of the twelfth century.[1] Metamorphoses were to be found everywhere in the natural world—the foodstuff in a seed becomes a tree; food "morphs" into blood and bile. And from Middle Eastern cultures came complex, often spiritual, operations for gradually changing matter that came to comprise alchemy.

Intellectual cross-fertilization was one benign by-product of a horrendous series of Crusades first launched by Pope Urban II in 1095 to remove Moslem control of the Christian shrine of the Holy Sepulchre in Jerusalem.[3] Jerusalem fell to the Crusaders in 1099, and they murdered its Moslem and Jewish inhabitants. Increasing control of the Holy Lands by the Crusaders continued until Zangi, a strong Moslem ruler, recaptured the city of Edessa (in Macedonia). A second Crusade was essentially defeated in 1154 by Zangi's successor Nureddin. By 1187, Nureddin's nephew Saladin had captured Jerusalem and virtually all of the Christian strongholds in the Holy Land. A third Crusade began in 1189 and achieved significant military successes. Although King Richard I (the Lion Heart) failed to reach Jerusalem, he obtained a peace treaty in 1192 with Saladin. However, this treaty quickly crumbled and more Crusades, including the pathetic Children's Crusade of 1212, continued until about 1270 with King Louis of France losing the eighth and final round.

Among the cultural artifacts that the Crusades brought back to Europe were the medicinal practices of Geber, Rhazes, and Avicenna and a cultural belief in the alchemical manifestation of metamorphosis—transmutation. "Geber" is actually a fourteenth-century name attributed to a number of works, some parts of which may well be ascribed to the eighth-century physician and alchemist Jābir ibn Hayyān (ca. 721–815), who was born in present-day Iraq and was educated in Arabia. The concept that all metals were a combination of mercury and sulfur is attributed by some to Jābir.[4] An influential thirteenth-century work, *Summa Perfectionis*, said to be authored by Geber (referred to by chemical historians as pseudo-Geber to avoid confusion) described procedures for characterizing and purifying metals.[5] Figure 42 is from one of the earliest printed books to employ figures derived from copper plates and purports to show Geber (pseudo-Geber or Jābir?) in the laboratory.[6] Avicenna (Latinized version for Abu Ali-al Hussin ibn Abdallah ibn Sina, 980–1037) was a Persian physician of great erudition. His name was also attributed to works of the early Renaissance, and chemical historians refer to the author as pseudo-Avicenna.[5] Figure 43 is from an eleventh-century manuscript said to show Avicenna amid his medicinal preparations.

Metamorphosis, according to Professor Bynum, involves a transformation from one form to another, while maintaining a common characteristic or aspect.[1] For example, Bynum relates[2] the poet Ovid's tale of Jove's punishment of King Lycaon, who was savage with his subjects and also attempted Jove's murder. Although Lycaon is thoroughly and bodily transformed into a wolf[2,7] ("*Lykos*" = "wolf" in Greek):

> He turns into a wolf, and yet retains some traces of his former shape. . . . There is the same grey hair, the same fierce face, the same gleaming eyes, the same picture of beastly savagery.

FIGURE 42. ■ An image of the eighth-century physician and alchemist Jābir ibn Hayyān ("Geber"), who was born in Arabia and educated in Iraq. This portrait is from Thevet's 1584 *Vies Des Hommes Illustres*. Numerous sixteenth- and seventeenth-century writings in alchemy and medicine were falsely attributed to Geber. To minimize confusion, modern historians credit these to a "pseudo-Geber" (or ψ-Geber).

FIGURE 43. ■ Painting in a fifteenth-century manuscript depicting the eleventh-century Persian physician Avicenna (Abu Ali-al Hussin ibn Abdallah ibn Sina, 980–1037) in an apothecary shop. See color plates. (© Archivo iconografico, S.A./CORBIS).

Bynum notes that "thirst for blood and delight in killing,"[2] what one might call the "essence of wolfishness," were characteristic of both King Lycaon and the wolf.

The notion that metals can be transformed into one another by metamorphosis is quite alien to we moderns. However, it is important to remember that, hundreds of years ago, there was no real concept of an element and metals were commonly found in various states of purity. Alloys such as bronze (copper and tin) and pewter (lead and tin is one formulation) offered a smooth continuity of metallic properties—almost a form of stop-motion metamorphosis or transmutation. The very nature of metals—luster, malleability, and thermal conductivity—argued for a common aspect (or substance)—an "essence of metallicity." And so, we have Johann Joachim Becher, in the mid-seventeenth century, believing that all metals contain quicksilver (mercury)[8] and another important chemist of that period, Johann Kunckel, reporting that he had extracted mercury from all metals.[9] And what magical stuff quicksilver is—the volatile, penetrating, very "essence of metallicity." Mercury dissolves gold and other metals, the nature and appearance of which are dramatically changed upon amalgamation. Heating an amalgam distills the mercury and returns the metal unscathed or perhaps even purer. One can imagine many samples of "pure" metals having some mercury contamination as the result of their history, and so, obtaining a trace of mercury from a combined sample of gold having a sordid history is fairly believable. Indeed, mercury is reputed to play male, female, and hermaphroditic roles. In the playful words of Professor Allison B. Kavey: "Mercury is a bit of a slut. . . ."[10]

Figure 44 is the beautiful frontispiece from the first English edition, published in 1658, of Giambattista della Porta's famous *Natural Magick.*[11] It was first published in Latin in four "books" in 1558, later expanded to twenty books in 1589, and published in numerous editions in Italian, French, and Dutch in addition to the English translation. Porta himself also claims versions in Spanish and Arabic.[12] His 1608 book *De Distillatione* begins with testimonials to him in Hebrew, Greek, Chaldee, Persian, Illyrian, and Armenian as well as a beautiful portrait of the author (see Figure 70). Figure 44 similarly honors Porta's ego, seemingly equating him with the four ancient elements, astrological cosmology, and the spirits of Art and Nature that form the basis of "natural magic." In fact, Porta (ca. 1535–1615) had wide-ranging interests in science, particularly physics. He is often credited with designing the *camera obscura* and produced a design for a steam engine. In addition, Porta wrote "some of the best Italian comedies of his age." However, much of Porta's *Natural Magick* is derived from the *Historia Naturalis* of the ancient Roman author Pliny, and he is almost totally credulous about "natural magic."[12]

Here are two brief excerpts from *The Fifth Book of Natural Magick: Which Treateth of Alchymy; Shewing How Metals May Be Altered and Transformed, One into Another:*[13]

CHAP. II.

Of Lead, and How It May Be Converted into Another Metal

> The Ancient Writers that have been conversant in the Natures of Metals, are wont to call Tinne by the name of white Lead; and Lead, by the name of black Tinne: insinuating thereby the affinity of the Natures of these two Metals, that they are very like each to another, and therefore may very easily be one of them transformed into the other. It is no hard matter therefore, as to change Tinne into Lead . . .

To Change Lead into Tinne

> It may be effected onely by bare washing of it: for if you bath or wash Lead often times, that is, if you melt it, so that the dull and earthy substance of it be abolished, it will become Tinne very easily: for the same quick-silver whereby the Lead was first made a subtil and pure substance, before it contracted that soil and earthiness which makes it so heavy, doth still remain in the Lead, as Gebrus hath observed; and this is it which causeth that creaking and gnashing sound, which Tinne is wont to yield, and whereby it is especially discerned from Lead: so that when the Lead hath lost its own earthy lumpish-

FIGURE 44. ■ Title page of the first English edition (1658) of Giambattista della Porta's *Magiae Naturalis* (first published in 4 books in 1558 and expanded to 20 books in 1589). Porta had a considerable knowledge of sixteenth-century chemical operations, has often been credited with invention of the *camera obscura,* and was a renowned playwright. This figure suggests that Porta was the veritable embodiment of order and logic—the antithesis of Chaos. (From The Roy G. Neville Historical Chemical Library, a collection in the Othmer Library, CHF.)

> ness, which is expelled by often melting; and when it is endued with the sound of Tinne, which the quick-silver doth easily work into it, there can be no difference put betwixt them, but that the Lead is become Tinne.

Note some interesting points here. Quicksilver (mercury) is common to both tin and lead. The removal of earthy impurities from lead is a continuum of metamorphosis. Thus, the lead-to-tin transmutation has the essential aspects of a metamorphosis between werewolf and human while conserving a common characteristic—the "essence of metallicity" imbued by quicksilver. Note too, the interesting point that the proof of metal identity is not density, melting point, or chemical reactivity but the sound obtained in its mechanical working!

---

1. C.W. Bynum, *Metamorphosis and Identity*, Zone Books, New York, 2001, pp. 15–36.
2. Bynum, op. cit., pp. 166–176.
3. *The New Encyclopedia Britannica*, Encyclopedia Britannica, Inc. Chicago, 1986, Vol. 16, pp. 880–892.
4. *The New Encyclopedia Britannica*, op. cit., Vol. 6, p. 451.
5. F.L. Holmes and T.H. Levere (eds.), *Instrumentation and Experimentation in the History of Chemistry*, The MIT Press, Cambridge, MA, 2000, pp. 44–49.
6. A. Thevet, *Vrais Portraits et Vies Des Hommes Illustres*, Paris, 1584, p. 73.
7. This is definitely not the two-legged Lon Chaney, Jr. *Wolfman*, B-movie fans.
8. J.R. Partington, *A History of Chemistry*, Vol. 2, MacMillan & Co. Ltd., London, p. 666.
9. Partington, op. cit., p. 362.
10. A.B. Kavey, "Mercurial Aspects: Gender and Sex in Popular English Alchemy," presented at The International Conference on the History of Alchemy and Chymistry, Chemical Heritage Foundation, Philadelphia, 19–22 July 2006.
11. J. Baptista Porta, *Natural Magick*, Thomas Young and Samuel Speed, London, 1658. I am grateful to The Roy G. Neville Historical Chemical Library (California) for supplying an image of the frontispiece of this book.
12. J. Baptista Porta, *Natural Magick* (reprint edited by Derek J. Price), Basic Books, Inc., New York, 1957.
13. Porta, op. cit., p. 163.

AN ALCHEMICAL BESTIARY

Symbols and metaphors allow us to represent phenomena we do not fully understand and thoughts having no rational translations. Four centuries ago, the wolf represented the "biting" behavior of antimony (or its sulfide) on "base" metals. At a much deeper, subconscious level we may employ sexual imagery to convey perceptions of the male and female nature of things. For millennia, these dualities were projected to explain properties of matter that could be understood only symbolically. It is no wonder that the psychologist Carl Jung wrote extensively on the symbolism of alchemy.[1]

THE WOLF AND THE IMPURE KING

In 1617 Michael Maier wrote a gloriously illustrated book titled *Atalanta Fugiens* (see Figure 82) for which he composed 50 fugues to accompany 50 illustrations (emblems) of the alchemical process.[2] Alluding to the three principles—sulfur, mercury, and salt—each of Maier's fugues was composed as an epigram in three verses for three voices. (In Brooklyn, this would translate as "three voices for three verses.")

Figure 45, Emblem 24 from *Atalanta Fugiens*, is a chemically astute depiction of a purification of gold.[3] The dead king symbolizes impure gold—say, gold contaminated with copper and other base metals. The wolf, often representing metallic antimony, here represents stibnite or antimony sulfide. Antimony ("not alone") is virtually always found in a combined state—hardly a "lone wolf." The wolf devours the impure king; that is to say, with application of heat, antimony loses its sulfur to copper and other base metals and the freed antimony alloys in a melt with gold. The sulfides of copper and the other base metals are dross easily separated from the melted alloy. The alloy is then placed in the fire, where chemically reactive antimony forms an oxide that sublimes off, leaving molten and chemically inert gold (the revivified king) (see page 57). Pierre Laszlo has provided evidence that the land and river in this image (and others in *Atalanta Fugiens*) implies a distinction between the dry way and wet way of chemical operations.[4] I wonder how this image would work as a question on a Chem I final?

LIONS AND DRAGONS AND SNAKES, OH MY!

Paired entities in struggle or passionate embrace (or both) represent the joining of the opposite principles [male–female; sophic ("philosophic") sulfur and sophic

FIGURE 45. ■ Here is a good question for the Chem I final exam: Write a description of the two simple chemical reactions depicted in this figure. (*Hint:* Think about the purification of gold.) Are you still "stumped"? Then see the accompanying text. (Figure from Maier's 1617 *Atalanta Fugiens*, from The Roy G. Neville Historical Chemical Library, a collection in the Othmer Library, CHF.)

mercury] thought to comprise all matter. Figure 46 (Emblem 16 from *Atalanta Fugiens*) depicts the struggle between two lions. The winged creature on the left is the Green Lion, representing the volatile (winged) sophic mercury: the female principle. The male is the Red Lion, symbolizing sophic sulfur, itself a symbol of fixity and combustibility.[3]

BLOOD OF THE DRAGON

In Maier's 1618 book *Viatorium* (Figure 47),[5] he describes how dragon's blood, another symbol for the philosopher's stone, is formed. An elephant engorges itself with water. In ambush lies a dragonlike serpent that attacks and wraps and tightens its coils about the elephant and drinks its blood (Figure 48). The weakened elephant eventually tumbles onto the serpent and crushes it to a bloody pulp. This dragon's blood, suffused with the matter of the elephant, is effectively a red tincture or philosopher's stone. The sexual imagery of this allegory should be quite obvious to anybody having a pulse.

SALAMANDER AS SPIRIT OF FIRE

The salamander is used to depict the "fiery masculine seed" that survives and is nourished by the fire.[6] The philosopher's stone is frequently likened to a seed

FIGURE 46. ■ The passionate struggle of the winged Green Lion (volatile female principle; "sophic" mercury) and the Red Lion (fixed male principle; "sophic sulfur") (from *Atalanta Fugiens*, fromThe Roy G. Neville Historical Chemical Library, a collection in the Othmer Library, CHF).

MICHAELIS
MAIERI
VIATORIVM,
hoc eſt,
DE MONTIBVS
Planetarum ſeptem,
ſeu
METALLORVM.
ROTHOMAGI,
Sumpt. IOANNIS
BERTHELIN.
1651,

FIGURE 47. ■ Title page from Maier's *Viatorium* (second edition, 1651; first edition, 1618) in which the seven ancient metals are represented (from top right clockwise): gold, silver, iron, copper, tin, lead, and mercury. The top center figure is the author Count Michael Maier, whom chemical historian John Read dubbed "a musical alchemist" referring to Maier's composition of fifty fugues for *Atalanta Fugiens*.

that may multiply metals. Sometimes the salamander simply represents fire or the spirit of fire.[7] Emblem 29 from *Atalanta Fugiens* (Figure 49) and its epigram help explain this mystical (and far from obvious to me) connection: "The Salamander cools the flame and goes."[2]

THE ONE AND ONLY FAMOUS, FABULOUS PHOENIX

One, and only one, phoenix[8] can exist in our world. This fabulous bird rises from the ashes of the penultimate phoenix, which self-immolated after 500 years of a lonely, sex-deprived existence. Closely associated with Egyptian mythology, the phoenix appears to have even more ancient oriental origins. Culturally, it is a symbol of rebirth and even life after death. In alchemical imagery, the phoenix is the last of four birds representing the successive color changes during The Great Work (see the top of Figure 50):

1. Crow, black, putrefaction
2. Swan, white, calcination
3. Peacock, yellow or rainbow colors signifying change
4. Phoenix, red, the red tincture or philosopher's stone

FIGURE 48. ■ Formula for "Dragon's Blood" (The Philosopher's Stone): (1) elephant gorges on water; (2) engorged elephant ambushed by huge serpent that tightens its coils and drinks the blood of its prey; (3) weakened elephant collapses and completely "smooshes" the serpent; *et voila*!; (4) "Dragon's Blood" fit for alchemical projection. (From Maier's *Viatorium*, second edition, 1651.)

Figure 51 shows a nineteenth-century Japanese fan (in black and white rather than the actual color) decorated with a watercolor painting of the phoenix. The figure bears a striking resemblance to a twentieth-century cultural icon—*Rodan*, the subject of a modern Japanese film genre that I will dub *Plastique Monstresque*.[9] Indeed, *Rodan* rises from the ashes of nuclear tests to menace the earth and teach us all a good lesson. It is a phoenix born on the funeral pyre of the atomic bomb.

The symbol of that august scientific body, the American Chemical Society (Figure 52), places a phoenix above Justus Liebig's nineteenth-century *kaliapparat*.[10] What do we make of *that*? The *kaliapparat* precisely measured carbon dioxide emitted from combustion of organic compounds, leading to accurate formulas and the scientific understanding of the vast "primeval forest"[11] of organic chemistry. In sharp counterpoint, the phoenix represents the culmination of the alchemical operation. Methinks my fellow chemists are hedging their bets—rational chemistry first but magic if it fails.

FIGURE 49. ▪ The salamander (from Maier's *Atalanta Fugiens*, from The Roy G. Neville Historical Chemical Library, a collection in the Othmer Library, CHF), which represents the resistance to fire attributed to the Philosopher's Stone. When Yale chemistry professor Benjamin Silliman visited the sweltering (115°F) laboratory of James Woodhouse of Philadelphia in the summer of 1802, he referred to "that salamander's home" (see page 392).

BEWARE THE AMOROUS BIRDS OF PREY[12]

Copulation is a major theme in alchemical imagery. The two snakes entwined about the rod of Mercury form the caduceus, the symbol of the medical profession (Figure 53).[13] These amorous serpents form three circles representing three cycles of separation and union of male and female principles.[14] Above the two serpents are two amorous birds of prey[15] also closely packed. These birds devour each other as they copulate, representing the process of chemical solution/combination and the loss of individual identities. Common decency prevents the depiction of the chemical wedding between man and woman (sol and luna).[16]

AND SHUN THE FRUMIOUS BASILISK[12]

Transmutation of "base metals" into silver or gold is ultimately achieved through "projection"—an unexplainable process that could occur at a distance—"the red tincture projected from the heart of the 'red king' (the red stone) onto his subjects, who personify the base metals."[17] The basilisk (or cockatrice) (Figure 54)[18] is a serpent of Roman mythology.[19] Some versions have it hatched by a serpent from an egg laid by a cock. Others view it as "a poisonous mixture of cock and

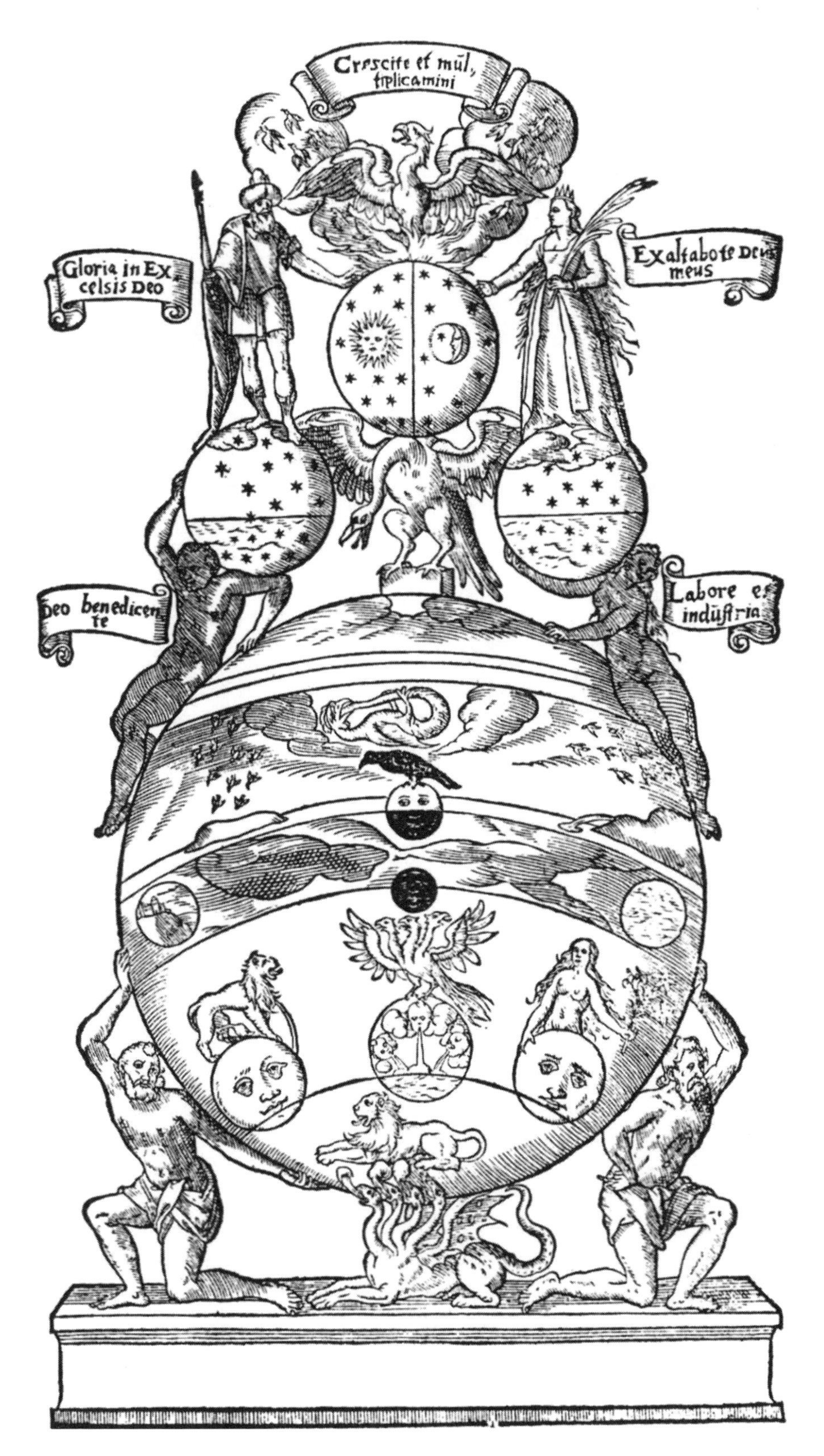

FIGURE 50. ■ Do you see the phoenix perched atop the Vase of Hermes (likened to a kind of Western mandala by alchemical interpreter Adam McLean, see p. 51)? (From Libavius' *Alchymia,* 1606.) The phoenix represents the last of the four color changes occurring during the Great Work. The work begins in darkness and abasement (black crow); the chemical mass whitens during calcination (swan), passes through bright color changes (peacock), and culminates with the rise of the red phoenix.

FIGURE 51. ■ Nineteenth-century Japanese watercolor on rice paper depicting the phoenix. Was the 1950s movie icon *Rodan* meant to represent a phoenix rising from the ashes of the nuclear bomb?

toad."[20] The mere glance of a basilisk (or distant exposure to its emanations) is deadly.[21]

> there is none that perisheth sooner than doth a man by the poison of a Cockatrice, for with his sight he killeth him, because the beams of the Cockatrices eyes, do corrupt the visible spirit of a man, which visible spirit corrupted, all the other spirits coming from the brain and life of the heart, are thereby corrupted, and so the man dyeth.

And it gets even scarier:[21]

> For it killeth, not only by his hissing and by his sight, (as is said of the Gorgons), but also by his touching, both immediately and mediately; that is to say; not only when a man toucheth the body it self, but also by touching a Weapon wherewith the body was slain, or any other beast slain by it; and there is a common fame, that a Horse-man taking a Spear in his hand, which had been thrust through a Cockatrice, did not only draw the poison of it into his own body and so dyed, but also killed his Horse thereby.

Clearly, it is a symbol for alchemical projection at a distance. Come to think of it, our newly named *Plastique Monstresque* film genre[9] affords us a 400-foot-tall basilisk called Godzilla whose breath vaporizes air force jets and army tanks hundreds of meters away.

FIGURE 52. ■ The symbol of the American Chemical Society, which includes the phoenix as well as Justus Liebig's early-nineteenth-century *kaliapparat* (see Figure 257), which revolutionized the analysis of organic compounds. This is a wonderful evocation of the mystical and rational roots of chemistry. (Used with permission from the American Chemical Society.)

FIGURE 53. ■ "Beware the amorous birds of prey . . . ." The amorous birds of prey are another representation of male-female (Sol-Luna; sulfur-mercury) duality. This image is from the 1755 *Medicinisch-Chymisch-und Alchemistisches Oraculum*.

DES VENINS. 105

DV BASILIC ROY DES SERPENS.

CHAPITRE XVIII.

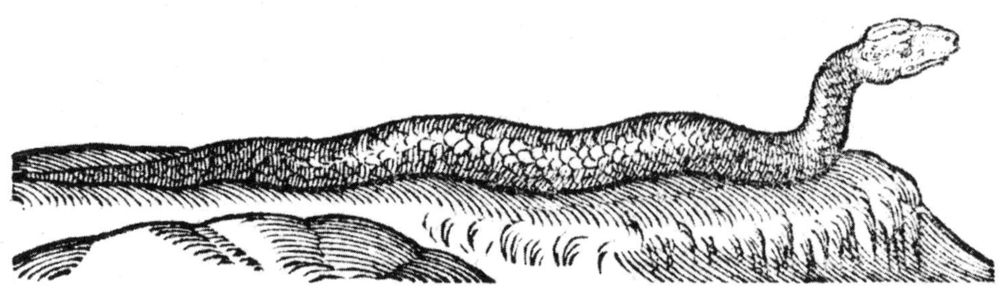

Βασιλίσκος, Baſiliſcus, Baſilic.

FIGURE 54. ■ "and shun the frumious basilisk." The basilisk (or cockatrice) is a symbol for "projection," the mysterious power of the Philosopher's Stone to transmute metals from a distance. Sometimes represented as a lizard, sometimes a combination of lizard and rooster, and sometimes as a serpent (a spitting cobra?). The mere glance of a basilisk is deadly. (From Grévin, *Deux Livres des Venins*, 1568, 1567.) (Courtesy of The New York Academy of Medicine and B & S Gventer: "Books"—each supplied copies of this image.) Was the Japanese film icon *Godzilla* a basilisk? Or perhaps another representation of the Philospher's Stone: the salamander, here as a survivor of the nuclear fire of atomic bonbs?

Actually, it is not hard to imagine cobras being the basis of basilisk mythology. They are among the world's most poisonous snakes; spitting cobras can cause blindness at a distance, and the sheer size of a king cobra (one reported to be 18 feet long[22]), coupled with this serpent's tall, vertical posture and hooded appearance, are the "stuff" of mythology.

THE OUROBOROS (OR KEKULÉ'S DREAM EXPLAINED?)

The ouroboros is a serpent constantly devouring itself as it regenerates. The concept surely flows from the molting of a snake to form a bright new skin. The circle represents unity and continuity. The ouroboros evokes the circular reflux distillation accomplished in a pelican (or even a modern reflux) apparatus. Did Auguste Kekulé dream about the ouroboros in imagining the benzene ring?[23] Figure 55, Emblem 14 from *Atalanta Fugiens*,[2] shows a variant on the ouroboros theme. Its Epigram translates as follows:[2]

The famished Polyps gnawed at their own legs,
    And hunger too, taught men to feast on men.
The Dragon bites its tail and swallows it,
    Taking for food a great part of itself.
Subdue it by hunger, prison, iron, until,
    It eats itself, vomits, dies, and is born.

Could the ouroboros also be a symbol for the law of the conservation of matter, far predating Lavoisier? Perhaps only Kekulé would have known for sure.

FIGURE 55. ■ The ouroboros—a symbol for completeness, cycle of life, and even the conservation of matter. The ouroboros continuously devours itself as it regenerates. Did August Kekulé actually dream about the ouroboros when he postulated that benzene was a cyclic compound? (From *Atalanta Fugiens*, from The Roy G. Neville Historical Chemical Library, a collection in the Othmer Library, CHF.)

---

1. N. Schwartz-Salant, *Encountering Jung On Alchemy*, Princeton University Press, Princeton, 1995.
2. M. Maier, *Atalanta Fugiens, hoc est, Emblemata Nova De Secretis Chymica, Accomodata partim oculis & intellectui, figures cupro incises, adjestisque sententiis, Epigrammatis & notis, partim auribus & recreatoni animi plus minus 50 Fugis Musicalibus trium Vocum*, . . . Oppenheimii Ex typographia Hieronymi Galleri, Sumptibus Joh. Theodori Bry, 1617. (I am grateful to The Roy G. Neville Historical Chemical Library, California, for supplying this image.) Second edition, 1618. A more recent translation of this book with English translation and commentary and a cassette audiotape of the fugues were published by Phanes Press, Grand Rapids in 1989.
3. J. Read, *Prelude to Chemistry*, The MacMillan Company, New York, 1937, pp. 236–254.
4. P. Laszlo, Aspects "de la tradition alchimique au XVIIe siecle," F. Greiner (ed.), *Chrysopoeia*, Vol. 4, pp. 278–285 (1998).
5. M. Maier, *Viatorium, hoc est, De Montibus Planetarum septem seu Metallorum* . . . , Oppenheimii, Ex typographia Hieronymi Galleri, sumptibus Joh. Theodori Bry, 1618. A second edition was published in Rouen in 1651.
6. L. Abraham, *A Dictionary of Alchemical Imagery*, Cambridge University Press, Cambridge, UK, 1998, p. 176.
7. Read, op. cit., pp. 128, 168, 244–245.
8. *The New Encyclopedia Britannica*, Vol. 9, Encyclopedia Britannica, Inc., Chicago, 1986, p. 393.

9. *Plastique Monstresque*: homage to the French who host the Caens Film Festival but also venerate Jerry Lewis, and of course to the Japanese who have given us this film genre.
10. The *Kaliapparat* is described in detail on pages 424–427 in this book.
11. The "primeval forest" was Friedrich Wöhler's description of organic chemistry.
12. With apologies to Lewis Carroll and his poem *Jabberwocky*. See also Read, op. cit., pp. 199–200. Apparently, Read actually knew what slithy toves and mimsy borogroves looked like.
13. *Medicinisch-Chymisch-und Alchemistisches Oraculum, darinnen man nicht nur alle Zeichen und Abkürzungen, welche sowohl in den Recepten und Büchern der Aerzte und Apotheker, als auch in den Schriften der Chemisten und Alchemisten verkommen, findet, sondern dem auch ein sehr rares Chymisches Manuscript eines gewissen Reichs***beygefüget*. Ulm und Memmingen, in der Gaumischen Handlung, 1755.
14. Abraham, op. cit., pp. 30–31.
15. Abraham, op. cit., pp. 23–25.
16. However, for a good time call ISBN 3–8228–8653-X. This is the Library of Congress call number for A. Roob, *The Hermetic Museum: Alchemy & Mysticism*, Taschen, Cologne, 1997. Pages 442–455 have lots of great pictures in color and black and white of the *Conjunctio*. This is soberly followed by pictures of the Rebis in which fusion seems to be permanent.
17. Abraham, op. cit., pp. 157–158.
18. J. Grévin, *Deux Livres des Venins, Ausquels il est amplement discouru des bestes venimeuses, thériaques, poisons & contrpoisons. Ensemble, Les oeuvres de Nicandre, Medecin & Poete Grec, traduictes en vers François*. Christopher Plantin, Antwerp, 1568, 1567. Both Bruce Gventer and The New York Academy of Medicine Library are gratefully acknowledged for providing images of this figure. I gratefully acknowledge helpful conversations with Miriam Mandelbaum of the Academy Library on images of the basilisk and other "fantastical" creatures.
19. *The New Encyclopedia Britannica*, op. cit., Vol. 3, pp. 420–421.
20. Roob, op. cit., pp. 354, 369.
21. E. Topsell, *The History of Four-Footed Beasts and Serpents, Volume 2. The History of Serpents Taken Principally from the Historiæ Animalium of Conrad Gesner* (reprint of 1658 London edition), Da Capo Press (Plenum), New York, 1967, pp. 677–681.
22. *The New Encyclopedia Britannica*, Encyclopedia Britannica, Inc., Chicago, 1986, Vol. 3, p. 415.
23. Read, op. cit., pp. 108, 117.

DRAGONS, SERPENTS, AND ORDER OUT OF CHAOS

The sometimes wildly allegorical depictions of alchemical relationships are well illustrated by Figures 56 and 57, which come from *Delia Tramutatione Metallica* (Brescia, 1599). It is a virtual reprint of the 1572 edition but with addition of the *Concordontia de filosifi:* a listing of alchemical works largely attributed to Arnold of Villanova.[1,2] The first edition (1564) contains a list of alchemists and alchemical works, which was expanded in the 1572 and 1599 editions.[2] The author, Giovanni Battista Nazari, is reported to have read widely in alchemy over a 40-year period but is blamed ". . . for describing spurious operations, which possibly helped ruin the people who tried them. . . ."[2] The book includes several dream sequences including one in which the author converses with Bernhardus Trevisanus (born 1406 in Padua), who, starting at age 14, devoted the remainder of his life to the study of alchemy.[4] The psychologist C.G. Jung had a lifelong interest in dreams and alchemy and owned a copy of the 1599 edition.[1]

FIGURE 56. ■ A depiction of the *tria prima* (Sophic Mercury, Sulfur, and Salt) from *Della Tramutatione Metallica* by Giovanni Battista Nazari (Brescia, 1599). A very similar figure, depicting Austrian physician Franz Anton Mesmer, appeared in an anti-Mesmer pamphlet published in 1784.[3]

Figure 56 represents the *tria prima*. Perhaps the old dragon is a representation of the ultimate source of these sophic elements—the *prima materia* or fundamental matter. Figure 57 is a depiction of the generation, starting from chaos, of the six lower metals (the six crowns) and ultimately gold (the King).

Figure 58 is a drawing, executed in 1999, by artist Rita L. Shumaker.[5] It depicts the male-female (gold–silver; sun–moon) relationship. The two entwined dragons also represent male and female (fixed and volatile) principles and, with the rod or central stem held by the male figure, form a caduceus—the familiar medical symbol. The central stem is said to consist of "the gold of the philosophers." The original form of the caduceus is said to have been a cross representing the four ancient elements.[6] The square in the background of Figure 58 represents these four elements. The drawing represents the *conjunctio*, the alchemical wedding of male and female, spirit and body. We encourage you, gentle reader, to find the "chymicall characters" (see Figure 6) in this figure. There are actually three dragons in this drawing representing the *tria prima* (salt, sulfur, mercury) as "metaphors for unconscious intuition and feeling, vital spirit or will, and the impulse to give creative form in matter."[5]

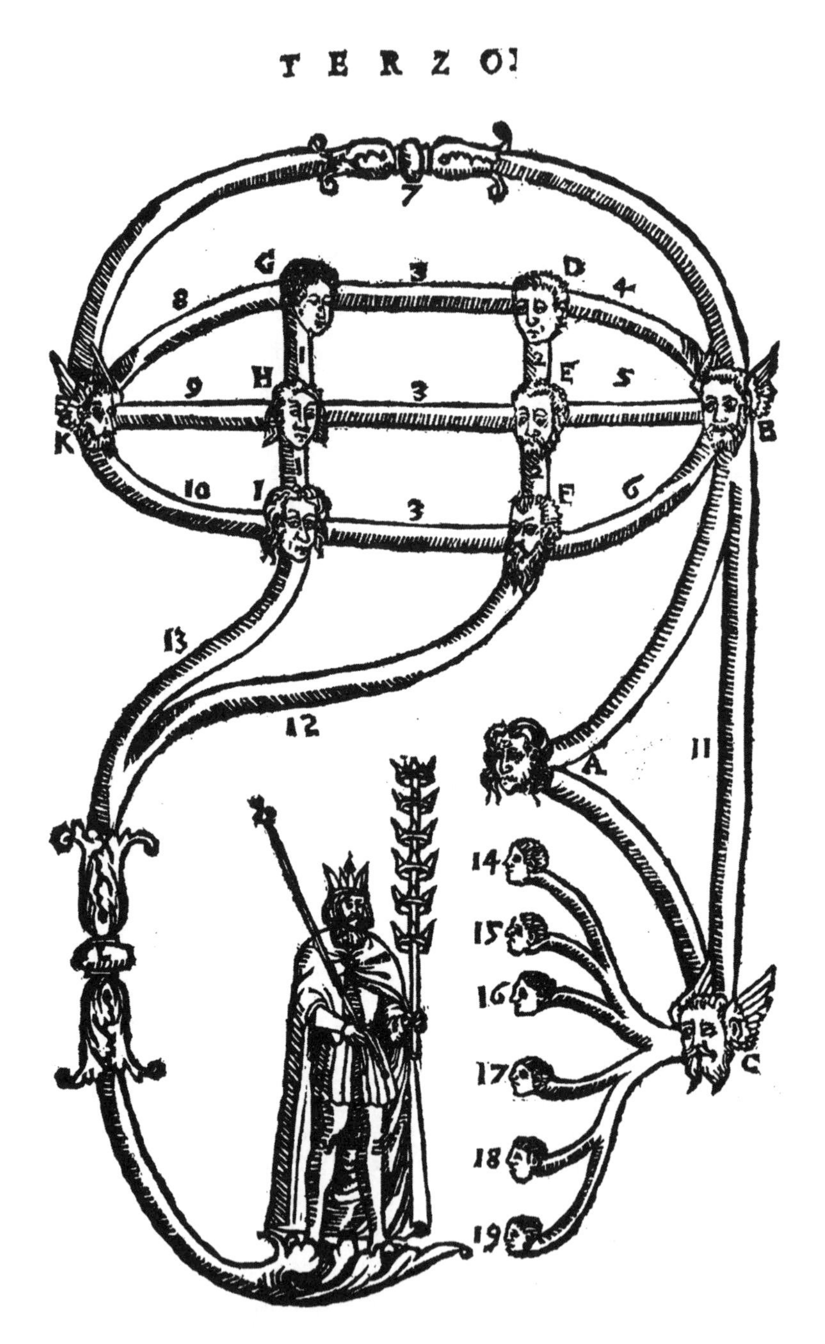

FIGURE 57. ▪ A depiction of the birth and evolution of the six lower metals (six crowns) and Gold (the King) starting from Chaos (Nazari, see Figure 56).

FIGURE 58. ■ Artist Rita L. Shumaker's rendition in 1999 of male and female allegorical images. The imagery of the caduceus is also evident in this drawing.

---

1. I. MacPhail, *Alchemy and the Occult*, Yale University Library, New Haven, 1968, pp. 178–181.
2. J. Ferguson, *Bibliotheca Chemica*, Derek Verschoyle, London, 1954 (reprint of 1906 ed.), Vol. II, pp. 131–132.
3. F. A. Pattie, *Mesmer and Animal Magnetism: A Chapter in the History of Medicine*, Edmonston, Hamilton, 1994. pp. 178–179.
4. J. Ferguson, *Bibliotheca Chemica*, Derek Verschoyle, London, 1954 (reprint of 1906 ed.), Vol. I, pp. 100–104.
5. The author thanks Ms. Rita L. Shumaker, a faculty member at the University of North Carolina at Charlotte, for this original drawing and its interpretation.
6. J. Read, *Prelude To Chemistry*, MacMillan, New York, 1937, pp. 105–116.

ALBERT THE GREAT AND "ALBERT THE PRETTY GOOD"

Toward the end of the Middle Ages (ca. 500–1450), European thinkers gathered the written lore of the ancients, combined it with knowledge acquired from Moslem cultures during the Crusades, and began to develop methods of inquiry that would begin to define modern science. One of the most important of these figures is Albertus Magnus (ca. 1200–1280).[1] He was born in Swabia (southwestern Germany) and educated at the University of Padua, where he was first exposed to and adopted Dominican beliefs. Ordained as a bishop, Albert was sent to the Dominican convent at the University of Paris some time before 1245. There he read deeply in the Aristotlean and Arabic tracts and began to interpret ancient physics and other sciences and write a summary of human knowledge. He was known as "Albert the Great" even during his own lifetime.[1] Albertus was canonized in 1931 and declared Patron Saint of the Natural Sciences by Papal decree in 1941. One of Albert's students at the University of Paris was St. Thomas Aquinas.[1]

Numerous books have been falsely attributed to Albert the Great, and only very few seem to be derived from his genuine writings.[2] Figures 59 and 60 are from a 1518 illustrated edition of one of his few authentic works on alchemy and mineralogy.[3,4] Figure 59 depicts an alchemist performing a distillation. Figure 60

Thiloninus Philyninus
Lectori.

QVi mirāda cupit populoq; indigna pphano
Noſcere, & in paruo diſcere magna libro.
Nos adeat, gēmas dabimus, gēmęq; colores
Quęq; ſit agnati patria concha ſoli
Si cupis, & venas æris, cauſaſq; metalli
Quas patitur propter vulnera tanta parens
Adde quod ex plumbo, ridebit barbar⁹, aurū
Soluere, natura vertere vera potes
Ars ea Philoſophis nomen Chemia pelaſgis
Accipit, areanis facta Magna notis.

FIGURE 59. ▪ Chemist at a still from the *Liber Mineralium Alberti Magni*, a 1518 text attributed to Albert The Great; note the poem above the figure (from The Roy G. Neville Historical Chemical Library, a collection in the Othmer Library, CHF).

FIGURE 60. ■ Note the six-line poem at the bottom of this figure from the 1518 *Liber Mineralium Albert Magni*, Oppenheim, 1518. (From The Roy G. Neville Historical Chemical Library, a collection in the Othmer Library, CHF.)

is from the last leaf of the book, often missing, and includes a six-line alchemical poem.[4]

Invoking the name Albertus Magnus or even providing tantalizing hints that implied a connection with this revered medieval genius was an effective way to sell books. In Figure 61(a) we see the title page from a nineteenth-century reprint of a "Marvelous Book of Natural Magic" from a "Petit Albert" or Albert Parvus and first published in 1668.[5,6] (I have adopted "Albert the Pretty Good" to avoid any possible confusion with Albert the Great and because it may be less insulting than "Albert the Little"). Although this book is said to be a "well-known collection of magical absurdities and impossibilities,"[6] the fact that it was reprinted for two centuries is certainly nothing to sniff at. How many university

(a)

LES SECRETS
MERVEILLEUX
DE LA MAGIE NATURELLE
DU
PETIT ALBERT
Tirés de l'ouvrage latin intitulé :
ALBERTI PARVI LUCII
Libellus de mirabilibus naturæ Arcanis
Et d'autres écrivains philosophes
ENRICHIS DE FIGURES MYSTÉRIEUSES, D'ASTROLOGIE, PHYSIONOMIE, ETC., ETC.
Nouvelle édition corrigée et augmentée

A LYON
Chez les Héritiers de BERINGOS fratres
A l'Enseigne d'Agrippa.

M. DC. LXVIII.

(b)

(c)

FIGURE 61. ■ (a) Title page from a nineteenth-century reprint of the 1668 "Marvelous Secrets" of "Little Albert" ["Albert The Pretty Good" (?)] as well as images of (b) Venus (copper) and (c) Jupiter (tin) in their triumphal chariots.

or commercial presses can claim such a best seller? Figures 61(b), 61(c), and 62(a–d) are depictions of gods and goddesses in triumphal chariots heralding the six ancient metals besides gold: Venus (copper), Jupiter (tin), Saturn (lead), Mercury, Luna (silver), and Mars (iron).

Here is Albert Parvus' recipe for Le Toothpaste:[7]

> Take blood of Dragon and Cinnamon three ounces, calcined alum two ounces; reduce all to a very fine powder, and polish your teeth twice each day.

Sound advice and a good formula—but where to get that first ingredient?

FIGURE 62. ■ Triumphal chariots of (a) Saturn (lead), (b) Mercury (quicksilver or mercury), (c) Luna (silver), and (d) Mars (iron) from "*Petit Albert*" (see Figure 61).

1. *The New Encyclopedia Britannica*, Encyclopedia Britannica Inc., Chicago, 1986, Vol. 1, pp. 218–219.
2. J. Ferguson, *Bibliotheca Chemica*, Derek Verschoyle, London, 1954, pp. 15–17.
3. Albertus Magnus, *Liber Mineralium Alberti Magni . . . Sequitur tractatus de lapidum et gemmarum material accidentibus . . . virtutibus ymaginibus, sigillis. De alchimicis speciebus, operationibus et utilitattibus. De metallorum origine et inventione, generatione . . . colore . . . virtute, transmutatione. Ad Emtores Thilonius*, Jacob Koebel, Oppenheim, 1518. I am grateful to The Roy G. Neville Historical Chemical Library for furnishing these images.
4. The Roy G. Neville Historical Chemical Library (California), catalog in preparation. I am grateful to Dr. Neville for helpful discussions.
5. A. Parvus, *Les Secrets Merveilleux de la Magie Naturelle du Petit Albert, Tirés de l'ouvrage latin intitulé: Alberti Parvi Lucii Libellus de mirabilibus naturæ Arcanis Et d'autres écrivains philosophes . . .* , Chez les Héritiers de Beringos fraters, Lyon, 1668 (The copy employed here is a nineteenth-century reprint.)
6. Ferguson, op. cit., p. 17.
7. Parvus, op. cit., p. 154.

A CANTERBURY TALE OF ALCHEMY

Was England's greatest poet a true Adept, or merely adept at rhyming verses? The *Canon's Yeoman's Tale* (or *CYT*) of Geoffrey Chaucer (ca. 1340–1400) implies such a detailed knowledge of alchemical operations[1] that Elias Ashmole[2] included this work in his *Theatrum Chemicum Britannicum*, published in 1652, among those of the other "Famous English Philosophers who have written the Hermetique Mysteries in their owne Ancient Language."[3] Figures 63 and 64 are illustrations from the *Theatrum*. In the first figure, the Master Adept bestows alchemical secrets on the young alchemist: "Receive the gift of God un-

FIGURE 63. ■ "Receive the gift of God under the sacred seal" sayeth the Master Adept to the young alchemist (from Ashmole, *Theatrum Chemicum Britannicum*, 1652, from The Roy G. Neville Historical Chemical Library, a collection in the Othmer Library, CHF).

FIGURE 64. ■ Here is a well-funded Renaissance research laboratory. Years later, the young alchemist in Figure 63 can now afford a research scientist on the GOld from Lead Discovery ("GOLD") program funded by the National Treasury. (Ashmole, *Theatrum Chemicum Britannicum*, from The Roy G. Neville Historical Chemical Library, a collection in the Othmer Library, CHF.)

der the sacred seal."[4] The next figure shows an active and well-funded laboratory suggesting that the young adept has indeed heeded good academic counsel. He has become a successful grantsman and is well on his way to tenure and promotion.

No less an expert than John Read suggested that "Chaucer himself had first-hand experience of the joys and sorrows of a 'labourer in the fire.'"[5] To these bits of circumstantial evidence, we now add the apparent "smoking gun"—sixteenth-century manuscripts in the library of Dublin's Trinity College titled *Galfridus Chauser his worke*, describing two alchemical procedures for obtaining the Philosopher's Stone followed by a poem concerning the Elixir.[1]

However, Gareth Dunleavy's careful research suggests that these manuscripts are pseudepigraphons falsely attributed to Chaucer.[1] False attributions to Geber, Albert the Great, Arnold of Villanova and Hermes himself were not uncommon attention-getting devices during the Renaissance. Dunleavy indicates that while Chaucer might have been familiar with general aspects of alchemy, the details in CYT closely resemble the writings of Arnold of Villanova.[1] Thus, he is skeptical about Chaucer the alchemist but notes that the manuscript itself may once have belonged to the library of John Dee, astrologer, mathematician, and alchemist to Elizabeth I.[1]

Now, back to *The Canterbury Tales*. The CYT Prologue sets the scene. The canon, a clergyman who, in this tale, is also an alchemist, is accompanied by his yeoman or assistant as they encounter a group of travelers on the road. The canon is dismissed by the group's host and the ash-darkened, poverty-stricken, indentured yeoman, who has been badly used by his master, tells a bitter and ironic tale of alchemical chicanery. The canon appears to be part "puffer" (earnest but misguided seeker of The Stone) and part charlatan.

The canon has offered to transmute a Priest's quicksilver into precious silver metal using a mysterious powder of projection. In reality, he has placed an ounce of pure silver into a hole drilled in a lump of coal and sealed the hole with blackened wax. The yeoman describes the canon's bait for the priest:

> For here shul ye se by experience,
> That this quicksylver I wol mortifye
> Right in your syght anon withouten lye,
> And make it as good Sylver and as fyne,
> As there is any in your purse or myne,

The canon produces his mysterious powder:

> I have a poudre that cost me deere,
> Shall make all good, for it is cause of all
> My connyng, which I you shewe shall

The greedy and gullible priest watches as the canon removes his own crucifix ("crosslet") and sets it in the fire. To this, the priest adds his quicksilver and the canon adds some of the powder. In the yeoman's bitter words:

> This Preest at this cursed Chanon's byddyng,
> Uppon the fyre anon set this thyng;
> And blewe the fyre and besyed him ful faste,
> And this Chanon into this crosslet caste
> A pouder, I not whereof it was,
> Ymade either of Chalke, Erthe, or Glasse
> Or somwhat els, was not worth a fly,

And now the Canon's trick:

> This false Chanon, the foule fende him fetche;
> Out of his bosome toke a bechen cole,
> In which ful subtelly was made an hole,
> And therein was put of Sylver lymayle,[6]
> An unce, and stopped was without fayle,
> The hole with waxe to kepe the Lymayle in.

The priest is industriously tending the fire and the canon distracts him by noting that the burning coals need to be rearranged and offering the priest a cloth to wipe his sweaty face. Whereupon the hollowed-out lump of coal is added, the fire is vigorously stirred up, and the canon then joins the priest for a hearty drink. Returning to the fire, the canon finds and recovers metallic silver for the delighted priest.

Now a second demonstration occurs in which the crafty canon actually leaves the priest to perform the transmutation on his own. He provides a hollowed-out stirring stick to the priest that . . . you guessed it . . . is filled with an ounce of silver secured with blackened wax. So now, completely outside the influence of the canon, powder of projection works for the priest himself. The canon works a final demonstration-transmuting copper to silver, leaving the priest in an ecstasy of joyful greed:

> This sotted Preest who was gladder than he,
> Was never Byrd gladder agenst the day,
> Ne Nightyngale agenst the ceason of May,
> Was never none, that lyft better to synge,
> Ne Lady lustier in Carolyng:

The priest pays the canon forty pounds, a vast sum, for his secret (including powder I suspect). Note Chaucer's distrust of a clergy so widely perceived during the Renaissance as corrupt. There are two clergyman here—one, poor but dishonest; the other, gullible yet incredibly wealthy.

One can just hear the canon as he bids the priest adieu—"A Fulle Moneybacke Guarantee! And if you're ever in Canterbury . . . try to find me."

---

1. G.W. Dunleavy, Ambix, Volume XIII, No.1, pp. 2–21 (1965). I am grateful to the late Professor Gareth Dunleavy for helpful discussions.
2. Elias Ashmole (1617–1692) was a gentleman of incredibly wide-ranging interests whose collection formed the basis for the first public museum in England, the Ashmolean Museum of Oxford University. A book published in 1650, titled *Fasiculus Chemicus: Or Chymical Collections . . .* was authored by a James Hasolle [an interesting, perchance smutty(?), anagram of Eljas Ashmole].
3. E. Ashmole, *Theatrum Chemicum Britannicum. Containing Severall Poeticall Pieces of our Famous English Philosophers, who have written the Hermetique Mysteries in their owne Ancient Language. Faithfully Collected into one Volume, with Annotations thereon.*, J. Grismond for Nath:Brooke, at the Angel in Cornhill, 1652. The images were supplied by The Roy G. Neville Historical Chemical Library (California). See also the facsimile reprint with a preface by C.H. Josten, Georg Olms Verlagsbuchhandlung, Hildesheim, 1968.

4. S. Klossowski De Rola, *The Golden Game. Alchemical Engravings of the Seventeenth Century*, Thames and Hudson, London, 1988, pp. 214–221.
5. J. Read, *The Alchemist in Life, Literature and Art,* Thomas Nelson and Sons Ltd., London, 1947, p. 29.
6. Powder or filings—"lymayle" rhymes better with "fayle."

THE SHIP OF FOOLS

In 1494, some 20 years prior to the Protestant Reformation, Sebastian Brant,[1,2] a German poet and humanist, published a long poetic satire titled *The Ship of Fools* (*Das Narrenschiff*). He has been termed "a man of deep religious convictions and of stern morality, even to the point of prudishness."[3] The book imagined a collection of "fools" reflecting mores and excesses that would have tickled the fancies of readers of the day by deflating recognizable character types. The ship, loaded with these fools, was bound for "Narragonia," the Land of Fools. The book's language was accessible, the woodcuts (some possibly by Albrecht Durer)[4] handsome and amusing. Six editions appeared during Brant's life (first English in 1509) with numerous additional authorized and pirated editions through 1629.[5] The book was "rediscovered" two centuries later and an edition published in 1839 with others following throughout the nineteenth century and into the early twentieth century.

The stern Brant did not have a very high regard for the sensual pleasures procurable in the streets of Basel. The prologue to his fiftieth poem, "Of Sensual Pleasure," expresses his self-righteous scorn:[6]

> The stupid oft by lust are felled
> And by their wings are firmly held:
> For many, this their end hath spelled.

Romantic "night music" was also verboten—according to the prologue to Number 62 "Of Serenading at Night":[7]

> The man who'd play the amorous wight
> And sing a serenade at night
> Invites the frost to sting and bite.

So, no sensual pleasures or serenading at night on the streets of Basel! You wouldn't expect Brant to be very open-minded or have a sense of humor about alchemy either, and he doesn't disappoint. Thus, we see in Figure 65[8] alchemists in dunce hats (Oh, the shame of it!) and a snippet from poem 102:[9] "Of Falsity and Deception" (Ha! There's a dead giveaway!):

> But let there not forgotten be
> Our quite deceptive alchemy:
> Pure gold and silver doth it yield
> But this in ladles was concealed.

FIGURE 65. ■ "Our quite deceptive" alchemist from *The Ship of Fools*. This figure is from the 1506 Basel edition, from The Roy G. Neville Historical Chemical Library, a collection in the Othmer Library, CHF.

Ah, the Canterbury Canon's old "gold-hidden-in-the-ladle-or-stirrer" trick. I wonder if Brant's was the voice of experience or whether he had only witnessed the bamboozling of wealthy priests and other easy marks.

1. *The Catholic Encyclopedia*, Vol. II, 1907, Robert Appleton Co.
2. S. Brant, *The Ship of Fools*, translated into rhyming couplets with introduction and commentary by Edwin H. Zeydel, Dover Publications, New York, 1962 (reprint of 1944 edition).
3. Brant, op. cit., p. 7.
4. Brant, op. cit., p. 20.
5. Brant, op. cit., pp. 21–24.
6. Brant, op. cit., pp. 178–180.
7. Brant, op. cit., pp. 206–208.
8. This figure is from the 1506 Basel edition translated by J. Locher into Latin and reinterpreted by Badius Ascensius, courtesy of The Roy G. Neville Historical Chemical Library (California), catalog in preparation. I am grateful to Dr. Neville for helpful discussions.
9. Brant, op. cit., pp. 327–330.

THE FIRST MODERN ENCYCLOPEDIA

The elegantly simple illustration of an alchemist tending his furnace, with distillation apparatus in the background, depicted in Figure 66 is found in the first edition of the *Margarita Philosophica*, published in 1503.[1,2] It is "the first modern encyclopedia of any importance"[3] and was printed less than fifty years after Johannes Gutenberg printed his first books in 1455. The *Margarita Philosophica* reflects the university curriculum at the end of the fifteenth century. It covers grammar, logic, rhetoric, mathematical topics, astronomy, music, childbirth, astrology, and hell.[4] Books 8 and 9 cover chemical topics, including transmutation.[3] The author, Gregorius Reisch, was the Prior of a Carthusian monastery at Freiburg and confessor of Maximilian I,[4] Holy Roman Emperor (1493–1519), who established the dominance in Europe of the Habsburg Family.[5] Figures such as 59, 60, 65, and 66 are elegant in their simplicity, and three more figures (Figures 67–69) are from the incunabula (pre-1501) and immediate post-incunabula periods.[6]

FIGURE 66. ■ An early-sixteenth-century alchemist from "the first modern encyclopedia of any importance" (Reisch, *Margarita Philosophica*, 1503, from The Roy G. Neville Historical Chemical Library, a collection in the Othmer Library, CHF).

FIGURE 67. ■ Woodcut from Bartholomaeus de Glanvilla, Anglicus. *De las Propriedades de las Cosas*, Toulouse, Henri Mayer, 1494.[6] This is a printed edition of a very famous encyclopedia of the Middle Ages. It depicts a visit by a physician and a consult with an apothecary.

FIGURE 68. ■ Woodcut from Hieronymus Brunschwig, *Buch der Vergift des Pestilenz*, Strassburg, Johann (Reinhard) Grüninger, 1500.[6] This book by the Strassburg surgeon Hieronymus Brunschwig concerns the plague. The woodcut depicts an apothecary preparing a draught.

FIGURE 69. ■ Woodcut from Hortus Sanitatis, *Le Jardin de Santé*, Paris, Philippe le Noir, ca 1510.[6] It depicts a doctor in his laboratory.

1. G. Reisch, *Margarita Philosophica* (*totius philosophiae rationalis, naturalis et moralis principia dialogice duodecim libris complectens*), Joannem Schott, Freiburg, 1503. The author is grateful to The Roy G. Neville Historical Chemical Library (California) for supplying a copy of the woodcut in Figure 16 and to Dr. Neville for helpful discussions.
2. J.R. Partington, A History of Chemistry, MacMillan & Co., Ltd., London, 1962, Vol. 2, p. 94.
3. D.I. Duveen, *Bibliotheca Alchemica et Chemica*, facsimile reprint, HES Publishers, Utrecht, 1986, p. 501.
4. Neville, Roy G., *The Roy G. Neville Historical Chemical Library: The Annotated Catalogue of*

Printed Books on Alchemy, Chemistry, Chemical Technology, and Related Subjects, Chemical Heritage Foundation, Philadelphia, 2006. I am grateful to Dr. Neville for helpful discussions.

5. *The New Encyclopedia Britannica*, Encyclopedia Britannica, Inc., Chicago, 1986, Vol. 7, p. 965.
6. Maggs Brothers, No. 520, Manuscripts and Books on Medicine, Alchemy, Astrology & Natural Sciences, London, 1929. See the following pages for: Figure 67 (pp. 43–44); Figure 68 (pp. 70–71); Figure 69 (pp. 86–87).

TODAY'S SPECIALS: OIL OF SCORPION AND LADY'S SPOT FADE-IN CREAM

Figure 70 is the frontispiece from the 1608 book *De Distillatione* depicting the author Giambattista Della Porta (1545–1615),[1] a polymath who authored books on plants, physiognomy, physics, chemistry, and mathematics, wrote "some of the best Italian comedies of his age," and published a design for a steam engine.[1,2] "This book is as rare as it is beautiful."[3] The dedications in the preface are set in Hebrew, Persian, Chaldaic, Illyrian, and Armenian typescripts attributed to the Vatican type foundry.[4]

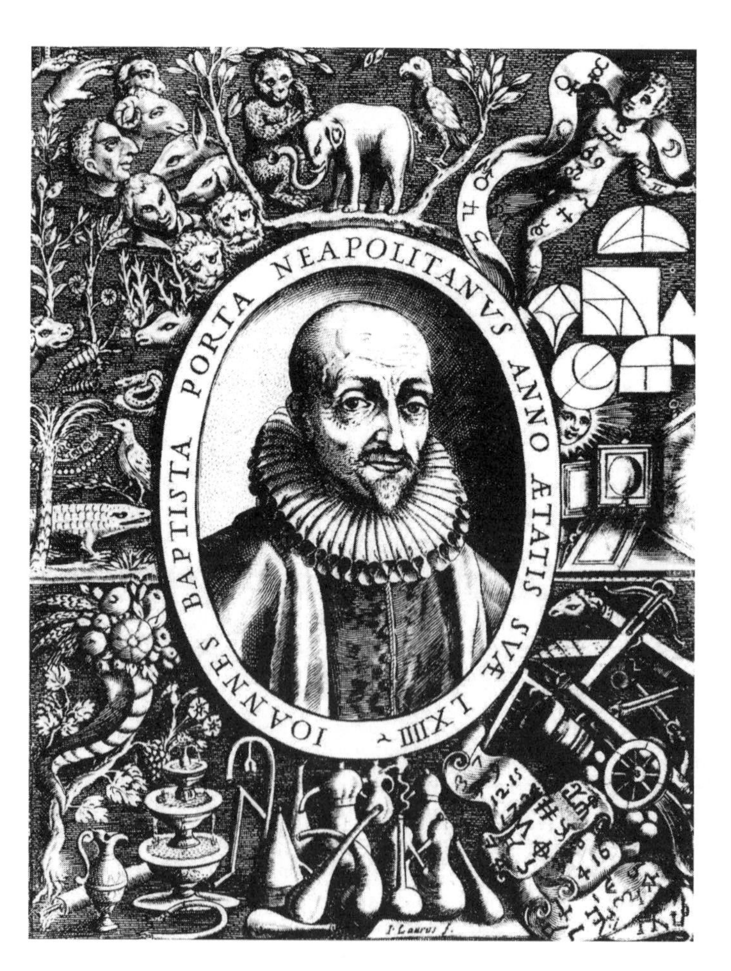

FIGURE 70. ■ Frontispiece depicting the polymath Giambattista Della Porta in his beautiful book *De Distillatione Lib. IX* (Rome, 1608).

Porta's book *Magia Naturalis*, first published in 1558, a compendium of popular science, was reprinted for over 100 years. A mixture of technical information and misinformation, it cites the procedure of the Greek physician and pharmacist Pedanius Dioscorides[1] (ca. 40–ca. 90 A.D.) for heating "antimony" [really stibnite—see Saturn and the wolf in Figure 38(a)] into lead despite the fact that sixteenth-century practitioners knew they were different and could not be so interconverted.[5] *Magia Naturalis* includes a preparation of a cosmetic that will produce spots (a kind of fade-in cream for women)—a bit of Renaissance fraternity house humor perhaps.

De Distillationibus also exemplifies the playful wit of the Renaissance, likening chemical glassware to animals. Figure 71(a) depicts a *matrass*[6,7]: it has a round bottom and long neck like an ostrich (phials for rectifying alcohol had a similar appearance) and is part of a distillation apparatus called an *alembic*, which has a distilling head that could be attached to a receiver (see Figures 72 and 73). The liquid to be distilled must be fairly volatile to make it to the top of the long neck. Figure 71(b) is a flat, stylized retort called a tortoise along with a rather stylized tortoise with a doglike head.

Could the hexagons with circles inside them on the tortoise's shell be a leap of about 330 "years into the future to our modern structure for benzene? We suspect not since benzene would not be discovered for another 200 years. However, when we discover that Kekulé claimed in the 1860s to have dreamed of benzene's structure formed from three snakes biting tails in a circle, perhaps a subliminal message from another reptile 260 years earlier might not seem quite so strange.

The distillation apparatus in Figure 72(a) places the alembic head on top of a wide-mouth flask (a kind of *cucurbit*, a more squat version of a matrass). This apparatus would be more useful for a less volatile liquid. Figure 72(b) is a one-piece *pelican*. Note how the bird's neck forms a curved arm as it bites its chest. When

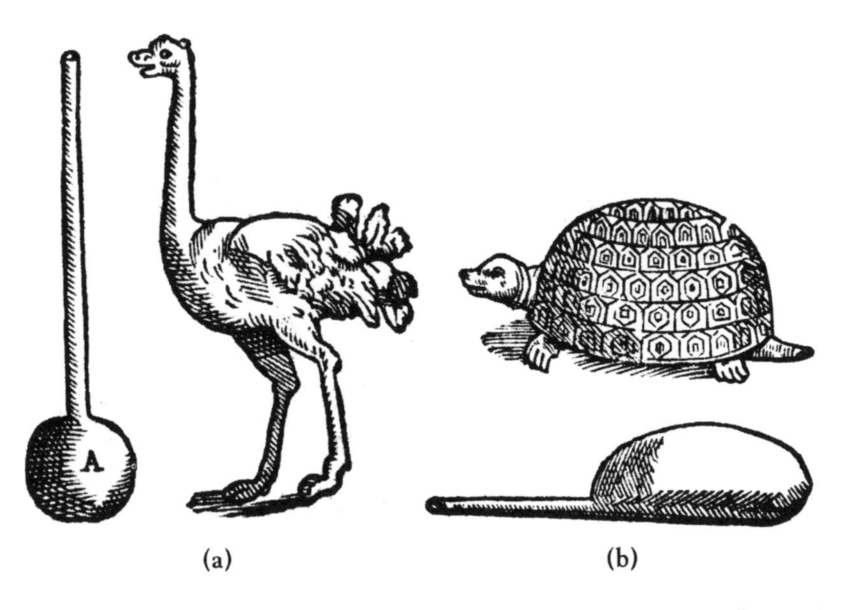

FIGURE 71. ■ Depictions of glassware and metaphors from Porta's *De Distillatione* (Figure 70): (a) Matrass and ostrich; (b) Flat retortlike "tortoise" and tortoise (somewhat "dog-headed" methinks; is that benzene on the shell?).

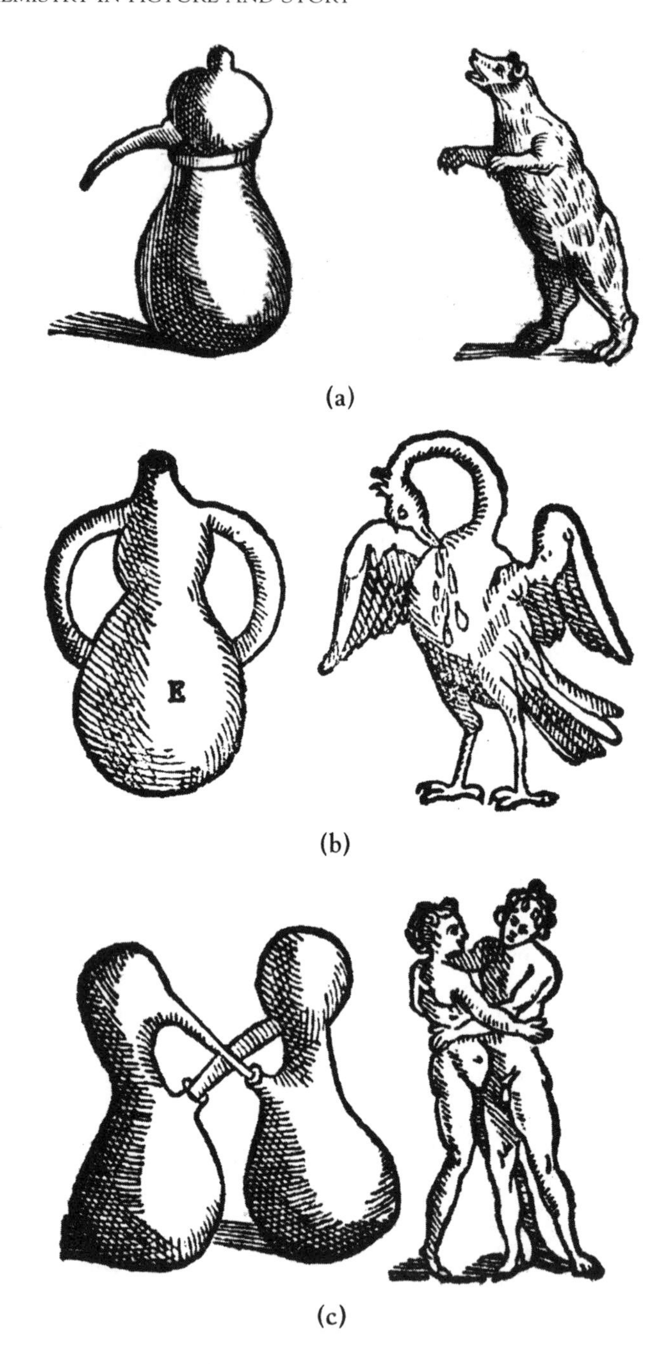

FIGURE 72. ■ Depictions of glassware and metaphors from Porta's *De Distillatione* (Figure 70): (a) Distillation apparatus employing a distillation head (alembic) atop a wide-mouth flask (or cucurbit) along with matching bear; (b) one-piece pelican for refluxing a liquid and the pelican itself biting its chest—considered a blood of Christ symbol; (c) A double pelican for prolonged exchange of hot fluids and an interesting metaphor.

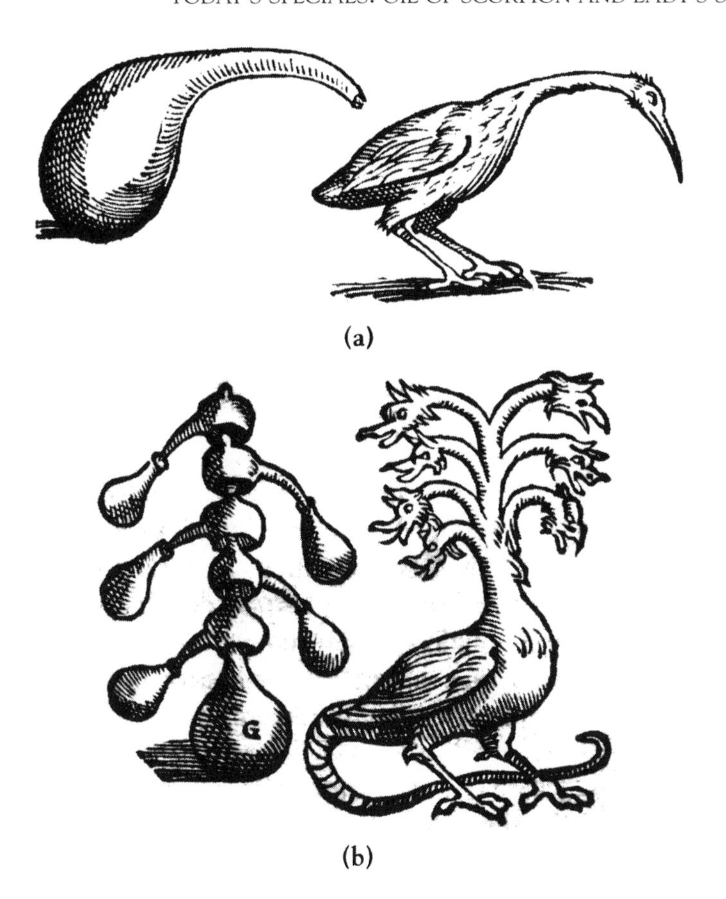

FIGURE 73. ■ Depictions of glassware and metaphors from Porta's *De Distillatione* (Figure 70): (a) Common retort and appropriate bird; (b) fractional distillation apparatus and depiction of a seven-headed beast (or perhaps the Organic Chemistry Laboratory Instructor).

closed at the top, the pelican was used for prolonged heating at the boiling point of the recirculating (refluxing) solvent. Figure 72(c) shows a *double pelican* in which the two wedded vessels exchange vapors and fluids for a prolonged period. We hesitate to provide further interpretation of the metaphor except to remind the reader that the book *was* printed in Rome seemingly with some degree of church assent.[4] Figure 73(a) shows a common retort. Figure 73(b) depicts a still capable of fractional distillation. The upper receivers are enriched in the more-volatile substances and the lower vessels are enriched in the less-volatile substances. Since fractional distillation is one of the first experiments in an introductory organic chemistry course, perhaps the seven-headed monster is a college sophomore's preconception of his or her laboratory instructor. Then again, perhaps not.

1. Also called Giambattista della Porta as well as Giovanni Battista Della Porta (see *Encyclopedia Brittanica,* 15th ed., 1986, Chicago, Vol. 9, p. 624, which lists his birthdate as "1535?").
2. J. Ferguson, *Bibliotheca Chemica*, Derek Verschboyle, London, 1954, Vol. II, p. 216.
3. D.I. Duveen, *Bibliotheca Alchemica et Chemica*, HES, Utrecht, 1986, p. 481.
4. I. MacPhail, *Alchemy and The Occult*, Yale University Library, New Haven, CT, 1968, Vol. 1, pp. 212–215.

5. J.M. Stillman, *The Story of Alchemy and Early Chemistry*, Dover, New York, 1960, pp. 349–352.
6. J. Ekiund, *The Incompleat Chemist—Being An Essay on the Eighteenth-Century Chemist in the Laboratory With a Dictionary Of Obsolete Chemical Terms of the Period*, Smithsonian Institution Press, Washington, D.C., 1975.
7. F. Ferchl and A. Sussenguth, *A Pictorial History of Chemistry*, William Heinemann, London, 1939, pp. 73–75, 105–108.

"VULGAR AND COMMON ERRORS"

Why would a knowledgeable scholar like Porta reinforce incorrect information such as that heating of antimony produces lead? Scientific experimentation was still only in its infancy. Early writers such as Pliny often turned folklore into fact. In his book *Pseudodoxia Epidemica: Or, Enquiries Into Very Many Received Tenents, and Commonly Presumed Truths*[1] the physician Thomas Browne notes on page 83:

> And first we hear it in every mans mouth, and in many good Authors we reade it, That a Diamond, which is the hardest of stones, and not yielding unto steele, Emery, or any thing, but its own powder, is yet made soft, or broke by the bloud of a Goat;

Goat's blood softens a diamond so that it can be shattered? Browne refers to this "vulgar and common error" and notes that, while some scholars accepted it, diamond cutters, whom we can presume as unscholarly, knew it was not true. He traces the misconception to the notion that in order to produce such potent blood, some scholars wrote that goats must be fed certain herbs that were said to dissolve kidney stones in humans. Since kidney stones are also extremely hard and can be "broken," why not diamonds?

Browne further noted that "glasse is poyson, according unto common conceit." Yet he pointed out that glass is made from sand, which is not poisonous. He had also fed finely ground glass to dogs: "a dram thereof, subtilly powdered in butter or paste, without any visible disturbance." The confusion arises from the common and successful practice of adding "glasse grossely or coursely powdered" to bait in order to "destroy myce and rats." Clearly, it is internal bleeding caused by the coarse glass rather than the chemical nature of glass that is deadly.

---

1. Thomas Browne, *Pseudodoxia Epidemica: Or, Enquiries into Very Many Received Tenents, and Commonly Presumed Truths*, T.H. for Edward Dod, London, 1646. I thank my daughter Rachel Greenberg for bringing goat's blood and diamonds to my attention.

WHAT IS WRONG WITH THIS PICTURE?

"Distillatio" (Figure 74) is an engraving[1] by Phillip Galle executed around 1580 from an oil by Stradanus (Jan van der Straet or Johannes Vander Straaten,

FIGURE 74. ■ *Distillatio*, an engraving (ca. 1580) by Phillip Galle after a painting by Stradanus. How many laboratory safety violations can you spot?

1523–1605), who painted other alchemical scenes as well.[2] This beehive of activity would cheer the heart of any modern-day research director or university professor. The alchemist is perhaps reading from the contemporary chemical literature and, in the manner of the recently born scientific method, trying to replicate a recent advance (Porta's 1558 work on the distillation of scorpion oil?).

What is wrong with this picture? Well, for starters, none of the graduate students, post-doctoral researchers, or technicians is wearing any eye protection. The professor's eyeglasses might pass muster through the middle twentieth century but not afterward as long as they lack protection on the sides. The ventilation system is antiquated to say the least; there is no evidence of fire extinguishers or a sprinkler system. A visit from the Fire Marshall should be in the offing. Admittedly, the large pestle hanging by an elastic band in the right front is a nice safety touch, although there should be a protective wire screen around it to keep it from swinging and conking an unsuspecting researcher.

The laboratory seems to be well equipped with the best the late sixteenth century has to offer. That large water bath with the multitude of stills shown in the center suggests a well-funded research operation. The two (!) hooded stills at the right front, one in operation and one idle (without hood), and the high-tech athanor, a furnace for incubating the Egg of the Philosophers, in the upper left, suggest that money is no object and that the alchemist is a good grantsman. And therein lies the greatest inconsistency. The professor is actually in "the trenches" doing science with his research group and not writing funding proposals, midsemester grade warnings, or explanations of low teaching evaluations by his students. Post-tenure reviews are still over 400 years away.

John Read describes this picture as a depiction of a late-sixteenth-century Italian laboratory bustling with "ordered and affluent activity."[2] This is in marked contrast to the poverty depicted in the 1558 "An Alchemist At Work" by Pieter Brueghel the Elder.[2] The sheaf of grain lying on the floor in the Stradanus "Distillatio" is said by Read to typify the "vital principle" although we would recognize it today as a fire hazard.

1. This engraving is from the author's private collection.
2. J. Read, *The Alchemist in Life, Literature and Art,* Thomas Nelson, London, 1947, pp. 66–68.

PROTECTING THE ROMAN EMPIRE'S CURRENCY FROM THE BLACK ART

Figure 75 is a whimsical eighteenth-century drawing[1] partly in the style of David Teniers, the Younger. There is a mysterious Arab, or possibly a Jew,[2] inappropriately garbed for a day in the laboratory. Then, there is the furtive figure peering in the doorway—a dark, sinister-looking cloaked character. From the expression of the alchemist, ancient magic is happening in the flask or perhaps there's a clue in the analysis of a woman's urine.[2]

Egyptian and Arabic cultures played crucial roles in the development of practical chemistry and alchemy.[3] Figures and ornaments of almost pure copper

FIGURE 75. ■ Pen-and-wash drawing of an alchemist (or a physician) signed F.P. Bush, 1769 from the author's personal collection (photograph by Dr. James Tait Goodrich, MD).

dating from around 4000 B.C. have been isolated from ancient Egyptian and Chaldean sites (Chaldea is now southern Iraq). Bronze alloys from Egypt dating to 2000 B.C., glass furnaces found at Tel-El Amarna dating back to 1400 B.C., evidence of pigments, cosmetics, and medicines all support the early and profound impact of these cultures.

While there is evidence suggesting the possibility of an even earlier origin in China,[3] the role of ancient Middle Eastern cultures in preserving western culture and in the beginnings of chemistry is undisputed. The origins of the word *alchemy* itself are very murky. *The Oxford English Dictionary*[3] cites the Arabic *alkimiya* derived from the Greek *chymeia*—itself related to a word meaning "to pour." A completely different origin[4] is also cited by this same august reference work referring to *Khem*, a word meaning "black," as the ancient name for Egypt due to the blackness of its soil. Alchemy could have a double meaning here: referring to its Egyptian origins or its place as "a black art." And even here, "black" could refer to alchemy's dark and secretive nature *or* to the first step on the pathway to the Philosopher's Stone—the *Nigredo*—or initial conversion of matter to blackness. The father of alchemy, Hermes Trismegistus ("Hermes Thrice Magisterial"), who was said to predate Jesus by 2500 years, was an invention. 'Tis a mystery. He is the reputed author of the mythical *Emerald Tablet*.

What is known is that the first verifiable person attached to an alchemical manuscript was Zosimos of Panopolis, who wrote in Alexandria, Egypt around 300 A.D. Alexandria was home to the greatest library of the classical world. Started in the third century B.C., it housed 400,000 to 500,000 books and manuscripts, mostly in Greek. The library was largely destroyed during civil wars toward the end of the third century A.D. and its "daughter library" sacked by Christians in 391 A.D.[5]

It is noteworthy that the Roman Emperor Diocletian,[6] who ruled from 285 to 305 A.D., was said to have ordered the destruction of alchemical books and manuscripts throughout the Roman Empire. As the story goes, he feared that transmutation of base metals to silver and gold would devalue the Empire's currency. (However, see the next essay, p. 135).

What manner of Emperor would destroy alchemical books and manuscripts merely to preserve the value of the Empire's currency? Being a bibliophile but not a scholar of antiquity, I tried to assess Diocletian as a politician from a *fin de siecle* (*actually, fin de millenium*) American perspective. Diocletian[6] stood for preservation of ancient virtues and the obligation of children to feed their parents in old age (no Social Security program here: a "social conservative"? a conservative Republican?). He also introduced a progressive income tax and the beginnings of the vast system of bureaucracy and technocracy that even today makes visits to state Departments of Motor Vehicles so memorable (a "tax-and-spend" liberal Democrat?). The coins he was trying to protect were inscribed *dominus et deus* ("ruler and god"). Does any reader out there know the Latin word for the Greek *hubris*[7].

In 1979, the Nobel Prize in Physics, awarded for a theory of unified weak and electromagnetic interactions between elementary particles, was shared by three scientists: Sheldon Glashow, Steven Weinberg, and Abdus Salam, two Jews born in New York and a Moslem born in Pakistan. Salam's Nobel Prize lecture[7] was particularly beautiful. He told of a young Scotsman named Michael who, almost eight centuries earlier, had traveled to study at the Arab Universities of Toledo and Cordova in Spain, centers for "the finest synthesis of Arabic,

Greek, Latin, and Hebrew scholarship" and home to the Hebrew scholar Maimonides. Salam notes that Sarton's *A History of Science* credits the period 750 to 1100 A.D. to an unbroken period of intellectual dominance by Middle Eastern cultures. In contrast to wealthy countries with flourishing schools of research such as Syria and Egypt, Scotland, a poor but developing land, had little to offer Michael upon his return, and Salam says: "At least one of his masters counseled young Michael the Scot to go back to clipping sheep and to the weaving of woolen cloth." But it was around this time that scientific superiority began to shift to the West, and Salam continues:

> And this brings us to this century when the cycle begun by Michael the Scot turns full circle, and it is we in the developing world who turn westward for science. As Al-Kindi wrote 1100 years ago: "It is fitting then for us not to be ashamed to acknowledge truth and to assimilate it from whatever source it comes to us. For him who scales the truth there is nothing of higher value than truth itself; it never cheapens or abases him."

---

1. This pen and wash drawing is signed "H.P. Bush, fecit 1769" and is from the author's personal collection. The author thanks Dr. James Tait Goodrich, M.D., James Tait Goodrich Antiquarian Books and Manuscripts for a photograph of this drawing.
2. The author thanks Dr. Alfred Bader, founder of Aldrich Chemical Company and renowned art collector, for his interpretation (personal correspondence). The stylized letters on the chemist's garb evoke both Arabic and Hebrew.
3. J.M. Stillman, *The Story of Alchemy and Early Chemistry*, Dover, New York, 1960, Chap. I.
4. *The Oxford English Dictionary*, Clarendon, Oxford, 1989, Vol. 1, p. 300.
5. *Encyclopedia Brittanica*, 15th ed., 1986, Vol. 1, Chicago, p. 251.
6. *Encyclopedia Brittanica*, 15th ed., 1986, Vol. 4, Chicago, pp. 105–106.
7. A. Salam, *Reviews of Modern Physics*, 52(3):525–526, 1980. I am grateful to Professor Joel F. Liebman for making me aware of and suggesting Salam's Nobel Prize lecture.

WHO IS ATHANASIUS KIRCHER, AND WHY ARE THEY SAYING THOSE TERRIBLE THINGS ABOUT HIM?

Figures 76–78 are from the 1665 book *Mundus Subterraneus*[1] authored by Athanasius Kircher (1602–1680), a Jesuit priest whose early professorial appointment was in Würzburg and last appointment was at the Jesuits College in Rome.[2] He was a person of both great learning and incredible credulity. It is worthwhile reminding the reader that the towering seventeenth-century scientists Boyle and Newton were both credulous about alchemy. I like to think of the mid-seventeenth-century figure Johann Baptist van Helmont as perhaps half scientist–half pseudoscientist (see p. 195). Kircher appears to be somewhat less of a scientist than van Helmont. He wrote voluminously. However, the science historian John Ferguson says of Kircher:[3]

> Kircher was a man of vast—almost cumbrous—erudition, of equal credulity, superstition, and confidence in his own opinion. His works in number, bulk, and uselessness are not surpassed in the whole field of learning.

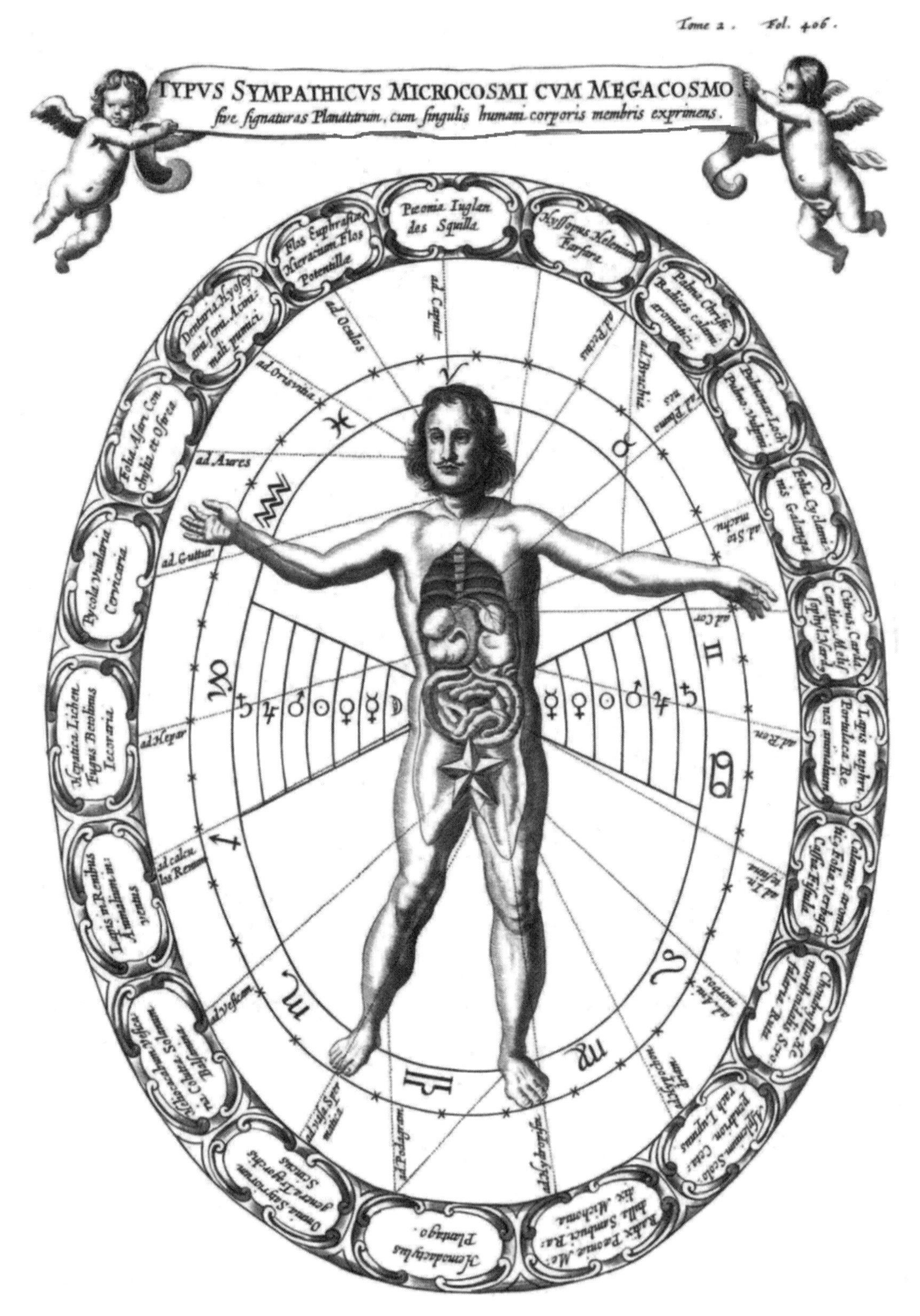

FIGURE 76. ■ Astrological unity of the microcosm and macrocosm in Athanasius Kircher's *Mundus Subterraneus* (1665). The sun is the human heart and the moon the human brain. (Courtesy J.F. Ptak Science Books.)

FIGURE 77. ■ A fabulous spagyrical (pharmaceutical) furnace from Kircher's *Mundus Subterraneus*. Said to be housed at the Jesuits College in Rome, what was its real purpose? (Courtesy J.F. Ptak Science Books.)

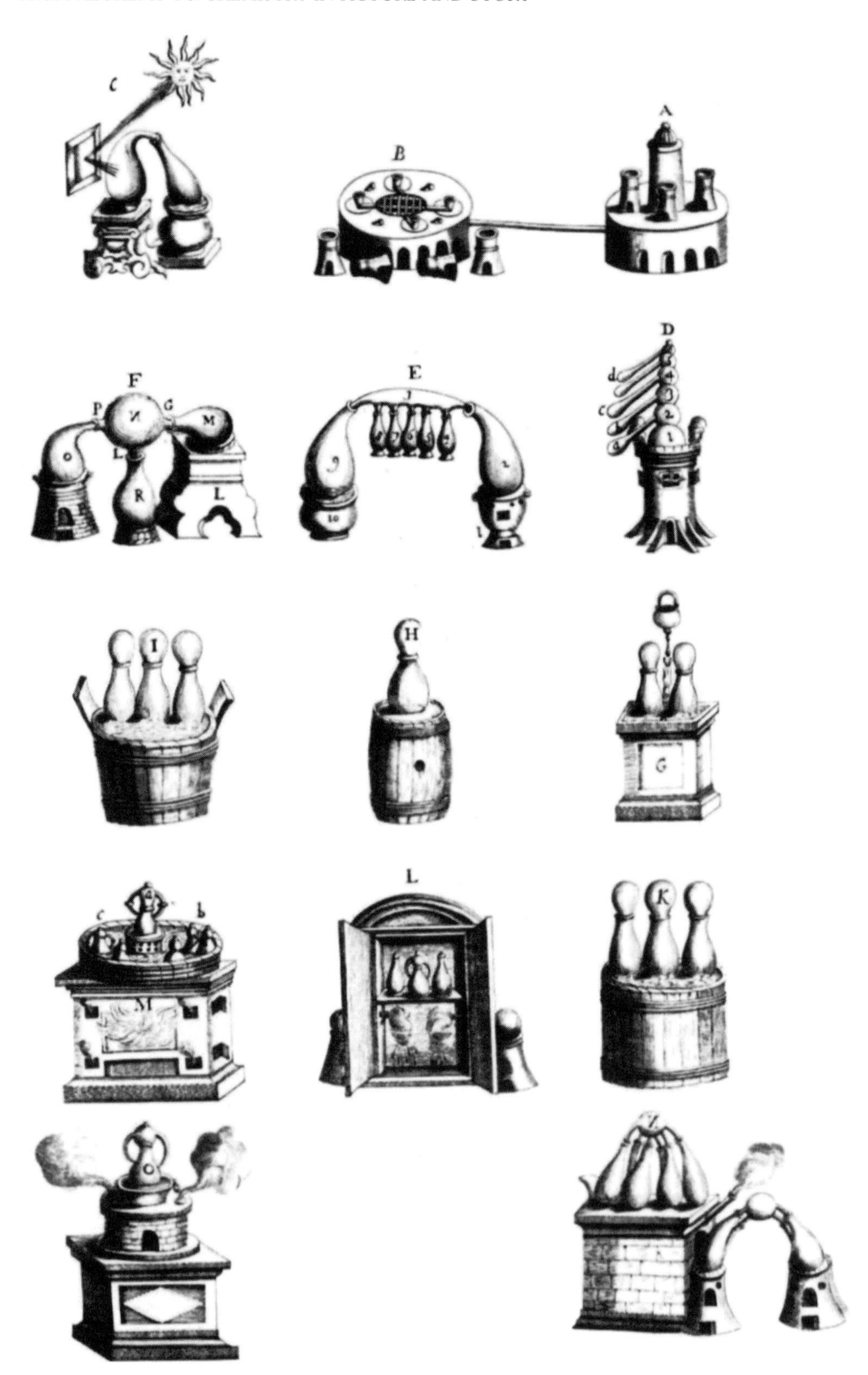

FIGURE 78. ■ It seems that the laboratory at the Jesuits College in Rome was equipped for all manner of distillation. Apparatus *E* evokes the wolf in the Romulus-and-Remus legend, and the tall pelican apparatus in M (left column, second from bottom) has the aspect of a stern cleric on a pulpit admonishing his congregants. (From Kircher's *Mundus Subterraneus*, courtesy J.F. Ptak Science Books.)

The *Mundus* included a vast array of descriptions of mining chemistry, metallurgical chemistry, and spagyrical (pharmaceutical) chemistry as well as chemistry useful to artists and artisans.[2] Most notable, from the historical perspective, is his disbelief in alchemy, which was expounded in the *Mundus*. Perhaps not surprisingly, Boyle "demurred at paying forty shillings for it."[2,4]

Figure 76 is a very astrological depiction of the microcosm–macrocosm description of the human body and its processes representing the larger universe. His relations[5] are as follows: sun = heart; moon = brain; Jupiter = liver; Saturn = spleen; Venus = kidneys; Mercury = lungs; Earth = stomach; veins = rivers; bladder = the sea; The seven major limbs represent the seven ancient metals.

Figure 77 is said to be a "pharmaceutical furnace" at the Jesuits College in Rome although Partington avers that it was "so named to disguise its real function."[2] One wonders whether this apparatus, that looks more like an alien loaded with hatchlings, ever truly existed. I confess that one of the stills in Figure 78 reminds me of the wolf who suckled Romulus and Remus—part of the mythology of ancient Rome.

---

1. A. Kircher, *Mundus Subterraneus, in XII. Libros digestus* . . . Joannem Janssonium & Elizeum Weyerstraten, Amsterdam, 1665. I thank John Ptak, J.F. Ptak Science Books, Washington, DC, for providing these three figures.
2. J.R. Partington, *A History of Chemistry*, Vol. 2, MacMillan & Co. Ltd., London, 1961, pp. 328–333.
3. J. Ferguson, *Bibliotheca Chemica*, Vol. 1, Derek Verschoyle, London, 1954, pp. 466–468.
4. The book is probably worth about $25,000 today.
5. A. Roos, *The Hermetic Museum: Alchemy & Mysticism*, Taschen, Cologne, 1997, p. 565.

ALCHEMISTS AS ARTISTS' SUBJECTS

The sixteenth, seventeenth, and eighteenth centuries witnessed the painting of numerous masterworks of European art depicting alchemists and physicians at work. Two prominent American collections including such artwork are the Fisher Collection of Alchemical and Historical Pictures (now kept by Duquesne University, Pittsburgh, PA) and the Isabel and Alfred Bader Collection (Milwaukee, WI). Although there are suggestions that Albrecht Durer (1471–1528) understood alchemical imagery, he apparently never engraved an alchemist or a laboratory.[1] Two early masters who represented medieval alchemists were Hans Weiditz ("An Alchemist and his Assistant at Work"—executed around 1520) and Pieter Brueghel the Elder (1525–1569).[1] Brueghel's 1558 "An Alchemist at Work" achieved widespread fame due to the contemporary engraving of it by Hieronimus Cock.[1] Some other noted artists depicting alchemical scenes during this period were Stradanus (see Figure 74), The De Bry (see Figure 34), Adriaen van Ostade, David Tenters the Younger, Jan Havickz Steen, Cornelis Pitersz Bega, Hendrik Heerschop, Charles Meer Webb, Matheus van Hellemont, Bathasar van den Bosch, Franz Christophe Janneck, Fernand Desmoulin, Thomas Wijck, Wenzel von Brozik, William Pether, and David Ryckaert. Most

FIGURE 79. ▪ Black-and-white reproduction of the color photograph of the 1671 oil painting, *The Alchemist*, by Hendrick Heerschop, in the Collection of Isabel and Alfred Bader. See color plates. The author expresses his gratitude to Dr. Bader for permission to reproduce the image and also for his helpful discussion of the Bush drawing (see Figure 75).

of these were in the Dutch–Flemish school.[1] A notable painting by Englishman Joseph Wright ("The Discovery of Phosphorus," 1771) and a piece by Richard Corbould near the start of the nineteenth century began to depict the science rather than the art. In the nineteenth century, carricaturists James Gillray, Thomas Rowlandson, and George Cruikshank (see Figures 126 and 127) took a shot at depicting chemical activity.[1]

Figure 79 was executed by Hendrick Heerschop in 1671 and is titled "The Alchemist." It is a black-and-white reproduction of a color photograph of a beautiful oil painting from the Isabel and Alfred Bader Collection.[2] The alchemist appears to smoke his pipe while watching a distillation. Hopefully, he is not distilling diethyl ether.

1. J. Read, *The Alchemist in Life, Literature and Art,* Thomas Nelson, London, 1947, pp. 56–91.
2. The author is grateful to Dr. Alfred Bader for making this photographic reproduction available and providing permission to reproduce it in black and white, as well as in color.

ALLEGORIES, MYTHS, AND METAPHORS

The sixteenth and seventeenth centuries witnessed the publication of many beautiful books that illustrated the alchemical art. For example, the title page of *Le Tableau des Riches Inventions* (Figure 80), authored by Francesco Colonna, translated by Beroalde de Verville, and published in Paris in 1600,[1] depicts The Great Work beginning in chaos and culminating with the Stone (rising of the phoenix).[2] The tree stump in this figure represents the putrefaction (*Nigredo*) or the debasement at the beginning of the process.[3] The figure on the lower left is a Philosophical Tree[3] representing the complete work as well as the aspect of multiplication of gold. Fire is the transforming element. The winged dragon and wingless serpent represent the union of Sophic Mercury (winged means volatile) and fixed (nonvolatile) Sophic Sulfur. Their symbols are also shown. The psychologist C.G. Jung owned a copy of this book and wrote extensively on dreams and alchemical imagery.[4,5]

Michael Maier (ca. 1568–1622), whom John Read calls "a musical alchemist," was a physician, philosopher, alchemist, and classical scholar. His extensive classical scholarship influenced his unification of alchemy and classical mythology.[6] Figure 81 is the title page of the 1618 book *Tripus Aureus* (The Golden Tripod).[7] It is a pun on the *tria prima* (mercury, sulfur, and salt), which "support" the synthesis of gold. The main objective is to present works by the "Three Possessors of the Philosopher's Stone." In Maier's own words:[7]

> Amiable reader, you behold three nurslings of the wealthy Art who by their studies have achieved the Stone. Cremerus in the middle, Norton himself on the left, Basil, lo, is seen on the right. Pray read their writings and search for the arms of Vulcan you who wish to pluck the apples of the Hesperian ground.

FIGURE 80. ■ Engraved title page from Francesco Colonna's *Le Tableau Des Riches Inventions* (Paris, 1600 [after 1610]) (courtesy of The Beinecke Rare Book and Manuscript Library, Yale University). The figure depicts the rising of the Phoenix beginning in chaos.

TRIPVS AVREVS,
Hoc eſt,
TRES TRACTATVS
CHYMICI SELECTISSIMI,
Nempe

I. BASILII VALENTINI, BENEDICTINI ORDInis monachi, Germani, PRACTICA vna cum 12. clauibus & appendice, ex Germanico;
II. THOMÆ NORTONI, ANGLI PHILOSOPHI CREDE MIHI ſeu ORDINALE, ante annos 140. ab authore ſcriptum, nunc ex Anglicano manuſcripto in Latinum translatum, phraſi cuiuſque authoris vt & ſententia retenta;
III. CREMERI CVIVSDAM ABBATIS WESTmonaſterienſis Angli Teſtamentum, hactenus nondum publicatum, nunc in diuerſarum nationum gratiam editi, & figuris cupro affabre inciſis ornati operâ & ſtudio
MICHAELIS MAIERI Phil. & Med. D. Com. P. &c.

FRANCOFVRTI
Ex Chalcographia Pauli Iacobi, impenſis LVCÆ IENNIS.
Anno M.DC. XVIII.

FIGURE 81. ▪ Engraved title page from Michael Maier's *Tripus Aureus* ("Golden Tripod"), published in Frankfurt in 1618. The Big Three of Alchemy are Basil Valentine, Thomas Norton, and John Cremer (courtesy of The Beinecke Rare Book and Manuscript Library, Yale University).

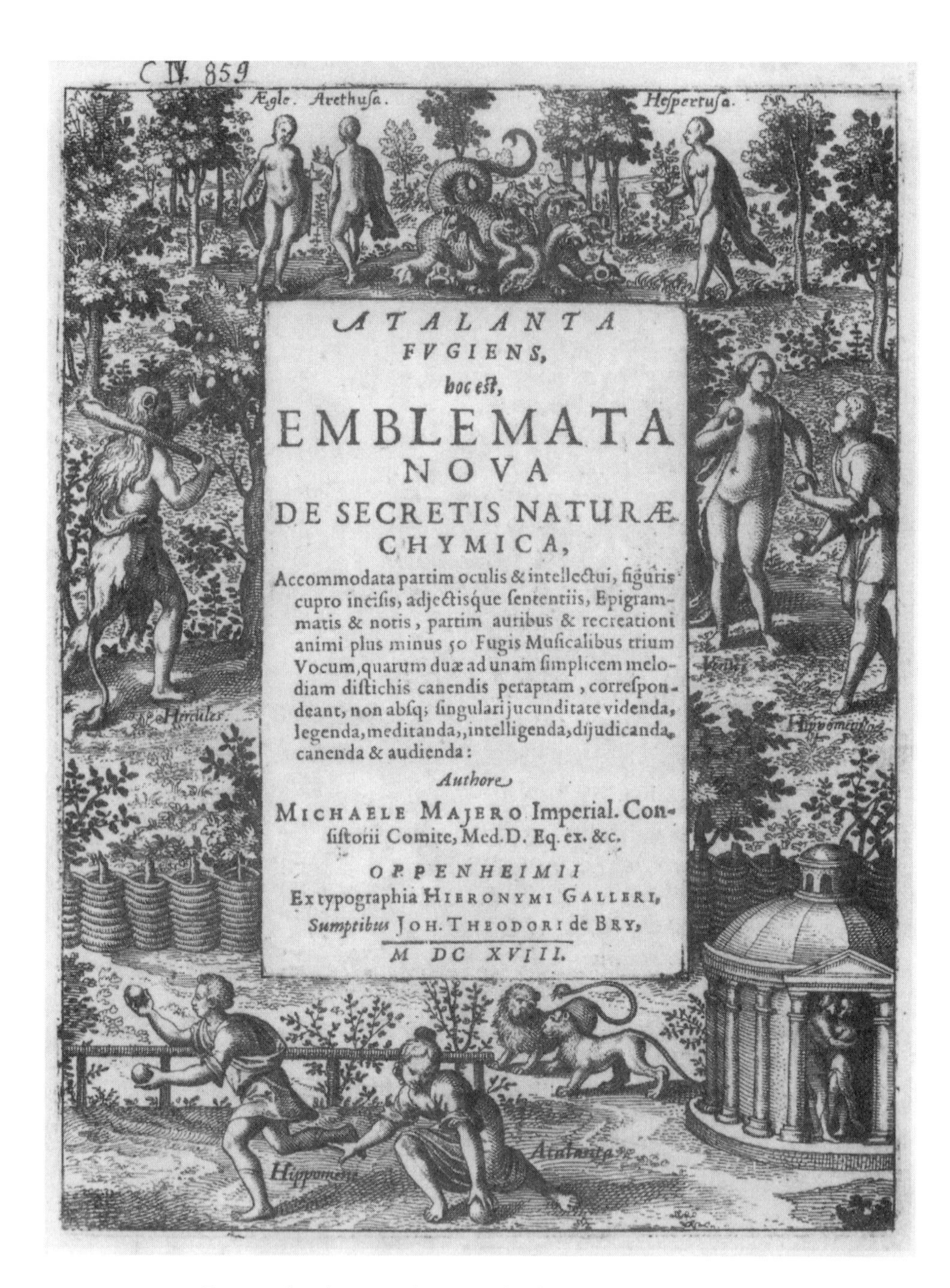

ATALANTA
FUGIENS,
hoc est,
EMBLEMATA
NOVA
DE SECRETIS NATURÆ
CHYMICA,
Accommodata partim oculis & intellectui, figuris cupro incisis, adjectisque sententiis, Epigrammatis & notis, partim auribus & recreationi animi plus minus 50 Fugis Musicalibus trium Vocum, quarum duæ ad unam simplicem melodiam distichis canendis peraptam, correspondeant, non absq; singulari jucunditate videnda, legenda, meditanda,, intelligenda, dijudicanda, canenda & audienda:
Authore
MICHAELE MAJERO Imperial. Consistorii Comite, Med.D. Eq. ex. &c.
OPPENHEIMII
Ex typographia HIERONYMI GALLERI,
Sumptibus JOH. THEODORI de BRY,
M DC XVIII.

FIGURE 82. ■ Engraved title page from Michael Maier's *Atalanta Fugiens* ("Atalanta Fleeing") published in Frankfurt in 1618. The engraving recounts the mythology of Atalanta, the fastest mortal, who challenged suitors to a foot race—losers were put to death. But Hippomenes used three golden apples (from Venus) to distract her and win the race and her hand (see discussion in text) (courtesy of The Beinecke Rare Book and Manuscript Library, Yale University). Maier composed fifty fugues for *Atalanta Fugiens*—an obvious pun since "fugues" and "fugiens" have the same root.

FIGURE 83. ■ Birth of Athena from the head of her father Zeus (no mother here) and a chemical *conjunctio* with her brother Apollo (from Michael Maier's *Atalanta Fugiens*). The *conjunctio*, or chemical marriage, weds Sol and Luna or Sophic Sulfur and Sophic Mercury (courtesy of The Beinecke Rare Book and Manuscript Library, Yale University).

John Cremer, a fourteenth-century abbot who lived in Westminster, was reputed to have joined Raymond Lully in alchemical works in Westminster Abbey and the Tower of London.[7] Thomas Norton, author of the *Ordinall of Akhimy*, began writing this famous work in 1477.[7] The reputed Basil Valentine is mentioned later in our *Chemical History Tour* and also by Read.[8] Vulcan is the god of fire. "Arms of Vulcan" refers to fire as an instrument of chemical change.[7] The picture depicts an immediate proximity between a chemical laboratory and a chemical library—the duality of practice and theory. Although the American Chemical Society recommends location of a university chemical research library in the chemistry laboratory building, it is unlikely that they have quite this closeness in mind.

Maier's 1618 book *Atalanta Fugiens* (Atalanta Fleeing) contains 50 beautiful engravings (see Figures 45, 46 and 49) and also includes music for about 50 three-voice (*tria prima*—get it?) fugues (a pun) he composed.[9,10] The title page (Figure 82) depicts the legend of Atalanta and the golden apples.[9] Atalanta, the fastest mortal, challenged any suitor to a foot race. If they lost, they died. She would wed the man who defeated her. The figure tells of Hercules picking the three golden apples of the Hesperides (guarded by Aegle, Arethusa, and Hesperia and their guard-dragon Ladon). The apples are presented by Venus to Hippomenes who drops them at opportune moments to distract Atalanta and thus wins the race and her hand. "Plucking the apples of the Hesperian ground" (see above) is a metaphor for achieving the Stone. Unfortunately, the couple later profanes the temple of Cybele (the act of profaning is shown in the lower right—clearly they are "fast company" albeit slightly incautious) and are changed into lions.

What is going on in Figure 83 (from *Atalanta Fugiens*)[7] And wouldn't this be a dandy question for a chemistry exam? In Greek mythology, the goddess Athena is born from the head of her father Zeus (she has no mother). One version has the Greek god of fire (Hephaestus, or Vulcan in Roman lore) splitting the head of Zeus. The god Apollo, born of Zeus and Leto, is also depicted. The incestuous *conjunctio*, presided over by Cupid, unites the male principle (sulfur) with the female principle (mercury). Lynn Abraham mentions the myth of Jupiter (Zeus) wherein he is transformed into an eagle, transporting Ganymede to heaven, and converted into a shower (distillation) of gold.[11]

On the other hand, perhaps this is merely an advertisement for taking a name brand of aspirin just prior to the chemistry final exam. I guess I'd have to give at least half credit for that answer.

The first public performance of alchemical music (examples of Maier's fugues) appears to have occurred at the Royal Institution of Great Britain on November 22, 1935. Student members of the St. Andrews University Choir ("The Chymic Choir") and the Music Department faculty "conspired" to give voice to the admirable scholarship of their colleague—Professor John Read.[12]

---

1. I. MacPhail, *Alchemy and the Occult*, Yale University Library, New Haven, CT, 1968, pp. 189–191.
2. C.G. Jung, *Psychology and Alchemy*, 2nd ed., translated by R.F.C. Hull, Princeton University Press, Princeton, NJ, 1968, p. 38.
3. L. Abraham, *A Dictionary of Alchemical Imagery*, Cambridge University Press, Cambridge, 1998, pp. 150–151, 205.

4. I. MacPhail, op. cit., pp. xv–xxxii (essay by A. Jaffe).
5. N. Schwartz-Salant, *Encountering Jung on Alchemy*, Princeton University Press, Princeton, NJ, 1995.
6. J. Read, *Prelude To Chemistry*, MacMillan, New York, 1937, pp. 228–236.
7. J. Read, op. cit., pp. 169–182.
8. J. Read, op. cit., pp. 183–211.
9. J. Read, op. cit., pp. 236–254.
10. J. Read, op. cit., pp. 281–289.
11. Abraham, op. cit., p. 110.
12. J. Read, op. cit., pp. xxiii–xxiv.

THE WORDLESS BOOK

One of the most beautiful books of the seventeenth century was titled *Mutus Liber* (wordless book), published in France in 1677 and authored by "Altus" a pseudonym representing the "Classic Elder" of alchemy. It was printed at the instigation of one Jacob Saulat and includes 13 folio-sized figures with only slight text in the title figure, which depict The Great Work.[1] In 1702, the *Mutus Liber* became more widely known due to its inclusion at the end of the first volume of Manget's *Bibliotheca Chemica Curiosa* and had 15 figures.[1] The figures are totally allegorical and there is no firm interpretation of them. It is interesting that the pictures depict a man and a woman (possibly husband and wife) apparently working as co-equals. This was a rather novel aspect of the book since women played virtually no significant role in chemistry for a long period. There was, however, an ancient Alexandrian woman called Mary Prophetissa,[2–4] sometimes equated to Miriam, the sister of Moses, who discovered hydrochloric acid and developed an early still called a kerotakis as well as the water bath, used for gentle heating. The water bath has survived into modern times and is termed the Bain Marie. Moreover, Mary Prophetissa is said to have originated the process of fusing lead–copper alloy with sulfur to make a blackish material.[4] Such black materials were often the starting points for transmutation and represent allegorical death preceding resurrection. This is one of the origins of the term "Black Arts" for alchemical practices.

The six figures shown here are selected from a 1914 Paris reissue of the *Mutus Liber*.[5] The title page (Figure 84) shows a picture believed to depict Jacob and the Ladder to Heaven and is totally spiritual.[1] His head rests on a rock some say represents the Philosopher's Stone. The translation is as follows:[4]

> The Wordless Book, in which nevertheless the whole of Hermetic Philosophy is set forth in heiroglyphic figures, sacred to God the merciful, thrice best and greatest, and dedicated to the sons of art only, the name of the author being Altus.

The last three lines are biblical references in reverse: *Genesis* 28:11, 12; *Genesis* 27:28, 39; *Deuteronomy* 33:18, 28.

Figure 85 is the second plate in the *Mutus Liber* and depicts the sun above two angels holding a vessel containing Sol and Luna at the sides of Neptune, who

FIGURE 84. ■ Engraved title page from *Mutus Liber* ("Silent Book"). This figure is from a 1914 Paris reproduction of the *Mutus Liber* in Manget's 1702 *Bibliotheca Chemica Curiosa* which reprinted the plates of the original 1677 book. The image depicts Jacob and the Ladder to Heaven.[5]

FIGURE 85. ■ The second plate (of 15 in Manget's printing) in *Mutus Liber* that depicts, in its spiritual upper section, the sun above two angels holding Sol and Luna in the presence of Neptune, representing the watery substance needed in the Great Work. In the earthly lower section, the male and female alchemists place the Philosophical Egg in the athanor where it is gently heated with a sand or water bath.[5]

FIGURE 86. ■ The fourth plate in *Mutus Liber* that depicts the collection of dew (a kind of *Prima Materia*).[5]

FIGURE 87. ■ The fifth plate in *Mutus Liber* depicts the two alchemists preparing the dew for distillation. The distillate is divided into four bottles and then heated (apparently for 40 days). The residue is spooned into a bottle and given to an old man (Saturn?).[5]

FIGURE 88. ■ In plate 14 in Manget's printing of *Mutus Liber* the man, the child and the woman are trimming the wicks and filling their lamps with oil. Equal parts of Lunar Tincture and Solar Tincture are ground together to provide Sophic Mercury.[1,5]

FIGURE 89. ■ The work is finished (plate 15) and the alchemists proclaim: "Given Eyes To See, Thou Seest."[5]

is considered to represent a watery or liquid substance needed in the Great Work as the two alchemists kneel at their furnace. The upper part represents the spiritual dimension of the work.[1] The lower section is the earthly part. In the furnace, the bottom section is flame, the middle funnel is a sand or waterbath for controlled heating of the Philosophical Egg.[1] (Any chemist who has tried a new reaction will appreciate the prayerful aspect of this picture.) Figure 86 is the fourth picture in *Mutus Liber* and shows the collection of dew in sheets spread in the pasture under the influence of the sun in Aries and the moon in Taurus (springtime). This illustrates the astrological dimension of the opus. Dew is considered to be a type of *Prima Materia* and the two alchemists wring it into a large collection plate.

In the book's next figure (Figure 87), the man and woman prepare the dew for distillation in an alembic. The man subsequently takes the distillate and pours it into four vessels that are heated, apparently for 40 days. The woman removes the residue from the distillation vessel and spoons it into a bottle that she gives to an old man, holding a child and bearing the mark of Luna. Some interpret the old man as Saturn.

In Figure 88 (Plate 14 in Manget's work) we see three furnaces and the man, child, and woman trimming the wicks on their furnance lamps. Equal parts of Lunar Tincture and Solar Tincture are ground together to make Sophic Mercury. The two alchemists seal their lips and the words read: "Pray, Read, Read, Read, Read again, Labor and Discover."[1] In Figure 89, the last picture in *Mutus Liber* is like the first (Figure 84) and is totally spiritual. The Ladder is no longer needed, a body, possibly Hercules (son of Zeus), lies at the bottom under the influence of Sol and Luna, the Zeus figure is being crowned with laurel wreaths by angels, and the two enlightened alchemists exclaim in unison:

> Given eyes To see, thou seest.

The Great Work is finished.

---

1. A. McLean, *A Commentary On The Mutus Liber*, Phanes, Grand Rapids, MI, 1991.
2. C.A. Burland, *The Art of The Alchemists*, MacMillan. New York, 1967, pp. 188–198. This book shows all 15 figures in a reasonably large format.
3. *Secrets of the Alchemists*, Time-Life Books, Alexandria, VA, 1990, pp. 70–77. This book depicts all 15 figures (in gold tint, no less!), significantly reduced in size, but with nice textual discussion.
4. J. Read, *Prelude To Chemistry*, MacMillan, New York, 1937, pp. 155–159.
5. *Mutus Liber—Le Livre d'Images sans Paroles, ou toutes les opérations de la Philosophie hermetique sont décribes et représentés. Reédite d'après l'original et precedé d'une Hypotypose explicative par MAGOPHUN.*, Librairie Critique, Emile Nourry, Paris, 1914.

STRANGE DOINGS IN AN ALCHEMIST'S FLASK

Johann Conrad Barchusen was born in Lippe, one of the former states of Germany, in 1666. He studied pharmacy, became a physician, and was appointed

Lector in Medicine at Utrecht in 1698 and Professor of Chemistry in 1703.[1] Figure 90 is from his 1698 book titled *Pyrosophia* and depicts Barchusen's chemical laboratory. Is that Barchusen himself carefully weighing reagents? And what efficacious liquids flowed from the caduceus-like still on the right side of his laboratory?

Although Barchusen admitted that he had never actually witnessed a transmutation,[1] he was a believer and apparently felt that a simple "how-to" laboratory manual could clarify synthesis of the Philosopher's Stone for the aspiring adept. His *Elementa Chemiae*, published in 1718, was the second edition of the *Pyrosophia* but also included a series of 78 emblems, copied from an early alchemical manuscript, and added his interpretations "to do a great favor to the adepts of gold-making."[2] Figures 91–94 depict 24 of these emblems; their interpretation here is essentially that of scholar Alexander Roob.[2] There are two different pathways to the Philosopher's Stone, the Dry Path and the Wet Path, and Barchusen has chosen to illustrate the Wet Path, which is longer and requires numerous distillations and sublimations.

The first 14 emblems in the *Elementa* represent a philosophical and mystical preface to the alchemical operations in the flask ("retort") that follow. The circles typically represent perfection and divide earthly from heavenly realms. Emblem 2 in Figure 91 invokes the presence of God and His privilege for the success of the Great Work. Emblem 5 is a Western-style alchemical mandala[3] depicting the four ancient elements and the hand of God. In Emblem 7, Philosophical ("Sophic") Mercury (not the familiar slippery silvery liquid) is depicted as playing a major role in these operations; the "mercurial spirit" (Azoth, a term taken from the first and last letters of the Greek, Hebrew, and Latin alphabets[4]) is its essence. The dove[5] represents the "mercurial spirit" in all metals (the large symbol above the dove is that of mercury enclosed within the sun) and the pathway to it is a well-kept secret (the locked trunk). The early view that "mercurial spirit" is present in and contributes luster and other common properties to all metals makes transmutation a much more reasonable operation than we understand it to be today (see "Natural Magick: Metamorphoses of Werewolves and Metals," page 64). Starting from the wing tip at the right side and moving clockwise are symbols for lead, tin, silver (Moon), gold (Sun), copper, and iron. The upward and downward triangles represent volatilizations and condensations and the symbol just under the wing at the right represents the operation of sublimation. At the bottom of Emblem 7 is a large symbol representing the Philosophical ("Sophic") Sulfur. The union of opposites [male Sulfur and female Mercury; male Gold (Sun) and female Silver (Moon)] (Emblem 9) is necessary in order to complete the Great Work but they must first be released by fire from the earthly materials that hold them (Emblem 11). The union of Sophic Mercury and Sophic Sulfur produces a homogeneous liquid (Emblem 13).

Before the lab work begins, gold (the lion) is eaten by the wolf (antimony) and placed into the fire for purification. Michael Maier's 1618 book *Atalanta Fugiens*, discussed earlier, also depicts the steps in the purification of gold by antimony but symbolizes gold by a king rather than a lion (Figure 45).

In Figure 92, Emblem 15 represents the dissolution of purified gold to form Sophic Sulfur in the alchemical flask. Placing the flask in a furnace (athanor) unites Sophic Sulfur with Sophic Mercury and the homogeneous mixture (Em-

FIGURE 90. ■ Is that author Johann Conrad Barchusen weighing the "midnight oil" in his Utrecht laboratory? (*Pyrosophia*, Utrecht, 1698).

FIGURE 91. ■ Emblems 2, 5, 7, 9, 11, and 13 (left to right in each row, starting from the top) from Johann Conrad Barchusen's, *Elementa Chemiae* (Leiden, 1718).

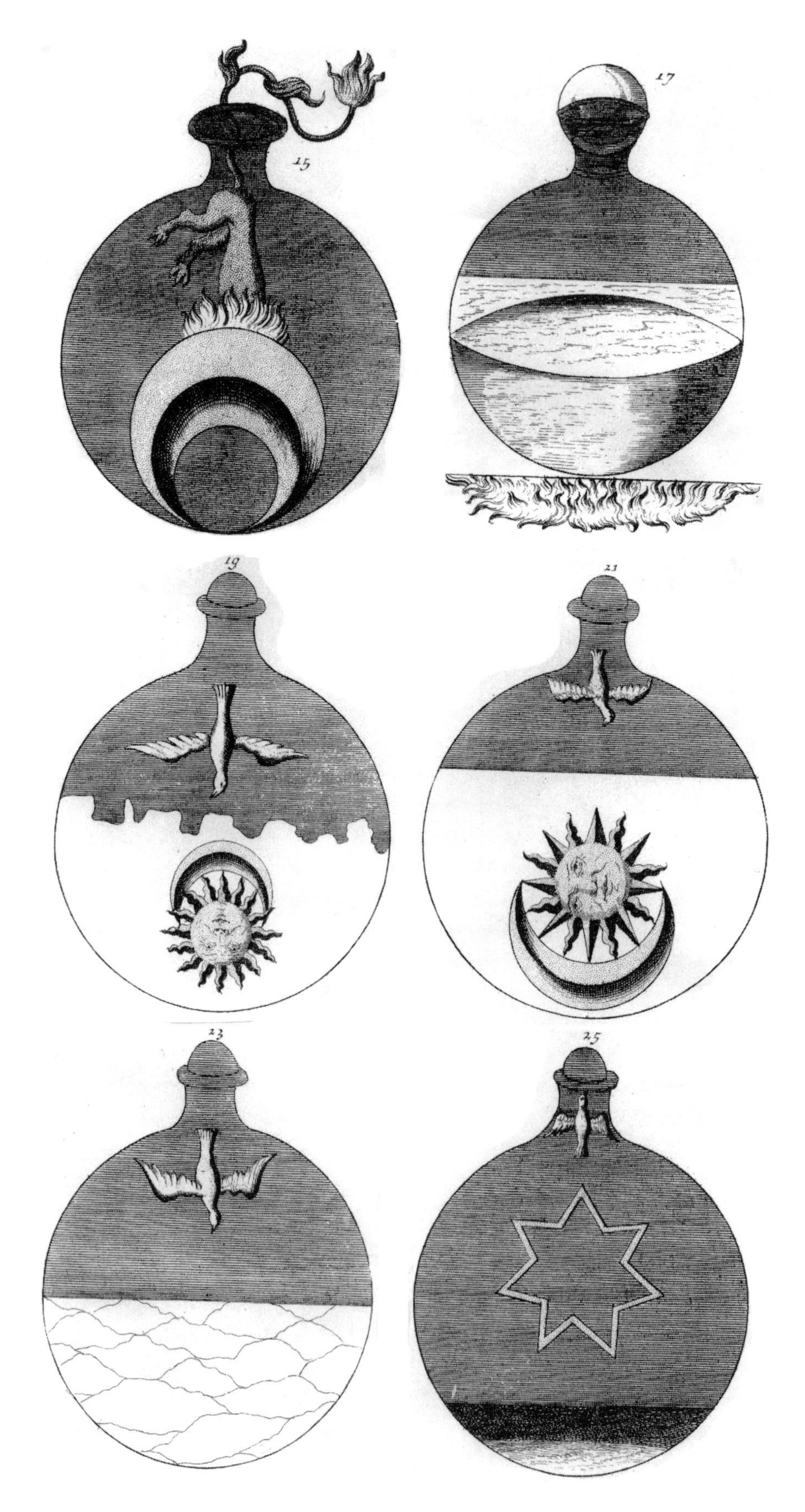

FIGURE 92. ■ Emblems 15, 17, 19, 21, 23, and 25 (left to right in each row, starting from the top) from Johann Conrad Barchusen's *Elementa Chemiae* (Leiden, 1718).

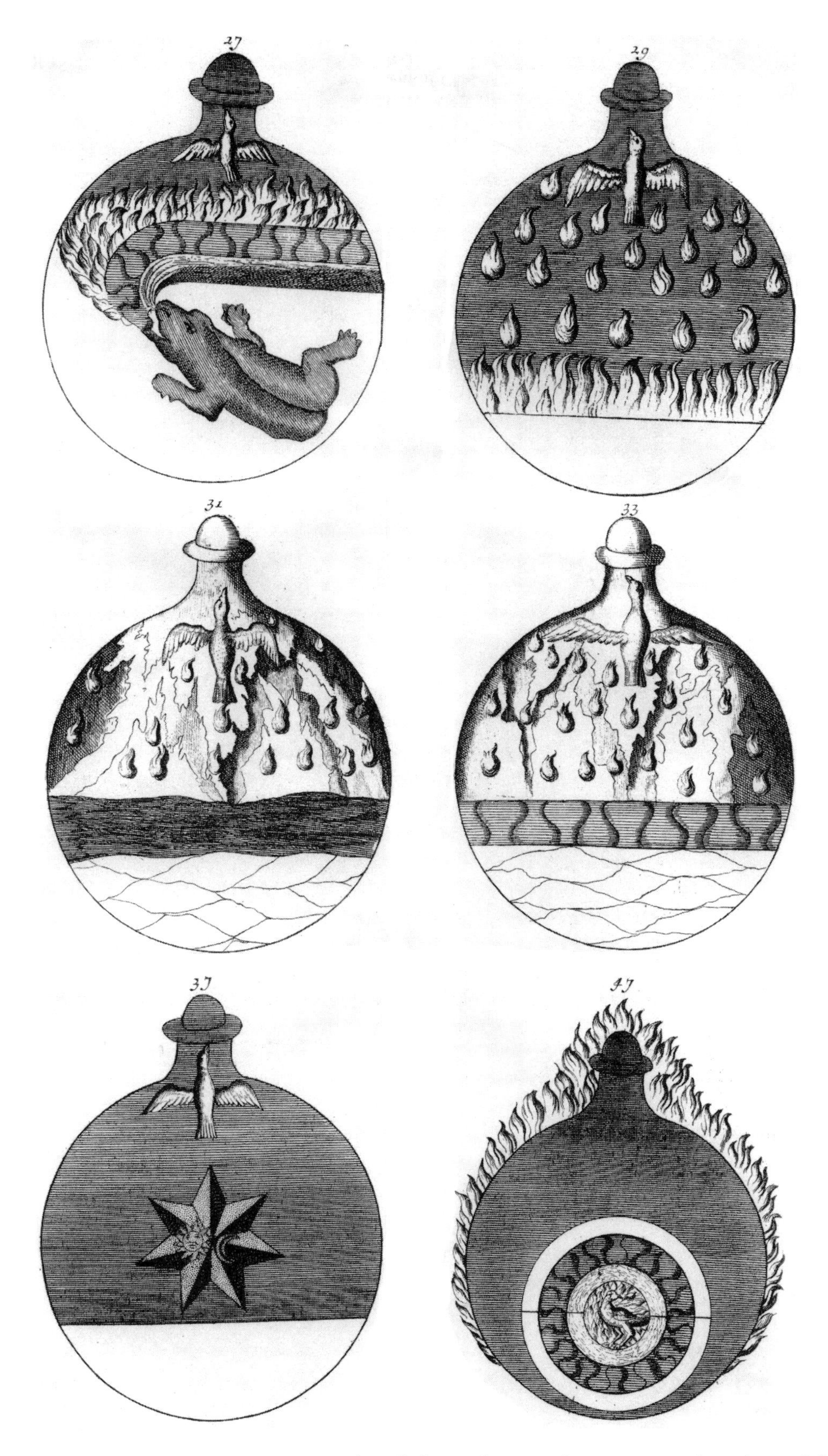

FIGURE 93. ■ Emblems 27, 29, 31, 33, 37, and 47 (left to right in each row, starting from the top) from Johann Conrad Barchusen's *Elementa Chemiae* (Leiden, 1718).

FIGURE 94. ■ Emblems 55, 57, 59, 65, 67, and 75 (left to right in each row starting, from the top) from Johann Conrad Barchusen's *Elementa Chemiae* (Leiden, 1718).

blem 17) is continuously heated and undergoes putrefaction (blackening), with separation of the elements gold, silver, mercury and sulfur. There follow multiple distillations: the dove flies upward (e.g., Emblem 25) and the resulting distillate is poured back into the contents of the flask [the dove flies downward (Emblems 19, 21, 23)] and the process is repeated multiple times. A degree of solidification occurs (Emblem 23) and the flask contains the seeds of the seven ancient metals that form a closed, mutually transmutable set (Emblem 25).

The black material (toad) turns white when distilled essence (Azote) is poured on it and great heat causes the toad to release liquids (Figure 93, Emblem 27). As intense heat continues to be applied, the elements separate and stratify. Heat and distillation continue (Emblems 29, 31, 33) and the collected distillate is poured back into the flask each time until the appearance of Sun and Moon (Emblem 37) indicate that success is near. Following the ninth and tenth distillations, the Philosopher's Stone begins to take form and rises as the Phoenix rises from the fire (Emblem 47) (see also Figure 52, the symbol of the American Chemical Society, evoking chemistry's ancient heritage).

And yet, the truly fussy adept is never finished and desires the most transparent, subtle, intensely red Stone (citrine or citrinish-red will *simply* not do!). Further sublimations are required. (Any modern organic chemist preparing an ultrapure pharmaceutical will understand such obsessive behavior.) Sophic Mercury (a serpent this time) is added and when the serpent swallows its own tail (the ouroboros image, Figure 55), numerous continuous (complete circle) sublimations (Figure 94, Emblem 55) lead to the Stone's final resolidification (Emblem 57). But a new dragon enters to intensify the fire (Emblem 59), "sweating out" the soul and burning the Stone for an extended period (Emblem 65). The Stone is remoistened again (Emblem 67, yuck!), redistilled, and "tortured by fire" until there is the perfection and resurrection (Emblem 75) that unifies Spirit, Soul, and Body. As long as God is with you, this should work every time, but there *must* be an abridged procedure somewhere!

---

1. Ferguson, J., *Bibliotheca Chemica*, Vol. 1, Derek Verschoyle, London, 1954, pp. 71–72.
2. A. Roob, *The Hermetic Museum: Alchemy & Mysticism*, Benedikt Taschen Verlag Gmbh, Cologne, 1997, pp. 126–145. This beautiful and amazingly affordable book includes hundreds of plates, a large number in color. It shows all 78 alchemical plates from Barchusen's *Elementa Chemiae* (Leiden, 1718) and provides explanations that I have employed, slightly adapted, in the present essay.
3. A. McLean, *The Alchemical Mandala*, Phanes Press, Grand Rapids, MI, 1989.
4. Roob, *op. cit.*, p. 308.
5. L. Abraham, *A Dictionary of Alchemical Imagery*, Cambridge University Press, Cambridge, 1998, pp. 58–59.

SECTION III
MEDICINES, PURGES, AND OINTMENTS

GEBER AND RHAZES: ALCHEMISTS FROM THE BIBLICAL LANDS

The Diocletian story (see p. 106) is a nice one. However, it seems that Arabic alchemy only reached the West (including Rome) around the eleventh century, so the story may be charitably termed "legendary."[1]

Most of our knowledge of Arabic alchemy derives from the writings of a mysterious eighth-century person named Jabir ibn Hayyan or Geber. Figure 95, the title page of *De Alchimia Libri Tres* (Of Alchemy in Three Books), published in Strasbourg in 1529, depicts a distillation furnace.[2] While alchemy also had origins in China and India, the cultures and languages of the biblical lands were

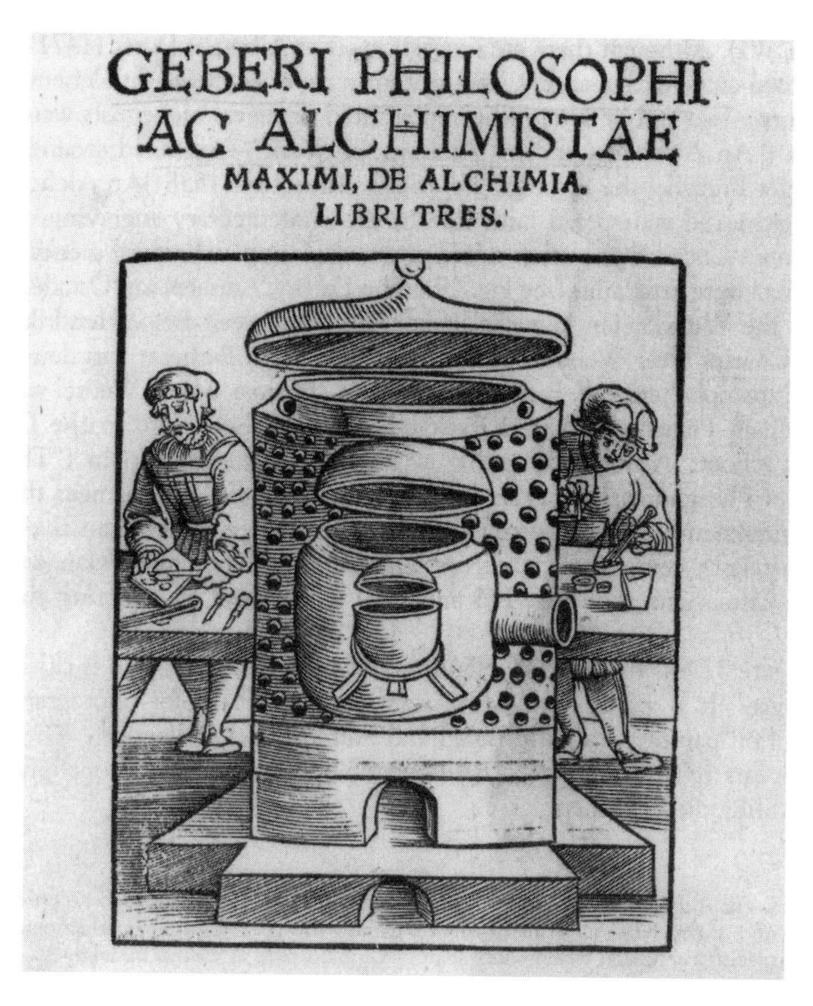

FIGURE 95. ■ Frontispiece from Geber, *De Alchimia Libri Tres* (Strasbourg, 1529) (courtesy of The Beinecke Rare Book and Manuscript Library, Yale University).

more accessible to the Europeans who, starting in the fifteenth century, produced the first printed books.

Al-Razi (850–ca. 923) or Rhazes, a Persian physician, produced the text *Secret of Secrets*. It included a great deal of practical and useful chemistry. Brock suggests that the preparation of pure hydrochloric, nitric, and sulfuric acids by Europeans in the thirteenth century depended crucially on the technology described by Rhazes.[1] These incredibly powerful "biting serpents" played critical roles in opening up new chemical reactions: for example, the ability to "release phlogiston" from metals or (as we have understood for over 200 years) to oxidize the metal to its calx while reducing an aqueous acid so as to release hydrogen gas.

---

1. W.H. Brock, *The Norton History of Chemistry*, Norton, New York, 1993, pp. 19–24.
2. I. MacPhail, *Alchemy and the Occult*, Yale University Library, New Haven, CT, 1968, pp. 32–34. We acknowledge the Beinecke Library, Yale University for this figure.

PARACELSUS

Theophrastus Bombast Von Hohenheim (1493–1541), who called himself Paracelsus, applied chemistry to effect medical cures and fathered a field called iatrochemistry. His break with the ancient medical doctrines of Galen was total and his tone intolerant and bombastic. He is recognized as having introduced experiment and observation into medical treatment.

Rather than search for Paracelsan quotes, we borrow from the novel by Evan S. Connell, *The Alchymist's Journal*[1] in order to gain insight into his mind and style:

> I have said that all metals labor with disease, except gold which enjoys perfect health by the grace of elixir vitae. I have taught Oporinus how this metal is sweet and exhibits such goodly luster that multitudes would look toward gold instead of the generous sun overhead. In fixity or permanence this substance cannot be exceeded and therefore it must gleam incorruptibly, being derived from an imperial correspondence of primary constituents which makes it capable of magnifying every subject, of vivifying lepers, of augmenting the heart. Conceived by our gracious Lord, it is a powerful medicament. False gold, which is a simulacrum boasting no remedial virtue, assaults internal organs and therefore it should be abjured, since the alchymic physician repudiates meretricious matter. We must not keep true gold beyond its measure but distribute what we hold, allegorically reminding each man of an earthly choice he is obliged to make between damnation and bliss.
>
> Pseudo-Alchymists that labor against quicksilver, sea salt, and sulfur dream of hermetic gold through transformation, yet they fail to grasp the natural course of development since what they employ are literal readings of receipts. Accordingly they bring baskets of gilded pebbles to sell, or drops of silver in cloudy alembics—futile panaceas meant for a charnel house. This is false magistry.

> Should it be God's will to instruct an alchymist at his art He will dispense understanding at the appropriate season. But if by this wisdom He concludes that any man was unfit or should He decide that irrevocable mischief would ensue, then that sanction is withheld.

The first novelized quotation indicates the imperfections in baser metals that are converted to gold (perfection) using the elixir vitae (or the Philosopher's Stone). True gold can be used as a medication. The second quotation indicates the hopeless quest of false alchemists, sometimes called "putters" after their furnace bellows, whose goal is solely gold making without an eye toward the unity of alchemy with nature. The last is perhaps most interesting: failure to duplicate an alchemical recipe is due to God's denial of the secret to the unworthy seeker rather than shortcomings in the original formula or method. It is, by the way, never clear how the Stone or the Elixir brings about its transformations.

We may obtain some feeling for the medicine of the period by visiting some of the cures attributed to Paracelsus in a book published in London in 1652 titled *THREE EXACT PIECES of LEONARD PHIORAVANT, Knight and Doctor in PHYSICK, viz. His RATIONAL SECRETS, and CHIRURGERY, Reviewed and Revived, Together with a Book of Excellent EXPERIMENTS and SECRETS, Collected out of the Practices of Severall Expert Men in both Faculties, Whereunto is Annexed PARACELSUS his One hundred and fourteen EXPERIMENTS: With certain Excellent Works of B.G. a Portu Aquitano, Also Isaac Hollandus his SECRETS concerning his Vegetall and Animal Work, With Quercetanus his Spagyrick Antidotary for GUN-SHOT.* (Nice to know what's in a book before you buy it):

- A certain woman was long sick of the Passion of the heart, which she called *Cardiaca*, who was cured by taking twice our *Mercuriall* vomit, which caused her to cast out a worm, commonly called *Theniam*, that was four cubits long.
- A boy of fifteen years old, falling down a stone staires, had his arme and leg benummed and voide of moving, whose neck with the hinder part of the head, and all the back bone I annointed with this unguent: a) Of the fat of a Fox; b) Oyle of the earthwormes; c) *Oleum Philosophorum*. I mixed them together, and annointed therewith, and in short space no wound nor swelling appeared in him so hurt.
- One that spit bloud, I cured by giving him one scruple of *Laudanum Precipitatum*,[2] in the water of Plantaine, and outwardly I applied a linnen cloth to his brest, dipped in the decoction of the bark of the roots of Henbane.
- One had two Pushes, as it were warts upon the yard, which he got by dealing with an unclean woman, so that for six moneths he was forsaken of all Physitians as uncureable, the which I cured by giving him *Essentia Mercurialis*, and then mixed the oyle of Vitriol with *Aqua Sophia*, and laid it on warm with a suppository four daies.
- A boy of eighteen years old had a tooth drawn, and three months after a certain black bladder appeared in the place of the tooth, the which I daily annointed with the oyle of Vitriol, and so the bladder was taken away, and the new tooth appeared.
- A fat drunken Taverner was in danger of his life by a surfet, who was restored to his health by letting of bloud.
- One who was troubled with paines in the stomack through weaknesse, who took *Oleum salis* in his drink, and caused him to have many seeges or stooles,

FIGURE 96. ▪ Title page of the *Basilica Chymica* (Frankfurt, 1611) by Oswald Croll, perhaps the major early source of Paracelsan chemical lore.

and so was restored to his health, as we have written on our book called *Parastenasticon*.

- A man that was troubled with the head-ach, I purged by the nostrills, casting in the juice of *Ciclaminus* with a siringe.
- A woman being almost dead of the Collick, I cured with the red oyle of Vitriol, drunk in Anniseed water, and a while after that potion, she voided a worm and was cured.
- To cause nurses to have abundance of milk, I have taken the fresh branches or tops of fennell, and boyled them in water or wine, and given it to drinke at dinner or supper, and at all times, for it greatly augmenteth the milk.
- A man being vehemently troubled a years space with pains in the head, I cured onely by opening of the skull, and in the same manner I cured the trembling of the brain, taking therewithall, *Oleum salis* in water of Basil.
- A Prince in Germany that was troubled with the Frenzie, by reason of a Sharp Fever, whom I cured by giving him five grains of *Laudanum nostrum*,[2] which expelled the Fever, and caused him to sleep six houres afterward.

Figure 96 is from the 1611 edition of *Basilica Chymica,* by Oswald Croll, which was printed in subsequent editions for 100 years. It is credited for passing the knowledge of Paracelsus and his followers into the seventeenth century.

The book's beautiful frontispiece depicts the *Alchemical Dream Team*:

Hermes Trismegistus, Egyptians
Geber, Arabs
Morienes, Romans
Roger Bacon, English
Ramon Lull, Spanish
Paracelsus, Germans

It's a Dream Team in another sense as well. Although the reputed author of the putative Emerald Tablet, the touchstone of alchemy, there is no evidence that a Hermes Trismegistus ever existed. The name of the reputed father of alchemy, Hermes-The-Thrice-Great, is a bit suspicious. In any case alchemy came to be called the "hermetic art." When we hermetically seal something, we protect it from air much as some alchemical experiments were sealed in glass and buried literally for years.

1. Evan S. Connell, *The Alchymist's Journal*, North Point, San Francisco, 1991.
2. J.R. Partington, *A History of Chemistry*, MacMillan, London, 1961, Vol. 2, p. 150, notes that opium had been employed by the Arabs in their medicine well before Paracelsus. But he also raises doubts over whether Paracelsus' laudanum ever had any opium. If not, then the above cures suggest effective placebos.

THE ALCHEMIST IN THE PIT OF MY STOMACH

Galen's views had dominated medicine for 1400 years. He believed that a balance of four bodily humours (phlegm, black bile, yellow bile, and blood) was required for sound health. Paracelsus was as bombastic as his family name (Theophrastus Bombast von Hohenheim) and joyfully trashed the Galenists and everybody else he disagreed with, leaving scores of enemies wherever he traveled.

Paracelsus believed that the purpose of alchemy was to create new medicines rather than gold.[1] He relied upon synthetic inorganic (metallic) compounds as medicines rather than the extracts of herbs used since ancient times.

FIGURE 97. ■ Artist Rita L. Shumaker's 1999 rendition of the *Archeus* believed by Paracelsus to inhabit the region near the stomach. It has a head and two hands and separates the nutritive part of food from the excrementous. If the Archeus was overwhelmed with excrementous matter, it would become ill as would the human inhabitant. Take calomel, sayeth Paracelsus, to unburden the Archeus.

As noted earlier, he believed in similitude and used poisons to kill poison—only, however, after sweetening or dulcifying them. Thus, liquid mercury metal dissolved in *aqua fortis* (nitric acid) followed by evaporation and calcination produced mercury oxide, which was employed against venereal diseases.[1] Similarly, metallic mercury dissolved in *aqua fortis* could be precipitated by adding salt water to produce solid calomel (Hg_2Cl_2).[1] This was an effective purgative (laxative) that could relieve gastric stress and remove intestinal worms.

Paracelsus' mysticism included a belief that the body is born completely healthy but that during life disease is received from food.[1] In the stomach he felt that there exists a vital force—the *Archeus*—a kind of Alchemist of Nature having a head and hands only. The *Archeus* separates the nutritive part of food from the poisonous part, the latter eliminated, in part, as excrement. Air is similarly digested to produce a nutritive part and a poisonous part ("excrementous air"?[2]). The *Archeus* can become sick if the separation is incomplete, and this results in the individual's illness. In this light, the laxative calomel clearly helps the *Archeus* remain healthy—a sound mind (and a sound *Archeus*) in a healthy body. Figure 97 is artist Rita L. Shumaker's depiction Of the *Archeus*.[3] The artist used tripe to model the human omentum, the inner membrane that connects the stomach with other organs and supporting blood vessels. The androgenous *Archeus* is depicted in the vicinity of the intestines, intimately part of the membranous fabric.

1. J.R. Partington, *A History of Chemistry*, MacMillan, London, 1961, Vol. 2, pp. 115–151.
2. "Thy Worst. I fart at thee" (said by *Subtle*, the second line in the 1610 play, *The Alchemist*, by Ben Jonson).
3. The author thanks Ms. Rita L. Shumaker, a faculty member at the University of North Carolina at Charlotte, for this original drawing and her imaginative evocation of the *Archeus*.

A SALTY CONVERSATION

We will speak later of Johann Baptist Van Helmont (Figure 138) in light of the Powder of Sympathy and his famous Tree Experiment. Although he was a disciple of Paracelsus and a believer in metallic medicines, he did not accept Paracelsus' *tria prima* nor the four elements of the ancients. Van Helmont believed in two elements, Water and Air, with only the first comprising matter. He was an independent thinker and drew the attention of the Spanish Inquisition (Spain occupied the Low Countries during parts of the sixteenth and seventeenth centuries). He spent the final 20 years of his life under house arrest.[1] Following his death in 1644, his son Franciscus Mercurius Van Helmont published his complete medical writings as *Ortus Medicinae* (1648). His work included recognition of the role of acid in digestion, the role of bile in digestion, and the role of acid in inflammation and the production of pus.[2] Van Helmont and Sylvius[1,3] (Francois Dubois, Franciscus de la Boe, 1614–1672) represented the golden age of iatrochemistry. Sylvius rejected the *Archeus* (see the next essay), which Van Helmont had merely modified. He recognized that although bile (for example, dog bile) tasted acidic (!), it was really alkaline. Aware that acidic substances and alkaline substances produced effervescence and/or heat upon mixing, Sylvius envisioned warfare between acid and alkali in living beings.! Sylvius' student Otto Tachenius promoted the acid–alkali theory of his master but added the unifying concept of the salt—the union of acid and alkali. This greatly improved the classification beyond the taste test, but it was Robert Boyle who discovered the quantitative test. In his *Rejections upon the Hypothesis of Akali and Acidium* (1675) Boyle defined acids as bodies that turn syrup of violets red and alkalies as bodies that turn this indicator green.[1]

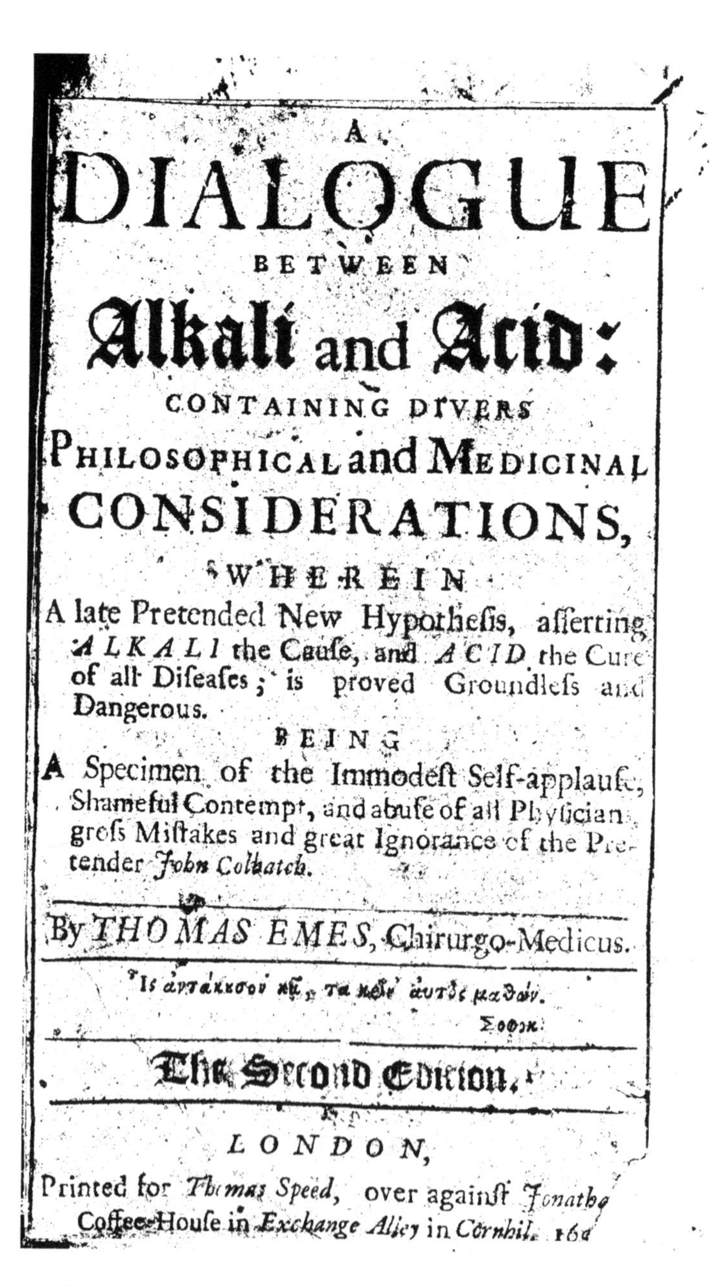

A
DIALOGUE
BETWEEN
Alkali and Acid:
CONTAINING DIVERS
PHILOSOPHICAL and MEDICINAL
CONSIDERATIONS,
WHEREIN
A late Pretended New Hypothesis, asserting *ALKALI* the Cause, and *ACID* the Cure of all Diseases; is proved Groundless and Dangerous.
BEING
A Specimen of the Immodest Self-applause, Shameful Contempt, and abuse of all Physicians, gross Mistakes and great Ignorance of the Pretender *John Colbatch*.

By *THOMAS EMES*, Chirurgo-Medicus.

The Second Edition.

LONDON,
Printed for *Thomas Speed*, over against *Jonath* Coffee-House in *Exchange Alley* in *Cornhil*. 16

FIGURE 98. ■ Title page of the second edition (1699) of physician Thomas Ernes' *Dialogue Between Alkali and Acid*. The question is: Which one is the cause of disease and which the cure?

A Dialogue Between Alkali and Acid (Figure 98) published by physician Thomas Ernes (2nd ed., 1699; 1st ed., 1698) is a wonderful example of invective directed against another physician, John Colbatch, who believed that the causes of diseases were alkaline and the cures acidic.[4] Ernes ends his 59-word title thus: *Being a Specimen of the Immodest Self-Applause, Shameful Contempt, and abuse of all Physician gross Mistakes and great Ignorance of the Pretender John Colbatch.* Now why won't my publisher let me use such a nifty title? And so the book begins:

Alkali: Well met Mr. Acid, whither are you hurrying so fast, to Some Heroe run through the Lungs, or the Heart?

Acid: I should hardly have to tell you Mr. Alkali, but that I am engag'd to oppose you where-ever we meet, you Principle of Death and Corruption, I am always provoked by you, you have done so much mischief in the World: And now to your farther reproach, I have a fresh instance of your badness, by a Messenger from my Lord Lazington, whom you have plagued with a fit of the Gout, and that a desperate one if I come not in time to his assistance, none can help him but I, and he thinks it 7 Years ere I come to him.

Alkali: You are very sharp Mr. Acid . . .

Stop! Enough! I hate that pun. Acid in the role of Scarlet Avenger? I suspect that when I start to hear test tubes of acid and alkali speaking to each other, it may be the moment to retire from chemistry and open that bookstore I've dreamed about.

---

1. W.H. Brock, *Norton History of Chemistry*, W.W. Norton, New York, 1993, pp. 41–63.
2. J.R. Partington, *A History of Chemistry*, MacMillan, London, 1961, Vol. 2, pp. 209–216.
3. J.R. Partington, op. cit., pp. 281–290.
4. J.R. Partington, op. cit., p. 290.

THE MAGIC OF DISTILLATION

Most of us who have been lucky enough to perform distillations know the thrill of winning a clear "spirit" from a dark and dingy solution, capturing a pure oil from a messy residue and even witnessing the collected distillate's abrupt solidification into white crystalline needles. A crude fermentation mixture will, upon distillation, yield an intoxicating "spirit of wine." Small wonder that a synonym for distillation is "rectification"—making things right. Indeed, distillation itself may almost be regarded as a religious act of ascent and descent:[1]

> Ascend with the greatest sagacity from the earth to heaven, and then again descend to the earth, and unite together the powers of things superior and things inferior. Thus you will obtain the glory of the whole world, and obscurity will fly away from you.

The archeological evidence suggests that distillation may have been performed as early as 5000 years ago.[2] We all know that the lid covering a pot of boiling broth will condense water vapor. It is not difficult to imagine fabricating lids with their edges turned up to form a gutter to collect condensed liquids.[2,3] Archeological research in the regions corresponding to ancient Mesopotamia has uncovered evidence of such "apparatus" dating from about 3500 years B.C.[2] Chemical historian Aaron Ihde provided a pictorial "Evolution of the 'Still'" and the major advance was the development, perhaps two thousand years ago, of the alembic or still head that transfers the distilled condensate down a "beak" into a separate receiver.[3] Medications up to the time of Paracelsus were derived from plants and animals as well as distillates from these natural sources.

Figure 99 is the frontispiece of the 1512 edition of the magnificent book by Heironymous Brunschwig, *Liber de Arte Distillandi*.[4] It depicts a double still allowing two separate operations likely to be rectifications of wine. The central unit is a tower containing cold water through which two "snakes" or condenser tubes pass. It is curiously reminiscent of the caduceus symbol wherein male and female snakes are entwined about the rod of Hermes and the four loops symbolize four "copulations." The hot alcohol vapors heat the tower's coolant, and the resulting warm water may be tapped off at the bottom as cool water is added from the top of the tower. The still operator on the left is comparing the temperature gradient between his cucurbit and the first condenser loop.[5] Figures 100–105 are also from Brunschwig's book. Figure 100 depicts the collection of distillate from rose extracts using four "rosenhuts" connected by beaks to receivers. The rosenhut was a form of air-cooled condenser. Figure 101 shows an efficient furnace having thirteen alembics for what appears to be a very profitable commercial operation.

Astrological influences are very much in evidence in Brunschwig's book. Figure 102 instructs that a particular distillation be performed under the influence of the Ram, corresponding to the Sun being in Aries (March 21–April 19), and thus, calling for mild heating.[6,7] The distillations in Figure 103 (Twins, Gemini, May 21–June 21) and Figure 104 (Crab, Cancer, June 22–July 22) are carried out under progressively warmer conditions with maximum heat (Figure 105) under the influence of Leo the Lion (July 23–August 22).[6,7]

Books of distillation enjoyed considerable popularity during the sixteenth and seventeenth centuries.[5] Figure 106 is from the 1528 edition of the *Coelum Philosophorum* ("Heaven of the Philosophers") of Philip Ulstad.[8] The two sets of distillation apparatus in Figure 107 are from the *Distillier Buch* of Walter H. Ryff.[9] Figure 108 is the frontispiece from the 1576 book by Conrad Gesner, *The newe Iewell of Health* . . . (109-word title!).[10] Although it is interesting to speculate over whether the woman depicted was an alchemist or chemist, it is more likely that she was simply an operator of the still since this was common sixteenth-century practice.

And what of the fruits of these countless distillations? Let us examine a few recipes. From the 1599 Gesner *The practice of the new and olde phisicke*:[11]

An oile or ointment sharpening the wit, and increasing memory
Take of Stœchas, of Rosemarie flowers, of Buglosse flowers, of Borage flowers, of Camomill flowers, of Maioram, of sage, of baulme, of violet flowers, of red rose leaves, and of bay leaves, of each one ounce and a half, al these

Liber de arte Distil
landi de Compositis.
Das büch der waren kunst zü distillieren die
Composita vñ simplicia/vnd dz Büch thesaurus pauperũ/Ein schatz d armẽ ge/
nãt Micariũ/die brösamlin gefallen võ dẽ büchern d Artzny/vnd durch Experimẽt
võ mir Jheronimo brüschwick vff geclubt vñ geoffenbart zü trost denẽ die es begerẽ.

FIGURE 99. ■ Title page (hand-colored; see color plates) from *Liber de Arte Distillandi* (Heironymous Brunschwig, 1512), depicting a double-still in which the central tower contains cooling water that is continuously replenished. Come to think of it, this apparatus evokes an image of the caduceus: two serpents entwined about the staff of Hermes where the loops symbolize "couplings." (From the Othmer Library, CHF.)

put by into a glasse bodie strongly luted, with foure pints either of Malmesie, Rennish wine, or Aqua vita, let these so stand to infuse for five daies, and distilled, add to it of the best Turpentine, one pound and a halfe, of Olibanu, of chosen Myrre, of Mastick, Bolelium, of gum jute, of each two ounces, of Vernicis integrae, one ounce, of Mellis anacardi, three ounces, all these brought to powder & infused for five dayes with the foresaid distillation, in a bodie with a head close luted, distill againe, adding to it of Cynamon, of Cloves, of Mace, of Nutmegs, of Cardamomum, of graines of Paradice, of the long and round Pepper, of Ginger, Xyloaloes, and of Cubebæ, of each one ounce, all these finelie brought to powder. To these adde of

FIGURE 100. ■ Distillation employing rosenhuts (air-cooled still heads) in Brunschwig's 1512 *Liber de Arte Distillandi* (from the Othmer Library, CHF).

FIGURE 101. ■ Thirteen alembics and a furnace [from Brunschwig's 1512 *Liber de Arte Distillandi* (from the Othmer Library, CHF)].

FIGURE 102. ■ Distillation using very mild heat under the influence of Aries (March 21–April 19) from Brunschwig's 1512 *Liber de Arte Distillandi* (from the Othmer Library, CHF).

FIGURE 103. ■ Distillation using moderate heat under the influence of Gemini (May 21–June 21) from Brunschwig's 1512 *Liber de Arte Distillandi* (from the Othmer Library, CHF).

FIGURE 104. ■ Distillation using moderately strong heat under the influence of Cancer (June 22–July 22) from Brunschwig's 1512 *Liber de Arte Distillandi* (from the Othmer Library, CHF).

FIGURE 105. ■ Distillation using conditions of strongest heat under the influence of Leo (July 23–August 22) from Brunschwig's 1512 *Liber de Arte Distillandi* (from the Othmer Library, CHF).

COELVM PHILOSO
PHORVM SEV DE SECRETIS
naturæ. Liber.

PHILIPPO VLSTADIO PATRICIO
Nierenbergensi. Authore.

FIGURE 106. ■ Title page from Philip Ulstad's "Heaven of the Philosophers," Nuremburg, 1528 (from The Roy G. Neville Historical Chemical Library, a collection in the Othmer Library, CHF).

> Muske & Amber gréece, of each two drams, all these mired together distill (after that these added & put into the former distillation have remained five dayes) the fire in the beginning soft, increase after by little and little unto the end of the work. The use of it, is, that the same may be applied in the winter time once in the wœke, but in the sommer time once in a month, the head before being washeth, the temples & hinder part part of the head anoint with it.

Now, if I can only remember this!

From John French's *The Art of Distillation*, published in 1653:[12]

> *How to turn Quick-Silver into a water without mixing any thing with it, and to make thereof a good Purgative and Diaphoretick medicine*
> Take an ounce of Quick-silver, not purified, put it into a bolt head of glass, which you must nip up, set it over a strong fire for the space of two months, and the Quick-silver will be turned into a red sparkling Precipitate. Take this

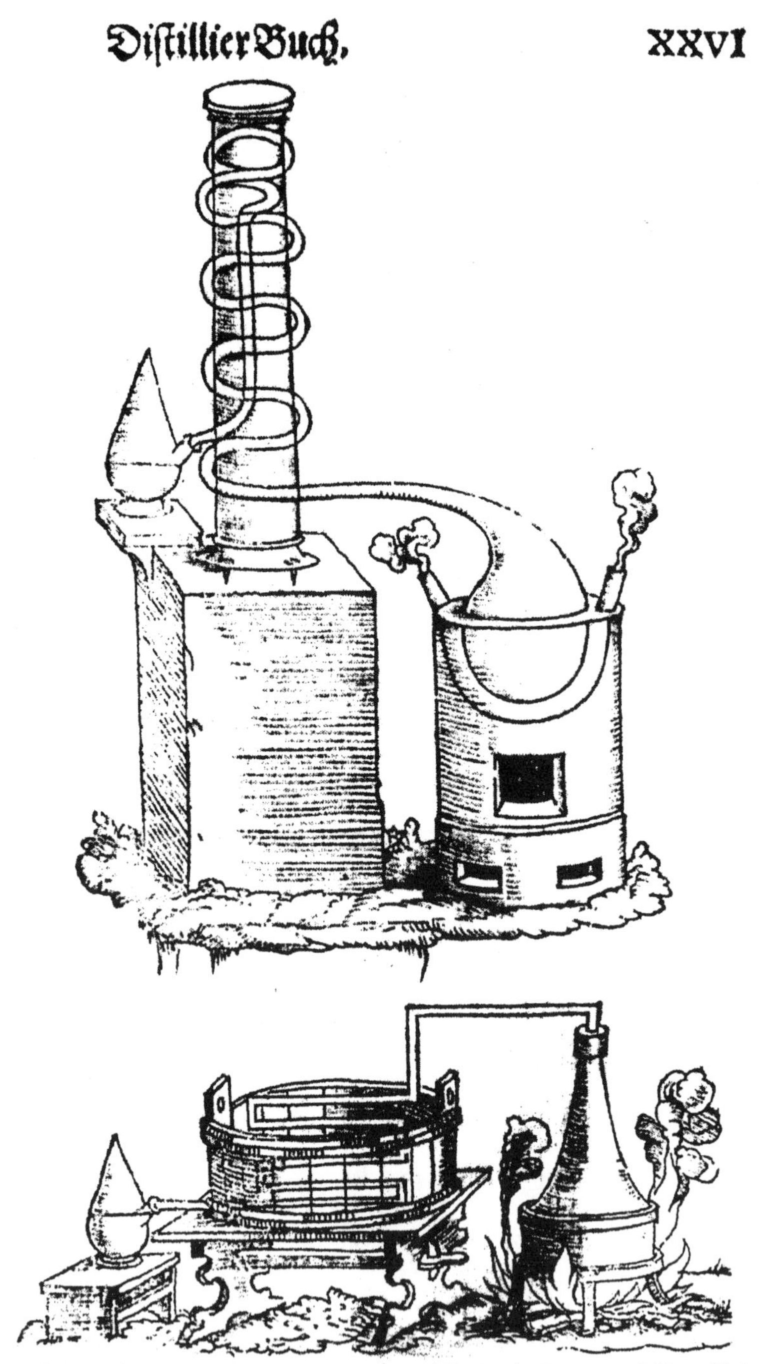
Distillier Buch. XXVI

Etliche bereydeen ein ofen wie obgemelt/in zimlicher grösse/also dz
sie auch mit holtz vnder fewern mögen/setzen ein irdin helm daruff/ a-
ber doppel/ aller maß als wir in der erste angezeyget haben/ durch ein
E 2

FIGURE 107. ▪ Two stills from Ryff's 1545 *Distillier-Büch* (from the Othmer Library, CHF).

FIGURE 108. ▪ Frontispiece from Conrad Gesner's 1576 *The New Iewell of Health* (from The Roy G. Neville Historical Chemical Library, a collection in the Othmer Library, CHF). Women commonly operated stills in the sixteenth century.

> powder, and lay it thin on a Marble in a cellar for the space of two months, and it will be turned into a water, which may be safely taken inwardly, it will work a little upward and downward, but chiefly by sweat.

Hmm—I recognize the calcination of mercury to its red oxide. The rest is a bit of a mystery. And let us take one more from Mr. French, Doctor of Physick:

> *A famous spirit made out of Cranium humanum.*
> Take of Cranium humanum as much as you please, break it into small pieces, which put into a glass Retort well coated, with a large Receiver well luted, then put a strong fire to it by degrees continuing of it till you see no more fumes come forth; and you shal have a yellowish spirit, a red oyl, and a volatile salt.
>
> Take this salt and the yellow spirit and digest them by circulation two or three months in Balneo, and thou shalt have a most excellent spirit.
>
> It helps the falling sickness, gout, dropsie, infirm stomache, and indeed strengthens all weak parts, and openeth all obstructions, and is a kinde of Panacea.

No limit apparently on the availability of *Cranium humanum*. I wonder who the supplier is. I'm a bit surprised that *Cranium humanum* was not employed in Gesner's *Memory and Wit Ointment*.

Harkening back to Gesner, let us try one additional medication made by bruising, rather than distilling, the cantharides beetle (*Spanish fly*):[14]

> Their virtue consists in burning the body, causing a crust, or . . . to corrode, cause exulceration, and provoke heat; and for that reason are used mingled with medicines that are to heat the Lepry, Tettars, and Cancerous sores.

But, despite its toxicity, the unique qualities of Spanish fly "are good for such as want erection, and do promote venery very much." And here is a more specific observation by Gesner:[14]

> When Anno 1579, I staid at Basil, a certain married man (it was that brazen bearded Apothecary that dwelt in the Apothecaries shop) he fearing that his stopple was too weak to drive forth his wifes chastity the first night, consulted one of the chief Physicians, who was most famous, that he might have some stifte prevalent Medicament, whereby he might the sooner dispatch his journey.

Propriety forces us to end here, but the outcome was both painful and fruitless.

---

1. R.G.W. Anderson, in F.L. Holmes and T.H. Levere (eds.), *Instruments and Experimentation in the History of Chemistry*, The MIT Press, Cambridge, MA, 2000, pp. 7–8.
2. Anderson, op. cit., pp. 5–34.
3. A.J. Ihde, *The Development of Modern Chemistry*, Harper & Row, New York, 1964, pp. 13–18.
4. H. Brunschwick, *Lieber de arte distillandi de composites. Das buch waren kunst zu distillieren die composita und simplicia und ds Buch thesaurus pauperum, ein Schatz der armen genant Micarium . . .*, Strassburg, 1512. I am grateful to Ms. Elizabeth Swan, Chemical Heritage Foundation for supplying an image of this hand-colored plate.
5. J.R. Partington, *A History of Chemistry*, MacMillan and Co. Ltd., London, 1961, Vol. 2, pp. 82–89.
6. A. Roob, *The Hermetic Museum: Alchemy & Mysticism*, Taschen, Cologne, 1997, p. 146.
7. *The New Encyclopedia Britannica*, Encyclopedia Britannica, Inc., Chicago, 1986, Vol. 12, p. 926.
8. P. Ulstadt, *Coelum Philosophorum seu de Secretis naturae. Liber*, Ioannis Grienynger, Strassburg, 1528. I am grateful to The Roy G. Neville Historical Chemical Library (California) for providing the image from this book.
9. W.F. Ryff, *New gross Distillier-Būch, wolgegrūndter kūnstlicher Distillation . . .*, Bei Christian Egenolffs Erben, Frankfort, 1545. I am grateful to Ms. Elizabeth Swan, Chemical Heritage Foundation, for supplying an image of this page.
10. C. Gesner, *The newe Iewell of Health, wherein is contained the most excellent Secretes of Phisicke and Philosophie, divided into fower Bookes. In the which are the best approved remedies for the diseases as well as inwarde as outwarde, of all the partes of mans bodie: treating very amplye of all Dystillations of Waters, of Oyles, Balmes, Quintessences, with the extraction of artificiall Saltes, the use and preparation of Antimonie, and Potable Gold. Gathered out of the best and most approved Authors, by that excellent Doctor Gesnerus. Also the Pictures, and maner to make the Vessels, Furnaces, and other Instruments thereunto belonging. Faithfully corrected and published in Englishe, by George Baker, Chirurgian*, Henrie Denham, London, 1576. The Roy G. Neville Historical Chemical Library.
11. C. Gesner, *The practice of the new and old phisicke, wherein is contained the most excellent Secrets of Phisicke and Philosophie, divided into foure Bookes, In the which are the best approved remedies for the diseases as well inward as outward, of al the parts of mans body: treating very amplie of al distillations of waters, of oyles, balmes, Quintessences, with the extraction of artificiall saltes, the use and preparation of Antimony, and potable Gold Gathered out of the best & most approved Authors, by that excel-*

lent Doctor Gesnerus. Also the pictures and maner to make the Vessels, Furnaces, and other Instrumentsd thereunto. Newly corrected and published in English, by George Baker, one of the Queenes Maiesties chiefe Chirurgians in ordinary, printed by Peter Shaw, London, 1599, p. 240 (i.e., p. 140).
12. J. French, *The Art of Distillation or, A Treatise of the Choicest Spagiricall Preparations Performed by way of Distillation. Together with the Description of the Chiefest Furnaces & Vessels Used by Ancient and Moderne Chymists, Also a Discourse of Divers Spagiricall Experiments and Curiosities: And the Anatomy of Gold and Silver, with the Chiefest Preparations and Curiosities thereof; together with their Vertues. All which are contained in VI. Bookes; Composed by John French Dr. of Physick*, E. Cotes, London, 1653, pp. 73–74.
13. French, op. cit., p. 91.
14. T. Muffet, *The History of Four-Footed Beasts and Serpents and Insects*, Vol. 3, *The Theatre of Insects* (reprint of 1658 London edition), Da Capo Press (Plenum), New York, 1967, pp. 1003–1005.

DISTILLATION BY FIRE, HOT WATER, SAND, OR STEAMED BOAR DUNG

Conrad Gesner (1516–1565) was born in Zurich into "the very poorest circumstances."[1,2] His early brilliance was noted by his father who sent him to his uncle,

The ſecond Booke of Diſtillations,
containing ſundrie excellent ſecret
remedies of Diſtilled
Waters.

FIGURE 109. ■ The title page of Book Two of Conrad Gesner's *The Practice of the New and Old Physicke, Wherein is Contained the Most Excellent Secrets of Phisicke and Philosophic, divided into foure Bookes. In the Which are The Best Approved Remedies for the Diseases as well Inward as Outward of al the Parts of Man's Body, etc.* (London, 1599). Now *that's* a title!

FIGURE 110. ■ Distillation apparatus from Gesner's treatise of 1599: (a), left: A furnace employing a water bath is termed the *Bain Marie* (*Balneum Marie* or bath of Marie); (b) heating samples in closed cucurbits using sand heated by the sun preferably in July or August; (c) A heating bath of boar dung freshly steamed. I suggest calling this the "Bane of Marie" and further advise outdoor use only.

who sold medicinal herb extracts, for further education. In that setting, Gesner developed a lifelong interest in plants and the medicines derived from them. His teachers sponsored Gesner's later education, despite his foolishness, at the age of 19, in marrying a bride with no dowry. He compiled a Greek–Latin dictionary and was appointed Professor of Greek in Lausanne Academy by the age of 21. This allowed Gesner to accumulate money, and he attended medical school for one year achieving the Doctorate in Medicine at the age of 25. The remainder of his life was spent as a physician in Zurich and a Lecturer in Aristotelian physics at the Collegium Carolinum. Gesner died of plague at the age of 49.

Figure 109 is from Gesner's *The Practice of the New and Old Physicke . . .* (102-word title!) published in London in 1599. The first edition of this book (*The Treasure of Evonymous . . .*—"evonymous" means anonymous) appeared in 1552.[1] Figure 109 is the title page of the second book of four in this volume. The sun and the moon represent the male (Sophic Sulfur) and female (Sophic Mercury) principles. In Figure 110(a) we see the *Bain Marie* (or *Balneum Marie; bain* is French and *balneum* is Latin for "bath"; Marie refers to the third-century al-

¶ **The fourth Booke of Dyſtillations,**
containing many ſingular ſecret
Remedies.

FIGURE 111. ■ The title page of Book Four of Gesner's 1599 treatise. The winged pet dragon represents Sophic Mercury; the Tree of Life flowers in cucurbits that produce Birds of Hermes signaling success of the Great Work.

chemist Mary Prophetissa). It is a furnace using a water bath to achieve a gentle and controlled distillation. Similar results can be achieved with the simpler apparatus on the right in this figure. A cucurbit (or retort) is fitted with an alembic (or limbeck) on top, having a beak to condense the vapors into a collecting retort.

Figure 110(b) depicts the heating of distillates in sealed cucurbits in a sand bath heated by the sun (Gesner advises July and August as the best times for this work, which may take periods as long as 40 days). Another technique for gentle distillation is to place cucurbits topped with alembics into a box of continuously steamed boar dung [Figure 110(c)]. I suggest "Bane of Marie" as the name for this apparatus. The operation is probably best done outdoors.

The title page for Book Four (Figure 111) is full of wonderful symbols. The sun and moon witness the growth of the Philosopher's Tree (or Tree of Life), representing the growth of The Great Work.[3] The pet dragon eating (eating what?!) from *her* bowl is winged and probably represents Sophic Mercury. The cucurbit, when sealed, can be considered to be a Philosopher's Egg.[3] (In this figure, we are one short of a dozen eggs.) A Bird of Hermes[3] ascends from each egg, symbolizing completion of The Great Work.

Figures 112 to 114 are from *The Art of Distillation* by John French (1653). The first [Figure 112(a)] represents a steam-distillation apparatus. Figure 112(b) depicts a *Bain Marie* made using a brass kettle and cover and heated in the cen-

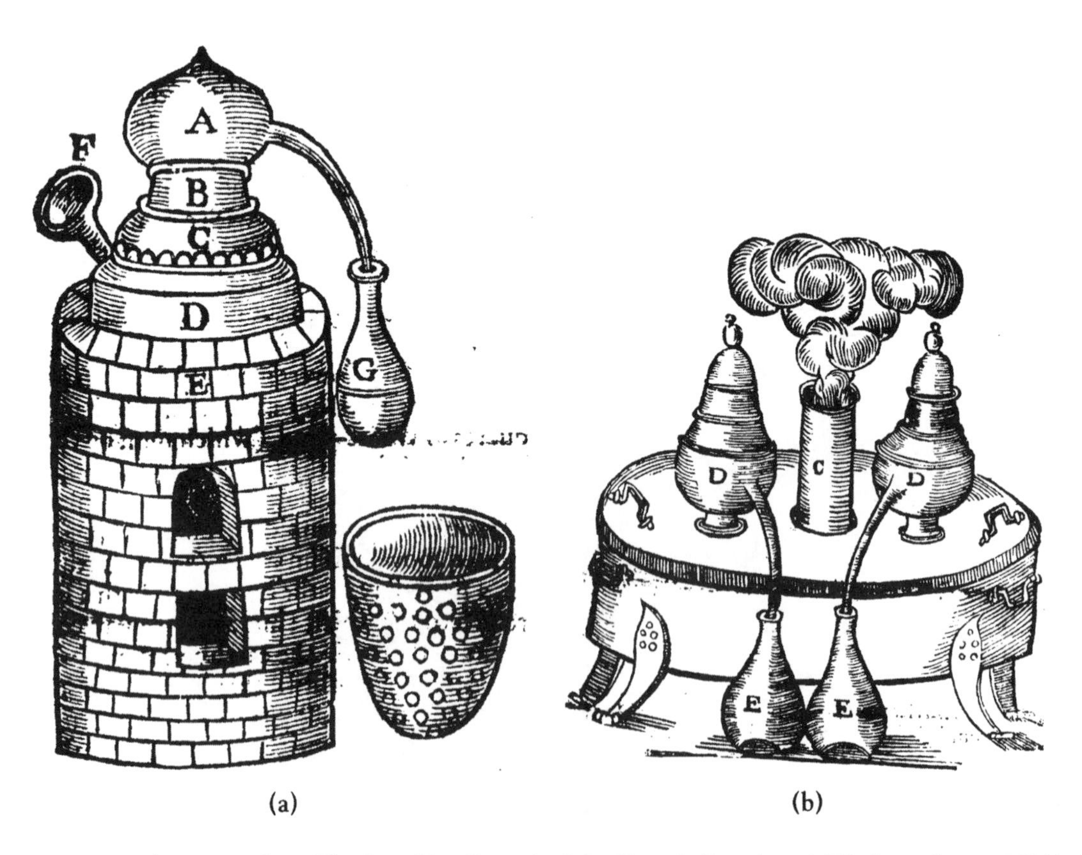

FIGURE 112. ■ Apparatus from *The Art of Distillation* by John French (London, 1653; first edition, 1651): (a) Apparatus for steam distillation; (b) A *Balneum Marie*.

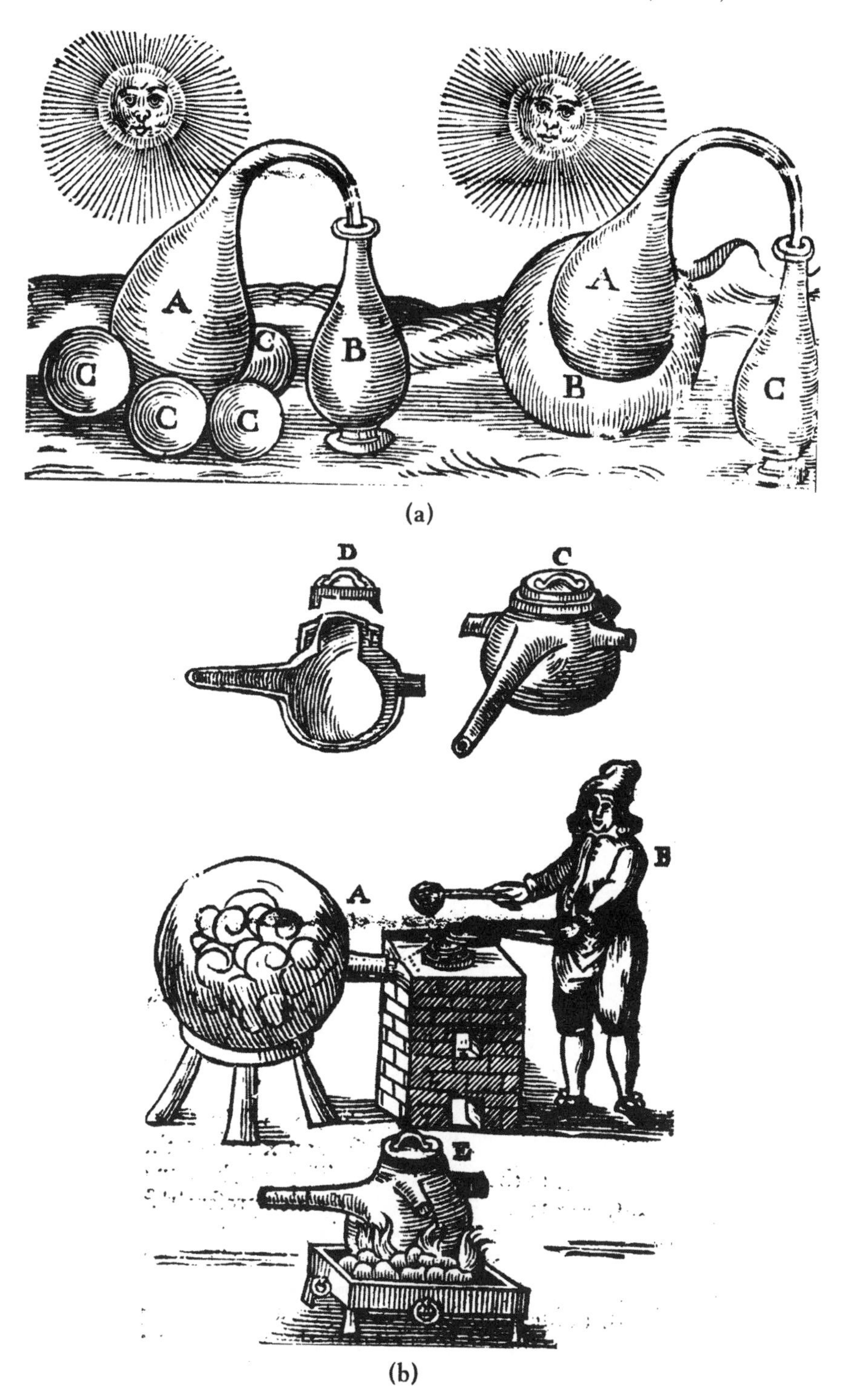

FIGURE 113. ■ From French's *The Art of Distillation:* (a), left, glass crystals heated by the sun as heat source; right, iron or marble mortar as the heat source; (b) heavy-duty furnace for distillation from large quantities of bones, horns, minerals, and vegetables.

ter by a stack oven. Figure 113(a) illustrates the use of sunlight for heating glass crystals or an iron (or marble) mortar as the heat source for distillation. The heavy-duty furnace in Figure 113(b) promises distillation of large quantities of spirits and oils from minerals, vegetables, bones, and horns in only 1 hour instead of the usual 24 ("time is money" even in 1653). Figure 114(a) depicts the distillation of spirit of salt (hydrochloric acid). Figure 114(b) depicts a still for volatile substances including condensers (one of these water-cooled) at the end: state-of-the-art, maintenance contract available for additional purchase.

FIGURE 114. ■ From French's *The Art of Distillation:* (a) Apparatus for distilling spirit of salt (hydrochloric acid), a very biting "serpent" indeed; (b) State-of-the-art still with water-cooled condenser for distilling volatile liquids. Similar apparatus are still found in the hills of Kentucky and West Virginia.

1. J. Ferguson, *Bibliotheca Chemica*, Derck Verschboyle, London, 1954, Vol. 1, pp. 315–316.
2. *Encyclopedia Brittanica*, Vol. 5, Encyclopedia Brittanica, Chicago, 1986, p. 225.
3. L. Abraham, *A Dictionary of Alchemical Imagery*, Cambridge University Press, Cambridge, 1998.

THE JOY OF SEXTODECIMO

It is self-indulgent, inappropriate, and simply rude to "kvell" or joyously brag about a book purchase in a "solemn" text such as this one. However, our text admits to being idiosyncratic, and self-indulgent idiocy is not a long stretch. Among the greatest joys of book collecting is "the hunt." A devoted collector will constantly be alert to quarry and prepared to pounce and feed at any opportune moment. Figure 115 displays the title page for the exceedingly rare true first English edition of the important seventeenth century text by Nicolas Le Fèvre (translated as Nicasius le Febure in the English editions). The original French edition was published in Paris in 1660, and the usual expert sources speak only of a first English edition of 1664 and a second of 1670.[1–3] While the British Museum[4] lists a copy of the 1662 edition[5] (which really constitutes the first half of the later editions), it appears to be almost unknown. My copy was purchased on a well-known World Wide Web auction site, and I stayed up three hours past my normal bedtime to win it. The English editions are in quarto (4to or 4o) format meaning each original sheet for printing has been folded twice to produce four leaves. Octavo (8vo or 8°), the most common modern book format, requires three folds and provides eight leaves, while the sextodecimo format (16mo or sixtodecimo for polite company) has sixteen leaves per original sheet.[6]

Le Fèvre had presented numerous well-regarded public lectures on chemistry in mid-seventeenth-century France and was appointed demonstrator in chemistry at the *Jardin du Roi* in 1650.[3] The mid-seventeenth century was a period of crisis throughout Europe. Failures of crops, famine, desperate poverty, frequent wars, and divided and shifting loyalties between nobles and kings, played out against a background of pervasive conflict between Catholics and Protestants.

The Reformation played out violently in sixteenth-century France and was punctuated by the Religious Wars during its latter half. A founding father of Protestant thought, Jean (John) Calvin, was born in France (1509) converted to Protestantism, and in 1534 emigrated to Geneva, where he established a model church and wrote numerous influential tracts. The Catholic king Henry II (1547–1559), ruled particularly harshly and this, in turn, influenced the ascendancy of less compromising Protestants, the French Calvinists or Huguenots. Ironically, the regency of Catherine de Médicis, the Queen Mother of Charles IX, tried to take a moderate approach, prompting a violent response from powerful Catholics and a counterresponse from the Huguenots. France was in danger of coming apart during the latter half of the sixteenth century, and a strong King, Henry IV, signed the Edict of Nantes in 1598, guaranteeing religious freedom to the Huguenots in designated parts of France and giving them the right to build fortresses (just in case).

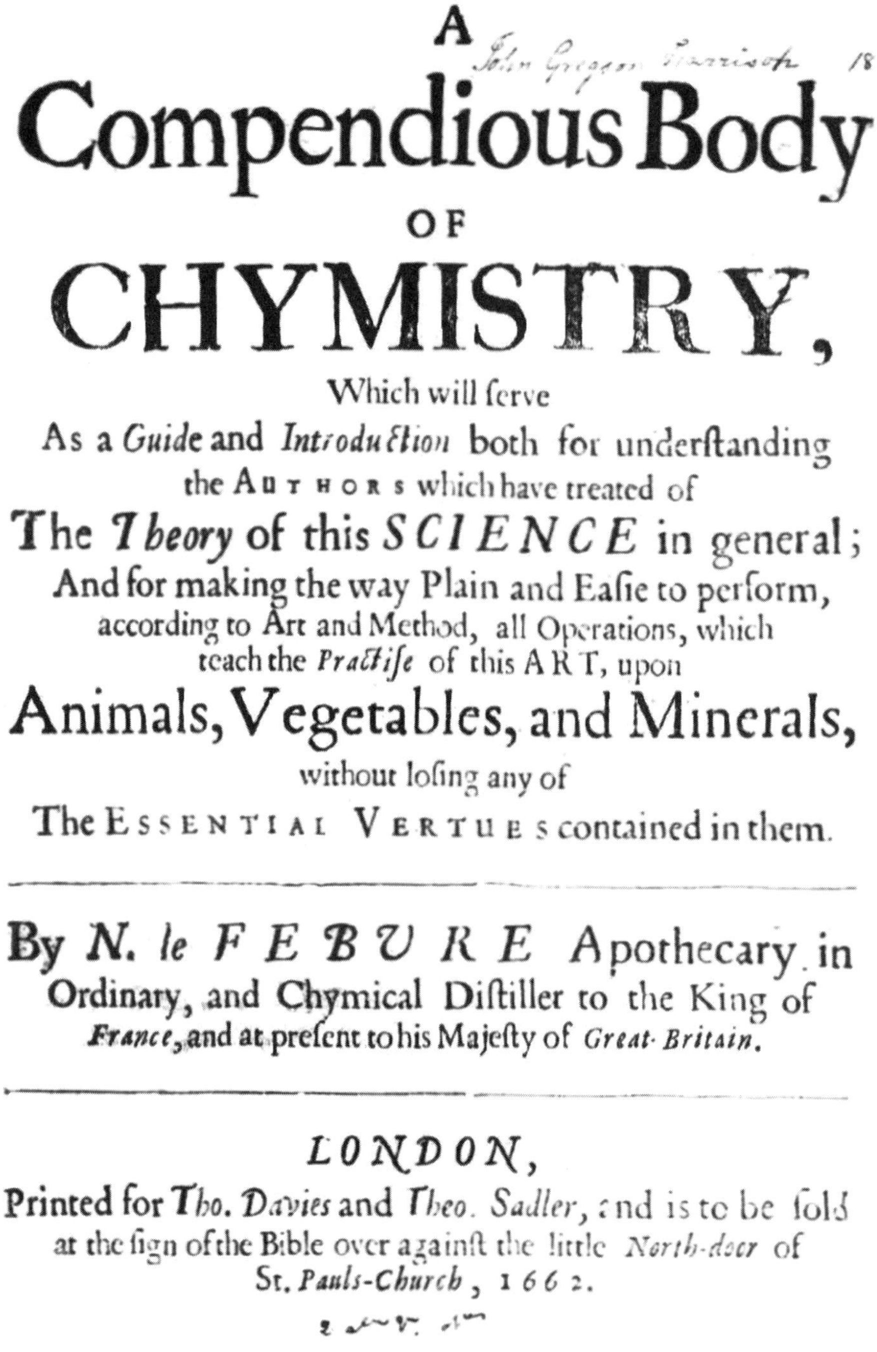

A
Compendious Body
OF
CHYMISTRY,
Which will ſerve
As a *Guide* and *Introduction* both for underſtanding
the AUTHORS which have treated of
The *Theory* of this *SCIENCE* in general;
And for making the way Plain and Eaſie to perform,
according to Art and Method, all Operations, which
teach the *Practiſe* of this ART, upon
Animals, Vegetables, and Minerals,
without loſing any of
The ESSENTIAL VERTUES contained in them.

By *N. le FEBURE* Apothecary in
Ordinary, and Chymical Diſtiller to the King of
France, and at preſent to his Majeſty of *Great-Britain*.

LONDON,
Printed for *Tho. Davies* and *Theo. Sadler*, and is to be ſold
at the ſign of the Bible over againſt the little *North-door* of
St. *Pauls-Church*, 1662.

FIGURE 115. ▪ Title page from the exceedingly rare 1662 English edition of Le Fèvre's famous text that is generally said to be first published in 1664. Such finds are cherished by rare-book collectors who will bore their unfortunate families, relatives, and friends to death with blow-by-blow accounts of successful book hunts.

During the first half of the seventeenth century threats to stability were constant during the early reign Louis XIII (1610–1643). Early in this reign, one of the great figures in French history, Cardinal Richelieu, came to the notice of the ruling house and by 1624 had become the King's principal minister. The powerful Richelieu, whose political acumen became legendary, dedicated himself to the consolidation of regal and religious authority. He died in 1642, and Louis XIII died in 1643. Louis XIV ("The Sun King") was not yet five years old when he assumed the throne and within a brief period another all-powerful Catholic leader, Cardinal Mazarin, became the ultimate authority in France. A series of rebellions started in 1648 and were crushed by 1653. In an environment of increasing intolerance, Le Fèvre moved to London in 1660. Others, including the surgeon Moyses Charas, also emigrated during this period. Louis XIV "viewed himself as God's representative on earth and considered all disobedience and rebellion to be sinful."[7] He declared himself absolute Monarch in 1661 and in 1685 revoked the Edict of Nantes, causing great dislocation and misfortune. During his lifetime Louis established the grandeur of France perhaps best symbolized by the palace he built at Versailles—its cost was estimated to equal that of a modern municipal airport.[7] His extravagant style and arrogance probably foreshadowed the fall of the French monarchy. The king died in 1715, and his body was carried in procession to the jeers of the populace.[7]

Le Fèvre entered England in 1660 at the beginning of the Restoration of the Monarchy. Religious fervor on the part of Protestants had overthrown the monarchy in the person of Charles I in 1649. This was the culmination of religious conflict that started with the split by King Henry VIII of the Church of England from Rome in 1534. The power of Protestantism advanced under King Edward VI. However, during the reign of Queen Mary (1553–1558), Catholics assumed power, and many Protestants were killed and others fled, some to Geneva, where they were influenced by Calvin. The accession of Elizabeth I in 1558 again placed Protestants in power, but her moderate treatment disappointed more radical sects, some of whom came to be called "Puritans." The Puritans sought to "purify" Protestantism from the last traces of Catholicism, and they placed the rulers of England under increasing pressure. Some of these Puritan groups emigrated to America, establishing communities in Virginia and New England. The pressure mounted during the reign of Charles I (1625–1649) and culminated in a military coup that overthrew the monarchy and turned power over to the military leader Oliver Cromwell. The Great Persecution occurred toward the end of this decade-long period and moderate Puritans finally helped to restore the monarchy and Charles II ascended the throne.

Starting around 1645, scholars from London and Oxford and other colleges began to meet in what came to be called the "Invisible College." This loose organization evolved into the Royal Society of London for the Promotion of Natural Knowledge, founded in 1660 and chartered by Charles II in 1662. It is not clear how enthusiastic the king was about his Royal Society, but it kept the Puritans who dominated the universities and their faculties occupied. In 1663 Le Fèvre became one of the initial members of the Royal Society.[3] England had become a beacon for learned men.

While on religious topics, it is amusing to read Le Fèvre's introduction in which he explores the antiquity of chemical knowledge:[8]

> and so we have upon record, that Moses took the Golden Calf, an Idol of the Israelites, did calcine it, and being by him reduced to powder, caused those Idolators to drink it, in a reproach and punishment of their sin. But no body, how little soever initiated in the mysteries of this Art, can be ignorant, that Gold is not to be reduced to powder by Calcination, unlesse it be performed either by immersion in Regal Waters, Amamulgation with Mercury or Projection; all of which three Operations are only obvious to those which are fully acquainted both with the Theorical and Practical part of Chimistry.

So, did Moses really calcine the Golden Calf? In the thirteenth century BCE it appears beyond reasonable doubt that *aqua regia* ("regal waters") was unknown. Alchemical projection would not be in vogue for at least another millennium or so. Amalgamation was a chemical possibility, but we assume that Moses wanted to punish his people, not poison them. The possibility that the calf was really made of marble (limestone) has been suggested.[9] Thus, a drink of the powdered calf would have been an excellent treatment for upset stomachs following prolonged hedonistic partying (see "There Is Truth in Chalk," p. 265).

---

1. J. Ferguson, *Bibliotheca Chemica*, Vol. II, Derek Verschoyle, London, 1954 (reprint of original edition of 1906), pp. 17–18.
2. D. Duveen, *Bibliotheca Alchemica et Chemica*, H&S Publishers, Utrecht, 1987 (reprint of original 1949 edition), pp. 345–346.
3. J. Read, *Humour and Humanism in Chemistry*, G. Bell and Sons Ltd., 1947, pp. 101–114.
4. British Museum Dept. of Printed Books. *General catalogue of printed books*. London Trustees, 1959–1966.
5. N. le Febure, *A Compendious Body of Chymistry Which will serve as a Guide and Introduction both for understanding the Authors which have treated of the Theory of this Science in general; And for making the way Plain and Easie to perform, according to Art and Method, all Operations, which teach the Practice of this ART, upon Animals, Vegetables, and Minerals, without losing any of the Essential Vertues contained in them*, Thos. Davies and Theo. Sadler, London, 1662.
6. J. Carter, *ABC for Book Collectors*, fifth edition, revised, Alfred A. Knopf, New York, 1987, pp. 100–101.
7. *The New Encyclopedia Brittanica*, Vol. 7, Encyclopedia Brittanica, Inc., Chicago, 1986, pp. 500–501.
8. le Febure, op. cit., p. 2.
9. *The New Encyclopedia Brittanica*, Vol. 24, Encyclopedia Brittanica, Inc., Chicago, 1986, p. 374.

THE COMPLEAT APOTHECARY

Le Fèvre's book *A Compendious Body of Chymistry* (see previous essay) was noteworthy for its clarity concerning the construction of apparatus and execution of chemical operations. In Figure 116(a) we see a "superdeluxe" philosopher's furnace or athanor with all of its accessories. Clearly, Charles II royally supported his "Royal Professor in Chymistry" and "Apothecary-in-Ordinary" to the royal household.[1] (One can only imagine the negotiations for "start-up monies" and moving expenses to bring this young professor from Paris. Was immediate tenure part of the package? Was a faculty committee involved or was tenure granted by Royal Decree?)

(a) (b)

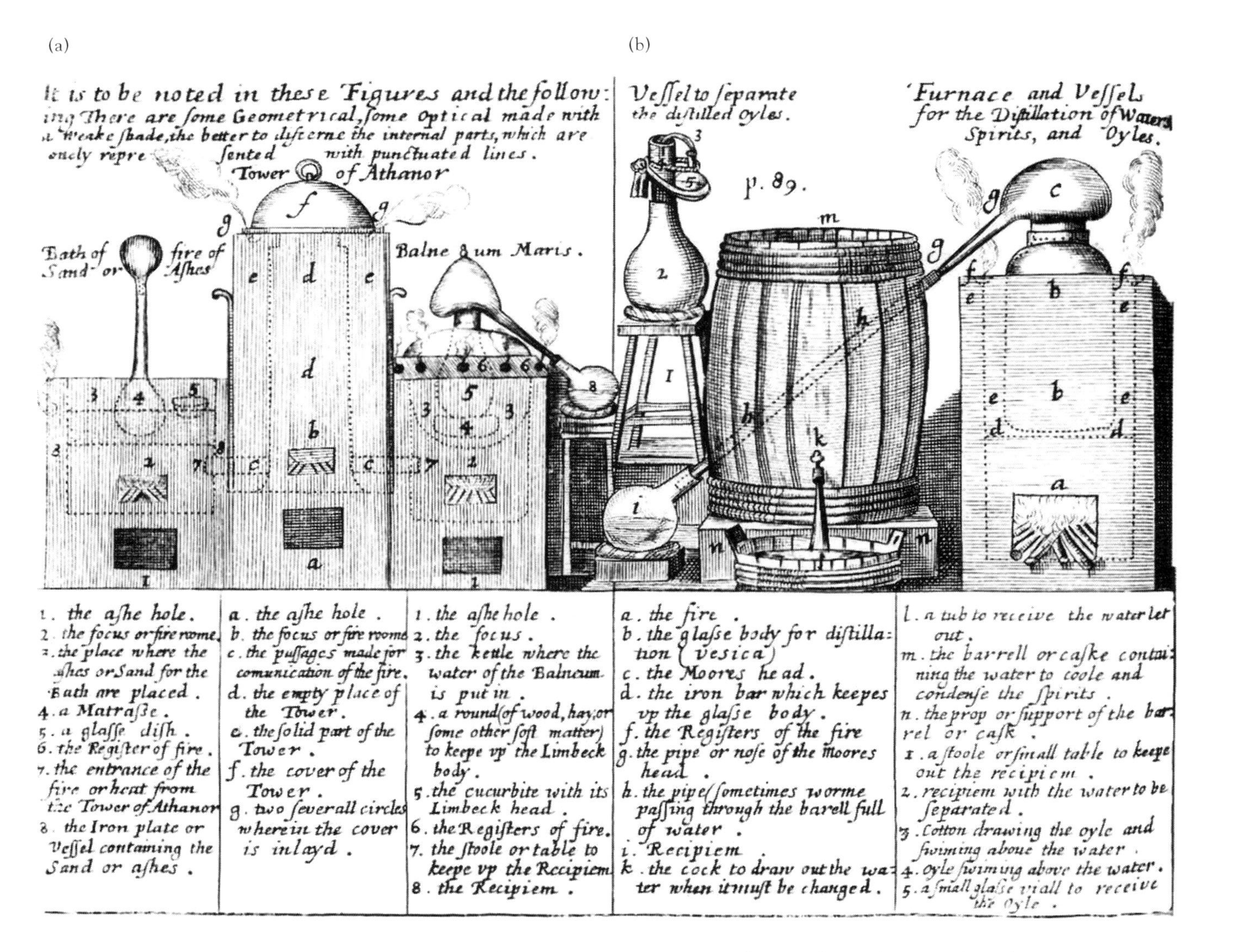

FIGURE 116. ■ Multipurpose stills in Le Fèvre's *A Compendious Body of Chymistry* (1662).

The heat from the athanor was communicated as needed to the *balneum maris* (*balneum marie* or *bain marie*—hot-water bath) accessory on the right and also to the sand bath accessory on the left in Figure 116a. Both the *balneum maris* and sand bath accessories had their own furnaces for specialized operations. The athanor itself was commonly used for operations involving a sealed vessel or philosopher's egg.

Figure 116b depicts an apparatus for distilling alcohol and other volatile spirits. The long, straight "worm" *h* descends from the "Moores head" c through a barrel filled with cold water in order to condense the distillate. When water was added, as a "menstruum," to crushed herbs, flowers, or animal parts, an oily substance, sometimes steam-distilled to form an upper layer over the water collected in receiver *i*. The oil was collected through capillary action by dipping cotton into the oil layer and having it drain into the small glass vial 5.

The lamp furnace (Figure 117a), "used by the most curious Artists for many Chymical Operations," was made of clay and designed to carefully control more modest degrees of heat. Control was performed through the screwdrive attached to lamp *b* as well as by the number of wicks burned simultaneously in the lamp. The most fascinating aspect of this figure is instrument n, the "Weatherglass, Thermometer, or Engin to judge of the quality or degrees of heat." Thermometry was in its infancy—the nature of heat was not understood. Boyle explained that an air current was cooler than standing air because it "drove away the 'warm streams of the body' that normally shielded the skin from the ambient cold" and also apparently penetrated the pores of the skin more than calm air.[2] Le Fèvre's thermometer contained some water in the lower (righthand) bulb, a strip of dyed water in the lower loop, and a hole in the upper bulb. The lower bulb would be inserted in the part to be sensed, and the heat of the water in the bulb would be transmitted to the air that would move the column of dyed water. The purpose was to improve the reproducibility of chemical operations. Some 60 years later, Herman Boerhaave would make the thermometer a standard part of chemical operations[2,3] (see Figure 169). The sublimatory furnace (Figure 117b) had a number of condensing vessels cooled by surrounding air. The least volatile sublimates would be largely collected in f and the most volatile in *k*.

The wind furnace (Figure 118a) was used for mineral and metallic fusions and vitrifications, and was particularly useful for obtaining the pure form ("regulus") of a metal. The crucible d was made of iron or clay. The alembic (limbeck) in Figure 118b was cooled in a cold-water bath rather than air as in Figure 116a. Figure 119a shows a shelf of glassware, including my personal favorite the double-pelican (5) suggestively symbolized by Porta in 1608 as a man and woman mutually circulating bodily fluids [Figure 72(c)]. Item *3* in Figure 119a was termed an "infernell glass" or "hell" since nothing introduced escaped. That is also the function of a philosopher's egg. Indeed, item 9 is LeFèvre's own *Ovum in Ova* ("egg within an egg")—an apparatus that also shares the function of *3*. We hesitate to call this Le Fèvre's "private little hell."

The apparatus in Figure 119b has a built-in efficiency of two collection vessels. This double alembic or distilling head can be made of iron if vegetables are distilled or steam-distilled. However, the distillation of oil of vitriol and other acidic substances (Figure 120)[3] requires tin or tin-lined vessels. Distillation of mercury, Le Fèvre notes, can never employ metallic vessels since amalgamation will occur. The diarist Samuel Pepys visited Le Fèvre's laboratory on January 15,

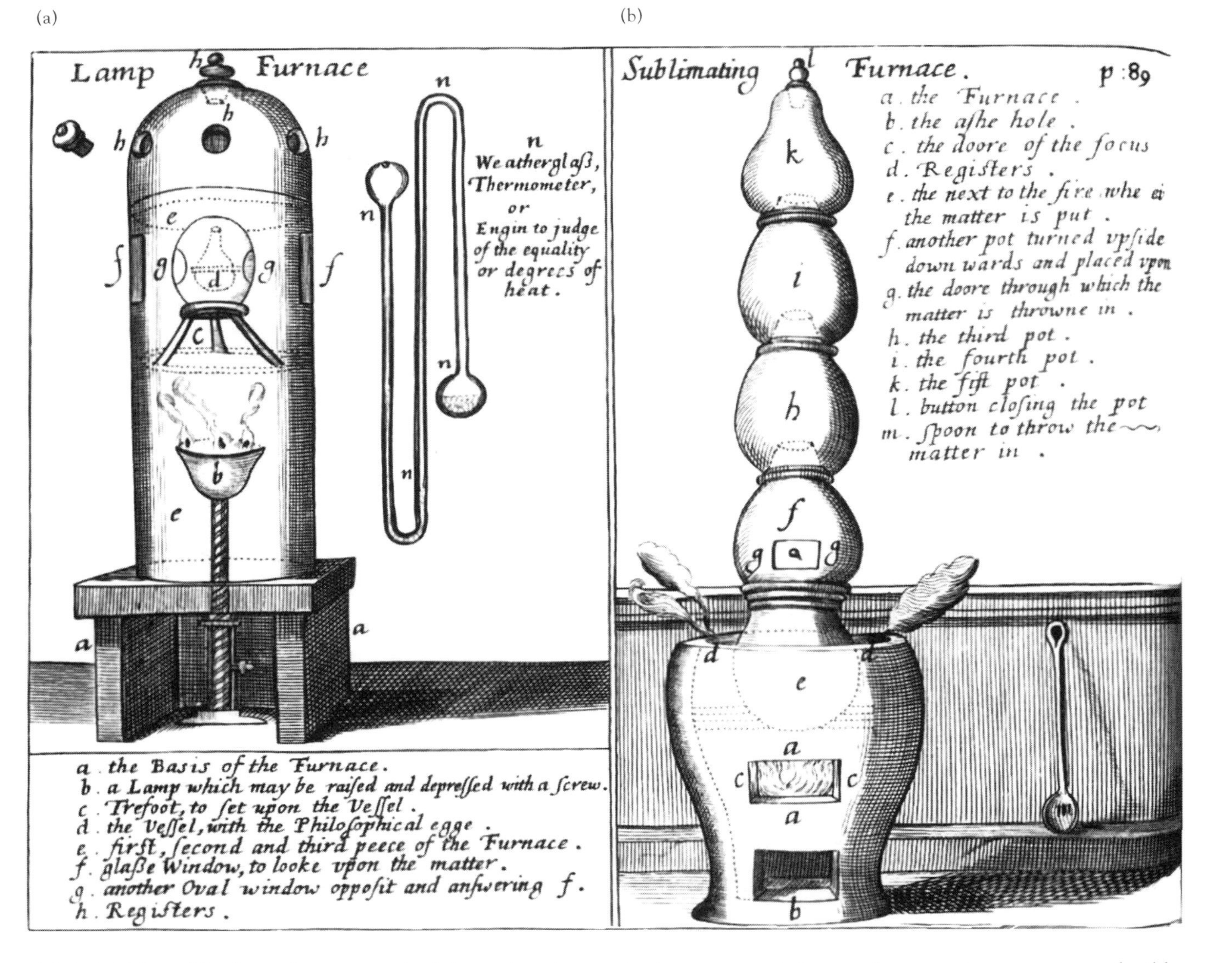

FIGURE 117. ■ The lamp furnace on the left employed an early thermometer (*n*) to render operations more reproducible. The sublimates obtained in the vessel on the right were condensed through air cooling. (From Le Fèvre's *A Compendious Body of Chymistry*.)

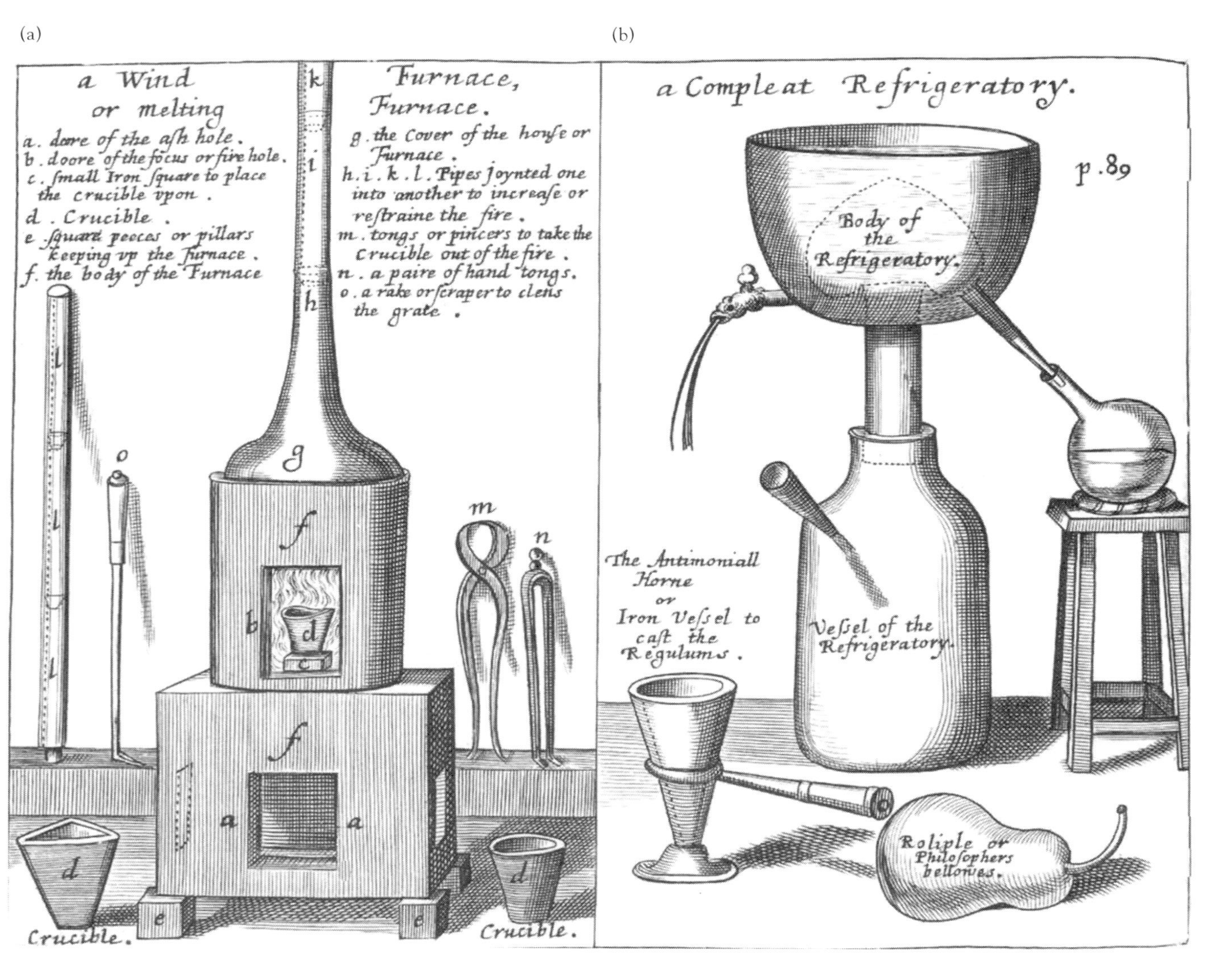

FIGURE 118. ■ The wind furnace on the left was used to obtain the regulus (purest form) of various metals; the alembic on the right used cold water rather than air to condense distilled "spirits" (from Le Fèvre's *A Compendious Body of Chymistry*).

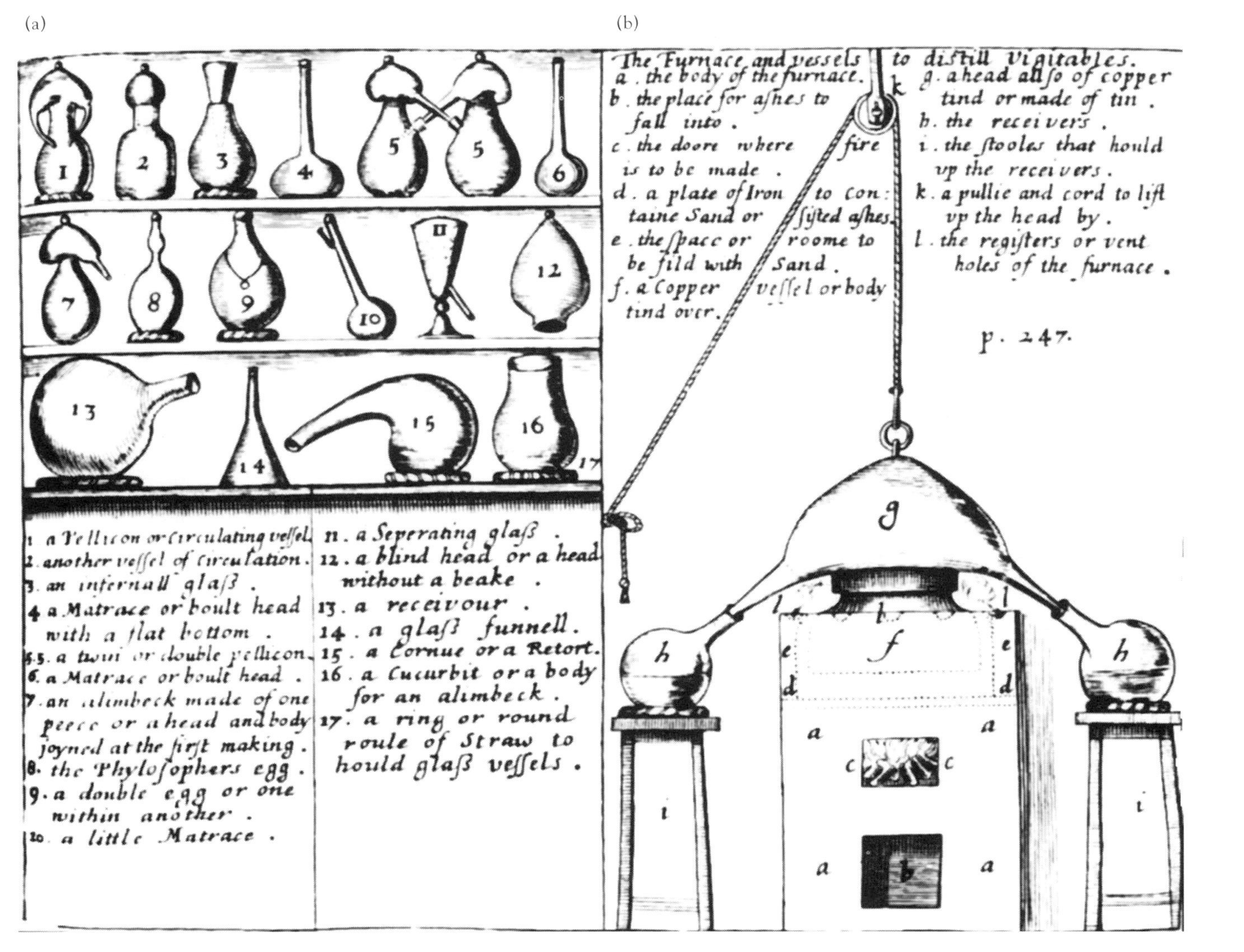

FIGURE 119. ■ A nice shelf of mid-seventeenth-century glassware and an efficient-looking double alembic (still head) that allows the entire apparatus to be lifted off of the furnace when such control is desired (from Le Fèvre's *A Compendious Body of Chymistry*).

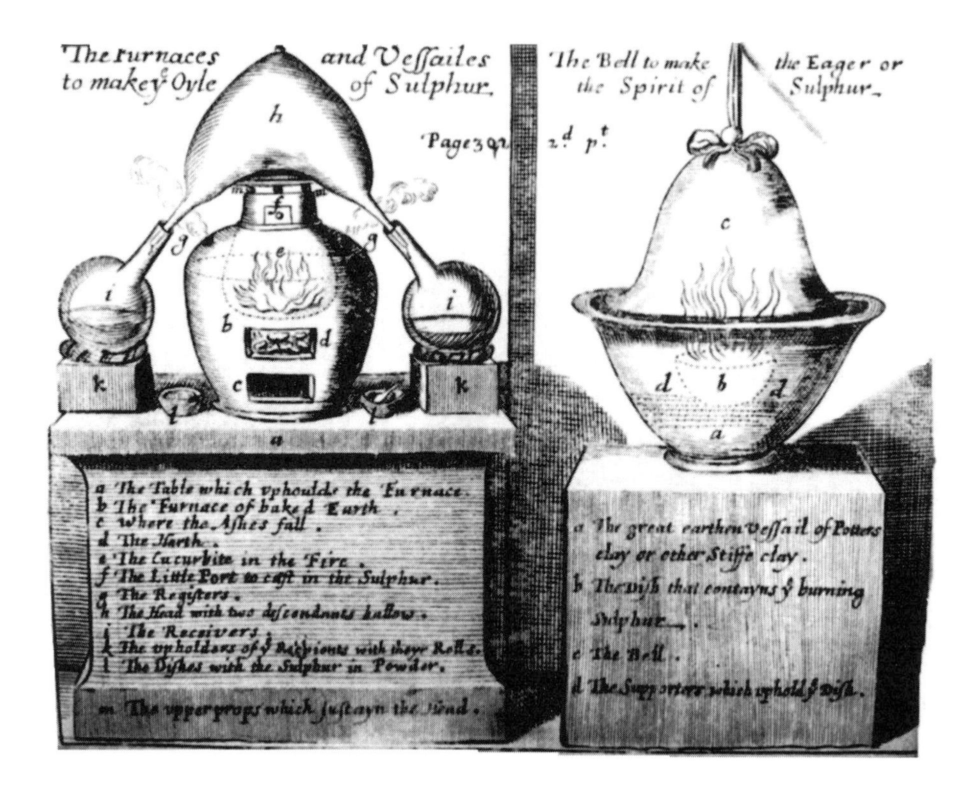

FIGURE 120. ▪ Distillation of oil of vitriol (sulfuric acid) in tin-lined vessels (from Le Fèvre's 1670 edition, *A Compleat Body of Chymistry*).

1669 "and there saw a great many chemical glasses and things, but understood none of them."[1]

---

1. J. Read, *Humour and Humanism in Chemistry,* G. Bell and Sons Ltd., London, 1947, pp. 101–114.
2. J. Golinski, In *Instruments and Experimentation in the History of Chemistry,* F.L. Holmes and T.H. Levere (eds.), The MIT Press, Cambridge, MA, 2000, pp. 185–210.
3. Figures 117–119 are common to all three of the English editions of Le Fèvre's book (1662, 1664, and 1670) while Figure 120 is not in the 1662 edition that is in reality the first half only of the latter two editions.

"RARE EFFECTS OF MAGICAL AND CELESTIAL FIRE"

Antimony is readily released from its ore stibnite (Sb_2S_3) by heating with iron and has been known for centuries. It may also be roasted and the oxide, thus formed, heated with charcoal to obtain the metal.[1] But it has also been "a puzzlement" for centuries. While it is a silvery-white, brittle metal, antimony has a number of allotropes (differing arrangements of atoms as in diamond and graphite—carbon allotropes) that differ in properties. Indeed, rapid condensation of antimony vapor produces a soft yellow, nonmetallic solid that changes spontaneously to the metal in sunlight. Moreover the oxides (calxes) of antimony also exhibit interesting properties. Antimony burns with a bright blue flame, producing a vapor that rapidly condenses into a white powder (Sb_2O_3). This oxide dissolves in both acids and bases—remember that oxides of nonmetals such as phosphorus and sulfur are acidic while oxides of metals are typically basic.[2] Metallic oxides such as "rust" are not volatile while Sb_2O_3 is. Do you remember the wolf in Maier's *Atalanta Fugiens* (see Figure 45)? It represents stibnite (sometimes antimony itself). The final purification of the king (gold) in the fire occurs because the antimony alloyed with gold burns to form its volatile oxide, which sublimes. Obviously, such a metal *must* have powerful medicinal properties. Indeed, tartar emetic (salt of potassium tartrate and antimony) is a potent purgative (as antimony itself is), although these substances also exhibit toxicity. Basil Valentine's *Triumphal Chariot of Antimony* celebrated the medicinal value of antimony (see p. 186).

Nicolas Le Fèvre performed careful quantitative work and discovered that calcinations involving sunlight amazingly increased the mass of antimony:[3]

> But those that are ignorant of the noble Works and rare Effects of Magical and Celestial Fire, drawn from the Rayes of the Sun, by the help of a Refracting or Burning-Glass, shall scarce believe that which we have to say, and are to demonstrate upon this Subject.

The "burning glass" is designed to be three to four feet in diameter (although that in Figure 121 is much smaller), made with two concave pieces of glass, filled with water and sealed with fish glue.[3] Le Fèvre was aware that calx of antimony could also be made using niter (e.g., by adding the metal to nitric acid and heating).[3] In this, he was anticipating Mayow's discovery a decade later (see later essay). However, the incompleteness of the reaction, which forms mixtures in any case, and losses during recovery were vastly inferior to the results achieved with divine sunlight. Crude burning of 12 grains of antimony is found to produce white smoke (said by Le Fèvre to be nothing more than "sublimated" antimony) and calx weighing only 6–7 grains. The calx so produced still has the emetic (nauseative) qualities of antimony powder, albeit somewhat reduced. On the other hand, absorption of "magic and celestial fire" produces 15 grains of calx from 12 grains of antimony.[3] Indeed, we know that conversion of 12 grains of antimony should produce about 14.5 grains of Sb_2O_3.

> but that which is yet more to be admired, and less conceivable, is, that these xv grains of white Powder, are neither vomitive nor purging, but contrariwise Diaphoretical and Cordial; which doth cast into admiration, not without reason, the most curious and intelligent searchers of Nature, and the wisest Physicians.

Diaphoretical substances induce perspiration—a more gentle purge than vomiting.

Le Fèvre was not the first to discover that calxes were heavier than the metals. Biringuccio noted it in 1540 (see p. 28) and Jean Rey had made this discov-

FIGURE 121. ▪ Calcination of antimony using "celestial fire" (from Le Fèvre's 1670, *A Compleat Body of Chymistry*). Le Fèvre was among the earliest chemists to discover that the calx was heavier than the corresponding pure metal.

ery almost a century later. Like Boyle, who also made this discovery, Le Fèvre assumed some kind of incorporation of fire (for Boyle, it was "igneous particles") to augment the weight of the metal (see p. 251).

---

1. F.A. Cotton and G. Wilkinson, *Advanced Inorganic Chemistry*, fifth edition, John Wiley and Sons, New York, 1988, pp. 387–388, 401.
2. T.L. Brown, H.E, LeMay, Jr., and B.E. Bursten, *Chemistry—the Central Science*, seventh edition, Prentice-Hall, Upper Saddle River, NJ, 1997, pp. 841–843, 847–848.
3. N. le Febure, *A Compleat Body of Chymistry: Wherein is contained whatsoever is necessary for the attaining of the Curious Knowledge of this Art; Comprehending in General the whole Practice thereof; and Teaching the most exact preparation of Animals, Vegetables and Minerals, so as to preserve their Essential Vertues. Laid open in two Books, and Dedicated to the Use of all APOTHECARIES, &c.*, O. Pulleyn Junior, London, 1670, pp. 212–217.

SECRETS OF A LADY ALCHEMIST

Very little is known of the life of Marie Meudrac,[1,2] but it appears that she was the first woman to author a printed book on chemistry. *La Chymie Charitable et*

Facile, en Faveur des Dames, was first published in 1656 followed by editions of 1674 and 1687.[3] The title page and frontispiece from the third (French) edition are shown in Figures 122 and 123.[4] There were also at least six German editions and one Italian edition.[3,5]

The Rayner-Canhams note that her work was based on the three alchemical principles, sulfur, mercury and salt, but presented clear discussions of useful chemical operations.[1] The book consists of six parts:[1,2]

Part 1—Principles and operations

Part 2—"Simples"—methods of preparation and treatment

Part 3—Animals

Part 4—Metals

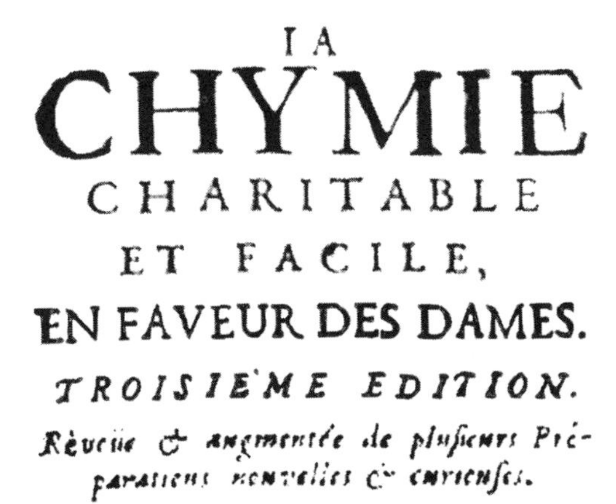

LA
CHYMIE
CHARITABLE
ET FACILE,
EN FAVEUR DES DAMES.
TROISIE'ME EDITION.
Reveüe & augmentée de plusieurs Préparations nouvelles & curieuses.

A PARIS,
Chez LAURENT D'HOURY, ruë
Saint Jacques, devant la Fontaine
S. Severin, au Saint Esprit.

M. DC. LXXXVII.
Avec Privilege & Approbation.

FIGURE 122. ▪ The third edition (1687) of the first chemistry book published by a woman, Marie Meurdrac. The first edition was published in 1656. This wonderful pocket-size book was dedicated to Madame La Comtesse De Guiche and includes a sonnet to Mademoiselle Meurdrac's book by a Mademoiselle D.I.

Part 5—Making compound medicines

Part 6—Preserving and increasing the beauty of ladies

Meudrac's doubts about publishing her work were summarized in her preface:[1,2]

> When I began this little treatise, it was solely for my own satisfaction and for the purpose of retaining the knowledge I have acquired through long work and oft-repeated experiments. I cannot conceal that upon seeing it completed better than I had dared hope, I was tempted to publish it: but if I had reasons for bringing it to light, I also had reasons for keeping it hidden and for not exposing it to general criticism.

Apparently Ms. Meudrac also kept chemistry apparatus mysteriously hidden behind her exquisite curtain, if the book's frontispiece (Figure 123) is taken as a clue. But further in her preface[1,2]

> On the other hand, I flattered myself that I am not the first lady to have something published; that minds have no sex and that if the minds of women were cultivated like those of men, and that if as much time and energy were used to instruct the minds of the former, they would equal those of the latter.

FIGURE 123. ■ The enchanting frontispiece from Marie Meudrac's 1687 chemistry book (see Figure 122) promising to disclose hitherto hidden secrets of chemistry.

This wonderfully assertive nonapology for daring to author a book precedes by about 140 years a similar nonapology by the brilliant Elizabeth Fulhame[6] introducing her own Essay on Combustion in 1794 (see Figure 220).

1. M. Rayner-Canham and G. Rayner-Canham, *Women in Chemistry—Their Changing Roles from Alchemical Times to the Mid-Twentieth Century,* American Chemical Society and Chemical Heritage Foundation, Washington, DC and Philadelphia, 1998, pp. 9–10.
2. L.O. Bishop and W.S. DeLoach, *Journal of Chemical Education*, Vol. 47, pp. 448–449 (1970).
3. D.I. Duveen, *Bibliochemica Alchemica et Chemica,* facsimile edition, HES Publishers, Utrecht, 1986, p. 401.
4. I am grateful to Ms. Elizabeth Swan, Chemical Heritage Foundation, for supplying these images.
5. J. Ferguson, *Bibliotheca Chemica,* facsimile edition, Vol. 2, Derek Verschoyle, London, 1954, pp. 92–93.
6. Rayner-Canham, op. cit., pp. 28–31.

"PRAY AND WORK"

Any chemist who has ever "punted" understands the poetic dictum of Saint Benedict of Montecassino:[1,2] "Ora et Labora" ("Pray and Work"). A series of well-planned, rational experiments may fail to yield an expected product, while a well-placed "punt"[3] ("going for broke"—with a one-step, less rational . . . vacuum sublimation, for example) sometimes works. The chemist depicted in Figure 124 may be trying just such a "punt" and pays homage to God whose all-encompassing wisdom is captured in his reaction vessel—a philosopher's egg nestled into an athanor (philosophical furnace). He even captures sunlight, as "philosophical fire," through a magnifying lens. The words passing between the sun and the chemist translate as "Without me you can do nothing" (possibly paraphrasing the word of the Lord).[1] And from the hand of God straight to the flask, we learn "The beginning of wisdom is the fear of God."[1] This delightful full-page woodcut comes from a German pamphlet published in 1755.[4]

This pamphlet, "Medical, Chemical and Alchemical Oraculum" (an "oraculum" is a "divine announcement"), consists of two parts. The second part is said to be a previously unpublished fourteenth-century manuscript.[5] The first part includes a 33-page table of chemical symbols (hieroglyphics?[6]), accompanied by Latin and German definitions (see Figures 125 and 126). For example, there are 9 symbols collected for *aqua regis* (*aqua regia*), the 3:1 solution of hydrochloric acid and nitric acid that "dissolves" gold (actually it oxidizes or "calcines" gold to $AuCl_4^-$: no oxygen is involved). Most of these symbols are representations of water (either a downward-pointing triangle or waves) appended to an "R." *Aqua vita* (nitric acid) has at least 20 symbols perhaps reflecting differences in chemical properties (it is an acid as well as an oxidizing agent), origins and uses. Figure 125 depicts 17 symbols for silver while the pamphlet also includes 34 discrete symbols for gold, 35 for arsenic, 40 for "quicksil-

FIGURE 124. ▪ "Work and pray"—helpful advice for chemists who are not atheists or agnostics (from the 1755 *Medicinisch-Chymisch-und Alchemistisches Oraculum*).

ver" (mercury), and no less than 54 for various preparations of the fabulous antimony!

Figure 126 depicts symbols for the "bezoardic forms" of the seven ancient metals. "Bezoar" may be defined as "a hard mass deposited around a foreign substance, found in the stomach or intestines of some animals and formerly thought to be a remedy for poisoning."[7] Apparently, ruminants were particularly prized because of the complicated digestions in their chambered stomachs. I suspect that, were I to be poisoned, I would be willing to spare the life of a goat, forgo a dose of bezoardic gold—as valuable as it is, and take my chances with a good, old-fashioned purge.

FIGURE 125. ■ Chemical hieroglyphics from the 1755 *Oraculum* (see Figure 124).

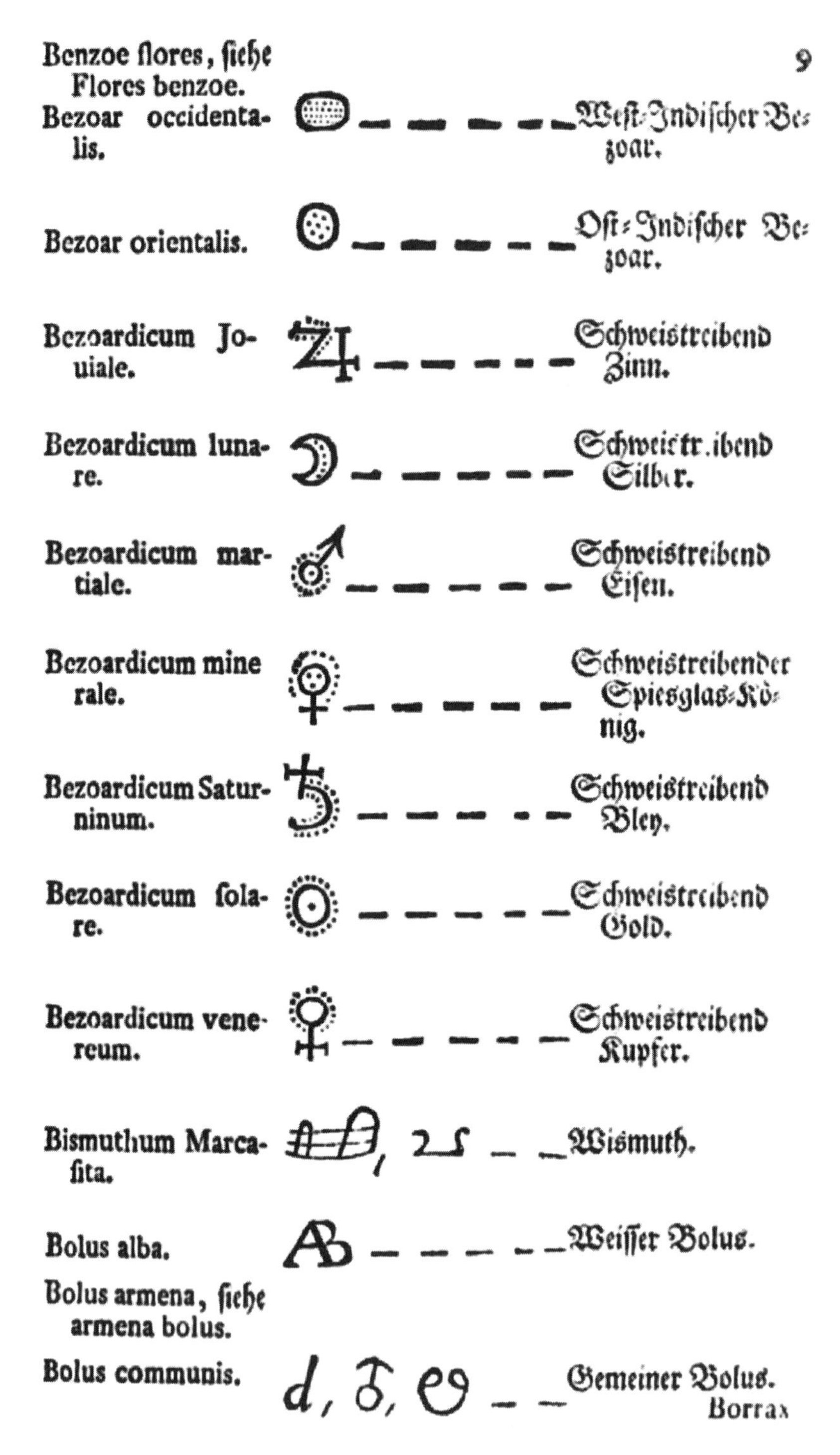

9

Benzoe flores, siehe Flores benzoe.

Bezoar occidentalis. — West-Indischer Bezoar.

Bezoar orientalis. — Ost-Indischer Bezoar.

Bezoardicum Jouiale. — Schweistreibend Zinn.

Bezoardicum lunare. — Schweistreibend Silber.

Bezoardicum martiale. — Schweistreibend Eisen.

Bezoardicum minerale. — Schweistreibender Spiesglas-König.

Bezoardicum Saturninum. — Schweistreibend Bley.

Bezoardicum solare. — Schweistreibend Gold.

Bezoardicum venereum. — Schweistreibend Kupfer.

Bismuthum Marcasita. — Wismuth.

Bolus alba. — Weisser Bolus.

Bolus armena, siehe armena bolus.

Bolus communis. — Gemeiner Bolus.

Borrax

FIGURE 126. ■ Additional chemical hieroglyphics from the 1755 *Oraculum* (see Figure 124). "Bezoar" may be defined as "hard masses deposited around foreign masses found in the walls of stomachs or intestines" of animals, especially ruminants.

1. I thank Professors Heinz D. Roth and Pierre Laszlo for help in the interpretation of Figure 124.
2. *The New Encyclopedia Britannica*, Vol. 2, Encyclopedia Britannica, Inc, Chicago, 1986, p. 97.
3. It is painful to confess, but over 25 years ago I found that rational, well-precedented syntheses failed to produce an exciting and previously unknown lactam. In desperation, I tried a thermal dehydration of the precursor amino acid under vacuum. Crystals of apparent product were obtained, and they were . . . the starting material only. Drat! Almost simultaneously, and completely independently, a research group at another university also found that the precedented reactions failed and these scientists also "punted" with a thermal dehydration under vacuum. They employed slightly different conditions than mine and were rewarded with a tiny yield of the desired compound (see H.K. Hall, Jr. and A. El-Shekeil, *Chemical Reviews*, Vol. 83, p. 549, 1983, which reviews this synthesis and their other outstanding work in this field). Perhaps I should have employed "philosophical" apparatus rather than simple glassware.
4. *Medicinisch-Chymisch-und Alchemistisches Oraculum, darinnen man nicht nur alle Zeichen und Abkürzungen, welche sowohl in den Recepten und Büchern der Aerzte und Apotheker, als auch in den Schriften der Chemisten und Alchemisten vorkommen, findet, sondern dem auch ein sehr rares Chymisches Manuscript eines gewissen Reichs*** beygefüget*, Ulm und Memmingen, in der Gaumischen Handlung, 1755.
5. *Bibliotheca Alchemica Et Chemica*, H&S Publishers, Utrecht, 1986, p. 440. This is a reprint of the book published in 1949 by E. Weil, London and supplemented by Catalogue 62, H.P. Kraus, originally printed in 1953 by H.P. Kraus, New York.
6. A. Roob, *The Hermetic Museum: Alchemy and Mysticism*, Taschen, Cologne, 1997, p. 600.
7. *Webster's New World Dictionary of the American Language—College Edition*, The World Publishing Co., Cleveland and New York, 1964, p 143.

A GOOD OLD-FASHIONED PURGE

Paracelsus revolutionized Renaissance medicine through his use of synthetic metallic drugs. For example, he employed calomel (Hg_2Cl_2)[1] as a purgative, and it is easy to imagine unburdening the *archeus* (p. 139) by ridding the body of "ill humours" as well as intestinal worms. The *London Pharmacopoeia*,[2] first authorized by the Royal College of Physicians in London and published in 1618, listed a synthesis of calomel in which mercury was dissolved in *aqua fortis* (nitric acid), sea salt added, and the precipitate collected and washed with water. Violent purging of the bowels was one means for cleansing the body. Another was through emetics, substances that induced repeated vomiting. Again, it is easy to imagine how medically helpful this would be for clearing the stomach of tainted food or poison. Perhaps the best known was tartar emetic, antimony tartrate. A preparation of this substance provided by Hadrian Mynsicht involved boiling cream of tartar (potassium hydrogen tartrate obtained from wine dregs) with antimony carbonate and allowing crystallization in a cool place.[3] The use of antimony for medicinal purposes stirred up great controversy in the mid-seventeenth century since antimony compounds were known to be quite toxic (p. 186). However, the "Chariot of Antimony" did indeed triumph, and antimony compounds were used as purgatives, emetics, and sudorifics ("sweating" agents). Books throughout the sixteenth and seventeenth centuries also referred to *Aurum Potabile* ("potable gold").[4] This was considered to be the universal medicine, tincture of gold, suitable for treating (and improving) animals, plants, and *minerals*. Logically speaking, it may have been a dilute solution of gold dissolved in *aqua regia* (hydrochloric acid–nitric acid, 3 : 1).

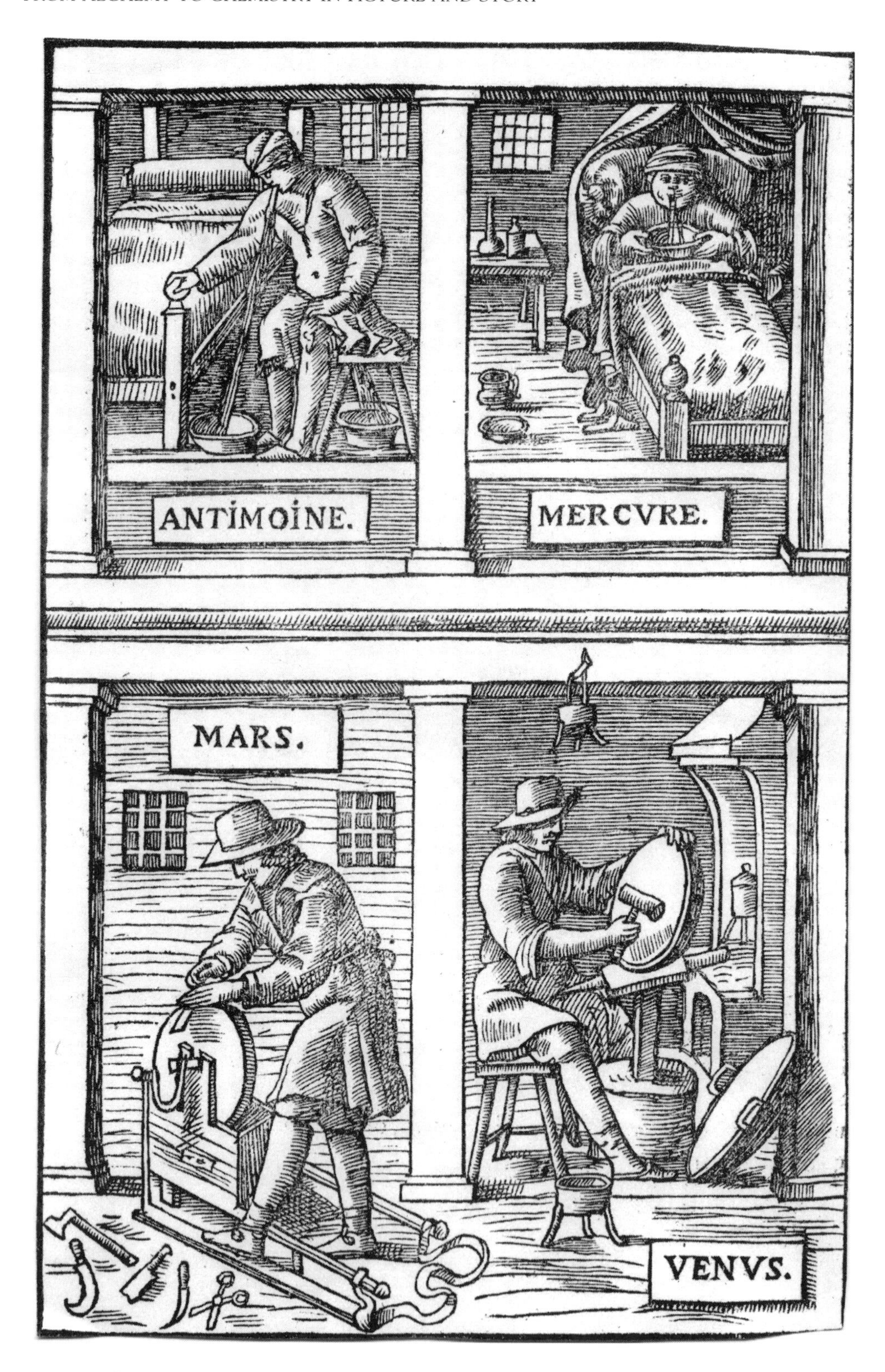

FIGURE 127. ▪ Illustration of uses for the metals antimony, mercury, iron ("Mars"), and copper ("Venus"). Clearly, both antimony and mercury compounds are effective emetics, but a closer look at this figure suggests that some antimony compounds are effective laxatives (purgatives) as well. (From Barlet's 1657 *Le Vray Méthodique Cours de la Physique*.)

FIGURE 128. ■ Applications for lead ("Saturne"), tin ("Jupiter"), silver ("Lune"—moon), and gold ("Soleil"—sun). (From Barlet's 1657 *Le Vray Méthodique Cours de la Physique.*)

FIGURE 129. ■ A seventeenth-century laboratory for processing animal products (from Barlet's 1657 *Le Vray Méthodique Cours de la Physique*).

FIGURE 130. ■ A seventeenth-century laboratory for processing plant products (from Barlet's 1657 *Le Vray Méthodique Cours de la Physique*).

FIGURE 131. ■ A seventeenth-century laboratory for processing minerals (from Barlet's 1657 *Le Vray Méthodique Cours de la Physique*).

FIGURE 132. ■ A seventeenth-century laboratory for processing metals (from Barlet's 1657 *Le Vray Méthodique Cours de la Physique*).

Figures 127 and 128 are from Annibal Barlet's 1657 *Le Vray et Méthodique Cours de la Physique*.[5] Little is left to the imagination concerning the efficacy of antimony compounds as violent emetics (as well as effective purgatives). Figure 127 also suggests mercury salts as all-purpose purges. The remaining six panels in Figures 127 and 128 depict the metallurgy of iron (Mars), copper (Venus), lead (Saturn), tin (Jupiter), silver (*Lune* or the moon), and gold (*Soleil* or the sun). Barlet apparently taught chemistry but made no contribution to the field.[6] Although the book was said by contemporaries to be of little value, its illustrations of laboratories treating animal extracts (Figure 129), plant extracts (Figure 130), mineral chemistry (Figure 131) and metallurgical chemistry (Figure 132) provide some insights into seventeenth-century chemical operations.

---

1. J.R. Partington, *A History of Chemistry*, MacMillan & Co. Ltd., London, Vol. 2, 1961, p. 145.
2. Partington, op. cit., p. 165.
3. Partington, op. cit., pp. 178–179.
4. J.R. Glauber, *The Works of the Highly Experienced and Famous Chymist, John Rudolph Glauber: Containing, Great Variety of Choice Secrets in Medicine and Alchymy in the Working of Metallick Mines, and the Separation of Metals: Also, Various Cheap and Easie Ways of making Salt-petre, and Improving of Barren-Land, and the Fruits of the Earth. Together with many other things very profitable for all the Lovers of Art and Industry*, London, printed by Thomas Milbourn, 1689, pp. 206–220.
5. (A.) Barlet, *Le Vray et Méthodique Cours de la Physique résolutive, uulgairement dite Chymie. Réprésenté par Figures générales & particulières. Pour connoistre la Théotechnie Ergocosmique. C'est à dire, l'Art de Diev, en l'ouvrage de l'univers. Seconde Édition. Avec l'indice des Matières de ce Volume, & quelques Additions*. Paris, Chez N. Charles, 1657. I am grateful to The Roy G. Neville Historical Chemical Library (California) for supplying these images.
6. J. Ferguson, *Bibliotheca Chemica*, Vol. 1, reprint edition, Derek Verschoyle, London, 1954, pp. 72–73.

"OPENING" METALS—THE ART OF CHYMISTRY

In 1667, a book with the very prosaic title *Cours de Chymie* was authored by a certain P. Thibaut, who announced himself "distiller in ordinary to the King."[1] John Starkey, the London publisher, who understood salesmanship and titled the 1668 English translation *The Art of Chymistry*, dubbed Pierre Thibaut "Chymist to the French King."[2] The King was Louis XIV (1638–1715), "The Sun King," who ruled from 1643 until his death. He expanded France's borders, made his country the dominant continental power, built the opulent palace at Versailles, and remains today the very symbol of absolute monarchy. He is alleged to have said "*L'État c'est Moi*" (I am the State) when approached about sharing some powers of government. Chymist to the Sun King sounds like a pretty nice job (as long as one keeps the flattery flowing): laboratory at Versailles, royal entertainments in the evening, venison and the best Bordeaux wines, and the favors of . . . but I digress.

Thibaut describes chymistry as a "liberal art" with a much higher calling than "vulgar pharmacy." "Vulgar" (that is to say "common") pharmacists prepared their remedies from plants or animals by only four techniques:[3]

> [F]or, either they pressed the Juice out of their Ingredients, or beat them to powder, or they boyled them, or else they infused them in Water, or some other Liquor, and so gave them to their Patients.

In contrast, Paracelsus and his followers discovered that the greatest remedies could be derived from metals if only they could be "opened" through chymistry:[3]

> [T]he Vertues of those excellent Remedies lay buried in their own Bodies, as in a Grave.

And here is an illustration of Thibaut's point—that only true chemical change will actually "open up" the medicinal virtues derivable from metals[3,4] (see also Figure 127):

> This may be clearly seen in Crude Antimony, of which a pound, either in Powder, or in Infusion, or Decoction, works no other effect in the Body, than if we had swallowed as much Saw-dust. But if you know how to open its Body by the Keys of Chymistry, then for interiour Remedies, you shall have an Emetick, a Purgative, a Sudorifick, A Diaphoretick, a Diuretick, and a Cordial, which you need only give in a different proportion of so many Grains. And for exteriour Applications, you may, out of the same Antimony, have a Deficcative, a Mundificative, a Consumptive, or Escarotick, with other rare Remedies, as we shall hereafter set down at large.

Antimony medications were extremely controversial during the seventeenth century since they were also known to be highly toxic. Mercury compounds (Figure 127) are also highly toxic. Although the metal itself has less obvious immediate (acute) effects, chronic exposure produces very serious impairment, including damage to the nervous system ("mad hatter" disease, for example) and it can certainly be lethal. Thus, readers should *never* attempt the rather disgusting and *dangerous* practice reported below by Thibaut:[3]

> If we consider Mercury, they that work in the Mines in Spain, teach us by their Theft, that more than a pound of it may be taken inwards without harm; for, a little before they give over working, they swallow a good quantity of it, which, when they are at liberty, they ease themselves of by Stool, and so keep it to sell in secret. And this the Overseers having discovered, do now force every Workman to stay there a considerable time, after his giving over working, that these Mercurial Thieves may be so forced to leave their theft behind them. But, if by the virtue of Chymical Dissolvants, you open the body of Quick-silver, it will produce in very small Doses, such various and wonderful Effects, that out of it alone, may be had Remedies to answer all the Indications of Physick.

Hi-Ho Quick-Silver!

It was not uncommon for Gallic enthusiasm to bubble up into an ode celebrating a friend's book (for example the sonnet by a Mademoiselle D. I. for Marie Meudrac's *La Chymie Charitable et Facile, en Faveur des Dames;* see Figure 122). Here we have a poem (in rhyme) by a certain Jacquet, an admirer of Thibaut:[1] (see reference 5 for an English translation):

Royal Distillateur et tous les Vegetaux,
Qui tire de leurs Corps l'eau, le sel, le Mercure
En les rendent si purs qu'ils changent la Nature,
Et prolongent la vie au Roy des animaux.

Ton Art produit au jour l'Ame des Mineraux,
Qui donne aux Clairs Cristaux l'admirable teinture,
La Lune avec Venus fait une Creature
Mariant sa Beauté au Prince des Metaux.

Grands Filous, Charlattans & Sousseurs d'Alchymie
De Qui la Verité est toujours ennemie,
Cassez vos Alembics, abbatez vos Fourneaux.

Thibaut nous fait tout voir par son experience,
Enseignant aux humains la divine Science,
De se render Immortels par l'eau de ses Tonneaux.

---

1. E.F. Smith, *Old Chemistries*, McGraw-Hill Book, New York, 1927, pp. 5–6.
2. P. Thibaut, *The Art of Chymistry: As it is Now Practiced,* John Starkey, London, 1668. Here I must make a confession: I was aware of the Thibaut title while completing my earlier book, *The Art of Chemistry.* Indeed, I wanted to use the 335-year-old title verbatim but was wisely advised that customers using database search engines would not search under "chymistry." Willing to compromise *Art* for *Gold,* I stayed with the modern spelling. I should have acknowledged the title of the 1668 book in my preface.
3. Thibaut, op. cit., The Preface.
4. Some of these terms are defined elsewhere in the present book and some are all too obvious. We note that a "mundificative" was an all-purpose healing or "cleansing" agent and an "escarotick" finds use in treating sores and lesions.
5. Translation of Jacquet's *Ode to Thibaut:*

Royal Distillers of all Vegetables,
Who draw from their Bodies water, salt, Mercury
In making them so pure that it alters Nature,
And prolongs the life of the King of animals.

Day-by-day, Your Art produces the Soul of Minerals,
That gives admirable hues to Clear Crystals,
The Moon together with Venus made a Creature
Marrying her Beauty to the Prince of Metals.

Great Swindlers, Charlatans and Low-Lifes of Alchemy
For whom Truth is always inimical,
Smash your Alembics, demolish your Furnaces.

Thibaut has made us see all through his experience,
Teaching the divine Science to humans,
Thus making them Immortals with water from his casks.

THE TRIUMPHAL CHARIOT OF ANTIMONY

The table of "Chimical Characters" shown in Figure 133 is from Nicholas Le Fèvre's *A Compleat Body of Chymistry* (second English edition, 1670), one of the

Explanation of the Chimical Characters p. 191

Steele iron or mars	celestial signe	Gumme	Crocus
Loadstone	Cancer	Hower	martis
Ayre	another 69	Oyle	Sagitari. a celestial sign
Lymbeck	Ashes	Day	Soap
Allom	Pot Ashes	Gemini a celestial signe II	Scorpi. a Celestial sign
Amalgama aaa	Calx C	Leo another signe	Salt alkali
Antimony	Quick lime	Stratū sup stratū or	Armoniac Salt
Aquarius a signe of	Cinnabar or	lay upon lay SSS	Comon Salt
the zodiack	Vermillion	Marcassite	Salgemme
Silver or Luna	Waxe	Precipitate of Quicksilv.	Brimsto or sulph.
Quicksilver or	Crucible	Sublimate	Black sulphur
Mercury	Calcinated copper	Moneth	Philosophers sulphur
Aries another	as ustū or crocus	Niter or Saltpeter	To sublimate
celestial signe	veneris	Night	Talck X
Arsenick		Gold or Sol	Tartar
Balneum B	Note of Distillation	Auripigmentū	Taur. a Celestial signe
Balneum	Water	Lead or Saturne	Earth
Maris MB	Aqua fortis	Pisces a Celestial signe	Caput Mortuū
Vaporous	aqua Regalis VR VR	Powder	Tuty
Bath VB	Spirit	To precipitate	Glasse
Libra another	SP.	To purify	Vert degrice, or flower of Copper
celestial signe	Spirit of Wyn	Quintessency QE	Vinegar
Borax	Tinne or Jupiter	Realgar	Distilled Vinegar
Bricks	Powder of Bricks	Retorte	Vitriol
capricornus another	Fire	Sand	Urine

FIGURE 133. ■ A table of Chimical Characters in Le Fèvre's 1670 edition of *A Compleat Body of Chymistry.*

important texts of the seventeenth century. This book also reported Le Fèvre's observation of the increase in weight as antimony is calcined (Figure 121).

Antimony was one of the nine elements known to the ancients.[1] It was found as the ore stibnite (Sb_2S_3), and this black sulfide was used by women as an eye cosmetic in biblical times. An early means for obtaining the metal was to roast the ore on charcoal heated to incandescence. Later methods involved heating stibnite with tartar and nitre or with iron. The resulting "lead" was used to fashion a Chaldean vase of pure antimony around 4000 B.C.[1]

Early chemical books show an amazing fascination with antimony far beyond our modern interest. Why? One reason was its preferred use for releasing gold from metallic impurities. Antimony has a fairly low affinity for sulfur (higher than gold, lower than silver—see Geoffroy's Table of Affinities [Figures 76 and 77]—pure antimony or Regulus of Antimony is represented by a three-pointed crown). Thus, its common ore will release sulfur to baser metals forming "scum" easily scooped from molten gold. It can separate silver from gold since silver captures sulfur from stibnite and the resulting liquid slag of silver sulfide and antimony sulfide is separable from gold antimonide. This last is burned to free the volatile antimony oxide, leaving pure gold.[2]

The wolf depicted in the First Key of Basil Valentine [see Figure 38(a)], represents antimony (sometimes called lupus metallorum or wolf of the metals by the alchemists). Another famous seventeenth-century picture depicts the wolf devouring a dead human (impure gold) with subsequent burning of the wolf (loss of volatile antimony oxide) to release the King (gold) (Figure 45).[3]

Now if metals could so effectively be purged of their impurities by antimony, should it not also be an effective human medicine—a purge (or emetic) to remove illness? Paracelsus first described antimony as a purge and set off a violent philosophical debate among physicians. The classical Galenical view was the use of a medicine with properties contrary to the disease. The Paracelsans argued for cure by *similitude* (i.e., fight poison with poison). The question of whether antimony was a medicine or a poison raged over centuries but was apparently settled by the cure of Louis XIV with *vin emetique* (emetic wine—yum) in 1658.[4] Le Fèvre was much taken with medicinal antimony and particularly with its purification and fixation (as the calx) by the sun.[4] He too noted the increase in weight upon calcination. The book *Triumphal Chariot of Antimony*, first published in 1604 and attributed to the legendary Benedictine Monk, Basil Valentine, used this flashy, Hollywood-like title to strike a blow for antimony in this long and passionate debate. For a modern encore, we eagerly await the movie version starring Charlton Heston as the chariot-driver.

It is worthwhile recognizing that modern anticancer agents "poison" normal cells, but are greater poisons to cancerous cells that multiply much more rapidly. Thus, the Paracelsian view is vindicated in this case but not in neutralizing stomach acid.

---

1. N.N. Greenwood and A. Earnshaw, *Chemistry of the Elements*, Pergamon, Oxford, 1984, p. 637.
2. F. Ferchi and A. Sussenguth, *A Pictorial History of Chemistry*, William Heinemann, London, 1939. p. 61.
3. J. Read, *Prelude To Chemistry*, MacMillan, New York, 1937, pp. 200–202; 240–241 [see Plate 47 in this book, which is taken from the book by Michael Maier (1687) titled *Secretions Naturae Scrutmium Chymicum*].
4. A.G. Debus, *The French Paracelsans*, Cambridge University Press, Cambridge, 1991, pp. 21–30, 95–99

SECTION IV
AN EMERGING SCIENCE

THE ANCIENT WAR OF THE KNIGHTS

A mysterious fable of a battle pitting Gold and Mercury, armed as knights, against the Philosopher's Stone first appeared in print (in German) in 1604, although manuscripts probably existed earlier.[1,2] Various versions were published throughout the seventeenth century in German and French. The definitive text (Figure 134)[3] was published in English in 1723 and compare both German and French sources. Nonetheless, the author's identity remains a mystery: was it Alexandre Toussaint De Limojon De Saint Disdier (nice name, but apparently not the true author[1]), or Johann Thölde, the original publisher, who might also have been the legendary Basil Valentine?

While the amusing aspect of this book is its use of allegory, the most fascinating aspect is the manner in which solid scientific reasoning, based upon experimentation, is employed to discredit some fundamental tenets of alchemical lore. While the "magistery" of alchemy wins the day ("The Hermetical Triumph"), early stirrings of the Scientific Revolution are quite audible.

In this fable, Gold is an arrogant and aggressive knight while Mercury, his subordinate knight, dutifully supports Gold. But let us hear Gold in his own bombastic voice as he confronts the Philosopher's Stone:[4]

> 'Tis God himself who has given me the Honour, the Reputation, and the glittering Brightness, which renders me so estimable, it is for that Reason that I am so searched for by every one. One of my greatest perfections is to be a Metal unchangeable in the Fire, and out of the Fire: So all the World loves me, and runs after me; but you, you are only a Fugitive, and a Cheat, that abuses all Men: This is seen in that, that you fly away and escape out of the Hands of those who work with you.

And here is the start of the Stone's measured, yet powerful response:[4]

> 'Tis true, my dear Gold, 'tis God who has given you the Honour, the Durability, and the Beauty, which makes you precious; 'tis for that Reason that you are obliged to return (eternal) Thanks (to the divine Bounty,) and not to despise others as you do; for I can tell you, that you are not that Gold, of which the Writings of the Philosophers make mention; but that Gold is hidden in my Bosom.

The Stone's point is that the substances that embrace in the legendary "chymical wedding" are not the two hopelessly naive metals who confront him but rather Philosophers (or "Sophic") Gold and "Sophic" Mercury each having a much more complex origin. And then, Gold makes the fundamental alchemical argument that the "wedding" (*conjunctio* to be more precise) between Gold and Mer-

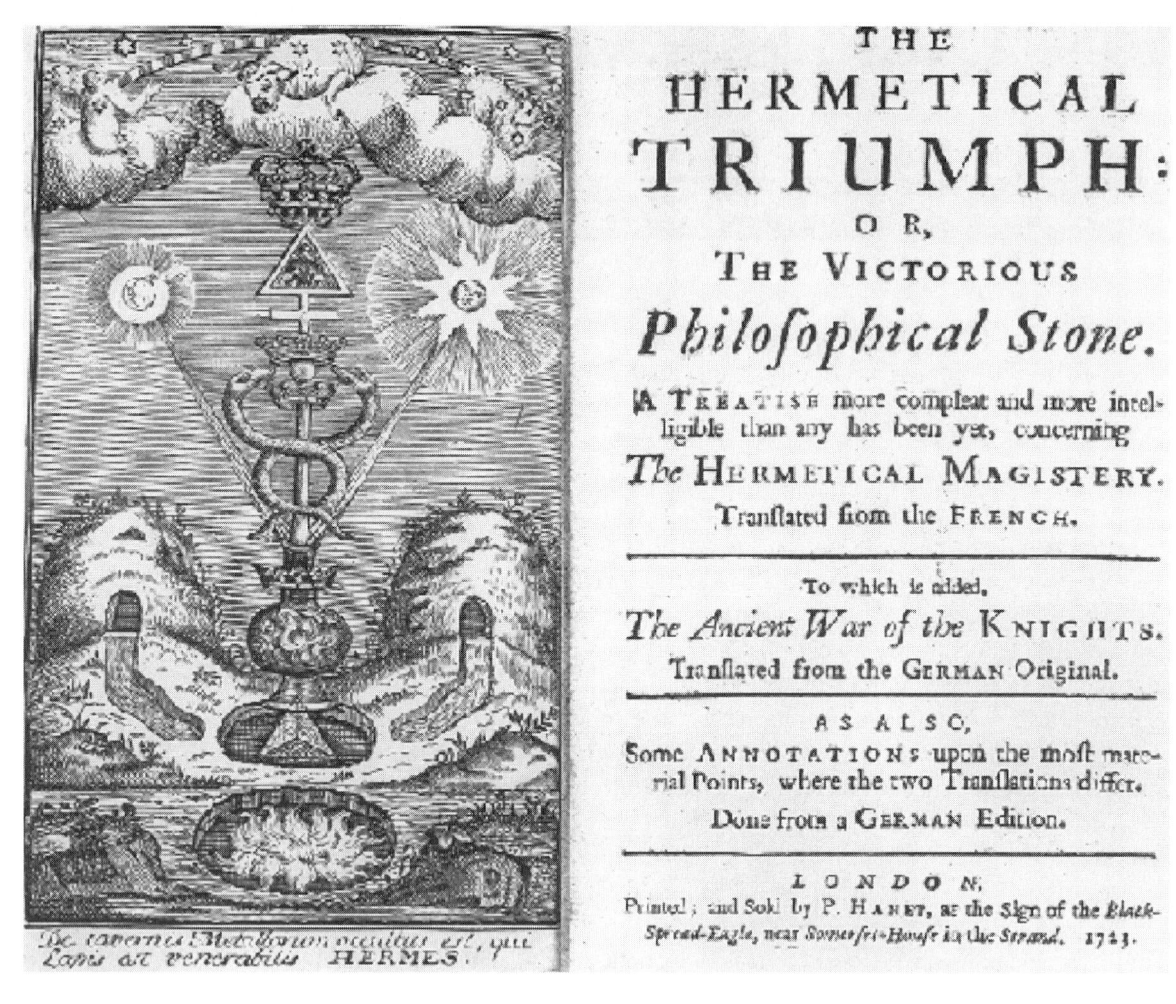

THE
HERMETICAL
TRIUMPH:
OR,
THE VICTORIOUS
Philosophical Stone.
A TREATISE more compleat and more intelligible than any has been yet, concerning
The HERMETICAL MAGISTERY.
Translated from the FRENCH.

To which is added,
The Ancient War of the KNIGHTS.
Translated from the GERMAN Original.

AS ALSO,
Some ANNOTATIONS upon the most material Points, where the two Translations differ.
Done from a GERMAN Edition.

LONDON,
Printed; and Sold by P. HANET, at the Sign of the *Black-Spread-Eagle*, near *Somerset-House* in the *Strand*. 1723.

FIGURE 134. ■ Frontispiece and title page for "The Hermetical Triumph" which includes the fable "The Ancient War of the Knights." This allegorical tale tells of combat between Gold and Mercury, girded for battle against the Philosopher's Stone.

cury is necessary in order to multiply gold, indeed, it is Nature's universal way for procreation:[5]

> I am not ignorant, that the Philosophers speak after this manner; yet this may be apply'd to my Brother Mercury, who is as yet imperfect; but if one join both of us together, he then receives from me the Perfection (which he wants). For he is of the Feminine Sex, and I am of the Masculine Sex; which makes the Philosophers say, that the Art is one quite homogeneal Thing. You see an Example here in the (the Procreation of) Men, for there can no Child be Born without (the Copulation of) Male and Female; that is to say, without the Conjunction of the one with the other. We have the like Example thereof in Animals, and in all living Beings.

Let us leave the sexual complexities of this statement to the psychologists and read the Stone's scientifically respectable response:[5]

> 'Tis true, your Brother Mercury is imperfect, and by consequence he is not Mercury of the wise. So though you should be join'd together, and one should keep you thus in the Fire during the Course of many Years, to endeavor to unite you perfectly to one another, there will always happen (the same Thing, namely,) that as soon as Mercury feels the Action, of Fire, it separates itself from you, it is sublimed, it flies away, and leaves you alone below. That if one dissolve you in Aqua-fortis, if one reduce you into one only (Mass), if one melt you, if one distill you, if one coagulate you, you will never produce any Thing but a Powder, and a red Precipitate: That if one make a Projection of this Powder on an imperfect Metal, it tinges it not; but one finds as much Gold as one put therein at the beginning, and your Brother Mercury quits you and flies away.

In other words, the union of the metals Gold and Mercury has changed nothing in any profound chemical way. Heat gold amalgam and pure mercury is distilled leaving pure gold behind. Alternatively, if the newlyweds were to bathe together in a (heart-shaped?) tub containing *aqua fortis* (nitric acid), mercury would separate as a red calx just as it does in the absence of gold. Enraged by his superior logic, Gold and Mercury violently attack the Philosophers Stone and are consumed, leaving no trace.

The Gold and Silver of this fable represent false alchemists: they suffer equally from ignorance and hubris. In contrast, the Philosopher's Stone is the True Adept—a natural philosopher pursuing truth and seeking the wisdom of God. He is the proto-scientist whose experimentation and reasoning will one day lead to a true chemical science. Or, does he somehow presage the birth of Robert Boyle, the "Sceptical Chymist" of our next essay?

---

1. J. Ferguson, *Bilbiotheca Chemica*, Vol. II, Derek Verschoyle, London, 1954, pp. 486–487.
2. L.I. Duveen, *Bilbiotheca Alchemica Et Chemica*, HES Publishers, Utrecht, 1986, p. 361.
3. Limojon De Saint Disdier, Alexandre Toussaint de, *The Hermetical Triumph;, or, The Victorious Stone. A Treatise more compleat and more intelligible than any has been yet, concerning The Hermetical Magistery. Translated from the French. To which is added, The Ancient War of the Knights. Translated from the German original. As also, some Annotations upon the most material Points, where the two Translations differ. Done from a German Edition*. P. Hanet, London, 1723.
4. Limojon De Saint Disdier, op. cit., pp. 4–5.
5. Limojon De Saint Disdier, op. cit., pp. 13–14.

THE FIRST TEN-POUND CHEMISTRY TEXT

The first systematic textbook of chemistry was the *Alchemia*, published in Frankfort in 1597 by Andreas Libavius (ca. 1540–1616).[1] The title page of the beautiful enlarged and illustrated second edition, the *Alchymia* (1606, Frankfurt), is shown in Figure 135. My copy of this book is bound in ornate, Italian-tooled vellum, measures about 9 inches by 13½ inches and weighs about 10 pounds. Libavius had a classical education and, in addition to obtaining the M.D. and serving as a physician, was Professor of History and Poetry at the University of Jena. In

D.O.M.A

ALCHYMIA
ANDREÆ
LIBAVII, RECOGNI-
TA, EMENDATA, ET
aucta, tum dogmatibus &
experimentis non-
nullis;

TVM COMMENTARIO
Medico Physico Chymico:

QVI EXORNATVS EST VA-
riis Instrumentorum Chymicorum picturis;
partim aliunde translatis, partim
planè nouis:

In gratiam eorum,

QVI ARCANORVM NATVRALIVM
cupidi, ea absq; inuolucris elementarium &
ænigmaticarum sordium, intueri
gaudent.

PRÆMISSA DEFENSIONE ARTIS:
opposita censuræ Parisianæ:

Cum Gratia & Priuilegio Cæsareo speciali ad decennium.

FIGURE 135. ■ The title page of the second edition (1606) of Libavius' *Alchymia*—the first chemistry textbook (the first edition, published in 1597, was smaller and not illustrated and said by Partington to be rarer than Newton's *Principia*, which itself "hammered down" at almost $400,000 at a 1998 book auction).

the manner of Paracelsus, Libavius employed metallic remedies including potable gold (gold dissolved in *aqua regia*) as well as calomel. However, his opinion of Paracelsus was stated thusly: "Paracelsus, as in many other matters he is stupid and uncertain, so also here writes like a madman."[1] While a believer in alchemy, Libavius performed much practical chemistry and noted that lead gains 8–10% in weight upon calcination.[1]

Alchymia describes the construction of a hypothetical chemical "house" (*Domus chymici*} (Figure 136) with detailed floor plans. The *Domus chymici* was to have a main laboratory, storeroom for chemicals, preparation room, a room for laboratory assistants, a room for crystallizing and freezing, a room for sand and water baths, a fuel room, a museum, gardens, walks, and . . . a wine cellar.[1,2] The book goes on to describe fume hoods, furnaces, glassware, luting material, mor-

FIGURE 136. ■ The *Domus chymici* ("house of Chemistry") in Libavius' *Alchymia* was never built. I suspect that zoning laws would have kept it out of a respectable residential neighborhood.

tars, forceps, chemical preparations, and everything else needed to be "state of the art" during the time of Shakespeare.

But Libavius means to cover all bases in the textbook market and concludes with an amply illustrated section on the *Lapidum Philosophorum* (Philosopher's Stone). Figures 137 and 50 are described by John Read as representing the Vase of Hermes heated at the bottom.[2] In Figure 137, we see a serpent, representing

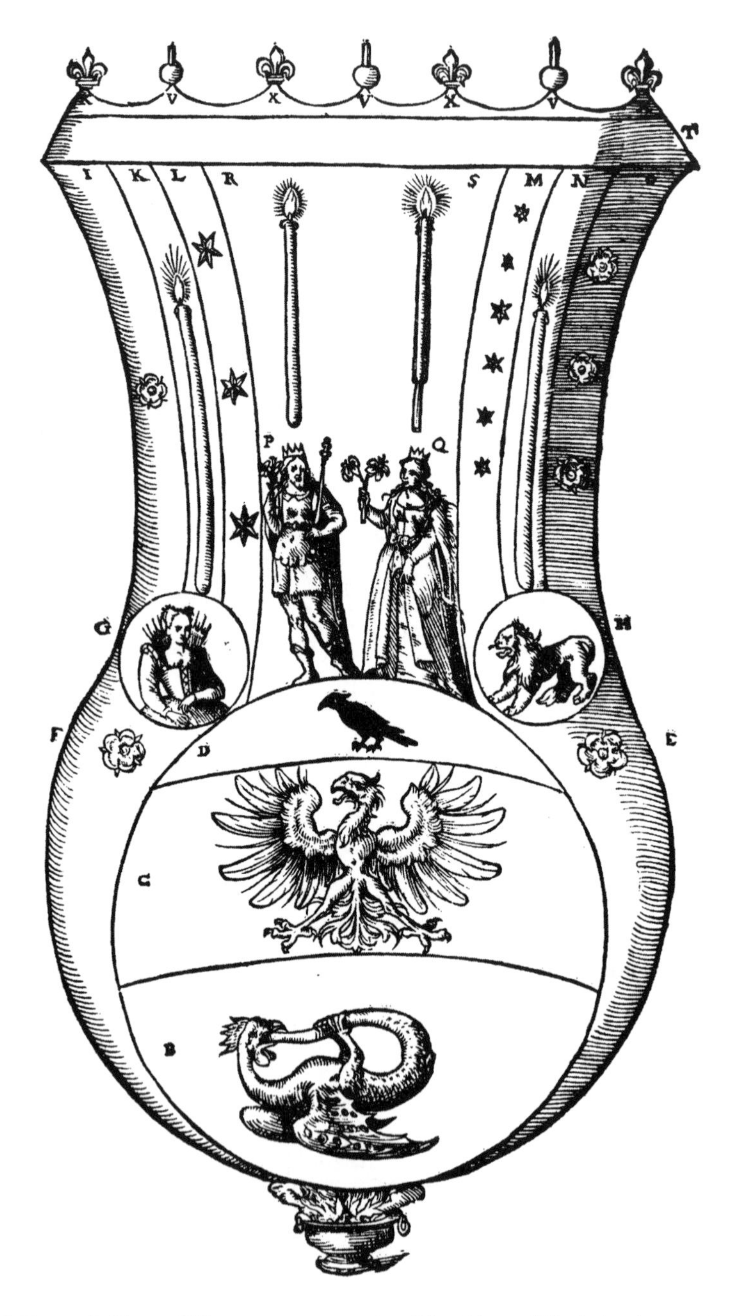

FIGURE 137. ■ A Vase of Hermes representing The Great Work in the section on The Philosopher's Stone in Libavius' 1606 *Alchymia*. Nice to see this schematic after all of the rational description of furnaces, flasks, lutes, forceps, and chemical preparations of the age.

Sophie Mercury, eating its tail—a representation of coagulation and fixity. The eagle has multicolored feathers representing color changes during fermentation. The black crow represents putrefaction. The maiden represents the moon or silver; the lion represents the sun or gold. The king and queen similarly represent male and female, sulfur and mercury.[2] Some highlights of Figure 50 include the base representing earthly foundation; two Atlases supporting the vessel; a four-headed dragon representing four stages of fire; the Green Lion representing mercurial liquid, the first matter of the stone; a three-headed silver eagle, a black crow representing putrefaction, the winged serpent biting its own tail again, a pretty nasty white swan between two globes, and a number of male–female, earth–moon, or sulfur–mercury images. Pretty obvious when you know what to look for (or have John Read's *Prelude to Chemistry* at your side).

---

1. J.R. Partington, *A History of Chemistry*, MacMillan, London, 1961, Vol. 2, pp. 244–267.
2. J. Read, *Prelude To Chemistry*, MacMillan, New York, 1937, pp. 212–221.

A TREE GROWS IN BRUSSELS

How ironic that Johann Baptist Van Helmont (1577–1644) refers to "Dame Nature" as the "Proto-Chymist,"[1] for if ever there was a human protochemist it was he. His writings navigate the borders between science, pseudoscience, and superstition. Van Helmont was bom in Brussels but travelled extensively. The picture of Van Helmont (Figure 138, left) is from the *Ortus Medicinae*, compiled by his son the alchemist and polymath Franz Mercurius (Figure 138, right) and first published in 1648.

At a time when measurement and experiment were just beginning to define science, Van Helmont performed his famous "tree experiment." He believed that there were only two true fundamental elements, water and air, and that trees were composed of the element water. To test this hypothesis, he weighed 200 pounds of dried earth, moistened it with distilled water and added the stem of a willow tree weighing 5 pounds. After five years of judicious watering he determined that the tree weighed 169 pounds, the soil, when separated and dried, still weighed 200 pounds and, thus, the extra 164 pounds could only come from addition of the element water.[2]

These conclusions were, of course, totally erroneous. We now know that the mass of the tree is comprised of cellulose and water. Cellulose is derived from photosynthesis (only discovered some 140 years later) involving carbon dioxide and water. And again, how ironic that the person who coined the term *gas* (from *chaos*) and effectively discovered carbon dioxide did not understand its role in his "tree experiment."

The law of conservation of matter is typically associated with the father of modern chemistry, Antoine Laurent Lavoiser, who worked in the late eighteenth century. Van Helmont's tree experiment demonstrates that this law was a tenable hypothesis over 120 years earlier. And about 150 years after the death

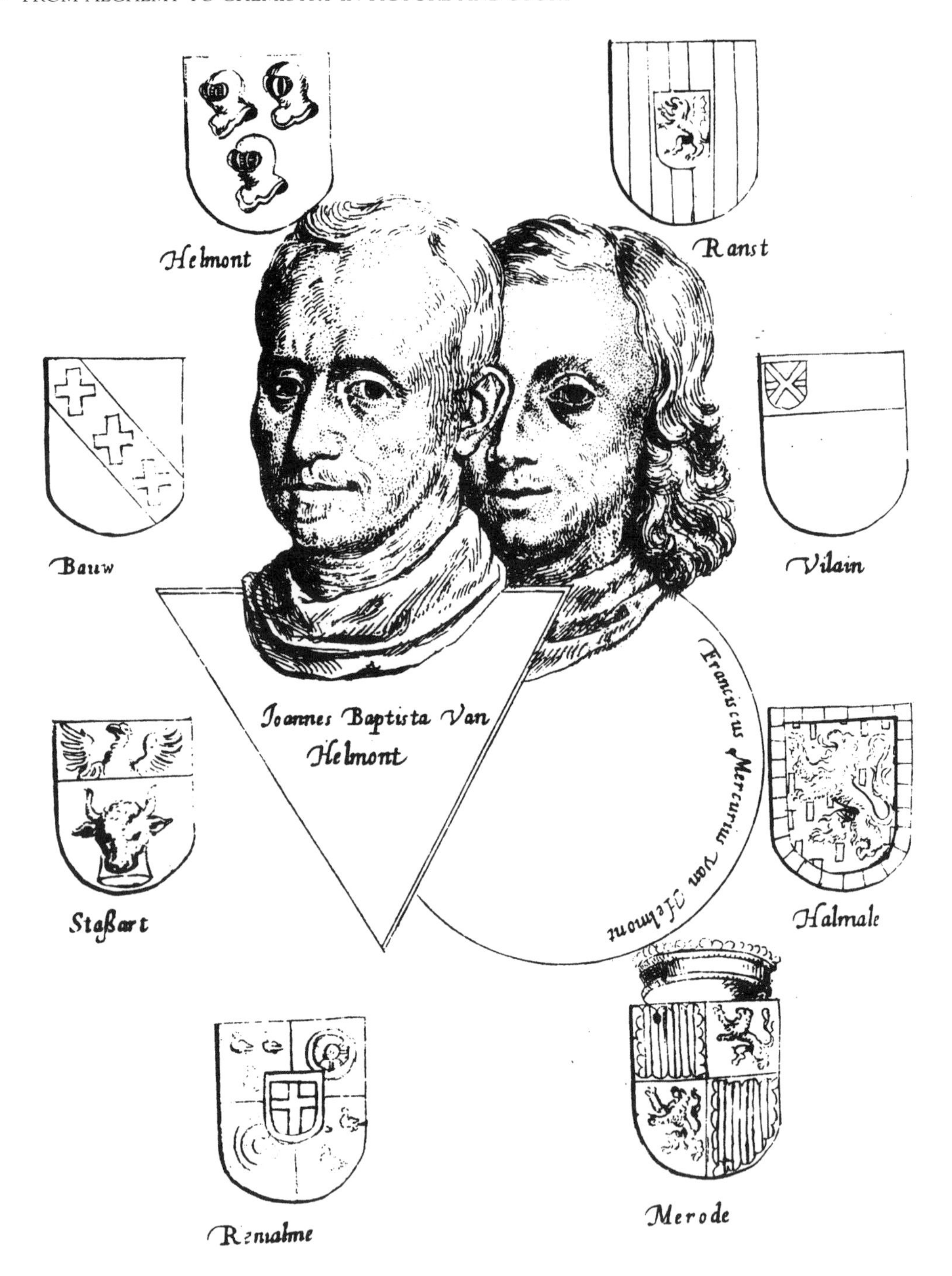

FIGURE 138. ■ Frontispiece from Johanne Baptist Von Helmont's *Ortus Medicinae* (Amsterdam, 1648) published by his alchemist son Franciscus Mercurius (right).

of Lavoisier, it evoked near-religious awe in Betty Smith's novel, *A Tree Grows in Brooklyn:*[3]

"Francie came away from her first chemistry lecture in a glow. In one hour she found out that everything was made up of atoms which were in continual motion. She grasped the idea that nothing was ever lost or destroyed. Even if something was burned up or left to rot away, it did not disappear from the face of the earth; it changed into something else—gases, liquids and powders. Everything, decided Francie after that first lecture, was vibrant with life and there was no death in chemistry. She was puzzled as to why learned people didn't adopt chemistry as a religion."

---

1. J.B. Van Helmont, *A Ternary of Paradoxes* (translated by Walter Charleton), London, 1650, p. 7.
2. H.M. Leicester and H.S. Klickstein, *A Sourcebook in Chemistry 1400–1900*, McGraw-Hill, New York, 1952, pp. 23–27.
3. B. Smith, *A Tree Grows in Brooklyn*, Harper & Brothers, New York, 1943, p. 389. (I thank Professor Susan Gardner for making me aware of this passage.)

CURING WOUNDS BY TREATING THE SWORD WITH POWDER OF SYMPATHY

Let us explore Van Helmont's beliefs a bit further. He was a fervent believer in the Powder of Sympathy. Rather than defining it, let us let Van Helmont describe its application[1]:

> . . . Mr. James Howel . . . interceding betwixt two Brothers of the sword, received a dangerous wound through the Arm: By the violent pain whereof, and other grievous accidents concommitant, he was suddenly dejected into extreem Debility and Danger. That in this forlorn plight, despairing to finde ease or benefit, by the fruitless continuance of Chirurgery, and fearing the speedy invasion of Gangraen, he consulted Sir K.D.,[2] who having procured a Garter cruentate, wherewith the hurt was first bound up, inspersed thereon, without the privacy of Master Howel, a convenient quantity of Roman Vitriol. That the powder no sooner touched upon the blood, in the Garter, then the patient cryed out, that he felt an intolerable shooting, and penetrative torment, in his Arm: which soon vanished upon the remove of all Emplasters and other Topical Applications, enjoyned by Sir K.D. That thence-forward, for three days, all former symptoms departed, the part recovered its pristine lively Colour, and manifest incarnation and consolidation ensued: but then Sir K.D. to compleat his experiment, dipt the garter in a fawcet of Vinegar, and placed it upon glowing coals; soon whereupon the Patient relapsed into an extream Agony, and all former evils instantly recurred. And finally, that having obtained this plenary satisfaction, of the Sympathy maintained between blood extravenated, and that conserved in the veins. . . he took again the Garter out from the Vinegar, gently dryed it, and freshly dressed it with the Powder, whereupon the Sanation proceeded with such admirable success, that within few days, there remained only a handsom Cicatrice, to witness there was once a wound.

In other words, treat the dressings that once covered the wound and are covered with blood with the Powder of Sympathy and the cure will be communicated to the blood still in the body. This could be, incidentally, regarded as a conceptual advance beyond Paracelsus' earlier doctrine wherein sprinkling Powder on the sword which caused the wound would heal the wound. Van Helmont argued that it is not the sword, but the blood on the sword, that communicates with the wound. The sympathy concept was the basis for the "wounded dog theory" tested by the Royal Navy in 1687.[3] A dog would be wounded and sent off to sea while its bandage remained in London. At noon, London time, powder of sympathy was sprinkled on the bandage, the dog was supposed to immediately cry and comparison with on-ship time was to provide longitude.

---

1. J.B. Van Helmont, *A Ternary of Paradoxes* (translated by Walter Charleton), London, 1650, Prologue.
2. Sir K.D. was Sir Kenelm Digby, scientist, physician, privateer, and gifted scoundrel whose Powder of Sympathy (a copper sulfate) was considered the best. He certainly was not faint of heart or capable of much sympathy himself.
3. D. Sova, *Longitude*, Penguin Books, New York, 1995. I am grateful to Professor Thomas W. Mattingly for bringing this to my attention.

DO ANONYMOUS PASSERSBY DEFECATE AT YOUR DOORSTEP? A SOLUTION

Here is another practical use of the Sympathy concept:[1]

> Hath any one with his excrements defiled the threshold of thy door, and thou intendest to prohibit that nastiness for the future, do but lay a red-hot iron upon the excrement, and the immodest sloven shall, in a very short space, grow scabby on his buttocks, the fire terrifying the excrement, and by dorsal magnetism driving the acrimony of the burning, into his impudent anus.

Think about this. First, it is not only blood that can "communicate" over long distances. Second, if only the blood of the wounded or the anus of the perpetrator is affected, then the cure and the punishment are "DNA-fingerprinted"—a major advance over late-twentieth-century medical care and forensic science.

---

1. J.B. Van Helmont, *A Ternary of Paradoxes* (translated by Walter Charleton), London, 1650, p. 13.

A HOUSE IS NOT A HOME WITHOUT A BATH TUB AND A STILL

Johann Rudolph Glauber (1604–1670) is widely considered the father of industrial chemistry and chemical engineering. Although he certainly believed in transmutation, Glauber made numerous important contributions to chemistry. He was the first to describe crystalline sodium sulfate (Na_2SO_4), commonly termed Glauber's salt, and its seemingly amazing medicinal properties[1]:

FIGURE 139. ■ These figures are from the folio-sized book by Johann Rudolph Glauber, *The Works of the Highly Experienced and Famous Chymist . . . Containing, Great Variety of Choice Secrets in Medicine and Alchymy . . .* (London, 1689). Some consider Glauber to be the first chemical engineer. The figures at the bottom show a common-man's bathtub and sauna designed by Glauber (see text).

> Externally adhibited, it cleanseth all fresh wounds, and open Ulcers and healeth them; neither doth it corrode, or excite pain, as other salts are wont to do. Within the body it exerciseth admirable virtues, especially being associated with such things whose virtues it increaseth, and which it conductith to those places to which it is necessary they should arrive . . .

He called sodium sulfate *Sal Mirabile* (Wonder Salt).

The three panels shown in Figure 139 are from *The Philosophical Furnaces*, a work reprinted in the beautiful 1689 folio *The Works of The Highly Experienced and Famous Chemist. . . .* Stills (used for making wine, beer, and medicines) and bathtubs of the period were usually made of copper and thus were extremely expensive and required special furnaces for each application. While not a problem for the wealthy, ownership of these household appliances was often out of reach for the less well off.

Glauber designed a copper globe, about the size of a human head, along with its own furnace that could be moved and plumbed into inexpensive wooden stills and baths. (Plumbing seals were made with "Oxe-bladders" or with starch and paper.) *Fig. I* (Figure 139) shows the furnace (A) and copper globe (B) and their attachment to distilling vessel C, itself attached to refrigeratory D (with "worm"—twisted copper tubing for condensation), which feeds to receiver E. *Fig. II* depicts a balneum (bath apparatus) with a cover having holes for glasses containing samples for gentle, controlled heating. *Fig. III* shows a wood bath tub as well as a wooden box for a dry bath (to provoke sweat with volatile spirits). The same furnace (A) and copper globe (B) could be used with each appliance.

Although Glauber notes that heat is supplied more slowly than would be the case for a copper appliance with its own customized furnace, the savings are worthwhile (except for the wealthy for whom time is always money). As for those foolish enough not to avail themselves of this innovation, Glauber says:

> Let him therefore keep to his copper vessels, who cannot understand me, for it concernes not me.

---

1. H.M. Leicester and H.S. Klickstein, *A Source Book in Chemistry 1400–1900*, McGraw-Hill, New York, 1952, pp. 30–33.

SKEPTICAL ABOUT "VULGAR CHYMICAL OPINIONS"

Robert Boyle (1627–1691) was born in Ireland to a wealthy family, educated at Eton, received further education on the Continent, and returned to England in 1645.[1] He began his scientific studies during the following decade and in 1656 moved to Oxford, where he secured the assistance of Robert Hooke. Hooke built a vacuum pump (Figure 145) for Boyle, who used it for numerous studies, including study of the relationship between volume and pressure of gas that now bears his name (see p. 210).[2] Boyle is generally considered to be the Father of Chemistry due in part to his gas law and other physical studies but also because of his classic book, *The Sceptical Chymist*, which included the first serious attempts to define chemical elements and atomistic concepts with experimental justification.

When The Honorable Robert Boyle published *The Sceptical Chymist* in 1661 (Figures 140 and 141),[2,3] two untested theories of matter dominated the protoscience of chemistry. The earliest of these "vulgar" (i.e., "common") "chymical opinions" was based upon the four elements (earth, fire, air, water) routinely attributed to Aristotle. Aristotelians were often referred to as "peripatetics" because of their master's style of teaching that included walking around.

THE
SCEPTICAL CHYMIST:
OR
CHYMICO-PHYSICAL
Doubts & Paradoxes,
Touching the
SPAGYRIST'S PRINCIPLES
Commonly call'd
HYPOSTATICAL;
As they are wont to be Propos'd and
Defended by the Generality of
ALCHYMISTS.
Whereunto is præmis'd Part of another Difcourfe
relating to the fame Subject.

BY
The Honourable *ROBERT BOYLE*, Efq;

LONDON,
Printed by *J. Cadwell* for *J. Crooke*, and are to be
Sold at the *Ship* in St. *Paul's* Church-Yard.
MDCLXI

FIGURE 140. ▪ The polite version of the title page of Robert Boyle's 1661 classic *The Sceptical Chymist*. This book is written in the form of a discussion among fictional characters, including Themistius, representing the "Peripateticks," defenders of the four ancient elements, Philoponus, who defends the three Paracelsian principles and Carneades, the voice of reason (i.e., Boyle), who, of course, gets all the best lines. (From The Roy G. Neville Historical Chemical Library, a collection in the Othmer Library, CHF.)

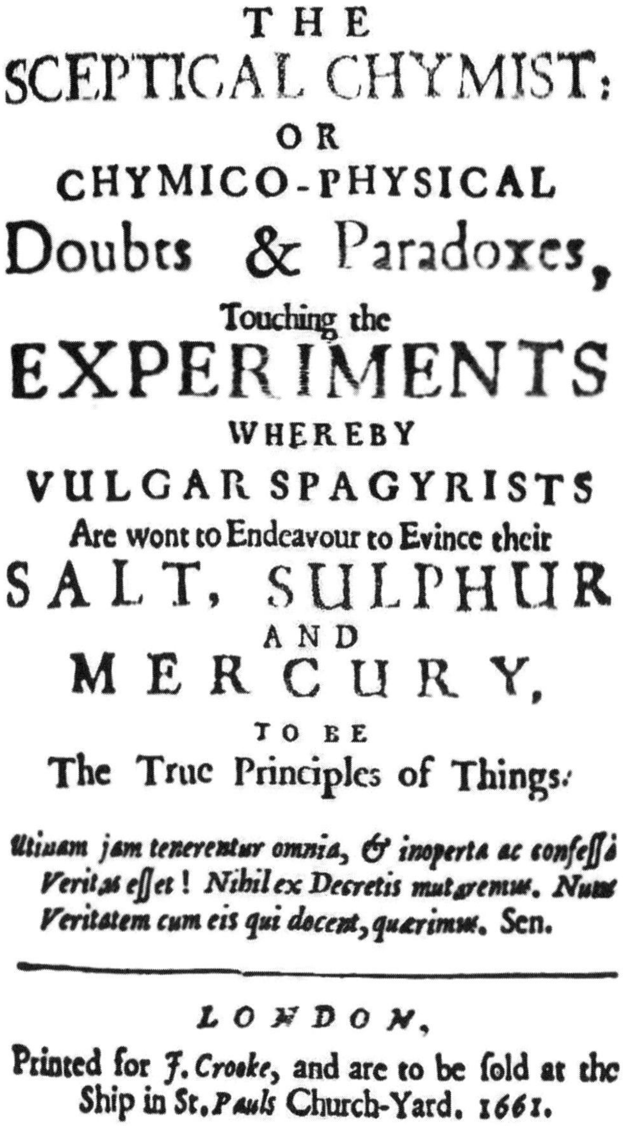

THE
SCEPTICAL CHYMIST:
OR
CHYMICO-PHYSICAL
Doubts & Paradoxes,
Touching the
EXPERIMENTS
WHEREBY
VULGAR SPAGYRISTS
Are wont to Endeavour to Evince their
SALT, SULPHUR
AND
MERCURY,
TO BE
The True Principles of Things.

Utinam jam tenerentur omnia, & inoperta ac confessa Veritas esset! Nihil ex Decretis mutaremus. Nunc Veritatem cum eis qui docent, quærimus. Sen.

LONDON,
Printed for *J. Crooke*, and are to be ſold at the Ship in St. *Pauls* Church-Yard. *1661.*

FIGURE 141. ▪ The less polite version of the title page also included in Boyle's 1661 *The Sceptical Chymist* (see Figure 140) (from The Roy G. Neville Historical Chemical Library, a collection in the Othmer Library, CHF).

The other prevailing vulgar opinion, dating from the time of Paracelsus (1493–1541), was based on the *tria prima* (sulfur, mercury, and salt). Boyle referred to the adherents of this theory as "chymists" (we would refer to them as "alchemists" and "iatrochemists"). There was no great honor in being a "chymist," although a "sceptical chymist" was, at least, capable of salvation. In addition to "chymists" and "Peripateticks," there were "hermetick philosophers" who believed that "fire ought to be esteemed the genuine and universal instrument of analyzing mixt bodies"; that is to say, the role of fire is chemical decomposition. And although Boyle believed in a corpuscular theory of matter, akin to the ancient atomic theory of Leucippus, he faulted the Greek philosophers for performing no experimentation:[4]

> And therefore we sent to invite the bold and acute Leucippus to lend us some light by his atomical paradox, upon which we expected such pregnant hints, that 'twas not without a great deal of trouble that we had lately word brought us that he was not to be found;

While Boyle's writing style makes for slow reading, the brief selection above illustrates some of the humor employed in *The Sceptical Chymist*. Moreover, he used an entertaining technique that would probably not work in today's scientific journals and monographs. Boyle set up imagined conversations between himself and convenient "straw men" whose arguments he could readily demolish. Although Boyle the narrator plays a passive role as he accompanies "the inquisitive Eleutherius" on a visit to "his friend Carneades," the latter is really Boyle's voice. Carneades is seated at a little round table in a garden with Themistius, who argues for the "Peripateticks," and Philoponus, who defends the Paracelsian view.

Here is Themistius' "proof" that green wood "disbands" into the four elements upon combustion:[5]

> The fire discovers itself in the flame by its own light; the smoke by ascending to the top of the chimney, and thereby vanishing into the air, like a river losing itself in the sea, sufficiently manifests to what element it belongs and gladly returnes. The water in its own form boiling and hissing at the ends of the burning wood betrays itself to more than one of our senses; and the ashes by their weight, their firiness, and their dryness, put it past doubt that they belong to the element of earth.

Boyle (oops, Carneades) responds that there is confusion here. First, it appears that the "element" fire must be applied to free the "element" fire from green wood. A second point, however, is the assumption that application of fire merely releases the four elements unchanged. Here is Carneades' excellent scientific counterargument:[6]

> When, for instance, a refiner mingles gold and lead, and exposing this mixture upon a cuppel to the violence of the fire, thereby separates it into pure and refulgent gold and lead (which driven off together with the dross of the gold, is thence called *lythargyrium auri*), can any man doubt that sees these two so differing substances separated from the mass, that they were existent in it before it was committed to the fire.

The point is that metallic lead and gold may be melted together to form an alloy. However, prolonged heating in air converts lead to its calx—the yellowish-red powder litharge, a pigment that we recognize today as lead oxide. The molten gold is chemically unchanged, and any impurities (dross) as well as the newly formed litharge will be absorbed into the cupel leaving pure gold. Litharge was clearly never present in the original alloy but was "released" by fire. Indeed, this is almost identical with the argument of the Philospher's Stone in the previous essay.

And here are some other problems with the four elements.[7] It appears to be impossible, even with the aid of fire or other agents, to draw earth, air, fire, or water out of gold. Indeed, if anything might appear to be a true element, it is gold. On the other hand, when blood is "analyzed" by fire, it yields "five distinct substances": phlegm, spirit, oil, salt, and earth.[7]

In answering Philopones, the Paracelsian Spagyrical Chymist, Carneades questions whether the nature of fire always requires that it produce "analysis" of substances or, at least, consistent "analysis."[8] Thus, combustion of "guajacum" wood produces soot and ash, while its distillation (by fire) in a retort yields "oil, spirit, vinegar, water and charcoal." Boyle (oops, Carneades) further argues that if the resulting charcoal is removed from the retort and burned openly, it becomes ash. Similarly, open combustion of camphor produces soot, which may be captured and examined.[8] This soot retains none of the properties of camphor. However, if camphor in a closed glass vessel is exposed gently to fire, a smoke rises that condenses as a white solid, retaining the characteristic penetrating camphor odor. We recognize, in this second case, that camphor has sublimed, unchanged by the "analytical knife" of fire.

In *The Sceptical Chymist*, Boyle offers four propositions[9] that define his views of the organization of matter. He is attempting to fundamentally describe both the physical organization of matter (originating as minute particles) and chemical organization (as elements that cannot be further simplified chemically):

Proposition I. It seems not absurd to conceive that at the first production of mixt bodies, the universal matter whereof they among other parts of the universe consisted, was actually divided into little particles of several sizes and shapes variously moved.

Proposition II. Neither is it possible that of these minute particles divers of the smallest and neighboring ones were here and there associated into minute masses or clusters, and did by their coalitions constitute great store of such little primary concretions or masses as were not easily dissipable into such particles as composed them.

Proposition III. I shall not peremptorily deny, that from most of such mixt bodies as partake either of animal or vegetable nature, there may by the help of the fire be actually obtained a determinate number (whether three, four, or five, or fewer or more) of substances, worthy of differing denominations.

Proposition IV. It may likewise be granted, that those distinct substances, which concretes generally either afford or are made up of, may without very much inconvenience be called elements or principles of them.

The first two propositions deal with the physical structure of matter.[10] There are two levels of organization—minutest particles, Boyle's "corpuscles," which may associate into "coalitions" of "minute masses or clusters." Microscopes, invented around the start of the seventeenth century, provided direct evidence of "the extream littleness of even the scarce sensible parts of concretes."[11] Boyle's associate Hooke published his masterpiece, *Micrographia* (see the next essay), just four years after *The Sceptical Chymist*. Boyle noted further that quicksilver could be distilled, dissolved in acids and filtered, and converted to amalgams that could be finely ground, but all finely divided forms could eventually be recovered as the shiny, metallic liquid. One of Boyle's most wonderful works is his *Effluviums* essay (1673) (see Figure 170), in which he imagines the smallest physically measurable "minute masses" (or "effluvia") of matter. For example; 1¼ grains of gold could be beaten into six 3¼-inch squares. Boyle's finest ruler (100 divisions per inch) could, in principle, produce $6 \times (3.25 \times 100)^2$ or 2,535,000 gold squares, each of which would weigh 0.000000032 gram.

The third and fourth propositions deal with Boyle's chemical concepts of the elements. Boyle believed in transmutation. Indeed, in an anonymous essay of 1678, Boyle, in the voice of a certain Aristander, recounts witnessing a "retro-transmutation" (gold degraded alchemically to a lesser metal) by a Pyrophilus. When the other witness, Simplicius, asks in effect "What's the point in degrading gold?", the sage Boyle (oops, Aristander) replies, in effect, "If you know how to transmute in one direction, you can transmute in the other as well." Perhaps this essay should have been titled "The Credulous Chymist." In any case, two things are abundantly clear—Boyle's concept of atoms and elements differed profoundly from the modern concepts because of his belief in transmutation, and his definition of elements suggested no scientific tests. In contrast, Lavoisier's defini-

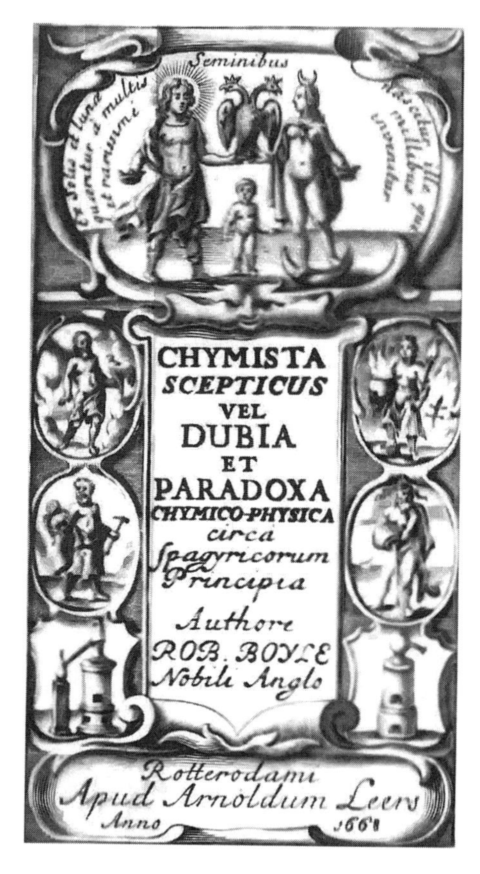

FIGURE 142. ▪ The title page for the 1668 continental edition of Boyle's *The Sceptical Chymist*. What is going on here?! Boyle has demolished the "contraries," the four elements, and the three principles, and here we see this "mumbo-jumbo" adorning the title page of this translation. One can only imagine publisher Arnold Leers' desire to sell a serious book using tabloid techniques. And one might also imagine Boyle's pained response upon receiving his gratis copy: "We are not amused," he might say, anticipating Queen Victoria by two centuries.

tion of elements ("simples") over a century later was testable—a substance was an element if it could not be further "simplified" chemically:[11]

> Thus, as chemistry advances towards perfection by dividing and subdividing, it is impossible to say where it is to end; and these things we at present suppose simple may soon be found quite otherwise. All we dare venture to affirm of any substance is, that it must be considered as simple in the present state of our knowledge, and so far as chemical analysis has hitherto been able to show.

Finally, I cannot resist the temptation to show the frontispiece (Figure 142) from the 1668 Latin translation of *The Sceptical Chymist* published in Rotterdam.[12] What was the publisher Arnold Leers thinking about?! The figures are the classical Sol–Luna (sulfur–mercury), Amorous Birds of Prey, et cetera that honor the dualities that Boyle demolished. Since it seems that Boyle did not have a sense of humor about things scientific, we can probably assume that Leers never consulted Boyle. *What* was Leers thinking about? Profits, no doubt. And one wonders whether the Right Honorable Robert "boyled" when he received his complimentary copy.

---

1. J.R. Partington, *A History of Chemistry*, MacMillan and Co. Ltd., London, Vol. 2, pp. 486–549.
2. R. Boyle, *The Sceptical Chymist: Or Chymico-Physical Doubts & Paradoxes, Teaching the Spagyrist's Principles Commonly call'd Hypostatical, As they are wont to be Propos'd and Defended by the Generality of Alchymists. Whereunto is præmis'd Part of another Discourse relating to the same subject,* F. Caldwell for F. Crooke, London, 1661. I am grateful to The Roy G. Neville Historical Chemical Library (California) for supplying these two images. Although the title page cited above (Figure 140) is commonly quoted, the original first title page appears to be that shown in Figure 141 (Dr. Neville, personal correspondence). It is both anonymous and a bit nasty ("Vulgar Spagyrists"), and perhaps Boyle (or the publisher) had some second thoughts.
3. E. Rhys (ed.), *The Sceptical Chymist by The Hon. Robert Boyle*, J.M. Dent & Sons, London; E.P. Dutton & Co., New York, 1944.
4. Rhys, op. cit., p. 13.
5. Rhys, op. cit., p. 21.
6. Rhys, op. cit., p. 24.
7. Rhys, op. cit., p. 27.
8. Rhys, op. cit., pp. 36–37.
9. Rhys, op. cit., pp. 30–34.
10. Brock, op. cit., pp. 54–70.
11. A. Lavoisier, *The Elements of Chemistry in a New Systematic Order Containing All the Modern Discoveries* (Robert Kerr, translator), William Creech, Edinburgh, 1790, p. 177.
12. R. Boyle, *Chymista Scepticus Vel Dubia Et Paradoxa Chymico-Physica circa Spagyricorum Principia,* Apud Arnoldum Leers, Rotterdam, 1668. This is the second Latin edition, the first edition was published in 1662 (Partington, op. cit.).

THE ATMOSPHERE IS MASSIVE

What is air? Paraphrasing David Abram[1]: We are immersed in the invisible air, but we barely even perceive it. We sense its effects—it is needed to support life—

but not its substance. Perception also rides upon windy drafts, which, in early times, might have been regarded as ethereal breaths of nature.

Why learn the gas laws in chemistry? We have known since the early nineteenth century that the gaseous state is where molecules roam as freely as individuals. This permits understanding of their physical and chemical behavior at the simplest levels. We also learned that two equal-sized balloons of hydrogen gas react totally and precisely with one equal-sized balloon of oxygen gas to produce water identical in mass to the two gases combined.

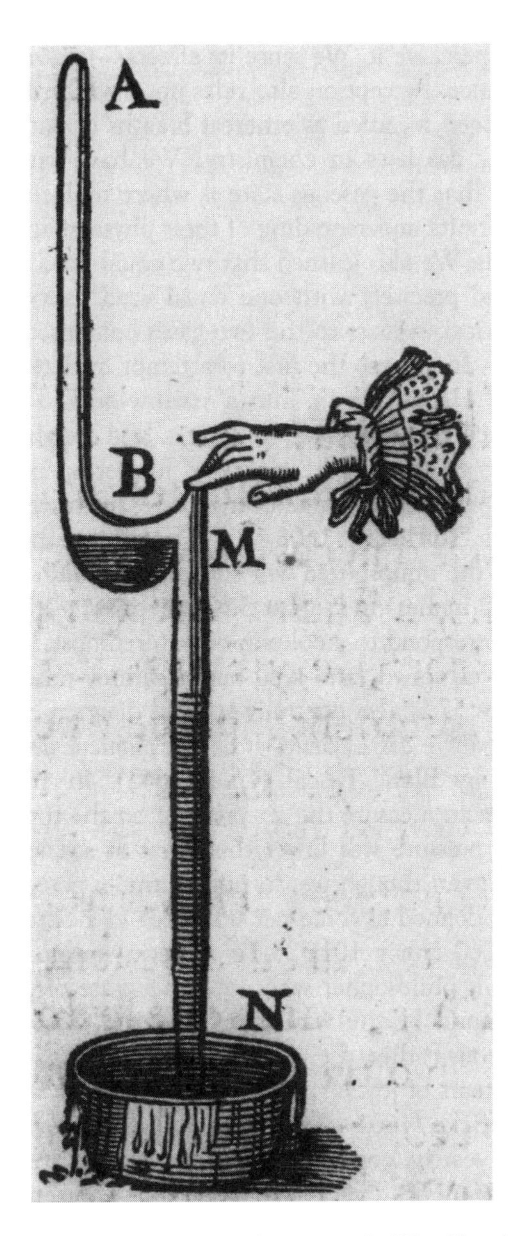

FIGURE 143. ▪ A figure from Blaise Pascal's *Traitez de l'Equilibre des Liqueurs, et de La Pesanteur de La Masse de L'Air* (Paris, 1663) depicting a highly stylized barometer. He sent his brother-in-law Perier to measure the atmospheric pressure on a mountain top (courtesy Edgar Fahs Smith Collection, Rare Book and Manuscript Library, University of Pennsylvania).

Galileo (1564–1642) was the first to attempt to determine the density of air (around 1638).[2] He forced air into a narrow-necked bottle, weighed the closed bottle, allowed the excess air to escape, and weighed the closed bottle again. (Galileo, who discovered the moons of Jupiter, spent the last eight years of his life under house arrest for teaching the Copernican view of the solar system.) Evangelista Torricelli (1608–1647) invented the barometer around 1643. At sea level, the atmosphere will support a column of mercury precisely 760 mm (roughly 30 inches) in height. Since mercury is 13.6 times denser than water, this would correspond to a column of water almost 34 ft high. This is the reason why an old-fashioned farm-type pump cannot raise water that is 34 ft deep or more. Figure 143 is a wonderfully stylized diagram of a barometer in the book *Traitez de l'Equilibre des Liqueurs et de la Pesanteur de la Masse de l'air . . .* published in 1663 by Blaise Pascal (1623–1662). In 1648, Pascal sent his brother-in-law Perier to measure the air pressure on the top of a mountain and confirmed that the pressure was lower than that at sea level—clearly the atmosphere has

FIGURE 144. ■ One of the greatest science demonstrations of all time: When von Guericke used his vacuum pump to remove the air from a sphere only 14 inches in diameter, teams of horses could not overcome the 2,260-pound (1.1-ton) force of atmospheric pressure pushing the hemispheres together [from Von Guericke's *Experimenta Nova* (1672)] (courtesy Edgar Fahs Smith Collection, Rare Book and Manuscript Library, University of Pennsylvania).

mass even though we do not routinely perceive it. [The modern unit of air pressure, defined as force per unit area (1 newton per square meter) is the pascal (Pa): 760 mm = 101,325 Pa]. The inventor of the first computer, Pascal was a religious philosopher who entered a state of grace late in his life: "He can only be found by the ways taught in the Gospel. Greatness of the human soul. 'Righteous Father, the world has not known thee, but I have known thee.' Joy, Joy, Joy, tears of joy."

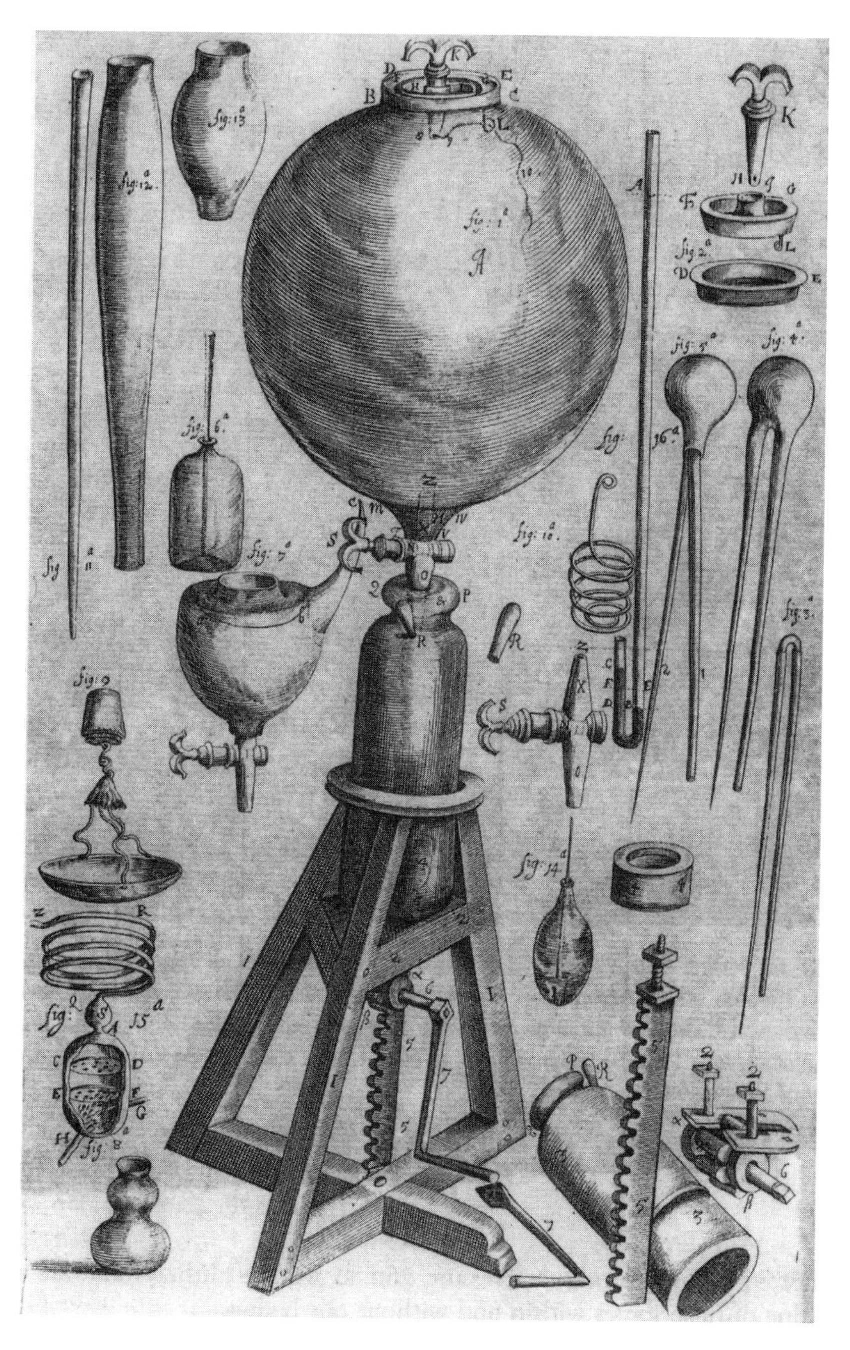

FIGURE 145. ■ The Boylean vacuum pump, built by Robert Boyle's assistant Robert Hooke (from *New Experiments Physico-Mechanical, Touching The Spring of the Air*, 2nd ed., London, 1662).

Otto von Guericke (1602–1686) invented the first vacuum pump around 1654.[2] During that year he conducted one of the greatest scientific demonstrations of all time. Figure 144 depicts the scene in Regensburg, Germany. In the presence of Emperor Ferdinand III, von Guericke used his pump to evacuate the air from a sphere assembled from two copper hemispheres. Although the sphere was only 14 inches in diameter, two teams of eight horses each could not pull the hemispheres apart. A 760 mm column of mercury with a base of one square inch weighs about 14.7 pounds. Thus, atmospheric pressure is about 14.7 pounds per square inch. Since the cross-sectional area of the evacuated sphere was about 154 square inches, the total force on it was equivalent to a weight of 2,260 pounds (*ca.* 1 ton). The total surface area of an adult human is much larger than that of the copper sphere, and thus the weight of the atmosphere upon us is much greater than a mere ton. Fortunately, we are not hermetically sealed. Our internal pressure equalizes the outside pressure and so we are blithely unaware of this matching of huge forces within and without our bodies.

In Figure 145, we see the Boylean vacuum pump, built by Boyle's youthful assistant Robert Hooke (1635–1703) in 1655. The large glass globe is sealed at the top with a brass rim and brass key. A stopcock (SN) connects the globe to brass cylinder P, which has a piston in it sealed with leather and run by a rack-and-pinion mechanism worked by hand crank. Plug R fits tightly into a hole in the cylinder. A vacuum is pumped as follows: With stopcock SN open and plug R in place, the piston is drawn down, removing air from the globe. The stopcock is closed, plug R removed, and the airtight piston raised to force out the collected air. The process is repeated.[2] In his 1665 book *Micrographia* Hooke first used the word cell to describe the honeycomb structure of cork visible by microscope.

---

1. D. Abram, *The Spell of the Sensuous*, Pantheon, New York, 1996, p. 260. I am grateful to Professor Susan Gardner for introducing me to this book and suggesting some of the themes of the present essay.
2. J.R. Parrington, *A History of Chemistry*, MacMillan, London, 1961, Vol. 2, pp. 512–519.
3. J. Steinmann, *Pascal* (translated by M. Turnell), Harcourt, Brace & World, New York, 1965, p. 80.

BOYLE'S LAW

The second edition of Boyle's first book, *New Experiments Physico-Mechanical Touching the Air*, was published in 1662 and contained a section titled "A Defense of Mr. Boyle's Explications of his Physico-mechanical Experiments, against Franciscus Linus." In this section, he disclosed the relationship between the pressure and the volume of a gas that we now call Boyle's Law—the first Ideal Gas Law. Why must all high school chemistry students learn this simple relationship? In part, because Boyle's Law and the other gas laws helped to establish the reality of atoms and molecules over 150 years later.

In the plate shown here (Figure 146), *Fig.* 5 depicts the J tube Boyle designed to test the pressure–volume relationship of the only gas he knew—air.

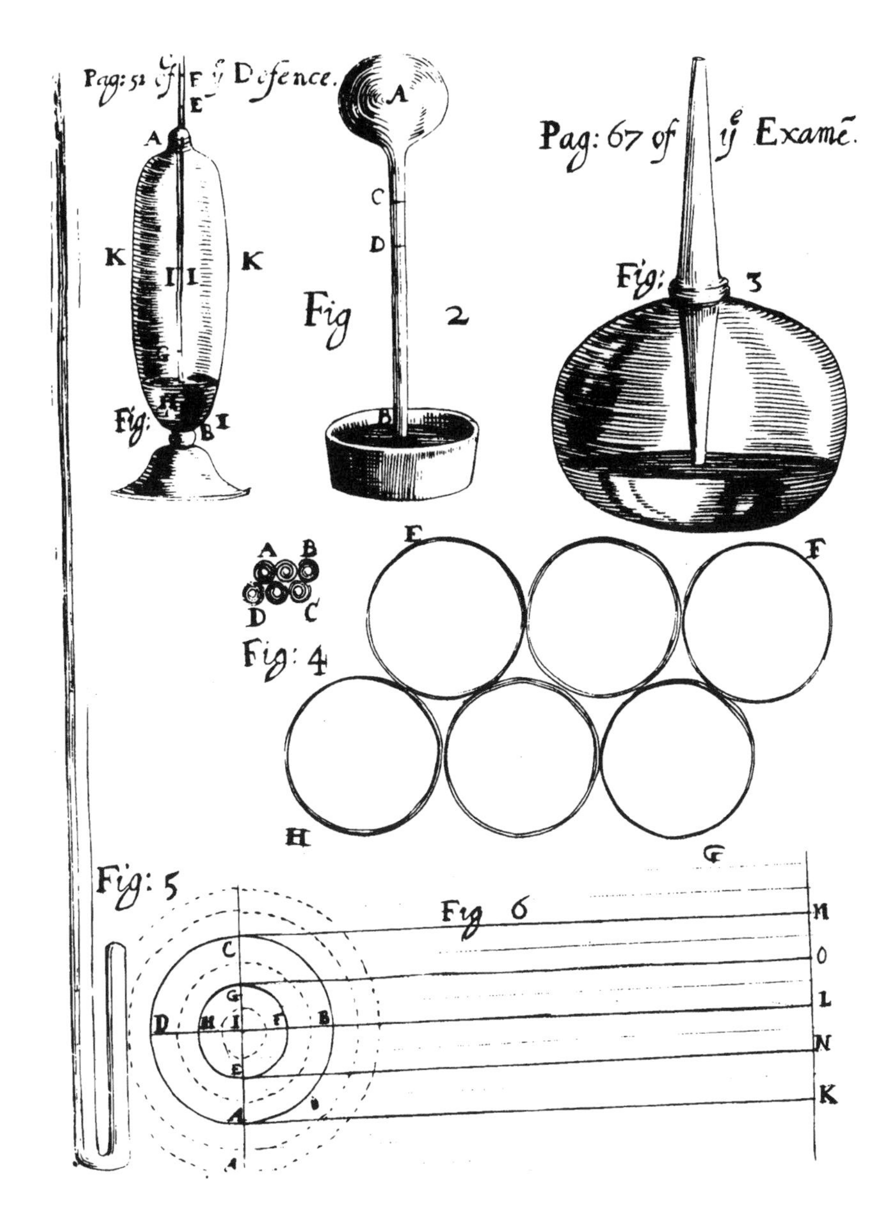

FIGURE 146. ■ In *Fig*. 5 we see Robert Boyle's famous J tube used to demonstrate that $PV = k$ (Boyle's Law). Air is trapped by mercury in the small arm of the J tube. As more mercury is added, the volume of the air decreases. (From *New Experiments Physico-Mechanical* . . . , 1662).

The experimental data described here are taken directly from Boyle's book. On the day he performed the experiments, the air pressure measured with a barometer was $29\frac{2}{16}$ inches of mercury (the pressure of atmospheric air supports a column of $29\frac{2}{16}$ inches against a vacuum). Boyle poured mercury into the open end of the J tube so as to trap a parcel of air, and he carefully adjusted the amount of mercury so as to have equal heights of mercury in both arms. This means that the pressure on the trapped air sample is $29\frac{2}{16}$ inches. (Since the two arms of the tube have the same cross-sectional area, the volume is directly related to height, in inches, which Boyle used as his measure of relative volume.) If enough mercury is added to compress the air "volume" to 9 inches ($^3/_4$ of the original volume), the total pressure is $39^4/_{16}$ inches ($29^2/_{16}$ + $10^2/_{16}$) or about $^4/_3$ of the original pressure. If sufficient additional mercury is added to compress the height of the trapped air to 6 inches from its original 12 inches, this air packet is supporting $29^{11}/_{16}$ inches of mercury in addition to the atmospheric $29^2/_{16}$ inches for a total of $58^{13}/_{16}$ inches: double the pressure, halve the volume. When enough mercury has been added to compress the air to 3 inches (one-fourth of original volume), the total pressure on the trapped air packet is $88^7/_{16}$ + $29^2/_{16}$ or $117^9/_{16}$ inches or four times the original pressure.

Thus, the form of Boyle's Law is:

$$PV = \text{constant} \qquad \text{or} \qquad P_1V_1 = P_2V_2 = P_3V_3 = P_4V_4 = \cdots$$

ENHANCING FRAIL HUMAN SENSES

We chemists seem to have almost surrendered Robert Hooke (1635–1703) to the physicists and biologists, if introductory textbooks are any indication. From middle school onward everybody learns that Hooke coined the term "cell" to describe the microscopic structure of cork. Those who take physics learn that springs, coiled or not, obey Hooke's law. We do know that not long after Otto von Guericke invented the vacuum pump (1654) (see Figure 144), Hooke, assisting Robert Boyle, constructed the "Boylean" vacuum pump (see Figure 145). However, Hooke would have termed himself a "natural philosopher," and his incredible scope of activity would have amply justified it. Trained at Oxford, he was appointed curator of experiments to the Royal Society, and was elected FRS in 1663 and professor of geometry of Gresham College in 1665.[1,2] Hooke was said to have "had poor health and slept badly," was something of a hypochondriac, and "For a few years before his death he is said never to have gone to bed or taken off his clothes."[5] This is easy to understand since "The dispersion of his effort seems to have been due at least in part to the varying interests of the Royal Society, which set Hooke to perform a variety of experiments without giving him time to finish any of them. The Society also asked him to repeat the same experiment over and over again, refusing to see the correct interpretation Hooke put upon it."[3]

Hooke's major published work was his 1665 folio *Micrographia*,[4] one of the most beautiful books in the history of science. It is overwhelmingly a book of microscopy, although the final two essays describe telescopic studies of the stars and the moon. Hooke's later sketches of Mars were employed in the nineteenth century to determine the planet's period of rotation.[1]

From the distant mirror of the seventeenth century, Hooke[5] assures us that we can "recover some degree of those former perfections" (lost upon Adam and Eve's expulsion from Eden) if

> The next care to be taken, in respect of the Senses, is a supplying of their infirmities with Instruments, and as it were, the adding of artificial Organs to the natural;

And while Hooke's microscopic tour-de-force is state-of-the-art in 1665, he avers:[5]

> 'Tis not unlikely, but that there may be yet invented several other helps for the eye, as much exceeding those already found, as those do the bare eye, such as by which we may perhaps be able to discover living Creatures in the Moon, or other Planets, the figures of the compounding Particles of matter, and the particular Schematisms and Textures of Bodies.

An ambitious agenda, indeed. But let us select a few micromorsels from *Micrographia*.

Observation XIII[6] explores the microscopic appearances of crystalline materials and offers the profound hypothesis that these regular, three-dimensional structures can be explained by (hexagonal) closest packing of spheres (Figure 147). Observation XIV[7] (*Of Several kindes of frozen* figures) depicts crystals of ice having different origins (Figure 148). Crystals observed on the surface of frozen urine are sometimes quite huge (especially those "observ'd in Ditches which have been full of foul water"). They have near sixfold symmetry (*Fig. i*). (*Note:* Italic figure numbers cited in text refer to the original figures shown collected in these composite illustrations.) What is the nature of the urine crystals? In Hooke's words, "Tasting several cleer pieces of this *Ice*, I could not find any *Urinous* taste in them, but those few I tasted, seem'd as *insipid* as water."[7]

Fig. 2 (in Figure 148) depicts snow flakes caught on a black cloth, the six arms in each flake identical but different from those of other flakes. *Fig. 3* is an enlargement of a single snowflake. *Fig. 4* and *Fig. 5* are ice crystals removed with a knife from the surface of a glass vessel filled with water and chilled. Although hexagonal symmetry is not obvious at first, the angles made with the "central stem" in *Fig. 4* are 60° and 120°. *Fig. 6* depicts the water surface just starting to freeze. Hooke referenced these observations to Observation XIII (Figure 147) and specifically to the hexagonal closest packing of spheres to form crystals. While he could assign no chemical meaning to these spheres, some 140 years later John Dalton used his atomic theory to explain the crystalline structure of ice with similar illustrations employing spheres to represent molecules of water (see Figure 232).

Chemically speaking, Observation XVI (*Of* Charcoal, or *burnt* Vegetables)[8] is the most exciting essay in *Micrographia*. Vegetable matter may be placed in a crucible, thoroughly surrounded and covered with sand and heated by fire. Once the heating is stopped and the sand is allowed to cool, charcoal may be recovered. However, if the sand is still hot (or even warm), the uncovered charcoal will burst into flames and be completely consumed. Other oxygen-deficient environments (including vacuum) did not support combustion of charcoal. However, charcoal heated in vacuo inflamed as soon as atmospheric air was introduced. Of course, gunpowder (charcoal, sulfur, and saltpetre) had been known for centuries. Heating charcoal with saltpeter produced very vigorous and complete combustion in a closed vessel (as well as under water). In contrast, combustion of charcoal in a closed vessel containing atmospheric air soon petered out. Similar observations were also made by Hooke using sulfur instead of charcoal. He posited that air is "a menstruum" capable of "dissolving" "sulphureous" (i.e., combustible) bodies. Furthermore:[8]

> the dissolution of sulphureous bodies is made by a substance inherent, and mixt with the Air, that is like, if not the very same, with that which is fixed in Salt-petre, which by multitudes of Experiments that may be made with Saltpetre, will, I think, most evidently be demonstrated.

We will shortly speak of Dr. John Mayow, a friend and companion of Hooke. Why is Mayow, rather than Hooke, generally credited with the discovery

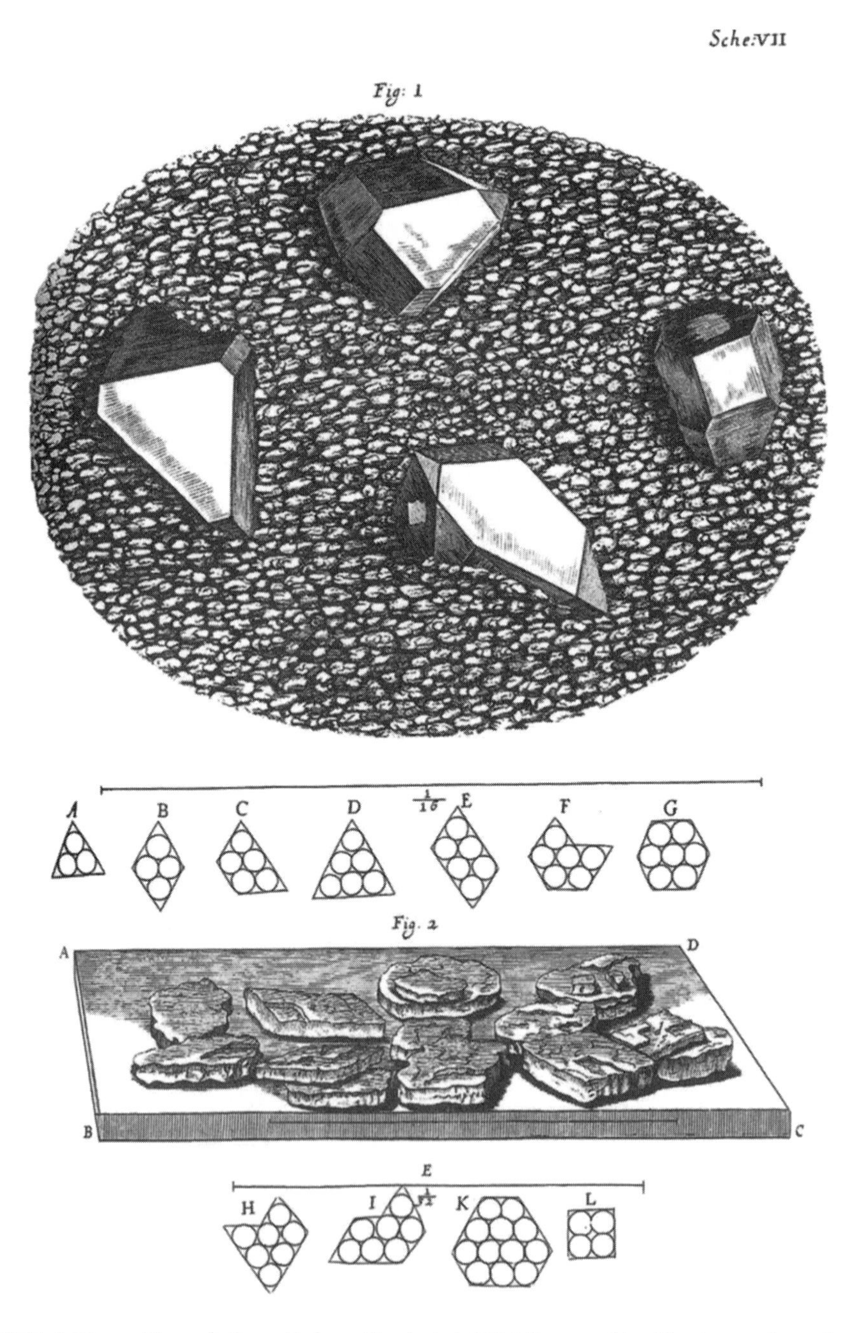

FIGURE 147. ■ Crystals from Robert Hooke's 1665 *Micrographia*. Hooke explained crystalline structures on the basis of close packing of spheres, an insightful anticipation of Dalton's explanation 140 years later (see Figure 232).

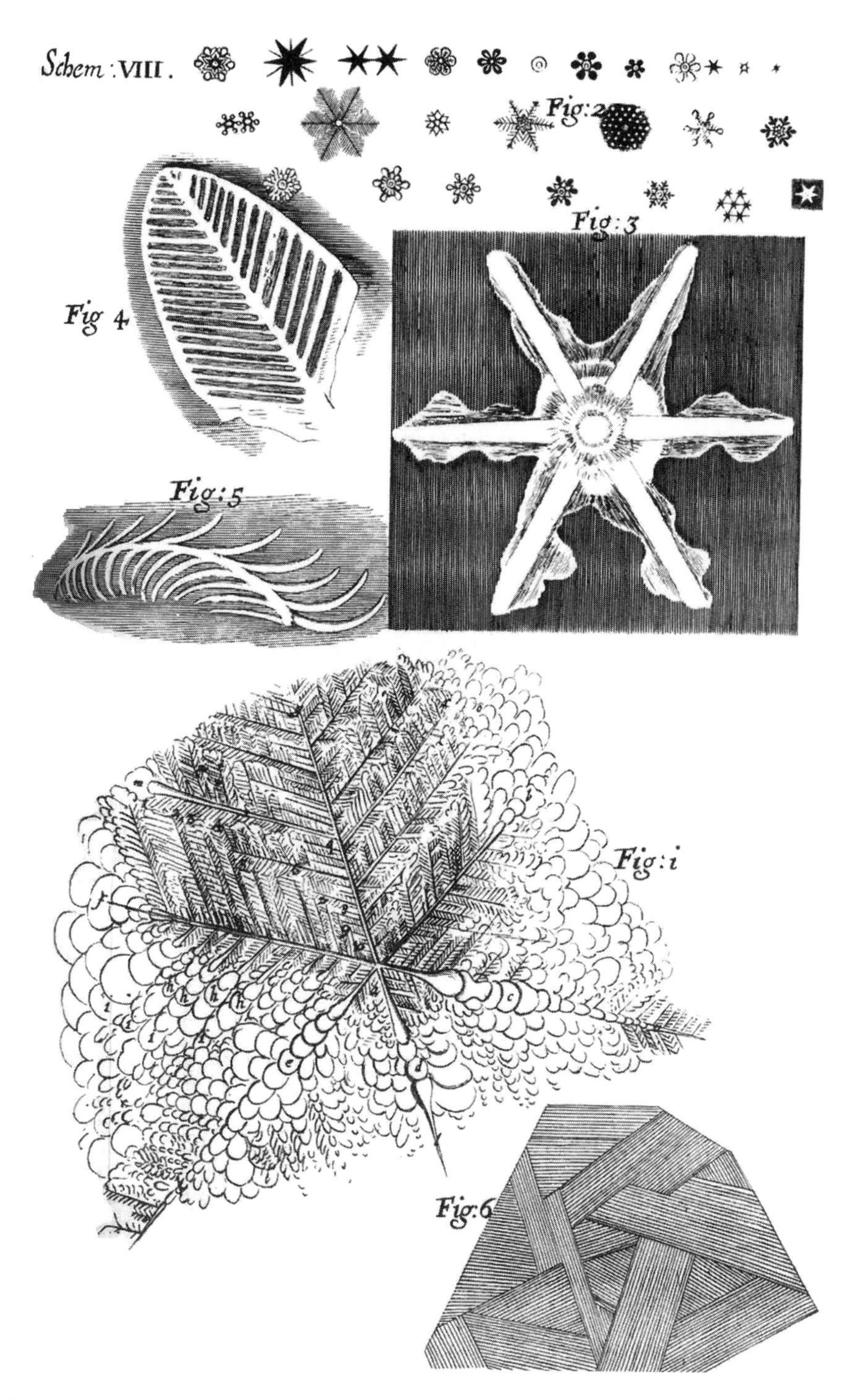

FIGURE 148. ■ Ice crystals viewed by Hooke under the microscope. He understood that their sixfold symmetry derived from packing of spheres (Figure 147). Hooke found that the ice crystals derived from urine were pure water (they lacked the "urinous" taste).

that a component of air supports both combustion and respiration? Partington[9] points out that Hooke postulated that the substance "mixt" in air was a "saline substance" (finely divided, suspended saltpetre or niter, perhaps?) that might be somehow "strained out." It was Mayow who correctly proved that the active substance was a gaseous component of atmospheric air.

Micrographia illustrated numerous small objects microscopically. The enlarged image of the stinger of a bee,[10] for example, actually provided very useful insights into its mode of action. However, for collectors of old tomes, far worse than fierce stinging bees and plague-carrying fleas is the fearsome bookworm (Figure 149)![11] Let us hear Hooke:

> And indeed, when I consider what a heap of Saw-dust or chips this little creature (which is one of the teeth of Time) conveys into its entrails, I cannot chuse but remember and admire the excellent contrivance of Nature, in placing in Animals such a fire, as it is continually nourished and supply'd by the materials convey'd into the stomach, and fomented by the bellows of the lungs; and in so contriving the most admirable fabrick of Animals, as to make the very spending and wasting of that fire, to be instrumental to the procur-

WANTED DEAD OR ALIVE

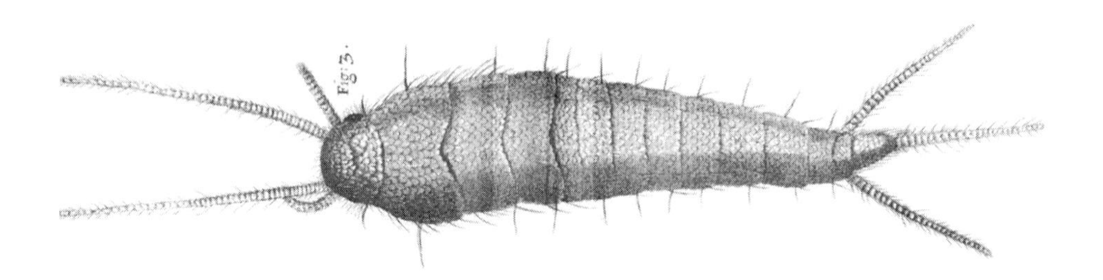

"The Bookworm"

aka "Silverfish"

aka *Tysanura*

Reward Offered by Antiquarian Book Collectors Anonymous

FIGURE 149. ■ No book collector or librarian will protest this "wanted" poster.

> ing and collecting more materials to augment and cherish it self, which indeed seems to be the principal end of all the contrivances observable in bruit animals.

And this some 120 years before Lavoisier proved with balance and calorimeter that respiration is combustion!

---

1. *The New Encyclopedia Britannica*, Encyclopedia Britannica Inc., Chicago, 1986, Vol. 6, p. 44.
2. J.R. Partington, *A History of Chemistry*, MacMillan & Co. Ltd., London, Vol. 2, 1961, pp. 550–570.
3. Partington, op. cit., pp. 551–552.
4. R. Hooke, *Micrographia: Or Some Physiological Descriptions of Minute Bodies Made by Magnifying Glasses. With Observations and Inquiries thereupon*, Jo. Martyn and Ja. Allestry, Printers to the Royal Society, London, 1665. See also the facsimile reprint published by Culture Et Civilisation, Brussels, 1966.
5. Hooke, op. cit., Preface
6. Hooke, op. cit., pp. 82–88.
7. Hooke, op. cit., pp. 88–93.
8. Hooke, op. cit., pp. 100–106.
9. Partington, op. cit., p. 558.
10. Hooke, op. cit., pp. 163–165.
10. Hooke, op. cit., pp. 208–210.

GUN POWDER, LIGHTNING, THUNDER, AND NITRO-AERIAL SPIRIT

Gunpowder is a mixture of saltpetre (KNO_3) or nitre ($NaNO_3$), sulfur, and carbon developed possibly as early as 1150 A.D. by the Chinese. Its explosive power is due to the exothermic reaction below in which a large volume of gas (carbon dioxide and nitrogen) is generated suddenly and violently along with a great deal of heat. Gunpowder burns under water or in a vacuum. In modern terms, we see saltpetre as the oxidizer (in place of gaseous oxygen), which con-verts charcoal to carbon dioxide. Thus, saltpetre and nitre are

$$2\ KNO_3 + 3\ C + S \rightarrow N_2 + 3\ CO_2 + K_2S$$

capable of supporting combustion. John Mayow (1641–1679)[1] came very close to "nipping" phlogiston theory "in the bud" almost immediately after Becher first proposed its original form in 1669. He first entered Oxford in 1658, was admitted as a scholar in 1659, and was elected a fellow of All Souls College in 1660. Mayow became a "profess'd physician" around 1670, although Partington could find no evidence for a formal medical degree.[1] It is not exactly clear when Mayow and Hooke met or whether Mayow ever met Boyle. He does appear to have been given access to the Boylean vacuum pump in Oxford during the 1660s.[2]

In 1668 Mayow published two tracts dealing with respiration and rickets. These were revised in 1674 and published with three additional tracts to constitute the *Tractatus Quinque Medico-Physici. . . .*[3] It is in this work that Mayow identifies his "*Spiritu Nitro-Aereo*" or "nitro-aerial spirit." Just as Hooke had done in his 1665 book *Micrographia*, Mayow identifies a substance in the air that is required to support the combustion of "sulphureous" matter just as niter or saltpetre was known to. Flammable substances were said to contain "sulphureous matter," and this was, of course, strongly related to phlogiston theory. Indeed, one can trace these concepts back to Paracelsus' three principles—sulfur, mercury, and salt. However, Mayow's powerful contributions[1] to chemistry were the realizations that (1) a component of air supported combustion; (2) this component of air had the same effect as nitre or saltpetre; (3) this component of air also supported respiration; and, most uniquely, (4) this was a specific gaseous component of air. Thus, "atmospheric air" contained a gaseous component capable of supporting combustion and respiration and another gaseous component that could not. Mayow's experiments are illustrated in Figure 150. He correctly explained a curious observation of Boyle's concerning gunpowder. It was known that the saltpetre in gunpowder provided a much greater amount of "nitro-aerial spirit" than did atmospheric air. Furthermore, the "fuel" components of gunpowder, carbon, and sulfur could each burn in a closed vessel up to a point and then would extin-

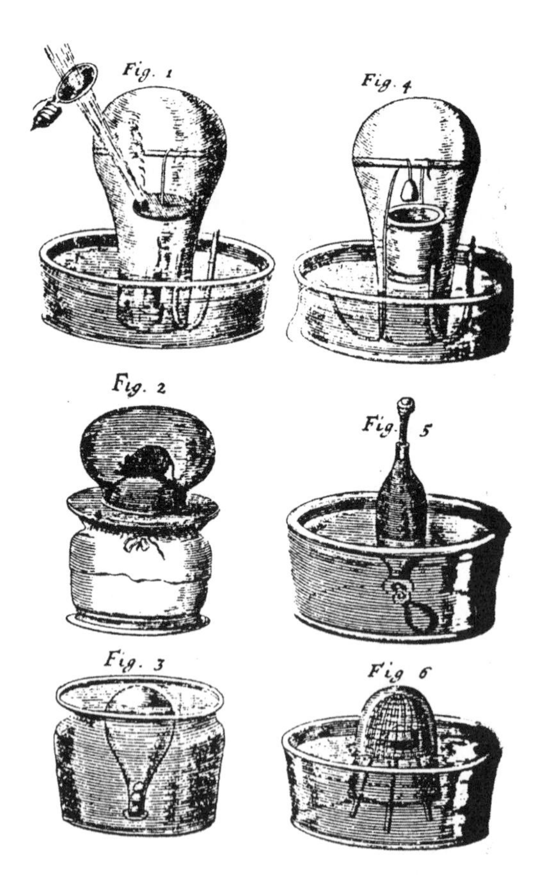

FIGURE 150. ▪ This plate is from John Mayow's *Tractus Quinque Medico Physici* (Oxford, 1674). It shows his experiments in which "nitro-aerial spirit" in saltpetre was transferred to antimony under a heating glass. In effect oxygen was transferred between the two substances (from the Dr. Roy G. Neville Historical Chemical Library, a collection in the Othmer Library, CHF).

guish. In contrast, a closed vessel containing carbon or sulfur combined with saltpetre would burn to completion just as gunpowder would under these circumstances. However, Boyle placed some gunpowder in a circle on a surface under vacuum.[4] Under a heating lens he observed slow, localized ignition of the particles of gunpowder directly exposed to the intense light. Removal of the lens stopped the burning. However, if the burning lens was trained on some crystals of gunpowder in the powder circle and the system was then opened to the atmosphere, full conflagration occurred instantly. Mayow correctly reasoned that "nitro-aerial" particles had to be in direct contact with charcoal or sulfur to produce combustion.[5]

Partington[6] remarks on Mayow's claim to have heated niter and collected the resulting nitric acid. He notes "if he had actually tried the experiment he could have discovered oxygen." However, as Partington and others remarked,

FIGURE 151. ■ A water spout depicted in John Mayow's *Tractus Quinque Medico Physici* (1681 edition, published in The Hague; the first edition was published in Oxford in 1674).

Mayow did not have the means to capture, manipulate, and study gases. These techniques awaited development by Stephen Hales in 1727 and subsequent improvements by William Brownrigg, Joseph Black, Henry Cavendish, and Joseph Priestley.

The *Tractatus Quinque* concerned itself with other scientific questions beyond the medical, physiological, and chemical. For example, Mayow discussed the origins of water spouts as due to air turbulence (see Figures 151 and 152; see also Benjamin Franklin's studies of these phenomena and Figure 119 later in this book). Mayow's explanation of lightning and thunder are reminiscent of those of Paracelsus[7] and imagine explosions between "nitro-aerial" spirit and "sulphureous" matter in the atmosphere.

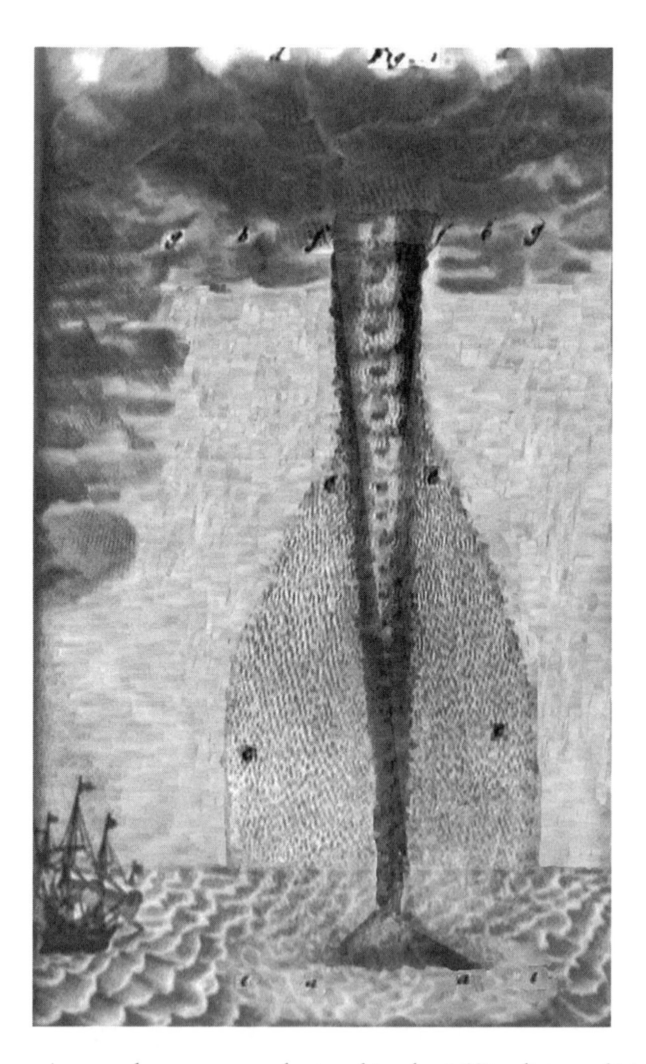

FIGURE 152. ▪ A second water spout depicted in the 1681 edition of Mayow's *Tractus Quinque Medico Physici*. Mayow came quite close to discovering that saltpetre contained oxygen, which could support combustion. Robert Boyle and his assistant Robert Hooke, who was friendly with Mayow, also investigated the ability of saltpetre to sustain combustion.

1. J.R. Partington, *A History of Chemistry*, MacMillan & Co. Ltd., London, 1961, Vol. 2, pp. 577–614.
2. Partington, op. cit., p. 604.
3. J. Mayow, *Tractatus Quinque Medico-Physici. Quorum primus agit de Sal-Nitro, et Spiritu Nitro-Aereo. Secondus de Respiratione. Tertius de Respiratione Foetus in Utero, et Ovo. Quartus de Motu Musculari, et Spiritibus Animalibus. Ultimus de Rhachitide*, Sheldonian Theatre, Oxford, 1674. (A second Latin edition was published in The Hague, 1681; an English translation was published by the Alembic Club, Edinburgh, in 1907.)
4. Partington, op. cit., p. 527.
5. Partington, op. cit., p. 589.
6. Partington, op. cit., p. 588.
7. Partington, op. cit., p. 133.

WHO WOULD *WANT* AN ANTI-ELIXIR?

A strange narrative indeed! Although *The Sceptical Chymist* rid chemistry of the Aristotlean Elements, Boyle was a believer in the possibility of transmutation (as was fellow member of the Royal Society Isaac Newton).

This pamphlet (Figure 153) is considered to be the rarest of Boyle's works. Of the first (anonymous) edition published in 1678 and this second, attributed edition of 1739, Duveen[1] accounted for only four known copies combined, although Ihde[2] suggests possibly four copies of each edition. Lawrence M. Principe has discovered that Boyle's "An Historical Account . . ." is really part of a much longer continuous text, *Dialogue on Transmutations*, developed in the late 1670s and then in the early 1680s.[3] In it Boyle narrates a series of one-time-only reverse transmutation experiments he witnessed in which the transmuting agent was a miniscule amount of solid substance. The claim tested was that the substance could transform gold into a baser metal. Why would anybody be interested in such an "anti-elixir"? Using very modern chemical logic, Boyle reasoned that if one learns how to transmute gold into a baser metal, then one would also gain the knowledge to perform the reverse operation.

The experiments narrated in this pamphlet gave tantalizing but inconclusive evidence for the chemical degradation of gold into a lesser metal, perhaps even a salt, but the world's known supply of anti-elixir was consumed—apparently never to be rediscovered. Ihde[2] speculated over whether the experiment was ever done at all, done incompetently, or was possibly a joke by Boyle. His firmer conclusions were that the experiment was, in all likelihood, actually carried out at Boyle's customarily high level of competence and that Boyle had no sense of humor, especially in regard to experimentation. Ihde's tentative conclusion: some sleight of hand by one of Boyle's laboratory assistants to give the chief his desired conclusion and help him recover from an earlier embarrassment at the hands of that young upstart Isaac Newton.[2] In 1676, Boyle had believed he had prepared a new mercury from quicksilver, but Newton argued that a purely physical, not an alchemical change, had occurred.[2] In contrast to Boyle, Newton believed that alchemical secrets were not to be shared, and his alchemical writings remained in the form of hundreds of manuscripts. In 1936 a great number of these manuscripts were sold at auction in London. Subsequently, the economist John Maynard Keynes acquired them from many of the buyers and studied them intensively. Keynes' collection passed to Kings College Cambridge upon his death in 1946. The fact that Newton[4,5] was credulous about alchemy plays a part in a novel[6] in which the "Aetherial Spirit" is embodied in the 9 lives of a Golden Cat born every 81st generation to parents (Feline Sol and Luna) whose *conjunctio* produces the quintessential cat.

AN

HISTORICAL ACCOUNT

OF A

DEGRADATION

OF

GOLD,

Made by an

ANTI-ELIXIR:

A STRANGE

CHYMICAL NARRATIVE.

By the HONOURABLE

ROBERT BOYLE, Efq;

The SECOND EDITION.

LONDON:

Printed for R. MONTAGU, at the *Book-Ware-Houfe*, in *Great Wilde-Street*, near *Lincoln's-Inn Fields*.

MDCCXXXIX.

FIGURE 153. ■ Title page from the second edition of Boyle's rarest work—his witnessing of a "reverse transmutation." The first edition, published in 1678, was anonymous.

1. D. Duveen, *Bibliotheca Alchemica Et Chemica*, HES, Dordrecht, 1986, p. 97.
2. A. Ihde, *Chymia*, No. 9, 47–57, 1964.
3. L.M. Principe, *The Aspiring Adept. Robert Boyle and His Alchemical Quest*, Princeton University Press, Princeton, 1998.
4. Partington, J.R., *A History of Chemistry*, Vol. 2, MacMillan & Co. Ltd, London, 1961, pp. 468–485.
5. B.J.T. Dobbs, *The Foundation of Newton's Alchemy: or, The Hunting of the Greene Lyon*, Cambridge University Press, Cambridge, 1975.
6. S.G. King, *The Wild Road*, Ballantyne, New York, 1997, pp. 328–329. I thank Ms. Susan Greenberg for bringing this to my attention.

A HARVARD-TRAINED ALCHYMIST

Renaissance alchemy conjures up back alleys in Prague and other Old World images. Harvard University is, of course, strictly New World in our minds—a cradle for progressive thought and the home of Nobel Laureates. How delightful that George Starkey (born in Bermuda in 1628, died in the London plague of 1665) provides us with a surprising conjunction of the Old and New Worlds.[1,2] Eirenaeus Philalethes ("A Peaceful Lover of Truth") was the pseudonym provided for his posthumous writings and Secrets Revealed [see Figure 154] is the English translation of his most influential book.

Starkey graduated from Harvard in 1646, one of a class of four[3] who received their lectures from President Dunster.[2] He shared a dorm room (measuring no more than 7 feet 9 inches by 5 feet 6 inches) with a John Allin.[2] Courses included "Logick," "Physicks," "Ethicks and Politicks," and "Arithmetick and Geometry."[2] The natural philosophy curriculum at Harvard reflected some of the finer points of the great debate between the Aristotleans and Cartesians (matter is continuous; there are no vacant spaces; "nature abhors a vacuum") and those who believed in corpusular matter, including Newton and Boyle. According to William R. Newman,[2] this division was not so clear-cut at Harvard where a late Aristotlean view, which allowed for finite particles, was in currency. In any case, following graduation Starkey rated the natural philosophy curriculum at Harvard as "totally rotten."[2]

On the basis of his examination of Harvard theses from the mid-seventeenth to late-eighteenth centuries, Newman[2] notes the following successfully defended positions:

1687	Is there a stone that makes gold? Yes.
1698, 1761	Is there a universal remedy? Yes in 1698, no in 1761.
1703	Can metals be changed into one another alternately? Yes.
1703, 1708, 1710	Is there a sympathetic powder? Yes.
1771	Can real gold be made by the art of chemistry? Yes.

As Newman notes,[2] "Obviously, Harvard was far from being an uncongenial place for the budding alchemist; as late as 1771, Harvard undergraduates were defending the powers of the philosopher's stone" (and these were not only the "New Age" people).

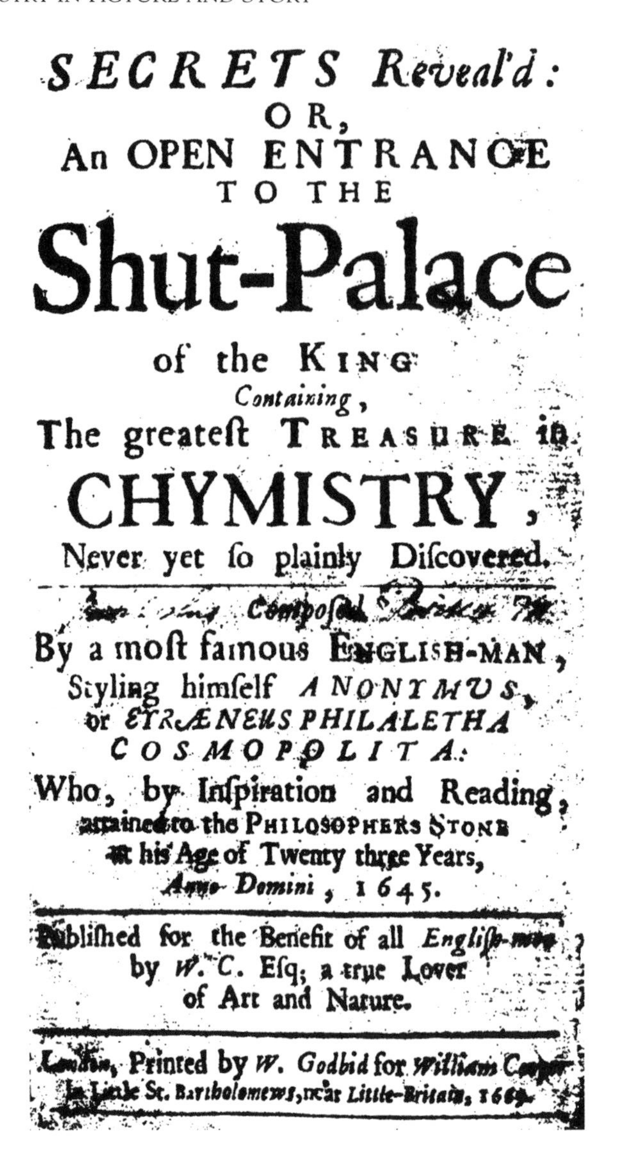

SECRETS Reveal'd:
OR,
An OPEN ENTRANCE
TO THE
Shut-Palace
of the KING:
Containing,
The greateſt TREASURE in
CHYMISTRY,
Never yet ſo plainly Diſcovered.
[illegible] Compoſed [illegible]
By a moſt famous ENGLISH-MAN,
Styling himſelf ANONYMUS,
or EYRÆNEUS PHILALETHA
COSMOPOLITA:
Who, by Inſpiration and Reading,
attained to the PHILOSOPHERS STONE
at his Age of Twenty three Years,
Anno Domini, 1645.

Publiſhed for the Benefit of all Engliſh-men
by W. C. Eſq; a true Lover
of Art and Nature.

London, Printed by W. Godbid for William Cooper
in Little St. Bartholomews, near Little-Britain, 1669.

FIGURE 154. ■ Eirenaeus Philalethes ("A Peaceful Lover of Truth") was, in reality, John Starkey, Harvard Class of 1646. His *Secrets Reveal'd* (spelling was not emphasized at Harvard) was cited extensively by Isaac Newton.

Moving to England in 1650, Starkey became an important exponent of Van Helmont's approach and worldview. Van Helmont did not support the Galenical view of medication (contraries) or the Paracelsian view (similitude). Instead he believed in cures that produced "healing ideas in the *Archeus*"—the inner architect or life spirit (Figure 97) located in a region between the stomach and spleen.[4] Van Helmont and Starkey shared a belief in the importance of pyrotechny (arts, such as distillation, involving fire) and the utility of practical, experimental work. Starkey had little use for the abstractions of mathematics. He referred to himself as a Philosopher By Fire, in sneering contrast to the safe academicians who eruditely cited published facts. Such fiery rhetoric made him few academic friends. However, he had important correspondence with Robert

Boyle, and Newman establishes that Isaac Newton, who seriously studied alchemy, cited Starkey's works far more often than any other alchemist of the period.[2] "Heady stuff" for a young man of modest means from the Colonies.

---

1. C.C. Gillespie (Editor-In-Chief), *Dictionary of Scientific Biography*, Charles Scribner, New York, 1975, Vol. XII, pp. 616–617.
2. W.R. Newman, *Gehennical Fire: The Lives of George Starkey, an American Alchemist in the Scientific Revolution*, Harvard University Press, Cambridge, MA, 1994.
3. The historical evidence is thus unambiguous: There was no freshman basketball club team when Starkey attended Harvard.
4. J.R. Partington, *A History of Chemistry*, MacMillan, London, 1961, Vol. 2, pp. 209–241.

LUCIFER'S ELEMENT AND KUNCKEL'S PILLS

Urine, the golden liquid endowed with vital and mystical properties, was used for centuries in thousands of alchemical preparations. Undoubtedly it has been distilled to dryness countless times. However, in 1669 a competent but obscure alchemist, Hennig Brand, boiled urine, concentrating it to a thick syrup, from which a red oil was distilled. The retort's black carbon residue was added to this oil and heated in an earthen retort. The scene is imagined by John Emsley in his wonderful book *The 13th Element*.[1] We "witness" Brand observe the distillation of a heavy glowing liquid that bursts into flames as it contacts air.[2] Once Brand isolates the liquid in his receiver, it solidifies but continues to glow. Imagine the wonder that this provoked—a glowing, fiery secret concealed in our own bodies and our excreta! Brand had discovered the element phosphorus ("bringing light"). The melting point (44°C) and boiling point (280°C) of white phosphorus are quite low. This explains its ease of distillation and the fact that impurities may lower the melting point sufficiently to make "liquid phosphorus".

Rather than publishing an epistle on urine and its transformation to phosphorus, Brand kept his work in Hamburg secret for about six years, hoping to make money as soon as he discovered what it was good for.[2] However, exhausting the fortune of his second wife on this fruitless search, he eventually publicized the discovery, and his work came to the attention of Johann Kunckel, recently retired unsuccessful gold-maker to the Elector of Saxony,[3] and now professor at the University of Wittenberg. Kunckel visited Hamburg but could learn no details from Brand. Kunckel, in turn, informed another alchemical colleague, Johann Daniel Kraft in Dresden, about the new substance. Sensing opportunity, Kraft hastened immediately to Hamburg and purchased all of Brand's phosphorus supply along with exclusive rights and a pledge by Brand to secrecy. This bit of business chicanery occurred right under Kunckel's nose, as it were, and the best the distinguished academician could get from Brand was the hint that urine was the source.[2]

Kunckel (1630–1703),[4] the son of an alchemist, had received no formal academic training but was an able and respected scientific investigator. With only the vague hint provided by Brand, Kunckel independently discovered the process and published on the properties of his *noctiluca constans* ("unending nightlight"), although not the method of its preparation, in 1678. Robert Boyle obtained samples from Kraft and Kunckel during this period and published work in 1680 on the preparation and properties of liquid phosphorus, the "Aerial Noctiluca" ("nightlight spirit"), and another in 1682 on solid phosphorus, the "Icy Noctiluca."[4] Even the eminent mathematician Gottfried Wilhelm Leibniz (1646–1716), who independently of Sir Isaac Newton developed the calculus, maintained a lifelong interest in alchemy, as did arch rival Newton, and wrote on the experimental investigations of phosphorus.[4,5] The magic imagery associated with phosphorus is apparent in Figure 155, the fanciful frontispiece to Johann Heinrich Cohausen's 1717 book[6] on phosphorus, where we

FIGURE 155. ■ The mystical title page of Johann Heinrich Cohausen's 1717 treatise on phosphorus. Hermes and the flying dragon are sources of fire and light—properties of white phosphorus.

FIGURE 156. ■ An illustration from Cohausen's 1717 *Lumen Novum Phosphoro Accensi*. How does one get the Holy Roman Empire phosphorus contract? Simple; devise a sign in which white phosphorus spells out the Emperor's name "LEOPOLDUS" so that it shines boldly in the dark.

see both Hermes and a flying dragon as sources of light and fire. Figure 156[7] apparently shows Johann Daniel Kraft "pitching" phosphorus to Leopold I early in his long reign (1658–1705) as Holy Roman Emperor. In this figure, *1* depicts flashing phosphorus (solid); *2*, liquid phosphorus giving off smoky fumes as it sits motionless; and *3*, a shine-in-the-dark barometer. Item 5 is a kind of "flashing sign"[8] with the Emperor's name printed neatly in solid phosphorus—a very modern bit of slick salesmanship.

Kunckel was certainly an enthusiastic alchemist, if unsuccessful gold-maker. Believing that mercury was the spirit of metallicity retained in the transmutation of metals, he reported extracting mercury from all metals.[9] However, his quantitative studies indicated that antimony gains weight upon calcination and also included measuring the strength of *aqua fortis* (nitric acid) by saturating it with silver, evaporating the solution to dryness, and weighing the remaining salt.[10] Kunckel also contributed to the art of glass-making and his 1679 book *Ars Vitraria Experimentalis*, included the 1612 work (in seven books) by Antonio Neri, updated with three of his own works.[11] Figure 157, from the 1679 book, depicts a glass-making furnace with workers fashioning bottles. In Figure 158 we see a pedal-operated bellows used to fabricate small glass toys.[4]

FIGURE 157. ■ A glass-making furnace depicted in Johann Kunckel's 1679 *Ars Vitraria Experimentalis* (from the Othmer Library, CHF). Kunckel was cheated by his "friend" Johann Daniel Kraft (he's the "pitchman" in Figure 156) who monopolized Hennig Brand's discovery of phosphorus. Nonetheless, the inadvertent hint from Brand that phosphorus came from urine was enough for the clever Kunckel to independently discover how to make it.

Emsley's book[1] traces the development and uses of phosphorus, which he suggests might be termed "The Devil's Element."[12] The basis for the destructive distillation of urine (or bone) to phosphorus is described. Organic matter, such as creatine, in urine decomposes under oxygen-poor conditions to form elemental carbon (e.g., charcoal). Under the high heat the carbon strips oxygen atoms from phosphate salts also present in the urine residue to form gaseous

FIGURE 158. ■ Fabricating small glass toys with the aid of a pedal-operated bellows (from Kunckel's 1679 *Ars Vitraria Experimentalis*, from the Othmer Library, CHF).

carbon monoxide. This is really not very different from the industrial process that produces white phosphorus from rock phosphate in the presence of coke and silica:[13]

$$2\,Ca_3(PO_4)_2 + 6\,SiO_2 + 10\,C \rightarrow P_4 + 6\,CaSiO_3 + 10\,CO$$

In his chapter[14] titled "The Toxic Tonic," Emsley notes the marketing of pills made of this exceedingly toxic element. Coated with a thin film of gold or silver for physical safety, these were marketed as "Kunckel's Pills" shortly after the famous chemist died. There are 60 pages covering the chemical, practical, business, and sociological history of matches, and this very interesting section almost reads like a novel. (Red phosphorus, the polymeric allotrope, was discovered in the nineteenth century by heating white phosphorus to 400°C in a closed vessel. It helped make the match industry safer.) Here, we first meet the late nineteenth-century English social reformer Annie Besant, who forms a union for women in the dangerous and exploitative match fabrication industry. We will meet her later in our book—some 20 years hence "divining" the internal structures of atoms (see p. 293). And Emsley describes the origin of the glow of white phosphorus—fully understood only in 1974. At the surface of white phosphorus, a solid composed of tetrahedral P_4 molecules, reaction with oxygen produces highly unstable molecules of HPO and P_2O_2, which luminesce close to the surface that formed them just before they "die."[15]

---

1. J. Emsley, *The 13th Element—the Sordid Tale of Murder, Fire, and Phosphorus*, John Wiley & Sons, Inc., New York, 2000. This book was first published in Great Britain as *The Shocking History of Phosphorus*, MacMillan Publishers Ltd. in 2000. It is a truly admirable book—a "page-turner," possibly a "barn-burner." The scope of the book can be imagined as author Emsley relates sadly and ironically that phosphorus was discovered in Hamburg and used in its horrific fire-bombing almost three centuries later. He notes that phosphorus was "the thirteenth chemical element to be isolated in its pure form." Aaron J. Ihde might have contested that since he lists zinc among the elements to have been isolated before 1600 (see A.J. Ihde, *The Development of Modern Chemistry*, Harper & Row, New York, 1964, p. 747). However, the separation of zinc from its oxide, a high-temperature process, was scientifically reported in the mideighteenth century, so Emsley's appellation appears to be "kosher."
2. Emsley, op. cit., pp. 3–24.
3. An "Elector" was a prince in the Holy Roman Empire who could participate in the election of an emperor.
4. J.R. Partington, *A History of Chemistry*, MacMillan & Co. Ltd., 1961, Vol. 2, pp. 361–377.
5. Partington, op. cit., p. 485.
6. J.H. Cohausen, *Lumen Novem Phosphoris Accensum, sive Exercitatio Physico-Chymica De causa lucis in Phosphoris tam naturalibus quam artificialibus*, Joannem Oosterwye, Amsterdam, 1717.
7. Cohausen, op. cit., p. 203.
8. Actually, the "*pomum*" was a globular hand warmer for clerics.
9. Partington, op. cit., p. 362.
10. Partington, op. cit., p. 375.
11. Partington, op. cit., p. 368.
12. Emsley, op. cit., pp. 299–302.
13. F.A. Cotton and G. Wilkinson, *Advanced Inorganic Chemistry*, fifth edition, John Wiley & Sons, New York, 1988, p. 386.
14. Emsley, op. cit., pp. 47–63.
15. Emsley, op. cit., p. 16.

THE EMPEROR'S MERCANTILE ALCHEMIST

Johann Joachim Becher (1635–1682)[1–3] has been gone for well over three centuries, and we usually think of him only as the *ur*-father of chemistry's first comprehensive theory: phlogiston. What we generally miss is that Becher may well have been the greatest mercantilist of the seventeenth-century Holy Roman Empire.[3] In this, he would share some common ground with Antoine Laurent Lavoisier, who, during the late eighteenth century, will become the Father of Modern Chemistry even as he functions as one of France's greatest economists.[4] In 1666, the 31-year-old Becher was appointed economic advisor to Leopold I and titled himself "Advisor on Commerce to His Majesty, the Emperor of the Holy Roman Empire."[5] Sixteen years later he would die in London and leave his family in such poor circumstances that one of his daughters was forced to become a domestic.[6]

The Holy Roman Empire lasted, in name at least, for over a thousand years following the conferral of the imperial title to Charlemagne by Pope Leo III in 800.[7] It consisted of a vast realm in central Europe with Germanic people at its core who furnished most of its traditional rulers. The Reformation in the sixteenth century created rebellious centers of power, notably among German princes who adopted Protestantism and rebelled against the Emperor. These religious tensions reinforced a bewildering "cat's cradle" of territorial conflicts and alliances leading to the start of the disastrous Thirty Years War (1618–1648).[8] By the time the war was settled with the Treaty of Westphalia in 1648, Spain had lost the Netherlands and its preeminence on the Continent, France had emerged as the major western European power, and many Germanic towns were ruined economically and the Empire irrevocably weakened. A century later, the famous French author and satirist Voltaire would quip that the Holy Roman Empire was "neither holy, nor Roman, nor an empire."[7] It would inauspiciously pass out of existence in 1806, two years after Napoleon declared himself emperor of France.

Becher came of age in the aftermath of the Thirty-Years War and like some other prominent chemists of the period, including Kunckel and Johann Rudolph Glauber, devoted himself to the prosperity of Germany. Becher was self-educated, developed an early interest in technology, and published his first work in 1654 on alchemy using the pseudonym Solinus Saltzthal.[1,2] By 1655 Becher had established himself as mathematical advisor to the Holy Roman Emperor Ferdinand III in Vienna and was advising him on alchemical processes.[9] Becher's first book on metal chemistry and iatrochemistry, *Natur-Kündigung der Metallen* (Figure 159), was published in 1660.[1] He believed in a "vitalist theory" in which minerals, as well as animals and plants, "have a sort of life and grow in the earth from seeds."[1] He obtained the M.D. degree at Mainz in 1661 and was appointed to the university's medical faculty in 1663 and became physician to the Elector at Mainz. Married in 1662 to a woman from a prosperous and well-respected family, the restless Becher moved to Munich in 1664 and became Medical and Mathematical Advisor to the Elector of Bavaria.[1–3] It was during this period that he became very much involved in commerce, organized the Eastern Trading Company and tried to establish for his patron a commercial colony in South America. As noted above, in 1666 he joined Leopold I in Vienna.

FIGURE 159. ■ Title page from the 1660 Frankfort edition of *Natur-Kündigung der Metallen* by Johann Joachim Becher. Becher is known to chemists as the father of phlogiston theory. However, he was at least as famous for his knowledge of economics and his status as Advisor on Commerce to Leopold I, Emperor of the Holy Roman Empire. (From The Roy G. Neville Historical Chemical Library, a collection in the Othmer Library, CHF.)

In his 1664 *Oedipus chimicus* (see Figure 162). Becher describes his early concepts of the elementary composition of matter.[2] His most famous work is his 1669 book commonly referred to as the *Physica subterranea*.[1–3,10] In this work, Becher argued[1,2] that air, water, and earth constituted the true elements with air being "an instrument of mixing." Metals and stones were said by Becher to be composed of three earths: *terra vitrescible* (glassy earth—the substance of subterranean matter), *terra pinguis* (fatty earth—combustibility) and *terra fluida* (odor, volatility, and other subtle properties). Becher concluded that all substances that

For full description and English translation, see page iv.

FIGURE 3. ■ For full description, see page 6.

FIGURE 36. ■ For full description, see page 52.

FIGURE 43. ■ For full description, see page 66.

FIGURE 79. ■ For full description, see page 112.

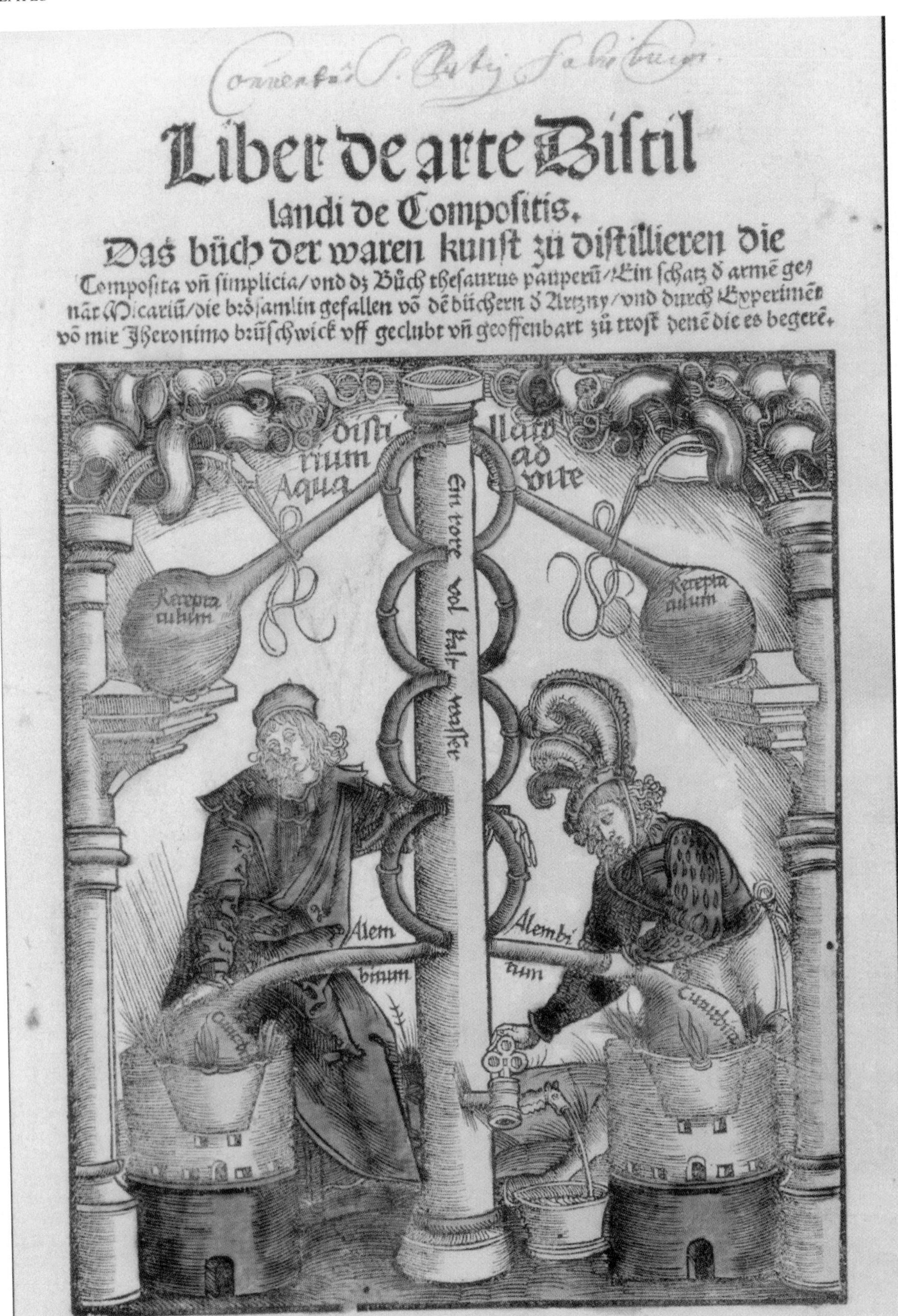

Liber de arte Distil
landi de Compositis.
Das büch der waren kunst zü distillieren die
Composita vñ simplicia/vnd dz Büch thesaurus pauperū/Ein schatz d armē ge/
nāt Micariū/die brösamlin gefallen vō dē büchern d Artzny/vnd durch Experimēt
vō mir Iheronimo brūschwick vff geclubt vñ geoffenbart zü trost denē die es begerē.

FIGURE 99. ■ For full description, see page 145.

FIGURE 164. ■ For full description, see page 242.

FIGURE 208. ■ For full description, see page 320.

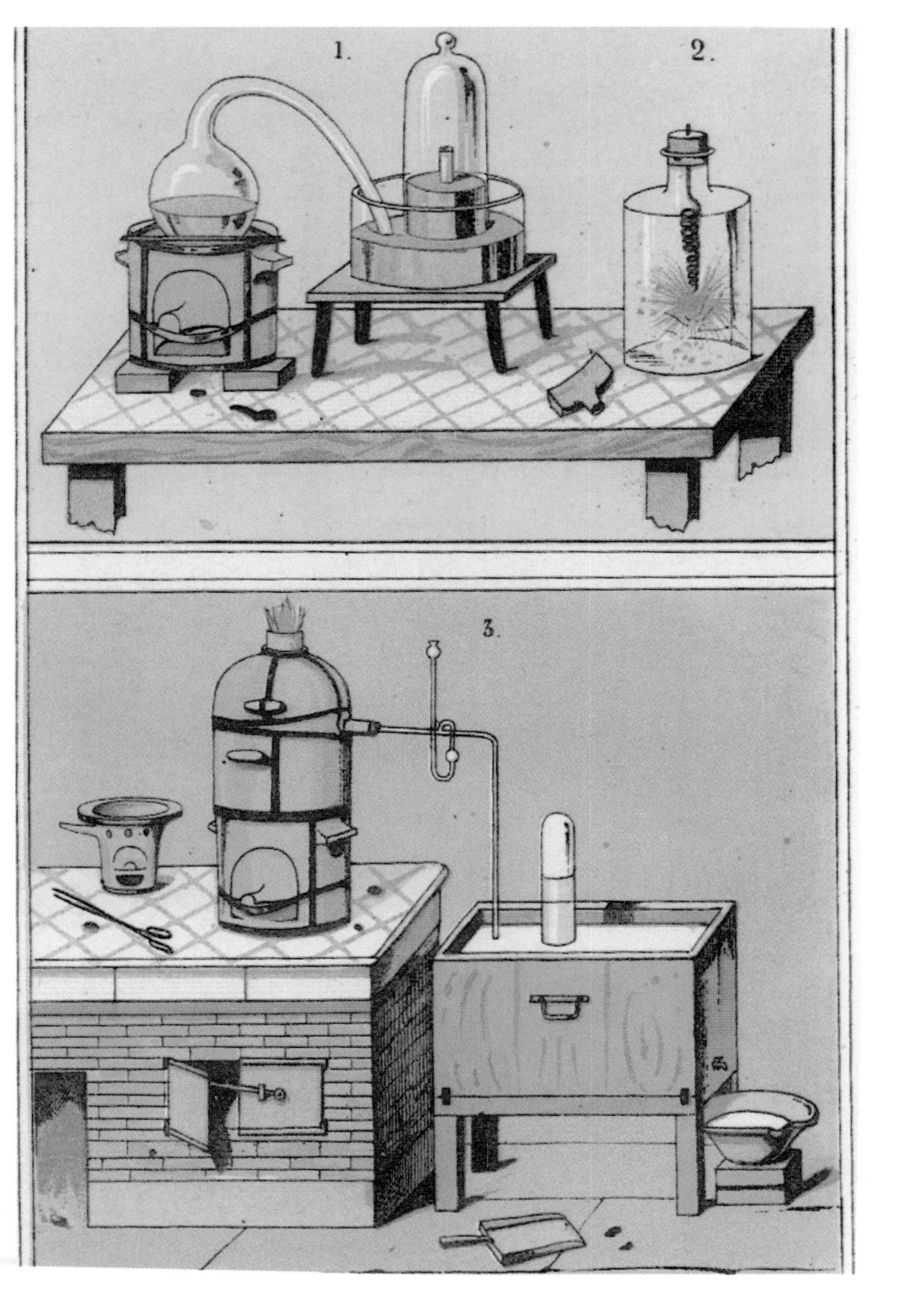

FIGURE 250. ■ For full description, see page 415.

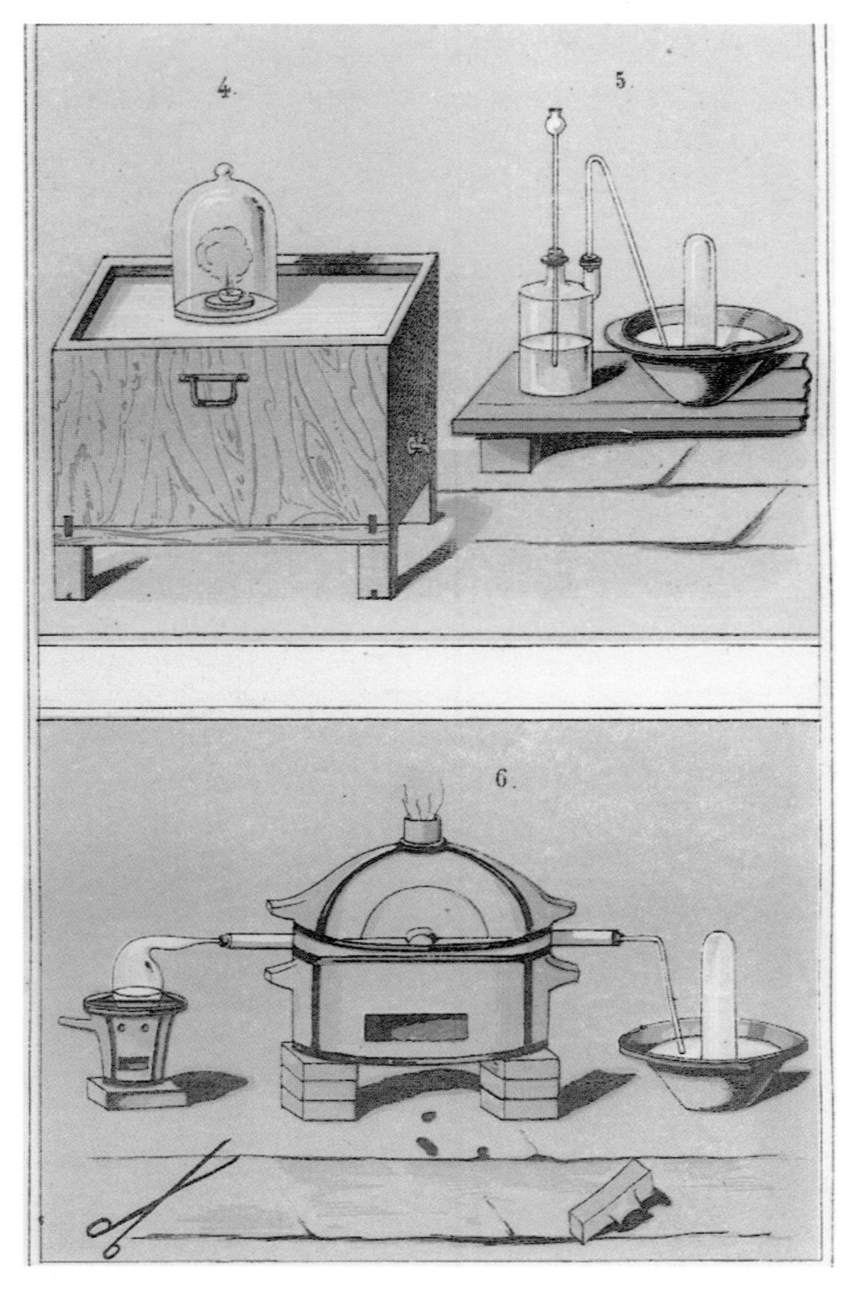

FIGURE 251. ■ For full description, see page 416.

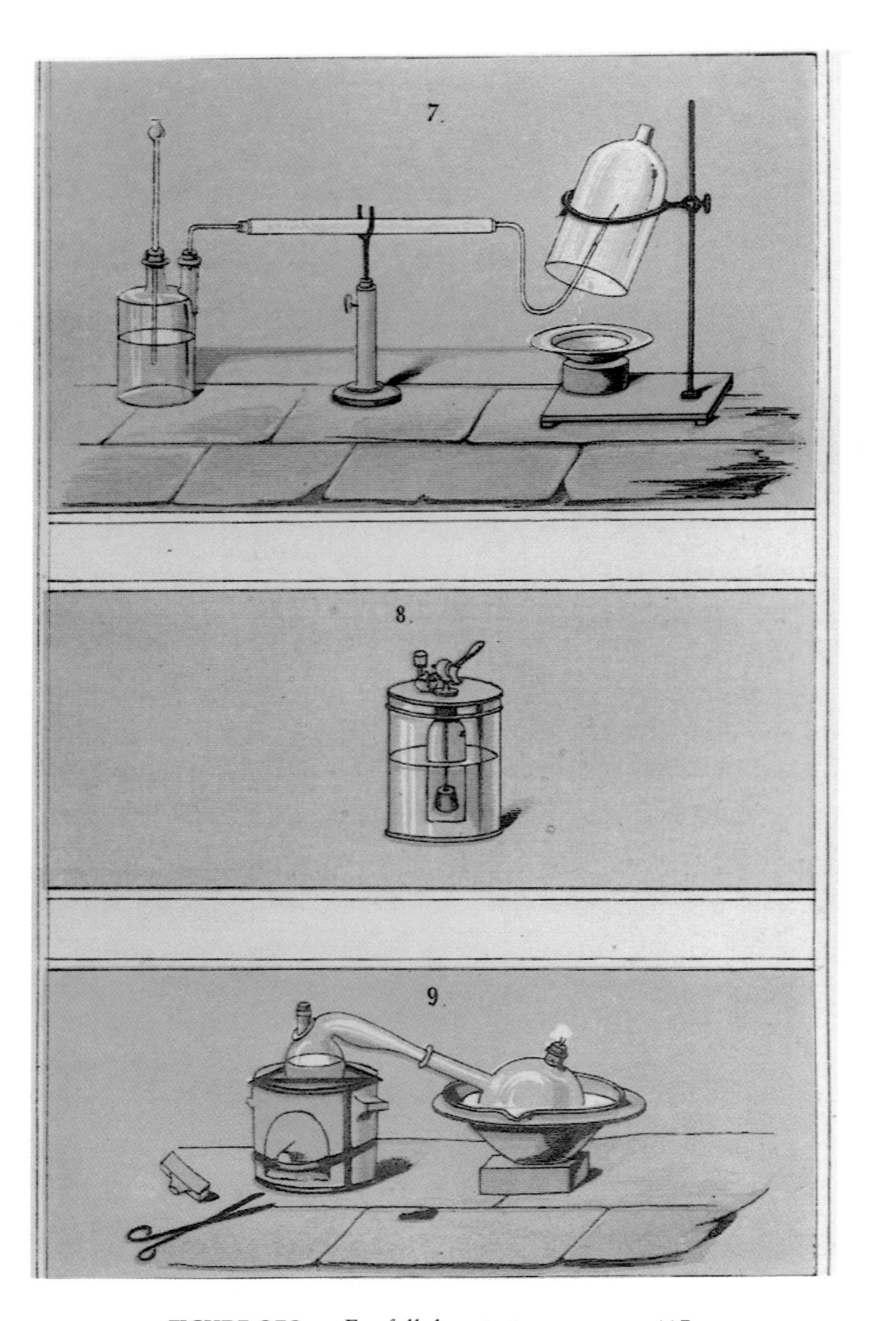

FIGURE 252. ■ For full description, see page 417.

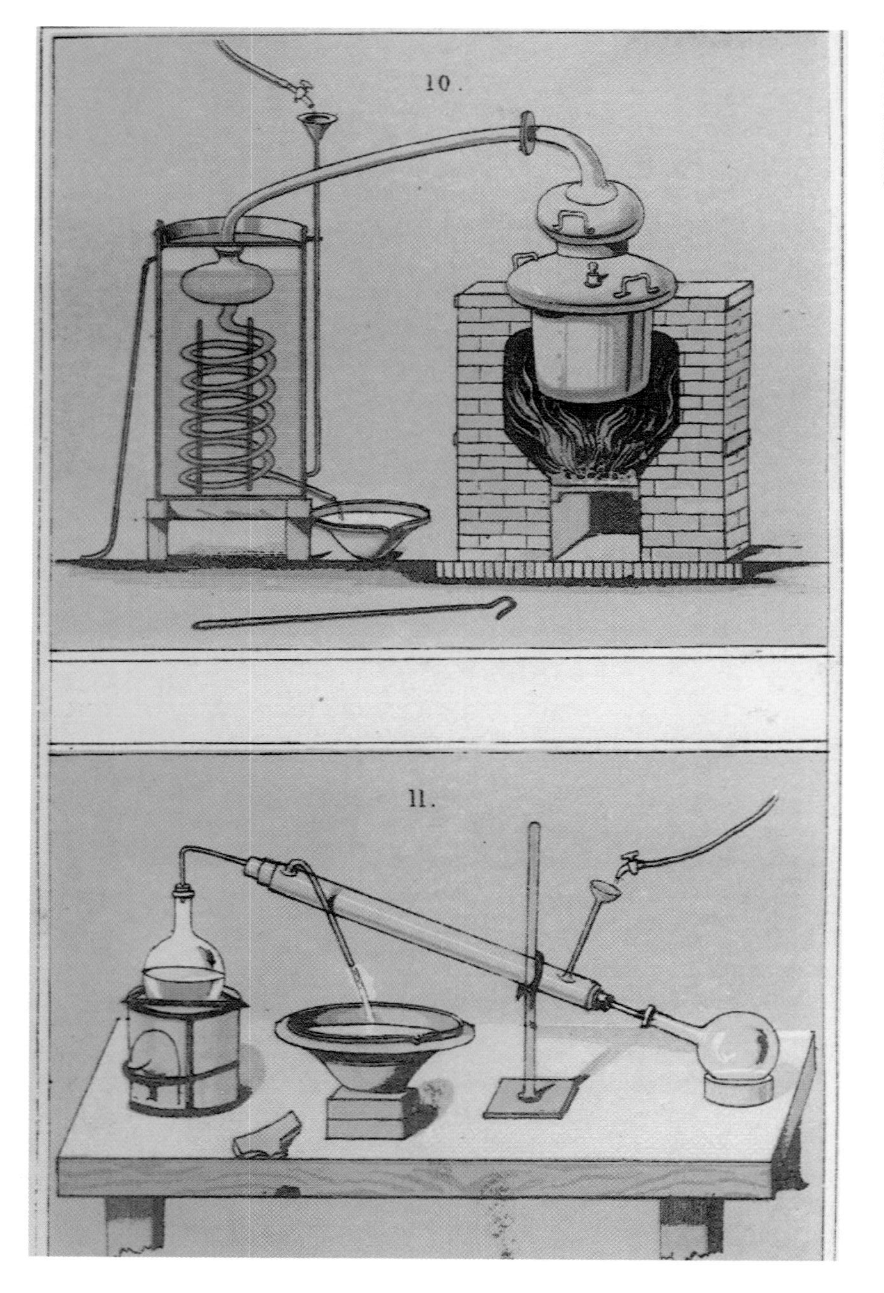

FIGURE 253. ■ For full description, see page 419.

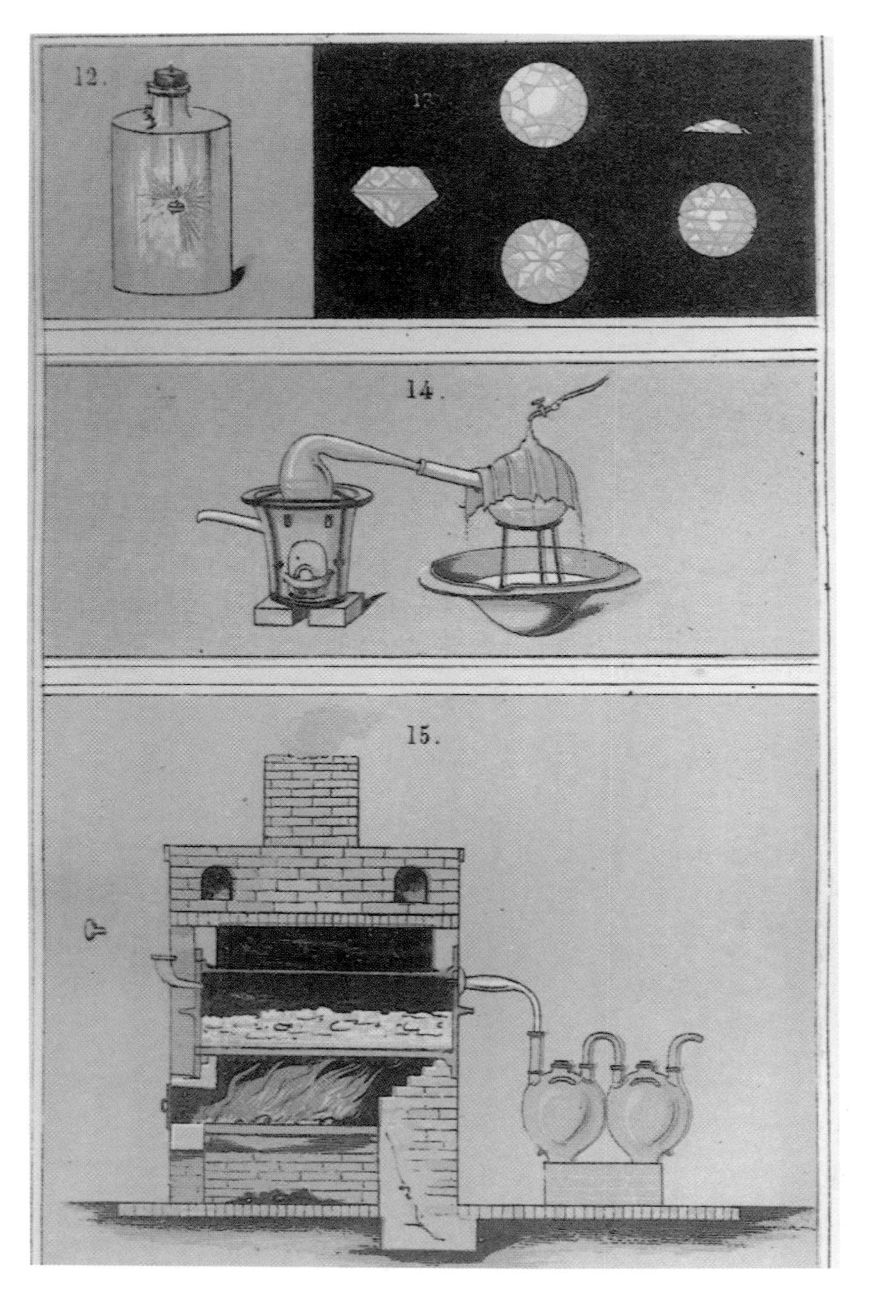

FIGURE 254. ■ For full description, see page 420.

FIGURE 255. ■ For full description, see page 421.

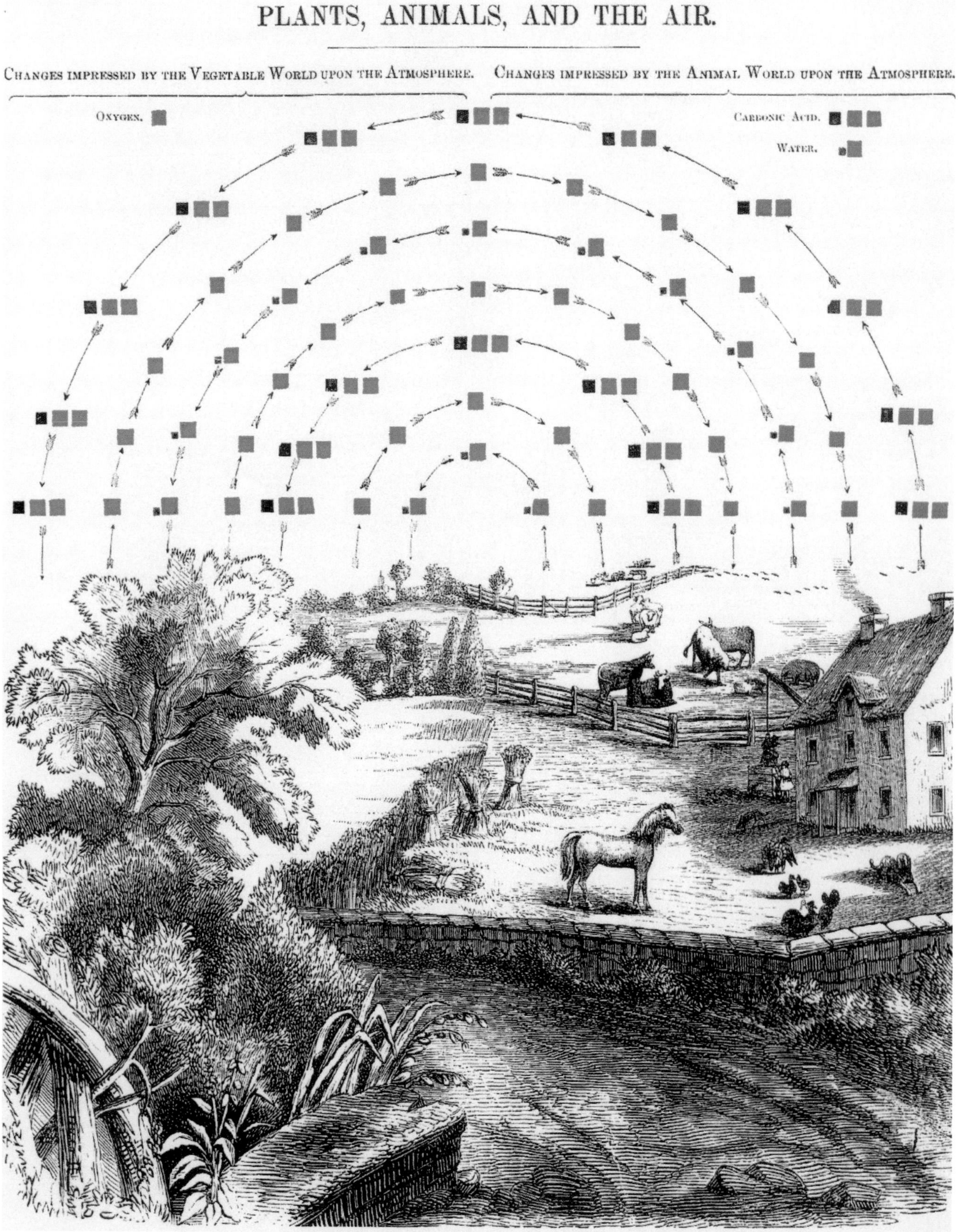

PLANTS, ANIMALS, AND THE AIR.

CHANGES IMPRESSED BY THE VEGETABLE WORLD UPON THE ATMOSPHERE. CHANGES IMPRESSED BY THE ANIMAL WORLD UPON THE ATMOSPHERE.

OXYGEN.

CARBONIC ACID.

WATER.

FIGURE 256. ■ For full description, see page 423.

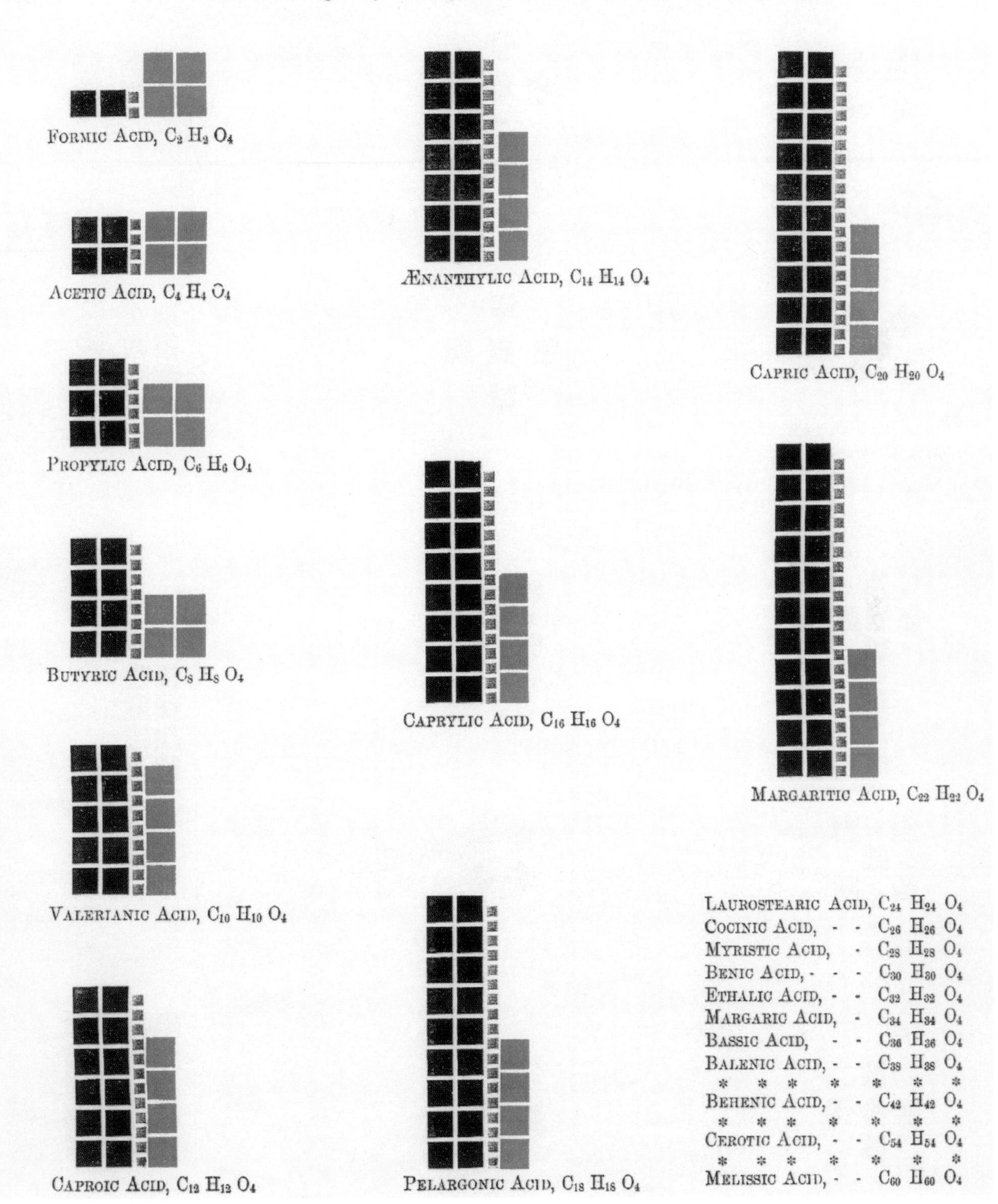

FIGURE 258. ■ For full description, see page 428.

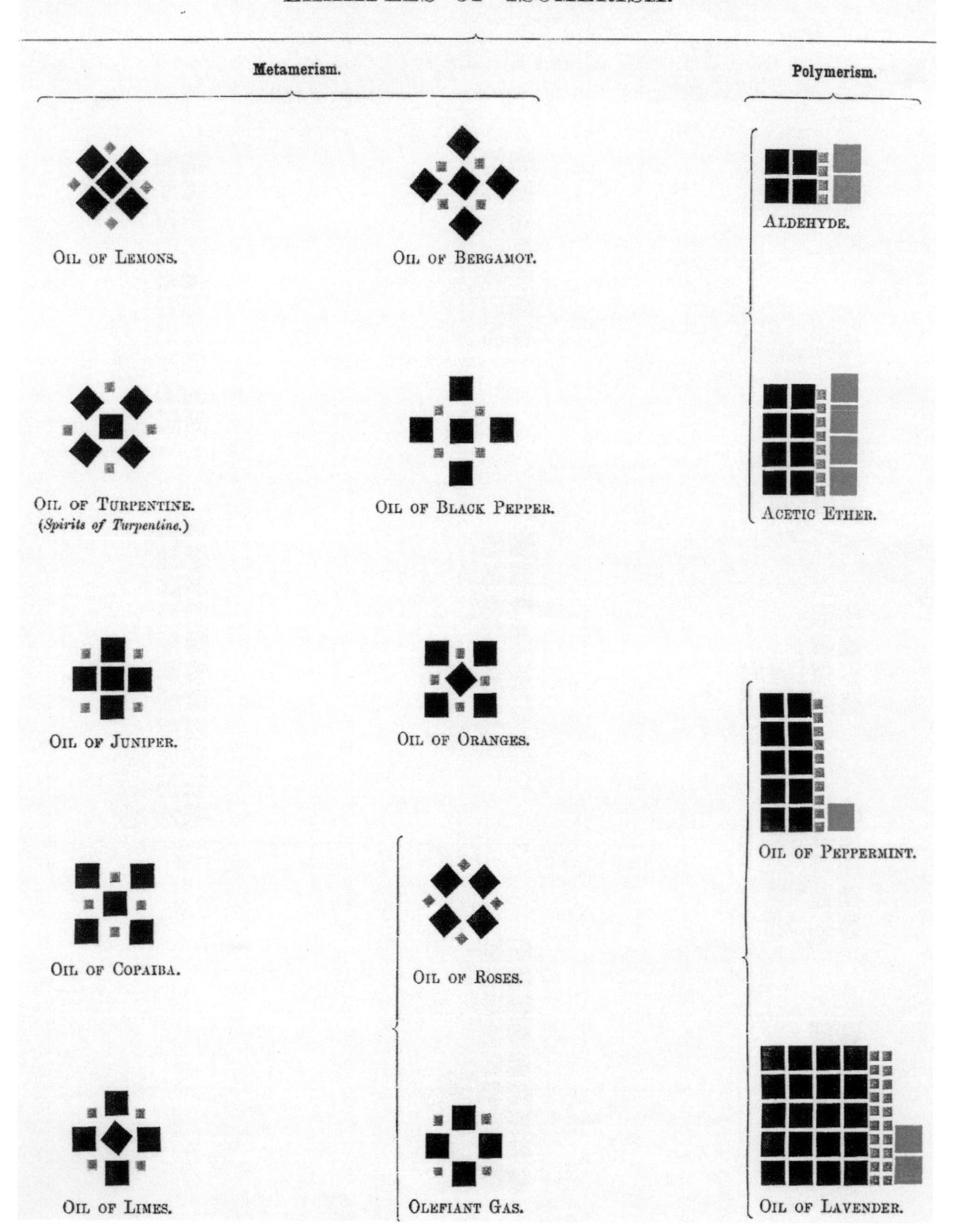

FIGURE 259. ■ For full description, see page 430.

ILLUSTRATION OF THE THEORY OF COMPOUND RADICALS.

ETHYLE.

OXIDE OF ETHYLE.
(Ether.)

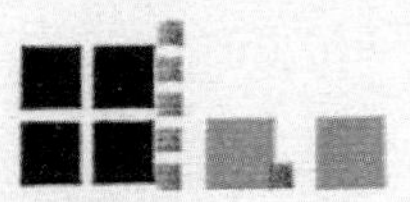

HYDRATED OXIDE OF ETHYLE.
(Alcohol.)

FORMYLE.

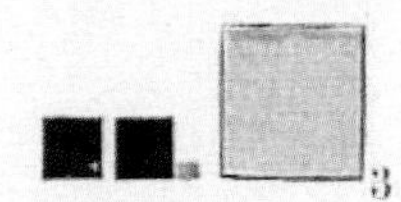

TERCHLORIDE OF FORMYLE.
(Chloroform.)

ACETYLE.

HYDRATED PROTOXIDE OF ACETYLE.
(Aldehyde.)

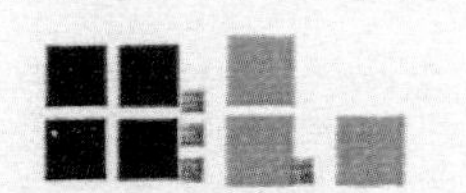

HYDRATED DEUTOXIDE OF ACETYLE.
(Acetylous Acid.)

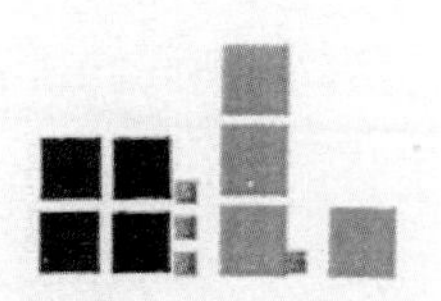

HYDRATED PEROXIDE OF ACETYLE.
(Acetylic Acid, Vinegar.)

THEORY OF CHEMICAL TYPES—DOCTRINE OF SUBSTITUTION.

AMMONIA.

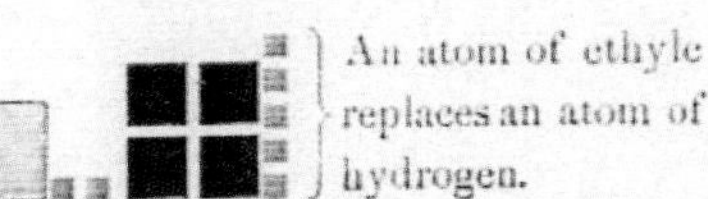

An atom of ethyle replaces an atom of hydrogen.

ETHYLAMINE.

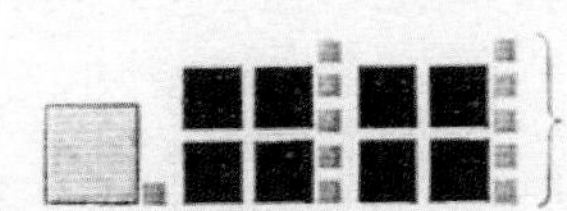

Two atoms of Ethyle replace two of hydrogen.

DIETHYLAMINE.

Three atoms of ethyle replace three of hydrogen.

TRIETHYLAMINE.

In this case, the hydrogen of the ammonia is replaced by three different compound radicals—ethyle, methyle, and propyle.

ETHYLO-METHYLO-PROPYLAMINE.

THEORY OF PAIRING—EXAMPLE OF COUPLED ACIDS.

BENZOYLE.
(Benzoic Acid.)

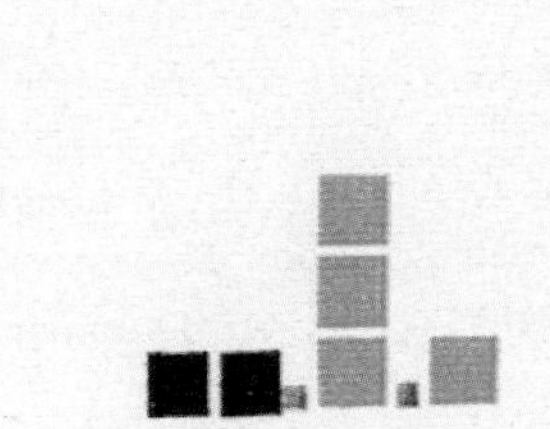

HYDRATED PEROXIDE OF FORMYLE.
(Formic Acid.)

FORMOBENZOIC ACID.

FIGURE 260. ■ For full description, see page 434.

FIGURE 278. ■ For full description, see page 463.

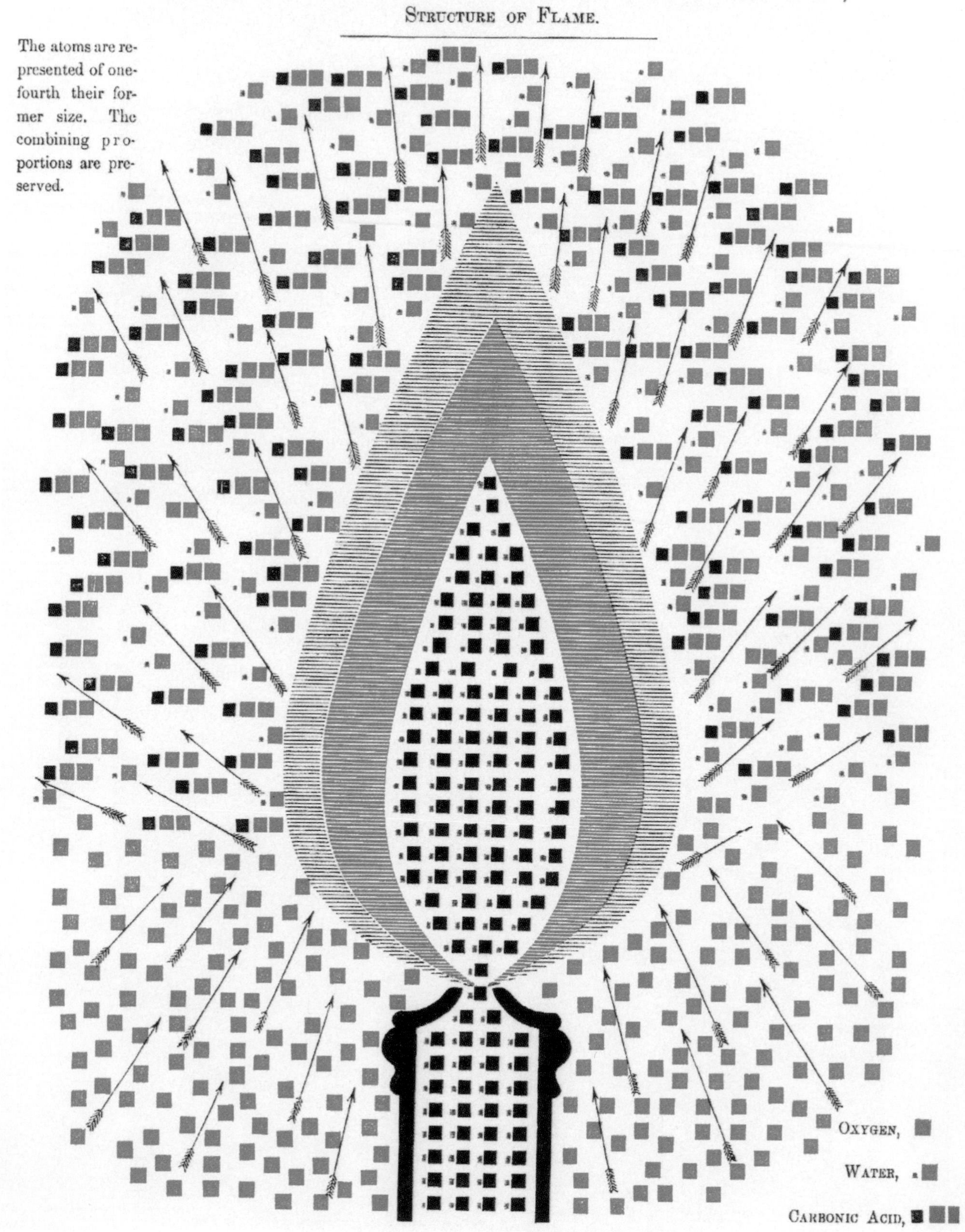

FIGURE 288. ■ For full description, see page 482.

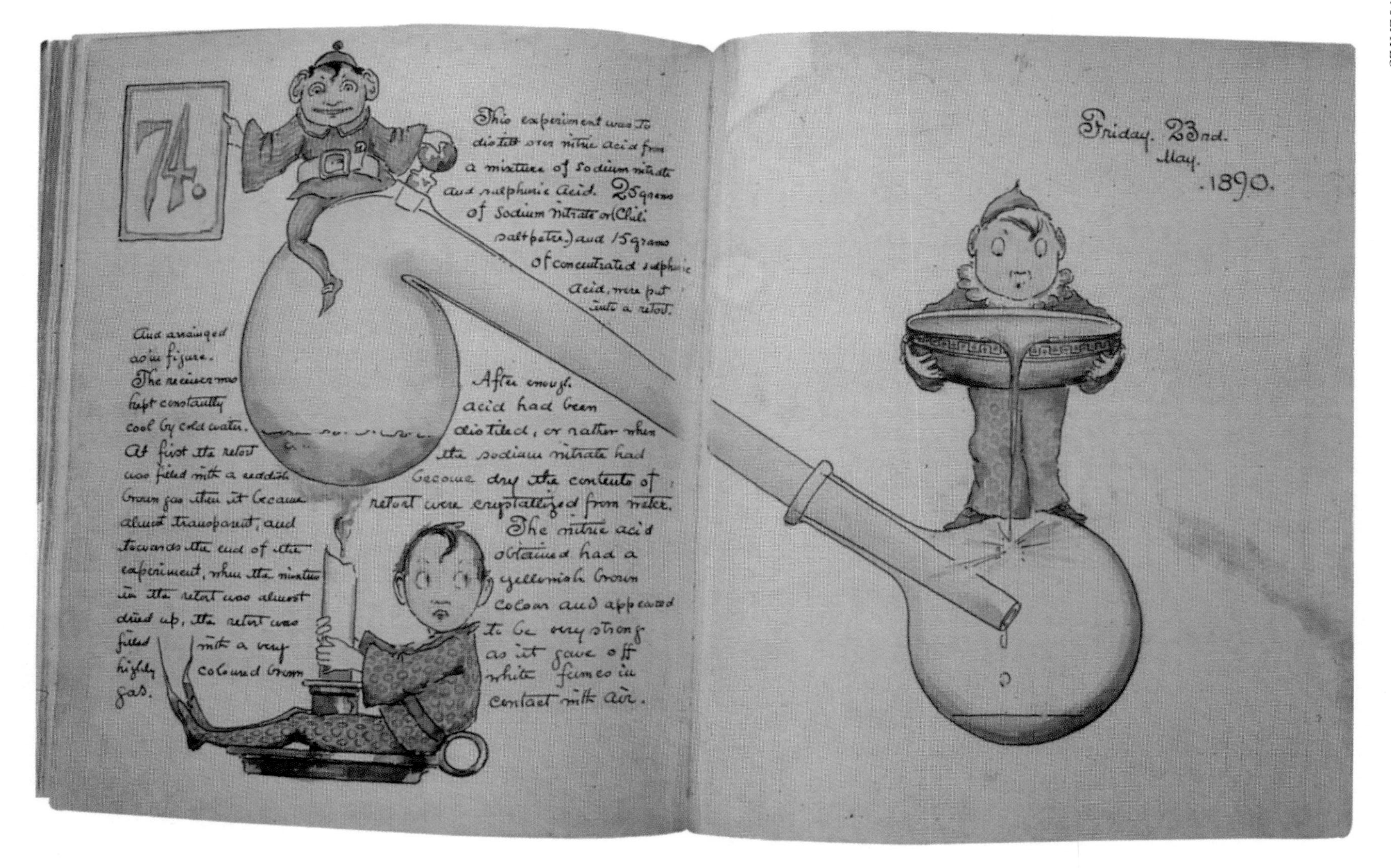

FIGURE 296. ■ For full description, see page 495.

FIGURE 304. ■ For full description, see page 521.

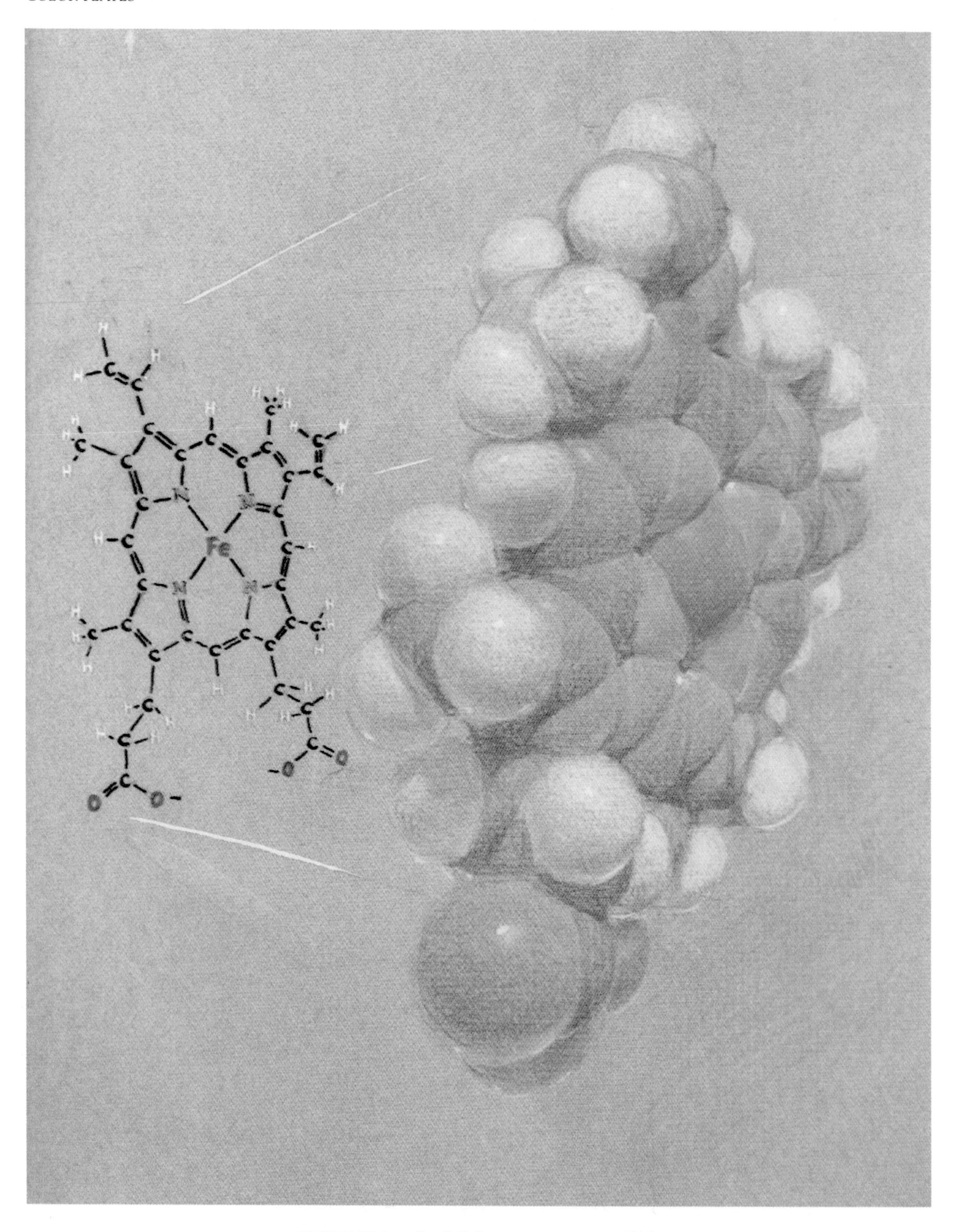

FIGURE 326. ■ For full description, see page 566.

Nanocar 1

FIGURE 341. ■ For full description, see page 594.

FIGURE 350. ■ For full description, see page 615.

FIGURE 351. ■ For full description, see page 616.

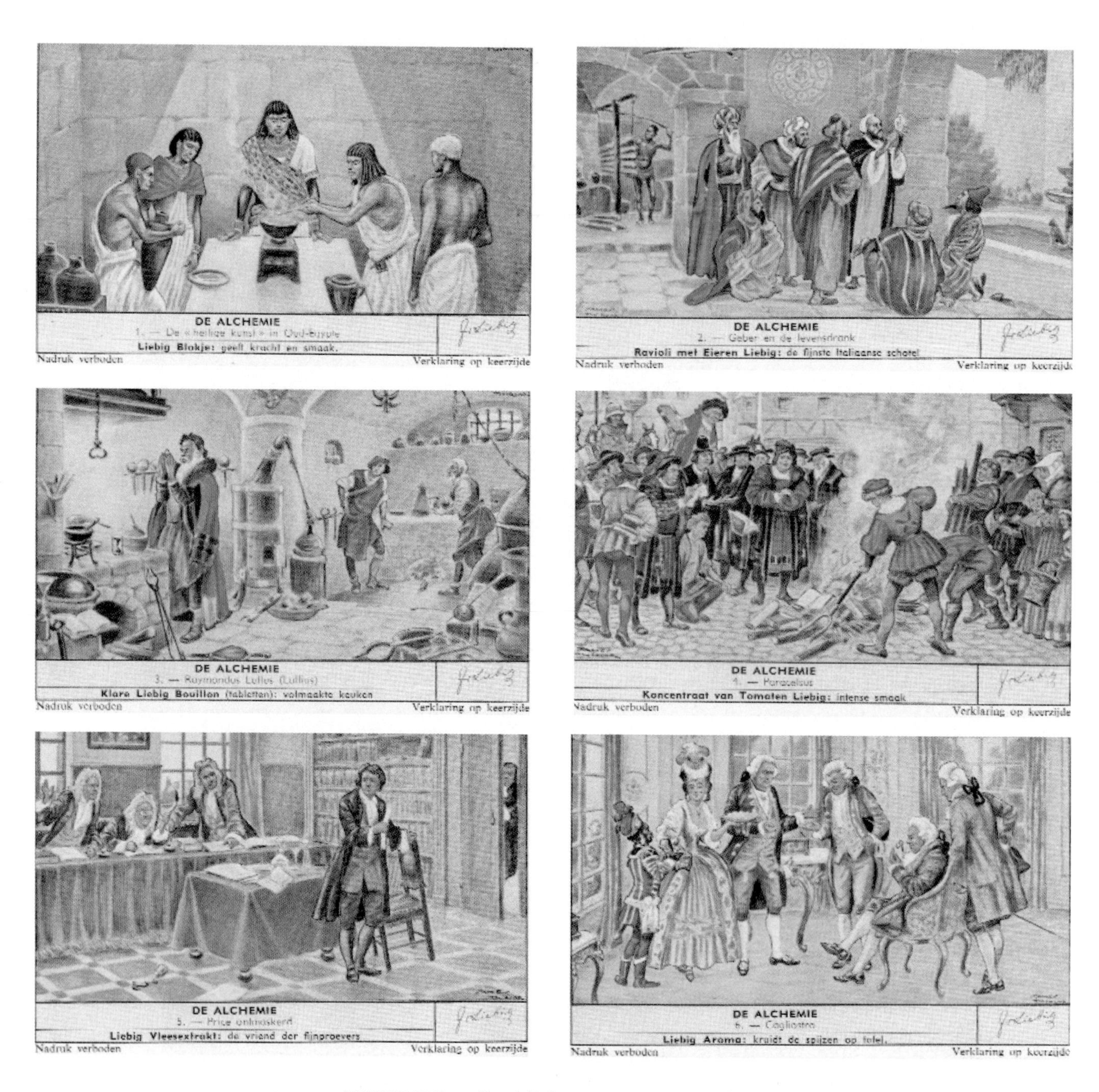

FIGURE 352. ■ For full description, see page 618.

burn, including metals such as tin and zinc, must contain *terra pinguis*, although Partington[1] notes that the fate of *terra pinguis* is never described by Becher. Indeed, Becher was well aware that metals *increased* in weight upon forming calxes. He attributed this to the accretion of fiery effluvia onto the metal as proposed earlier by Boyle.[11]

Becher's transformation[12] from a "purveyor of alchemical secrets" before 1655 into the trusted technical advisor to nobles and emperors over the next 15 years relied on his mastery of mechanics and science, amplified by his ability to market himself as the expert to consult in a world full of unscrupulous pretenders. Here is a segment from a letter to Emperor Leopold authored during the 1670s:[12]

> Above all, however, because Your Imperial Majesty has a desire to have some trials made in these things, it would be necessary to take into service a loyal, honest, and knowledgeable subject, whom Your Imperial Majesty could trust with the processes of such worthless vagabonds, and who, privately and secretly, could in silence faithfully work out the processes and report on them to Your Majesty. If this does not happen, Your Imperial Majesty will never get to the bottom of this, nor understand the nature of these things, but instead will always be duped by these scoundrels.

Now, who does Doktor Becher have in mind as the Emeror's expert? Becher biographer Pamela H. Smith notes ironically that "Becher's portrayal of the selfish and gain-seeking projector resembles remarkably his own situation ten years earlier."[12]

Becher's involvement with German mercantile interests led to his design of a factory for manufacture of glassware and crafts, complete with laboratory and library. His edicts, in 1677, against French imports into southern Germany failed and led to his brief imprisonment in 1678.[2] In 1678, he was also involved in an unsuccessful attempt to commercialize Henning Brand's technology for phosphorus manufacture.[13] However, another "syndicate," headed by Gottfried Wilhelm Leibniz, the famous mathematician who was "in cahoots" with the shadowy industrial spy Johann Daniel Kraft, succeeded in bringing Brand and his technology to Hanover. You will remember our earlier discussion, gentle reader, of Kraft's swift appropriation of Kunckel's hint about Brand's discovery of phosphorus, followed by his attempt to "corner the market" and "shut Kunckel out." Brand and his wife Margaretha, themselves, were not above using the threat of joining Becher to extort additional funds from the Leibniz "syndicate." Frau Brand's letter to Leibniz is not very subtle: "Dr. Becher is ever so honest and four weeks ago, as he left Hamburg for Amsterdam, he honored my husband with ninety-four Reichsthaler."[13]

Anticipating the likely failure of a large-scale demonstration of his process for extraction of gold from sea sand scheduled in Holland in March, 1680,[2] Becher abruptly left for London, without his family. Although Robert Boyle was one his patrons in England, Becher was unsuccessful in his entreaties to the Royal Society for election to its membership. He did, however, sell three of his portable furnaces (Figures 160 and 161) at 12 pounds each. One of these was purchased by Boyle. I confess that I would love to read a play or short story reconstructing the interplay between the aristocratic Englishman Boyle and the eight-year-younger Becher, very possibly "burned out" from his Continental intrigues, close scrapes

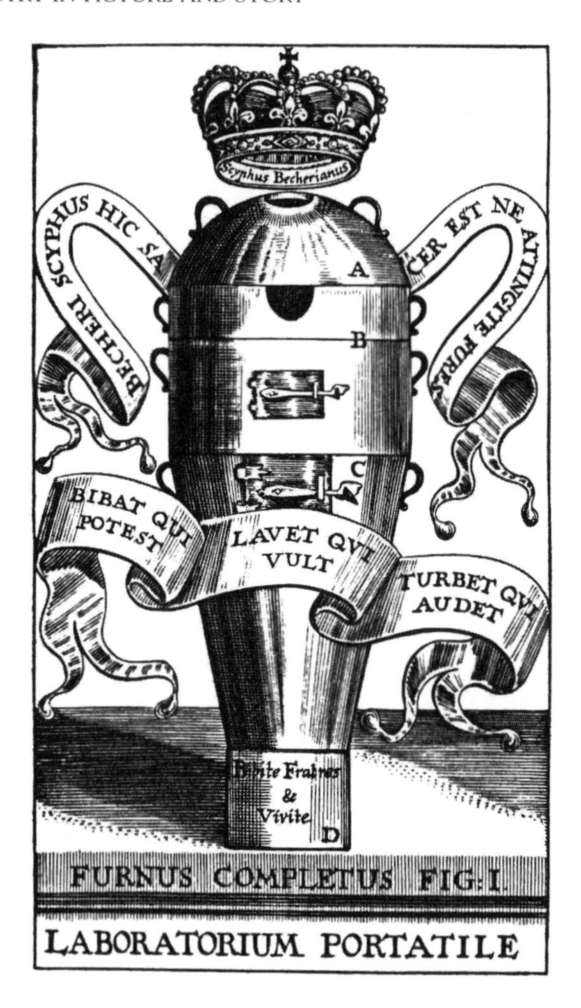

FIGURE 160. ■ Becher was nothing if not a venture capitalist, and here is the portable furnace he invented and marketed. One of these was purchased for twelve pounds by Robert Boyle. (From Becher's 1660 *Natur-Kündigung der Metallen;* from The Roy G. Neville Historical Chemical Library, a collection in the Othmer Library, CHF.)

with the law, and abandonment of his family, in the final two years of his brief, adventurous life.

Becher's theory was largely unrecognized in his time and was embraced some three decades later by the famous physician Georg Ernst Stahl (1660–1734).[14,15] Although one frequently reads that Stahl was a "student" or a "disciple" of Becher, it is worthwhile to note explicitly that Stahl had just attained the age of 22 and was studying medicine in Jena (Germany) when Becher died in London in 1682. Although Stahl's interest in chemistry started very early, there is no mention of the two having ever met. Nonetheless, Stahl's reading of Becher's work and adoption of his theory led him to republish Becher's *Physica subterranea* in 1703. It is Stahl who coined the term "phlogiston" and developed the concept that this essence of fire was lost to the surroundings during combustion and calcinations.[14] Partington notes that "Stahl was proud, morose, atrabilious, . . . quarreled with his senior colleague Hoffmann, to whom he owed his ap-

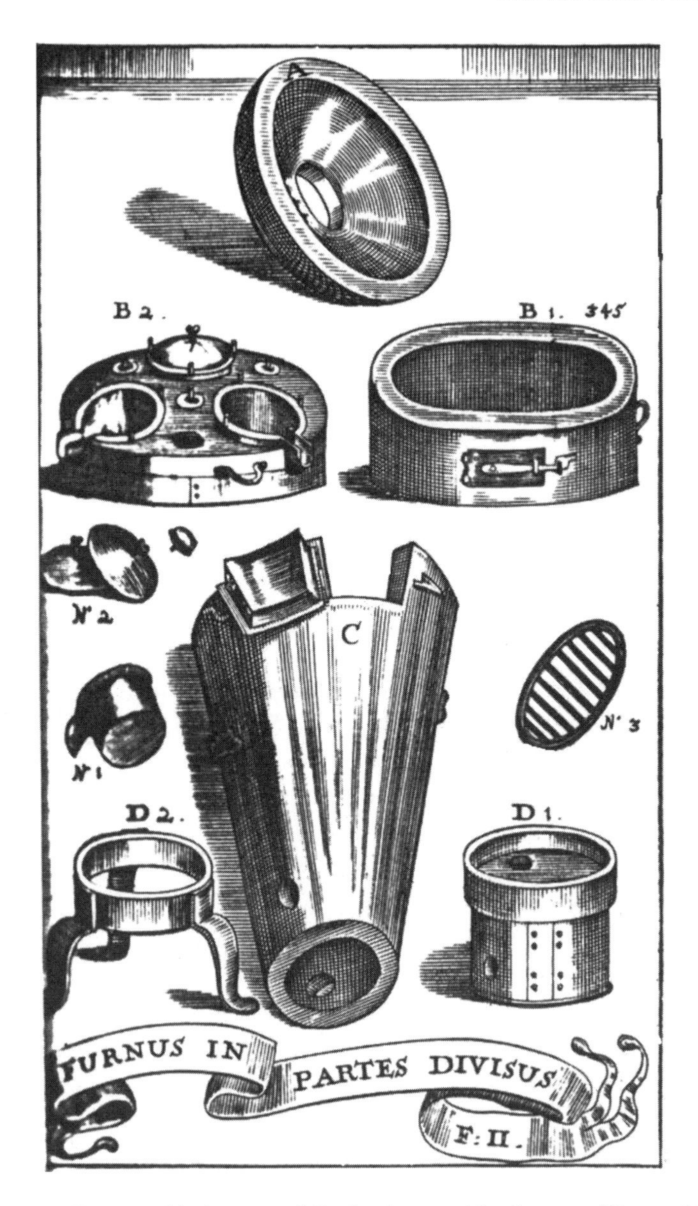

FIGURE 161. ■ Disassembled view of Becher's portable furnace (Figure 160). (From Becher's 1660 *Natur-Kündigung der Metallen*, fromThe Roy G. Neville Historical Chemical Library, a collection in the Othmer Library, CHF).

pointment at Halle, . . . rarely answered letters, . . . showed contempt for all who differed from his views and reacted violently to criticism. These qualities . . . greatly enhanced his reputation."[14,15]

1. J.R. Partington, A *History of Chemistry*, MacMillan and Co. Inc., London, 1962, Vol. 2, pp. 637–652. I am grateful to The Roy G. Neville Historical Chemical Library (California) for supplying the three figures shown from Becher's 1660 *Kündigung der Metallen*.
2. C.C. Gillispie (ed.), *Dictionary of Scientific Biography*, Charles Scribner's Sons, New York, 1970, Vol. I, pp. 548–551.
3. P.H. Smith, *The Business of Alchemy—Science and Culture in the Holy Roman Empire*, Princeton

University Press, Princeton, 1994. It is interesting that the international financier George Soros has written a book titled *The Alchemy of Finance*, Simon & Schuster, New York, 1987. Soros, a protégé of philosopher of science Karl Popper, employs finance and philanthropy to promote open societies.
4. J.-P. Poirier, *Lavoisier—Chemist, Biologist, Economist*, University of Pennsylvania Press, Philadelphia, 1996.
5. Smith, op. cit., p. 18.
6. Partington, op. cit., p. 638.
7. *The New Encyclopedia Britannica*, Encyclopedia Britannica, Inc., Chicago, 1986, Vol. 6, pp. 21–22.
8. *The New Encyclopedia Britannica*, op. cit., Vol. 11, p. 711.
9. Smith, op. cit., pp. 16–17.
10. Partington (see above) also cites an earlier 1667 version of this work.
11. Partington, op. cit., p. 650.
12. Smith, op. cit., pp. 76–80.
13. Smith, op. cit., pp. 248–255.
14. Partington, op. cit., pp. 653–686.
15. C.C. Gillispie, op. cit., 1975, Vol. XII, pp. 599–606.

PHLOGISTON: CHEMISTRY'S FIRST COMPREHENSIVE SCIENTIFIC THEORY

The initial concept of phlogiston was due to Johann Joachim Becher (1635–1682) and has clear alchemical roots.[1] For Becher, the important elements were Water and three Earthy Principles. (He regarded Air and Fire to be agents of chemical change rather than elements in the chemical sense). His three Earthy Principles corresponded very roughly to the Paracelsian "salt," "mercury," and "sulfur." This last "sulfur-like" Earthy Principle was termed *Terra Pinguis* (fatty earth) by Becher and was said to be present in combustible matter and released upon combustion. It was this principle that Georg Ernst Stahl (1660–1734) later equated to his phlogiston.

Becher was aware, as was Boyle (see effluviums discussion on pp. 248–251), that calxes were heavier than the corresponding metals. He too attributed these observations to igneous ("fiery") particles small enough to move through glass and join the metal inside.

Becher was an argumentative man who described himself as follows[2]:

> . . . one to whom neither a gorgeous home, nor security of occupation, nor Fame nor health appeal, for me rather my chemicals amid the smoke, soot and flame of coals blown by the bellows. Stronger than Hercules, I work forever in an Augean stable, blind almost from the furnace glare, my breathing affected by the vapour of mercury. I am another Mithridates, saturated with poison. Deprived of the esteem and company of others, a beggar in things material, in things of the mind I am a Croesus. Yet among all these evils I seem to live so happily that I would die rather than change places with a Persian King.

Clearly, Becher was a truly "hard-core," "gung-ho" chemist. Happily, we modern chemists do not have to recite this pledge as our professional oath.

OEDIPVS
CHIMICUS.

ŒDIPUS.

SPHINX.

FRANCOFVRTI,
1664.

(a)

FIGURE 162. ■ (a) The title page of *Oedipus Chimicus* (perhaps only the Sphinx knows the riddle of the Stone) (courtesy of Jeremy Norman & Co., from Catalogue 5, 1978). (b) On the next page is the title page of Johann Joachim Becher's *Physicae Subterranae Libri Duo* (Frankfurt, 1681). The first edition, published in 1669, contained Becher's view of matter—the Phlogiston Theory, later modified by Georg Ernst Stahl [see Fig. 1 for the frontis from the final (1738) edition of this book] (courtesy of The Beinecke Rare Book and Manuscript Library, Yale University). *Illustration continued on following page.*

DOCT. JOH. JOACHIMI BECH
ERI PISICA SVBTERRANEA
TRIPLEX NATURÆ OFFICINA
GLOBUS TERR-AQUA-AEREUS
AER
Vegetabilis
Animalis
AQUA
TERRA
Mineralis

(b)

Figure 162(a) is from the book *Oedipus Chymicus*[3] (1664) and it depicts Oedipus solving the riddle posed by the Sphinx. It is thought to represent the chemist solving the alchemical riddle and is consistent with Becher's firm belief in transmutation. Of course, once Oedipus relieved Thebes of the dreaded Sphinx he was made King but other disasters followed. Perhaps personal disaster would have also afflicted the discoverer of the Philosopher's Stone or the Elixir: King Midas comes to mind. Figure 162(b) is from the 1681 edition[4] of this book. The last edition was published by Stahl in 1738 (see Figure 1).

---

1. H.M. Leicester and H.S. Klickstein, *A Source Book In Chemistry 1400–1900*, McGraw-Hill, New York, 1952, pp. 55–58.
2. J.R. Partington, *A History of Chemistry*, MacMillan, London, 1961, Vol. 2, pp. 637–652 (quotation on p. 639).
3. This figure, from the book *Oedipus Chymicus*, *is from Science and Technology: Catalog 5* Jeremy Norman, San Francisco, 1978, p. 13. Courtesy of Jeremy Norman & Company, Inc.
4. I. MacPhail, *Alchemy and the Occult*, Yale University Library, New Haven, 1968, Vol. 2, pp. 472–476.

THE "MODERN" PHLOGISTON CONCEPT

The Phlogiston concept was chemistry's first truly unifying theory and was developed in its useful form early in the eighteenth century by Georg Ernst Stahl (1660–1734), an irrascible, egotistical, and rather unpleasant chemist and physician. It was said of him that "Stahl seems to have regarded his ideas at least in part due to divine inspiration and the common herd could have no inkling of them."[1] ". . . [H]is lectures were dry and intentionally difficult; few of his students understood them."[1] Stahl attacked adversaries vehemently and while he clearly acknowledged his debt to Becher (Stahl reissued Becher's *Physicae Subterranae*; see Figure 1), he also found much to criticize.

Figure 163 is the title page of Stahl's famous 1723 textbook. It summarizes Stahl's views as early as 1684. Over a half-century later this book was ceremoniously burned by Madame Lavoisier dressed in the outfit of a Priestess (see our later discussion). In the sixteenth century Paracelsus was said to have burned texts of Galen and Avicenna—an earlier act in the theatre of invective.

Phlogiston was postulated to be present in substances that could burn as well as in metals, which were known to form calxes. The concept works like this:

Charcoal (has Phlogiston) ⟶ Residue + Phlogiston

Metal (has Phlogiston) ⟶ Calx + Phlogiston

Aside from relating these two seemingly very different kinds of chemistry, it explained the well-known ability to convert calxes into metals by heating with charcoal:

Metal calx + Charcoal ⟶ Metal + Ashy residue

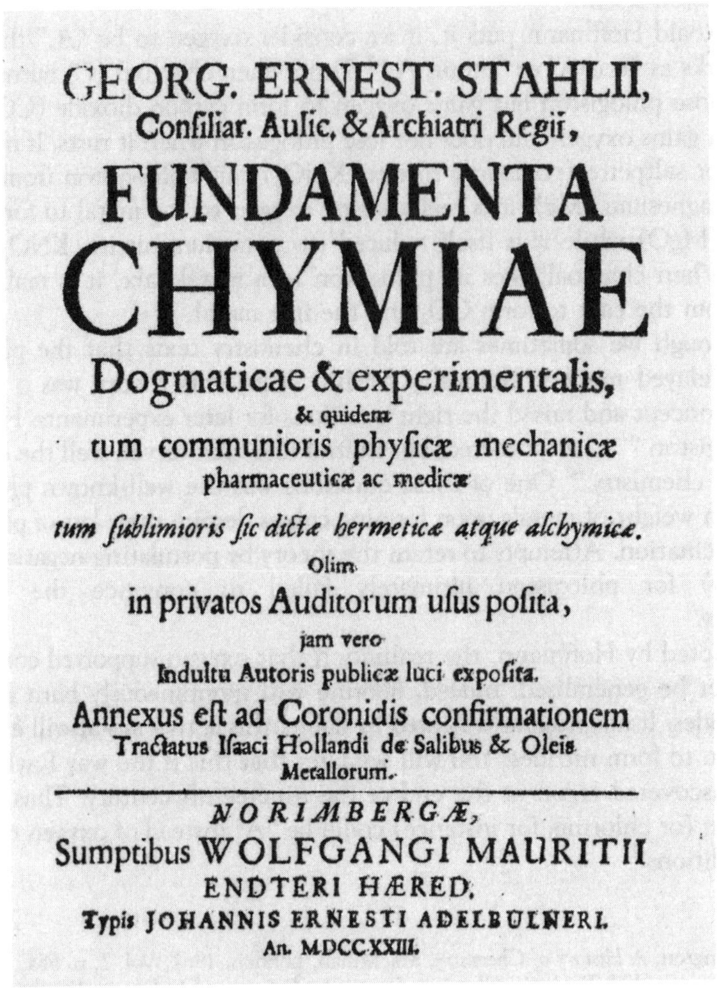

GEORG. ERNEST. STAHLII,
Confiliar. Aulic. & Archiatri Regii,
FUNDAMENTA
CHYMIAE
Dogmaticae & experimentalis,
& quidem
tum communioris phyficæ mechanicæ
pharmaceuticæ ac medicæ
tum fublimioris fic dictæ hermeticæ atque alchymicæ.
Olim
in privatos Auditorum ufus pofita,
jam vero
Indultu Autoris publicæ luci expofita.
Annexus eft ad Coronidis confirmationem
Tractatus Ifaaci Hollandi de Salibus & Oleis
Metallorum.

NORIMBERGÆ,
Sumptibus WOLFGANGI MAURITII
ENDTERI HÆRED.
Typis JOHANNIS ERNESTI ADELBULNERI.
An. MDCCXXIII.

FIGURE 163. ■ Title page of text by Georg Ernst Stahl who formulated the "modern" phlogiston theory. Madame Lavoisier, dressed as a Priestess, ceremoniously burned this book to mark the publication of Lavoisier's *Traité Elémentaire de Chimie* in 1789.

where charcoal and metal have phlogiston. Similarly, combustion of phosphorus in air formed phosphoric acid; and of sulfur, sulfuric acid. Heating of these acids with charcoal produced elemental phosphorus and sulfur, respectively.

This powerful and conceptually useful theory held sway for about a century. When Priestley discovered oxygen in 1774, he called it *dephlogisticated air* since it supported combustion, which allowed it to attract phlogiston vigorously from substances such as charcoal or iron. Nitrogen was initially called *phlogisticated air* because it did not support combustion and was obviously saturated with phlogiston. When Cavendish discovered hydrogen gas in 1766 and found that its density was less than one-tenth that of air, he thought this flammable gas was phlogiston itself.

As Roald Hoffmann puts it, if we consider oxygen to be "A," then phlogiston works as "not A" or "minus A."[2] Thus, when charcoal (C) burns, carbon does not lose phlogiston but gains oxygen to form carbon dioxide (CO_2). Similarly, iron gains oxygen and does not lose phlogiston when it rusts. If nitric acid (HNO_3) or saltpetre (potassium nitrate, KNO_3) gains phlogiston from a metal such as magnesium (Mg), it is really losing oxygen to the metal to form a calx or oxide (MgO) while it is itself reduced (to potassium nitrite, KNO_2, for example). When charcoal loses its phlogiston to a metal calx, it is really taking oxygen from the calx to form CO_2 and the free metal.

Although we sometimes are told in chemistry texts that the phlogiston concept delayed modern chemistry by 100 years, this theory was a powerful unifying concept and raised the right questions for later experiments. Hoffmann calls phlogiston ". . . an incorrect but fruitful idea that served well the emerging science of chemistry."[2] One of these questions was the well-known problem of the gain in weight of metals upon forming calxes despite their *loss* of phlogiston during calcination. Attempts to retain the theory by postulating negative weight (buoyancy) for phlogiston ultimately failed to convince the scientific community.

As noted by Hoffmann, the realization that oxygen supported combustion would later be generalized. Indeed, fluorine will spontaneously burn metals to form fluorides. If magnesium is heated by flame, this active metal will even burn in nitrogen to form nitrides. You will see later that this is the way Rayleigh and Ramsay discovered argon at the end of the nineteenth century. Thus, fluorine or nitrogen (or chlorine, for instance) could be "A" instead of oxygen under the right conditions.[2]

---

1. J.R. Partington, *A History of Chemistry*, MacMillan, London, 1961, Vol. 2, p. 655.
2. R. Hoffmann and V. Torrence, *Chemistry Imagined—Reflections On Science*, Smithsonian Institution Press, Washington, D.C., 1993, pp. 82–85.

THE HUMBLE GIFT OF CHARCOAL

Charcoal is hardly an awe-inspiring substance, yet it has played a critical role in human history. As noted in an earlier essay, charcoal's ability to strip phosphate of its oxygens at high temperature provided the surprised Brant with elemental (white) phosphorus. In modern terms we understand that the driving forces are thermodynamic. Carbon monoxide has the strongest covalent chemical bond in nature.[1] Energetically, the creation of strong bonds at the sacrifice of weaker bonds is a powerful driver of chemical reactions. Moreover, we understand that entropy (the degree of disorder in a system) can be a strong driving force as well. Production of a gas (carbon monoxide), which increases disorder and therefore entropy, is a potent driving force in reductions by charcoal.[2] Escape of the gas into the open atmosphere prevents recombination of carbon monoxide's lone oxygen to re-form its "parent" substance, and this further drives the reaction. Through the ages charcoal has been heated with various metal calxes (powdery oxides) to reduce them to the corresponding pure metals.[3] The by-product, carbon monoxide, simply disappears into thin air. What a magical effect! Indeed, freshly made activated charcoal is also an incredibly powerful absorbent that can remove odors, decolorize liquids, and even make red wine look like water. What a *terrible* effect!

Figure 164 shows an oil painting on porcelain by an L. Sturm and given the title "The Alchemist."[4] The artist is likely to be a porcelain painter in Bamberg and Munich named Ludwig Sturm (1844–1926).[5] The title *could* be an apt one in the sense that the central figure appears to be performing an operation for the benefit of two wealthy clients in eighteenth-century dress. However, alchemy had reached its apex during the midseventeenth century, and by the mid-eighteenth century chemistry was firmly on its way to becoming a precise science.

FIGURE 164. ■ Oil-on-porcelain painting by artist L. Sturm, very likely the porcelain painter Ludwig Sturm (source: Dr. Alfred Bader). Although the painting is titled "The Alchemist," rational chemistry is occurring. The key to the picture is the pan of charcoal. It is likely that a metal oxide is being reduced to the metal by charcoal in the red-hot crucible. See color plates. (I am grateful to the Art Museum of the State University of New York at Binghamton for permission to use this image.)

The gullible wealthy had also "wised up" by this period. What *is* happening in this painting? Clearly, the chemist's assistant has provided a red-hot crucible using his furnace and bellows and the chemist is adding a powdery calx to the crucible. The key to the figure is the pan of powdered charcoal that appears just in front of the wealthy client on the right. One might imagine adding the charcoal to the crucible just prior to addition of the calx. Were the calx black oxide of copper, the result would be particularly thrilling—a hissing of gas from the dark mass and appearance of a reddish, golden drop of liquid metal that would soon solidify into copper.

A phlogistonist's view would be somewhat different. Charcoal is "fatty" and loaded with phlogiston—the essence of fire. The copper calx would actually be copper devoid of phlogiston. The chemical operation shown would reduce the calx back to the metal by returning its full complement of phlogiston. The ashes remaining in the crucible would then be charcoal devoid of its phlogiston. Now was the central figure an alchemist ("puffer" or charlatan?) seducing two well-heeled investors who wished to multiply their fortunes? I suspect not. He was probably a competent early chemist or metallurgist seeking support from some wealthy investors.

---

1. Professor Joel F. Liebman, personal communication.
2. The thermodynamics of this reaction can be calculated using standard enthalpy and entropy of formation data in M.W. Chase, Jr., *NIST-JANAF Thermochemical Tables*, fourth edition, *Journal of Physical and Chemical Reference Data, Monograph* 9, 1998 (see also the NIST Website at *http://nist.gov*). For a metal having weaker affinity for oxygen, such as in mercury(II) oxide, both enthalpy and entropy favor this reaction. For a metal having somewhat stronger affinity, as in copper(II) oxide, enthalpy disfavors the reaction but is overwhelmed by entropy.
3. The most important pyrometallurgical operation is the reduction of the iron ores hematite (Fe_2O_3) and magnetite (Fe_3O_4). Although carbon (in the form of coke) is added to these ores at high heat, the chemistry is more complex than simply passing oxygen from iron to carbon. The blast furnace housing this operation provides hot air. Oxygen in the air forms carbon monoxide from the coke and it is the carbon monoxide that reduces the iron ores by stripping them of oxygen to produce carbon dioxide. This is complemented by water also present in the blast air that likewise converts coke to carbon monoxide. Water's by-product, hydrogen, similarly strips the iron ore of oxygen to produce water. (See T.L. Brown, H.E. LeMay, Jr., and B.E. Bursten, *Chemistry—the Central Science*, seventh edition, Prentice-Hall, Upper Saddle River, NJ, 1997, pp. 872–875.
4. I am grateful to Dr. Lynn Gamwell, Director, University Art Museum, State University of New York at Binghamton for kindly providing this image from the museum collection.
5. H. Vollmer (ed.), *Allgemeines Lexikon Der Bilden den Künster von der Antike Bis Zur Gegenwart*, Verlag Von E.A. Seeman, Leipzig, 1938, Vol. XXXII, p. 257. I am grateful to Dr. Alfred Bader for commenting on this painting and making me aware of the artist.

BEAUTIFUL SEVENTEENTH-CENTURY CHEMISTRY TEXTS

Following Libavius' *Alchymia*, a series of useful and beautifully illustrated textbooks appeared throughout the seventeenth century.[1–3] Let us begin the Beguin—Jean Beguin's *Tyrocinium Chymicum* ("The Chemical Beginner") was first published in 1610. It went through more than 50 editions before the last in 1669. Figure 165 shows the title page depicting an alchemical Cupid for the 1660 edition. Nicolas Le Fèvre first published his *Traicté de la Chimie* in 1660. The second French edition appeared in 1669, English editions in 1664 and 1670, and the last German edition in 1688. Figures 121 and 133 are from the 1670 edition of Le Fèvre. Christophe Glaser first published his *Traité de la Chymie* in 1663. An English edition was published in 1677. German editions (see Figure 166) were published through 1710. Nicolas Lemery published an incredibly successful text (Lemery was Glaser's student). The first French edition of *Cours de Chimie* was published in Paris in 1675 (Figure 167 is from the 1686 Paris edition). The final French edition was published in 1756—an incredible 81-year run! (See discussions of early glassware in figures.) Figure 168 is from Moses Charas' *Royal Pharmacopoea* (London, 1678).

Figure 169 is from the first English edition of Herman Boerhaave's *Elements of Chemistry* (London, 1735). Boerhaave was a renowned physician and teacher of chemistry.[4] His lectures were so excellent that a pirated edition was published by his students in 1724 (translated into English in 1727). Although he was not a significant primary contributor to chemical science, he was rigorous and skeptical about the phlogiston concept. The first authorized edition of Boerhaave's Elements was published in 1732 (Leiden). He signed each copy of the huge tome as

FIGURE 165. ■ The first edition of Jean Beguin's *Tyrocinium Chymicum* ("The Chemical Beginner") was published in 1610. This 1660 edition used a Chemical Cupid to entice readers to love chemistry.

Chimischer
Wegweiser/
das ist/
Sichere Anweisung zur Chi-
mischen Kunst/darinnen durch einen kur-
tzen Weg und leichte Handgriffe gewiesen
wird/wie man allerley Artzneyen durch
die Chimie bereiten kan.
Erstlich in Frantzösischer Sprach beschrieben
von
CHRISTOPHORO GLASER,
Ordinar - Apothecker des Königes und
Ihrer Durchleuchtigkeit des Hertzogs
von Orleans in Paris;
Anitzo aber auf Begehren in unsere Teut-
sche Sprache übersetzet von einem
Philochimico.
Jena/verlegts Joh. Jac. Bauhofer/1684.

FIGURE 166. ■ The 1684 German edition of Christophe Glaser's 1663 *Traité de la Chymie*.

verification of its legitimacy. Boerhaave's *Elements* included perhaps the first really comprehensive history of chemistry. Boerhaave was the first great exponent of clinical teaching and he made the medical college at Leiden one of the best in Europe. Following his death, Dr. Samuel Johnson wrote a piece titled "Life of Herman Boerhaave" in the *Gentleman's Magazine* (1739). Johnson's biographer Boswell wrote that Johnson then "dicovered the love of chymistry which never 'forsook him.'"[4] At least twenty years of Johnson's life were spent in his own chymical laboratory.[4]

In Figure 169 we see (*Fig. I*) a thermometer designed to be free standing so that the bulb *AB* can sit in the vessel *PQ* into which liquids can be poured or mixed. In *Fig. II* we see Fahrenheit's first thermometer meant to be filled with alcohol containing red dye. *Fig. III* shows Fahrenheit's second thermometer, this to be filled with mercury. *Fig. IV* shows Fahrenheit's third thermometer, to be used to measure "the Heat of the Human Body." This one can use mercury or alcohol and red dye. It is placed in a hermetically sealed glass chamber. The thermometer is to be used under the arm, upon the breast under one's clothes, or in the mouth . . . whew!

FIGURE 167. ■ Glassware in the 1686 edition of Nicolas Lemery's *Cours de Chimie*. This book was first published in Paris in 1675; the last edition was also published in Paris in the year 1756.

FIGURE 168. ▪ Seventeenth-century glassware in Moses Charas' *The Royal Pharmacopoea* (London, 1678). Note the double pelican (*KK*) and alembics O and *E*.

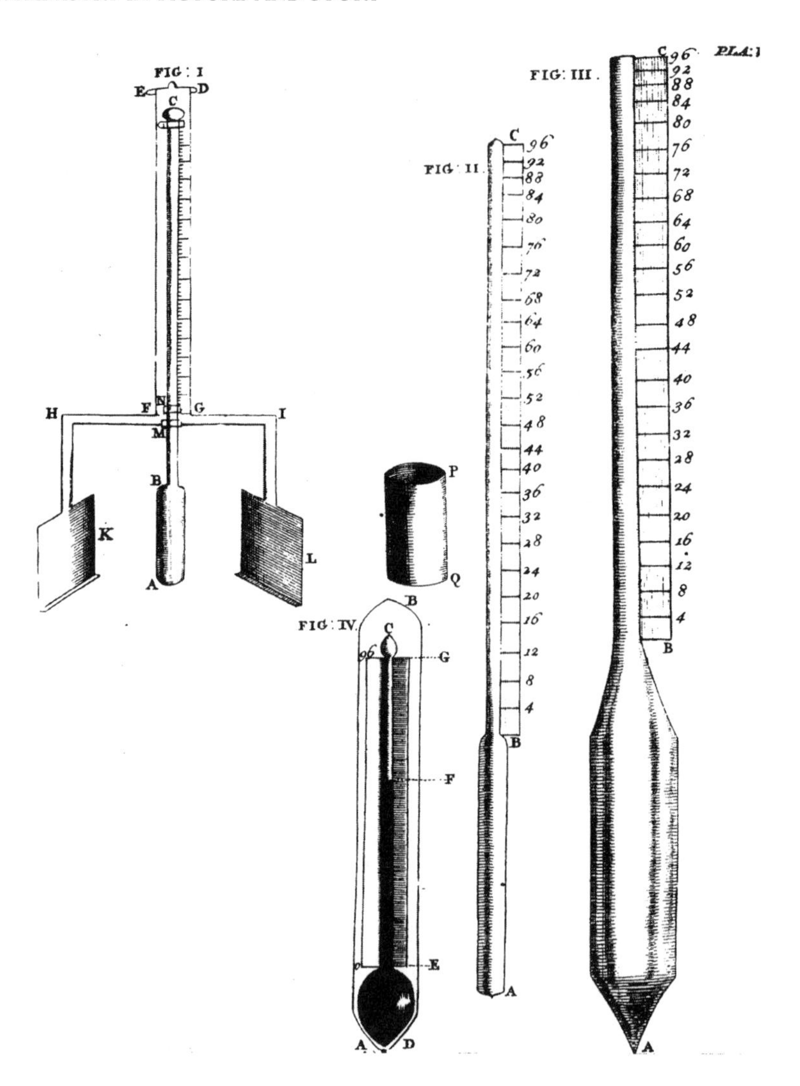

FIGURE 169. ■ These thermometers (see text) are found in the 1735 English edition of Herman Boerhaave's *Elements of Chemistry*. Boerhaave was not a distinguished chemist who made primary discoveries but rather a distinguished teacher of chemistry and medicine who helped introduce clinical teaching into medical school curricula.

1. J. Ferguson, *Bibliotheca Chemica*, Derek Verschoyle, London, 1954 (reprint of 1906 edition).
2. D. Duveen, *Bibliotheca Alchemica et Chemica*, HES, Utrecht, 1986 (reprint of 1949 edition).
3. J. Read, *Humour and Humanism in Chemistry*, G. Bell, London, 1947, pp. 79–123.
4. J. Read, op. cit., pp. 128–153.

WHAT ARE "EFFLUVIUMS"?

Effluvium: Now there's a rare word! *Websters's New World Dictionary of the American Language* (College Edition) defines it as follows: 1. A real or supposed outflow of a vapor or stream of invisible particles; aura. 2. A disagreeable or noxious vapor or odor (plural: effluvia).

Boyle believed in a corpuscular theory of matter—something of a fore-bearer to atomic theory. In this pretty little *Effluviums* book (Figure 170) he conducts *gedanken* (thinking) experiments to calculate the upper limits to the measurable masses of effluvia. But before we illustrate some of these, let Boyle define the contemporary debate:

> Whether we suppose with the Ancient and Modern ATOMISTS, that all sensible Bodies are made up of Corpuscles, not only insensible, but indivisible, or whether we think with the CARTESIANS, and (as many of that Party teach us) with ARISTOTLE, that Matter, like Quantity, is indefinitely, if not infinitely divisible: It will be consonant enough to either Doctrine, that the EFFLUVIA of Bodies may consist of Particles EXTREMELY SMALL. For

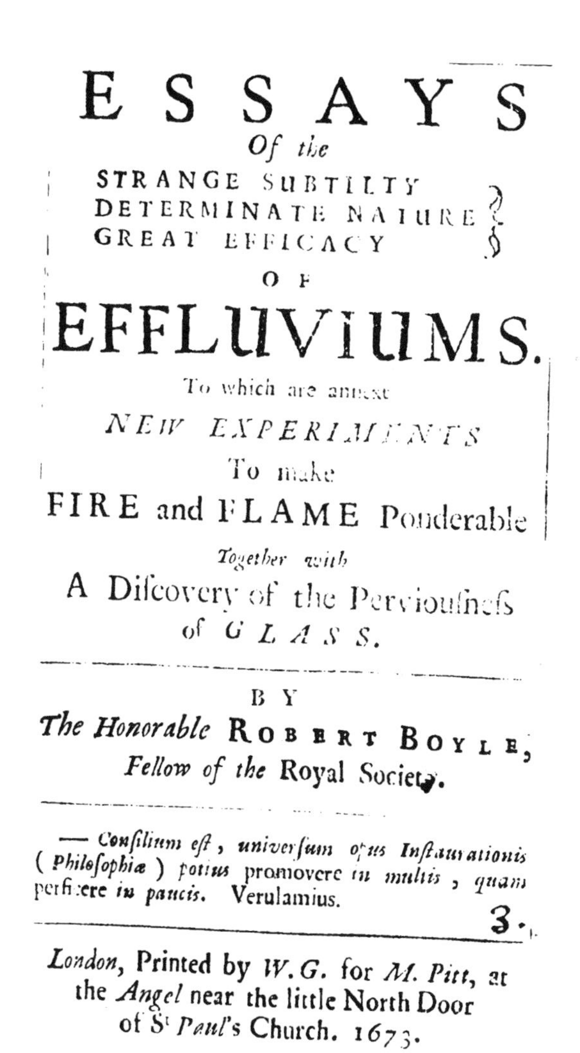

ESSAYS
Of the
STRANGE SUBTILTY
DETERMINATE NATURE
GREAT EFFICACY
OF
EFFLUVIUMS.
To which are annext
NEW EXPERIMENTS
To make
FIRE and FLAME Ponderable
Together with
A Diſcovery of the Perviouſneſs of *GLASS*.

BY
The Honorable ROBERT BOYLE,
Fellow of the Royal Society.

— *Conſilium eſt, univerſum opus Inſtaurationis (Philoſophiæ) potius* promovere *in multis, quam* perficere *in paucis.* Verulamius.

London, Printed by *W.G.* for *M. Pitt*, at the *Angel* near the little North Door of St *Paul's* Church. 1673.

FIGURE 170. ■ Title page of Robert Boyle's wonderful essay in which he estimates the mass of the smallest measurable "effluvia" of silver, gold, silk, and alcohol vapors. In the Denis I. Duveen collection there is a copy of this book autographed by Robert Boyle for Isaac Newton (kind of like the Old Testament autographed by Abraham for Jesus).

> if we embrace the OPINION OF ARISTOTLE or DES-CARTES, there is no stop to be put to the subdivision of Matter, into Fragments, still lesser and lesser. And though the EPICUREAN Hypothesis admit not of such an INTERMINATE division of Matter, but will have it stop at certain solid Corpuscles, which for their not being further divisible are called ATOMS ("atomos") yet the Assertors of these do justly think themselves injured, when they are charged with taking the MOTES or small Dust, that fly up and down in the Sun-Beams, for their Atoms, since, according to these Philosophers, one of those little grains of Dust, that is visible only when it plays in the Sun-Beams, may be composed of a multitude of Atoms, and may exceed many thousands of them in bulk.

(*Modern*) *English Translation:* Do not think, for a moment, that I am so foolish, as to assume that the effluvia whose masses I will estimate are the same things as my version of atoms (corpuscles). I am estimating an upper limit for the masses of effluvia each of which are composed of many thousands of corpuscles. In any case, stay tuned and see how effluvia explain my observations of metals and their calxes.

Here are some thought experiments by Boyle:

1. One grain (0.0648 grams or g) of silver has been drawn by a master silversmith into a wire 27 ft long. Boyle had a special ruler subdivided into 200 divisions per inch. Therefore, the wire can be subdivided into $27 \times 12 \times 200 =$ 64,800 silver "cylinders" each weighing 0.0000010 gram (1.0×10^{-6} g).
2. If it were possible to gild this silver wire, the mass of the gold sheath would be even much less per "cylinder of sheath."
3. "An Ingenious Gentlewoman of my Acquaintance, Wife to a Learned Physician" drew 300 yards of silk gently from the mouth of a silkworm. The silk strand weighed 2.5 grains. The division of the silk gave $300 \times 3 \times 12 \times 200 =$ 2,160,000 silk "cylinders" each of mass 0.000000075 g (7.5×10^{-8} g).
4. Six minute pieces of gold were each beaten into squares with 3¼ inch sides. The total mass of the six square leaves of gold was 1¼ grains. Therefore the six square leaves could be subdivided into a total of $6 \times (3.25 \times 200)^2 =$ 2.535,000 squares of gold each weighing 0.000000032 g (3.2×10^{-8} g).

It is most wonderful to note that some 240 years later, the 1926 Nobel Laureate in Physics, Jean Perrin, performed similar calculations in his 1913 book *Les Atomes* (English, 1916): "an upper limit to molecular size."[1] Gold leaf of 0.1 micron (10^{-5} cm) thickness implies that, at a maximum, gold atoms occupy cubes of 10^{-15} cm^3. Using gold's density, this means a mass of 10^{-14} g per gold atom. Since a hydrogen atom is $^1/_{197}$ the mass of a gold atom, its mass can be given an upper limit of 5×10^{-17} g. Actually, we have known for about 100 years that 1 mole of gold has a mass of 197.0 g and is comprised of 6.02×10^{23} atoms (Avogadro's number). Therefore, an individual gold atom weighs 3.27×10^{-22} g—about 100 trillion times less than Boyle's tangible gold effluvium and 100,000 times less than Perrin's upper limit. However, neither Boyle nor Perrin claimed to be weighing individual corpuscles or atoms. Derived from the experimental studies of Daniel Sennert (1572–1637), Boyle's corpuscles were only superficially similar to the modern concept of atoms. However, without a firm understanding of a chemical element, transmutation was considered to be reasonable.[2] This would imply that a corpuscle of lead could become a corpuscle of gold. This is fundamentally different from the atomic theory of Dalton developed at the start of the nineteenth century.

And why this interest in effluviums? Toward the end of the book Boyle describes his accurate measurement of the increase in weight upon heating a metal, such as iron, in air to form a calx such as rust. In 1673 this was an extremely important observation (also noted by Rey, Becher, Stahl, Mayow, and Le Fèvre among others). Boyle's explanation: minute effluviums in the flame (Becher's igneous particles) penetrate the pores of the sealed glass vessel containing metal and air and "adhere" to the metal, thus forming a calx weighing more than the metal. This was a near avoidance of the Phlogiston Theory that was already in its embryonic stage.

It is also worth noting that Van Helmont explained the sympathy concept as well as magnetic phenomena as arising from contact between effluvia (for example, between the blood on a sword and the blood in the wounded person's body; see pp. 197–198).

Toward the end of the *Effluviums* book Boyle explicitly raised the possible health issue of the effect of effluviums from fire landing on cooked meat and being consumed. We now know that when the fats from meat drip onto hot coals during charbroiling, they pyrolyze to form carcinogenic polycyclic aromatic hydrocarbons which rise up from the grill and deposit on the surface of the meat. Thus, in this regard, Boyle anticipated the human exposure health specialists by 300 years. It is also worth noting that William Penn corresponded with Boyle from Pennsylvania and during the early 1680s sent him samples of ore and medicinal plants from the New World.[3]

---

1. J. Perrin, *Atoms*, Constable, London, 1916, pp. 48–52.
2. W. R. Newman, *Atoms and Alchemy*, The University of Chicago Press, Chicago and London, 2006.
3. C. Owens Peare, William Penn—A Biography, Dobson Books, London, 1959, p. 268. (I thank Professor Susan Gardner for calling this to my attention).

THE SURPRISING *CHEMICAL* TAXONOMIES OF MINERALS AND MOLLUSKS

Eighteenth-century Sweden[1] enjoyed a powerful mining and metallurgy industry, becoming the main source of iron for the rest of Europe. Its other abundant natural resource—virgin forests—made Sweden a center for the lumber industry and furniture manufacture. This may explain why detailed scientific classification of both the plant and mineral kingdoms originated in Sweden during this period. On the other hand, it may simply have been the Lutheran yearning for order and harmony.

In 1753, Carolus Linnaeus (1707–1778)[2] published the *Species Plantarum*, providing the first systematic taxonomy of flowering plants and ferns. It was based largely on the external structures (morphologies) of flower parts. External appearance also played a major role in mineral classification. For example, gemstones such as diamonds and rubies were assumed to be closely related. In 1758, another Swede, Axel Frederic Cronstedt [1722 (or 1702)–1765],[3] published (anonymously) his *Försök till Mineralogie*. In this book, Cronstedt placed all minerals into four chemical groups: earths, salts, bitumens, and metals.[3] However, this crude classification was published only two years after Black reported "fixed air" and predated the chemical revolution by about two decades.

In order to illustrate the chemical confusion prevalent in eighteenth-century mineralogy, a couple of examples will suffice. Plumbago or Flanders stone is a slippery grayish mineral that darkens the hands.[4] It was also called "black lead," due to the superficial resemblance to the grayish, soft metal. Plumbago was employed in seventeenth-century pencils;[5] hence its more modern name *graphite*. Of course we still commonly employ "lead" pencils.[6] Johan Gottschalk Wallerius (1709–1785), another Swede, classified graphite as a kind of talc.[7] In 1779, the great Swedish chemist Carl Wilhem Scheele oxidized graphite with niter, collected the "fixed air" produced, and concluded that the mineral consisted of pure carbon.[8] Diamond, on the other hand, appears to be as distinct from graphite, as one can imagine. It is crystalline, clear and harder than rock. Indeed, diamond would appear to be a close relative of other rare, beautiful gems, including ruby, garnet, and sapphire. However, rumors about experiments in which diamonds were burned gained increasing credibility during the seventeenth and eighteenth centuries.[9] In 1760, François I, Emperor of Austria, described an experiment wherein diamonds and rubies were burned for 24 hours in crucibles.[9] Upon opening the crucibles, the rubies remained unchanged but the diamonds had disappeared without a trace. In 1797, Smithson Tennant oxidized diamond with nitre and proved conclusively that diamonds are also pure carbon.[10]

The blowpipe was an effective early instrument for analyzing the composition and chemistry of minerals. Although its origins are ancient, blowpipes were perfected in Sweden during the eighteenth century.[11] Scheele and Torbern Bergman used the instrument extensively. Figure 171 is from Bergman's essays and describes in detail the construction and use of a solid silver blowpipe.[12] Section A (upper right), starts with the mouthpiece and tapers to a tight fit with unit B, which forces the breath to make a 90° turn and collects in a trough the mist and droplets of water in the operator's exhalations. Unit B attaches to pipe C that ends in a smooth, small round orifice g that trains the exhalation on the candle flame. The flame is blown horizontally and its reducing or oxidizing regions may be focused on samples as desired. Tiny mineral samples may be exposed to the flame while sitting in spoon E, made of silver or gold, or, if the mineral is nonflammable, seated in an indentation in a piece of charcoal. Hammer F pounds pieces of minerals within metal ring H on metal plate G. Pieces of mineral are handled using forceps I.

Now here's the tricky part. The flow of exhaled air onto the flame must be smooth and continuous, sometimes for minutes at a time. Try *that*. However, Bergman[12] assures us that with practice the technique can be mastered—fill your cheeks with air and as you inhale and exhale through your nostrils, keep your cheeks replenished with air, and keep squeezing them steadily with the fingers of one hand so that the exhalation remains steady and continuous. Esteemed reader, you have my leave to take the day off and practice.

There are several protocols to follow. First, expose the sample to the outer (fuel-deficient), oxidizing part of the flame. If the sample survives, expose it to

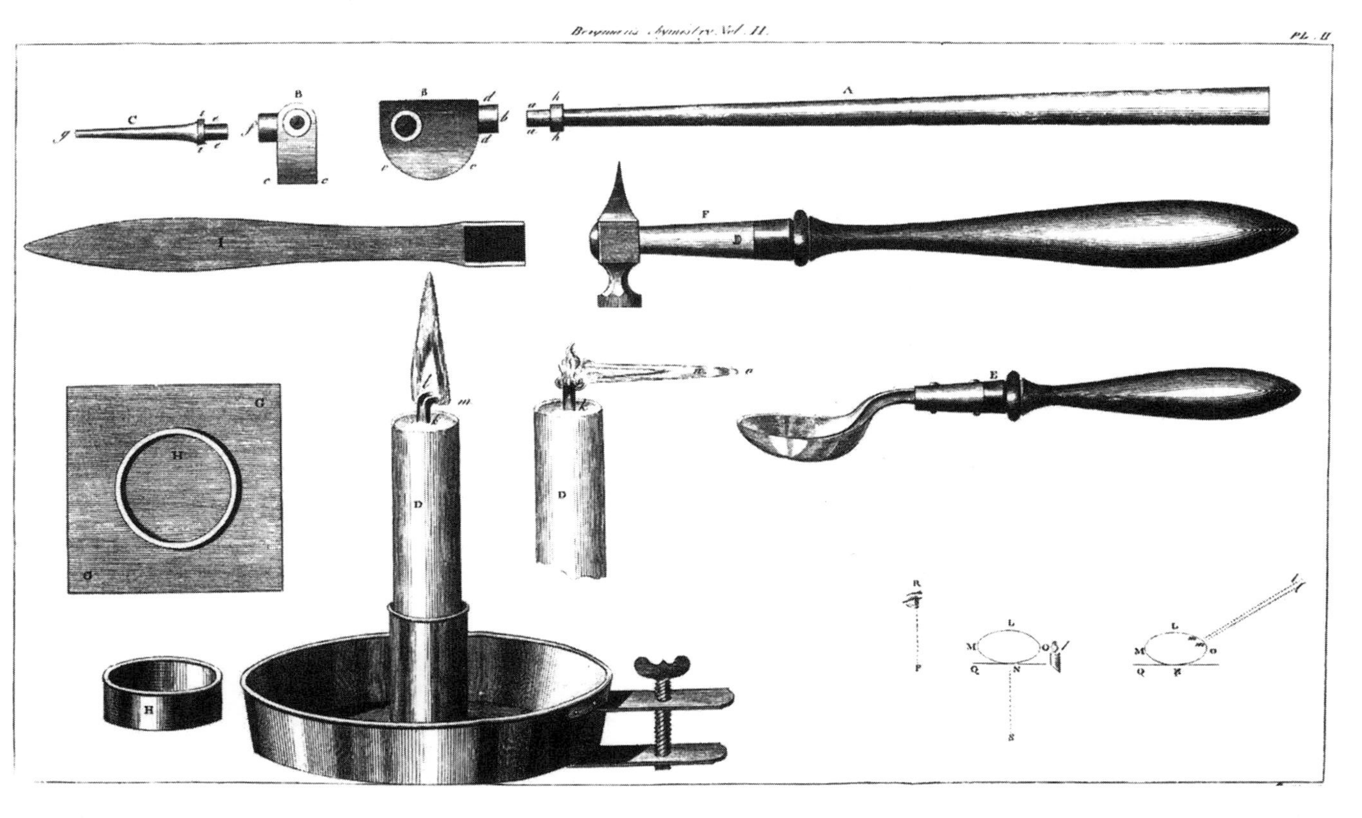

FIGURE 171. ■ Apparatus employed in blowpipe analysis of minerals was developed to a high state of technology in eighteenth-century Sweden (from Torbern Bergman's 1788 *Physical and Chemical Essays*).

the tip of the blue, reducing, hotter section of the flame. If the sample will not melt in the flame, then Bergman offers three "fluxes," substances that aid fusion (melting) of samples: an acidic phosphate salt, an alkaline salt (sodium hydroxide), and a neutral salt (borax). The flux is melted; then a finely ground mineral is added and brought to reaction with the flux, and the results are recorded. Such blowpipe analyses were both highly specific and extremely sensitive.

As the eighteenth century came to a close, advances in mineral analyses and chemical theory merged to set the stage for a revolution in mineral taxonomy. Figure 172 is from the 1801 *Traité de Mineralogie* by abbé René Just Haüy (1743–1822).[13] In this book the amorphous yellow crystals of sulfur and the clear crystalline diamond, as dissimilar as they are, appear together in Plate LXII as

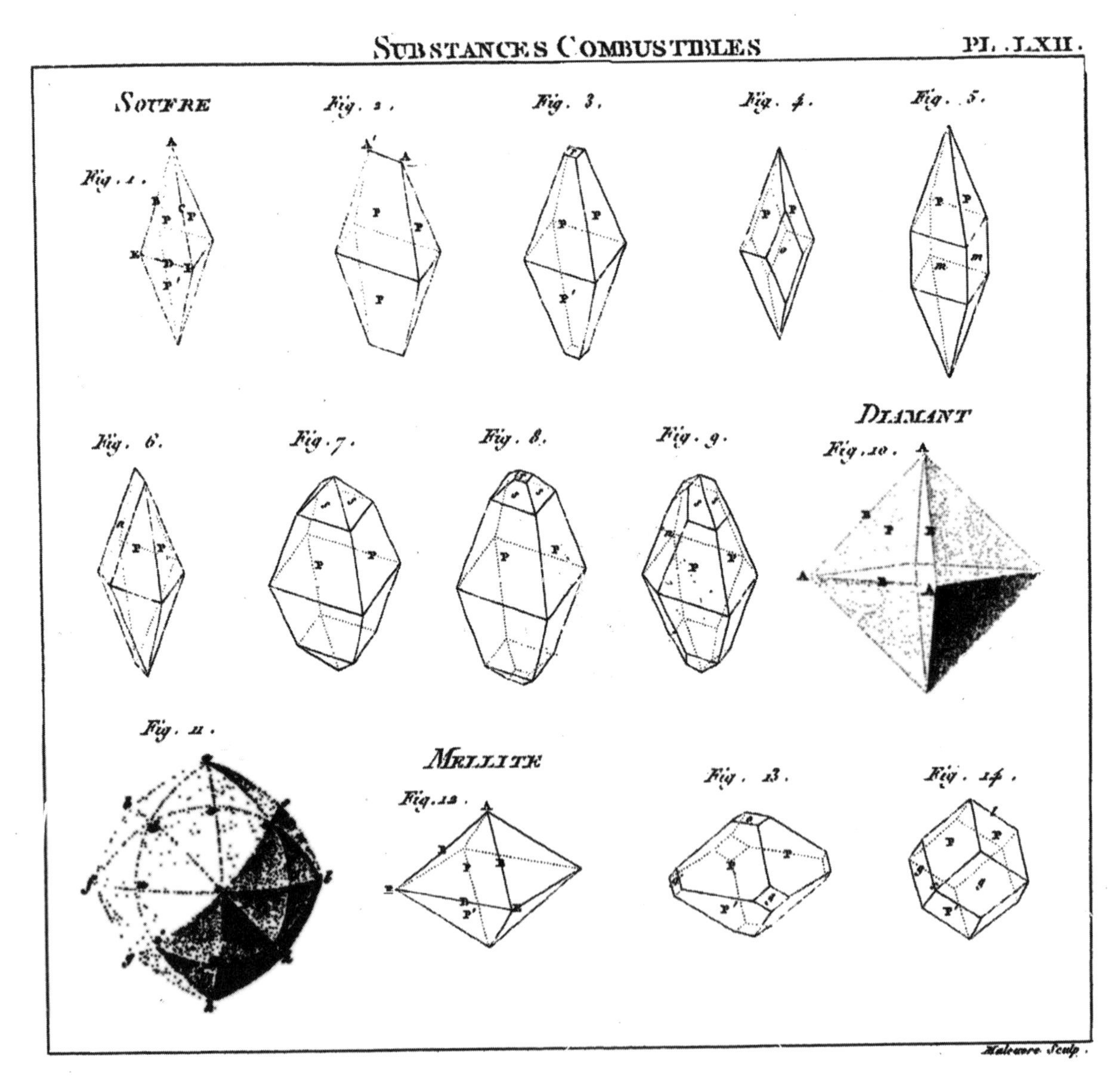

FIGURE 172. ■ Chemical classification of minerals illustrated in René Haüy, *Traité de Minéralogie*, Paris, 1801 (from the Othmer Library, CHF).

flammable chemical elements. Haüy was the first to understand that crystals cleaved along specific faces fixed by underlying crystal symmetries. He is perhaps the principle founder of crystallography.[14] William Hyde Wollaston (1766–1828), carefully measured the angles of crystalline faces, unified the ideas of Hooke (see Figure 147) and Dalton (see Figure 232) about crystal packing and applied them to understanding crystal symmetry and cleavage (see Figure 173).[14] The mineralogical studies started in Sweden combined with the "new chemistry" enabled American mineralogist James Dwight Dana to propose, in 1837, the chemical classification of minerals that remains functional today.[15]

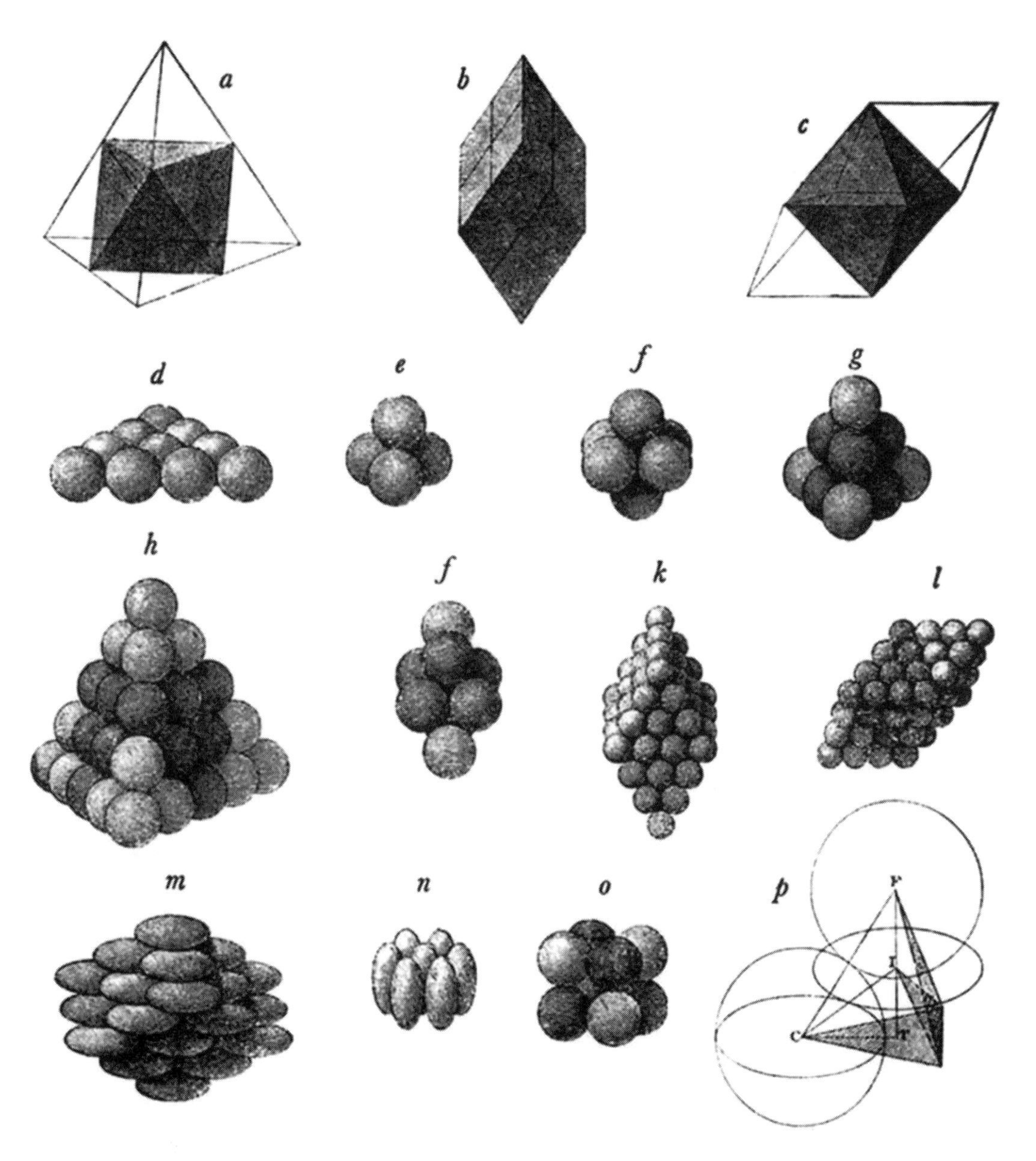

FIGURE 173. ■ William Hyde Wollaston's explanation of crystalline faces (from C. Singer, *The Earliest Chemical Industry*, The Folio Society, with permission from The Folio Society).

Now, how do these discussions of minerals relate to mollusks? The systematics of the Linnaeus taxonomy were based on external structure (morphology). Developed a century before Darwin's discovery of evolution, it lacked the insights derived from the theory of natural selection. Thus, we now better understand the external similarity between sharks and dolphins (frequently confused by nervous swimmers) by understanding the parallel evolutionary paths that allow them to occupy similar environments and niches. Morphology here is very misleading—the shark, a fish, and the dolphin, a mammal, are about as different as a ruby and a diamond. In contrast, dolphins and horses seem to be as morphologically unrelated as graphite and diamond. However, they both are warm-blooded, give birth to live young free of eggs, and nurse their young. The hidden chemical reality of graphite and diamond is that they are both pure carbon.

There are six species of clams shown schematically in Figure 174.[16] Four of these are species in the genus *Calyptogena*, and the remaining two are from genera *Vesicomya* and *Ectenagena*. These classifications have been based largely on the morphologies of the clams' shells since this is the portion that survives once the clam dies and also permits linkages with fossil ancestors (see the essay on Lamarck later in this book, see p. 346). However, during the latter part of the twentieth century, new chemical tools, including protein sequencing and later DNA sequencing, were developed to study the hidden dimensions in such phylogenetic relationships. Specifically, each characteristic protein in an organism is coded for by a gene. The hemoglobin of a horse more closely resembles that of a

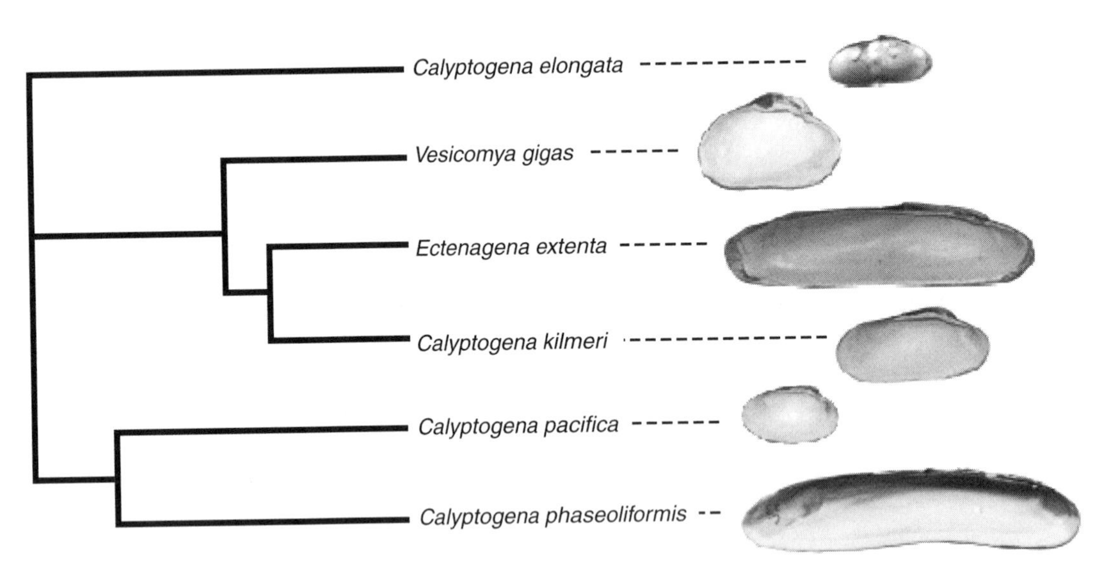

FIGURE 174. ■ Modern cladistic display of phylogenetic relationships among selected clams. These classifications are based on chemical criteria (DNA sequences) rather than shell structure (morphology). These biochemical studies indicate that *C. kilmeri* is genetically more similar to *V. gigas* and *E. extenta* than it is to the more morphologically similar *C. elongata*, *C. pacifica*, and *C. phaseoliformis*. Mineralogy underwent a similar evolution from morphological to chemical taxonomy more than two centuries ago. Prior to 1770, diamond was thought to be closely related to ruby and totally unrelated to plumbago (graphite). However, chemistry showed that diamond and plumbago are each pure carbon and are both totally unrelated to ruby. (I am grateful to Professor Robert Vrijenhoek for introducing me to cladistic concepts, discussing his studies on clams, and furnishing the figures shown.)

cow than that of a mouse. In this case, at least, the obvious morphological relationship reflects the underlying genetic differences. A new field termed "genomics" arose toward the end of the twentieth century as advances in chemical analysis, automation, and bioinformatics allowed direct comparisons of the huge genetic sequences of different organisms. The results are occasionally quite surprising.

Figure 174 depicts the systematic relationships between the six clam species explored based not on the morphologies of the shells but rather on the DNA sequences pertaining to mitochondrial oxidase subunit I.[16] The phylogenetic systematics shown are displayed as a cladogram[17] in which each branch in the tree represents a distinct modification to form a new species. Genomically, the relationships between the clams are quite different from the classification based on morphology (Figure 174). The hidden chemical reality is considerable different from the conclusions based on structure. Thus, on the basis of this chemical classification, *V. gigas*, *E. extenta*, and *C. kilmeri* are representatives of three different genera but could be assigned to one genus since they are "monophyletic" (trace to a common ancestor). The cladogram indicates that *C. pacifica* and *C. kilmeri*, currently classified in the same genus, have less in common than do *C. kilmeri* and *E. extenta* (*E. extenta* appears to be morphologically distant from *C. kilmeri*).[17] Professor Robert Vrijenhoek discovered that clams in Monterey Bay called *Calyptogena pacifica* in fact represent three morphologically similar but genetically distinct species occupying different depths. The true *Calyptogena pacifica* is not even found in Monterey Bay but in the vicinity of Washington State.[16]

As Genomics passes through its infancy it begins to raise many difficult questions. For example, not all proteins reflect identical phylogenetic relationships between closely related species. The use of the specific DNA segment noted above to determine the phylogenetic relationships between the six clams depicted in Figure 174 is not the only possibility. As more knowledge is gained in this revolutionary field, decisions on most appropriate sequences as well as weighting factors may emerge. Does this return us to the subjectivity of morphological classification, or can certain "critical" genes be accepted as the determining indices? These are certainly much more complex questions than those raised two centuries earlier about minerals that are chemically simpler, lack functional tissues and organs, do not metabolize other minerals, and, so far as we know, neither mate nor evolve.

---

1. *The New Encyclopedia Britannica*, Encyclopedia Britannica Inc., Chicago, 1986, Vol. 28, pp. 332–350.
2. *The New Encyclopedia Britannica*, op. cit., Vol. 7, pp. 379–380.
3. J.R. Partington, *A History of Chemistry*, MacMillan & Co., Ltd., 1961, Vol. 2, pp. 173–175.
4. Partington, op. cit., p. 91.
5. Partington, op. cit., p. 104.
6. Long-time friend Professor Joel F. Liebman is a theoretician who has not done an experiment for at least 35 years. He quips that the only chemical research hazard he faces is "lead poisoning" from an inadvertent jab from his own pencil.
7. Partington, *A History of Chemistry*, MacMillan & Co., Ltd., London, 1962, Vol. 3, p. 170.
8. Partington (1962), op. cit., pp. 216–217.
9. J.-P. Poirier, *Lavoisier—Chemist, Biologist, Economist*, Engl. transl., University of Pennsylvania Press, Philadelphia, 1996, pp. 47–50.

10. Partington (1962), op. cit., pp. 703–705.
11. J.J. Berzelius, *The Use of the Blowpipe in Chemical Analysis, and in the Examination of Minerals* (transl. J.G. Children), London, 1822.
12. T. Bergman, *Physical and Chemical Essays* (transl. E. Cullen, J. Murray, London, 1788, pp. 471–529.
13. R. Haöy, *Traité de Minéralogie*, Louis, Paris, 1801.
14. C. Singer, *The Earliest Chemical Industry*, The Folio Society, London, 1948, pp. 291–307. We are grateful to the Folio Society for permission to reproduce this figure.
15. *The New Encyclopedia Britannica*, op. cit., Vol. 24, pp. 129–138.
16. I am grateful to Professor Robert Vrijenhoek, who helped inspire this essay and who supplied the drawings in Figure 174 as well as extremely helpful discussions. I also wish to acknowledge helpful conversations with Professors Judith Weis and Will Clyde.
16. I.J. Kitching, P.L. Forey and D.M. Williams, in *Encyclopedia of Diversity*, S.A. Levin (ed.), Vol. 1, pp. 677–707.

CHEMICAL AFFINITY

Figures 175 and 176 depict two halves of the first logically organized tables of the properties of chemical substances. It was composed by Étienne-François Geoffroy (1672–1731) in 1718.[1] The top horizontal row depicts 16 substances (elements and compounds), classes of substances, and even mixtures in a fairly arbitrary order from left to right. Each column rank orders substances according to *affinity*. Those substances closest to the top have the highest affinities for the substance at the top (the *header*), while those toward the bottom have the lowest affinities.

Let's examine some brief illustrative examples. In Column 16, we see water as the header with alcohol above salt. This means that alcohol has a greater affinity for water than salt. Thus, if you added alcohol (say, ethanol or 200 proof vodka) to salt water (saline) the liquids would mix and the salt would precipitate, forming a filterable solid. Alcohol has displaced salt from water. In contrast, if you took a 50:50 alcohol–water mixture (100 proof vodka), you could not dissolve salt in it since water has greater affinity for alcohol.

Column 1 shows the chemical affinities of substances for acids. Most metals react chemically with acids and release hydrogen gas—they appear to "dissolve" and release "air." However, if we first mix an alkali (base) such as potassium carbonate (K_2CO_3) with the acid and neutralize it, the solution will no longer dissolve metals. If a metal is dissolved in acid and alkali is added, a solid will precipitate (actually the insoluble metal carbonate or hydroxide). Thus, the alkalis have higher affinity for acids than do metallic substances.

But wait a minute, dear readers. Those of you who have had some high-school chemistry realize that solubilities of alcohol and salt in water are physical properties while "solubilities" of metals in acids are chemical properties. You would not have received a good grade from me for confusing the two. Clearly, the differences were not yet fully clear to early eighteenth-century scientists.

Human history is writ large in Column 9! Let us look at the affinities of sulfur. Of the metals shown, iron has the highest affinity, with tin and copper (which form the alloy bronze) having lower affinities. Tin and copper ores (com-

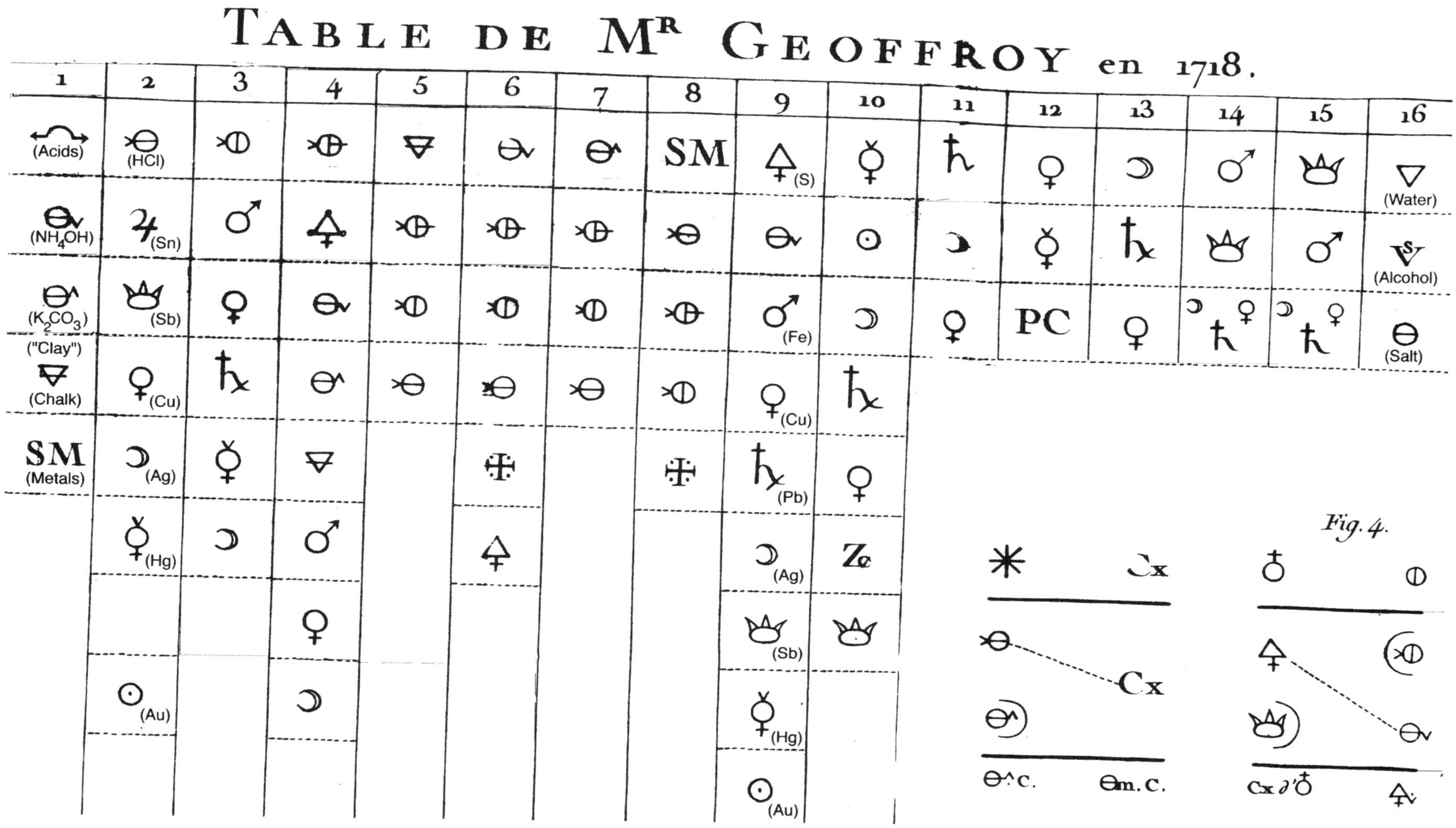

TABLE DE MR GEOFFROY en 1718.

1	2	3	4	5	6	7	8	9	10	11	12	13	14	15	16
[symbol] (Acids)	[symbol] (HCl)	[symbol]	[symbol]	🜃	[symbol]	[symbol]	SM	🜍 (S)	☿	♄	♀	☽	♂	♁	🜄 (Water)
[symbol] (NH_4OH)	♃ (Sn)	♂	🜍	[symbol]	[symbol]	[symbol]	[symbol]	[symbol]	☉	☽	☿	♄	♁	♂	🜈 (Alcohol)
[symbol] (K_2CO_3)	♁ (Sb)	♀	[symbol]	[symbol]	[symbol]	[symbol]	[symbol]	♂ (Fe)	☽	♀	PC	♀	☽ ♀ ♄	☽ ♀ ♄	⊖ (Salt)
("Clay") 🜃 (Chalk)	♀ (Cu)	♄	[symbol]	[symbol]	[symbol]	[symbol]	[symbol]	♀ (Cu)	♄						
SM (Metals)	☽ (Ag)	☿	🜃		[symbol]		[symbol]	♄ (Pb)	♀						
	☿ (Hg)	☽	♂		🜍			☽ (Ag)	Zc						
			♀					♁ (Sb)	♁						
	☉ (Au)		☽					☿ (Hg)							
								☉ (Au)							

Fig. 4.

Cx

Cx

[symbol] c. [symbol] m. c.

Cx ∂ [symbol]

FIGURE 175. ■ Top half of Étienne Geoffroy's 1718 Table of Affinities taken from *Recueil de Dissertations Physico-Chimiques Presentés a Differentes Académies* (Jacques Francois Demachy, Paris, 1781).

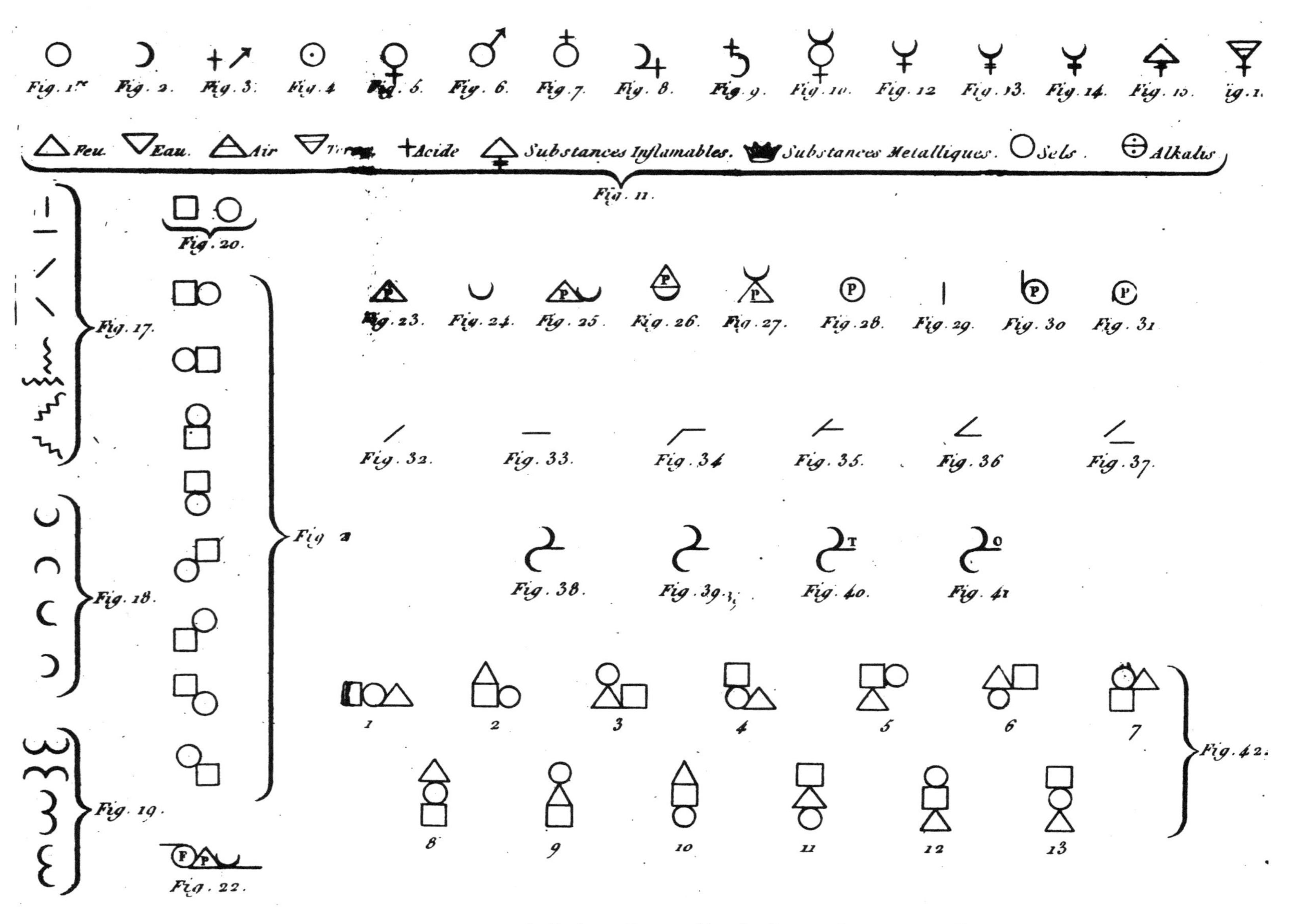

FIGURE 176. ■ Bottom half of Geoffrey's Table of Affinities (see Figure 175).

monly sulfides) could be smelted relatively easily, and the Bronze Age thus began around 3000 B.C. Higher temperatures and therefore more modern furnace technology were required to win iron from its sulfides and the Iron Age only began around 1200 B.C.

Notice gold at the bottom of Column 9. This noble metal has little affinity for sulfur and can often be found in nature as shiny nuggets or granules in an uncombined state.

Geoffroy's table, a somewhat arbitrary conglomeration of chemical and physical properties, elements, compounds, classes of substances and mixtures, is, nevertheless, a distant relative of the Periodic Table, formulated some 150 years later.

Following Vannucio Biringuccio in the sixteenth century and Albaro Alonso Barba in the seventeenth century, Christlieb Ehregott Gellert was one of the first to employ amalgamation to separate gold from its ores. In 1750 he extended Geoffroy's table to 28 columns (Figure 177).[3] Let us take a brief glance at the organization of this table of "solubilities" (actually chemical *and* physical affinities or rapports). Most of the symbols can also be found in the first four figures from the previous essay. In contrast to Geoffroy's table, Gellert placed the substance having *greatest affinity* for the column header at the bottom of the column. (Substances having no affinity are listed below the main table.) Let us examine Gellert's organization—the column headers are grouped as follows:[4]

Earths (columns 1–5)

1. Vitrifiable (fusible under fire) or siliceous earth (silicon dioxide)
2. Fluors or fusible earths (low-melting minerals)
3. Clay (possibly alumina)
4. Gypsum (calcium sulfate dihydrate or $CaSO_4 \cdot 2H_2O$)
5. Calcareous earth [strong—$Ca(OH)_2$ or mild—$CaCO_3$)

Alkalis (columns 6 and 7)

6. Fixed alkali (potassium carbonate—K_2CO_3)
7. Volatile alkali (aqueous ammonia solution—NH_4OH)

Acids (columns 8–12)

8. Distilled vinegar (acetic acid—$HC_2H_3O_2$)
9. Marine acid (hydrochloric acid—HCl)
10. Nitrous acid (actually nitric acid—HNO_3)
11. Vitriolic acid (sulfuric acid—H_2SO_4)
12. *Aqua regia* (HCl/HNO_3—3:1)

A salt (column 13)

13. Nitre (sodium nitrate—$NaNO_3$)

Nonmetals and metals (columns 14–27)

14. Sulfur
15. Liver of sulfur
16. Cobalt
17. Arsenic
18. Regulus of antimony
19. Glass of antimony (possibly antimony oxysulfate—$Sb_2O_2SO_4$)

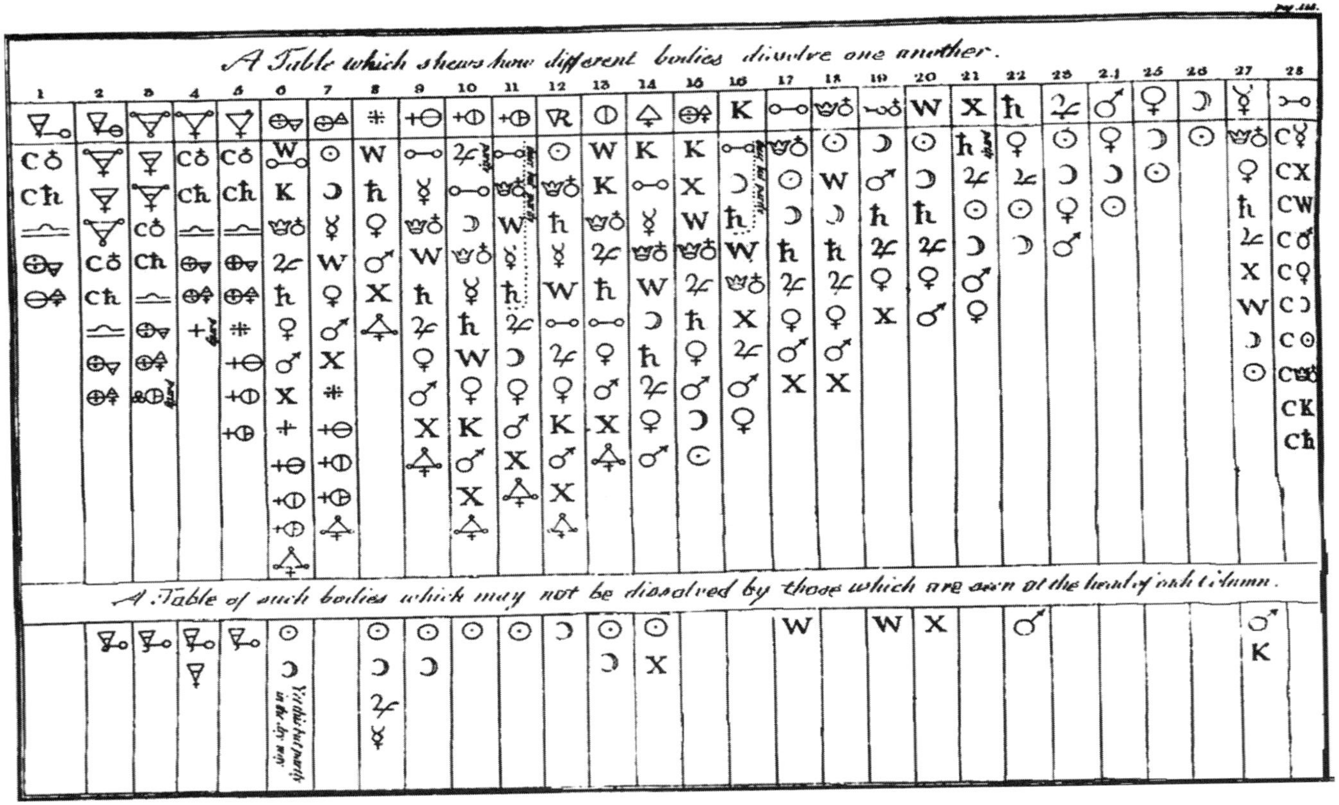

A Table which shews how different bodies dissolve one another.

A Table of such bodies which may not be dissolved by those which are seen at the head of each column.

FIGURE 177. ■ Table of Affinities developed by Christlieb Ehregott Gellert in 1750. In this table, the highest affinity for the substance at the top of the table (the "header") is shown by the lowest substance in the table. The lowest affinity is exhibited by the substance just below the header. Substances in the lower table are unreactive with the header. Thus, in column 10 (nitric acid), phlogiston is seen to have the highest affinity because nitric acid is "dephlogisticated"—it supports calcination of metals. The reactivity of metals for nitric acid is as follows: zinc > iron > cobalt > copper>lead > mercury > silver > tin. Gold is the least reactive since it is actually unreactive with nitric acid. (From C.E. Gellert, *Metallurgic Chemistry*, 1776; from the Othmer Library, CHF.)

20. Bismuth
21. Zinc
22. Lead
23. Tin
24. Iron
25. Copper
26. Silver
27. Mercury

Glass (column 28—fused form of any substance—"C" connotes calx, so that the first symbol below the header is calx of mercury).

The unreactive nature of gold is readily apparent in its frequent appearance in the small bottom table of unreactive substances. Gold is not attacked by any acids (columns 8—11) except for *aqua regia* (column 12), where it is still seen to be the least reactive of the 12 substances listed. However, gold is the last of eight substances in column 27, thus indicating its high affinity for mercury—the very basis for Gellert's amalgamation process for isolating gold from its ores. Phlogiston appears as the last symbol in columns 8—12, indicating the highest affinity with acids. This derives, in part, from the fact that metals readily "dissolve" in acids, apparently losing their phlogiston to the surroundings and form calxes. Interestingly, phlogiston also has the highest affinity for the alkalis (columns 6 and 7) that, in turn, also have high affinities for acids. Note that distilled vinegar (acetic acid) is, as expected, the weakest of the acids since it "dissolves" the fewest substances. The highest affinity for niter (column 13) is also assigned to phlogiston since niter (as well as saltpetre) readily converts metals to calxes upon addition of heat. A century earlier, Boyle, Hooke, and Mayow had noted the ability of saltpetre to substitute for air in the combustion of sulfur and carbon as well as in the calcinations of metals. In the mid—1770s, oxygen was considered to be dephlogisticated air "hungry for phlogiston." In this sense, niter and saltpetre, which chemists soon would learn supplies oxygen, could also be considered "dephlogisticated."

The orders of affinities of metals for acids in Gellert's table very much reflect the modern Metal Activity Series (which indicates ease of oxidation). Here is one such comparison:

Gellert Column 10 (Nitric Acid)	Metal Activity Series[5]
Zinc	Zinc
Iron	Iron
Cobalt	Cobalt
Copper	Tin
Lead	Lead
Mercury	Copper
Silver	Silver
Tin	Mercury
Gold	Gold

If one takes silver metal and adds it to an aqueous solution of nitric acid, the metal slowly "dissolves," releasing bubbles of "air." (Some 15 years after

Gellert's table was published, Henry Cavendish would capture these bubbles and determine that they were superlight and flammable and believe that his flammable air was phlogiston itself). We now recognize that silver has been oxidized and that the solution contains silver nitrate ($AgNO_3$) rather than the metal. If the more active metal copper is added (say, by placing a copper wire into the solution), the solution will begin to turn as blue as certain copper salts and silver will deposit on the wire. Thus, copper appears to have higher affinity for nitric acid than silver does, just as alcohol has a higher affinity for water than salt. In fact, there is considerable confusion here. For example, the metal oxidations are truly chemical changes; displacement of salt by alcohol is a purely physical change.

The organization of Geoffroy's or Gellert's tables into earths, alkalis, acids, salts, nonmetals, and metals has some hints of the future periodic table. The periodic table gathers metals, metalloids, and nonmetals in clusters and regions. Rankings within a chemical family reflect reactivity—fluorine is much more reactive than the other halogens; lithium is the least reactive of the alkali metals. The tables of affinities mix chemical and physical properties explicitly, while the periodic table is explicitly chemical in nature, although trends in physical properties also emerge. The periodic table includes elements only, while tables of affinities include elements and compounds. However, here it is well to remember that oxides of metals (left-hand side of the periodic table) are alkalis while oxides of nonmetals (right-hand side of the periodic table) are acids. What is chiefly missing from the tables of affinities is periodicity itself. Periodic behavior could occur only if there were some measurable scalar increase in a property such as atomic mass (early nineteenth century) or atomic number (early twentieth century). Nonetheless, these affinity tables forced chemists to explicitly imagine the systematic organization of their field—perhaps the first true step in becoming a modern science.

---

1. This version of Geoffroy's table is from M. De Machy, *Recueil de Dissertationes Physico-Chimiques*, Paris, 1781 (Plate 1). Also, see the discussion in H.M. Leicester and H. S. Klickstein, *A Source Book in Chemistry 1400–1900*, McGraw-Hill, New York, 1952, pp. 67–75.
2. J.R. Partington, *A History of Chemistry*, MacMillan and Co. Ltd., London, Vol. 3, 1962, pp. 49–55.
3. C.E. Gellert, *Metallurgic Chymistry. Being a system of Mineralogy in General, and of all the arts arising from the science. To the great improvement of manufacturers, and the most capital branches of Trade and Commerce. Theoretical and Practical. In two parts, Translated from the original German of C.E. Gellert*. By I.S. London and T. Becker, 1776. (This is the English translation of the original German edition (1751—1755.) The author is grateful to Ms. Elizabeth Swan, Chemical Heritage Foundation, for supplying this image.
4. J. Eklund, *The Incompleat Chymist—Being an Essay on the Eighteenth-Century Chemist in His Laboratory, with a Dictionary of Obsolete Chemical Terms of the Period*, Smithsonian Institution Press, Washington, DC, 1975.
5. T.L. Brown, H.E. LeMay, Jr., and B.E. Bursten, *Chemistry—The Central Science*, seventh edition, Prentice-Hall, Upper Saddle River, NJ, 1997, pp. 128–132.

DOUBLE-BOTTOM CUPELS, HOLLOW STIRRING RODS, AND OTHER FRAUDS

During the period from 1596 through 1601, Duke Friedrich I of Württenberg (1557–1608) hanged three unsuccessful alchemists from a special gallows in Stuttgart with the inscription: "He shall learn to make gold better."[1] It is interesting to note that although alchemy was essentially dead by the end of the seventeenth century, there remained popular interest well into the eighteenth century. Gullible savants and the just plain greedy were prey for "alchemists," and the venerable *Journal des Savans* continued to publish occasional papers on transmutation. Amazingly, a mainstream scientist, the aforementioned Geoffroy, was moved to publish a paper in *Histoire de l'Academie Royale des Sciences* in 1722 warning against such gullibility.[1] Among the frauds he warned against were the following:[1]

1. Double-bottom cupels
2. Hollow stirring rods
3. Amalgams concealing precious metals
4. Acids containing dissolved gold and silver
5. Filter papers with minute amounts of concealed gold or silver to be recovered upon ashing of the paper

Now, ladies, gentlemen, and children of all ages—watch as I scratch this little black crystal of samarium oxide (SmS). Presto—it changes to gold (colorwise, that is).[2] Who will be the first to purchase some of this "black gold"?

Apparently, even today there are alchemists busily working away in France earnestly trying to discover the Stone of the Philosophers.[3] Eureka! There may yet be a customer to buy the famous *Pont de Brooklyn* from me.

1. T. Nummedal, "Fraud and the Problem of Authority in Early Modern Alchemy," The International Conference on the History of Alchemy and Chymistry, Chemical Heritage Foundation, Philadelphia, 19–22 July 2006.
2. A. Debus, in *Hermeticism and the Renaissance*, I. Merkel and A.G. Debus (eds.), Associated University Press, Cranbury, NJ, 1988, pp. 231–250.
3. H. Rossotti, *Diverse Atoms: Profiles of the Chemical Elements*, Oxford University Press, Oxford, 1998, p. 439.
4. A. McLean, *A Commentary on the Mutus Liber*, Phanes, Grand Rapids, MI, 1991, pp. 8–10.

THERE IS TRUTH IN CHALK

In vino veritas ("there is truth in wine") implies that, suitably "lubricated," a person may be more likely to confess all. However, the German apothecary Johann Friedrich Meyer (1705–1765)[1] might well have said "*In calcis veritas*" ("there is truth in chalk")[2] since it is through calcium carbonate (chalk; limestone) that he discovered his *acidum pingue*—said to be the general principle innate in all bodies, the principle in fire, and the component of all acids (see Figure 178).[3] And

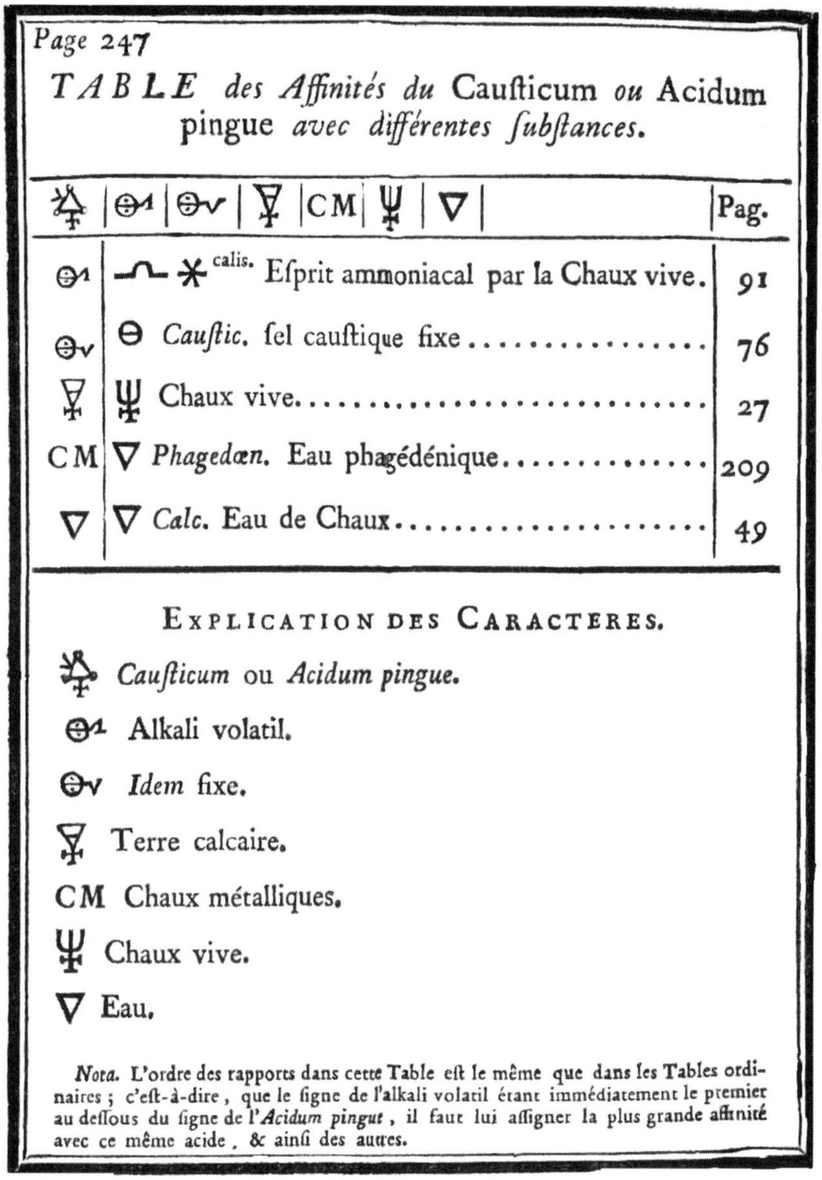

Page 247

TABLE des Affinités du Cauſticum *ou* Acidum pingue *avec différentes ſubſtances.*

🜍	⊖ᐟ	⊖v	🜄	CM	♆	∇	Pag.

		Pag.
⊖ᐟ	⏜ ✱ calis. Eſprit ammoniacal par la Chaux vive.	91
⊖v	⊖ *Cauſtic.* ſel cauſtique fixe................	76
🜄	♆ Chaux vive.........................	27
CM	∇ *Phagedæn.* Eau phagédénique..............	209
∇	∇ *Calc.* Eau de Chaux....................	49

EXPLICATION DES CARACTERES.

🜍 *Cauſticum* ou *Acidum pingue.*

⊖ᐟ Alkali volatil.

⊖v *Idem* fixe.

🜄 Terre calcaire.

CM Chaux métalliques.

♆ Chaux vive.

∇ Eau.

Nota. L'ordre des rapports dans cette Table eſt le même que dans les Tables ordinaires ; c'eſt-à-dire, que le ſigne de l'alkali volatil étant immédiatement le premier au deſſous du ſigne de l'*Acidum pingue*, il faut lui aſſigner la plus grande affinité avec ce même acide, & ainſi des autres.

FIGURE 178. ■ A table from Johann Friedrich Meyer's 1766 *Essais de Chymie*. He believed in an *acidum pingue* ("fatty" or "oily" acid) present in strong ("caustic") alkali (e.g., KOH) and absent in mild alkali (e.g., K_2CO_3). Addition of *acidum pingue* actually corresponded to loss of CO_2 much as the addition of phlogiston corresponded to the loss of another invisible gas—oxygen.

well might he have said it since he confessed to consuming 1200 pounds of chalk over eight years to cure his own violent stomach acidity.[4]

Until 1756, the only known gas or "air" was indeed common air. In that year, Dr. Joseph Black published his paper on the isolation and properties of "fixed air" (carbon dioxide or CO_2).[5] He had used the pneumatic techniques of Stephen Hales and William Brownrigg to capture the "air" that was "fixed" in chalk ($CaCO_3$). Although Van Helmont had worked with this "air" over 100 years earlier, neither he nor other contemporaries truly characterized it.

In Black's time only three major alkalis were recognized:[5] vegetable, marine, and volatile. Each of these was found in "mild" and "caustic" forms. Black's careful quantitative studies correctly convinced him that loss of "fixed air" is what converted mild alkalis to caustic alkalis. In modern terms, this can be summarized as follows:

Alkali Family	"Mild" Alkali	$\underset{+H_2O}{\xrightarrow{-CO_2}}$	"Caustic" Alkali
Vegetable	K_2CO_3		KOH
Marine	Na_2CO_3		NaOH
Volatile	$(NH_4)_2CO_3$		NH_4OH

Meyer's theory was essentially the reverse of Black's.[1] Meyer's *acidum pingue* ("fatty" or "oily acid") was said to be a component of all acids. When the mild alkalis (which Black understood to be carbonates) were reacted with acids, the effervescence indicated absorption of the *acidum pingue* found in all acids. Caustic alkalis were, as noted above, saturated with *acidum pingue* and thus did not effervesce when reacted with acids. The slippery feeling of caustic alkalies arose from the "oily acid" saturating them. Figure 178 depicts a table of affinities with *acidum pingue*.

At this point, parallels with phlogiston theory become all too apparent.[1] Metal calxes were said to gain phlogiston and become metals when heated with charcoal, a substance laden with phlogiston. Lavoisier established that, to the contrary, the calxes actually *lost* oxygen to the charcoal to form CO_2. According to Meyer, mild alkalis were said to gain *acidum pingue* from the fire to become strong alkalis. Black established that, in fact, they *lost* CO_2 in these transformations.

To add to this confusion, however, conversion of a metal such as calcium to its calx in the fire would require adding *acidum pingue* from the fire. The calx is indeed heavier. However, addition of *acidum pingue* to mild alkalis to form caustic alkalis would suggest that the latter should also be heavier. They are, however, lighter because of loss of carbon dioxide during the transformation. The source for this "confusion within confusion" is that phlogiston (*terra pinguis* or "fatty earth") is the essence of fire internal to a combustible substance. In contrast, *acidum pinguis* is a component of fire that acts as an external agent to the combustible material. Black and Lavoisier "trimmed the fat" from both *acidum pingue* and *terra pinguis* or, perhaps, emulsified these theories.

1. J.R. Partington, *A History of Chemistry*, Vol. 3, MacMillan & Co. Ltd., London, 1962, pp. 145–146.
2. "There is truth in chalk" is self-evident to any teacher who has not been fully converted to the computerized classroom.
3. (J.)F. Meyer, *Essais de Chymie, Sur La Chaux Vive, La Matiere Elastique et Electricque, Le Feu, Et L'Acide Universel Primitif, Avec un Supplement Sur Les élements* (M.P.F. Dreux, transl.), Vol. 1, Chez G. Cavalier, Paris, 1766, p. xv.
4. J. Ferguson, *Bibliotheca Chemica*, Vol. II, Derek Verschoyle Academic and Bibliographical Publications Ltd., London, 1954, p. 93.
5. Partington, op. cit., pp. 135–143.

SECTION V
THE CHEMICAL REVOLUTION

PEAS PRODUCE LOTS OF GAS

Stephan Hales (1677–1761) studied theology in Cambridge and became an active priest but preferred scientific pursuits.[1] He performed important studies on the hydrostatics of fluids in plants (*Vegetable Staticks* . . . , London, 1727) and blood flow (*Statical Essays: containing Haemastaticks*. . . , London, 1733—"a gruesome book"[1]). His studies on "airs" were performed between 1710 and 1727. Hales is considered to be the originator of pneumatic chemistry—the collection and manipulation of isolated gases.[1] The distinguishing characteristic of his apparatus was the separation of the collected gases from their sources.

In Figure 179 (see *Figure 33*) retort *r*, holding matter for distillation, is joined to the large long-necked flask *ab* using cement (tobacco pipe clay and bean flour well mixed with some hair) covered by a bladder. The large hole in the bottom of *ab* is for insertion of a glass siphon reaching to *z* inside the flask with the other end extending above the surface of the water container (*xx*) holding *ab*. Flask *ab*, while attached to *r*, is first immersed in a large bucket of water up to level *z* as excess air is pushed out through the siphon. Flask *ab* is then immersed into *xx*, which is filled with water. Heating of the vegetable matter in *r* produces "airs" that press the water level down from *z* to a new level *y*, which is carefully marked on the flask. The apparatus is allowed to cool to room temperature, r disconnected, and the top of *ab* corked. Inverted flask *ab*, first emptied of water, is filled to *z* and the mass of water determined. It is then filled to *y* and the total mass determined. The difference is the mass of water and therefore the volume of the gas generated. (Sometimes after cooling, there is actually net uptake of gas by the matter remaining in *r*.)

In Figure 180 (see *Figure 36*) we see a "strong Hungary-water bottle" having mercury at the bottom and otherwise filled with peas soaking in water. An evacuated glass column closed at the top and extending below the mercury pool at the bottom is tightly sealed at the top of the bottle. The peas absorb all of the water, and after two or three days, the gas they produce supports a column nearly 80 inches high (about 2.5 atmospheres of pressure). In Figure 180 (see Figure 37) we see a strong iron vessel *abcd* that is 2.5 inches in diameter and 5 inches deep, filled with peas soaking in water over a pool of mercury. In this homely but clever apparatus, a glass tube inside a concentric iron cylinder (for protection) has a drop of honey (*x*) at the bottom. The iron cover, closely fitted and sealed to the vessel with leather, is held closed with a cider press. After a few days, the press is loosened, pressure released, and the cover removed. Although the mercury column has fallen back to zero, a little dab of honey marks the spot (*z*) it arose to. The pressure was, again, about 2.5 atmospheres and corresponded to a force of about 189 pounds against the iron cover.

In Figure 181 (see *Figure 38*) we see the very famous Hales pneumatic apparatus in which various materials are heated in an iron gun barrel. Gases are collected in the inverted, suspended flask, which is initially filled with water. Only

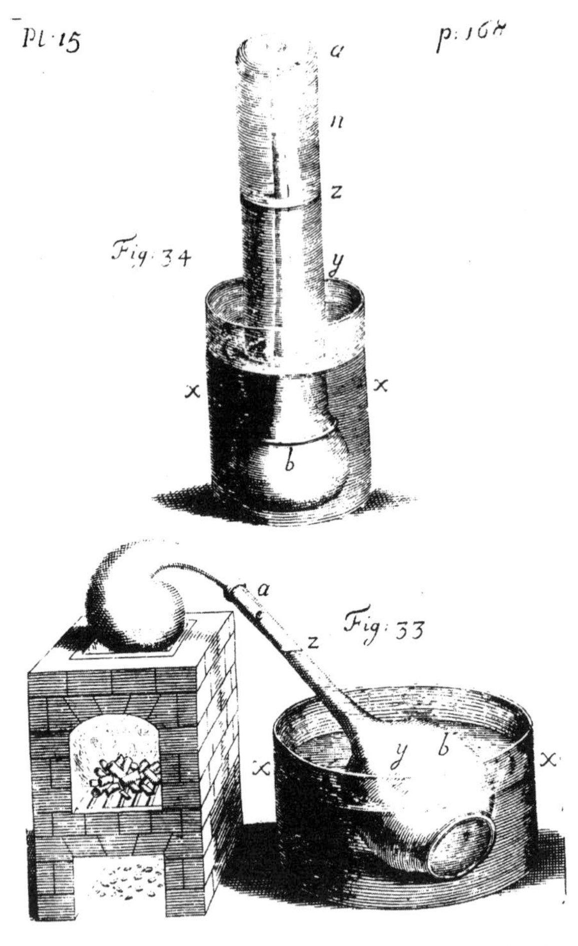

FIGURE 179. ■ Early pneumatic apparatus for measuring "airs" derived from distillation of vegetable matter [from the second edition of Stephan Hales' *Vegetable Staticks* (London, 1731); first edition, 1727].

water-insoluble gases can be collected in this manner. Water-soluble gases were subsequently collected over mercury or water coated with an oil layer.

Note the fascinating apparatus at the top of Figure 181 (*Figure* 39). Hales (*his* face?) breathes air from the sealed sieve bag by sucking through wooden soffet *ab*. At the bottom of *ab* (see *ib*) there is a valve that opens upon inhalation. A similar valve at *x*, entering the bag, is closed upon inhalation. The two valves switch roles upon exhalation. Hales found that he could perform inhalation–exhalation cycles for about 1.5 minutes using an empty bag. When the bag contained four flannel diaphragms [dipped in salt of tartar (K_2CO_3) solution and dried—this absorbs CO_2], he could breathe for 5 minutes. If the salt of tartar had been well-calcined (slightly basic due to some loss of CO_2 to form K_2O), he could breathe for 8.5 minutes.

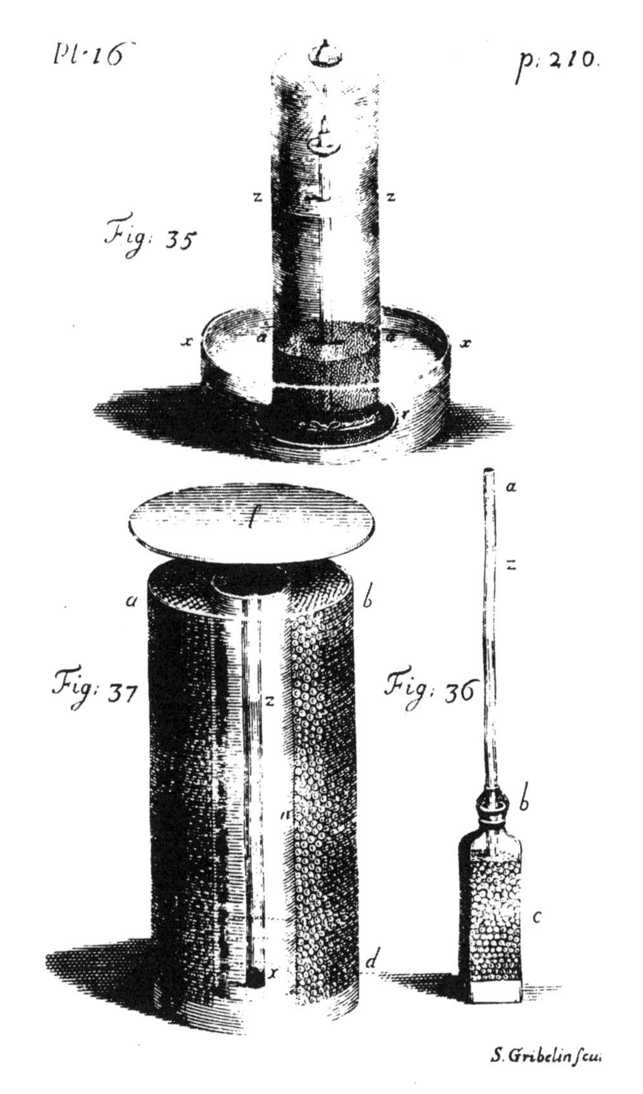

FIGURE 180. ■ Experiments measuring gases derived from peas (from Hales' 1731 edition of *Vegetable Staticks*).

1. J.R. Partington, *A History of Chemistry*, MacMillan, London, 1962, Vol. 2, pp. 112–123.

BLACK'S MAGIC

While Stephan Hales devised techniques of pneumatic chemistry to separate "airs" from their sources, he did not explore their differences in great detail. However, in 1756, in a continuation of his M.D. thesis (1754), Dr. Joseph Black (1728–1799) described the generation of an "air" that had been "fixed" in magnesium alba ($MgCO_3$) and released upon heating.[1] Moreover, he tested this "fixed air" and found that its properties were very different from those of every-

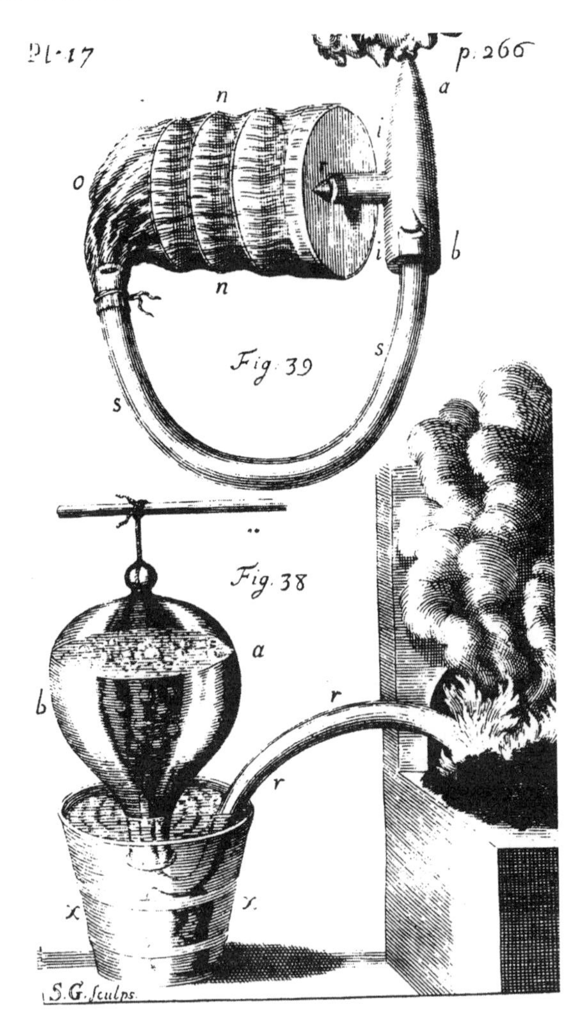

FIGURE 181. ■ Hales' early work collecting gases (1731 edition of *Vegetable Staticks*). The bottom figure shows gases collected from substances decomposed in the barrel of a gun and collected over water. This is the forerunner of the pneumatic troughs used by Scheele, Priestley, and Lavoisier to ignite the chemical revolution. The top figure depicts a bellows for collecting and recycling exhaled air. When the four diaphragms in the bellows were imbued with alkaline potassium carbonate, the breathing cycles would continue over longer periods due to removal of acidic carbon dioxide.

day air. For example, it extinguished flames rather than supporting them. The same "fixed air" is also generated when chalk ($CaCO_3$) is dissolved in acid. This "fixed air," when diffused into lime (CaO) water, would turn it cloudy by forming insoluble chalk. It is often said that until 1756 the only gas known was common air and that Black's discovery was the first of a pure gas. Actually, Van Helmont's studies in the seventeenth century involved discoveries of other gases that he recognized as different from common air, usually CO_2 often mixed with others, and he performed some characterizations. For example, Van Helmont knew that the poisonous gas that collects in mines (CO_2 with some CO) extinguishes flames.[2] However, his studies were not readily controllable and generally involved different mixtures of gases depending upon the source.

Black was a gifted teacher and his classic text *Lectures on the Elements of Chemistry* was published posthumously (Edinburgh, 1803; Philadelphia, 1807). He undoubtedly delighted and puzzled audiences by pouring "fixed air" (which is denser than common air) out of a glass to extinguish a candle flame. Black also showed that the same gas was generated by fermentation as well as by respiration since these emissions also turned lime water milky and were therefore CO_2.

Sometime during 1767–1768, Black filled a small balloon with hydrogen gas (newly discovered by Cavendish) and showed that it rose to the ceiling, surprising his audience who suspected that it was secretly raised by a black thread.[1] However, he argued against using hydrogen for manned balloons. In fact, the first hydrogen-filled balloon was flown by Jacques Alexandre Cesar Charles [yes, the ($V = kT$)-Charles-Law Charles] in 1783, the English Channel was crossed in 1785, and military balloons were flown as early as 1796.[1] Of course helium's discovery was about 100 years into the future. The explosion of the zeppelin Hindenburg over Lakehurst, New Jersey in 1937, with the loss of 36 lives, ultimately proved Dr. Black correct.

---

1. J.R. Partington, *A History of Chemistry*, MacMillan, London, 1962, Vol. 3, pp. 130–143.
2. J.R. Partington, *A History of Chemistry*, MacMillan, London, 1961, Vol. 2, pp. 229–231.

CAVENDISH WEIGHED THE EARTH BUT THOUGHT HE HAD CAPTURED PHLOGISTON IN A BOTTLE

Although we modern chemists go to some lengths to let the public know that we play tennis, like fast cars and stylish clothes, and are down-to-earth social-mixer types, we must admit that our passion for smelly, smoky mixtures will likely get us booted from most respectable country clubs. Henry Cavendish (1731–1810) was definitely an unworldly type. He lived with his father until the latter died in 1783, did not marry, communicated with his housekeeper using daily notes, and dressed in shabby, outdated clothing despite inheriting a fortune when he was 40.[1] The French physicist Jean-Baptiste Biot described him as "the richest of all learned men, and very likely also the most learned of all the rich."[1]

In our modern era when university tenure decisions are sometimes based upon the sheer poundage of publications, it is interesting to note that Cavendish published 18 papers in the *Philosophical Transactions of the Royal Society* (and no books).[2] He left many unpublished works and unstylishly referred to them in his published works.

But what works they were! In his first paper, published in 1766, Cavendish employed the pneumatic studies of Stephen Hales and Joseph Black to isolate hydrogen gas by pouring acids on metals such as zinc, copper, and tin. Indeed, the well-known affinities of these baser (more active) metals (see Geoffroy's Table of Affinities in Figures 175 and 176) for acids were long known to produce calxes. Moreover, the amount of gas collected did not depend on the identity of the acid (hydrochloric or sulfuric) or its amount but only on the quantity of metal. Thus,

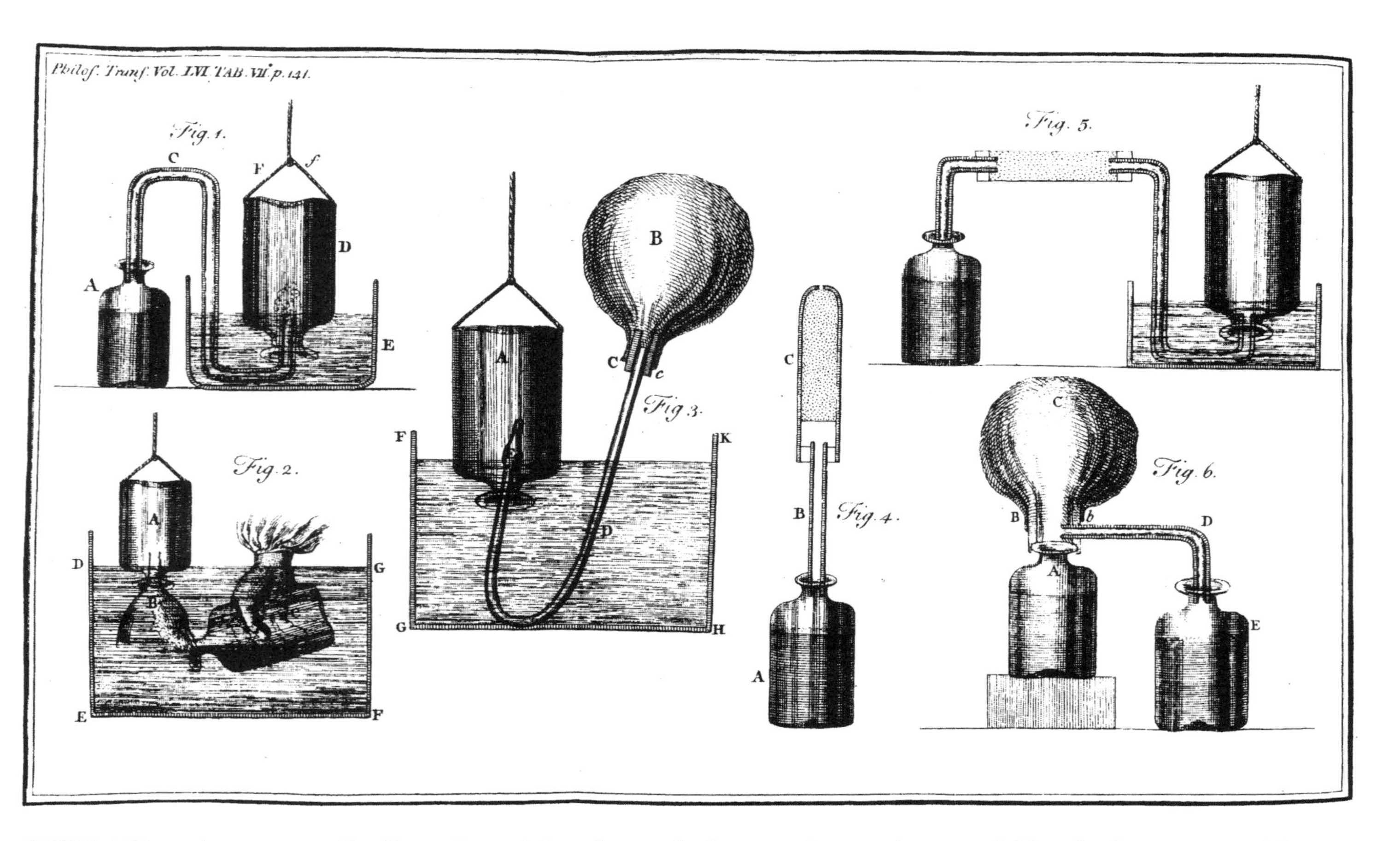

FIGURE 182. ■ Apparatus used by Henry Cavendish to discover hydrogen and manipulate gases [*Philosophical Transactions of the Royal Society (London)*, **LVI:** 141, 1766]. He thought that he had isolated phlogiston itself.

the metals were believed to lose their phlogiston to the air. The ignitable gas collected, which appeared to escape from the metal, was named "inflammable air" by Cavendish. It was less than one-tenth the density of atmospheric air and for a period Cavendish felt that phlogiston itself had been isolated. Figure 182 is taken from the 1766 work ("Three papers containing experiments on factitious Airs")[2,3] and shows in the panel labeled *Fig. 1* the collection of "inflammable air" over water; gases are transferred through funnels under water (*Fig. 2*). In *Fig. 3* we see transfer of gas into a bladder (a bit of wax is fixed to the end of the pewter siphon tube and then scraped off against the inside of the upper part of the bottle so as to keep water out of the tube). By pushing vessel *A* completely below the surface of the water in trough *FGHK*, all of the gas is pushed into bladder *B*, which is tied tightly around wood collar *Cc*, itself forming an airtight connection with the siphon tube with the aid of lute (almond powder made into a patty with glue). *Figure 4* shows the gas-generation vessel A filled with acid with metal added; glass tube *B* connects to C, which is filled with pearl ash (dry K_2CO_3, for removing aqueous acidic aerosols) and has a small opening at the top. The apparatus in *Fig. 4* allows determination of the weight of hydrogen lost from the top of C. *Fig. 5* shows collection of a gas through a drying tube (containing pearl ash) for probably the first time, and *Fig. 6* is an apparatus used to investigate the water solubility of "fixed air" (carbon dioxide).[2]

In 1784 Cavendish published his work on the composition of water based upon his experiments igniting hydrogen in air. (Primacy for the discovery that, once and for all, water is a compound and not an element was later given to James Watt). At this time he also noted that absorption of all of the oxygen (dephlogisticated air) and nitrogen (phlogisticated air) by chemical reaction left a tiny, but reproducible trace of unreactive gas. The apparatus is shown in Figure 183 ("Experiments on Air").[2] In *Fig. 1* of Figure 183 we see the apparatus used by Cavendish for conducting the experiment. Tube M is initially filled with mercury as are the two glasses. Gases are collected using the j tube in *Fig. 2* from a glass containing nitrogen or oxygen that is inverted in water. Exact gas volumes are cleverly introduced through tip A into tube M. Liquid containing the base and litmus indicator is also similarly introduced into tube M. Mercury serves as the container and electrical conductor for the sparking of known amounts of the two gases in the upper part of tube M. *Fig. 3* depicts an apparatus for repeated introduction of quantities of gas into tube M through tip A. In this work, Cavendish anticipated the discovery by Rayleigh and Ramsay of the inert gases (e.g., argon) 110 years later. With great admiration and respect, they quoted him extensively in their own prize-winning report.

In 1798, Cavendish applied Newton's gravitation law to an experiment involving two lead balls and two smaller spheres. In so doing, he accurately determined the mass of the earth.

Let's examine his tenure file: On the one hand, he had only 18 published papers and no books. On the other hand, he discovered hydrogen, was a vital contributor to understanding the composition of water, discovered nitrogen and the composition of the atmosphere, separated the inert gases from atmospheric air, and weighed the planet. His student evaluations indicate that they disapprove his choice of clothing and that they don't "identify with" him. He also has a low profile on campus and seems to avoid committee work Looks like this will be a difficult tenure decision.

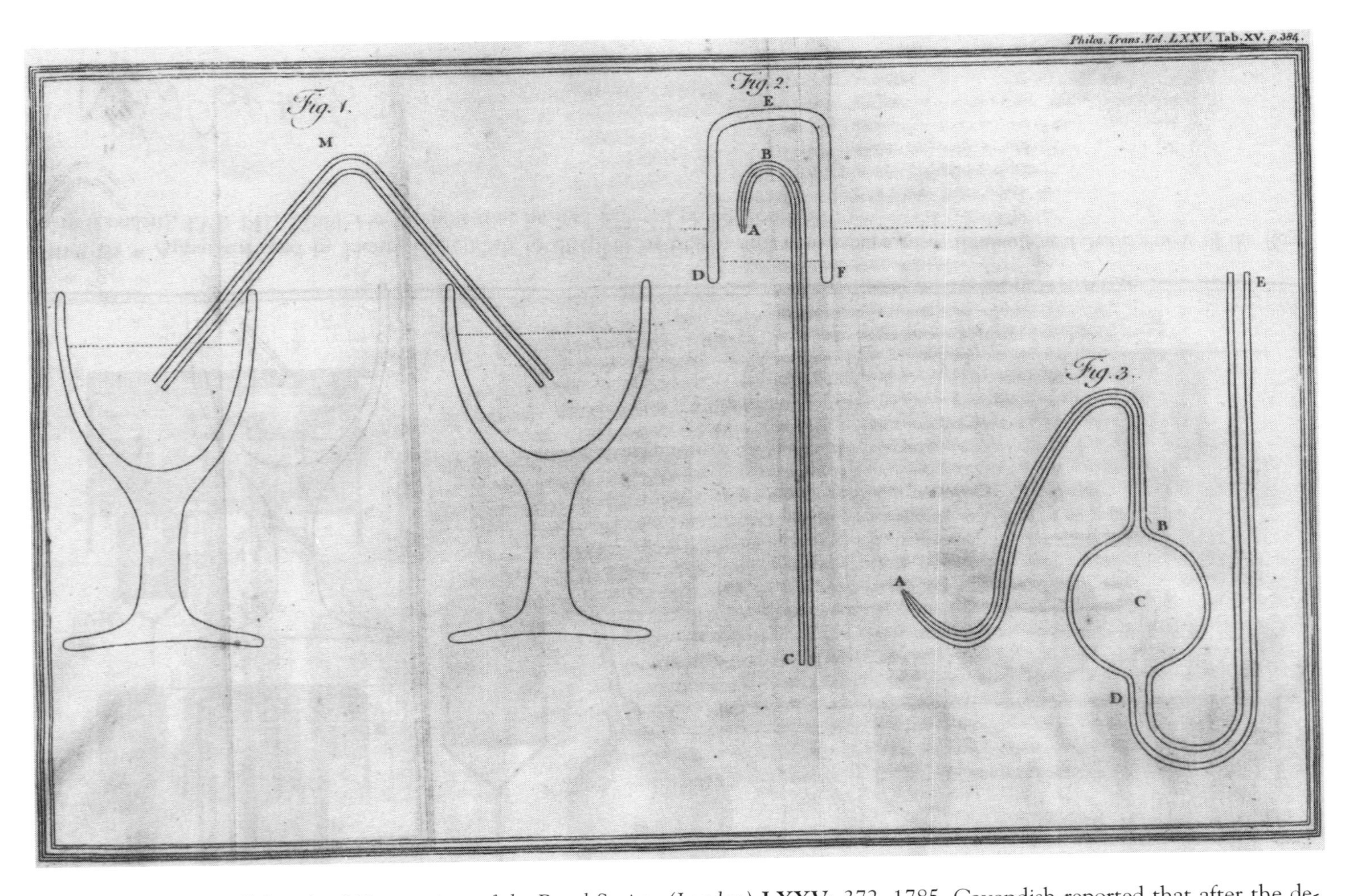

FIGURE 183. ■ In *Philosophical Transactions of the Royal Society (London)* **LXXV:** 372, 1785, Cavendish reported that after the dephlogisticated air (oxygen) was removed from atmospheric air, the remaining phlogisticated air (nitrogen) could be sparked with oxygen introduced into the vessel and the gas produced yields nitric acid when combined with water. However, total reaction of all of the nitrogen left a tiny bubble of unreactive residue. Cavendish had isolated the rare gases (mostly argon). Over 100 years later Rayleigh and Ramsay, who discovered argon, would express their great admiration for Cavendish's amazingly accurate work.

1. *Encyclopedia Brittanica*, 15th ed., Chicago, 1986, Vol. 2, pp. 974–975.
2. J.R. Partington, *A History of Chemistry*, MacMillan, London, 1962, Vol. 3, pp. 302–362.
3. "Factitious air" refers to gases derived from heating or other chemical actions on solids. Thus, hydrogen appears to be "liberated" from an active metal upon addition of acid and is, therefore, a "factitious air."

IN THE EARLY MORNING HOURS OF THE CHEMICAL REVOLUTION

Revolutions usually begin quietly, build momentum, and then explode with a cataclysmic event such as the storming of the Bastille. So, too, did the chemical revolution begin with early rumbles of discontent with phlogiston theory. The observed increases in weight as metals *lost* phlogiston to become their calxes were explicable only if chemists repressed all their experiences and common sense and accepted negative mass. Often, early acts of rebellion happen quite innocently, and the consequences are understood much later. Stephan Hales' publication in 1727 of his methods for collecting gases led to Joseph Black's report of carbon dioxide in 1756, the isolation of hydrogen gas by Henry Cavendish in 1766, and ultimately the isolation of oxygen gas by Carl Wilhelm Scheele in 1771 or 1772 and Joseph Priestley in 1774. Cavendish initially thought that he had actually captured the phlogiston released from metals by the aqueous acids that converted them to calxes. He could not possibly have understood that the gas actually came from the "element" water and not the metal. Scheele and Priestley imagined the "air" they captured as strongly attracting phlogiston, the essence of fire, from combustibles and metals. Perhaps *the* cataclysmic event in the chemical revolution was the full realization in 1783 that Cavendish's phlogiston and Priestley's dephlogisticated air combined to give water and did not simply disappear with a "poof" into "air" or simply *nothing*.

Figures 184–193 are from the fabulous 35-volume, folio encyclopedia of philosopher Denis Diderot (1713–1784) and mathematician Jean Le Rond D'Alembert (1717–1783), published between 1751 and 1772, with later supplements (1776–1780).[1] Diderot's progressive philosophy had earlier run afoul of the reactionary French Church and State and he spent three months in prison during 1749.[2] The encyclopedia, suffused with the more progressive thinking of the Enlightenment, is still highly treasured for its exquisite printing as well as content. The elegant symbols in Figures 184–187 illustrate the "Babel" of chemical nomenclature that existed prior to publication by De Morveau, Lavoisier, and their colleagues of the *Nomenclature Chimique* in 1787. These figures combine elements, alloys and other mixtures, compounds, chemical operations, quantities, glassware, and other apparatus. We will very briefly describe only a few of these.

The ancients likened the seven known metals to the sun, the moon and the five known planets (Mercury, Venus, Mars, Jupiter, and Saturn)[3]—see Figures 127 and 128. The first symbol in column 1 of Figure 184 represents the alloy steel (*Acier*), and its resemblance to iron (*Fer*—Figure 185, column 1) is obvious. *Limaille d'Acier* (Figure 185, column 2) refers to steel shavings or powder. The shield and spear of the god of war Mars symbolized the metal widely used in

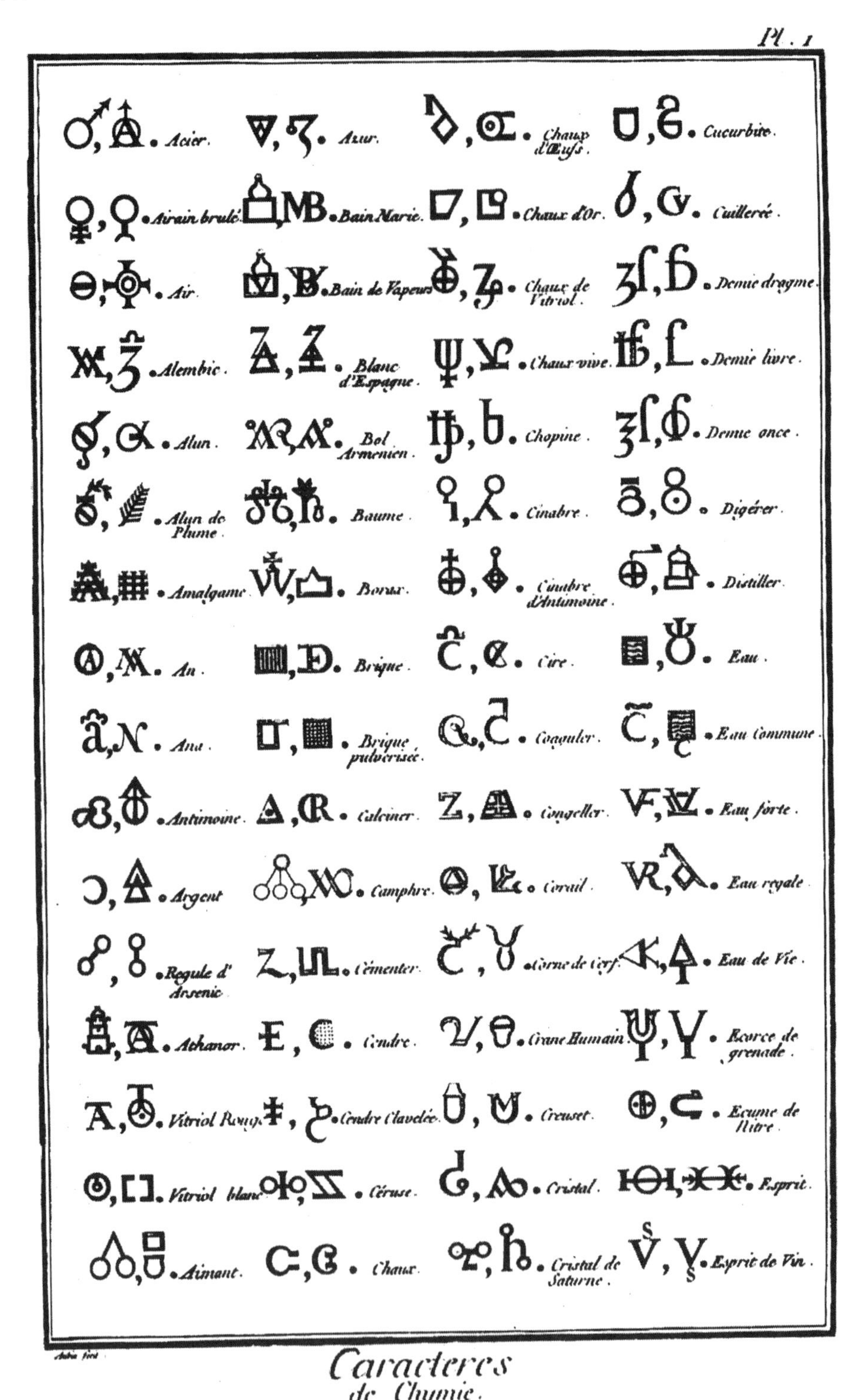

FIGURE 184. ■ Chemical symbols (see text) from the eighteenth-century encyclopedia published by philosopher Denis Diderot and mathematician Jean Le Rond D'Alembert.

Caracteres de Chymie.

FIGURE 185. ■ Chemical symbols (see text) from the eighteenth-century encyclopedia published by Diderot and D'Alembert.

Pl. III.

Pinte.	Safran de Venus.	Soufre noir.	Tutie.
Plomb.	Salpêtre.	Soufre des Philosophes.	Tutie Sublimée.
Poudre.	Sandarac.	Soufre vif.	Verd de gris.
Précipiter.	Sang de Dragon.	Stratifier.	Verre.
Purifier.	Savon.	Soufre des Prophetes.	Verre d'anti.
Putréfier.	Sel Alembroch.	Sublimé.	Vin.
Quarteron.	Sel alkali.	Sublimé de Mercure.	Vinaigre.
Quinte Essence.	Sel Armoniac.	Sublimé fait avec Soufre.	Vinaigre blanc.
Réalgar.	Sel commun.	Sublimer.	Vinaigre distilé.
Régule d'Antimoine.	Sel gemme.	Talc.	Vinaigre rouge.
Retorte.	Sel des Pélerins.	Talc noir.	Vin blanc.
Sable.	Sel de Plomb.	Thérébentine.	Vin rouge.
Safran.	Sel de Tartre.	Terre.	Vitriol.
Safran magistral.	Sel d'Urine.	Teste morte.	Vitriol blanc.
Safran de Mars.	Soude.	Sigillum.	Vitriol bleu.
Saffraner.	Soufre commun.	Tartre.	Urine.

Caracteres de Chymie.

FIGURE 186. ■ Chemical symbols (see text) from the eighteenth-century encyclopedia published by Diderot and D'Alembert.

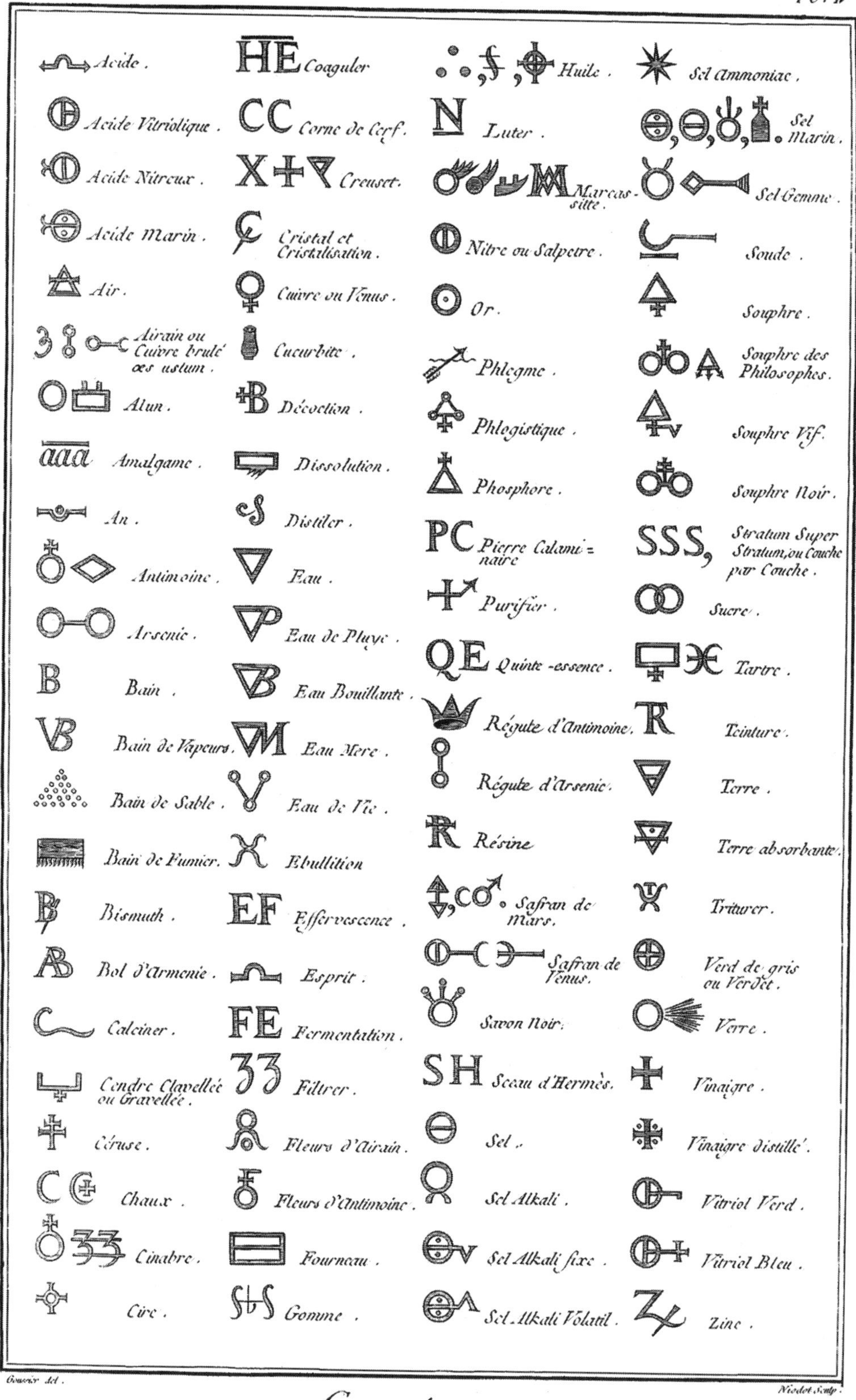

FIGURE 187. ■ Chemical symbols (see text) from the eighteenth-century encyclopedia published by Diderot and D'Alembert.

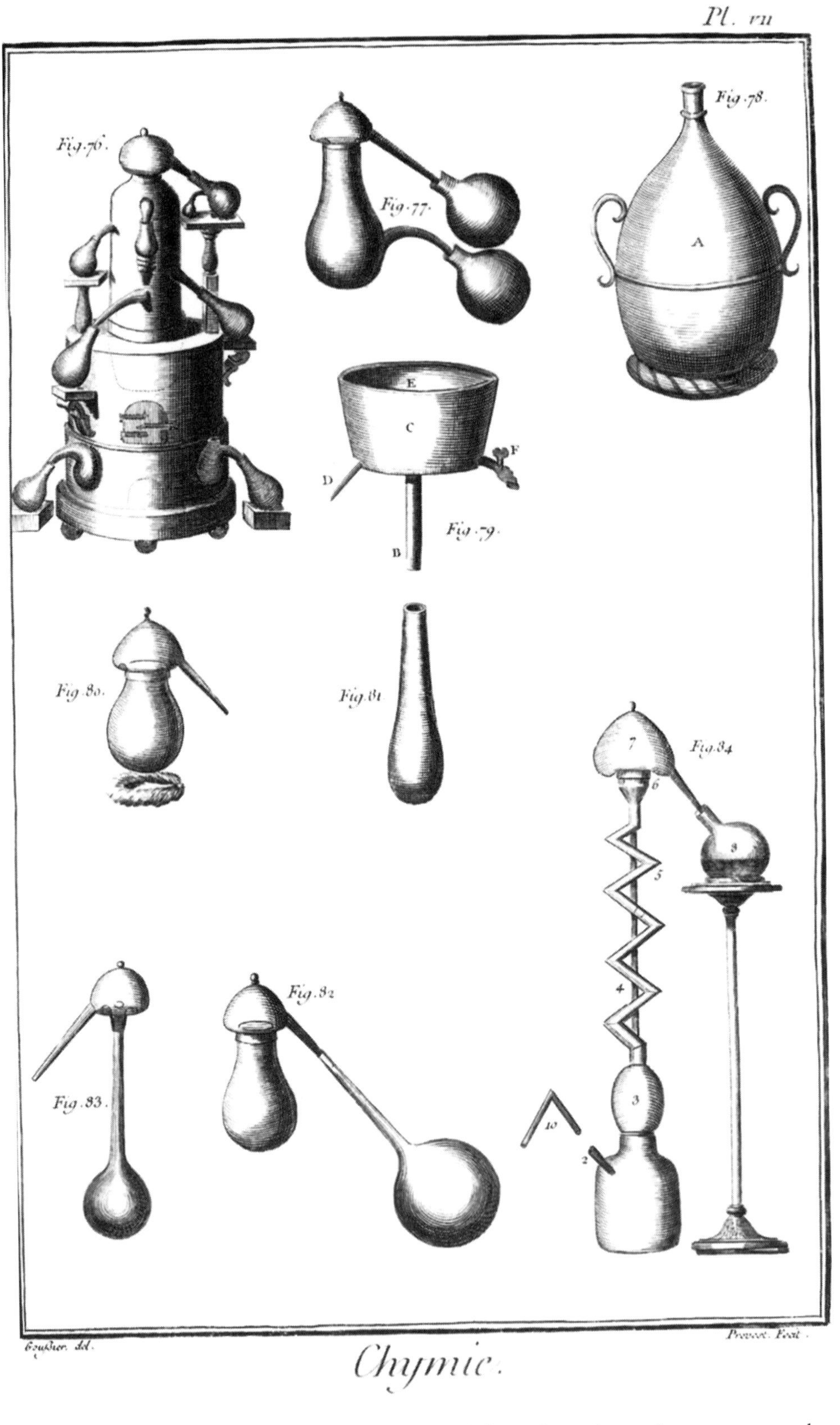

FIGURE 188. ■ Distillation apparatus (see text) from the eighteenth-century encyclopedia published by Diderot and D'Alembert.

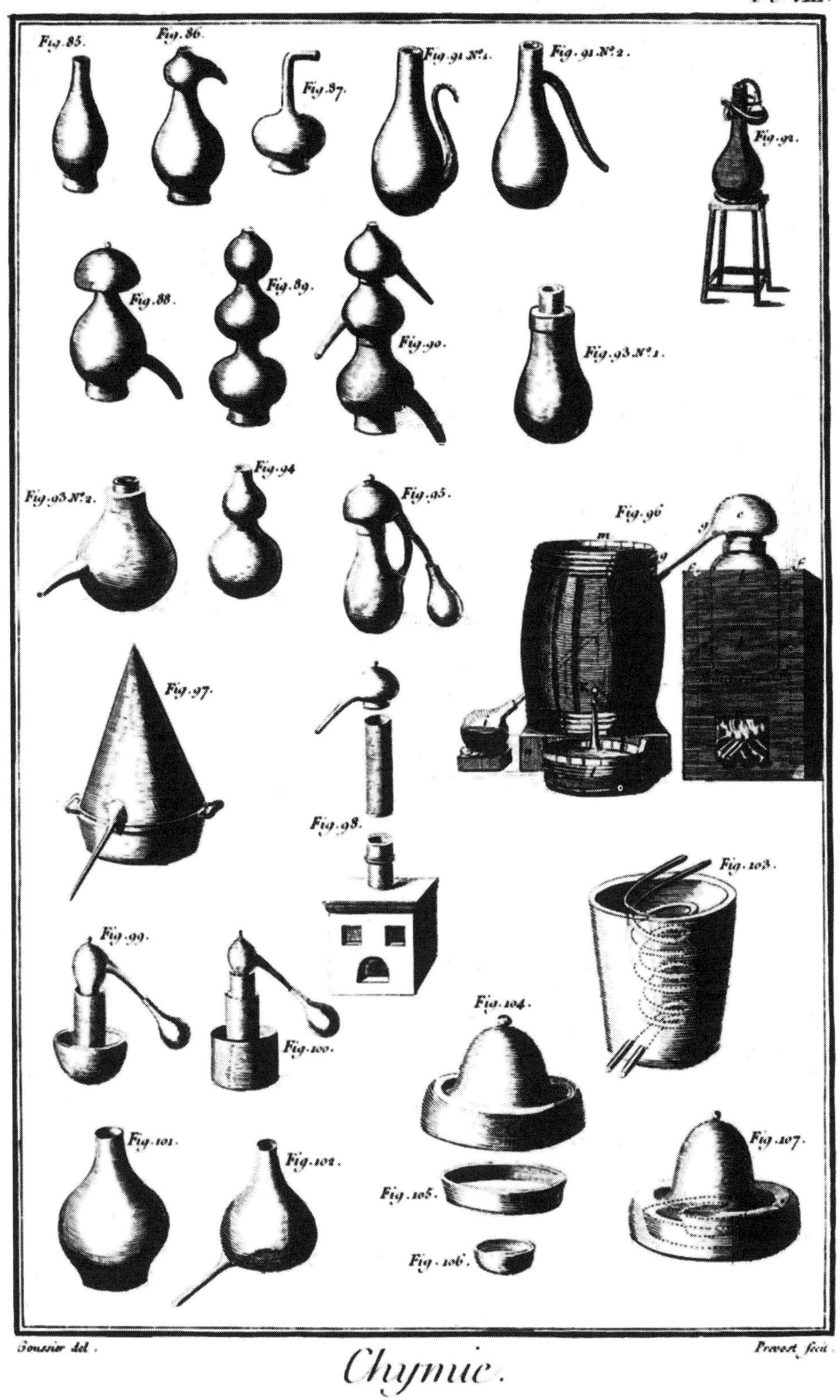

FIGURE 189. ■ Distillation apparatus (see text) from the eighteenth-century encyclopedia published by Diderot and D'Alembert.

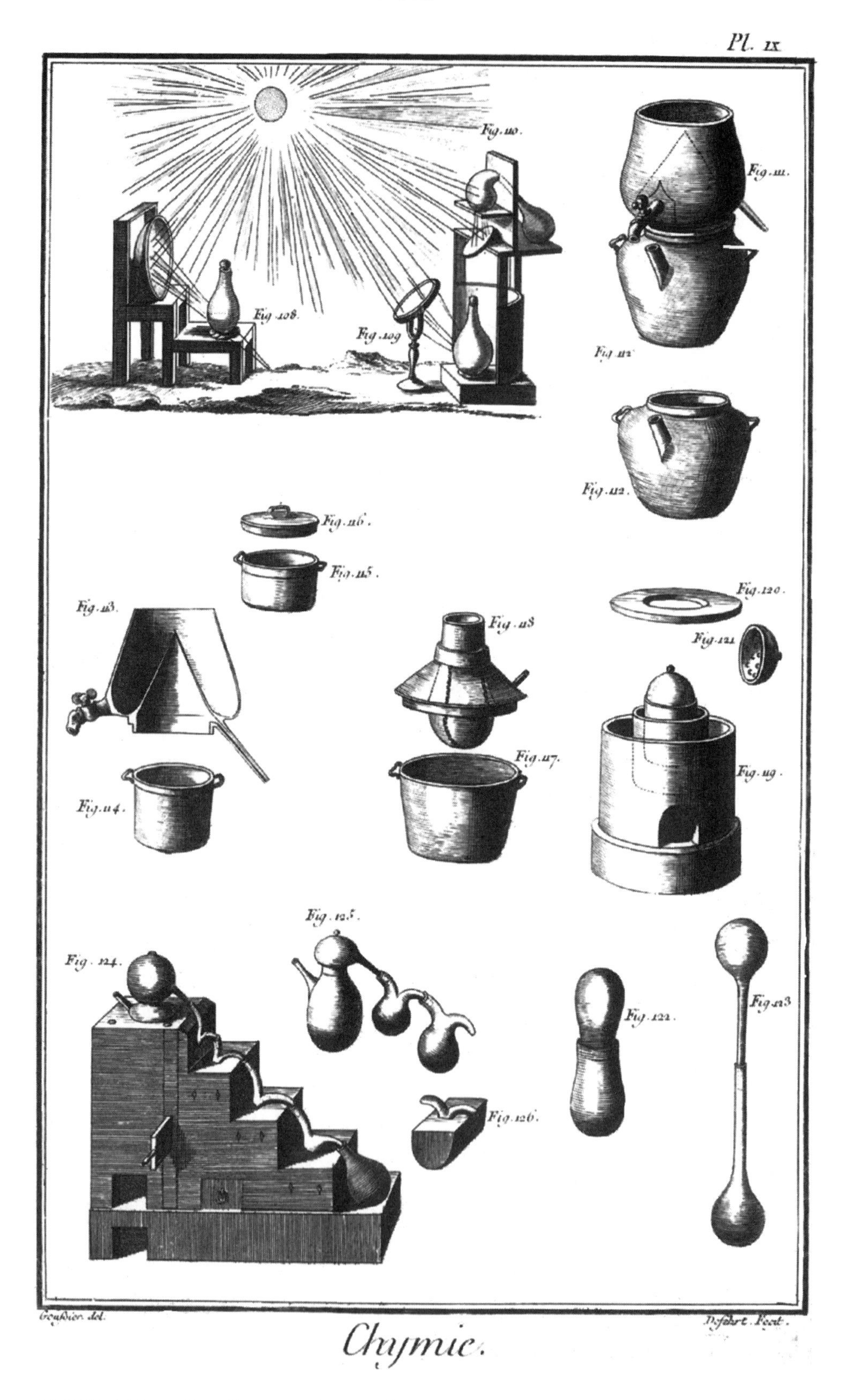

FIGURE 190. ■ Distillation and recirculation apparatus (see text) from the eighteenth-century encyclopedia published by Diderot and D'Alembert.

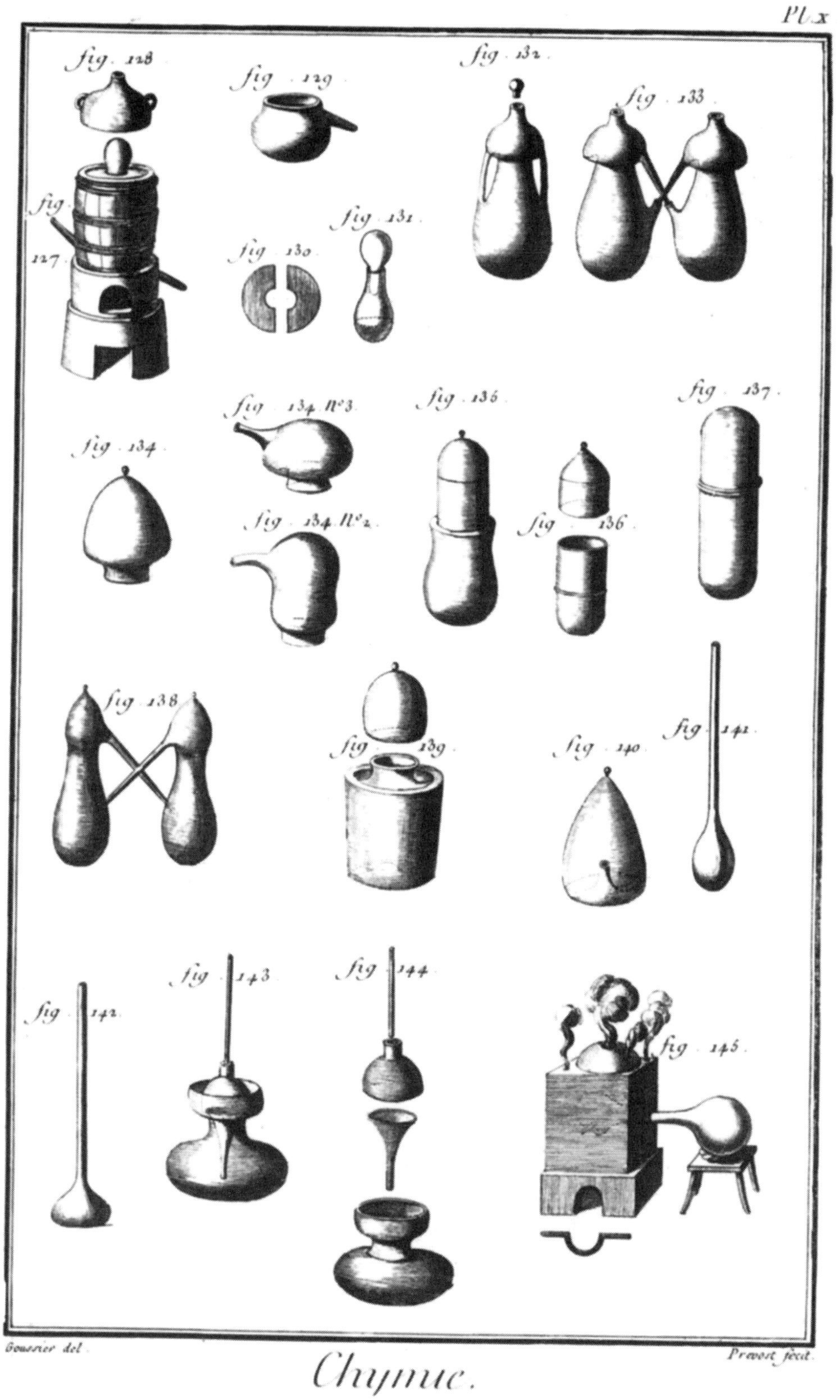

FIGURE 191. ■ Various types of recirculation apparatus (see text) from the eighteenth-century encyclopedia published by Diderot and D'Alembert.

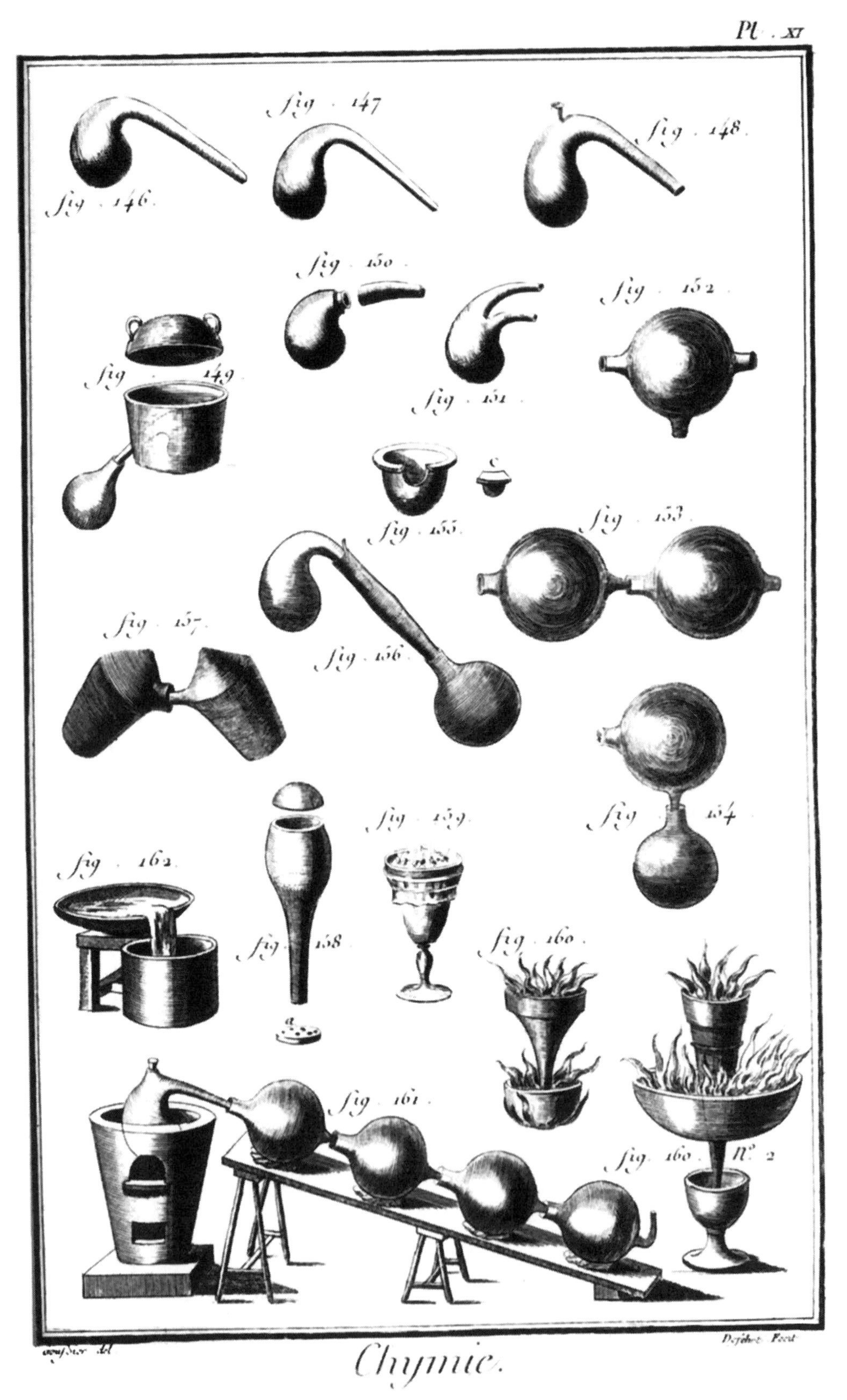

FIGURE 192. ■ Various retorts and "balloon" apparatus (see text) from the eighteenth-century encyclopedia published by Diderot and D'Alembert.

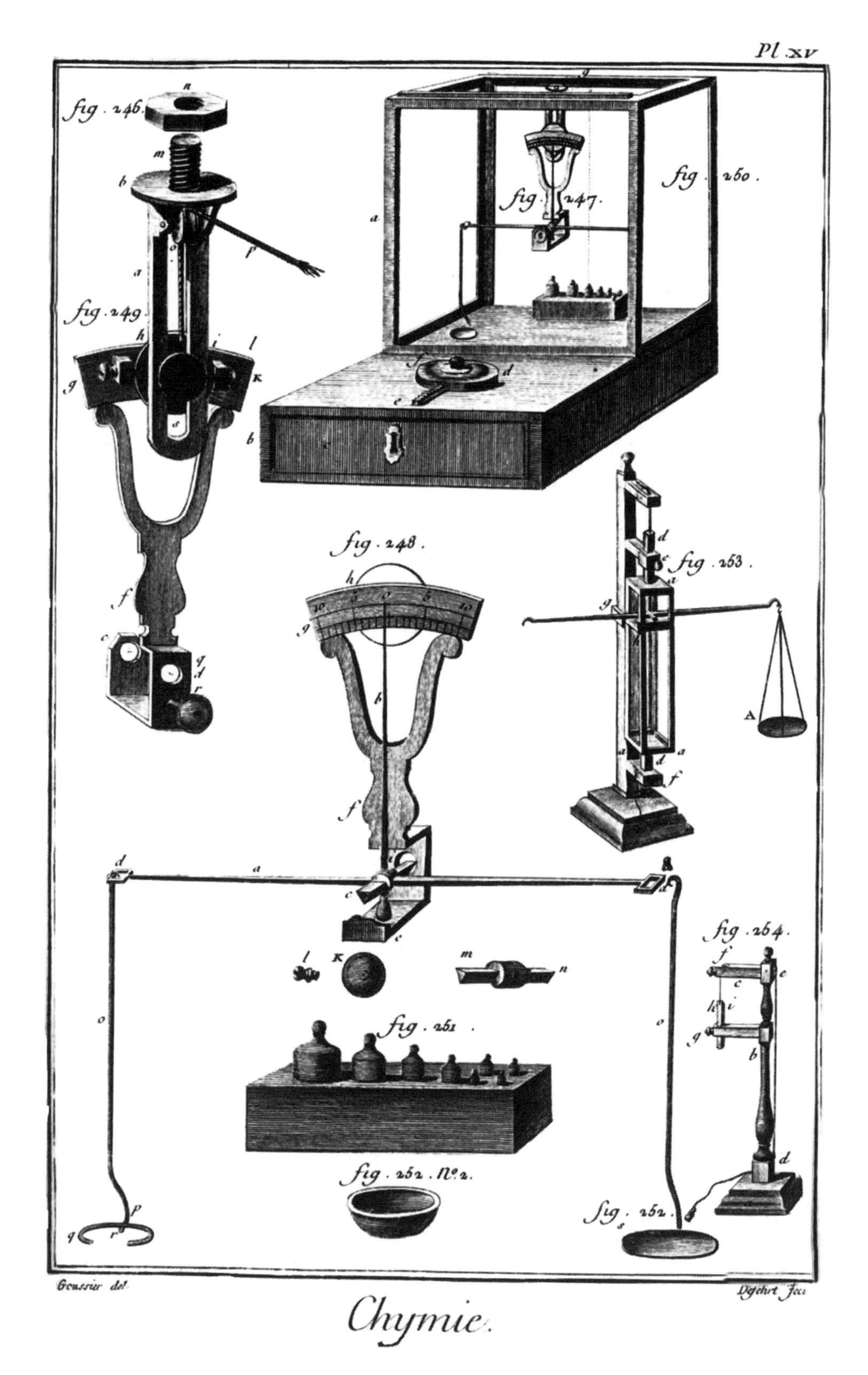

FIGURE 193. ■ Fully assembled chemical balance belonging to Guillaume François Rouelle, demonstrator at the *Jardin du Roi* in Paris, depicted in the eighteenth-century encyclopedia published by Diderot and D'Alembert. G.F. Rouelle's demonstrations inspired Antoine Lavoisier to enter chemistry. Rouelle was a firm phlogistonist, and in less than two decades, his former student would demolish phlogiston with the modern theory of oxidation.

weaponry, whose red calx suggested the red planet. *Airain brulé* (Figure 184, column 1) is roasted brass (or bronze), an alloy of copper (*Cuivre* or Venus, Figure 187, column 2—the mirror of Venus) and tin (*Estain*, Figure 185, column 1—Jupiter—the symbol may be derived from the Arabic four possibly because Jupiter is the fourth most distant planet from earth[3]). The outermost planet visible to the naked eye was Saturn—the slowest-moving according to earth-bound observers. Dark, dense lead (*Plomb*, Figure 186, column 1) or Saturn is symbolized by the scythe of the slow-moving god of sowing or seed. The caduceus of Mercury is itself a symbol consisting of male and female snakes (Figure 185, column 3) entwined about the god's staff. Silver (*Argent*) is the moon (Figure 184, column 1) and gold (Or) the sun (Figure 185, column 4).

Antimoine (Figure 184, column 1) actually represents stibnite (Sb_2S_3), while "regulus" (the refined state) of antimony (Figure 186, column 1) is actually the pure element.[4] "Flowers of antimony" (Figure 187, column 2, symbol 21) refers to what we now recognize as antimony oxide (Sb_2O_3) ("flowers" symbolize volatile salts that may be sublimed).[4] In Figure 186 there are no less than five listings for sulfur (*soufre*): common, black (*noir*), living or natural (*vif*), "philosophical sulfur" (one of Paracelsus' three principles "actually" derived alchemically from gold—but never mind) and "Sulfur of the Prophets." "Spanish white" (Figure 184, column 2, symbol 4)[4] can be bismuth oxychloride or bismuth oxynitrate but I see no resemblance to the symbol for bismuth itself (Figure 187, column 1). The *Bain Marie* (Figure 184, column 2, symbol 2) is the hot-water bath developed in ancient times by Maria Prophetessa (Mary the Jewess—a genuine historical alchemist[5]) for gentle controlled heating of chemical vessels. The following symbol represents a steam bath. *Aimant* (Figure 184, column 1) is a magnet. Phlogiston, of course, occupies an honored place (Figure 187, column 3). Incidentally, Phlegme, just above phlogiston, refers to an aqueous distillation fraction and not to the "yucchy" stuff the Brits call "catarrh." *Cornede Cerf* (Figure 184, column 3) is deerhorn. "Blood of the Dragon" (often symbolizing the Philosopher's Stone[6]—see Figure 48) is listed in Figure 186 (column 2, symbol 4).

Some of the glassware and other apparatus that preceded and ushered in the chemical revolution are depicted in Figures 188–193. Many of these predated the revolution by hundreds of years. In Figure 189 a complete alembic (*Fig.* 86) is accompanied by its cucurbit (*Fig.* 85) minus still head. *Fig.* 89 and *Fig.* 90 each depict stacks of three aludels, often used for sublimation. *Fig.* 97 is referred to as a "chapel of the ancients used to distill their rose water"; it is often referred to as a "rosenhut," which acts as a primitive air condenser.[7] *Fig.* 96 shows a complete distillation apparatus for obtaining spirit of wine, aromatic spirits, and essential oils. A condenser pipe passes through a barrel filled with ice water. An even more effective condenser system is the "double serpent" winding through a chamber filled with ice water. One serpent can be used to condense spirit of wine, the other for aromatic spirits or essential oils—drink and perfume, separate and not co-mingled. *Fig. 107* depicts an eighteenth-century apparatus for distillation using the heat of the sun. The inner glass bowl containing the sample to be distilled is placed in a larger earthen bowl covered by a glass bell jar. The apparatus is set in a base, and volatiles from the inner vessel collect on the inside of the bell jar, flow down the sides, and collect in the larger bowl.

The top of Figure 190 depicts three different means for using the sun to provide heat for distillation. *Figs. 122* and *123* show a double cucurbit and a double matras joined so as to allow reflux or "cohabation" (redistillation or "circulation"). The wonderful apparatus in *Fig. 124* also serves for cohobation. In Figure 191, the pelican (*Fig. 132*) and double pelicans (*Figs. 133 and 138*) also serve the purpose of cohobation. *Fig. 127* (with *Fig. 128*) is an apparatus designed to provide a smoke bath for a reaction vessel. *Fig. 143* depicts the "Enfer de Boyle" or Robert Boyle's "hell"—a piece of apparatus for fully sealing airtight a chemical mixture that will be strongly heated by fire.

Figure 192 shows five styles of retorts at the top. The glass balloons depicted in *Fig. 153* are combined to form an apparatus (*Fig. 161*) for distillation of vapors of antimony powder used in a canon powder also including charcoal and saltpetre. *Fig. 159* is a glass *descensum* apparatus for filtration of melts. If a melt tends to resolidify before purification, modified, heated apparatus such as *160* and *160 No. 2* may be employed.

Figure 193 shows details as well as the fully assembled assaying balance belonging to Guillaume François Rouelle (1703–1770).[8] Rouelle was demonstrator in chemistry from 1742 through 1768 at the *Jardin du Roi* in Paris. He was, not surprisingly, a firm believer in phlogiston. His dynamic lecture style has been described thusly:[8] "On entering the laboratory for his lecture he was correctly dressed in velvet and with a powdered wig, holding his three-cornered hat under his arm. As he warmed to his subject he dispensed with hat, wig, coat, and waistcoat in turn." One of Rouelle's entranced students[8] was a certain Antoine Laurent Lavoisier, who would eventually become the "master of the chemical balance" and, in so doing, overthrow the phlogiston theory that captivated his teacher.

1. *The Haskell F. Norman Library of Science and Medicine*, Part II, Christie's, New York, 1998, pp. 124–125.
2. *The New Encyclopedia Britannica*, Encyclopedia Britannica, Inc., Chicago, 1986, pp. 79–81.
3. J. Read, *Prelude to Chemistry*, The MacMillan Co., New York, 1937, pp. 88–89.
4. J. Eklund, *The Incompleat Chymist—Being An Essay on the Eighteenth-Century Chemist in His Laboratory, with a Dictionary of Obsolete Chemical Terms of the Period*, Smithsonian Institution Press, Washington, DC, 1975.
5. Read, op. cit., p. 128.
6. Read, op. cit., pp. 15, 195.
7. Read, op. cit., pp. 76–77.
8. J.R. Partington, *A History of Chemistry*, MacMillan & Co. Ltd., Vol. 3, 1962, pp. 73–76.

MAKING SODA POP

Joseph Priestley (1733–1804) began a religious odyssey at an early age and is now recognized as one of the founders of Unitarianism.[1,2] At 19 he entered the Dissenting Academy of Daventry to study for the Nonconformist Ministry, refecting the early influence of his aunt. By the age of 28 he taught languages (including Hebrew), history, law, logic, and anatomy at the highly regarded Dissenting Academy at Warrington. His scientific interests were well under way by this time—he had earlier purchased an air pump and an electrical machine. His scholarship was recognized with an LL.D. from Edinburgh in 1765, and, with

Benjamin Franklin's encouragement, Dr. Priestley published his *History of Electricity* in 1767 (he published a similar book on vision, light, and colors in 1772).

Priestley began his studies on "airs" (he disliked Van Helmont's term *gas*) around 1770. His home in Leeds was next to a brewery and Priestley collected "fixed air" (CO_2) directly from the surface of the brewing mixtures and investigated its properties. He also obtained this gas by heating natural mineral waters and recommended it for revitalizing flat beer.[1] His 1772 pamphlet (Figure 194) was addressed to the Right Honourable John Earl of Sandwich, First Lord Commissioner of the Admiralty. Any modern grantsman will recognize a final report for a Department of Defense contract in it. By pouring dilute oil of vitriol (sulfuric acid) on chalk (calcium carbonate) Priestley generated "fixed air" and impregnated water with it. This articial soda was more readily available and cheaper than the carbonated waters from spas so many of which were, unfortunately, located in the borders of the hated enemy France.

Carbonated water had long been reputed to prevent "the sea scurvy" on long voyages and to slow the putrefaction of water. In addition, it settled upset stomachs and acted as something of a substitute for the fresh vegetables that aid

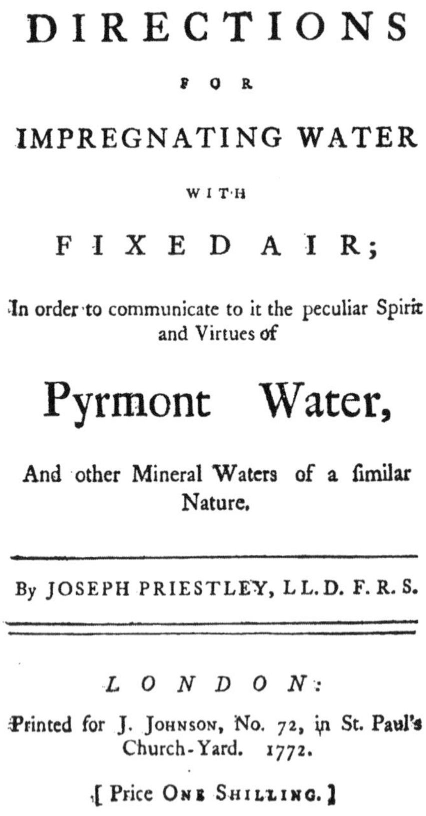

DIRECTIONS

FOR

IMPREGNATING WATER

WITH

FIXED AIR;

In order to communicate to it the peculiar Spirit and Virtues of

Pyrmont Water,

And other Mineral Waters of a ſimilar Nature.

By JOSEPH PRIESTLEY, LL.D. F.R.S.

LONDON:

Printed for J. JOHNSON, No. 72, in St. Paul's Church-Yard. 1772.

[Price ONE SHILLING.]

FIGURE 194. ■ Joseph Priestley made artificial soda by pressurizing water with chemically generated carbon dioxide. The work was vital to the strategic interests of the Royal Navy since carbonated water remained fresh longer than untreated water and was useful for treating upset stomachs. Fearful of a strategic "soda-pop gap," French spies reported this scientific advance and research was ordered. The young researcher? Antoine Laurent Lavoisier, who commenced the chemical revolution.

digestion. Priestley thus helped Brittania to "rule the waves." There is nothing like soda pop to help sailors down a ship's store of salt pork. Whether or not the meat was pressed between slices of bread is not clear from this slim pamphlet.

During this period of time, a Portuguese monk named Joao Jacinto de Magalhaens[3] (Magellan for short, a descendant of the famous navigator) was employed in England as a spy for France. He recognized the importance of a potential treatment for scurvy on the high seas and sent a copy of Priestley's pamphlet to France. Clearly, a strategic "soda-pop gap" between England and France was intolerable. The person in France who was requested (i.e., assigned) to study this chemistry? Antoine Laurent Lavoisier. It was the start of his pneumatic researches that ultimately revolutionized chemistry.[3]

1. J.R. Partington, *A History of Chemistry*, MacMillan, London, 1962, Vol. 3, pp. 237–297.
2. A.J. Ihde, *The Development of Modern Chemistry*, Harper & Row, New York, 1964, pp. 40–50.
3. J.-P. Poirier, *Lavoisier—Chemist, Biologist, Economist*, R. Balinski (translator), University of Pennsylvania Press, Philadelphia, 1996, pp. 51–54.

FIRE AIR (OXYGEN): WHO KNEW WHAT AND WHEN DID THEY KNOW IT?

Carl Wilhelm Scheele (1742–1786)[1,2] was born in Stralsund, Sweden, a Baltic Sea port that was ceded to Prussia in 1815 and is now in the northeastern part of Germany. He was the seventh child of eleven in a Swedish family and raised in very modest circumstances. The school that provided Scheele's solid elementary education did not offer a Gymnasium[2] (university-preparatory secondary education) course. Stimulated by his older brother's training, young Scheele was sent to the same pharmacy in Göteberg, under the supervision of Martin Bauch. Bauch was sophisticated and current in his chemistry, and Scheele began to bloom as a young chemist as he read the great texts of Johann Kunckel and Caspar Neumann and tried their experiments. When the pharmacy was sold in 1765, Scheele moved to Malmö, where he worked as apprentice in a pharmacy and met Anders Retzius, lecturer at the University of Lund.[1,2] At Malmö, his Master Kjellstrom described the young Scheele's reactions as he pored through texts: "that maybe; that is wrong; I will try that."[1] He was given considerable freedom to continue his experimental interests at Malmö. Tempted by the scientific sophistication of Stockholm, in 1768 Scheele accepted an assistant's position in a pharmacy but found his duties limited to preparing prescriptions.[2] In 1770, he joined a pharmacy in Uppsala, where he was finally given his own workbench.[2] In Uppsala he developed a friendship with young Johan Gottlieb Gahn (1745–1818), who would discover manganese four years later. Gahn introduced Scheele to Sweden's foremost chemist Tobern Bergman (1735–1784).

The Uppsala period (1770–1775) was one of incredible productivity for Scheele. He made numerous important discoveries, but his discovery of oxygen in 1771 or 1772 (possibly even earlier) is considered his single greatest work.[1] It is important to remark that Joseph Priestley independently discovered oxygen in

1774 and immediately published his findings. Scheele was apparently totally unaware of Priestley's work and submitted his book manuscript *Chemische Abhandlung von der Luft und dem Feuer* (eventual English title *Chemical Observations and Experiments on Air and Fire*[3]) to the printer in Uppsala in December 1775 as well as to Bergman in early 1776.[1] He first learned of Priestley's discovery of oxygen in August 1776, from a letter written to him by Bergman.[1] He was by that time already very frustrated with the slowness of Bergman's review of his manuscript as well as the printer's delay in publishing it.[1] Scheele's work, reporting his discovery of oxygen some six years earlier, was finally published in 1777 (Figure 195).[4] It would take over a century before Scheele's primacy in the discovery of oxygen would be widely accepted.[1,2]

Carl Wilhelm Scheele's
d. Königl. Schwed. Acad. d. Wissenschaft. Mitgliedes,
Chemische Abhandlung
von der
Luft und dem Feuer.
Nebst einem Vorberichte
von
Torbern Bergman,
Chem. und Pharm. Prof. und Ritter; verschied. Societ. Mitglied.

Upsala und Leipzig,
Verlegt von Magn. Swederus, Buchhändler;
zu finden bey S. L. Crusius.
1777.

FIGURE 195. ▪ Title page from the first (Upsala and Leipzig) edition of Carl Wilhelm Scheele's monumental work on air and fire. Although Scheele first discovered oxygen in 1771 or 1772, long delays in review by Bergman and publication allowed Joseph Priestley to be first in publishing oxygen's discovery (1774). (From the Roy G. Neville Historical Chemical Library, a collection in the Othmer Library, CHF)

Figure 196 is from the *Opuscula Chemica et Physica* (collected works in Chemistry and Physics, published in Leipzig in 1788–9).[5] In the upper left of Figure 196 (see *Fig. 1*) is shown a vial in which three teaspoonfuls of steel shavings and one ounce of water were placed and then followed by half an ounce of sulfuric acid. A cork with a hole containing a long, hollow glass tube was inserted tightly in the vial (*A*). The vial was placed in a vessel of hot water (*BB*, in order to accelerate solution). Flammable air (hydrogen) emerging from the top of the tube was then ignited by a candle and the flame positioned into the center of the 20-ounce retort C. When the fire self-extinguished, water from the vessel had climbed to level *D* in the retort. The space up to level *D* corresponded to four ounces of water. Thus, the volume of the original air in the retort had diminished

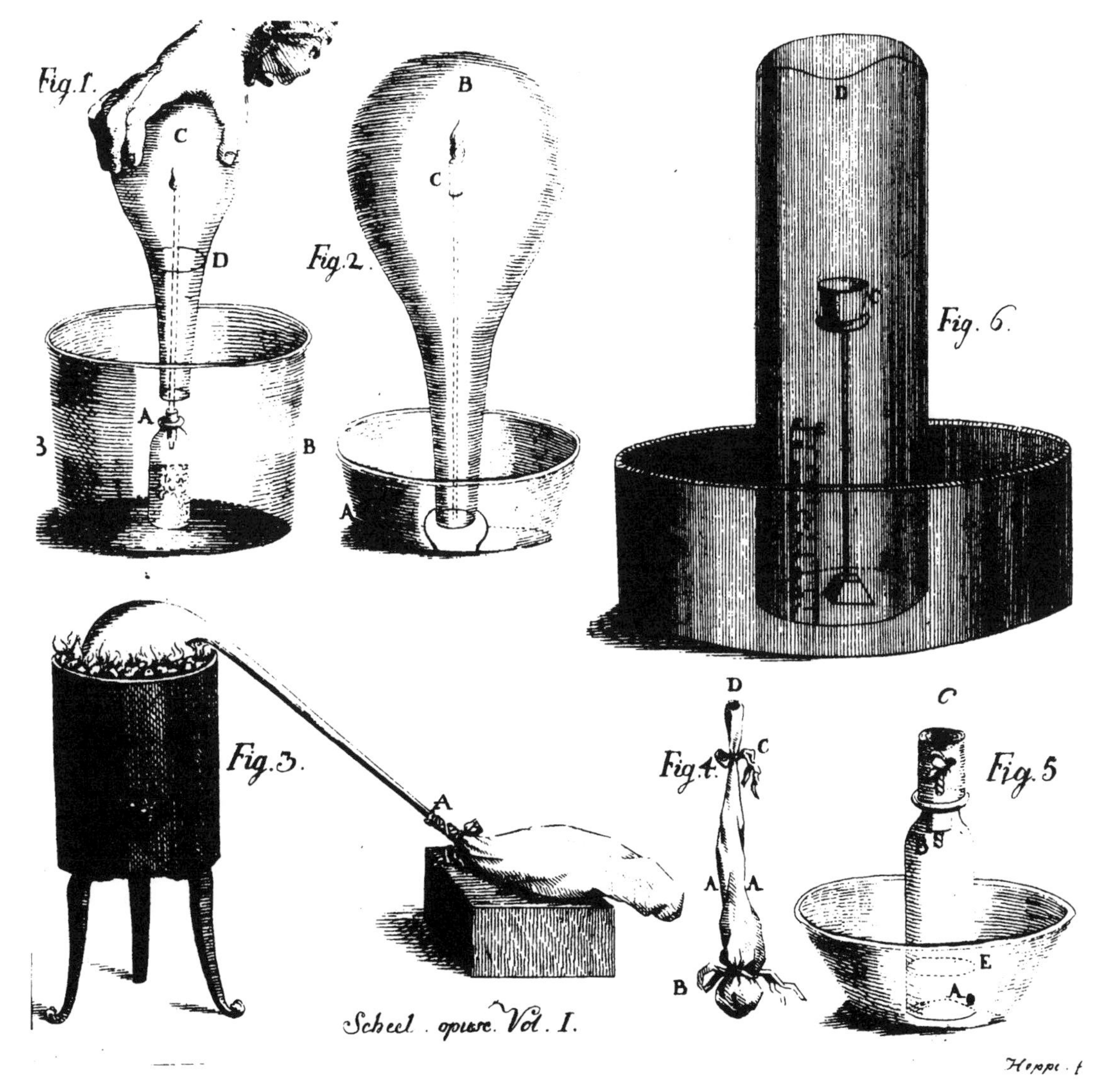

FIGURE 196. ■ Pneumatic experiments in which Carl Wilhelm Scheele was the first to discover oxygen ("fire-air"). These findings were first published in the exceedingly rare 1777 *Chemische Abhandlung von den Luft und dem Feuer* (Leipzig); first English edition (*Chemical Observations and Experiments on Air and Fire*, London, 1780). This figure is derived from *Opuscula Chemica et Physica* (Leipzig, 1788–89).

by one-fifth. Scheele added limewater to the retort and noted that no fixed air (carbon dioxide) was produced from this combustion. He did not, unfortunately, detect the small quantity of water vapor formed. Scheele concluded that there was a distinct gaseous component in atmospheric air that supports combustion. Scheele believed that phlogiston was absorbed from the flammable substance by this gaseous substance. But where did it go? He postulated that the imponderable phlogiston and the new gas combined to form a substance that escaped by penetrating the pores of the glass vessel.

Fig. (see Figure 196) depicts a similar experiment involving a candle. A tough waxy mass is placed in the bottom of dish A. A strong iron wire is planted into this mass at one end and pressed into the bottom of a candle at the other end C. The candle is lit, and an inverted retort is placed over it and fastened tightly onto the waxy mass below. Water is added to the dish. By the time the candle had extinguished, only 2 of the original 160 ounces of the retort's capacity were occupied by the water. Scheele found that carbon dioxide was created in this combustion and correctly reasoned that production of fixed air would soon extinguish the flame. He could not, at the time, understand that for every molecule of gaseous oxygen lost, one molecule of gaseous carbon dioxide would be formed, although this gas also has significant water solubility.

Scheele believed in the Phlogiston Theory and continued to do so throughout his life, as did Joseph Priestley. He felt that heat was a combination of phlogiston and what he called "fire air." His theoretical basis for this belief is nicely and succinctly described by Ihde.[6] Scheele reasoned that when a substance burned, it lost phlogiston, which combined with air, to some extent, increasing its mass and decreasing its volume. However, he found that the remaining "foul air" ("mephitic air" or nitrogen) was less dense rather than more dense than air. Thus, he reasoned, there was a component of common air he termed "fire air" that combined with phlogiston to produce heat, a kind of ethereal fluid, which escaped through the glass vessel. Scheele then decided to isolate "fire air" from heat by capturing the phlogiston using nitre. [Remember Mayow's experiments published almost 100 years earlier and depicted in Figure 150—nitre or salt-petre dephlogisticate (burn) charcoal or sulfur to produce the respective "acids."] Scheele's investigation involved heating saltpetre ("fixed" nitric acid) and capturing the "fire air":[6]

$$\text{Heat} + \text{Nitric acid} \rightarrow \text{Fire air} + \text{Red fumes}$$

Fig. 3 in Figure 196 shows the bladder employed to collect oxygen. Oil of vitriol (sulfuric acid) was added to saltpetre (KNO_3), and the gases produced included oxygen as well as red vapors of NO_2. These were collected in a bladder filled with milk of lime [$Ca(OH)_2$] solution that "fixed" the NO_2 vapors and left pure oxygen. *Figure 4* is another bladder apparatus. Its use can be illustrated as follows. Place chalk ($CaCO_3$) in the bottom of the bladder and tightly tie a knot at point *B*, sealing off the bottom. Add acid solution into the main part of the bladder AA, squeeze out residual air and tie this section off at C. Then fasten the opening *D* tightly around an inverted bottle stoppered tightly with a cork. When ready, the knot at *B* can be untied, that at *D* untied and the cork (still inside the bladder) removed and gaseous products (fixed air in this case) collected. Scheele employed a number of different reactions to obtain oxygen, including heating saltpetre, red calx of mercury (HgO), and heating "black manganese" (MnO_2) with sulfuric acid or phosphoric acid. Scheele's studies of respiration of oxygen included bees (*Fig.* 5). *Fig.* 6 depicts a supplementary experiment, conducted by Scheele in 1778 and published in 1779, to measure the volume of fire air in atmospherical air. One part of powdered sulfur and two parts of iron filings are placed in the cup C on top of a glass column anchored by a lead base B sitting in a glass vessel A filled with water. The free air volume in a glass cylinder D is 33 ounces. After placing this vessel over the apparatus and standing the open end in water, several hours pass and the volume (measured by E) of water entering the cylinder is 9 ounces. Thus the volume of fire air constitutes about $9/33$ or 27 percent of the air.

Scheele made a number of other extremely valuable contributions to chemistry.[1] He added oil of vitriol to fluorspar (CaF_2) and distilled it. He found extensive corrosion of the retort containing the reaction mixture; all the luting used to seal the apparatus was now brittle and friable, and a white deposit was formed on all interior surfaces of the apparatus. Scheele had formed hydrofluoric acid (HF), which attacked the glass vessel (silicon dioxide) and formed gaseous SiF_4. When he duplicated the experiment but added water to the receiver, he was surprised to find a layer of gelatinous silica on top of the water.

Scheele was, of course, well aware of the source of "urinous phosphorus," as it had been originally been reported by Henning Brand a century earlier (see the earlier essay in this book, p. 225). His friend Gahn had discovered calcium phosphate in bone and horn around 1769 or 1770.[1] Scheele treated burned hartshorn with nitric acid, precipitated gypsum with sulfuric acid, concentrated the filtrate and distilled the resulting phosphoric acid over charcoal, and collected the resulting phosphorus.[1]

Scheele discovered chlorine when he treated "black manganese" with hydrochloric acid and isolated a suffocating, greenish-yellow gas.[1] The bleaching properties of chlorine were readily apparent. Scheele's discovery that plumbago (graphite) burned completely to yield fixed air allowed him to conclude that "black lead" was, in reality, pure carbon, just like diamond.[1]

In February 1775, Scheele was still an apothecary's assistant (*studiosus pharmaciae*), but his fame was such that he was elected as a member of the Royal Academy of Sciences of Sweden.[1] At this point, he decided to finally achieve the stature of a full apothecary. Herman Pohl, who owned the pharmacy privilege that permitted him to own an apothecary shop in Köping, died in 1775, leaving his widow, Sara Margaretha Sonneman Pohl, searching for a buyer.[2] Scheele reached an agreement with her that was almost subverted by another buyer. However, his widespread and justified fame forced Fru Pohl to obey public sentiment and stick by her initial agreement.[2] Although not much seems to be known about their relationship, Roald Hoffmann and Carl Djerassi employ some literary license to speculate sensitively about this relationship in their play *Oxygen*.[7] A letter that Scheele sent to Lavoisier in October 1774, which was never answered,[8] plays a vital role in the play. We do know that on his deathbed, Scheele married Fru Pohl so that she could inherit the pharmacy business from him.[2]

Partington[1] notes that Scheele's contribution to chemistry are "astonishing both in number and importance" and quotes the great nineteenth-century chemist, Humphry Davy: "nothing could damp the ardour of his mind or chill the fire of his genius: with very small means he accomplished very great things."

He died at the age of 44 from a complication of disorders, including rheumatism contracted by work in unfavorable circumstances.[1]

1. J.R. Partington, *A History of Chemistry*, MacMillan and Co. Ltd., London, 1962, Vol. 3, pp. 205–234.
2. C.C. Gillispie, *Dictionary of Scientific Biography*, Charles Scribners Sons, New York, Vol. XII , 1975, pp. 143–150.
3. C.W. Scheele, *Chemical Observations and Experiments on Air and Fire*, London, printed for J. Johnson, 1780.
4. C.W. Scheele, *Chemische Abhandlung von der Luft und Feuer, nebst einer Vorbericht von Torbern Bergman*, verlegt von Magnus Swederus, Uppsala und Leipzig, 1777. This is one of the rarest, most desired books in the field of rare chemistry book collecting. I am grateful to The Roy G. Neville Historical Chemical Library (California) for providing this image.
5. Scheele, C.W., *Opuscula Chemica et Physica, Latin Vertit G.H. Schaefer*, Leipzig, I. Godofr. Mulleriana, 1788–9.
6. A.J. Ihde, *The Development of Modern Chemistry*, Harper and Row, New York, 1964, pp. 50–53.
7. C. Djerassi and R. Hoffmann, *Oxygen*, VCH-Wiley, Weinheim, 2001.
8. J.-P. Poirier, *Lavoisier—Chemist, Biologist, Economist*, R. Balinski (transl.), University of Pennsylvania Press, Philadelphia, 1996, pp. 76–83.

NICE TO HIS MICE

Priestley's first original scientific paper (1770) was on charcoal and had a number of errors.[1,2] However his 1772 paper "Observations on Different Kinds of Air"[3] was a "powerhouse" and was the start of his six volumes published between 1774 and 1786. Priestley's pneumatic trough[4] (Figure 197) evolved from Hales' apparatus (Figure 181) through modifications by William Brownrigg. Priestley capitalized on Cavendish's technique for collecting water-soluble gases such as carbon dioxide over mercury instead of water.[1,2]

In the landmark 1772 paper Priestley describes the isolation and properties of gases first observed by others but not so systematically. He described carbon dioxide ("*fixed air*"—sometimes termed *mephitic air*), nitrogen (the air remaining after a candle had burned out in common air and following CO_2 precipitation in lime water—he termed it "phlogisticated air," often also termed by others "mephitic air"), hydrogen (Cavendish's "inflammable air"—sometimes confused by Priestley with carbon monoxide), hydrogen chloride ("acid air"—later "marine air"), and nitric oxide (NO—"nitrous air").

Nitrous air was generated by exposure of brass, iron, copper, tin, silver, mercury, bismuth, or zinc to nitric acid.[5] Priestley discovered that it reacted instantly with common air to produce a reddish-brown gas (NO_2), which dissolved in water to produce nitric acid. After his own discovery of oxygen in 1774, two to three years after Scheele (Priestley was scrupulously honest and unaware of Scheele's work), Priestley realized that he had discovered a simple and reliable technique for testing the "goodness" of air: "every person of feeling will rejoice with me in the discovery of nitrous air, which supersedes many experiments with the respiration of animals."[6] Although the inverted beer glass in Figure 197 (see

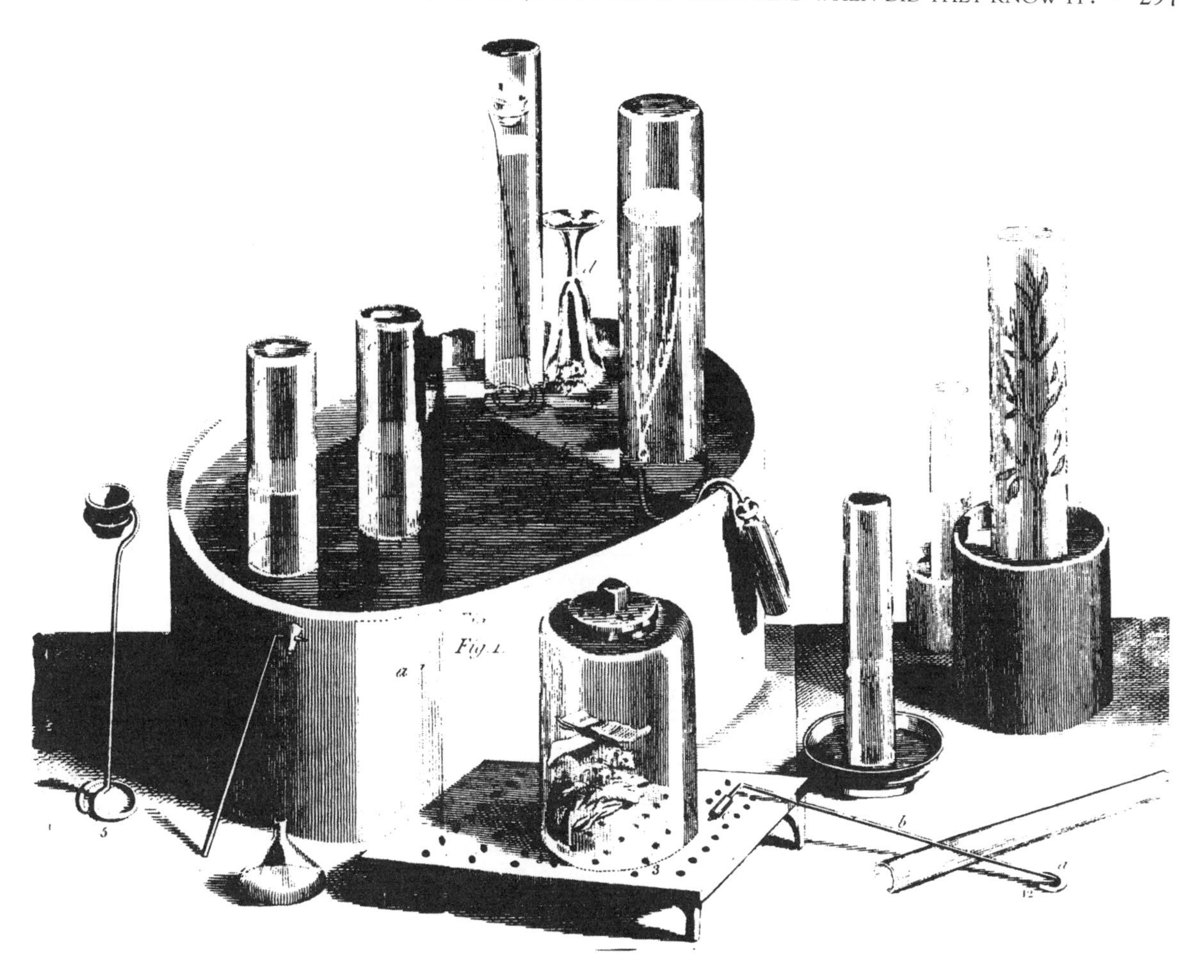

FIGURE 197. ■ Joseph Priestley's pneumatic trough for isolation of "factitious airs" (gases derived from solids). Although Scheele was the first to discover oxygen, Priestley published first (1774). He was gentle to his experimental mice (from the later abridged edition, *Experiments and Observations on Different Kinds of Air*, Birmingham, 1790).

Fig. 1, part d, and also *Fig. 3*) depicts an experimental mouse, Partington[6] notes that Priestley "always took pains to keep his mice warm and comfortable."

Priestley's discovery of oxygen on August 1, 1774 was made by heating red HgO (*mercurious cakinatus*), itself obtained by heating mercury in air or by reaction of mercury with nitric acid (remember Mayow's work, Figure 150). A firm believer in Stahl's Phlogiston Theory to the end of his life, Priestley called the amazing new air, which supported combustion and respiration, "dephlogisticated air." The idea is that a burning candle, for example, loses its phlogiston to something "dephlogisticated" that hungrily grabs it. Indeed, Priestley also found that exposure of "nitrous air" (NO) to iron filings produced a new gas, capable of supporting a brilliant flame, which he called "dephlogisticated nitrous air."[1,2] This was actually nitrous oxide (N_2O, or "laughing gas") which had apparently first been made prior to 1756 by Joseph Black, who heated ammonium nitrate and found vapors whose "effect on breathing and sensation was very far from being unpleasant."[7] Other gases explored by Priestley included ammonia (NH_3, "alkaline air"), sulfur dioxide (SO_2, "vitriolic acid air"), and silicon tetrafluoride ("fluor acid air").

The politically liberal Priestley was sympathetic to the aspirations of the American Colonies and was a regular correspondent of Franklin. In a climate of fear and conservative backlash to the American and French Revolutions and on July 14 (Bastille Day) of 1791 a wild Birmingham mob burned Priestley's meeting house to the ground (the family had fled earlier). In his entertaining book *Crucibles* Jaffe seems to have discovered an eighteenth-century video recording or a "fly on the Church wall" that overheard one rioter yell: "Let's shake some powder out of Priestley's wig."[8] Even the more cosmopolitan London was no longer friendly. In 1794 Priestley moved to the United States, modestly declining a Professorship at the University of Pennsylvania and the charge of a Unitarian chapel in New York, for the peace of living and writing in Northumberland, Pennsylvania.

---

1. J.R. Partington, *A History of Chemistry*, MacMillan, London, 1962, Vol. 3, pp. 237–297.
2. AJ. Ihde, *The Development of Modern Chemistry*, Harper & Row, New York, 1964. pp. 40–50.
3. J. Priestley, *Philosophical Transactions of the Royal Society*, 62:147–267 (1772).
4. This figure is from the 1790 edition (*Experiments and Observations on Different Kinds of Air and other Branches of Natural Philosophy*, three volumes, Birmingham, 1790) of the six books published between 1774 and 1786. The same figure appeared in Volume 1 of that series.
5. It is important to note that nitric acid is different from an acid such as hydrochloric in that the nitrate part (NO_3^- is a stronger oxidizing agent than aqueous hydronium ions (H_3O^+). Thus, copper and iron, which have more positive (more favorable) reduction potentials than H_3O^+, are not oxidized readily in HC1 to produce Hz gas. However, they are oxidized by the powerful NO_3^-, which has a very high reduction potential and is therefore readily reduced to NO. Magnesium, which is very easily oxidized (very hard to reduce), will produce H_2 gas in both hydrochloric and nitric acids.
6. J.R. Partington, op. cit., p. 253.
7. J.R. Partington, op. cit., p. 142.
8. B. Jaffe, *Crucibles*, Simon and Schuster, New York, 1930, p. 52.

LAUGHING GAS OR SIMPLY "SEMI-PHLOGISTICATED NITROUS AIR"

Pneumatic chemistry, the collection and handling of gases ("airs"), began in the 1720s thanks to the work of Stephan Hales (Figure 181) and later improvements by William Brownrigg.[1] Thus, prior to the 1770s the only gases obtained in pure form and characterized to some degree were fixed air (carbon dioxide) and inflammable air (hydrogen; Figure 182); both were termed "factitious airs" using the nomenclature of Henry Cavendish.[2] In the early 1770s, improving the techniques of pneumatic chemistry, Dr. Joseph Priestley isolated and characterized a series of new "airs" that included, among others, oxygen ("dephlogisticated air," O_2), nitrogen ("phogisticated air," N_2), nitric oxide ("nitrous air," NO), and nitrous oxide ("dephlogisticated nitrous air," N_2O).[3] Unbeknownst to Priestley, the Swedish apothecary Carl Wilhelm Scheele was performing a parallel series of experiments and had isolated and characterized oxygen ("fire air") in 1772 (at least two years earlier than Priestley) and possibly even in 1771.[4] Both Priestley and

Scheele were firmly anchored in phlogiston theory. Figure 198 depicts some of Priestley's pneumatic apparatus including his pneumatic tub (*Fig. 1*).

Some of the confusions engendered by the phlogiston theory can be briefly illustrated by the nomenclature described above. For example, "factitious airs" referred to gases derived from solids. Heating chalk (calcium carbonate, $CaCO_3$—both "chalk" and "calcium" are related to the Latin *calx or calcis*) produces "fixed air" (carbon dioxide, CO_2). Figure 198 (see *Fig. 11*)[5] depicts Priestley's apparatus for collecting factitious airs from solids heated in a gun barrel placed partly in a hearth. The gas is collected in a tube that is filled with mercury and inverted into a dish also containing mercury (*Fig. 11* in Figure 198). Oxygen can be obtained by heating solid calxes including mercuric oxide (HgO) or manganese dioxide (MnO_2). Carbon dioxide and oxygen are, thus, both "factitious airs." So far, so good. However, adding zinc to sulfuric acid ("oil of vitriol") produces another "factitious air"—"inflammable air" (hydrogen, H_2)—"clearly" obtained from the solid metal. Indeed, Cavendish thought that inflammable air *was* phlogiston when he isolated it in 1766. In fact, the hydrogen comes from the aqueous acid solution since water is a compound of hydrogen and oxygen. But this understanding was achieved only in the early 1780s.

Priestley found that when the "nitrous air" (NO), obtained by treating copper with *aqua fortis* (nitric acid, HNO_3), was stored over iron filings—a new gas was produced in which a candle burned more intensely than in either common air or his "nitrous air." Clearly, this new "air" wanted phlogiston just as oxygen ("dephlogisticated air") did and hence was also dephlogisticated. Since it was superior to "nitrous air" in supporting combustion, Priestley logically named it "dephlogisticated nitrous air."

Now recall the device of considering "phlogisticated" to mean "minus oxygen" and "dephlogisticated" to mean "plus oxygen."[6] We now know that the new gas was nitrous oxide (N_2O)—laughing gas—a story for another day (see Figure 224). In fact, it has *less* oxygen relative to "nitrous air" and, on the basis of oxygen content, should have been called "phlogisticated nitrous air" (or even worse, "semiphlogisticated nitrous air"). However, the degree of dephlogistication ("hunger" for the essence of fire) did not have a straightforward relationship to oxygen content. Indeed, chlorine, discovered by Scheele,[4] could ignite hydrogen. Even Lavoisier believed that chlorine must contain oxygen—but it does not. In fact, the observed chemistry of nitrous oxide reflected its structure and reactivity, factors that could not possibly be understood in the late eighteenth century, rather than oxygen content. The confusion was understandable.

---

1. J.R. Partington, *A History of Chemistry*, Vol. 3, MacMillan & Co., Ltd, London, 1962, pp. 112–127.
2. Partington, op. cit., pp. 302–319.
3. Partington, op. cit., pp. 237–268.
4. Partington, op. cit., pp. 205–229.
5. This is an early nineteenth-century engraving depicting Priestley's pneumatic apparatus. The collection of carbon dioxide from the gun barrel over a dish of mercury is to be found in J. Priestley, *Experiments and Observations on Different Kinds of Air, and Other Branches of Natural Philosophy*, Vol. III, Thomas Pearson, Birmingham, 1790, Plate II, as well as earlier editions.
6. R. Hoffmann and V. Torrence, *Chemistry Imagined—Reflections on Science*, Smithsonian Institution Press, Washington, DC, 1993, pp. 82–85.

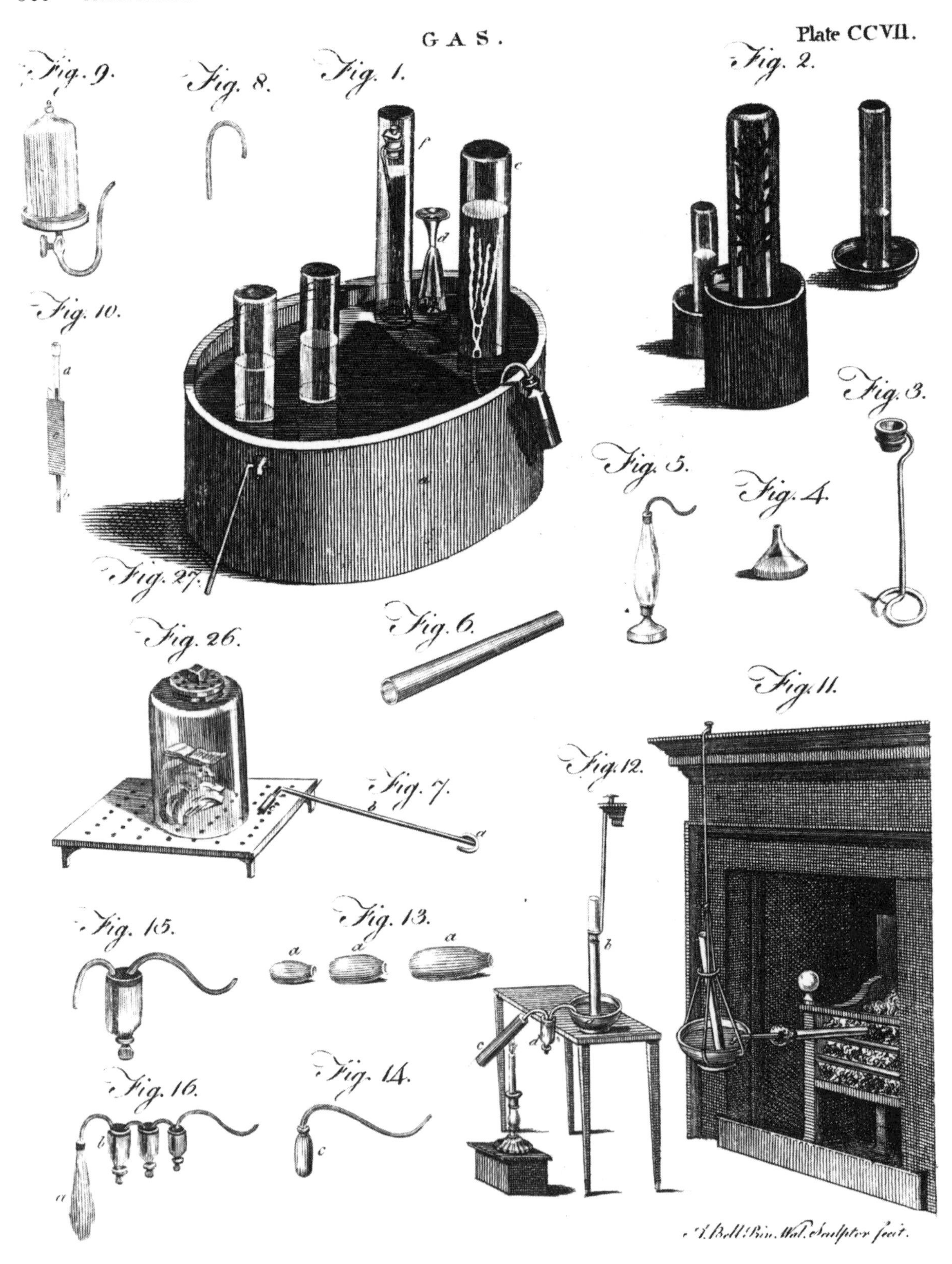

FIGURE 198. ■ Collage of Joseph Priestley's pneumatic apparatus. A gun barrel is used as reaction chamber and placed in a fireplace. The resulting gases, if water-soluble, are collected over mercury (*Fig. 11*).

EULOGY FOR EUDIOMETRY

Eudiometers,[1,2] at their simplest, were nothing more than inverted graduated tubes for measuring volumes of gases collected over water (or above mercury if the gases were water-soluble; see *Fig. 11* in Figure 198). The name is derived from the Greek for "fine weather measure."[2] The real power of eudiometry derived from the ability to measure volume changes when different gases reacted. Joseph Priestley employed very simple eudiometers for his studies of factitious airs such as "nitrous air" (NO or nitric oxide, isolated in 1772) and "dephlogisticated air" (O_2 or oxygen, isolated by him in 1774). Figure 199 depicts such a eudiometer (see g)[1] and glassware for handling gases in a eudiometry experiment. The three vessels labeled *f* in Figure 199 are typically one-, two-, and four-ounce measures for gases, initially filled with water, but inverted in a pneumatic trough and filled with the gas in question. For example, one measure of a gas, "common air," could be transferred to a "reaction chamber"—a glass jar filled with water and inverted in a pneumatic trough. When an equal measure of "nitrous air" is added, a reddish gas forms immediately then disappears. The reaction would be allowed to continue for about two minutes and the gaseous contents transferred into the eudiometer. For "common air," the typical result was that 1.36 measures, of the 2.00 mixed, remained in the eudiometer.[2] Eudiometry also indicated that one volume of pure "dephlogisticated air" reacted completely with two volumes of "nitrous air." Some 35 years later (combining the discoveries of Dalton, Gay-Lussac, and Cannizzaro), chemists would understand the gaseous reaction to be

$$2\ NO + O_2 \rightarrow 2\ NO_2 \text{ (reddish gas)}$$

We know also that nitrogen dioxide dissolves in water to form nitric acid:

$$3\ NO_2 + H_2O \rightarrow 2\ HNO_3 + NO$$

The NO regenerated would recycle in the reaction vessel, and eventually virtually all of it would react with oxygen and become nitric acid. Actually, these complications, combined with the fact that NO is considerably more water-soluble than oxygen, led to lots of inconsistencies depending on which gas was added first, the extent of shaking of the reaction vessel, and how long the gases were allowed to react. Eventually, Cavendish introduced consistency into the technique.[2] The most accurate eudiometer was considered to be that constructed by Felice Fontana (1730–1805).[2] Fontana was the first to discover the ability of charcoal to absorb gases.

It is now easy to understand the eudiometry results for mixing of "common air" and "nitrous air." We know that air is roughly 20% (by volume) oxygen. Thus, from 1.0 measure of "common air," an 0.2 measure of oxygen will disappear upon reaction with an 0.4 measure (from the original 1.0) of "nitrous air." The remaining volume of gas in the eudiometer is 1.4 measures—the total volume lost is 0.6 measure. In reality, 1.36 measures would typically remain (see above); take the volume lost (0.64 measure), divided by 3, and the result is a bit over 0.21. Thus,

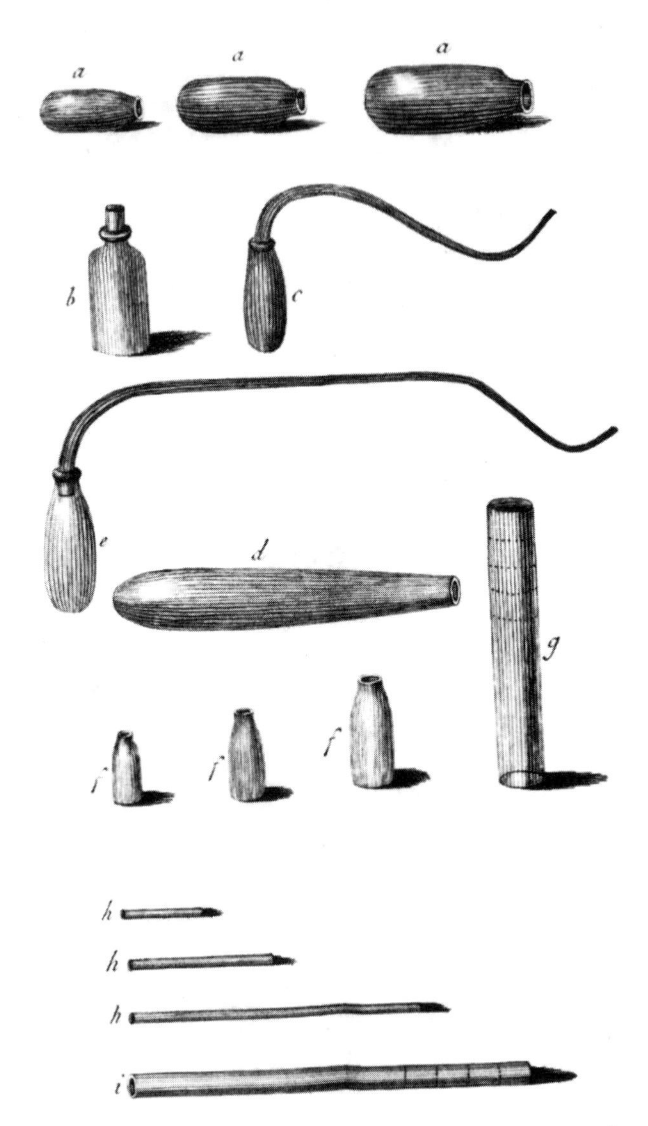

FIGURE 199. ■ Drawing g depicts the simplest possible eudiometer (a graduated tube sealed at one end). Measured volumes of gas are introduced and changes in final volume measured. Priestley's earliest eudiometers simply measured the "goodness" (oxygen content) of air by reaction with freshly generated "nitrous air" (actually nitric oxide, NO). (From Priestley, *Experiments and Observations on Different Kinds of Air*, 1790 edition.)

roughly 21% of "common air" is oxygen. If a sample of "common air" had been reduced to 10% oxygen (through calcination of a metal or introduction of a mouse into the container), mixing of 1.0 measure each of "common air" and "nitrous air" would leave a volume of 1.7 measures in the eudiometer. The "goodness" of this sample was only 1.7, compared to the "goodness" of common air (1.36).[2] Priestley was delighted that he no longer had to kill mice to test the "goodness of air" (see *Fig. 26* in Figure 198).[2] He and others tested the hypothesis that city air had less oxygen than country air. By this measure at least, these two "types" of "common air" had equal "goodness." However, Priestley (and Scheele independently) discovered that the air dissolved in water had slightly more oxygen than that found in "common air" (since nitrogen is a bit less water-soluble than oxygen).[2]

Throughout the latter part of the eighteenth and the early part of the nineteenth centuries, a variety of modifications were devised in which the eudiometer served as both volumetric cylinder *and* reaction chamber.[3] Sulfur powder or red phosphorus could be suspended in a dish surrounded by the air in question and situated above the water in the eudiometer. A powerful magnifying glass would ignite the solids, ultimately forming water-soluble acids, and diminishing the air volume accordingly. Although hydrogen ("inflammable air") and oxygen had been ignited in different admixtures starting in the 1770s—two volumes of hydrogen and one volume of oxygen gave the loudest "pop"—the eudiometer of Volta (see eudiometer *b* with electrodes *f* in *16* at the bottom center of Figure 200[4]) efficiently and accurately employed an electric spark to ignite the two gases and measure the remaining volume.

If we think of the eudiometer in its original incarnation as an apparatus employed to measure oxygen in air, it has long been as extinct as the dodo. Unlike the dodo, we can bring the eudiometer back to life for student demonstration purposes. In their PBS (Public Broadcasting Service) televison series, Philip and Phylis Morrison did just that.[5] They assembled a Volta-style eudiometer and added two "thimblefuls" each of hydrogen and oxygen gas. When they ignited the mixture with an electric spark, the gases expanded [the "slow-mo" (slow-motion) camera indicated no loss of gas bubbles from the bottom of the eudiometer] and left behind one "thimbleful" of gas. An afternoon of repetitions did not change the result. In their words: "The residue was simply the amount of oxygen that could not be taken up into water, always one volume of the total of four. Again, we learned H_2O. Something deep within water appears to know simple arithmetic."[5]

1. J. Priestley, *Experiments and Observations on Different Kinds of Air, and Other Branches of Natural Philosophy*, Thomas Pearson, Birmingham, 1790, Vol. 1, pp. 20–30; Vol. 3, Plate IV.
2. J.R. Partington, *A History of Chemistry*, Vol. 3, MacMillan and Co, Ltd., London, 1962, pp. 252–263; pp. 321–328.
3. [F. Accum] (i.e., "A Practical Chemist"), *Explanatory Dictionary of the Apparatus and Instruments Employed in the Various Operations of Philosophical and Experimental Chemistry with Seventeen Quarto Copper-Plates*, Thomas Boys, London, 1824, pp. 100–110, which describes 10 eudiometers (Priestley, Pepy, Scheele, De Marti, Humbolt, Hope, Seguin, Bertholet, Davy, Volta).
4. Accum, op. cit., Plate 2), figure *16*.
5. P. Morrison and P. Morrison, *The Ring of Truth—an Inquiry into How We Know What We Know*, Random House, New York, 1987, pp. 191–193.

WHERE IS THE INVECTIVE OF YESTERYEAR?

Invective was employed as an art form in scientific discourse centuries ago. A wonderful example is from the Preface to the 1776 edition of a book called *Phosphori*,[2] written by Benjamin Wilson (1721–1788). Like Priestley, he believed in phlogiston and held that the glow of phosphorescence was visible evidence of phlogiston, the fire trapped in many types of matter.

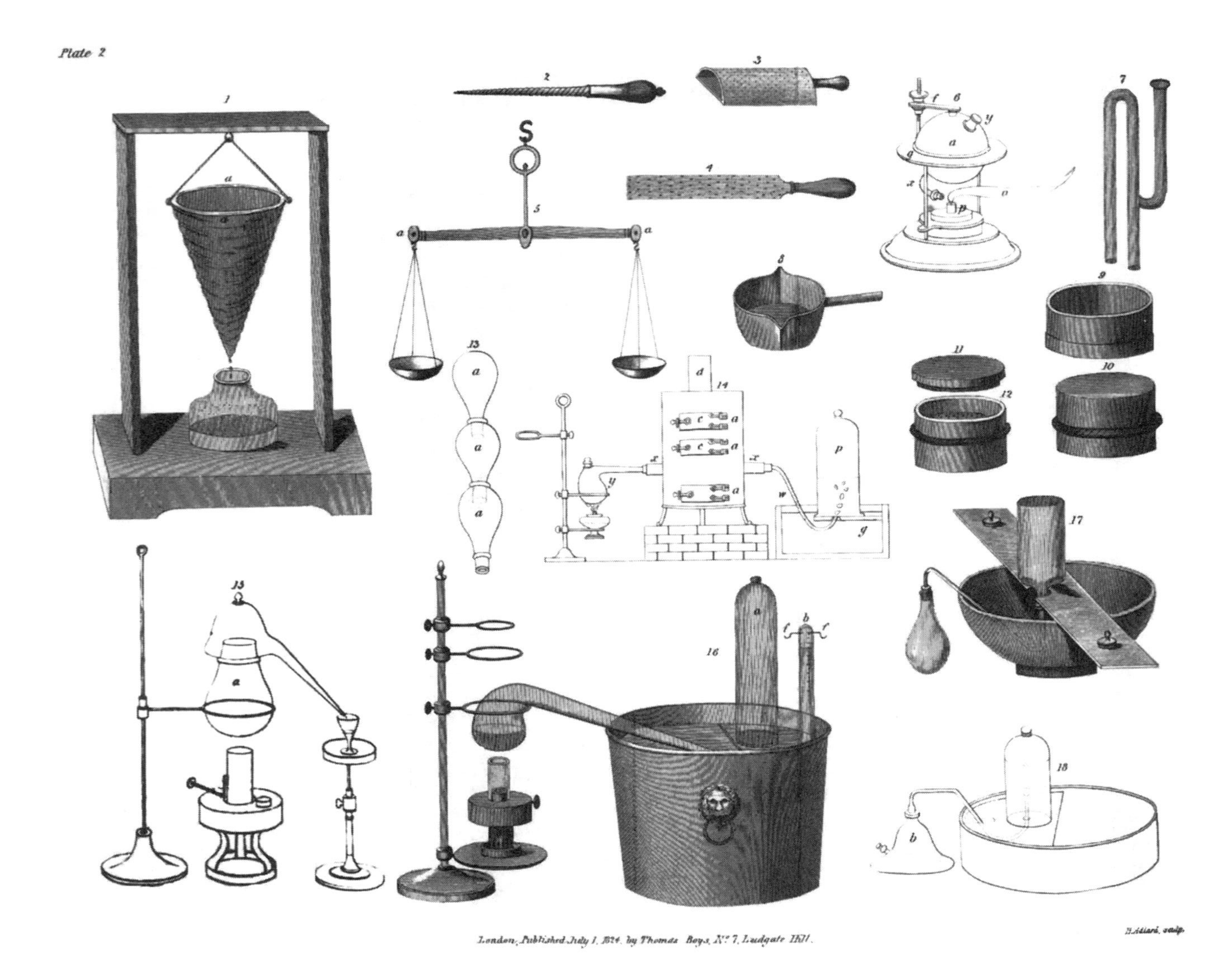

FIGURE 200. ■ Alessandro Volta's eudiometer (see bottom center figure, *16*—tube standing on the right in the tub). Combination of hydrogen and oxygen in this eudiometer could be ignited and the change in volume measured. (From Accum's 1824 *Explanatory Dictionary of the Apparatus and Instruments*. . . .)

Wilson[2] endured his family's poverty until not yet 20, worked in poor circumstances, commenced artistic studies in these circumstances, and started to enjoy some success in his 40s, being appointed by the Duke of York to succeed William Hogarth as Sergeant-Painter in 1764. He speculated in stocks and was declared a defaulter on the Stock Exchange in 1766. During the 1740s he also developed an interest in electricity and later engaged in a highly charged public debate with Benjamin Franklin on the shape of lightning rods. (Wilson had painted a portrait of Franklin in 1759.)[3] Franklin argued for a sharp point, and Wilson correctly argued for a rounded point that would not actually attract lightning. He won the debate but his arguments were so excessive that he received the following criticism in the *Philosophical Transactions:*[2]

> But he has been chiefly distinguished as the ostensible person whose perverse conduct in the affair of the conductors of lightning produced such shameful discord and dissensions in the Royal Society, as continued for many years after, to the great detriment of science.

The scorn so evident in the Preface of *Phosphori* is generally missing in scientific discourse. After all, Dr. X may eventually review Dr. Y's research grant proposal. In reading this excerpt one should note that Doctor Priestley was a painfully honest English clergyman and a friend of Franklin (and sympathetic to the American Colonies fight for independence) who had immense standing in the scientific community and had criticized Wilson's experiments:

> Now why may not such a plain philosopher (with the good Doctor's gracious leave) be supposed capable of, at least stumbling upon discoveries, which had escaped the observation of preceding philosophers, even of the highest and most respectable characters? For it is well known, that it is not always men of "vast and comprehensive understandings," that have been favoured by Providence with making discoveries sometimes the greatest, and most useful to the world: but on the contrary (to allude to the words of an eminent writer with whom Dr. Priestley is intimately acquainted), the Great Author of Nature hath frequently chosen "weak things," in the philosophical, as well as the spiritual world, to confound the mighty, and things that are not, to bring nought the things that are.

1. B. Wilson, *Phosphori*,2nd ed., London, 1776.
2. *Dictionary of Scientific Biography*, Charles Scribner, New York, 1976, Vol. XIV, pp. 418–419.
3. B.B. Fortune and D.J. Warner, *Franklin and His Friends—Portraying the Man of Science in Eighteenth-Century America*, Smithsonian National Portrait Gallery and University of Pennsylvania Press, Washington, D.C. and Philadelphia, 1999, pp. 74–77.

LA REVOLUTION CHIMIQUE COMMENCE

Antoine Laurent Lavoisier[1–3] (1743–1794) is justifiably said to be the father of modern chemistry. His greatest single contribution is the recognition that both combustion and calcination arise from combination of atmospheric oxygen with inflammable substances and metals rather than from loss of phlogiston from these substances. His greatest published work, *Traité Élémentaire de Chimie* (Paris, 1789; London, 1790; Philadelphia, 1796) is clearly a modern textbook. His contributions to chemistry, including its first systematic nomenclature, are far too numerous to mention in our brief *Chemical History Tour*. He was born to wealth, married additional wealth, lived a stylish and aristocratic life, and died by the guillotine on May 8, 1794 during the Reign of Terror. Before he died, Lavoisier experienced an angry fist-shaking crowd and perhaps stared across the River Seine to see the College Mazarin where he received his early education.[4] Twenty-eight members of the *Ferme Générale*, including Lavoisier, were executed in 35 minutes on that day.[4] The heads were placed in a wicker basket, the 28 bodies stacked on wagons and buried in large common graves dug in a wasteland named Errands ("maimed person").[4] On May 9, the famous mathematician Joseph Louis Lagrange commented: "It took them only an instant to cut off that head but it is unlikely that a hundred years will suffice to reproduce a similar one."[4]

The young Lavoisier showed a precocious interest in chemistry sparked by the lively demonstrations of Guillaume François Rouelle. His brilliance and his wealth gained him entry into the Academic Royale des Sciences in 1768 at the age of 25 and full membership in 1769. In 1768 he purchased a privilege to collect taxes in the Ferme Générale. Jacques Pauize, a senior member of the Ferme, had a beautiful and gifted young daughter named Marie-Anne Pierrette who was attracting unsuitably aged suitors. He introduced Antoine and Marie and they were married in 1771—just short of her fourteenth birthday. Intellectually, they were well met and Marie learned sufficient chemistry to be an effective and critical translator of texts in other languages including English, thus opening the wider chemical literature to Antoine. Her artistic talents also found some expression in the drawings used to illustrate his texts.

Lavoisier's earliest studies showed a respect for precise measurement. He demonstrated that diamonds decompose in strong heat (Boyle had proven this a century earlier) but showed that air was necessary and that the decomposition product turned lime water milky and was thus fixed air (CO_2). In 1772 his studies extended to the combustion of phosphorus and sulfur, which, like carbon, produced "acid airs" that weighed more than the solids that produced them. Similarly, he verified the observation by Jean Rey in 1630, also noted by Boyle and others, that the calxes formed by heating metals were heavier than the metals themselves. In his first great book (*Opuscules Chimiques et Physiques*, Paris, 1774; *Essays Physical and Chemical*, London, 1776), Lavoisier first offered the idea that these processes involved absorption of some "elastic fluid" present in air rather than loss of phlogiston to the air. In this book he confused this elastic fluid with fixed air.[1,2]

Figure 201 is from the 1776 *Essays Physical and Chemical*. In *Fig*. 8 (Figure 201) we see an apparatus for measuring the "air" absorbed during calcination of lead or tin under a powerful magnifying lens ("heating glass"). The inverted bell jar sits over a vessel filled with water. In the middle is a glass column with a cup-like indentation on the top. Some lead or tin is placed into a porcelain dish placed on top of the glass column. Siphon MN is placed under the bell jar and air withdrawn until the water level rises to the desired level. Heating of the metal should produce calx with the loss of some "aerial fluid" and a rise in the water

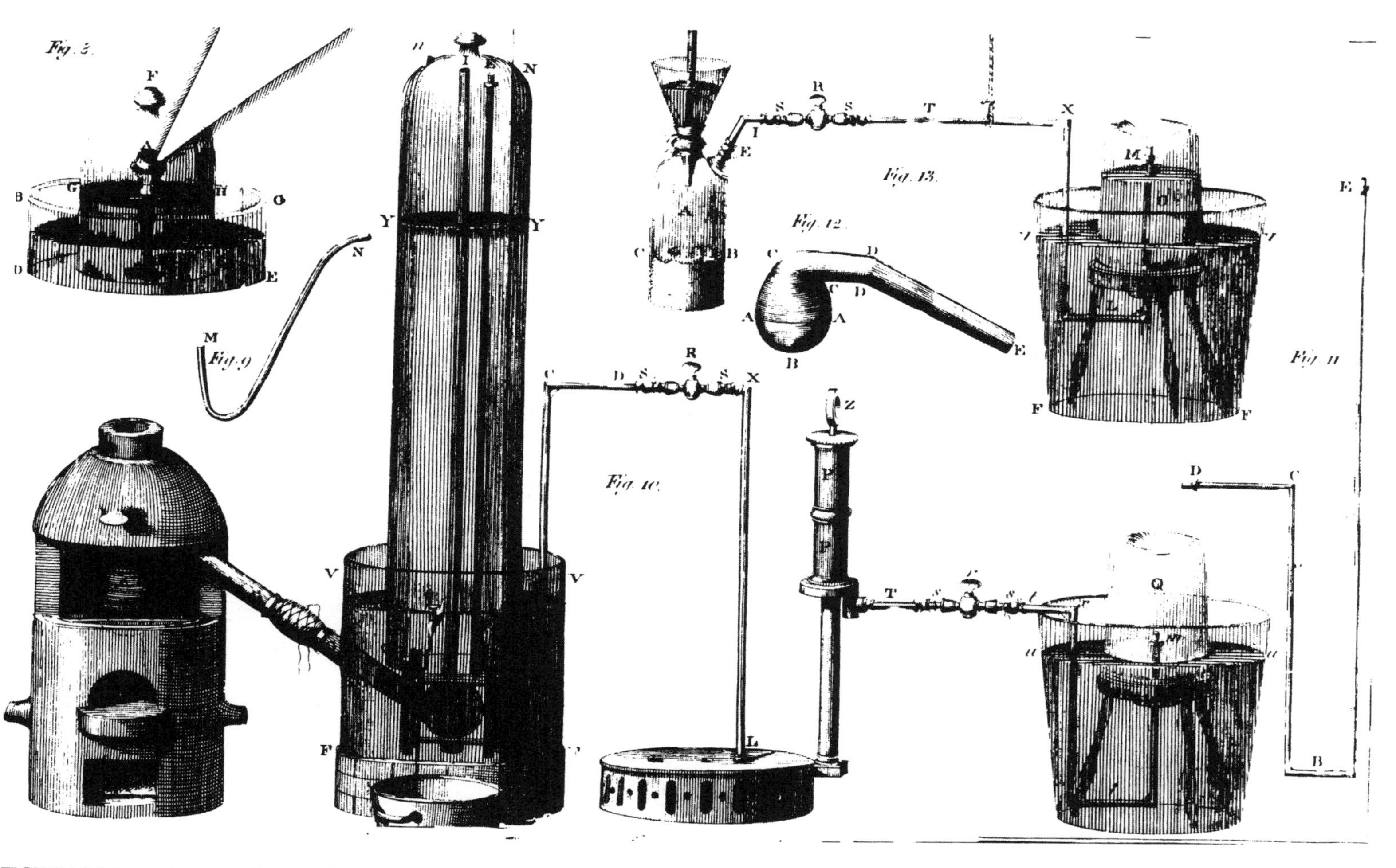

FIGURE 201. ■ Antoine Laurent Lavoisier's pneumatic experiments in *Essays Physical and Chemical*, London, 1774; the French edition was published in 1774; see text.

column. Unfortunately, the heating glass was too powerful and molten metal evaporated and splashed onto the sides of the bell jar giving inconclusive results.

Fig. 10 in Figure 201 shows an apparatus for measurement of the gas (CO_2) released when minium (red-lead or litharge, Pb_3O_4) mixed with charcoal is heated in a furnace. Glass retorts were attacked by this chemical mixture, so Lavoisier fabricated an iron retort (*Fig. 12*). The tall inverted bell jar *nNoo* sits in a wooden or iron trough filled with water. A siphon inserted at n raises the water to YY. Alternatively, hand-pump *P* can connected using siphon *EBCD* (*Fig. 11*) and used to raise the column fairly high. The top of the water in jar *nNoo* is coated with a thin layer of oil. This is another way to collect a water-soluble gas such as CO_2 rather than by using mercury. To the right in *Fig. 10* we see an apparatus for transfer of the gas collected in jar *N* to glass bottle *Q*. This important experiment demonstrated the release of an "aerial fluid" upon heating red-lead.

Fig. 13 in Figure 201 depicts an apparatus for generating CO_2 by adding dilute oil of vitriol onto powdered chalk. The water-soluble gas is collected over water having a layer of oil on top.

1. J.R. Partington, *A History of Chemistry*, MacMillan, London, 1962, Vol. 3, pp. 363–495.
2. A.J. Ihde, *The Development of Modern Chemistry*, Harper & Row, New York, 1964, pp. 57–88.
3. J.-P. Poirier, *Lavoisier—Chemist, Biologist, Economist*, R. Balinski, (translator), University of Pennsylvania Press, Philadelphia, 1996.
4. J.-P. Poirier, op. cit., pp. 381–382.

SIMPLIFYING THE CHEMICAL BABEL

Peter Bruegel The Elder depicted The Tower of Babel in 1563. This huge city reaching into the clouds was a human conceit and according to *Genesis* 11:9: "Therefore its name was called Babel, because there the Lord confused the language of all the earth." As the chemical edifice was erected through the eighteenth century a form of chemical babble arose in a confusing nomenclature. This was due in part to different degrees of purities of substances as well as uncontrolled neologisms. A look at the literature of the time shows that the term "mephitic air" (*mephitic* means "pestilential exhalation"), while most often used for carbon dioxide, was sometimes employed for the nitrogen that remained after "vital air" was totally consumed from common air. Eklund's useful work[1] is a helpful guide for understanding eighteenth-century nomenclature. In 1787, Lavoisier, de Morveau, Berthollet, Fourcroy, Hassenfratz, and Adet collaborated on the book *Méthode de Nomenclature Chimique* (Paris and London, 1788). Figures 202 and 203 are derived from this work but are actually taken from the second English edition (1793) of Lavoisier's *Traité*.

The work was of immense importance to the field, but let's note some interesting little flaws that prove that even Lavoisier was not infallible. First, he names "vital air" as oxygene, which means "acid maker." This was reasonable to Lavoisier since combustion of carbon, sulfur, and phosphorus in pure oxygen each produced acids. His oxygen theory of acids was well accepted. This included

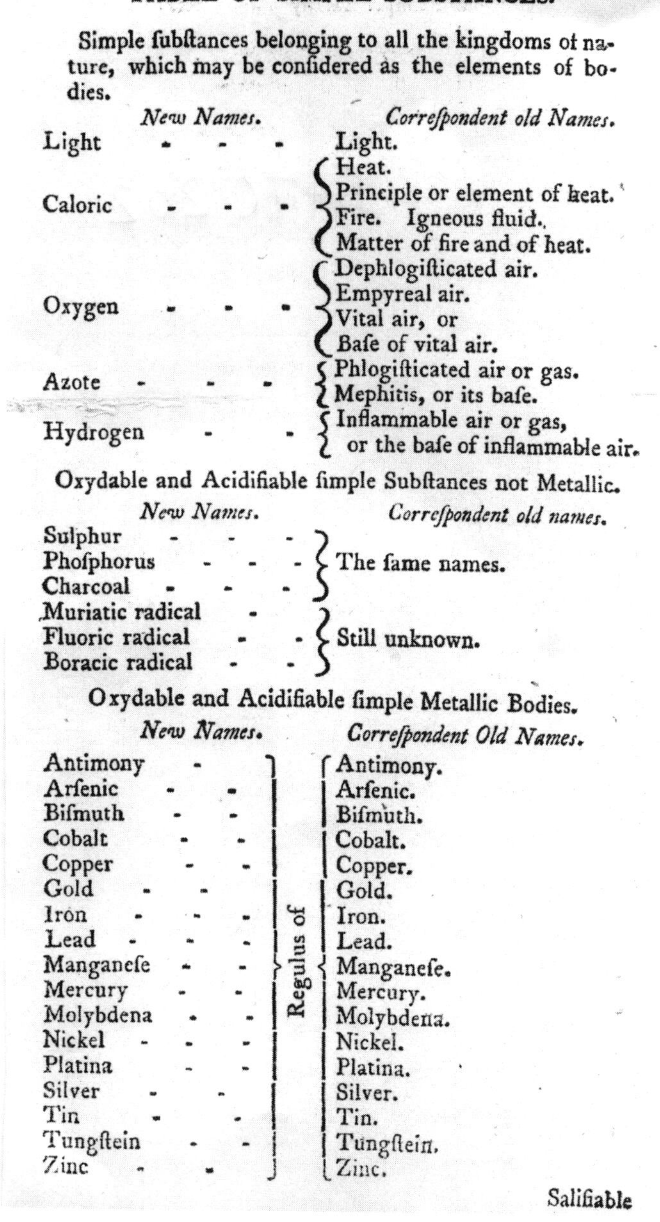

TABLE OF SIMPLE SUBSTANCES.

Simple ſubſtances belonging to all the kingdoms of nature, which may be conſidered as the elements of bodies.

New Names.	*Correſpondent old Names.*
Light	Light.
Caloric	Heat. Principle or element of heat. Fire. Igneous fluid. Matter of fire and of heat.
Oxygen	Dephlogiſticated air. Empyreal air. Vital air, or Baſe of vital air.
Azote	Phlogiſticated air or gas. Mephitis, or its baſe.
Hydrogen	Inflammable air or gas, or the baſe of inflammable air.

Oxydable and Acidifiable ſimple Subſtances not Metallic.

New Names.	*Correſpondent old names.*
Sulphur, Phoſphorus, Charcoal	The ſame names.
Muriatic radical, Fluoric radical, Boracic radical	Still unknown.

Oxydable and Acidifiable ſimple Metallic Bodies.

New Names.		*Correſpondent Old Names.*
Antimony	Regulus of	Antimony.
Arſenic		Arſenic.
Biſmuth		Biſmuth.
Cobalt		Cobalt.
Copper		Copper.
Gold		Gold.
Iron		Iron.
Lead		Lead.
Manganeſe		Manganeſe.
Mercury		Mercury.
Molybdena		Molybdena.
Nickel		Nickel.
Platina		Platina.
Silver		Silver.
Tin		Tin.
Tungſtein		Tungſtein.
Zinc		Zinc.

Salifiable

FIGURE 202. ■ Here is Lavoisier's list of "simple substances" (i.e., elements) in the first English edition *Elements of Chemistry* (London, 1790) of his monumental *Traité Elémentaire de Chimie* (Paris, 1789). Note caloric as an element. Count Rumford would disprove caloric about 10 years later and also marry Lavoisier's widow Marie-Anne Pauize Lavoisier.

the belief that hydrochloric acid (HC1) contained oxygen because its precursor chlorine (isolated by Scheele in 1774) must have contained oxygen. This was disproven by Humphrey Davy some 20 years after the *Nomenclature* was published.

A second problem was Lavoisier's postulation of the element "caloric"—a kind of imponderable heat fluid. In certain ways, caloric was a substitute for the

	Names of the ſimple ſubſtances.	Firſt degree of oxygenation.		Second degree of oxygenation.	
		New Names.	Ancient Names.	New Names.	Ancient Names.
Combinations of oxygen with ſimple non-metallic ſubſtances.	Caloric .	Oxygen gas . . .	Vital or dephlogiſticated air		
	Hydrogen .	Water*.			
	Azote .	Nitrous oxyd, or baſe of nitrous gas . . .	Nitrous gas or air . .	Nitrous acid . . .	Smoaking nitrous acid .
	Charcoal .	Oxyd of charcoal, or carbonic oxyd . . .	Unknown	Carbonous acid . . .	Unknown . . . .
	Sulphur . .	Oxyd of ſulphur . .	Soft ſulphur	Sulphurous acid . .	Sulphureous acid . .
	Phoſphorus	Oxyd of phoſphorus . .	Reſiduum from the combuſtion of phoſphorus	Phoſphorous acid . .	Volatile acid of phoſphorus . . . . . .
	Muriatic radical . .	Muriatic oxyd . . .	Unknown	Muriatous acid . . .	Unknown . . . .
	Fluoric radical . .	Fluoric oxyd . : .	Unknown	Fluorous acid . . .	Unknown . . . .
	Boracic radical . .	Boracic oxyd . . .	Unknown	Boracous acid . . .	Unknown . . . .
Combinations of oxygen with the ſimple metallic ſubſtances.	Antimony .	Grey oxyd of antimony	Grey calx of antimony	White oxyd of antimony	White calx of antimony, diaphoretic antimony
	Silver . .	Oxyd of ſilver . . .	Calx of ſilver . . .	. . . .	. . . .
	Arſenic . .	Grey oxyd of arſenic .	Grey calx of arſenic .	White oxyd of arſenic	White calx of arſenic .
	Biſmuth .	Grey oxyd of biſmuth .	Grey calx of biſmuth .	White oxyd of biſmuth	White calx of biſmuth
	Cobalt . .	Grey oxyd of cobalt .	Grey calx of cobalt .	. . . .	. . . .
	Copper . .	Brown oxyd of copper .	Brown calx of copper .	Blue and green oxyds of copper	Blue and green calces of copper . . . .
	Tin . .	Grey oxyd of tin . .	Grey calx of tin . .	White oxyd of tin . .	White calx of tin, or putty of tin . . .
	Iron . .	Black oxyd of iron .	Martial ethiops . .	Yellow and red oxyds of iron	Ochre and ruſt of iron .
	Manganeſe	Black oxyd of manganeſe	Black calx of manganeſe	White oxyd of manganeſe	White calx of manganeſe
	Mercury .	Black oxyd of mercury	Ethiops mineral † . .	Yellow and red oxyds of mercury . . .	Turbith mineral, red precipitate, calcined mercury, precipitate *per ſe*
	Molybdena	Oxyd of molybdena .	Calx of molybdena . .	. . . .	. . . .
	Nickel . .	Oxyd of nickel . . .	Calx of nickel . . .	. . . .	. . . .
	Gold . .	Yellow oxyd of gold .	Yellow calx of gold . .	Red oxyd of gold . .	Red calx of gold, purple precipitate of caſſius
	Platina . .	Yellow oxyd of platina	Yellow calx of platina .	. . . .	. . . .
	Lead . .	Grey oxyd of lead . .	Grey calx of lead . .	Yellow and red oxyds of lead	Maſſicot and minium .
	Tungſtein .	Oxyd of Tungſtein . .	Calx of Tungſtein . .	. . . .	. . . .
	Zinc . .	Grey oxyd of zinc . .	Grey calx of zinc . .	White oxyd of zinc . .	White calx of zinc, pompholix . . .

* Only one degree of oxygenation of hydrogen is hitherto known —A † Ethiops mineral is the ſulphuret of m

FIGURE 203. ■ This table is also from the first English edition of Lavoisier's 1789 *Traité*. Note that oxygen gas is said to be a combination of oxygen and caloric. When a substance burns or calcines it combines with oxygen and releases caloric as heat. This has a bit of the flavor of phlogiston in it. Dephlogisticated air (oxygen) was to absorb phlogiston from burning or calcining substances.

phlogiston Lavoisier demolished. According to this view, gaseous oxygen contains caloric (which helps to keep it in a rarified state). When a substance burns or forms a metallic oxide, it combines with ("fixes") oxygen (thus, increasing its weight) and, in the process, frees the caloric as heat. In Figure 202, we see caloric listed as an element ("simple substance"). In Figure 203, we see that just as hydrogen combines with oxygen to produce water, so does caloric combine with oxygen to produce oxygen gas. Ironically (or perhaps not), it is Madame Lavoisier's second husband, Count Rumford, who eventually disproves the existence of caloric.

---

1. J. Ekiund, *The Incompleat Chymist*, Smithsonian Institution Press, Washington, D.C., 1975.

WATER WILL NOT "FLOAT" PHLOGISTON

We have imbibed, sweated, and excreted water since time immemorial—so it might be nice to know what it's made of. Water boils, freezes, and is recovered unchanged from salts and other "earths." It is absolutely elemental to our very existence, as is air. One of the four Aristotelian elements, water can be transmuted to "air" by adding heat, to "earth" by removing wetness, and it "neutralizes" its contrary element—fire. Its status as an element survived Robert Boyle's scathing criticism of the ancients in his 1661 classic *The Sceptical Chymist*. As late as 1747, Ambrose Godfrey, Boyle's very capable assistant, reported the chemical conversion of water to earth,[1] an experimental conclusion once and for all time refuted in 1770 by Antoine Lavoisier.[2] So, when and how did we learn the true nature of water, or how to get "From H to eau," as Philip Ball so wittily phrases it?[3]

In 1766, Cavendish reported releasing a "flammable air" (hydrogen) that he equated to phlogiston, the very essence of fire, from metals through reaction with aqueous acids. Eight years later, Priestley reported isolating "dephlogisticated air" (oxygen). It must have been something of a disappointment when the two were mixed and didn't immediately consume each other[4] to give . . . to give *what?* . . . "dephlogisticated phlogiston"?—effectively nothing? Priestley first ignited this mixture in 1775 in a $1\frac{1}{2}$-inch bottle with a $\frac{1}{4}$ inch opening and later "amused himself by carrying these corked or stoppered bottles about and exploding them."[5] Formation of water, in the combustion of hydrogen in common air, was first noted in 1776/77 by Pierre Joseph Macquer, who observed dew collected on a cold porcelain dish above the flame. Similar observations were made in England by John Warltire, who burned hydrogen and oxygen in various combinations of volumes; the best proportion was 2 : 1.[5] But seeing residual water appear in chemical reaction vessels was commonplace. One explanation was that water had to be squeezed out of air before it could become fully phlogisticated. Listen carefully, dear reader; phlogiston theory was creaking loudly now and beginning to crumble. In fact, gases occupy huge volumes compared to liquids—one liter of the two gases properly mixed will yield about 10 drops of water—rather easy to dismiss.

Suffice it to say, we will not solve "The Water Controversy"[6] here. The key figures are Henry Cavendish, James Watt, and Antoine Lavoisier. Volta's eudiometer (see Figure 200) stimulated Priestley, Cavendish, and others to use this technique. Although James Watt (1736–1819), inventor in 1765 of a vastly improved steam engine, made a strong case that he first recognized that water is a compound rather than an element, it was Cavendish who is accorded the discovery. He was the first to experimentally combine the exact proportions of hydrogen and oxygen and find their mass conserved in liquid water. However, anchored in phlogiston theory, Cavendish's interpretation was that "flammable air" (hydrogen) was really water plus phlogiston (Φ) while dephlogisticated air (oxygen) was really water lacking phlogiston:[5,7]

$$\begin{array}{ccccc} \text{Hydrogen} & + & \text{oxygen} & \rightarrow & \text{water} \\ (\text{Water} + \Phi) & & (\text{water} - \Phi) & \rightarrow & \text{water} \end{array}$$

It would seem logical that if hydrogen were pure phlogiston and oxygen were "dephlogisticated air," then their "marriage" might produce, if anything, "phlogisticated air" or "azote" (nitrogen). Instead, water is found—it has body and considerable density. It must have been there all along, right?

It was Lavoisier who finally provided absolutely conclusive evidence for water's composition. Water played an important role in his career. His first paper, published at the age of 22, established one of Lavoisier's *leitmotifs*.[8] He calcined (heated) gypsum (calcium sulfate), collected and weighed the water of crystallization, and then reconstituted the original substance by adding the water collected to the anhydrous salt. As noted above, he disproved the claim by others that water was converted to earth upon heating. His claim for primacy in the discovery of the composition of water bears striking resemblance to his claim for the discovery of oxygen; the historical record indicates in each case, that he (1) did *not* make the initial discovery, (2) withheld some information from rivals, but (3) completely and correctly explained the breakthroughs with his antiphlogistic theory of oxidation.[9,10]

In Figure 204 we see Lavoisier's apparatus for *decomposing water* (*Fig. 11*)[11] in which an annealed glass tube (*EF*, surrounded by clay mixed with crushed stoneware and reinforced with an iron bar) is placed into furnace *EFCD*; connected at one end to retort A, which produces steam; and at the other to condenser SS, which drips unreacted water into *H*. Gas from the tube leaves through *KK* and is appropriately purified and collected. The first experiment found that the quantity of steam lost from A, and passing through the empty red-hot tube, precisely equaled the quantity of water collected in *H*. In the second experiment, 28 grams of charcoal were placed in the tube—these were gone following prolonged exposure to steam in the red-hot tube and produced 100 grams of carbonic acid gas (carbon dioxide), and 13.7 grams of "a very light gas . . . that takes fire" (hydrogen) with a loss of 85.7 grams of water. It was known that 100 grams of carbonic acid gas contained 72 grams of oxygen and 28 grams of carbon (*remember, no atoms or formulas yet*). Thus, 85.7 grams of water decomposed into 13.7 grams of hydrogen and 72 grams of oxygen—the *Ferme Générale*, in which Lavoisier was a shareholder, would have been well satisfied with this neat accounting. In the third experiment, 274 grams of soft iron, in thin plates rolled up into spirals, are exposed to steam in the red-hot tube. No carbonic acid gas is found this time. Instead, the iron is now a black oxide (the same as produced by combustion of iron in oxygen) and its weight augmented by 85 grams of oxygen. The amount of hydrogen collected was 15 grams; the amount of water lost, 100 grams. From these two very different experiments, the same result—water is 85% oxygen and 15% hydrogen.

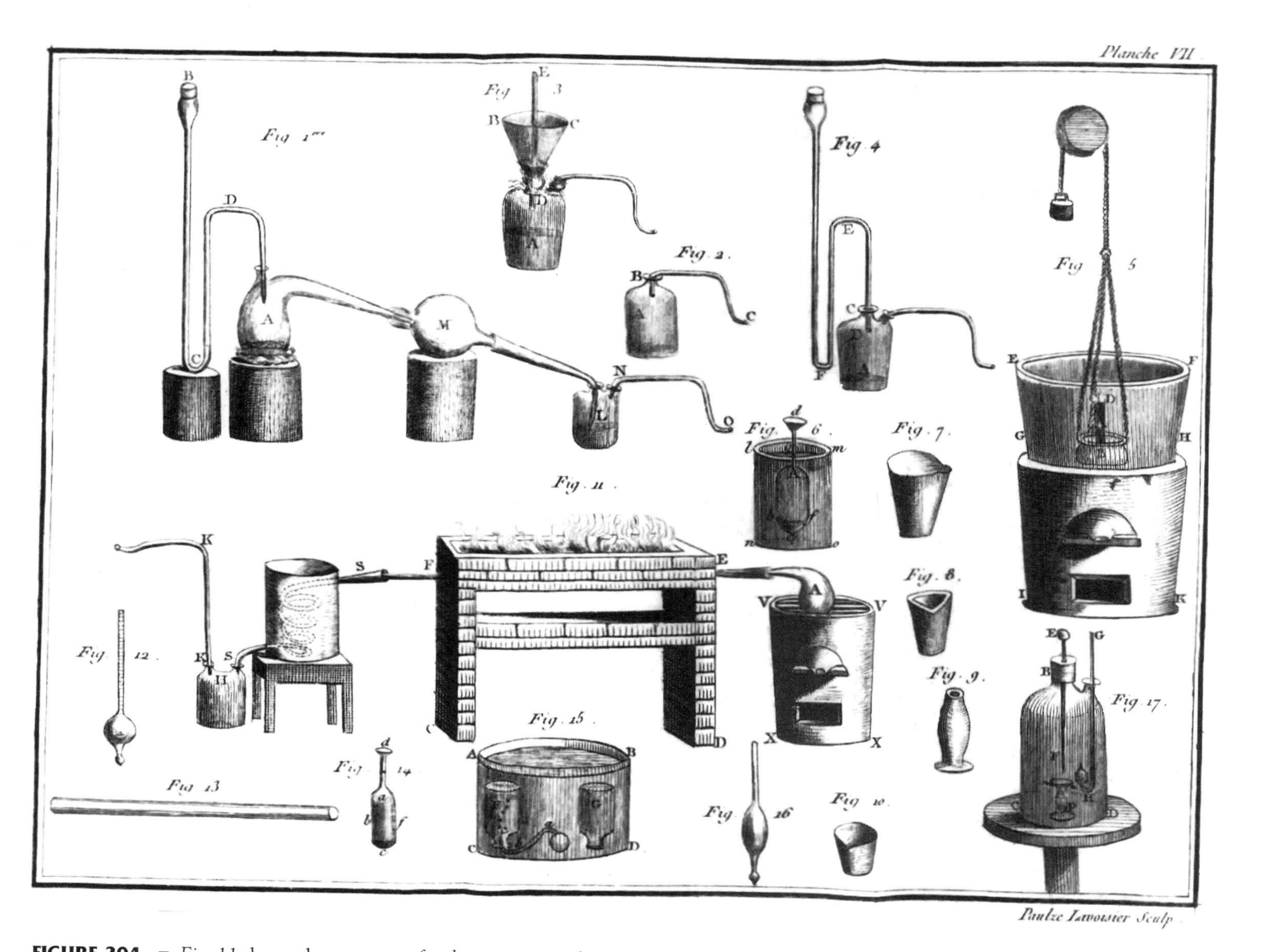

FIGURE 204. ■ *Fig. 11* shows the apparatus for *decomposition* of water into hydrogen and oxygen. Charcoal is placed in an annealed glass tube in the furnace. Over the fire, charcoal is oxidized to CO_2 that is collected while hydrogen is collected separately. Alternatively, iron plates could be placed in a tube and heated red hot, forming iron oxide and releasing hydrogen in the presence of steam. (From Lavoisier, *Traité Elémentaire de Chimie, seconde édition*, 1793.)

Figure 205 is from Lavoisier's *Traité*. In *Fig. 1* we see a very sophisticated distillation apparatus designed to trap and weigh everything generated. The sample is heated in retort A, volatile and semivolatile liquids are collected in the preweighed globe C; the first preweighed three-necked bottle after C contains water and the remaining three bottles contain potash (KOH) solutions (all preweighed) to trap the acidic gases. The remaining water-insoluble, nonacidic gases (e.g., oxygen) are delivered to a bottle in a pneumatic trough or similar collection device. The tall tubes luted into the center opening of each bottle have small openings—they reach to the bottom of the liquids and will only leak if there is a pressure buildup. If a vacuum is created, the mass of air introduced is negligible compared to the mass of the glassware and their contents. Lavoisier notes that if the masses in all vessels, including residue in the retort, do not total to that of the starting material, the experiment must be redone. Figure 206 (*Fig. 1*) shows an apparatus for separation of the gaseous components arising from fermentation or putrefaction. The matrass A is connected by brass tubing and valves to glass bulb B. If frothing in A exceeds the capacity of the matrass, excess froth is collected in B and drained periodically into bottle C. Water vapor is removed in glass tube h, which contains a drying agent such as calcium chloride. Carbon dioxide from fermentation is collected in potash solutions in bottles D and E. Putrefaction sometimes produces hydrogen, collected in bell jar F in pneumatic trough GHIK.

Fig. 2 in Figure 205 depicts the famous apparatus for heating metallic mercury in the presence of air. Lavoisier heated 4 ounces of mercury in the retort A.[11] After twelve days he stopped the heating and weighed the red calx (HgO) that had formed on the surface of the mercury. Its mass was 45 grains. The air volume had decreased from 50 cubic inches to 42 cubic inches (about 16%). The air remaining was "mephitic" (actually nitrogen). When the *mercurius calcinatus per se* was transferred to a small retort and heated it produced 8 to 9 cubic inches of "highly respirable air" and 41.5 grains of mercury. This was the gas that Scheele called "fire air" or "empyreal air," Priestley termed "dephlogisticated air," and Lavoisier later called "vital air" and eventually oxygen. When this oxygen was added to the "mephitic" air, normal air was reformed. The interesting apparatus in *Fig. 10* is a customized matrass [see Figure 26(a); you may ignore the ostrich]. Its bulb has been heated in flame and flattened. The flat bottom contains mercury, which can be heated on a sand bath. The small opening at the top permits slow circulation of atmospheric air but minimizes loss of mercury vapor. Over several months, good yields of red HgO are obtained. The retort-and-bladder apparatus (*Fig. 12*) is similarly used to heat mercury in the presence of a half-bladderful of oxygen—only small amounts of red calx were formed.

In *Fig. 3* we see a small apparatus for igniting iron in a porcelain dish in a bell jar filled with oxygen over mercury. Lavoisier siphons out some air in order to raise the mercury level. He uses a red-hot iron wire (*Fig. 16*) to touch off a piece of phosphorus attached to tinder attached to the iron wire sample. *Figure 17* (upper right) depicts a fine iron wire attached to a stopper and twisted into a spiral with a small piece of tinder at point C. With the stopper and wire out, the tinder is lit and the wire lowered into the oxygen-containing bottle. As it burns, iron forms a calx that falls to the bottom, is collected, powderized, and weighed.

Fig. 4 depicts a large vessel for combustion of phosphorus in oxygen (the opening at the top has a diameter of three inches). Phosphorus is placed in the

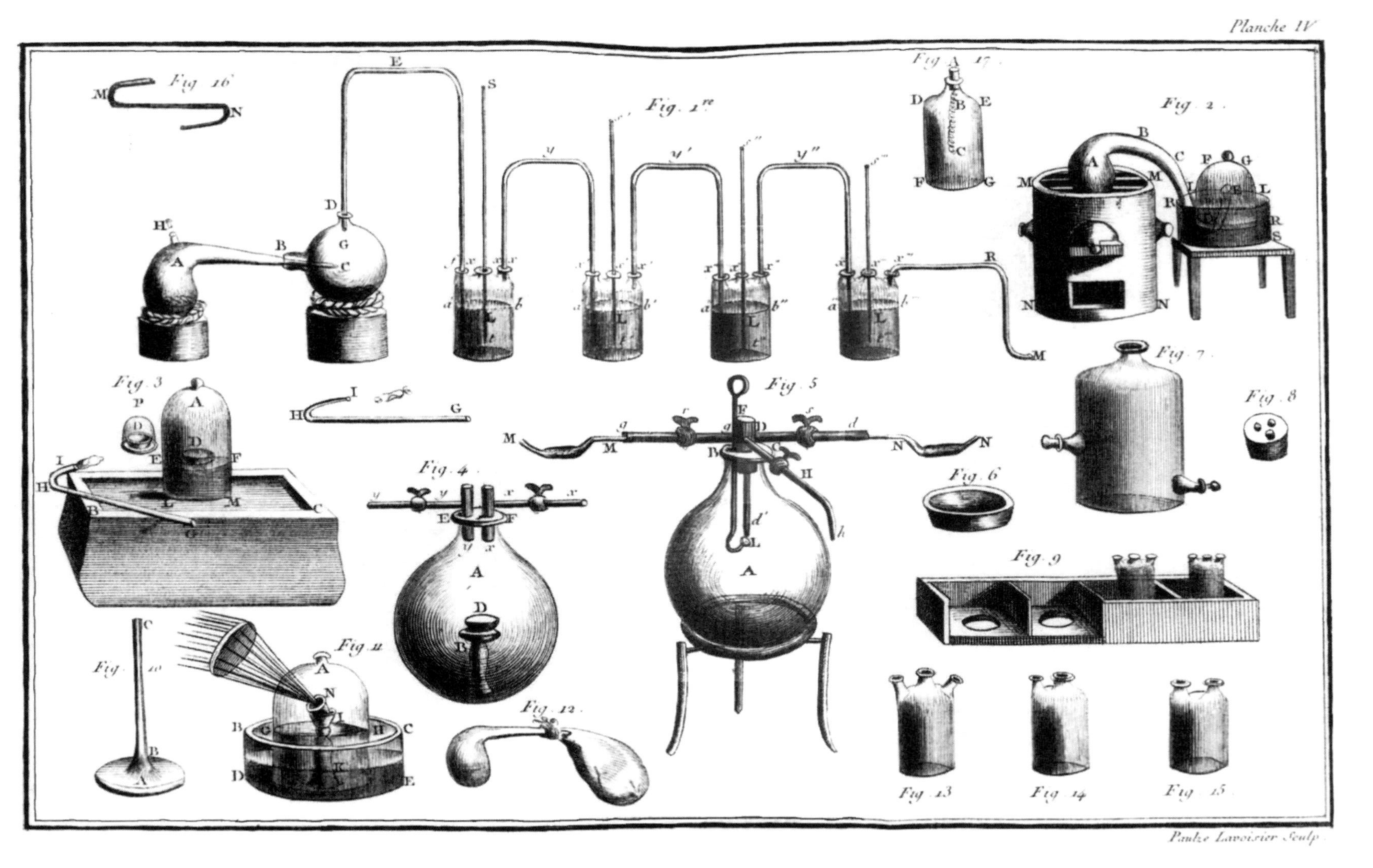

FIGURE 205. ■ *Fig.* 5 shows the apparatus for *synthesis* of water from hydrogen and oxygen. Once the correct amounts are added, a spark is set off at L and quantitative reaction occurs. (From Lavoisier, *Traité Elémentaire de Chimie*, *seconde édition*, 1793.)

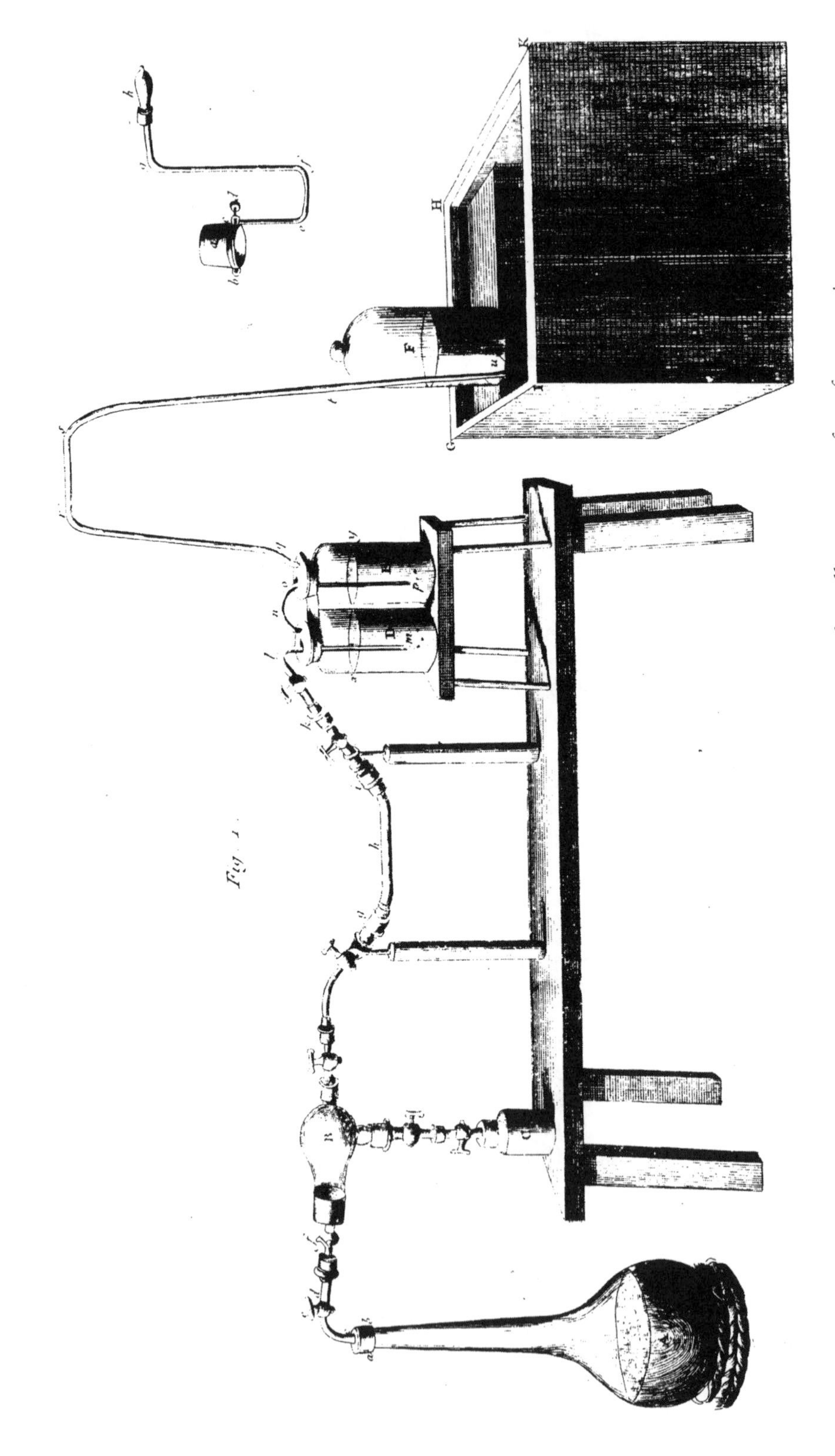

FIGURE 206. ■ From Lavoisier's *Elements*, an apparatus for collection gases from fermentations.

porcelain dish *D*. Air is evacuated through one stopcock and oxygen added through the other. Combustion is started with a burning glass. In phosphorus combustion, white flakes of phosphorus pentoxide[6] (actually P_4O_{10}, which sublimes at 360°C) coat the vessel wall and interfere somewhat with the efficacy of the burning glass. This solid is extremely hydroscopic ($P_4O_{10} + 6H_2O \rightarrow 4H_3PO_4$, phosphoric acid).

In Figure 205 we see Lavoisier's apparatus for *synthesizing water* (*Fig. 5*).[12] The 30-pint glass balloon A conducts pure oxygen from the left (through drying tube MM filled with powdered calcium chloride or similar) and pure hydrogen from the right (through drying tube *NN*). An electric spark will be supplied in the vicinity *Ld*. On the basis of the results of water decomposition above, 85 grams of oxygen and 15 grams of hydrogen are added slowly with periodic sparking. *Et voilá*—100 grams of liquid water!

Lavoisier did not invent the law of conservation of matter. It was a firm assumption in the minds of contemporary and earlier scientists. However, his careful trapping of gases in preweighed liquids and his requirement that all matter in a chemical reaction must be accounted for brought chemistry to a new level as a science—some even called it physics. If the mass at the start of a reaction and at the end could not be matched then there was not much point in analyzing the chemistry. It was as if the *Ferme Générale* were conducting an audit.

The clarity and authority of the *Traité Élémentaire de Chimie* (Paris, 1789) spelled the end for the phlogiston theory. Irish chemist Richard Kirwan[14] (1733–1812) published a book, *An Essay on Phlogiston and the Constitution of Acids* (London, 1787), that made the case forcefully for phlogiston—the English view. Madame Lavoisier translated the book into French (Paris, 1788) and it became the focus for the anti-phlogistic arguments of the French school. A second edition of Kirwan's book, including the appended 1788 essays of the French chemists, was published in London in 1789. However, by 1792 Kirwan had accepted the anti-phlogistic theory and wrote to Berthollet: "At last I lay down my arms and abandon Phlogiston."[15]

To celebrate the victorious *Traité*, Madame Lavoisier, dressed as a priestess, ceremoniously burned Stahl's works (an *auto-da-fe* of Phlogiston).[16,17] She had earlier asked one of the members of the Arsenal Laboratory, Jean Henri Hassenfratz, for suggestions for celebration of this success. In a letter dated February 20, 1788, he suggested three possibilities: a portrait of the Lavoisiers, a play involving the combat between phlogiston and oxygen, and a totally allegorical presentation about the chemical revolution.[17] The portrait was painted by the artist Jacques Louis David (who was also Mme. Lavoisier's art instructor). Hassenfratz suggested two possibilities for the play. One involved a grand battle. Oxygen's troops included carbonates, phosphates, sulfates, etc., against the allies of Phlogiston, *acidum pingue* and *acide igne*. The other was a confrontation between handsome Oxygen, with his brother-in-arms Hydrogen at his side, and the deformed Phlogiston already missing an arm. At Phlogiston's side is *acidum pingue*, already dead, and *acid igne*, ashen, defeated and dying of fear. Oxygen is poised to lop off Phlogiston's remaining arm. A play was apparently performed and reported to Crell's journal *Chemische Annalen* by a Dr. von E**. Phlogiston was placed on trial, weakly defended by Stahl, and then burned at the stake.[17] If you think about this, you will realize that if combustion releases phlogiston, then combustion of phlogiston leaves nothing.

Phlogiston had "met its Waterloo." Water's decomposition into the elements hydrogen and oxygen and its reconstitution from these elements sounded a death knell for phlogiston theory. The same result was achieved by electrolysis of water in 1789.[13] The light gaseous "essence of fire" released from iron by aqueous acids was phlogiston, according to Cavendish in 1766. How ironic to learn that "flammable air" comes from the aqueous acid, not the metal, and that it would be Cavendish who effectively discovers it! But confidentially, friends, at the true end of the second millennium, we still do not fully understand why hydrogen and oxygen don't immediately consume each other when mixed and form water without the aid of a spark.[4]

---

1. Godfrey (Ambrose and John), *A Curious Research into the Element Water; Containing Many Noble and Useful Experiments on that Fluid Body*, T. Gardener, London, 1747.
2. J.R. Partington, *A History of Chemistry*, Vol. 3, MacMillan & Co, Ltd., London, 1962, pp. 379–381.
3. P. Ball, *Life's Matrix: A Biography of Water*, Farrar, Straus and Giroux, New York, 2000, pp. 141–147.
4. A very high level of quantum chemical calculations seemingly clarifies some, but not all, of the mystery of hydrogen's very slow reaction with oxygen (see M. Filatov, W. Reckien, S.D. Peyerimhoff, and S. Shaik, *Journal of Physical Chemistry* A, Vol. 104, p. 12014 (2000). I thank Professor Joel F. Liebman for making me aware of this article.
5. Partington, op. cit., pp. 325–338.
6. Partington, op. cit., pp. 344–362.
7. W.H. Brock, *The Norton History of Chemistry*, W.W. Norton & Co., New York, 1993, pp. 109–111.
8. J.-P. Poirier, *Lavoisier—Chemist, Biologist, Economist* (transl. R. Balinski), University of Pennsylvania Press, Philadelphia, 1993, pp. 13–16.
9. Partington, op. cit., pp. 402–410; 436–453.
10. J.-P. Poirier, op. cit., pp. 76–83.
11. Partington, op. cit., p. 417.
12. A. Lavoisier, *Elements of Chemistry in a New Systematic Order, Containing All the Modern Discoveries*, second edition (transl. R. Kerr), London, 1793, pp. 135–149. See Plates VII and IV, respectively, for the apparatus for decomposition and synthesis of water.
13. Partington, op. cit., p. 457.
14. J.R. Partington, op. cit., pp. 660–671.
15. A.J. Ihde, *The Development of Modern Chemistry*, Harper and Row, New York, 1964, p.81.
16. J.R. Partington, op. cit., o. 491.
17. The author is grateful to Dr. Jean-Pierre Poirier for his correspondence, including a transcript of Hassenfratz's letter and to Professor Roald Hoffmann for making me aware of Poirier's findings concerning the Lavoisiers' play. Some discussion of the play is to be found in Poirier's book (see Ref. 8).

BEN FRANKLIN—*DIPLOMATE EXTRAORDINAIRE*

Here is a problem worthy of Benjamin Franklin, America's greatest diplomat: How do you retain the friendships of two close friends who disagree absolutely and fundamentally on chemical theory? Joseph has been your friend since the 1760s, took your advice and published *The History and Present State of Electricity*.[1] At considerable risk, this English clergyman steadfastly supported the independence of your beloved America. You met Antoine in 1772. He is brilliant, dashing, and wealthy, and his fellow countrymen venerate you (Figure 207).[2] His beautiful, gifted, and wealthy wife Marie Anne speaks to you in French, English, or Latin, as you wish, and has painted your portrait in oil (Figure 208).[3] The Lavoisiers host "smashing" parties and stimulating salons. In contrast, Joseph and Mary Priestley live an ascetic life and preside over sober teas. Joseph published the first paper describing the discovery of what he calls "dephlogisticated air." But he is firmly anchored in the century-old phlogiston theory. Antoine calls Joseph's air "oxigene" and believes that this theory is as dead as the dodo,[4] whose extinction roughly coincided with

FIGURE 207. ■ Portrait of Benjamin Franklin in the *Ouevres De M. Franklin*, Paris, 1773. The poem is translated in footnote 2 of this essay. (This book belonged to Dr. Werner Heisenberg, courtesy of his son Professor Jochen Heisenberg.)

FIGURE 208. ■ Madame Lavoisier was instructed in painting by the famous artist Jacques Louis David. This is a photo of the oil portrait she painted of her close friend Benjamin Franklin. See color plates. (Courtesy of a relative of Benjamin Franklin.)

the birth of phlogiston. So what is a great diplomat to do? The answer? Read both schools of thought in depth and leave no record whatsoever about where you stood on the greatest chemical controversy of the Enlightenment.[5]

Neutral as he was on phlogiston, Franklin nonetheless made some highly original and insightful chemical speculations. One of the most fascinating is his statement of the conservation of matter in a 1752 letter[6] composed when Lavoisier was but nine years old:[5,6]

> The action of fire only *separates* the particles of matter, it does not *annihilate* them. Water, by heat raised in vapour, returns to the earth in rain; and if we collect all the particles of burning matter that go off in smoke, perhaps they might, with the ashes, weigh as much as the body before it was fired: And if we could put them in the same position with regard to each other, the mass would be the same as before, and might be burnt over again.

Although the law of conservation matter is strongly and quite properly associated with Lavoisier, its chemical consequences were stated explicitly at least a century earlier and, indeed, the concept dates back to antiquity.[7] Nonetheless, Franklin's views on this matter are not widely known and his statement even suggests a specific experiment to verify the law. Franklin also reports witnessing the flammability of swamp gas (methane), in New Jersey, no less, over a decade before it was isolated by Alessandro Volta:[5]

> choose a shallow place, where the bottom could be reached by a walking-stick, and was muddy: the mud was first to be stirred with the stick, and when a number of bubbles began to arise from it, the candle was applied. The flame was so sudden and so strong, that it catched his ruffle and spoiled it, as I saw.

Dudley Herschbach notes that Franklin's observations over the course of six decades "convinced him of the danger of lead poisoning."[8] According to Herschbach, "he attributed a sudden attack of 'Dry Bellyach' that had beset a family to drinking rainwater collected from a leaded roof. He noted that trees planted around the house years before had grown tall enough to shed leaves on the roof, thereby creating acid that corroded the lead and 'furnished the water . . . with baneful particles'."[8]

Franklin and Lavoisier shared an interest in gunpowder, specifically the major (75%) component: saltpetre. Around 1775, as the American Revolution began to heat up, the English forbade shipments of gunpowder from Europe.[9] The Royal Navy was a convincing argument. In the same year the Continental Congress authorized publication of the pamphlet *Several methods of making salt-petre; recommended to the inhabitants of the United Colonies, by their representatives in Congress*. Sections were authored by Franklin and Dr. Benjamin Rush, who subsequently became signers of the Declaration of Independence.[9] Barn stalls and household chamber pots became vital sources of saltpetre. The French were more than eager to sell saltpetre and gunpowder to the American Colonies in order to weaken their long-time foes the English. Franklin dealt very successfully with his *bon ami* Lavoisier, who was an official in the *Régie de Poudres*, the government organization governing the production and quality of gunpowder. Figure 209 is from a report principally authored by Lavoisier on the fabrication of saltpetre.[10] It shows a factory in which earths and ashes rich in manure are washed, the water evaporated, and the remaining liquors cooled to allowed crystallization of saltpetre. In the lower left is a figure of a hydrometer (Lavoisier terms it an "aerometer") used to measure densities of the remaining liquors to assess when they are ready to crystallize.

Ballooning had its origins in France. The Montgolfier brothers gave the first public demonstration of a hot-air balloon on June 5, 1783. On August 27, 1783, J.-A.-C. Charles (of $V = kT$ fame)[11] made the first ascent in a balloon filled with "inflammable air" (hydrogen). Franklin was enthusiastic about ballooning and was said to quip that Montgolfier was the father of the balloon and Charles its wet nurse.[9] Both Franklin and Lavoisier served on committees that evaluated and exchanged information on ballooning. Strikingly, Lavoisier and Franklin also served on a committee to investigate the phenomenon of "human magnetism" pioneered by Franz Anton Mesmer. Although Franklin corresponded with Mesmer, both he and Lavoisier were skeptical of the claimed phenomenon, and the Committee's 1784 report was negative. We now know that Mesmer had, during his investigations, discovered hypnosis and the power of suggestion. Here is Franklin concluding an otherwise skeptical letter about Mesmer's work:[9]

> There are in every great rich city a number of persons who are never in health, because they are fond of medicines and always taking them, and hurt their constitutions. If these people can be persuaded to forbear their drugs in expectation of being cured by only the physicians finger or an iron rod pointing at them, they may possibly find good effects tho' they mistake the cause.

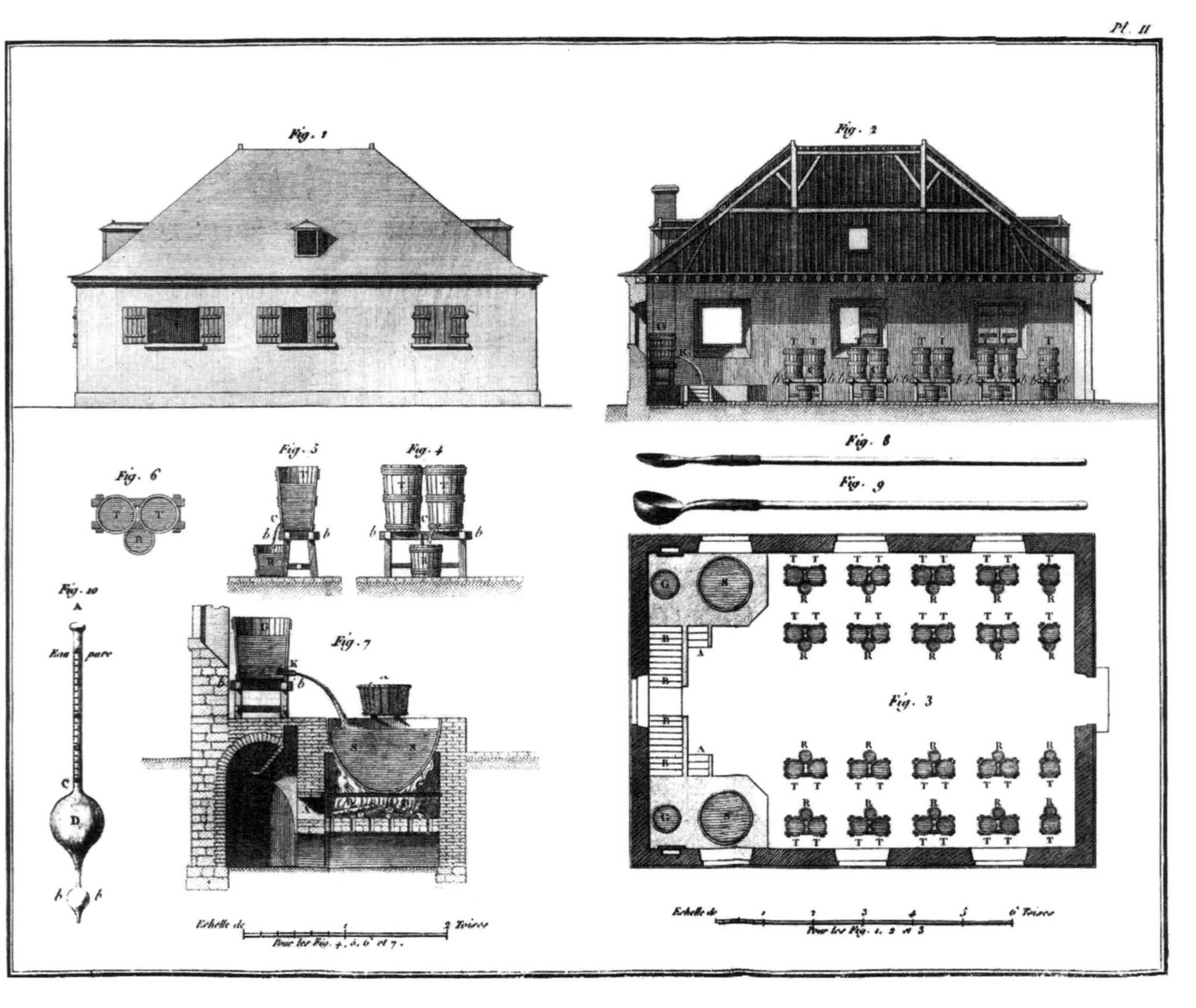

FIGURE 209. ■ A saltpetre factory described in a 1779 report (the figure is from the 1794 reprint) authored by Lavoisier and others on saltpetre manufacture.

Franklin is, of course, best known for his studies of electricity, and Herschbach compares their revolutionary importance with those of Newton or Watson and Crick.[12] Franklin believed electricity to be a fluid—*excess* corresponding to "positive" electricity and *deficit* to "negative" electricity. Indeed, his invention of the lightning rod around 1772 protected gunpowder storehouses and Franklin happily used his "thunderhouse demo" to show the efficacy of this invention.[12] However, his wide-ranging interests also led him to explain why the winds of a "noreaster" actually come from the southwest and, similarly, how the Gulf Stream affects climate in the northeastern United States. Figure 210 depicts Franklin's explanation of a waterspout in the ocean.[2] Clearly, he can be considered one of the fathers of earth science. Franklin invented the glass harmonica and *may* have composed a string quartet (in the key of F—what else?)—a slightly

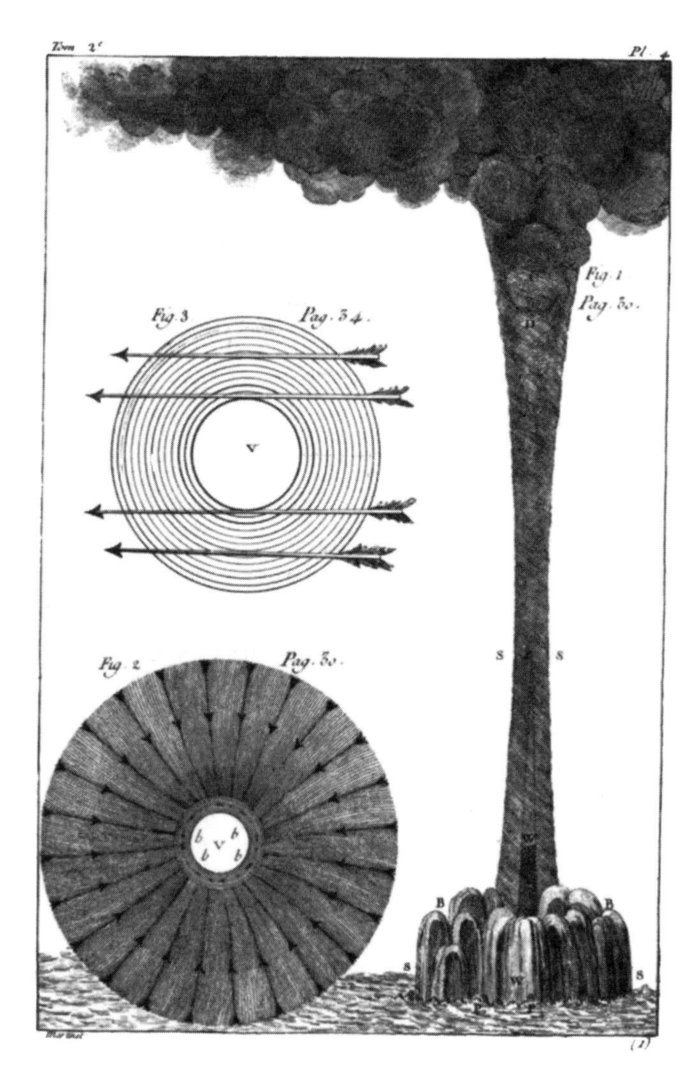

FIGURE 210. ■ Franklin was one of the earliest earth scientists. This figure shows a water spout and explains its origin. Franklin explained the effects of the Gulf Stream on weather in the northeastern United States. (From *Ouevres De M. Franklin*, courtesy of Professor Jochen Heisenberg.)

mischievous opus since the left hands of the surprised musicians remain idle throughout.[12]

---

1. J.R. Partington, *A History of Chemistry*, Vol. 3, MacMillan & Co., Ltd, London, 1962, pp. 237–256.
2. M. Barbeu Dubourg (transl.), *Oevres de M. Franklin*, Chez Quillau, Paris, 1773. I am grateful to Professor Jochen Heisenberg for providing for review the copy belonging to his father, Dr. Werner Heisenberg. The poem under the frontispiece portrait of Franklin was translated by my colleague Professor Jean Benoit as follows:

   He has conquered Heaven's fire
   He has helped the arts to blossom in wild climates
   America places him at the head of the sages
   Greece would have placed him amongst their gods

   It is abundantly clear that the French lionized Franklin. The "wild climates" referred to is a French Enlightenment view of the cultural milieu (or lack thereof) in the New World.
3. I am grateful to a relative of Benjamin Franklin for supplying a photograph of the oil-on-canvas portrait of Franklin by Mme. Lavoisier that is in his possession.
4. *The New Encyclopedia Britannica*, Encyclopedia Britannica, Inc. Chicago, 1986, Vol. 4, p. 148.
5. D.I. Duveen and H.S. Klickstein, *Annals of Science*, Vol. 11, No. 2, pp. 103–128 (1955).
6. I am grateful to Professor Dudley Herschbach for making me aware of this letter and its importance.
7. Partington, op. cit., pp. 377–378.
8. D. Herschbach, *Environmental Encyclopedia*. I thank Professor Herschbach for making me aware of this aspect of Franklin's work.
9. D.I. Duveen and H.S. Klickstein, *Annals of Science*, Vol. 11, No. 4, pp. 271–308 (1955); Vol. 13, No. 1, pp. 30–46 (1957).
10. [A. Lavoisier et al.], *Instruction sur l'Establissement des Nitrières, et sur la Fabrication de Salpêtre*, Cuchet, Paris (1794) (original edition 1779).
11. Charles' law for ideal gases—the volume of a gas is directly proportional to its absolute temperature.
12. D. Herschbach, *Harvard Magazine*, Cambridge, UK, Nov.–Dec. 1995, pp. 36–46.

***MON CHER* PHLOGISTON, "YOU'RE SPEAKING LIKE AN ASS!"**

The jubilant revolutionaries who overthrew the monarchy in August 1792 were convinced that France had been born anew. On 8 *frimaire An* II (Year 2), that is, November 28, 1793, Antoine Laurent Lavoisier and his father-in-law Jacques Paulze, reported for internment to the Port-Libre prison.[1] Lavoisier's situation had become increasingly perilous as the revolution radicalized and closed in all around him, stripping him of offices, colleagues, and the ability to travel, and after some days of hiding in Paris, his very freedom. He would lose his life in late spring 1794.

But in 1788 Lavoisier was at the height of his prestige and authority. As one of 40 wealthy partners in the *Ferme Générale*, he was a shareholder in the company empowered to collect taxes on all imports. This included the very salt that sustained people's lives. The General Farm also exercised some control of the flow of this revenue into the Royal Treasury. It therefore wielded considerable influence on France's fiscal policy. Lavoisier himself had, as a partner in the Farm, become a member of the board of directors of the Discount Bank, the "banker's bank," which advanced money to the Royal Treasury and supplied the gold and silver for minting into coins.[2] A world-class economist, he would soon become its president. Poirier nicely states this situation—a "private company was controlling the loans made to the government by a private bank."[2]

During most of 1788 Lavoisier wrote his masterpiece *Traité Élémentaire de Chimie* and it was published in early 1789 (see Figure 211).[3,4] The project had started as an attempt to provide an accessible introduction to chemistry, evolved to update the 1787 *Méthode de nomenclature* chimique [3,4] and became the most important treatise in the history of chemistry. It included the first modern list of chemical elements (33 in number, including the "imponderables" light and

TRAITÉ
ÉLÉMENTAIRE
DE CHIMIE,
PRÉSENTÉ DANS UN ORDRE NOUVEAU
ET D'APRÈS LES DÉCOUVERTES MODERNES;
Avec Figures:

Par M. LAVOISIER, de l'Académie des Sciences, de la Société Royale de Médecine, des Sociétés d'Agriculture de Paris & d'Orléans, de la Société Royale de Londres, de l'Institut de Bologne, de la Société Helvétique de Basle, de celles de Philadelphie, Harlem, Manchester, Padoue, &c.

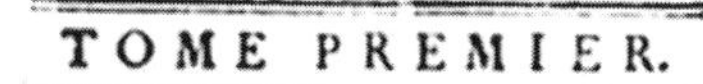

TOME PREMIER.

A PARIS,
Chez CUCHET, Libraire, rue & hôtel Serpente.

M. DCC. LXXXIX.

Sous le Privilège de l'Académie des Sciences & de la Société Royale de Médecine.

FIGURE 211. ■ Title page from the first edition (1789) of Lavoisier's 1789 masterpiece, *Traité Elémentaire De Chimie*, the first modern textbook of chemistry.

caloric).[3–5] Lavoisier was very much concerned with public education[6] and pedagogy, and this is reflected in his textbook. As late as September 22, 1793, he advocated to the Convention the education of a technologically proficient populace. However, in the prevailing climate of the "Cultural Revolution,"[7] Robespierre and the Jacobins wanted a more ideological education. This debate came to an abrupt halt with the Reign of Terror in October 1793.[6]

The year 1788 was a triumphant one for the Lavoisiers, even as the winds of revolution were stirring. Madame Lavoisier wrote to Jean Henri Hassenfratz, Director of the Arsenal, seeking suggestions for celebrating the victory of their chemical revolution. He suggested a portrait of the Lavoisiers and an allegorical play in which oxygen defeats phlogiston.[8,9] The portrait of the Lavoisiers was completed in 1789 by Jacques Louis David, for a fee of 7000 livres ($280,000 in current money)[9] and now sits in the Metropolitan Museum of Art. The only tangible evidence of a brief satirical play or masque is found in a letter by a Dr. von E** published in Crell's journal *Chemische Annalen*.[8,9]

Another, earlier masque is, however, creatively imagined in the play *Oxygen*, authored by two distinguished modern chemists, Carl Djerassi and Roald Hoffmann.[10] The year 2001 marks the hundredth anniversary of the Nobel Prize. In modern Stockholm the 2001 chemistry Nobel committee is informed, in secret, that they will also choose the first "retro-Nobel Prize." The committee quickly reaches consensus that the discovery of oxygen and its role in chemistry and respiration merits the first "retro—Nobel." Should it go to the Swede Carl Wilhelm Scheele who first isolated "fire air" (oxygen) in 1771 (or 1772) but did not publish the work until 1777? Should it go to Joseph Priestley, who independently discovered "dephlogisticated air" (oxygen) in 1774 and promptly published his discovery in the same year? Both Scheele and Priestley erroneously believed that their "air" drew phlogiston from burning or rusting substances. Or should it go to Antoine Laurent Lavoisier, whose intellectual synthesis fully explained the role of oxygen in combustion, calcination, and respiration? *Oxygen* actually begins in the Stockholm of 1777, where we meet Marie Anne Paulze Lavoisier, Mary Priestley, and Sara Margaretha Pohl, assistant and companion to Scheele, in a sauna. Their husbands have been summoned to Sweden to perform experiments before King Gustav III, who will render *The Judgement of Stockholm*. This play-within-a-play might somehow help the 2001 committee to settle its (or at least the play-goers') predicament. On the evening before the royal command chemistry demonstrations, the Lavoisiers perform their brief masque before King Gutav III and the assembled company. The Priestleys, Scheele, and Fru Pohl become increasingly uncomfortable and, finally, quite upset upon its conclusion. Antoine plays the vanquished "Phlogiston" and Marie Anne the victorious "Oxygène" in this play-within-a-play-within-a-play. Leading up to their masque's conclusion we have Madame Oxygène addressing Monsieur Phlogistique:[11]

> *Mon cher* monsieur, you're speaking like an ass!
> You know there's no such thing—negative mass!
> A revolution is about to dawn
> In chemistry, as Oxygen is born.
> Phlogiston is a notion of the past,
> Disproved and set aside, indeed, surpassed.

Madame Lavoisier was surely one of the most fascinating figures in the history of chemistry.[12] She plays the central role in *Oxygen*, and a mysterious note long hidden in her *nécessaire*[12] solves, in the play at least, a chemical riddle more than two centuries old. As to who wins the first "retro-Nobel"—that is for you, gentle reader, to guess—but first read the play.

Widespread resentment of Lavoisier, which accompanied admiration of this awesome polymath, predated the French Revolution. As a member of the Academy of Sciences, his investigation, along with Franklin and others, that dismissed mesmerism as bad science, was resented by a populace that wanted to believe in it.[13] Another longstanding grievance was the Farm's "watering of tobacco" prior to distribution to distributors.[14] More serious, however, was Lavoisier's role in the General Farm's collection of taxes. Imagine a powerful holding company comprised of 40 individuals, the purpose of which was to zealously collect taxes for the Royal Treasury but not before they had taken a very tidy profit. The salt tax was widely despised—salt was a staple for preserving meat—and indeed, is a very substance of life.[15,16] The salt tax was one of Lavoisier's specific responsibilities for the General Farm. Another was the tax upon imports into Paris. Lavoisier, using the genius in accountancy that he applied to chemistry, had realized late in the 1770s that only four-fifths of the goods needed to supply the inhabitants of Paris were actually reported and taxed.[17] The remaining fifth was being smuggled with a loss to the Royal Treasury (and, incidentally, the General Farm). His solution, a wall with toll gates surrounding Paris, was accepted in 1787 and built at a cost of 30 million livres ($1.2 billion).[17] Once again, try to imagine a private company owned by 40 of the wealthiest individuals in the United States, walling in New York City and building palatial toll gates at taxpayer expense for use by the Internal Revenue Service. One of the accusations leveled against Lavoisier years later was that the wall built around Paris confined the city's air to the detriment of its citizens' health.

But the ground was beginning to shake and soon Lavoisier would experience a foreshadowing of his own fate. As Director of the Gunpowder Administration, he had authority over gunpowder shipments from the Arsenal. Not long after the storming of the Bastille on July 14, 1789, mysterious transfers of gunpowder from the Arsenal were observed by citizens who concluded that a Royalist counterattack was in the offing.[18] Lavoisier was seized and held briefly in custody—some members of the crowd gathered along his route of transport demanded summary execution. However, he explained the shipments in detail, was declared innocent, and released. On March 20, 1791, the National Assembly abolished the General Farm.[19] In the aftermath, Lavoisier's business affairs were found to be aboveboard. The learned academies were abolished in August, 1793.[20] Academicians who had not abandoned their "elitist" views and wholeheartedly joined the people were now in danger. The "Reign of Terror" fully radicalized the Revolution, and Lavoisier, Paulze, and 26 other members of the *Ferme Générale* were guillotined over the course of 35 minutes on May 8, 1794.[21]

---

1. J.-P. Poirier, *Lavoisier—Chemist, Biologist, Economist* (R. Balinski, transl.), University of Pennsylvania Press, Philadelphia, 1993, pp. 346–369.
2. Poirier, op. cit., pp. 220–221.
3. Poirier, op. cit., pp. 192–197.

4. J.R. Partington, *A History of Chemistry,* MacMillan & Co. Ltd., London, Vol. 3, 1962, pp. 484–487.
5. A. Greenberg, *A Chemical History Tour,* John Wiley & Sons, New York, 2000, pp. 143–146.
6. Poirier, op. cit., pp. 336–345.
7. Poirier, op. cit., pp. 328–335.
8. The author is grateful to Dr. Jean-Pierre Poirier for supplying a copy of Hassenfratz's letter.
9. Poirier, op. cit., pp. 1–3.
10. C. Djerassi and R. Hoffmann, Oxygen, Wiley-VCH, Weinheim, 2001. I am grateful for permission from Professor Djerassi and Professor Hoffmann to use this section from their play and for their helpful comments.
11. Djerassi, op. cit., pp. 42–45.
12. R. Hoffmann, *American Scientist,* Vol. 90, No. 1 (Jan.–Feb. 2002), pp. 22–24.
13. Poirier, op. cit., pp. 154–159.
14. Poirier, op. cit., pp. 23–28, 115–116, 166–170.
15. Poirier, op. cit., p. 120.
16. P. Laszlo, *Salt: Grain of Life,* Columbia University Press, New York, 2001.
17. Poirier, op. cit., pp. 170–173.
18. Poirier, op. cit., pp. 241–245.
19. Poirier, op. cit., pp. 272–273.
20. Poirier, op. cit., pp. 333–335.
21. Poirier, op. cit., pp. 381–382.

"LAVOISIER IN LOVE"

Draft for a screenplay: "Lavoisier In Love." Noting the great critical (and commercial) success of the 1998 film "Shakespeare In Love," we1 feel that the roughly 30-year period between 1772 and 1805 that witnessed the chemical revolution could furnish a blockbuster.[2] Although we offer the idea later in a humorous vein, we honestly feel that an epic of more appropriate title could really be quite good. We see Kenneth Branagh directing the screenplay and playing Antoine; Gwyneth Paltrow as the young Marie; Judi Dench as Marie in her later years. Are there any financial backers out there?

We are not even wedded to the above title—"Antoine and Marie, The Tax Collector's Daughter" is another possibility.[3] Where "Shakespeare in Love" offers mere swordplay, we offer certified gunpowder and real pyrotechnics in the laboratory, in the streets and on the high seas. The film will be a period piece centered around the lives of Antoine Laurent Lavoisier, the father of modern chemistry, and his wife Marie-Anne Pierette Paulze-Lavoisier, one of the most sophisticated and alluring women of the age. The voice-over narration[4] is that of Madame Paulze-Lavoisier, the nexus of our drama. The background is the American Revolution, the French Revolution, and the fearful and violent reaction in England during and after the loss of the jewel in the crown of its North American empire. ("The Madness of King George" did pretty well in 1995 but it was aimed at the brie-and-merlot crowd, not the masses.) There's sex, violence, adultery, abandonment, lechery, espionage, treason—it will make "Les Liaisons Dangereuses" seem like "Sesame Street."

It is 1766 and phlogiston, a last remnant of alchemical thought, has held sway for nearly 100 years. In England, Henry Cavendish, an eccentric genius millionaire, thinks he has isolated the elusive phlogiston but has really made explosive hydrogen. (FLASH-FORWARD 20 years—French exploring air travel in hydrogen-filled balloons—a fiery disaster occurs). CUT TO Birmingham, England in the early 1770s: Joseph Priestley, a rather stiff-necked Unitarian clergyman discovers oxygen, finding that it sustains animals five times as long as regular air. He is a friend and correspondent of Benjamin Franklin as the American Revolution commences. JUMP CUT TO Fall, 1775; Ben advises:[5]

> "Joseph, Britain, at the expense of three millions, has killed one hundred and fifty Yankees this campaign, which is twenty thousand pounds a head . . . During the same time sixty thousand children have been born in America."

Enter the wealthy and brilliant 28-year-old Antoine to rescue 14-year-old Marie-Anne from the doddering lechers who work with him and Monsieur Pauize at La Ferme Generale. (La Ferme was a private finance company employed by the government to collect taxes—more on that later.) However, there is a foreshadowing as the shadow of a blade falls on a sausage for the Company Christmas party *choucroutes garnie*.[6] Marie and Antoine marry in 1771 and form a partnership that any couple would envy. Antoine begins his scientific studies in their residence. Marie's facility with languages brings Antoine access to the foreign chemical literature. He doesn't like what he reads and decides to change *everything*. Marie learns enough chemistry to be able to translate and comment critically on foreign texts. A gifted artist, also she engraves the plates for his monumental *Traité Élémentaire de Chimie* and paints a portrait of Franklin that Ben treasures highly. Saturdays are spent in the Salon with Antoine and Marie discussing the week's experiments with the *cogniscenti*.

Enter Pierre du Pont and his son Eleuthère Irenee who will later find a small measure of success with a start-up chemical company in wild and remote Delaware. Pierre is dashing and ebullient. Antoine is analytical and totally devoted to his beakers, flasks, and balances. Pierre and Marie begin an affair starting in 1781 that will last over 10 years without damaging the friendship between Pierre and Antoine (or, for that matter, Marie and Antoine)—ah, the French. From a scene:

Antoine: There are 26 of Pierre's robes in my armoire. I can't seem to find my lab coat!

Marie: It's in the laboratory with your clean underwear. I'll see you next Saturday in the Salon, Cheri.

FLASHBACK TO Benjamin Thompson, born to a family of modest means in the Colony of Massachusetts, who marries at the age of 19 a wealthy widow some 14 years his senior. During the American Revolution he spies for the British, is almost caught, abandons his wife, takes a fortune, and flees to England where he is knighted by George III in 1784. He will return.

Meanwhile, the English and French have been fighting for global dominance directly and by proxy for over 100 years. The phlogiston controversy gives them a fresh field for rivalry. Volleys of rhetoric fly back and forth across the Channel. Richard Kirwan, a viriolic Irishman, attacks Antoine (in print). Marie translates Kirwan's work—it provides Antoine with just the ammunition he needs and he appends his own notes to Marie's published translation. Hoist on his own petard and mortally wounded, Kirwan abandons phlogiston. Priestley holds fast—he never abandons phlogiston! The American Revolution triumphs. England is full of fear and anger. The excesses of the French Revolution add to this fear. Lavoisier, the tax collector, is guillotined (remember the foreshadowing?). Priestley escapes England as an angry rabble vows to "shake the powder from his wig" and burns his church to the ground. (Future portraits will show Priestley *sans* wig.)[7]

After Lavoisier is executed, the most eligible, wealthy, and brilliant suitors in Europe court Marie. REENTER Benjamin Thompson, now Count Rumford of the Holy Roman Empire, retired Head of the Bavarian Army, and vanquisher of Antoine's caloric theory, who emerges from the pack. The happily unmarried couple tour Europe together for four years. He is once again willing to endure the consequences of marrying a wealthy woman. They marry in 1805, but the marriage is on the rocks in two months—it seems that Rumford locked the front gates on Marie's guests one day and she responded by pouring boiling water on his prize flowers (lots of doctoral theses by students of cinema on the symbolism of these two actions). To this interesting cast we can add the laughing-gas-sniffing parties of Humphry Davy and his artistic friends.

In our advertising trailer—COMING TO A THEATRE NEAR YOU:

SEE! Franklin Cruise The Salons of Paris!
SEE! Cavendish Make Water!
SEE! Marie Paulze-Lavoisier Scald Rumford's Flowers!

---

1. Ideas were contributed to this essay by Professor Susan Gardner, University of North Carolina at Charlotte.
2. Professor Roald Hoffmann was kind enough to share with me his earlier ideas about dramatizing the Lavoisiers. This was, in part, developed into a play: Carl Djerassi and Roald Hoffmann, *Oxygen*, Wiley-VCH, Weinheim, 2001. A DVD is available from the University of Wisconsin.
3. In the movie, Shakespeare's original title appears to be: "Romeo And Ethel, The Pirate's Daughter."
4. Professor Susan Gardner proposed Mme. Lavoisier as the narrative voice.
5. E. Wright, *Franklin of Philadelphia*, Belknap Press of Harvard University Press, Cambridge, 1986, p. 239.
6. Homage to Jane Campion's movie "The Piano."
7. Well, at least Rembrandt Peale's 1801 oil-on-canvas portrait is wigless. See B.B. Fortune and D.J. Warner, *Franklin and His Friends: Portraying The Man of Science in Eighteenth-Century America*, Smithsonian Portrait Gallery and University of Pennsylvania, Washington, D.C. and Philadelphia, 1999, p. 151.

REQUIEM FOR A LIGHTWEIGHT

Although Phlogiston Theory was vanquished during the 1780s, it is worthwhile to summarize some of the definitions and arguments for and against phlogiston.[1] These will be limited almost entirely to coverage in the present book and is not meant to be a thorough treatment.

1. *What Were the Origins of Phlogiston Theory?* Diverse cultures had ancient beliefs in dualities (male–female; yin–yang; Sol–Luna; sulfur–mercury). This was modified by Paracelsus and others to the *tria prima:* sophic sufur, sophic mercury, and sophic salt, which constitute matter in various proportions. Becher (in the seventeenth century) recognized three "earths." One of these, *terra pinguis* or "fatty earth," was said to be present in combustible and metallic substances. It is analogous to sophic sulfur. Becher's theory was modified by Stahl (early eighteenth century) who coined the term "phlogiston" (Φ) to replace *terra pinguis.*

2. *What Is the Nature of Phlogiston (Φ)?* Phlogiston (Φ) is frequently defined as "the essence of fire." Sometimes Φ is simply identified as the fire released from a burning substance. Phosphorescence of a substance was considered to be a visual manifestation of the Φ stored in that substance. White phosphorus is loaded with Φ since it phosphoresces and can also ignite spontaneously. Phlogiston was usually considered to be an imponderable (superlight or even lacking mass) substance (often a fluid). However, fire did not always accompany release of Φ and, thus, might be merely one manifestation of its release.

3. *What Chemical Phenomena Did Phlogiston (Φ) Explain?* Most powerfully, it was a unifying theory for combustion of matter *and* for formation of calxes (often what we call oxides). This was by no means obvious. It is important to note that until the mid-eighteenth century, gases arising from combustion, such as carbon dioxide, were simply seen as "airs" and not collected.

Charcoal (contains Φ) + heat → ash + Φ

Copper (contains Φ) + heat → copper calx + Φ

Charcoal (contains Φ) + copper calx + heat → ash + copper (contains Φ)

Cavendish collected the "flammable air" derived from "dissolving" metals in aqueous acids. What remained upon evaporation of the solution was the calx. The "flammable air" he collected had only 7% of the density of "atmospherical air." It appeared that the superlight, superflammable gas "obviously" released from the metal might well be Φ itself.

Copper (contains Φ) + sulfuric acid → copper calx + Φ? ("flammable air")

Other "airs" could also remove Φ from metals:

Copper (contains Φ) + nitric acid → copper calx + "nitrous air" (contains Φ)

What Lavoisier called *oxygen* today was termed "dephlogisticated air" by Priestley. It comprised one fifth of the atmosphere and had great affinity for Φ. "Nitrous air" and "flammable air" each carry the same amount of Φ since one volume of each will lose all of its Φ to one-half volume of "dephlogisticated air." Atmospheric air that absorbs Φ is "wounded" and when fully phlogisticated is "deadly" or "mephitic." What remains is "mephitic" or "phlogisticated" air, or nitrogen, which had earlier been named "azote" ("without life").

Food contains Φ; fatty food is particularly rich in Φ

4. *What Were Phlogiston's Failures?* Most notably, increases in weight upon loss of Φ:

$$\begin{array}{ccc} \text{Copper (contains } \Phi) + \text{heat} \rightarrow & \text{copper calx} + \Phi \\ \text{(63.5 grams)} & \text{(79.5 grams)} \end{array}$$

This had been noted since the sixteenth century. If the law of conservation of matter holds, then Φ has negative mass (–16.0 grams in this case). When gases were collected, starting in the mid-eighteenth century, the results of exhaustive combustion of charcoal would be

$$\begin{array}{ccc} \text{Charcoal (contains } \Phi) + \text{heat} \rightarrow & \text{ash} + \text{“fixed air”} + \Phi \\ \text{(60 grams)} & (\ll 1 \text{ gram}) \text{ (ca 220 grams)} \end{array}$$

The mass of the gas generated was quite consequential and inconsistent with loss of Φ unless it had negative mass (~–160 grams in the present case). Also of immense significance was the problem of the composition of water. Water generally was unnoticed (or remarked upon) as a product of combustion. Combustion of "flammable air" (Φ?) by combination with "dephlogisticated air" might be expected to give "dephlogisticated phlogiston" (simply *nothing*?) or perhaps simply "air" devoid of Φ—possibly nitrogen? Instead, the product was water. Water could likewise be split chemically to give hydrogen ('flammable air") and oxygen ("dephlogisticated air"). It was now obvious that the Φ derived from "dissolving" copper in sulfuric acid came from the acid solution (in the form of "flammable air" or hydrogen) rather than from the metal itself. Similarly, the Φ derived from "dissolving" copper in nitric acid also came from the acid solution (in the form of "nitrous air" or nitric oxide, NO) rather than from the metal itself. There were other very fundamental questions, including "Where did Φ go once it was lost?" Why did the volume of "atmospherical air" decrease by one-fifth when Φ was gained? Did the "dephlogisticated air" lose its "elasticity" due to "injury"? There were countless other, more subtle, quantitative problems. Here is one—if 63.5 grams of copper is "dissolved" in sulfuric acid, all of its Φ is released in one volume of"flammable air" with total conversion of the metal to its calx. If 63.5 grams of copper is "dissolved" in nitric acid, all of its Φ is released in two-thirds volume of "nitrous air" with total conversion of the metal to its calx. In the first case, all of the Φ will be removed from the 1 volume of "flammable air" by one-half volume of "dephlogisticated air." In the second case, all of the Φ will be removed from the two-thirds volume of "nitrous air" by one-third volume of the very same "dephlogisticated air." The numbers simply do not add up.

5. *Consolidation.* Combustion and calx formation are both examples of chemical combinations with the oxygen in the air. That is why calxes are heavier than their metals and why the products of combustion weigh more than the combustibles (when oxygen is neglected). A useful strategy is to employ Roald Hoffmann's suggestion of considering Φ as "minus oxygen."[2] Thus, oxygen is lost from the atmosphere in combustion rather than the atmosphere gaining Φ. Metals gain oxygen, rather than losing Φ, when they form calxes. Substances are oxidized by oxygen, which is, of course, an oxidizing agent. Thus, gain of Φ corresponds to reduction; Φ would be a reducing agent. Try it. It's fun.

1. My father, Murray Greenberg, suggested this essay.
2. R. Hoffmann and V. Torrence, *Chemistry Imagined—Reflections on Science*, Smithsonian Institution Press, Washington, DC, 1993, pp. 82–85.

OKAY, I NOW KNOW WHAT "OXIDATION" MEANS, BUT WHAT IS "REDUCTION"?

Thanks to Lavoisier's work toward the end of the eighteenth century, we understand that metals add oxygen to form oxides and that combustion of organic matter adds oxygen to both carbon and hydrogen. Thus, propane forms carbon dioxide and water as it lights our grills. Oxidation of nonmetals such as nitrogen, phosphorus, and sulfur form oxides that behave as acids while metal oxides behave as bases. The oxidation concept was extended in the nineteenth century—the lower-valence form of copper could be oxidized to the higher-valence form (even if no oxygen were involved). For example, yellow cuprous oxide (Cu_2O, which contains 11.1% oxygen) adds oxygen to form black cupric oxide (CuO, which contains 20.1% oxygen), and cuprous chloride (CuCl) may be oxidized to cupric chloride ($CuCl_2$) in the absence of oxygen. Indeed, that means that chlorine is also an oxidizing agent, as is iodine.

Now it so happens that the "reduction" concept, the opposite of "oxidation," is centuries older than its "contrary." What is the origin of the older term? The first instinct is to posit that "reduction" might refer to the fact that conversion of a calx, such as CuO, to the metal is accompanied by weight reduction.[1] However, despite the fact that *some* seventeenth-century chemists (Rey, Boyle, Le Fèvre, etc.) first reported that metals are lighter than their calxes, the "reduction" concept was effectively in use much earlier.[2]

Happily, perusal of a fat, old dictionary provides the needed insight. Although a popular two-pound dictionary,[3] offers "to lessen in any way" as the first of 15 definitions of the word "reduce," an older weightier tome, a 20-pound dictionary,[4] offers its primary definition as "to bring back; to lead to a former place or condition; restore" in full agreement with the Latin root *reducere*, "to lead back." And so there amiable reader is, I believe, the crucial insight. The ancient artisans assumed the pure metal to be the reference state and the action of reversion to the metal (from its calx, for example, by heating with charcoal) was understood to be "reduction" (reversion to its pristine state).[2] And this is also interesting since for most metals (but certainly not gold), the true former (earthly) state is not the metal but a salt, often a sulfide that must be roasted to obtain the metal. John Read[5] noted that the process of "putrefaction" (*Fourth Key* of Basil Valentine; see Figure 38(d)) involves blackening of the initial alchemical mixture accompanied by the "death" of metals baser than gold. This is actually the roasting and oxidation of metals and their sulfide ores. The reverse is their "resurrection" (*Eighth Key* of Basil Valentine; Figure 40(a)), the restoration of the original metals, and reunion with their souls through the chemical process we call "reduction."

It has also long been known that heating a calx in a stream of hydrogen gas causes reduction of the calx to the lighter pure metal. (Oh! Did I mention that water is also formed?) So hydrogen gas is a reducing agent. Reduction by hydrogen of unsaturated fats, however, produces saturated fats, which weigh *more* than the corresponding unsaturated fats. The reduced, saturated fat contains more calories than does the unsaturated fat so . . . oh, never mind! In any case, the modern definition of "reduction," which unifies all of these diverse operations, is "a process in which a substance gains one or more electrons" (oxidation is the "contrary").[6]

Well, darn it, as a chemist I know what I mean when I say "reduction." To paraphrase Popeye The Sailor—"I yam what I yam and that's all [*spit*] what I yam!"[7]

---

1. C. Cobb and H. Goldwhite, *Creations of Fire*, Plenum Press, New York, 1995, p. 8.
2. J.R. Partington, *A History of Chemistry*, Vol. 2, MacMillan & Co. Ltd., London, 1961, p. 19.
3. *College Edition—Webster's New World Dictionary of the American Language*, The World Publishing Company, Cleveland and New York, 1964, p. 1219.
4. *Webster's New Twentieth Century Dictionary of the English Language Unabridged*, second edition, The World Publishing Company, Cleveland and New York, 1956, p. 1514. This is also the primary definition in the Oxford English Dictionary.
5. J. Read, *From Alchemy to Chemistry*, Dover Publications, Inc., New York, 1995, p. 33.
6. T.L. Brown, H.E. LeMay, Jr., and B.E. Bursten, *Chemistry—the Central Science*, Prentice Hall, Upper Saddle River, NJ, 1997, pp. G–11, G–13. Pauling notes a failed attempt by Professor E.C. Franklin of Stanford University to remove this confusion by coining the words "de-electronation" (for oxidation) and "electronation" (for reduction) [see L. Pauling, *General Chemistry*, privately printed (Edwards Brothers, Inc. Lithographers—Ann Arbor), Pasadena, 1944, p. 65].
7. Popeye's neologisms and puns ("vitalicky"; "I know what rough is, but what's roughined?") have outlasted those of Professor E.C. Franklin (see note 6 above).

THE GUINEA PIG AS INTERNAL COMBUSTION ENGINE

Since *calorique* was a simple substance (an element), albeit imponderable, naturally Lavoisier wanted to measure it. Figure 212 shows the ice calorimeter designed by Lavoisier and Laplace. The fully assembled calorimeter is shown in *Fig. 1* and the cutout view in *Fig. 3*. The basket *ffff*, with opening *LM*, is made of iron wire mesh and can be covered with lid *GH*. This basket holds the caloric-generating sample: hot metal, hot liquid, or chemical reaction via mixing (inside a suitable container), combustion sample, or live guinea pig. Crushed ice is placed in chamber *bbbb* as well as in jacket *aaaa*. Chamber *aaaa* insulates the apparatus—water may be tapped conveniently through *sT*. The ice in chamber *bbbb*, supported by screen *mm* and sieve *nn*, absorbs the heat from basket *ffff*. The resulting water is tapped through *xy* and weighed. Prior to the experiment, crushed ice is tightly packed into chambers *aaaa* and *bbbb*, into lid *GH*, and the apparatus cover (*Fig. 7*) and allowed to attain equilibrium. These experiments are best done in rooms not much warmer than 50°F and definitely not colder than 32°F (since ice must be at this temperature and not colder). A large sample is placed in a metallic bucket equipped with a thermometer (*Fig. 8*; a corrosive liquid would be placed in a glass vessel equipped with a thermometer, *Fig. 9*). The bucket or glass vessel is placed in a bath of boiling water. Just prior to transfer, the

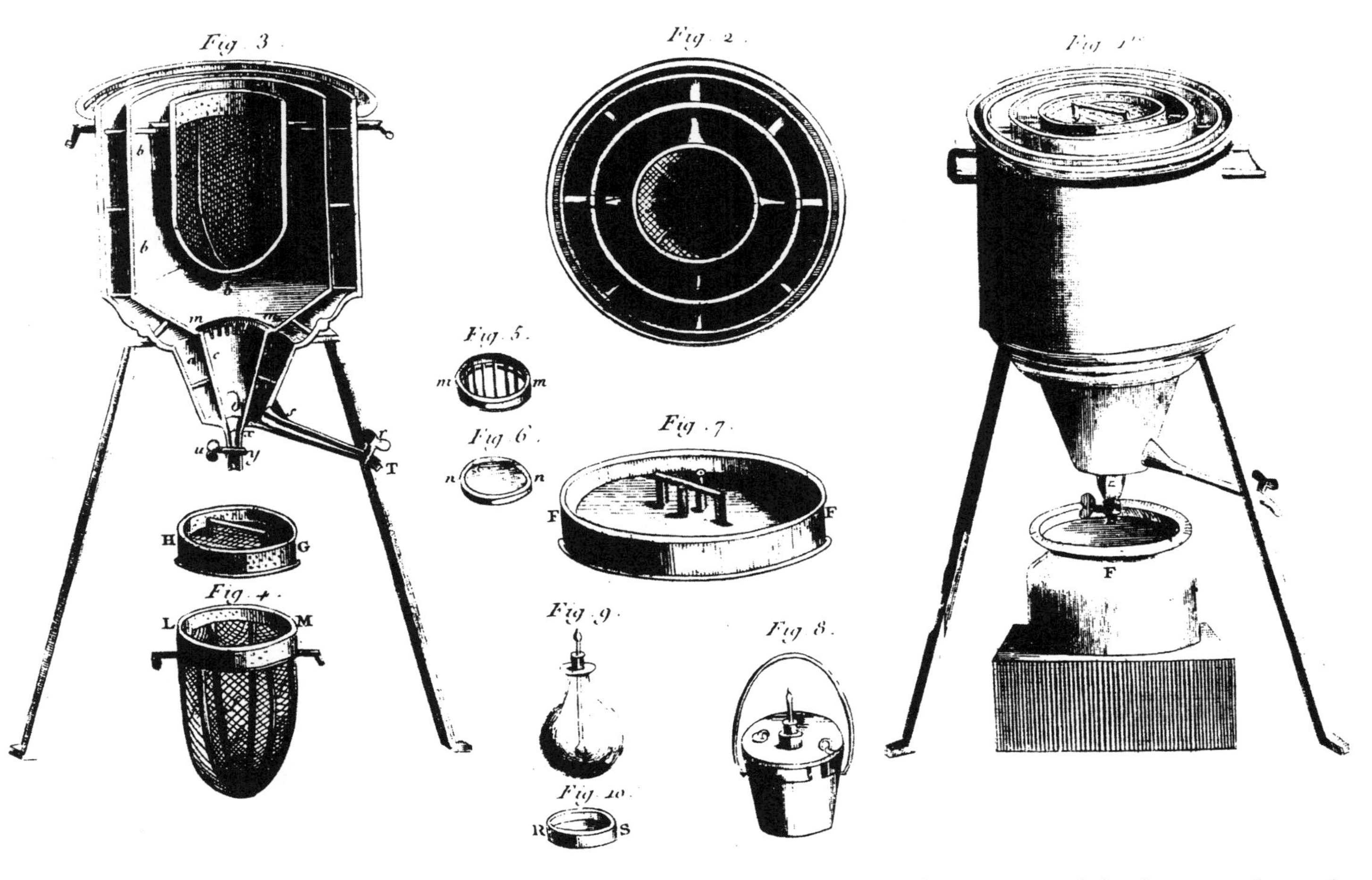

FIGURE 212. ■ Ice calorimeter, designed by Lavoisier and the famous mathematician Laplace. Heat was defined in units of ice melted. The idea that metabolism was similar to combustion derived from the knowledge that oxygen was required, carbon dioxide and water produced and heat generated by animals. Thus, Lavoisier realized that combustion, calcination and metabolism were all related in the sense that each involved combination with oxygen.

last drops of water are tapped through *xy* and discarded. Quick transfer of the hot sample is performed. It takes typically 10 to 12 hours for the entire internal calorimeter to return to 32°F. The water from chamber *bbbb* is then tapped and carefully weighed. Lavoisier and Laplace realized that there had to be heat losses that limited the accuracy of their determinations.

Lavoisier and Laplace defined their heat unit as the quantity required to melt one pound of ice (at 32°F). They demonstrated that it requires one pound of water starting at 167°F and cooling by 135°F (to 32°F) to melt this ice. Thus, they took 7.707 pounds of iron strips heated in a boiling water bath to 207.5°F and added the metal quickly into basket *ffff* and closed the calorimeter. After eleven hours, 1.109795 pounds of ice had melted. The iron had thus cooled by 175.5°F. Using the ratio $175.5/1.109975 = 135/x$, they found that $x = 0.85384$. Dividing this by 7.707, the quotient 0.1109 is the quantity of ice melted by one pound of iron cooling through 135°F. Other caloric-generating processes could be placed on this arbitrary scale.

Guinea pigs "thrive" in ice calorimeters better than mice. The air entering and leaving basket *ffff* had to, of course, pass through tubing immersed in the crushed ice. The realization gained over the previous decade that both respiration and combustion required oxygen "jelled" with the earlier observations that both processes produced carbon dioxide. It was thus a relatively small creative leap to equate the two and try to measure the slow internal combustion recognized as animal heat.

Imagining a mouse shivering inside the ice calorimeter (Figure 212), enduring "mephitic air" in the apparatus of Priestley (Figure 197), Scheele, Lavoisier or Mayow (Figure 150) inspires respect for the role this hardy and courageous mammal has played and continues to play in science. And while we previously noted that Priestley was "nice to his mice" (p. 296), Franklin wrote to him suggesting, in effect, that he ". . . repent of having murdered in mephitic air so many honest, harmless mice . . .".[1] Perhaps a statue should be erected honoring the mouse at the Royal Institute in Stockholm.

1. W. C. Bruce, *Benjamin Franklin Self-Revealed,* Second Revised Edition, Vol. I, Putnam, New York, 1923, pp. 106–107. I thank Professor Roald Hoffmann for bringing this material to my attention and Professor Susan Gardner for suggesting homage to mice.

THE MAN IN THE RUBBER SUIT

Antoine Lavoisier "buried" phlogiston theory and, in so doing, explained the basis of combustion and calcinations such as the rusting of iron. However, it is less widely appreciated that it was Lavoisier who first demonstrated that metabolism is simply a very slow combustion process. Where this metabolism actually occurred, heart, lungs, or both places, remained a mystery to him.

It was apparent to John Mayow as early as 1674 that respiration removed something from atmospheric air and the remaining depleted air could not support life or combustion.[1] Mayow's observations were strengthened a century later by the work of Scheele, Priestley, and Lavoisier, who isolated, manipulated, and characterized the "airs" they studied. When a mouse was confined in a bell jar containing atmospheric air, the air soon became "mephitic" or "deadly" because its full complement of "vital air" (oxygen) was depleted.[2] If the "mephitic air" (~99% nitrogen) were recharged with oxygen in the correct 4 : 1 proportion, a mouse would live equally long as it would in atmospheric air. The isolation and characterization of the mouse's exhalation gas, "fixed air" (carbon dioxide), was only first reported in 1756 by Joseph Black.[3] In 1777, Lavoisier concluded that animal metabolism combines carbon and oxygen to produce carbon dioxide, just as carbon unites with oxygen during combustion.[4,5]

The ice calorimeter, designed by Pierre Simon de Laplace, was first employed by Lavoisier during winter, 1782/83.[6,7] Heat from the reaction vessel was measured by the quantity of ice in the surrounding metal jacket that melted and was collected as water. Lavoisier and Laplace measured the heat given off by many chemical processes, including the combustion of charcoal. They also measured the heat produced by a living guinea pig.

By burning charcoal and measuring the "fixed air" produced, Laplace and Lavoisier equated formation of 1 ounce of fixed air to melting 26.692 ounce of ice.[6] Over a 10-hour period the amount of fixed air collected from a guinea pig's exhalations (224 grains, where 576 grains = 1 ounce) equated, on this basis, to melting 10.14 ounces of ice. The actual heat given off by the guinea pig over ten hours was greater, melting 13 ounces of ice. Although there were errors in the determination of the heat of combustion of charcoal, the main discrepancy was that the guinea pig "burned" not only carbon but also the hydrogen in its "fuel" to form water, thus releasing additional heat.[8] However, Lavoisier did not understand the true nature of water in early 1783 or that it was a product of respiration.

The discovery that water was not an element, but a compound of hydrogen and oxygen was absolutely critical to Lavoisier's growing understanding of respiration. In 1774, he ignited "flammable air" (hydrogen), isolated eight years earlier by Cavendish, in the presence of "vital air" and tried to collect the resulting "air" over water.[9] Naturally, the small quantity of water vapor generated went unnoticed. Although credit for the discovery of water's composition remains somewhat controversial, most chemical historians attribute it to Cavendish in 1783.[10] However, it was Lavoisier who determined the precise composition both in its decomposition into the elements and its synthesis from the elements (see Figures 204 and 205) and he reported these findings in early 1784. Water was not an element but a compound and a combustion product of compounds containing hydrogen. Thus, he realized that the water exhaled and sweated by animals was likely to be a product of respiration. The remaining problem for the "master of the chemical balance" was that a complete accounting of masses, input versus output, had not yet been demonstrated. Unfortunately, no human being had yet been hermetically sealed in a closed flask for precise studies of mass balance.

However, in 1790 Lavoisier and his assistant Armand Seguin conducted studies in which Seguin was sealed completely in a suit made of elastic-rubber-coated taffeta.[5,11] He breathed pure oxygen through a tube glued airtight around his lips and exhaled through this tube. The scene was depicted in a drawing by Madame Lavoisier (Figure 213)[12]—Seguin is seated at the left, while Lavoisier is

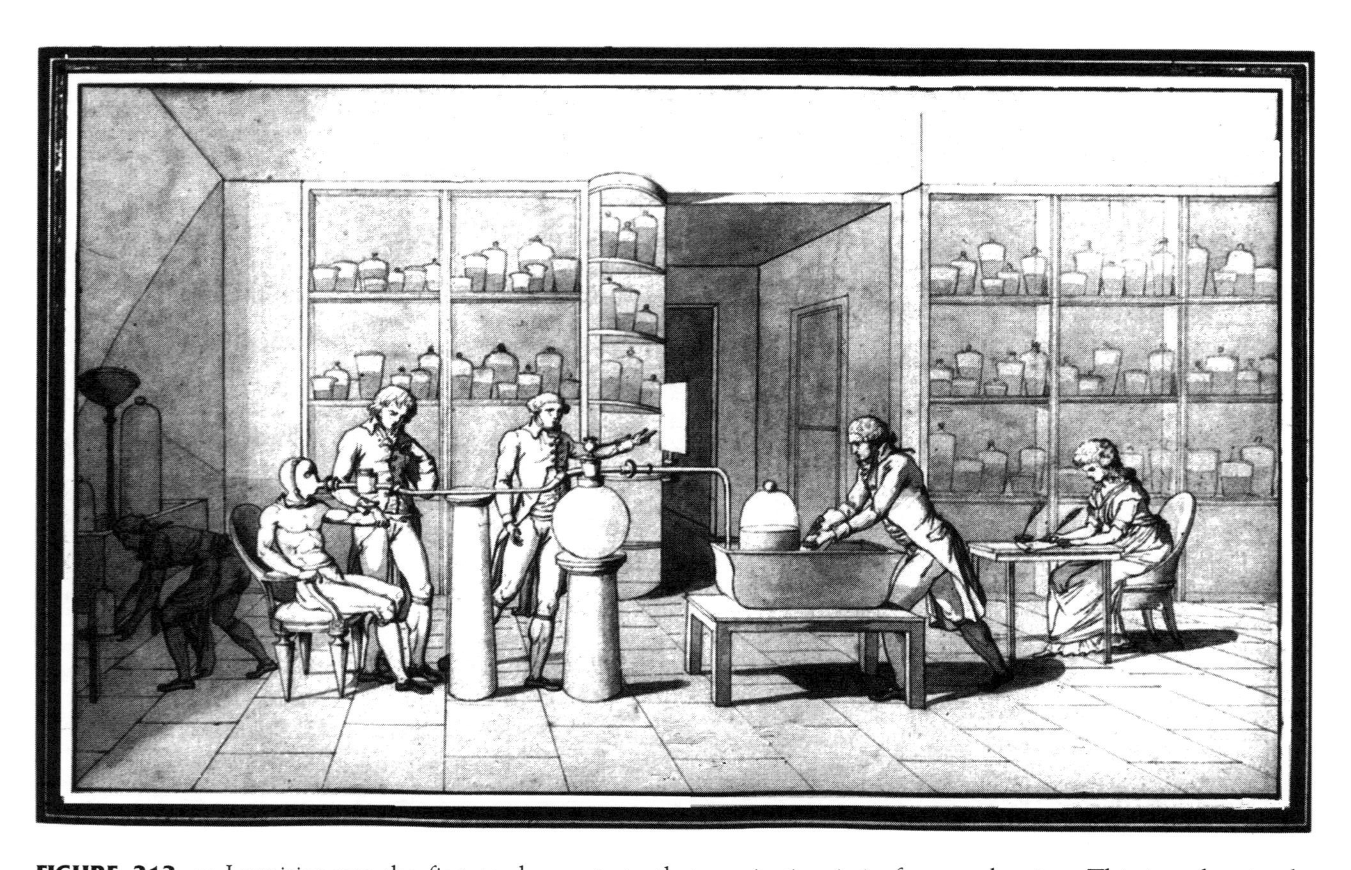

FIGURE 213. ■ Lavoisier was the first to demonstrate that respiration is in fact combustion. This is a drawing by Madame Lavoisier (depicted at right) of her husband conducting respiration experiments on his assistant Armand Seguin, completely enveloped in a rubber suit. Seguin survived and eventually became extremely wealthy as an army contractor. (Courtesy Professor Marco Beretta.)

FIGURE 214. ■ Another drawing of Lavoisier's respiration studies on his assistant Seguin-the man in the rubber suit. Madame Lavoisier depicts herself drawing this scene. (Courtesy Professor Marco Beretta.)

standing center right and providing oxygen and Madame Lavoisier is taking notes. The disappearance of oxygen was carefully measured and the exhaled carbon dioxide and water vapor collected and measured. The amount of oxygen inhaled closely matched the quantity exhaled in the forms of carbon dioxide and water. In order to fully account for the mass balance of water, the mass of the man in the rubber suit was measured carefully before and after experimental periods. Perspiration, and other "effluvia," trapped in the suit were measured to an amazing accuracy of 18 grains in 135 pounds[5]—an expensive balance, indeed! When it came to Lavoisier's apparatus, money was no object. Figure 214[12] depicts the same experiment but Seguin exerting himself by peddling a treadle. The uptake of oxygen was considerably greater. On November 17, 1790, Seguin and Lavoisier presented a memoir that stated in part: [5]

> Respiration is only a slow combustion of carbon and hydrogen, similar in every way to that which takes place in a lamp or lighted candle and, from this viewpoint, breathing animals are actual combustible bodies that are burning and wasting away.

---

1. J.R. Partington, *A History of Chemistry*, MacMillan and Co. Ltd., London, 1961, Vol. 2, pp. 577–614.
2. Partington, op. cit., 1962, Vol. 3, pp. 205–234; 237–297.
3. Partington 1962, op. cit., pp. 130–143.
4. Partington 1962, op. cit., pp. 471–479.
5. J.-P. Poirier, *Lavoisier—Chemist, Biologist, Economist* (transl. R. Balinski), University of Pennsylvania Press, Philadelphia, 1996, pp. 300–309.
6. Partington 1962, op. cit., pp. 426–434.
7. Poirier, op. cit., pp. 135–140.
8. Modern calorimetric data indicate that combustion of carbon (graphite) sufficient to produce exactly 1 ounce of carbon dioxide would melt 26.86 ounces of ice. If sufficient glucose ($C_6H_{12}O_6$) were burned to collect the same 1 ounce of CO_2, one might naively have expected in 1783 that 26.86 ounces of ice would be melted. However, we know that formation of 0.41 ounces of H_2O would accompany the 1 ounce of CO_2 formed in glucose combustion. The extra heat from formation of water added to the heat from formation of carbon dioxide would melt 31.89 ounces of ice.
9. Poirier, op. cit., pp. 140–144.
10. Partington 1962, op. cit., pp. 325–338.
11. Partington 1962, op. cit., pp. 471–479.
12. I am grateful to Professor Marco Beretta for supplying these images.

"POOR OLD MARAT"? I THINK NOT!

Jean-Paul Marat is considered today to have been a minor scientist and was so judged by the *Académie des Sciences* over two centuries ago. He remains, however, famous and infamous as an impassioned and uncompromising "Friend of the People"—a major actor in the triumphs, excesses, and tragedies of the French Revolution. Although Marat himself was murdered on July 13, 1793, some 10 months before the execution of Lavoisier, he certainly helped to inflame passions and create the atmosphere that led the brilliant aristocrat to the guillotine on May 8,

1794.

Born in the Swiss canton of Neuchâtel in 1743, Marat left home in 1759, spent 6 years in France and 11 in England and Scotland, writing philosophical tracts that gained some international notice.[1–4] One of these, *The Chains of Slavery* (1774), was said to have first advanced his idea of the "aristocratic" plot.[3] Marat started to attend medical classes in France around 1760, then moved to England and practiced medicine beginning in 1765. Following 10 years of successful practice, he was awarded the honorary degree of Doctor of Medicine at the College of St. Andrews in Scotland in 1775.[5] Although Samuel Johnson was critical of St. Andrews' practice of selling degrees (he said that the college would "grow richer by degrees"), the two medical faculty members recommending him were highly regarded.[5] Marat then resettled in France, proceeded to ingratiate himself with the aristocracy, and was appointed physician to the personal guards of Comte D'Artois, the youngest brother of King Louis XVI. He charged his aristocratic clients[4,6] almost 1 louis (24 livres) per consultation, or roughly $1000. today, and was considered a very accomplished physician.

Beginning in 1778, Marat began a series of scientific investigations of the "imponderables" light, heat, fire and electricity and ". . . began to lay siege to the Academy of Science."[4] Figure 215 is from Marat's 1780 book[7] in which he explained the nature of heat and fire. He identified a *fluide igné*[8] or fiery fluid that, in some ways, anticipated Lavoisier's *calorique* (caloric). He posited that when a hot object contacts a cold object, *fluide igné* is passed from the warmer object to the colder until the contents are equal. Marat viewed heat and fire as two closely related effects having the same origin. Heat is produced when energy input is only moderate and fire when energy input is high.[8] This physical interpretation of a continuum from heat to fire neglected the dramatic chemistry of fire. According to Marat, all substances must contain *fluide igné*, or they could not reach the temperature of their surroundings. It was the movement of this fluid, not its mere presence, that generated heat and fire.[8] Among a list of substances commonly recognized as flammable (carbon, camphor, naphtha, essential oils, alcohol, phosphorus), which Marat describes as "highly impregnated with *fluide igné*," he also lists nonflammable "fixed alkali" (or sodium carbonate, Na_2CO_3 today).[8] The confusion probably derives from release of "fixed air" (carbon dioxide) from heating this salt just as "fixed air" is "released" when carbon-containing substances are burned. Marat explained the fact that a flame will soon be extinguished in an enclosure as follows: "the air, violently expanded by the flame and unable to escape, dramatically compresses and smothers it."[4]

The shadowy figures in Figure 215 were said by Marat to be images of actual *fluide igné* escaping from a lighted candle (*Fig. 1*), a burning piece of charcoal (*Fig. 2*), and a red-hot cannonball (*Fig. 3*). These images were obtained using Marat's "solar microscope," a dark room with a tiny hole allowing entrance of a very thin beam of light that passes through the lens of a microscope and onto a screen. The small visible cone of a candle flame, for example, is imaged, as is the surrounding column of *fluide igné*. America's ambassador to France, Benjamin Franklin, attended one of Marat's demonstrations. He exposed his bald pate to the solar microscope, and it was duly observed that "we see it surrounded by undulating vapors that all end in a spiral. They look like those flames through which painters symbolize Genius."[4]

Marat's attempt for recognition by the French Academy of Science was re-

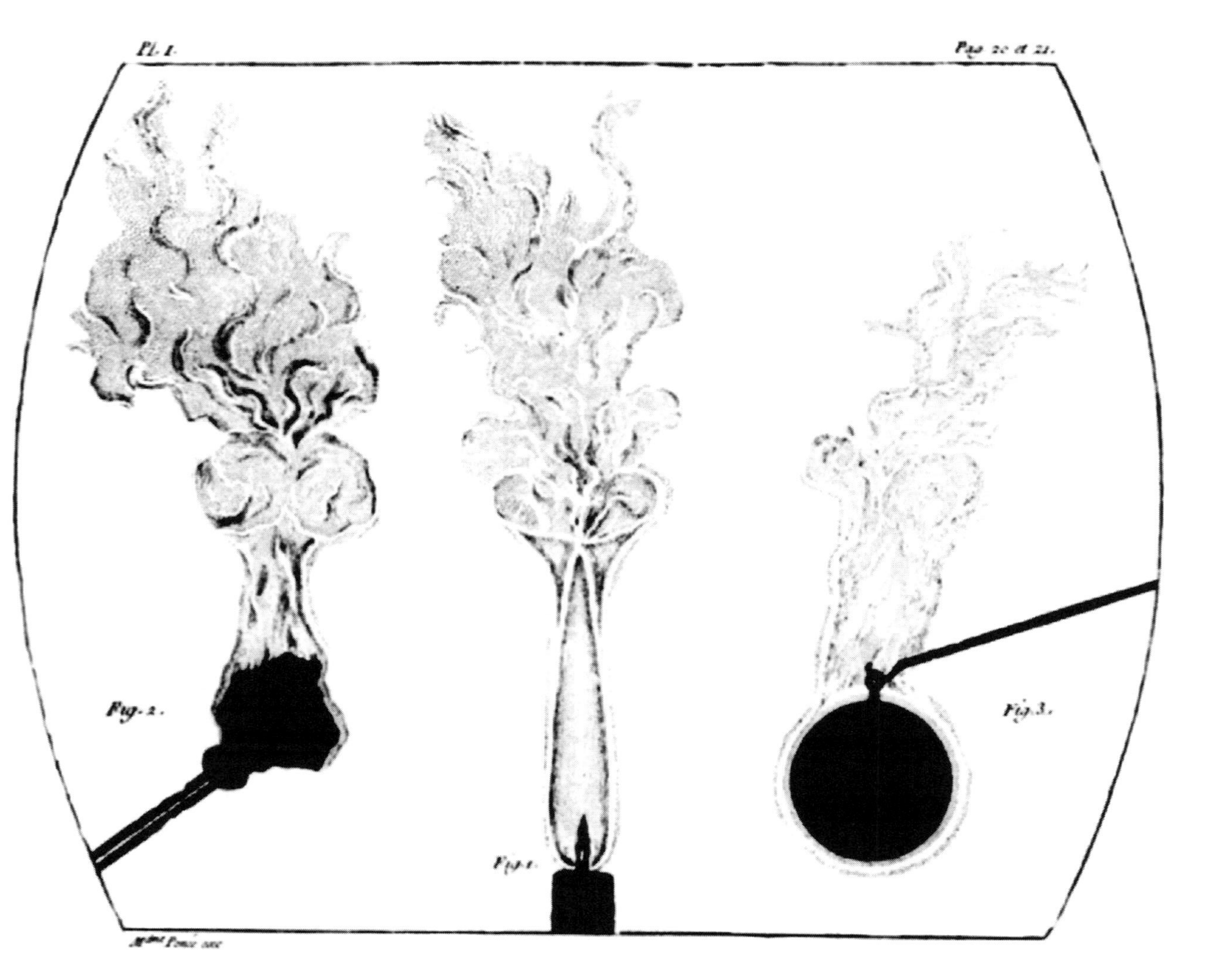

FIGURE 215. ■ Shadowy figures were obtained using Jean Paul Marat's "solar microscope" in a dark room. Marat postulated a *fluide igné* that could escape from heated or burning materials (he makes no real distinction between physical and chemical processes). Here he claims to have observed this imponderable substance escaping from (a) a lighted candle, (b) a burning piece of charcoal, and (c) a red-hot cannonball. Lavoisier played a major role in denying Marat's application for membership in the Academy of Sciences and thus earned his undying hatred, (From Marat's 1780 *Recherches Physiques sur le Feu.*)

jected in May 1779. In June 1780 Lavoisier called the Academy's attention to a paper by Marat that implied that the Academy had endorsed his *fluide igné* findings.[4] The Academy rebutted this assertion, and Marat now included the Academy and Lavoisier in particular on a lifelong enemies list. It should be noted that the wealthy and brilliant Lavoisier had become a member of the Academy at the age of 25. His rejection may have been the beginning of what Gottschalk refers to as Marat's "martyr complex."[9]

French society underwent enormous change throughout the eighteenth century with considerable stresses and cracks developing in the social fabric.[10] The authority of the French monarch remained absolute. In the hands of an inspirational king, Louis XIV—"The Sun King," the French remained reasonably quiescent. However, the next two rulers of the *Ancien Régime*, Louis XV and Louis XVI, were relatively ineffectual. The revolution began stirring largely in urban areas during 1788, but its most dramatic expression, the storming of the Bastille, occurred on July 14, 1789. Not long afterward, Marat, whose political activism dramatically increased during the 1780s, began to edit a newspaper, *Ami du Peuple* ("Friend of the People"), to raise and incite revolutionary fervor.[1–4] Marat initially supported a constitutional monarchy but quickly embraced the views of the rabid antimonarchists who forcibly brought the royal family from Versailles to Paris in October 1789. However, the monarchy remained intact, and Marat's criticism of the king's Finance Minister caused the revolutionary to flee to England briefly during 1790.

Poirier uses "Cultural Revolution"[11] to describe France's revolt against the intellectual authority of knowledge similar, by implication, to that seen in China during the 1960s. The Academy's 1784 criticism of mesmerism, in which both Lavoisier and Franklin played lead roles, was now attacked as elitist.[12] Two excerpts from Marat's 1791 pamphlet *Les Charlatans Modernes* illustrate the vulnerability of the academicians:[13]

> At the head of them all would have to come Lavoisier, the putative father of all the discoveries that have made such a splash. Because he has no ideas of his own, he makes do with those of others.

Recall that this was after Lavoisier had totally revolutionized chemistry. Marat's description of the Academy was no less demagogic:[14]

> A collection of vain men, very proud to meet twice a week to chatter idly about fleurs-de-lys; they are like automatons accustomed to following certain formulas and applying them blindly. . . . What a pleasure it is to see the mathematicians yawn, cough, spit, and snigger when a chemistry paper is being read, and the chemists snigger, spit, cough, and yawn when a mathematics paper is presented.

The French Revolution became increasingly radicalized and bloodthirsty during the next few years.[10] The more conservative of the revolutionary factions, the Girondins, supported the constitutional monarchy. However, it became ever clearer that the king would never abandon the arrogant and uncompromising aristocracy. A brief war with neighboring Prussia was instigated by the king, who had tried unsuccessfully to flee France, in June 1791. The hope was that a foreign war would quell the civil war. The plot backfired, and a "second revolution" overthrew the monarchy in August 1792. The National Convention, comprising relatively conservative Girondins and more radical Montagnards, was established to develop a new constitution. The king was convicted in December and executed in January 1793. Queen Marie Antoinette was also imprisoned and ultimately guillotined in October 1793.

By that time, the Montagnards had defeated the Girondins.[10] Indeed, the most radical factions of the Montagnards defeated their own bourgeoisie factions. Marat was, with Robespierre and Danton, among these most radical factions. Extended periods of hiding in cellars and sewers apparently contributed to a painful and disfiguring skin condition that Marat relieved with frequent baths. On July 13, 1793, Charlotte Corday, regarded now as an innocent, possibly brainwashed, tool of the Girondins, gained entry to Marat's home and stabbed him to death in his bathtub. The "Death of Marat" was memorialized by artist Jacques Louis David in a dramatic painting (Figure 216). David himself seems to have been a fashionable radical who had no problem "tacking with the prevailing wind." He had been Madame Lavoisier's art teacher and also collected a fee, roughly equivalent to $300,000 today, for painting the Lavoisiers' portrait in 1789.[15] Nevertheless, David attacked the academicians during the revolution, and was commissioned to paint the Marat portrait. David dressed Marat's body in Roman style for the funeral.[16] The "Law of Suspects," which began the Reign of Terror in October 1793, as a response to Marat's murder, devoured Marie Antoinette and half a year later Lavoisier and his father-in-law in its ravenous maw. Ultimately,

FIGURE 216. ■ Marat was trained as a physician and had a very profitable practice. He became radicalized throughout the 1780s and was a formidable member of the most radical groups after 1789. Extended periods of hiding in the sewers of Paris may have provoked a painful skin disease that he treated with countless baths. It was in the bath that he was stabbed to death by one Charlotte Corday on July 13, 1793. The scene (shown here in black and white) was painted by Jacques Louis David and is exhibited in the Musée Royaux des Beaux-Arts in Brussels. The events leading up to and including Marat's assassination were dramatized in 1964 by playwright Peter Weiss.

popular revulsion with this bloodbath caused some moderation in the later years of the decade, but France did not truly stabilize until Napolean Bonaparte imposed a military dictatorship in late 1799.

Historical viewpoints change with the times. The nineteenth century was unkind to Marat. However, in 1964 the German playwright Peter Weiss published a fascinating play that is often referred to by the abbreviated title "Marat/Sade."[17] Marat is portrayed more sympathetically in the play within this play. It is 1808 and the Marquis de Sade, a fallen nobleman and writer, is interred in the Asylum of Charenton. He is directing a play that dramatizes the murder of Marat. A Chorus intones "Poor Old Marat" during sections of the play (singer Judy Collins popularized the lyrics during the 1960s). The play dramatizes the murder of Marat. Charlotte Corday is depicted as an automaton—a sort of doomsday machine for Marat. Lavoisier makes a brief cameo appearance in the play. The play's critical juxtaposition is between the committed but fanatical Marat and the Marquis who has lived a life both intellectual and debauched testing the boundaries of human nature. Sade has sympathy for the common people and their revolution, but he is cynical, world-weary and deeply frightened of fanatic crusaders like Marat. In the play, there is this exchange between them:[18]

Sade: I don't believe in idealists
who charge down blind alleys
I don't believe in any of the sacrifices
that have been made for any cause
I believe only in myself

Marat: [*turning violently* to SADE]
I believe only in that thing which you betray
We've overthrown our wealthy rabble of rulers
disarmed many of them though
many escaped
But now those rulers have been replaced by others
who used to carry torches and banners with us
and now long for the good old days
It becomes clear
that the Revolution was fought
for merchants and shopkeepers
the bourgeoisie
a new victorious class
and underneath them
ourselves
who always lose the lottery

Historically, Marat and Sade never actually conversed.[17]

---

1. L.R. Gottschalk, *Jean Paul Marat—a Study in Radicalism*, Benjamin Blom, New York, 1927.
2. C.D. Conner, *Jean Paul Marat—Scientist and Revolutionary*, Humanity Books, Amherst, 1998.
3. *The New Encyclopedia Britannica*, Encyclopedia Britannica, Inc., Chicago, 1986, Vol. 7, pp. 813–814.
4. J.-P. Poirier, *Lavoisier—Chemist, Biologist, Economist*, University of Pennsylvania Press, Philadelphia, 1993, pp. 110–112.

5. Gottschalk, op. cit., pp. 4–5. See also Conner's spirited defense of Marat's medical training—Conner, op. cit., pp. 33–34.
6. Poirier, op. cit., p. 428.
7. [J.P.] Marat, *Recherches Physiques sur le Feu*, chez C. Ant. Jombert, Paris, 1780
8. Marat, op. cit., pp. 17–21.
9. Gottschalk, op. cit., pp. 1–31.
10. *The New Encyclopedia Britannica*, op. cit., Vol. 19, pp. 483–502.
11. Poirier, op. cit., pp. 328–333.
12. Poirier, op. cit., p. 159.
13. Poirier, op. cit., p. 196.
14. Poirier, op. cit., p. 329.
15. Poirier, op. cit., p. 1.
16. Poirier, op. cit., p. 330.
17. P. Weiss, *The Persecution and Assassination of Jean-Paul Marat as Performed by the Inmates of the Asylum of Charenton Under the Direction of the Marquis De Sade*, English version By Geoffrey Skelton, Atheneum, New York, 1965.
18. Weiss, op. cit., pp. 41–42. Copyright Suhrkamp Verlag Frankfurt am Main 1964. Permission to reprint English version (ISBN 0-7145-0361-4) courtesy Marian Boyars Publishers, London (UK) 1965.

POOR OLD LAMARCK

It is sad that the only thing we learn in school about Jean Baptiste Lamarck (1744–1829)[1] is that he explained the long limbs and necks of giraffes by noting that they must continuously stretch and extend themselves, thus strengthening and slightly elongating their necks and legs during their lifetimes, *and* that these acquired improvements are inherited by their offspring. Successive generations would continue to "improve" in this manner—we might now say "evolve." This explanation was offered almost 60 years before the publication of *The Origin of Species* by Charles Darwin in 1859, which presented evolution as an observed fact and offered natural selection as its mechanism. It would be another 6 years before Gregor Mendel would present his observations on hybrid peas, a further 35 years before the significance would be fathomed, and another 50 years before Watson and Crick would explain it. Nonetheless, we remember Lamarck for his wrong theory just as Brooklyn Dodger pitcher Ralph Branca is forever remembered for the homerun ball he threw to Bobby Thompson on October 3, 1951 that allowed the New York Giants to snatch the pennant from its rightful owners. Branca, who proudly wore number 13 on his uniform, had a very respectable lifetime record (88 wins; 68 losses;[2] "You could look it up"—C. Stengel[3]), but he will always be remembered for that gray, infamous Manhattan afternoon. Lamarck could just as well have worn number 13. He was one of 11 children born into the "semi-impoverished lesser nobility of Northern France."[1] He married three (possibly four) times—his three known wives dying early of illness, his total of eight children including one deaf son, one insane son, and all but one of his children were consigned to lives of poverty. In order to pay for his funeral in 1829, his family had to sell his books and scientific collection at public auction and appeal to the *Académie des Sciences* for funds.[1]

In fact, Lamarck made numerous important contributions to science in the late eighteenth and early nineteenth centuries. He coined the terms "biology" and "invertebrate" and developed a taxonomy system that was, in some respects, easier to use than that of Linnaeus.[1] He was widely recognized as one of the leading experts on invertebrate biology and was one of the first paleontologists to relate fossils to living creatures and to try to account for the differences between, say, fossil clams and their living relatives. This naturally led him to try to explain the sources of these differences. While he never used the term "evolution," he was certainly a protoevolutionist.[1] Lamarck had a unified view of Nature that was almost mystical in nature. Only living organisms could make "organic matter." These organisms could change (read "evolve") through the generations by strengthening and improving themselves—a very alchemical idea—strengthening of human characteristics leading to human perfection—the removal of impurities gradually perfecting and transmuting base metals into gold. When organisms died, the decomposing organic matter would become mineral matter. Indeed, he resisted the growing reductionism (he termed them "small facts") in science. Moreover, Lamarck, as a professor and curator at the *Muséum National d'Histoire Naturelle*, arranged the invertebrates according to systematic classification rather than the random "cabinet of curiosities" typical of earlier such museums. We see Lamarck's pioneering methods in the halls of dinosaurs and mammals in modern museums and, indeed, in the evolutionary organization within these halls.

Unfortunately, Lamarck's resistance to the "small ideas" developing in chemistry fixed him firmly in the pre-Boyleian seventeenth century. He retained his belief in the four elements and was particularly fascinated by the different forms assumed by the element fire. Lamarck believed that "matter of fire" and "matter of electricity" were essentially one and the same. This is not so surprising. If one is a great distance from a large fire, it is still possible to see the sky light up as the fire intensifies or as a new fuel source inflames. The appearance is not very different from the diffuse appearance of lightning obscured by clouds. Lamarck was well aware that Benjamin Franklin had demonstrated the electrical nature of lightning. Moreover, it was also apparent by the end of the eighteenth century that both fire and electricity caused chemical change.

Figure 217 is from Lamarck's first and most important chemical work,[1] which was published in 1794.[4] The top part of the figure shows two cork balls suspended over a hook by a silk thread. The light cork balls have been electrified through friction, and they separate (see *Fig. B* at the top). The reason offered by Lamarck is that electrification causes each sphere to be surrounded by superlight "matter of electricity." The pressure of the air pushes in on all sides equally (the thread causes a bit of distortion), thus separating the two electrified spheres. If the charged cork balls were forced to touch (*Fig. A*), the regions of electric matter between them would overlap and the overall shape of the electric matter would be oval (not the optimal spherical) with a small gap at the top. Air pressure would seep in and force these balls apart (*Fig. B*). In the bottom of Figure 217 we see a vase full of water placed over a fire. Again, air pressure keeps the flame concentrated under the vase, and the easiest path for the fire, according to Lamarck, is through the vessel and into the water. As more fire is absorbed, the water molecules are surrounded by ever larger coats of superlight "matter of fire"—the water gets warmer (and less dense). Eventually, the low density and high energy of these particles, forced by the downward pressure of the atmos-

FIGURE 217. ■ Jean Baptiste Lamarck was an important biologist who, unfortunately, is widely remembered for his incorrect theory of acquired traits. His chemistry, however, was very outmoded, and in his one chemistry text Lamarck tried to describe the repulsion between corpuscles of matter as they absorbed heat that caused them to expand and repel. (From Lamarck's 1794 *Recherches Sur Les Causes Des Principaux Faits Physiques*.)

phere, cause them to vaporize and carry away "matter of fire" with them. (Come to think of it fellow teachers—what an interesting way to represent the latent heat in molecules of water vapor!)

And so, we wish we could say some nice things about Lamarck's chemistry. However, his early contributions to biology and its museum display for the public were of great value and do honor to his memory.

---

1. C.C. Gillispie (Editor-in-Chief), *Dictionary of Scientific Biography*, Vol. VII, Charles Scribner's Sons, New York, 1973, pp. 584–593. His full name, for the record, was Jean Baptiste Pierre Antoine de Moncet de Lamarck.
2. D.S. Neff and R.M. Cohen, *The Sports Encyclopedia: Baseball*, St. Martin's Press, New York, 1989.
3. P. Dickson, *Baseball's Greatest Quotations*, HarperCollins Publishers, New York, 1991, p. 427.
4. J.B. Lamarck, *Recherches sur les Causes des Principaux Faits Physiques*, Tome Premier, Chez Maradan, 1794, pp. 198–204.

ELECTIVE ATTRACTIONS

The Swedish chemist Torbern Bergman (1735–1784)[1] systemized chemical affinities and displacements (single or double)[2] in the "wet way" or "dry way" in his book *A Dissertation on Elective Attractions*.[2] See Figure 218 and the enlargement of item 20 in Figure 219(a). In a single elective attraction, calcium sulfide [CaS or (1)] will be decomposed by sulfuric acid (H_2SO_4 or (2)] in water (3) to produce elemental sulfur (4), which precipitates (downward half-bracket) and calcium sulfate [gypsum, $CaSO_4$, or (5)], which also precipitates (downward bracket). Thus, sulfuric acid (2) has a higher affinity for pure calcareous lime (6)—really the source of calcium in (1)—than does sulfur (4).

In Figure 218 and the enlargement of item 26 in Figure 219(b) is depicted a double elective attraction. Silver nitrate (1) and sodium chloride [(2), table salt] decompose each other in water (3) to produce silver chloride (4), which precipitates (downward bracket), and sodium nitrate (5), which remains in solution (upward bracket).

---

1. J.R. Partington, *A History of Chemistry*, MacMillan, London, 1962, Vol. 3, pp. 179–199.
2. T. Bergman, *A Dissertation on Elective Attractions*, Edinburgh, 1785.

THE PHOENIX IS A "HER"?

Mrs. Elizabeth Fulhame, whom Laidler calls a "forgotten genius,"[1,2] authored a remarkable book (Figure 220) published in 1794 (German translation, 1798; American edition, 1810; see Figure 221). Women were not only not encouraged,

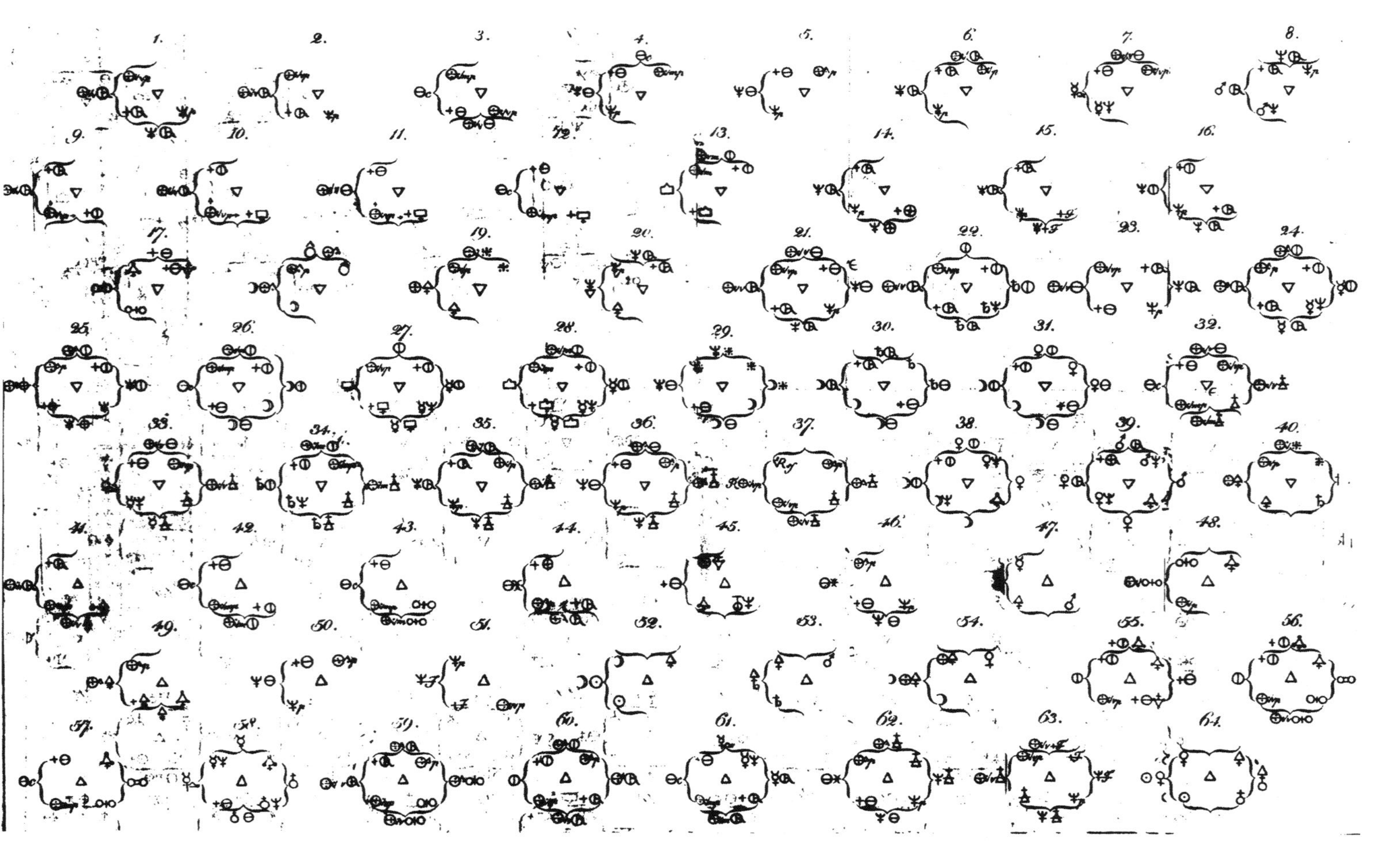

FIGURE 218. ■ This is a table of chemical affinities from Torbern Bergman's *A Dissertation on Elective Attractions* (London, 1785).

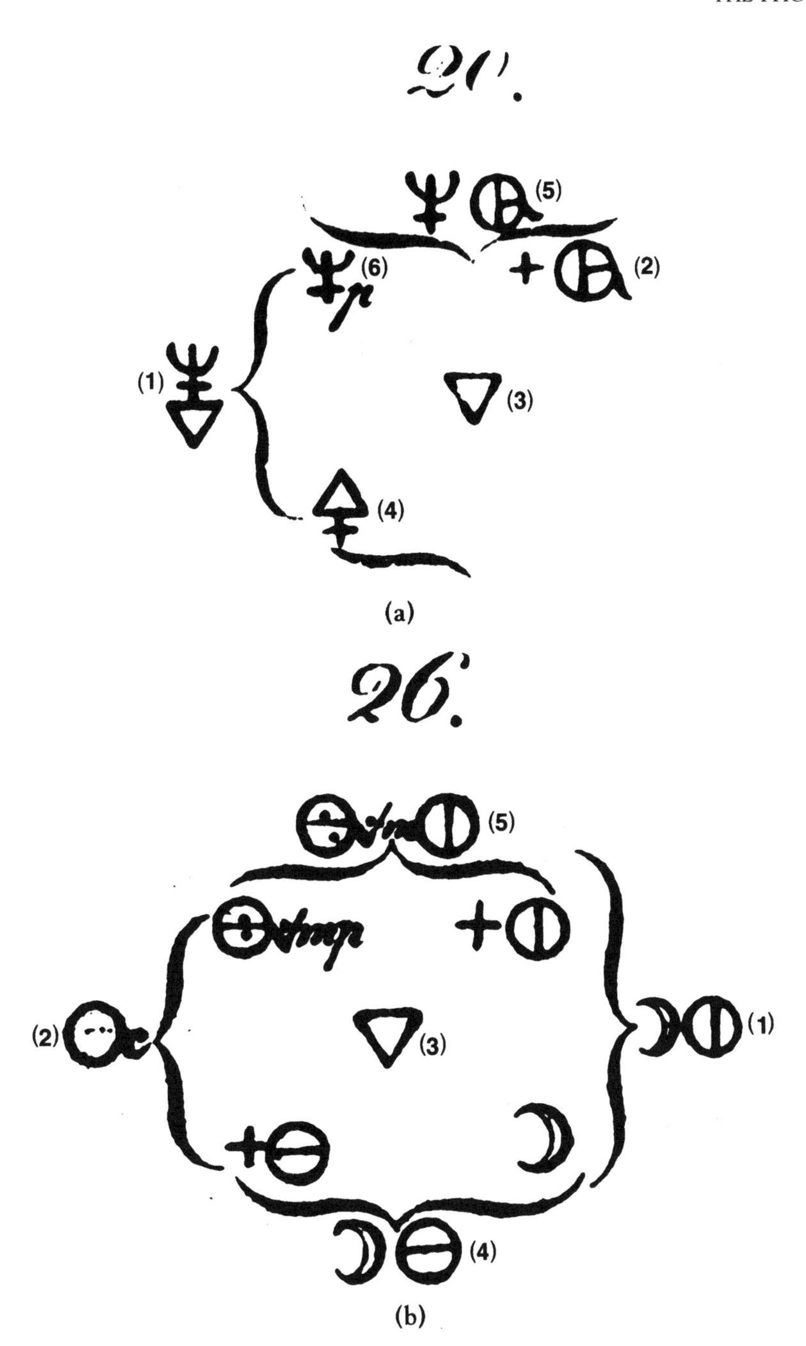

FIGURE 219. ■ (a) Single-elective attraction (decomposition) of calcium sulfide from Bergman's tables (see Figure 218). (b) Double elective attraction between silver nitrate and sodium chloride (see text).

they were actively discouraged from pursuing scientific interests. The 1794 edition was published privately for the author, presumably with the support of her husband Dr. Thomas Fulhame. From her preface to this book:[3]

> It may appear presuming to *some*, that I should engage in pursuits of this nature; but averse from indolence, and having much leisure, my mind led

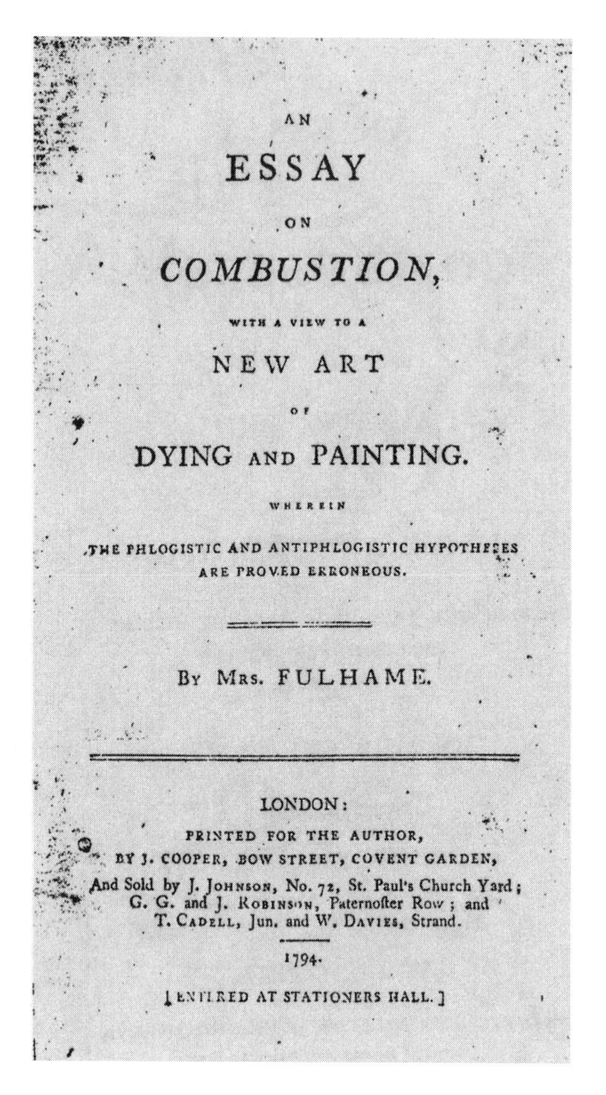

AN

ESSAY

ON

COMBUSTION,

WITH A VIEW TO A

NEW ART

OF

DYING AND PAINTING.

WHEREIN

THE PHLOGISTIC AND ANTIPHLOGISTIC HYPOTHESES
ARE PROVED ERRONEOUS.

BY MRS. FULHAME.

LONDON:

PRINTED FOR THE AUTHOR,
BY J. COOPER, BOW STREET, COVENT GARDEN,
And Sold by J. JOHNSON, No. 72, St. Paul's Church Yard;
G. G. and J. ROBINSON, Paternofter Row; and
T. CADELL, Jun. and W. DAVIES, Strand.

1794.

[ENTERED AT STATIONERS HALL.]

FIGURE 220. ■ Title page of Mrs. Elizabeth Fulhame's book on the theory of combustion. Laidler calls her a "forgotten genius" who first demonstrated photoimaging and may rightly be called the "mother of mechanistic chemistry (from The Roy G. Neville Historical Chemical Library, a collection in the Othmer Library, CHF).

> me to this mode of amusement, which I found entertaining and will I hope be thought inoffensive by the liberal and the learned. But censure is perhaps inevitable; for some are so ignorant, that they grow sullen and silent, and are chilled with horror at the sight of anything that bears the semblance of learning, in whatever shape it may appear; and should the *spectre* appear in the shape of a *woman*, the pangs which they suffer are truly dismal.

Mrs. Fulhame made two, probably three, great discoveries. She was the first to demonstrate photoimaging and used salts of gold and other metals. The famous Count Rumford (Benjamin Thompson—see pp 356–359) differed with her chemical interpretation as opposed to a purely physical one.[4] He was wrong — the photochemical reduction of gold or silver ions to the respective metals is

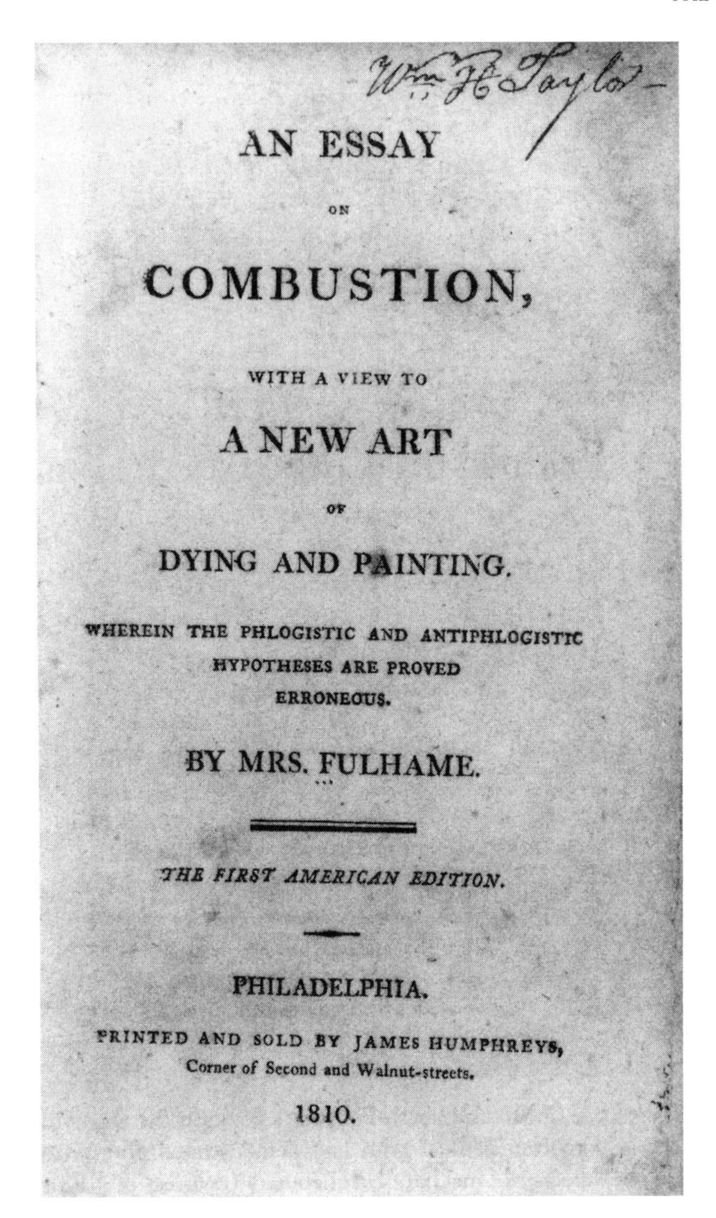
AN ESSAY

ON

COMBUSTION,

WITH A VIEW TO

A NEW ART

OF

DYING AND PAINTING.

WHEREIN THE PHLOGISTIC AND ANTIPHLOGISTIC HYPOTHESES ARE PROVED ERRONEOUS.

BY MRS. FULHAME.

THE FIRST AMERICAN EDITION.

PHILADELPHIA.

PRINTED AND SOLD BY JAMES HUMPHREYS, Corner of Second and Walnut-streets.

1810.

FIGURE 221. ■ Title page of first American edition of Mrs. Fulhame's Treatise. She was an early member of the Philadelphia Chemical Society, possibly nominated by Joseph Priestley with whom she differed about phlogiston (courtesy of Edgar Fahs Smith Collection, Rare Book & Manuscript Library, University of Pennsylvania).

considered to be the first demonstration that ambient aqueous chemistry can accomplish the work of high-temperature smelting.[3]

Her work on the participation of water as a catalyst in the oxidation of charcoal to carbon dioxide, later proven,[3] was of great importance and anticipated the concept of *catalysis* (term introduced by Berzelius in 1836—"wholly loosening" from the Greek[2]). Implicit in this is also the modern concept of the chemical mechanism: a stepwise, "blow-by-blow" account of a chemical reaction. We will illustrate this briefly with the rusting of iron—it was the Irish chemist William Higgins who first discovered the role of water in this process and he accused Mrs. Fulhame of plagiarism (but he also accused John Dalton of plagiarizing the Atomic Theory from him).[4] Mrs. Fulhame's concept was more general; she clearly overextended it.[2,4]

Although the rusting of iron involves reaction of the metal with oxygen to form red-brown iron(III) oxide (Fe_2O_3), we know iron doesn't just rust in the open air if it is kept dry. Water plays the roles of electrolytic solvent and catalyst. If iron is wet and exposed to an ample supply of oxygen, the following reactions occur.[5]

Reaction 1:

$$4Fe(s) + 4H_2O(l) + 2O_2(g) \rightarrow 4Fe(OH)_2(s)$$

Reaction 2:

$$4Fe(OH)_2(s) + O_2(g) \rightarrow 2Fe_2O_3 \cdot H_2O(s) + 2H_2O(1)$$

Net Reaction:

$$4Fe(s) + 3O_2(g) + 2H_2O(1) \rightarrow 2Fe_2O_3 \cdot H_2O(s)$$

It is clear that two of the water molecules in Reaction 1 were regenerated in Reaction 2 and therefore do not appear in the net reaction. They were "temporarily tied-up" in the $Fe(OH)_2$ *intermediate* but then regenerated when the intermediate reacted.

Although Mrs. Fulhame was much more anti-phlogistonist than phlogistonist and thus closer to Lavoisier, Laidler speculates that it may have been Priestley who nominated her for the Philadelphia Chemical Society.[2] Mrs. Fulhame ends her book with an ebullient reference to the Phoenix,[4] a majestic bird symbolizing renewal or rebirth from the ashes—it adorns the symbol for the American Chemical Society:

> This view of combustion may serve to show how nature is always the same, and maintains her equilibrium by preserving the same quantities of air and water on the surface of our Globe: for as fast as these are consumed in the various Processes of combustion, equal quantities are formed, and Regenerated like the Phoenix from her ashes.

Partington,[4] who was a great authority if ever there was one, cautiously avers: "The phoenix, it may be noted, was a fabulous bird regarded as sexless."

---

1. K.J. Laidler, *Accounts of Chemical Research*, 1995, Vol. 28, pp. 187–192.
2. K.J. Laidler, *The World of Physical Chemistry*, *Oxford University Press*, Oxford, 1993, pp. 250–252; 277–278.
3. M. Rayner-Canham and G. Rayner-Canham, *Women in Chemistry: Their Changing Roles From Alchemical Times to the Mid-Twentieth Century*, American Chemical Society and Chemical Heritage Foundation, Washington, D.C. and Philadelphia, 1998, pp. 28–31.
4. J.R. Partington, *A History of Chemistry*, MacMillan, London, 1962, Vol. 3, pp. 708–709.
5. J.C. Kotz and P. Treichel, Jr., *Chemistry and Chemical Reactivity*, 3rd ed., Saunders, Fort Worth, 1996, pp. 980–983.

CHEMISTRY IN THE BARREL OF A GUN

In the wrong circumstances, charcoal can be dangerous. Just ask Johann Baptist Von Helmont, who coined the term gas and then almost "gassed" himself by burning charcoal indoors.[1]

Carbon monoxide (CO) is formed as a by-product of combustion in oxygen-poor environments. In oxygen-rich environments, typical flammable materials burn (oxidize) completely to form carbon dioxide (CO_2) and water (H_2O). Under these conditions, CO is a short-lived *intermediate* that reacts as quickly as it is formed. [$Fe(OH)_2$, described in the previous essay, is a longer-lived intermediate en route to rust.] Mrs. Fulhame correctly concluded that water accelerates charcoal[2] combustion. The reason is, once again, clarified by *parts* of the very complex reaction mechanism for combustion.[3] In the absence of hydrogen-containing substances, the key chain-initiating reaction is

$$CO + O_2 \rightarrow CO_2 + O \tag{1}$$

This is followed by many other reactions that keep the chain going. One is thought to be (where M is a molecule or atom for collision):

$$M + CO + O \rightarrow CO_2 + M \tag{2}$$

However, where sources of hydrogen are present [water, methane (CH_4), etc.], a very different and even faster chemistry occurs:[3]

$$CO + H_2O \rightarrow CO_2 + H_2 \tag{3}$$

$$H_2 + O_2 \rightarrow H_2O_2 \tag{4}$$

$$H_2O_2 \rightarrow 2OH \tag{5}$$

$$OH + CO \rightarrow CO_2 + H \tag{6}$$

$$H + O_2 \rightarrow OH + O \tag{7}$$

Charcoal does not burn rapidly because of its solid structure and the absence of sources of hydrogen; the latter also explains the relative abundance of carbon monoxide in its emissions. To trap and observe CO, it needs to be made outside of normal combustion conditions. Imagine newborn guppies amidst a tank of voracious fish. Now imagine the same newborn guppies deposited by their mother directly into an incubator tank so they may be studied.

Carbon monoxide was a "puzzlement" during the late eighteenth century when it was discovered independently by Torbern Bergman, Joseph Marie François de Lassone, and Joseph Priestley.[4] Steam passed over red-hot charcoal produces "water gas," which is useful for combustion energy but highly toxic. (We understand today that it is a mixture of CO, H_2, and CO_2). Priestley observed that when chalk ($CaCO_3$) was heated in a red-hot gun barrel, the result was "inflammable air" of a relatively "heavy" nature that burned with a blue flame to form "fixed air" (CO_2). When slaked lime [$Ca(OH)_2$] was heated in a red-hot gun barrel, the result was "light inflammable air" that burned explosively. Reactions (8) and (9) correspond to the first case, wherein CO is formed. Reactions (10) and (11) correspond to the second case wherein H_2 is formed. Of course, to add to the confusion, water gas contained both "light" and "heavy inflammable airs."

$$CaCO_3 \rightarrow CaO + CO_2 \quad (8)$$

$$3CO_2 + 2Fe \rightarrow Fe_2O_3 + 3CO \quad (9)$$

Here the gaseous product is "heavy inflammable air."

$$Ca(OH)_2 \rightarrow CaO + H_2O \quad (10)$$

$$3H_2O + 2Fe \rightarrow Fe_2O_3 + 3H_2 \quad (11)$$

Here the gaseous product is "light inflammable air."

Now, if you are Joseph Priestley and firmly wedded to the phlogiston theory, the conclusion is obvious—both "fixed air" (CO_2) and steam are releasing the phlogiston from iron in the gun barrel although to different extents. In 1801, William Cruickshank, Ordnance Chemist, Lecturer in Chemistry in the Royal Artillery Academy, Surgeon of Artillery and Surgeon to the Ordnance Metal Department finally succeeded in differentiating hydrogen from carbon monoxide.[4] As we will soon see, Count Rumford also used artillery to do science. Perhaps England's eighteenth-century wartime economy produced a surplus of weapons to be later exploited as scientific apparatus.

---

1. J.R. Partington, *A History of Chemistry*, MacMillan, London, 1962, Vol. 2, p. 229.
2. Charcoal is formed from slowly heating wood to rather high temperatures. The result is a mass of about 75 carbon, 20 volatiles (boiled away in red-hot charcoal), and 5 ash.
3. K.K. Kuo. *Principles of Combustion*, Wiley, New York, 1986, pp. 148–149.
4. J.R. Partington, op. cit., Vol. 3, pp. 271–276.

A BORING EXPERIMENT

Even as Lavoisier demolished phlogiston, he postulated a new gaseous "simple substance" or "element" called *caloric*—the element of heat (see Lavoisier's Table of Elements, Figure 202). Lavoisier had fully explained mass transformation in chemical reactions. The nature of *energy* transformations remained a mystery. Caloric could be transferred from a warmer body to a cooler body without chemical change. However, Lavoisier also posited that oxygen the element was a component of calxes, while oxygen gas contained caloric, released as heat and light when a substance burned (see Figure 203). The similarity between the caloric concept and the phlogiston concept is almost obvious.

Figure 222 is from the *Philosophical Transactions of the Royal Society of London* by Benjamin Thompson (Count Rumford) (1753–1814). A solid iron pipe (*Fig. 1*), normally bored to make a cannon, was machined to leave a short cylinder (9.8 inches in length and 7.75 inches in diameter) attached by a thin

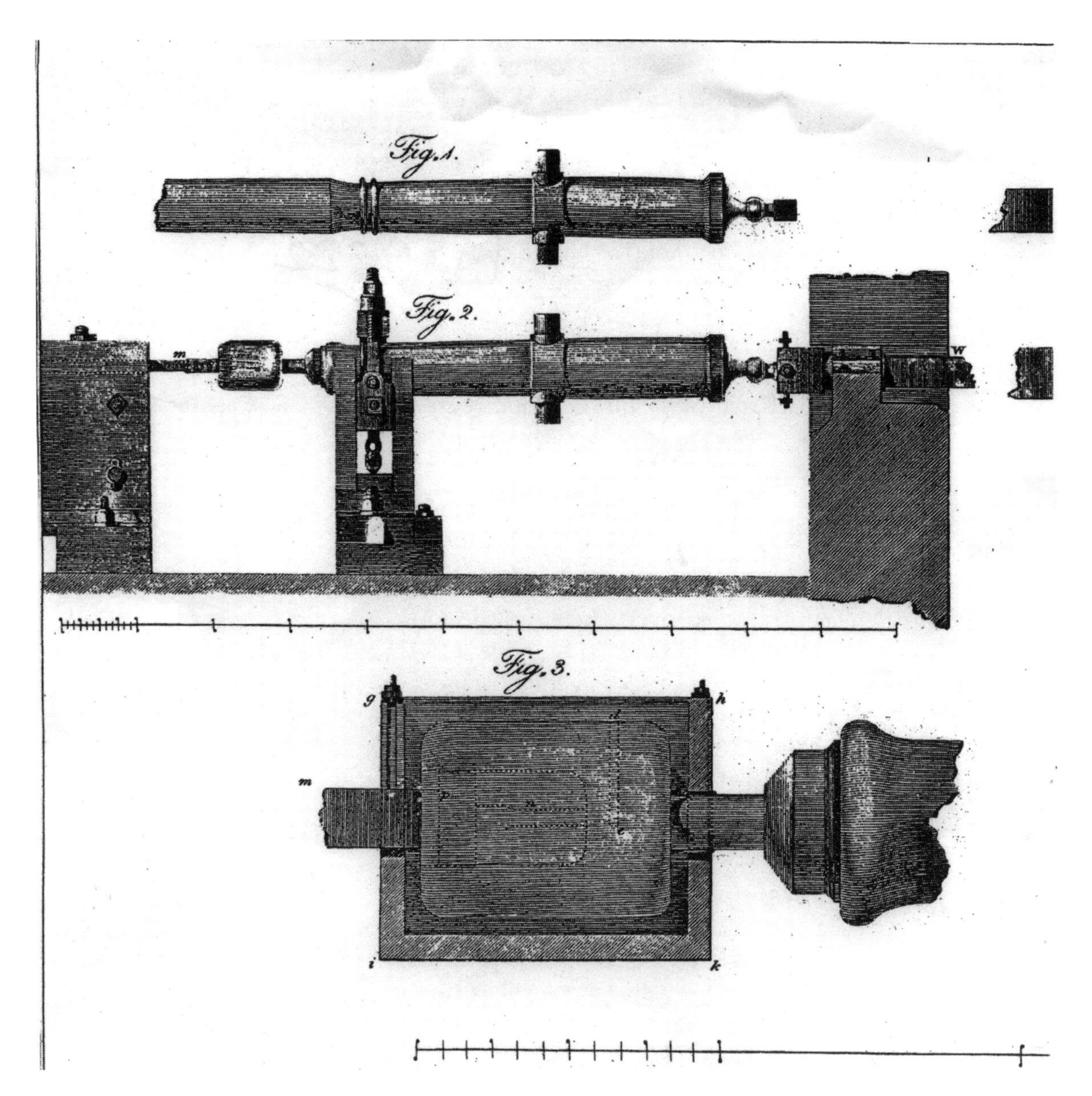

FIGURE 222. ■ The cannon-boring experiment of Benjamin Thompson (Count Rumford), which disproved the caloric theory (from *Philosophical Transactions of the Royal Society of London*, Vol. 88 (1798) pp. 80–102).

iron cylinder to the main cannon pipe (*Fig. 2*). In this short cylinder a tube 7.2 inches in length and 3.7 inches in diameter was bored. In addition, a small round channel was drilled for periodic removal and replacement of the thermometer as needed for temperature readings (*Fig. 3*). In operation, a steel borer was pressed by an iron bar (*m*) with great pressure (1000 pounds) into the end of the tubular channel drilled into the cylinder. The cannon (and cylinder) were turned, by horse power, on their common axis 32 times per minute. After 30 minutes, the temperature of the 113-pound cylinder rose from 60 °F to 130 °F. Rumford also demonstrated that the mechanical work involved in boring a brass cannon was sufficient to boil 5 pounds of ice-cold water. The heat capacity of the chips produced by boring remained the same as when these chips were part of the cannon.[1] One would have expected a loss in caloric to be manifested in a loss of mass and/or heating capacity. In effect, Rumford showed that there was no limit to the amount of caloric that could be released as the result of mechanical friction. This was, of course, impossible. He also carefully established that there is no change in mass upon freezing water. At the time, Rumford's work had little impact: Explanations offered were that the quantity of caloric present in the cannon was incredibly large and hardly any had been released in Rumford's experiments and that caloric was exceedingly light. However, the idea of an infinite quantity of immeasurably light caloric made little sense.

This study was a first quantitative step toward establishing the *First Law of Thermodynamics* in terms of the mechanical equivalent of heat:

$$\text{Energy}_{\text{System}} = (\text{Heat added})_{\text{System}} - (\text{Work on surroundings})_{\text{System}}$$

In the boring experiment, work is done by the surroundings on the system (the brass cannon), the energy of the system rises and heat is also released to the surroundings (water bath). The First Law of Thermodynamics and the mechanical equivalent of heat (1 calorie = 4.184 joule) were established in 1843 by James Prescott Joule (1818–89). In order to raise the temperature of 1 gram of water by 1 °C (1 calorie), 4.184 joule of mechanical work, such as spinning paddles in water (Joule's experiment), is required.

Benjamin Thompson was born to a modest farming family in Woburn in the Colony of Massachusetts in 1753.[2] He received little formal education, was largely self-taught, moved to Concord, New Hampshire to teach school and married a wealthy widow 14 years his senior when he was 19. New Hampshire was part of Massachusetts until it became a Crown Colony in 1679. Concord, New Hampshire was originally Rumford in the Colony of Massachusetts. The town was determined to be part of the Colony of New Hampshire in 1741. This decision was bitterly disputed and finally settled in 1762 and the town was renamed Concord (after the peaceful agreement) in 1765 and incorporated into New Hampshire. They separated permanently in 1775 as the American Revolution began and Thompson worked as a spy for the English, fleeing to England in 1776. He retired from the British Army and was Knighted by George III in 1784 and moved to Germany, became head of the Bavarian Army, and was appointed Count Rumford of the Holy Roman Empire in 1793. The early thermodynamics studies grew out of this military experience in Germany. Count Rumford returned to England in 1798, helped found the Royal Institution in 1799, and he appointed Humphry Davy Lecturer in Chemistry in 1801 following publication of his work on laughing gas.

Count Rumford successfully courted Madame Marie-Paulze Lavoisier over a four-year period and they were married in 1805. However, according to the Rayner-Canham's, "he was a rather conceited, boring individual, who was expecting to live well on Paulze-Lavoisier's finances, while pursuing his researches alone" and their marriage apparently deteriorated in two months with separation occurring in 1809.[3]

---

1. W. Kauzmann, *Thermodynamics and Statistics: With Applications To Gases*, Vol. II of *Thermal Properties of Matter*, Benjamin, New York, 1967, pp. 34–35.
2. C.C. Gillispie (Editor-in-Chief), *Dictionary of Scientific Biography*, Scribner, New York, 1970, Vol. 13.
3. M. Rayner-Canham and G. Rayner-Canham, *Women in Chemistry: Their Changing Roles from Alchemical Times to the Mid-Twentieth Century*, American Chemical Society and the Chemical Heritage Foundation, Washington, D.C. and Philadelphia, 1998, pp. 17–22.

LAUGHING GAS FOR EVERYBODY!

Humphry Davy[1] (1778–1829) was apprenticed to a surgeon in Penzance in 1795 but started reading Lavoisier's *Elements of Chemistry* and Nicholson's *Dictionary of Chemistry*, which still retained some phlogistic influences. His early investigations caught the attention of Thomas Beddoes and he was appointed to Beddoes's Pneumatic Institution in 1798. The Institution's purpose was to use inhalable gases to cure diseases.

Priestley's work on different kinds of air in 1772 produced impure nitrous oxide (N_2O). In 1799 Davy heated ammonium nitrate in the retort depicted in Figure 223 (*Fig. 2*) and obtained the pure gas, collected over water. His experimental and physiological studies were published in *Researches, Chemical and Philosophical, Chiefly Concerning Nitrous Oxide, Or Dephlogisticated Nitrous Air And Its Respiration* (1800). The plate shown is the frontispiece from the 1839 reprint of this exceedingly rare book and depicts a gas holder and breathing apparatus. Davy's reckless breathing of the newly discovered gases of the period were, for once, rewarded with nitrous oxide (laughing gas):

> On April 16th, Dr. Kinglake being accidentally present, I breathed three quarts of nitrous oxide from and into a silk bag for more than half a minute, without previously closing my nose or exhausting my lungs. The first inspirations occasioned a slight degree of giddiness. This was succeeded by an uncommon sense of fulness of the head, accompanied by loss of distinct sensation and voluntary power, a feeling analogous to that produced in the first stage of intoxication; but unattended by pleasurable sensation. Dr. Kinglake, who felt my pulse, informed that it was rendered quicker and fuller.

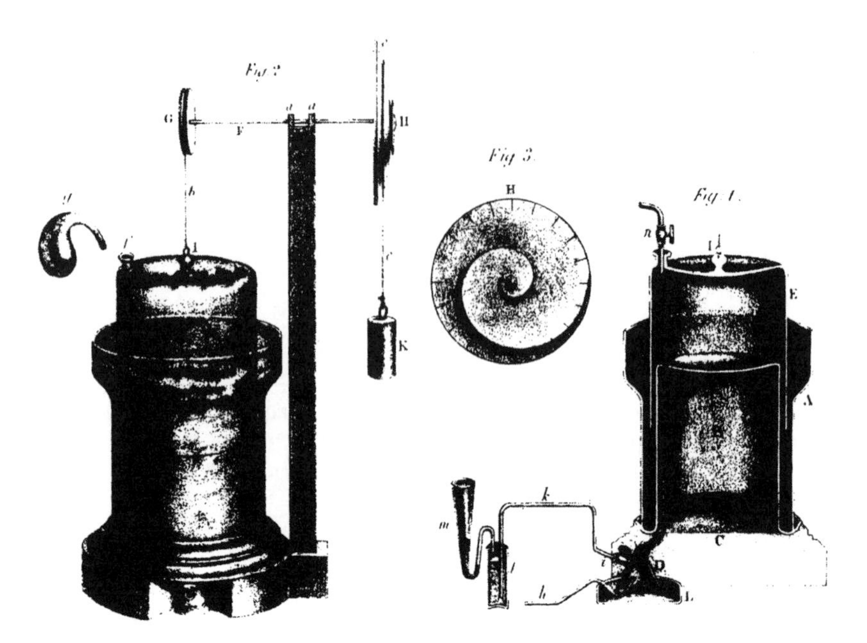

FIGURE 223. ■ Diagram of Humphry Davy's apparatus for storing and breathing nitrous oxide (from *Researches, Chemical and Philosophical, chiefly concerning Nitrous Oxide, or Dephlogisticated Nitrous Air, and its Respiration;* the original edition, published in 1800, is of very great rarity). Davy's artistic circle of friends included poet Samuel Taylor Coleridge and Dr. Mark Roget (*Thesaurus* fame) who sampled laughing gas with Davy and recorded their scientific observations.

Davy, who wrote good poetry and was an avid fisherman,[1] had a wide variety of friends and correspondents who sampled nitrous oxide: these included Dr. Peter Mark Roget, future physician and author of the *Thesaurus*, but only 20 years old at the time, and Samuel Taylor Coleridge, one year after composing *The Rime of the Ancient Mariner*. Coleridge's description is the more poetic:

> The first time I inspired the nitrous oxide, I felt a highly pleasurable sensation of warmth over my whole frame, resembling that which I remember once to have experienced after returning from a walk in the snow into a warm room. The only motion which I felt inclined to make, was that of laughing at those who were looking at me. My eyes felt distended, and towards the last, my heart beat as if it were leaping up and down. On removing the mouth-piece, the whole sensation went off almost instantly.

Nitrous oxide was first used as an anesthetic in 1846 but not before it had caused a stir in college dorms of the period. And James Gillray's 1802 carricature (Figure 224) shows Davy holding bellows and assisting a lecture-hall laughing gas demonstration. And at the right standing, that is Count Rumford smiling with approval.[2]

1. J.R. Partington, *A History of Chemistry*, MacMillan, London, 1964, Vol. 4, pp. 29–73.
2. J. Read, *Humour and Humanism in Chemistry*, G. Bell, London, 1947, p. 207.

FIGURE 224. ■ Artist James Gillray depicted a Royal Institution Lecture in 1802 in which Humphry Davy, holding the bladder, delights as a subject receives a dose of laughing gas (see *Chemical Heritage*, Vol. 17, No. 2. p. 45, 1999). The tall standing figure on the right is Count Rumford. (Courtesy of Edgar Fahs Smith Collection, Rare Book & Manuscript Library, University of Pennsylvania.)

SOME LAST-MINUTE GLITSCHES BEFORE THE DAWN OF THE ATOMIC THEORY

Introductory chemistry books paint a fairly neat picture of the orderly march toward Dalton's atomic theory: Discovery of the Laws of Conservation of Matter, Definite Composition and Multiple Proportions, and thence Atomic Theory. It was never quite so neat.

Chemists who preceded Lavoisier for decades if not centuries implicitly assumed that matter could not be created nor destroyed.[1] Why else would they postulate the addition of effluviums of fire (see Becher, Boyle, or Freind) to explain the increase in mass when metals form calxes, or the need to postulate buoyancy (or negative mass) for phlogiston, to explain the same phenomena? However, Lavoisier's careful work with chemical balances and pneumatic chemistry established the Law of Conservation of Matter on firm scientific ground.[1] Similarly, the Law of Definite Composition had long been assumed—that the back oxide of copper, for example, would always be 80% by weight copper and 20% by weight oxygen no matter the country, chemist, or method of origin. The studies of Joseph Louis Proust (1754–1826) established this and helped to solidify the principles of chemical composition (*stoichiometry*).

However, Claude Louis Berthollet (1748–1822), one of the great collaborators with Lavoisier on the *Nomenclature Chimiques*, raised some difficult questions in his book *Essai de Statique—Chimique* published in 1803 (Figure 225).[2] Although there was some confusion about mixtures and compounds, he noted that there were some crystalline compounds having indefinite and varying compositions. He was correct. For example, the iron ore wustite is typically given the formula FeO although it really ranges from $Fe_{0.95}O$ (76.8% iron) to $Fe_{0.85}O$ (74.8% iron) depending, as we know today, on the balance between Fe^{2+} and Fe^{3+} ions to balance the O^{2-} ions in the ionic salt.[3] Since two Fe^{3+} ions will be equivalent to three Fe^{2+} ions in neutralizing three O^{2-} ions, replacement of Fe^{2+} by Fe^{3+} ions will produce gaps in the crystalline lattice and cause the Fe/O ratio to be less than 1:1 and slightly variable. Wustite is an example of a *nonstoichiometric compound* and such compounds are sometimes called *berthollides*.

Even more serious was Berthollet's finding that in some cases the products obtained in a chemical reaction depended upon reaction conditions. For example, a well-known laboratory chemical reaction is:

$$CaCl_2 + Na_2CO_3 \rightarrow CaCO_3 + 2NaCl$$

where $CaCl_2$ is a muriate of lime, Na_2CO_3 is soda, $CaCO_3$ is limestone, and NaCl is salt. The precipitation of solid limestone drives this "double elective attraction." However, accompanying Napolean on a trip to Egypt in 1798, Berthollet was surprised to discover deposits of soda on the shores of the salt lakes.[4] He reasoned that high concentrations of salt in the lakes could reverse the normal affinities, and thus the products of the reaction depended upon conditions. In fact, he had discovered the reversibility of chemical reactions and the Law of Mass Action, but this was only understood later.

$$CaCl_2 + Na_2CO_3 \rightleftarrows CaCO_3 + 2NaCl$$

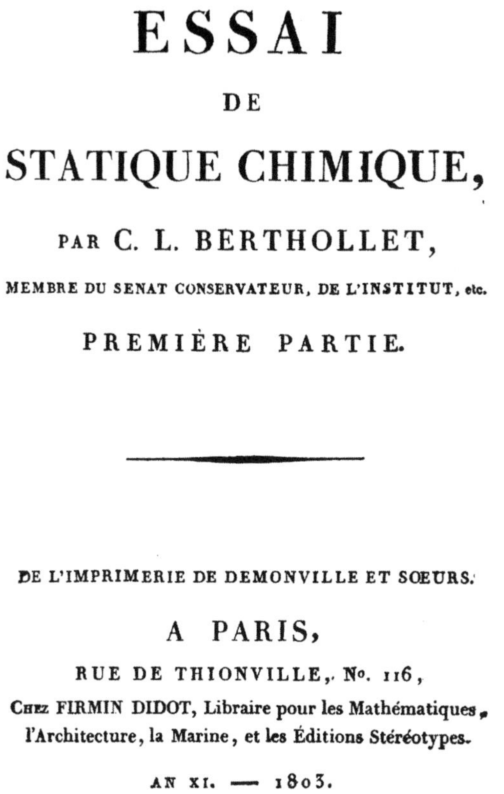

ESSAI

DE

STATIQUE CHIMIQUE,

PAR C. L. BERTHOLLET,

MEMBRE DU SENAT CONSERVATEUR, DE L'INSTITUT, etc.

PREMIÈRE PARTIE.

DE L'IMPRIMERIE DE DEMONVILLE ET SOEURS.

A PARIS,

RUE DE THIONVILLE, N°. 116,

CHEZ FIRMIN DIDOT, Libraire pour les Mathématiques, l'Architecture, la Marine, et les Éditions Stéréotypes.

AN XI. — 1803.

FIGURE 225. ■ Title page of Claude Louis Berthollet's book, published just before Dalton's Atomic Theory. Berthollet discovered that chemical compositions were not always "definite" but often depended upon reaction conditions. He had really discovered the law of mass action.

There is something here for us to learn about the Scientific Method. To borrow the oft-cited example given by the philosopher Karl Popper: *If* one observes only white swans for decades, then the hypothesis "All swans are white" appears reasonable and as it continues to be verified over decades it assumes the status of a confirmed theory and possibly even a law. It can never be proven true since all possible future cases cannot be tested. However, the confirmed scientific observation of a black swan will overturn the theory. Now Berthollet's scientific observations might have been taken as invalidating the Law of Definite Composition and seriously undermining the Atomic Theory. However, rather than tossing them out due to the observation of a few "black swans," chemists retained these explanations, correctly anticipating that the inconsistencies would be explained in the future.

1. F.L. Holmes, *Chemical and Engineering News*, **72** (37): 38–45, 1994.
2. H.M. Leicester and H.S. Klickstein, *A Source Book in Chemistry 1400–1900*, McGraw-Hill, New York, 1952, pp. 192–201.
3. D.W. Oxtoby and N.H. Nachtreib, *Principles of Modern Chemistry*, 3rd ed., Saunders College Publishing, Fort Worth, 1996. p. 9.
4. W.H. Brock, *The Norton History of Chemistry*, Norton, New York, 1993, p. 144.

ATMOSPHERIC WATER MOLECULES AND THE MORNING DEW

John Dalton[1] recorded atmospheric measurements throughout his long scientific life and at 6 A.M. on July 27, 1844 he made his final diary entry—"little rain this day"—in a feeble hand just before he died.[2] This lifelong interest led him to try to understand the occurrence of water vapor in the air (why doesn't it simply condense?). Another puzzler was why air was completely homogeneous—why didn't the denser oxygen component settle out from the more abundant, lighter nitrogen gas? Could it be that the two gases formed a weak 1 : 4 compound susceptible to displacement of nitrogen by more reactive substances, such as metals or hydrogen, having higher affinities for oxygen?

A decade before he enunciated his atomic theory, Dalton had found that the amount (pressure) of water vapor in air or introduced into a vacuum depended solely on the temperature of the liquid water in equilibrium.[3] (Dalton also developed the concept of the dewpoint).[3] This suggested that water vapor did not form a chemical compound with air (else, why would it enter an evacuated vessel?). It also suggested that water's vapor pressure and very existence were completely independent of other gases in the air. In 1801, his experimental studies permitted a more general statement of what we now call Dalton's law of partial pressures:[4]

> When two elastic fluids, denoted by A and B, are mixed together, there is no mutual repulsion amongst their particles; that is, the particles of A do not repel those of B, as they do one another. Consequently, the pressure or whole weight upon any one particle arises solely from those of its own kind.

In modern terms, we could write this as

$$P_{\text{total}} = P_{\text{oxygen}} + P_{\text{nitrogen}} + P_{\text{water vapor}}$$

This is an interesting, if not very straightforward notion. Why should "particles" of "A" (e.g. nitrogen) repel other "A" particles but not "particles" of "B," such as oxygen, toward which they remain indifferent?

Dalton's first atomic theory was a "physical one." From his 1801 presentation[5] we see his depiction of the four atmospheric gases (water, oxygen, nitrogen, and carbonic acid). Separately, each gas repels like "atoms" (top of Figure 226), but mixed "atoms" of different gases do not repel or attract (bottom of Figure 226). Dalton, modest Quaker that he was, nonetheless compared his theory to Newton's law of universal gravitation.[5] This comparison was *not* immodest. A few years later, Dalton would realize that his theory explained chemistry as well as physics.

Dalton's explanation mixes Lavoisier's caloric theory with Newton's mechanical repulsion theory and then adds a dash of his own special ingredient. First, it is important to recall that when combined in a metal oxide or calx, the element oxygen is in its "fixed" state. Thus, according to Lavoisier, "oxygen gas" is actually "oxygenated caloric" since heat was required in order to free the ele-

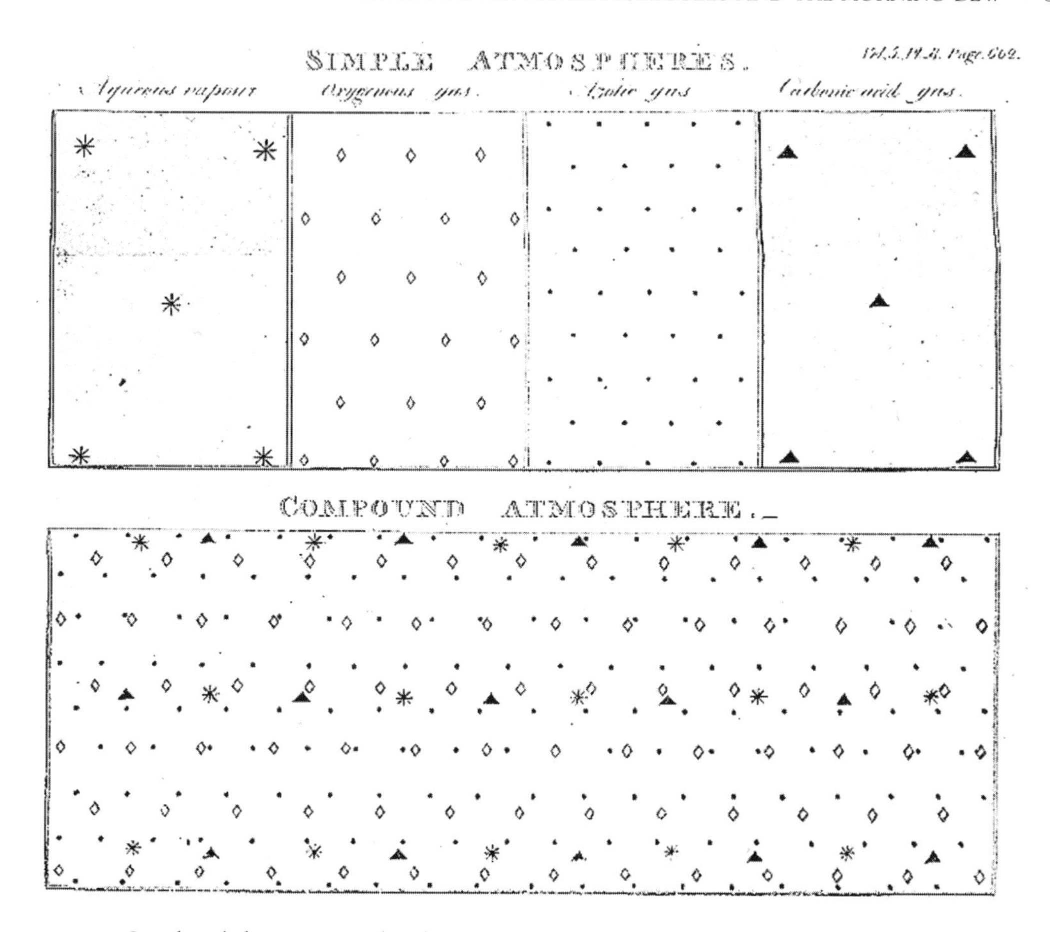

FIGURE 226. ■ Graphical description of Dalton's law of practical pressures printed in the 1802 *Memoirs of the Literary and Philosophical Society of Manchester.* According to Dalton's theory, the corpuscles of a single atmospheric gas (e.g., water, oxygen, nitrogen, carbon dioxide) repel each other but not corpuscles of other gases. Thus, the gases may be mixed ("superimposed: as in the bottom of this figure) without any interaction; hence, the atmosphere is freely mixed, not layered. See Figure 227, which depicts Dalton's attempt at explaining these phenomena.

ment from its calx. Similarly, boiling water adds caloric to form vapor. Note, in an earlier essay, how Lamarck's contemporary picture (Figure 217) of evaporation of water places "jackets" of caloric around corpuscles of water so as to increase the space between them causing these corpuscles to move into the gas phase.

Dalton's clever "take" on this problem of repulsion amongst like molecules was diagrammed in Part II of his *Chemical Philosophy* (Figure 227).[6] At the top part of this figure are represented "jackets" of caloric surrounding gaseous molecules. (*Note:* Hydrogen gas is thought to be monoatomic.) The lower part of the figure demonstrates why like molecules, such as *azote* (nitrogen), repel each other while different types of gas molecules are mutually indifferent. Since at a given temperature, the sizes and "jackets" of caloric surrounding all azote gas molecules must be equal in dimension, lines of force are totally aligned and repulsion occurs.[7] The same is true for repulsion between hydrogen molecules. However, such perfect alignment does not occur between hydrogen and *azote*, and they have totally independent existences and make their own independent contributions to

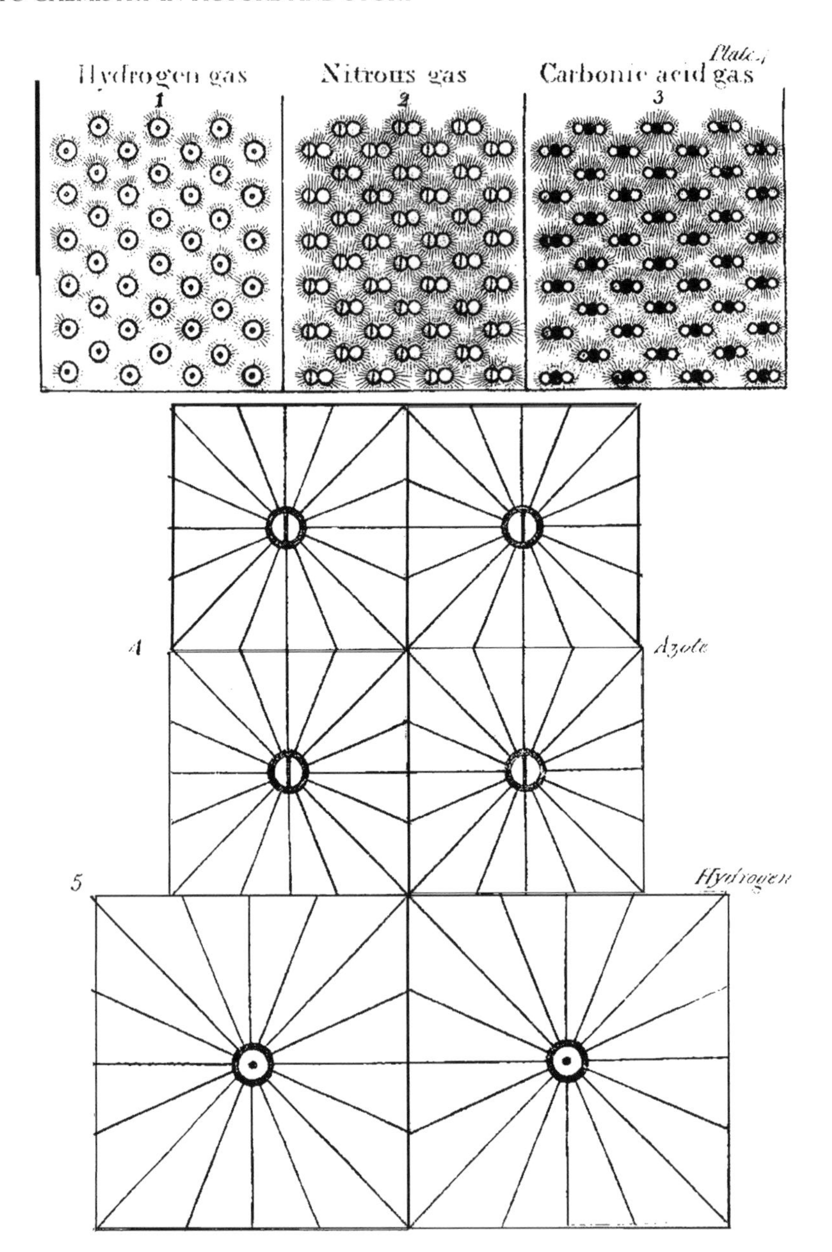

FIGURE 227. ■ Figures of jackets of caloric and "lines of force" postulated by Dalton to explain why like molecules of gas must repel each other so that gases of different densities remain mixed rather than layered in the atmosphere (from Dalton, *A New System of Chemical Philosophy*, Part I, Manchester, 1808).

the total pressure of a mixture. It was vital for Dalton that like "atoms" repelled and different "atoms" did not, since this prevented separation and layering of bulk gases according to gas densities.

Not only is air uniformly mixed at sea level, rather than consisting completely or mostly of the denser gas, oxygen, but the same mixture persists high into the atmosphere. This was known at the time Dalton was formulating his atomic theory. In 1804, Joseph Louis Gay-Lussac piloted a balloon some 23,000

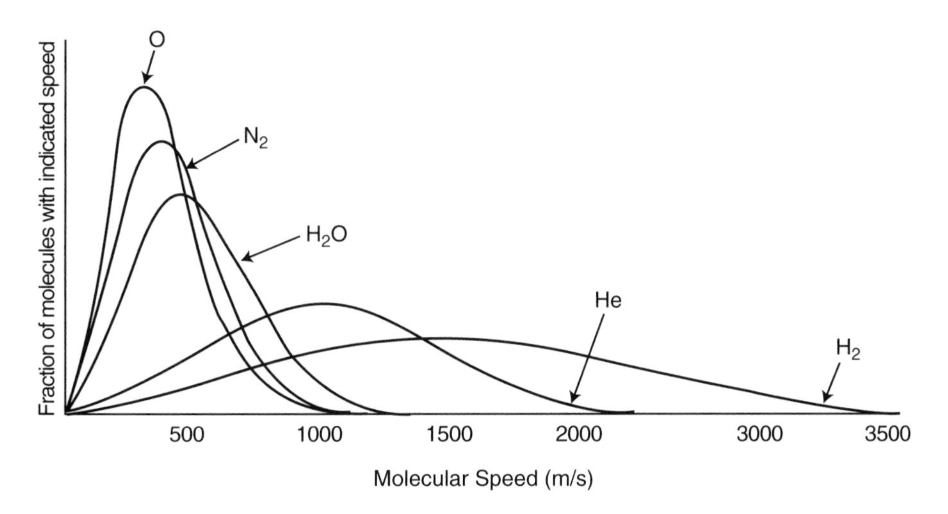

FIGURE 228. ■ The distribution of velocities of gas molecules at 25°C. Note how much greater the velocities are for ultralight hydrogen molecules and helium atoms. These substances thus escape the earth's atmosphere, unlike the heavier gases. (Adapted from Brown et al., *Chemistry—the Central Science*).

feet above Paris, and collected two air samples, which were shown to have the same composition as air at sea level.[8]

We now know that the troposphere, stratosphere and mesosphere extending up to about 60–70 miles above the earth have essentially the same uniform chemical composition.[9] Oxygen, nitrogen, water vapor, and the other atmospheric gases move at velocities[10] very far below the velocity needed to escape the earth's gravitational pull [11 kilometers per second (km/s) or 22,000 miles per hour (mph)].[11] The fact that these gaseous molecules have negligibly weak attractions for each other, the natural tendency to maximize disorder (increase entropy) and constant mixing by winds driven by the earth's rotation guarantees total mixing of the atmospheric gases. Figure 228 depicts the velocity distributions of simple gaseous molecules.[10] The two lightest gases, hydrogen and helium, show drastically different distributions relative to the other gases in this figure. Even though their average velocities are well below the earth's escape velocity, the most rapidly moving "outlier molecules" will escape into space and over time these gases are lost from the earth's atmosphere. Thus, although hydrogen is the most abundant element in the universe, only the minutest traces are found in the earth's atmosphere as the result of continuing high-energy processes that split water molecules. Similarly, the minute traces of helium in the atmosphere are due to fresh outgassing of radioactive materials that emit *alpha* particles (helium nuclei).

---

1. J.R. Partington, *A History of Chemistry*, MacMillan & Co. Ltd., London, 1962, Vol. 3, pp. 755–822.
2. Partington, op. cit., p. 760.
3. Partington, op. cit., pp. 762–765.
4. Partington, op. cit., pp. 765–767.
5. J. Dalton, *Memoirs of the Literary and Philosophical Society of Manchester*, Vol. V, Part II, Codell and Davies, London, 1802, pp. 535–602.

6. J. Dalton, *A New System of Chemical Philosophy*, Part I, Manchester, 1808; Part II, Manchester, 1810 (see Plate 7).
7. Partington, op. cit., pp. 778–781.
8. J.R. Partington, *A History of Chemistry*, MacMillan & Co. Ltd., London, 1964, Vol. 4, p. 78.
9. *The New Encyclopedia Britannica*, Encyclopedia Britannica, Inc., Chicago, 1986, Vol. 14, p. 311.
10. T.L. Brown, H.E. LeMay, Jr., and B.E. Bursten, *Chemistry—the Central Science*, seventh edition, Prentice-Hall, Upper Saddle River, NJ, 1997, pp. 364–368.
11. S. Mitton (ed.), *The Cambridge Encyclopedia of Astronomy*, Crown Publishers Inc., New York, 1977, pp. 193–195.

EXCLUSIVE! FIRST PRINTED PICTURES OF DALTON'S MOLECULES

> there are things which exist with solid and everlasting body, which we show to be the seed of things and their first-beginnings, out of which the whole sum of things now stands created.[1]
>
> There is then a void, mere space untouchable and empty. For if there were not, by no means could things move; for that which is the office of body, to offend and hinder, would at every moment be present to all things; nothing, therefore, could advance, since nothing could give the example of yielding place.[2]
>
> But as it is, because the fastenings of the first-elements are variously put together, and their substance is everlasting . . . No single thing then passes back to nothing, but all by dissolution pass back into the first-bodies of matter,[3]

Thus, does the Latin poet Lucretius speak to us, from a distance of 2000 years, in *De rerum natura* (*The Nature of Things*) "justifying"

1. That atoms are the ultimate and indestructible "seeds" of matter
2. The existence, indeed requirement, of empty space (void or vacuum)
3. The law of conservation of matter

Before Lucretius is awarded a share of the "*retro*–Nobel Prizes"[4] in chemistry, physics, *and* literature, we must admit that these were purely philosophical premises—no scientific hypotheses were tested experimentally. Lucretius' epic poem summarized views of earlier Greek philosophers including Democritus, Leucippus, and Epicurus (see Robert Boyle's ironic comment, pp. 202–203).

An early near-scientific theory of corpuscles or atoms was published by Daniel Sennert (1574–1637), Professor of Medicine at the University of Wittenberg as early as 1618.[5] The French Philosopher René Descartes (1596–1650) fathered analytical geometry, but his contributions to physics and chemistry were not significant.[6] He believed in atomlike ultimate particles that packed together such that the universe contained no voids ("Nature abhors a vacuum"). All motion in the universe had to be coordinated in a form of "cosmic gridlock." In contrast were the views of Pierre Gassendi (1592–1655),[7] a classicist who studied Epicurus and adopted the Epicurean concept of atoms and voids rather than the

Cartesian continuum. Gassendi found a firm scientific argument for the true existence of void and vacuum in Torricelli's barometer, invented in 1643.

Robert Boyle and Isaac Newton, both strong adherents of alchemy, advanced a corpuscular theory of matter. It is worth remembering that, since they believed that lead could be transmuted to gold, there could be no "uniquely" gold or lead corpuscles. Their views were influenced by Gassendi.[8]

Boyle's law (1662) had demonstrated that if a gas expanded, for example, to eight times its volume, its pressure would decrease by a factor of 8. The density was also found to decrease by 8. Now, one might imagine that "thinning" of a "cartesian fluid" could decrease its density by expanding its continuum of "atoms." However, it is even harder to rationalize the reduction in pressure using this model. Instead, a model imagining the gas to be composed of individual "corpuscles," separated in space, might explain such behaviors in terms of Newtonian physics. The rarified gas would have, on average, even greater space between corpuscles. In 1687, Newton attempted to explain Boyle's law by postulating repulsion between hard corpuscles in the gaseous fluid.[9] Relating the repulsion to centrifugal force, he predicted that it would be inversely proportional to the distance between the centers of the atoms. Thus, reduced pressure in the gas was the result of reduced repulsion between corpuscles now further separated. Just as Newton did not attempt to explain the nature of the gravitational force that drew objects to the earth, he also did not attempt to understand the source of this mysterious repulsive force between atoms.[9]

John Dalton's earliest atomic theory originated in 1801 and was purely physical in nature.[10] Its basis was Boyle's law and his own law of partial pressures (see next essay). But his truly fundamental breakthrough, which occurred in 1803, was to produce the modern paradigm that ties together everything we know today about chemistry. Dalton's atomic law was the culmination of the chemical revolution that had occurred during the preceding three decades.[11]

Lucretius' poem suggests that the Law of Conservation of Matter has been assumed for at least two millennia. It was certainly a fundamental scientific assumption during the scientific revolution. However, it was Lavoisier who propounded the view that, unless all material mass could be accounted for in a chemical reaction, one could not even try to understand it. Critical, too, were Richter's establishment of tables of equivalents and his concept of stoichiometry and Proust's law of definite composition, that successfully survived his debate with Berthollet. Dalton's notebook entry on September 6, 1803 (his thirty-seventh birthday) includes the first symbolic drawings and relative weights of his atoms.[11]

Thomas Thomson[12] received his M.D. degree at Edinburgh in 1799, where he was inspired by Joseph Black. Starting in 1800 he lectured on chemistry at Edinburgh and published the first edition of his comprehensive *A System of Chemistry* in 1802. Thomson visited Dalton in 1804 and enthusiastically adopted his atomic theory. Interestingly, the first published statement of Dalton's theory appeared in the third edition (1807) of Thomson's five-volume chemical treatise.[13] Dalton's *Chemical Philosophy* was published the next year.[11] It is thrilling to read Thomson's polite and tentative remarks and view the first printed pictures of atoms as they appeared in his book (Figure 229):[12]

> We have no direct means of ascertaining the density of the atoms of bodies; but Mr. Dalton, to whose uncommon ingenuity and sagacity the philosophic

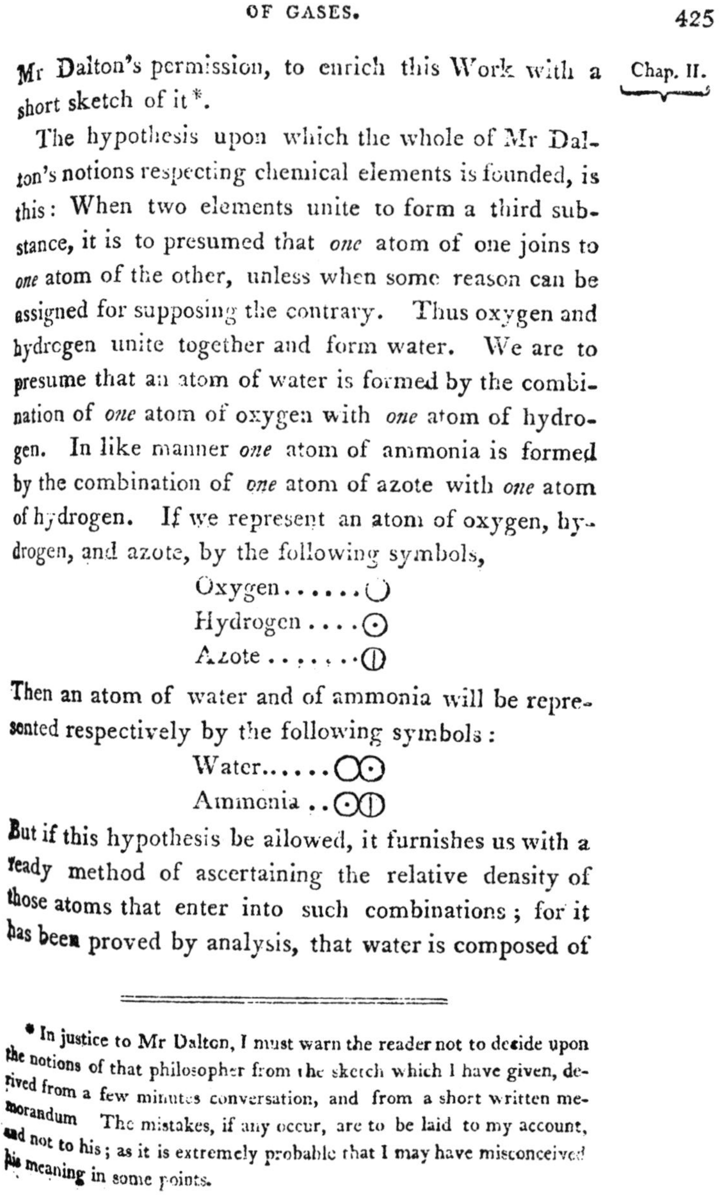

OF GASES. 425

Mr Dalton's permission, to enrich this Work with a short sketch of it*.

Chap. II.

The hypothesis upon which the whole of Mr Dalton's notions respecting chemical elements is founded, is this: When two elements unite to form a third substance, it is to presumed that *one* atom of one joins to *one* atom of the other, unless when some reason can be assigned for supposing the contrary. Thus oxygen and hydrogen unite together and form water. We are to presume that an atom of water is formed by the combination of *one* atom of oxygen with *one* atom of hydrogen. In like manner *one* atom of ammonia is formed by the combination of *one* atom of azote with *one* atom of hydrogen. If we represent an atom of oxygen, hydrogen, and azote, by the following symbols,

Oxygen......○
Hydrogen....⊙
Azote.......⦶

Then an atom of water and of ammonia will be represented respectively by the following symbols:

Water......○⊙
Ammonia ..⊙⦶

But if this hypothesis be allowed, it furnishes us with a ready method of ascertaining the relative density of those atoms that enter into such combinations; for it has been proved by analysis, that water is composed of

* In justice to Mr Dalton, I must warn the reader not to decide upon the notions of that philosopher from the sketch which I have given, derived from a few minutes conversation, and from a short written memorandum The mistakes, if any occur, are to be laid to my account, and not to his; as it is extremely probable that I may have misconceived his meaning in some points.

FIGURE 229. ■ Although John Dalton developed a physical atomic theory in 1801 and extended it to chemistry in 1803, he did not publish his theory until 1808. However, Thomas Thomson at the University of Edinburgh was an early advocate of atomic theory and, with the Dalton's permission, published its first printed discussion in 1807 (see *A System of Chemistry*, 3rd ed., London, 1807).

> world is no stranger, has lately contrived an hypothesis which, if it prove correct, will furnish us with a very simple method of ascertaining that density with great precision.

The Quaker Dalton postulated a principle of "greatest simplicity" and thus assumed, for example, that water was comprised of one atom each of hydrogen and oxygen and ammonia comprised of one atom each of nitrogen and hydrogen (see Figure 229). This led to values of atomic weights in 1803 that we would now see as anomalies[11] (e.g., "*azot*" or nitrogen = 4.2; oxygen = 5.5; where hydrogen assumed = 1.0). Dalton was, however, also aware that "carbonic acid" contained twice the weight of oxygen per weight of carbon than did the newly discovered "carbonic oxide."[14] Similar findings were extant for oxides of nitrogen.[11] Thus, this law of multiple proportions (e.g., CO_2 vs. CO), developed by Dalton, was a clear corollary of his atomic theory.

---

1. C. Bailey (transl.), *Lucretius on the Nature of Things*, Oxford University Press, Oxford, 1910, p. 43.
2. Bailey, op. cit., p. 38.
3. Bailey, op. cit., p. 35.
4. The idea of a "*retro*–Nobel Prize" forms the premise for the inventive play Oxygen by Carl Djerassi and Roald Hoffmann; see C. Djerassi and R. Hoffmann, Oxygen, Wiley-VCH, Weinheim, 2001.
5. J.R. Partington, *A History of Chemistry,* MacMillan & Co. Ltd., London, 1961,Vol. 2, pp. 271–276.
6. Partington, op. cit., pp. 430–441.
7. Partington, op. cit., pp. 458–467.
8. Partington, op. cit., pp. 502–507.
9. Partington, op. cit., pp. 474–477.
10. Partington, *A History of Chemistry,* MacMillan & Co., Ltd., London, 1962, Vol. 3, pp. 765–782.
11. Partington (1962), op. cit., pp. 782–786.
12. Partington (1962), op. cit., pp. 716–721.
13. T. Thomson, *A System of Chemistry in Five Volumes,* third edition, Bell & Bradfute, and E. Balfour, London, 1807, Vol. III, pp. 424–431.
14. Partington (1962), op. cit., pp. 271–276.

THE ATOMIC PARADIGM

Paradigm is a much overused word. However, the existence of atoms is so fundamental to the very fabric of chemical understanding that we can say virtually nothing scientifically sensible without it. *This* is a paradigm! Figures 101 and 102 are derived from John Dalton's 1808–1810 *A New System of Chemical Philosophy* (Vol. I, Parts I and II; the third volume, Vol. II, Part I appeared in 1827; although less important than the earlier volumes it is of utmost rarity and pricy indeed). Dalton (1766–1844), born to Quaker parents of modest means,[1] was largely self-educated. He taught school at the age of 12 and in 1793 moved to Manchester where he was for a period Professor of Mathematics and Philosophy at New College. This college moved from Manchester in 1803 and, after a variety of incarnations, became Manchester College in Oxford in 1889.[1] Dalton, however, remained in Manchester where he earned a modest living tutoring, lecturing, and consulting while performing his research.[1] Partington conjectures that the "robust and muscular" Dalton inherited his nature largely from his "energetic and lively" mother.[1] He never married but was attracted briefly to a widow of "great intellectual ability and personal charm": "During my captivity, which lasted about one week, I lost my appetite and had other symptoms of bondage about me as incoherent discourse, etc., but have now happily regained my freedom."[1]

Dalton had a lifelong interest in meteorology.[1,2] He published a book on this topic in 1793. However, his studies of the composition of the atmosphere gave him the first clues leading to his atomic theory. Dalton realized that the composition of air was independent of altitude. Although oxygen and nitrogen differed in density, they did not form layers. His thoughts at this time included the idea that individual atoms were surrounded by envelopes (atmospheres) of caloric that repelled like atoms and attracted different atoms, thus explaining atmospheric mixing. During the period 1799–1801 he defined the vapor pressure of water and realized that when water was added to dry air, the total pressure was the sum of the dry air pressure and water's vapor pressure—the gases mixed yet acted totally independently (Dalton's Law of Partial Pressures). He also showed, as had Charles earlier, that air expands its volume linearly upon heating.

Although evidence suggests that chemists of the late eighteenth century assumed that specific substances had definite compositions,[2] Berthollet's studies (see Figure 225) showed that compositions of "substances" often depended upon starting conditions. We now understand that Berthollet was observing mixtures whose proportions changed with conditions prevailing at equilibrium. Joseph Louis Proust (1754–1826) was educated in Paris but moved to Madrid where he held academic positions.[2] He engaged in a respectful debate with Berthollet over the course of a number of years and eventually prevailed. Proust demonstrated that there were two distinct oxides of tin and two distinct sulfides of iron—each with their own definite composition. Previous uncertainties were the result of mixtures of each pair of binary compounds.[2]

Dalton applied the Laws of Conservation of Mass and Definite Composition to explain his Atomic Theory.[3] He developed the chemical theory in 1803 and told Thomas Thomson, University of Edinburgh, about it in 1804. The third edition of Thomson's multi-volume A *System of Chemistry* (Edinburgh, 1807) was the first book to include Dalton's Atomic Theory (Figure 229) and atams and molecules were represented in Dalton's new book (Figure 230). Dalton developed a third law, the Law of Multiple Proportions, to explain different formulas for binary compounds. For example, look at the binary compounds (*41* to *45*) formed between oxygen and nitrogen (Figure 231). In comparing nitric oxide (*41*, NO) with nitrous oxide (*42*, N_2O), we see that the mass of nitrogen combining with a mass of oxygen in compound *42* is twice that in *41*. Dalton's atoms were real, indestructible, and unique for each of Lavoisier's ponderable elements. They were a total denial of alchemical transmutation. Dalton even built molecular models.

Dalton, a Quaker in science style as well as lifestyle, also assumed a "rule of greatest simplicity."[2] He originated the concept of atomic weights but could neither measure them nor even understand their basis. He chose to assign a relative weight of 1 to hydrogen, the lightest element, and to assume that combinations were the simplest possible. For example, we know water to be H_2O, ammonia to be NH_3, and methane to be CH_4. Dalton assumed they were HO, NH, and CH, respectively. Based upon the chemical analyses of 1803, which were good but far from perfect, he derived atomic weights as follows: hydrogen, 1.0 (assumed); oxygen, 5.5; nitrogen, 4.2; carbon, 4.3. By 1808, he had modified the values to include the latest data and rounded them off to whole numbers: hydrogen, 1 (assumed); nitrogen, 5; carbon, 5; oxygen, 7.[3,4] These assumptions would continue

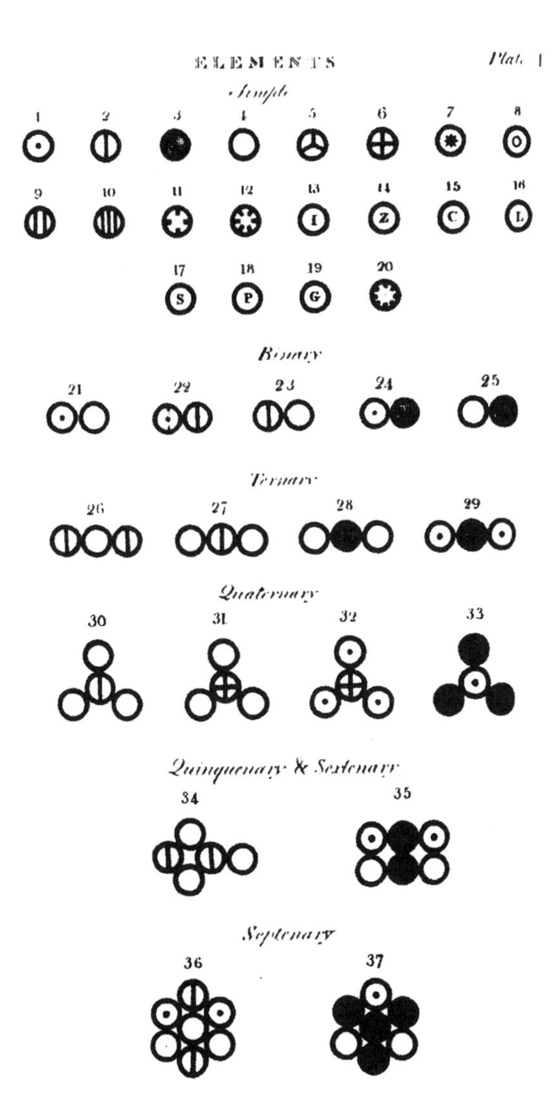

FIGURE 230. ■ Plate depicting atoms in Dalton's *A New System of Chemical Philosophy* (Manchester, 1808–1810). Dalton's "rule of greatest simplicity" (perhaps a Quaker style of science as well as lifestyle) has the formula of water as HO rather than H_2O (hydrogen peroxide, H_2O_2, would be discovered by Gay-Lussac and Thenard in 1815).

to cause confusion for decades. The assumption of 1.0 for hydrogen appears to be prescient although there was no basis at all for its assumed "oneness." We now know that hydrogen's "oneness" derives from its nucleus that has one proton only. Although hydrogen gas is actually H_2, oxygen (O_2), nitrogen (N_2), and some others are also diatomic and their relative densities are direct reflections of atomic masses. It remained for physician William Prout (1785–1850) to hypothesize in 1815–1816 that atomic weights are whole-number multiples of the atomic weight of hydrogen.[3,4]

Partington[1] notes that "Dalton never pretended that his teaching work interfered with his research, saying that 'teaching was a kind of recreation, and if richer he would not probably spend more time in investigation than he was accustomed to do.'" This is food for thought for those occasional self-important

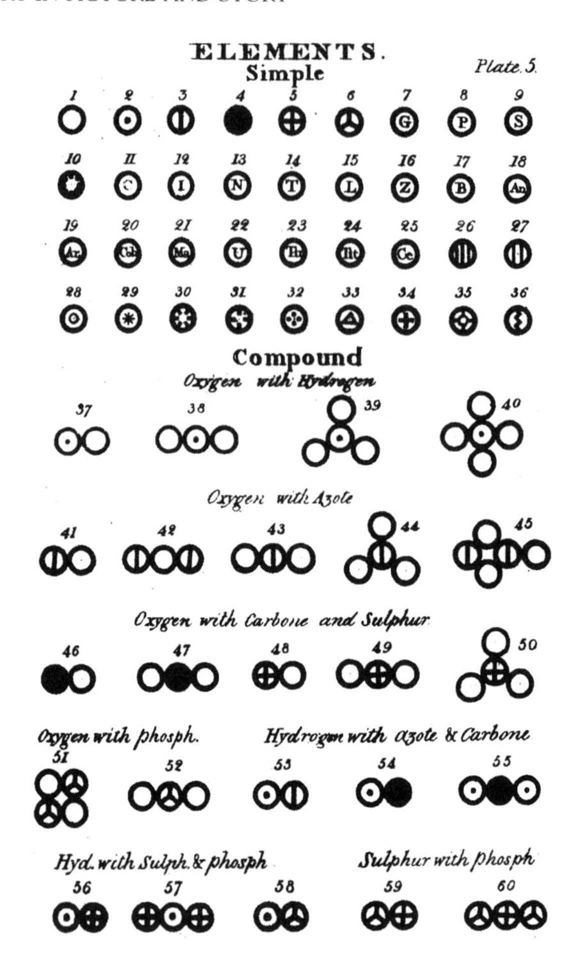

FIGURE 231. ■ Plate depicting atoms in Dalton's *A New System of Chemical Philosophy* (Manchester, 1808–1810).

professors whose accomplishments do not include discovering their discipline's paradigm. Of course, one of Dalton's students was the renowned physicist James Prescott Joule—now there is a "career student" for any dedicated teacher!

1. J.R. Partington, *A History of Chemistry*, MacMillan, London, 1962, Vol. 3, pp. 755–822.
2. A.J. Ihde, *The Development of Modern Chemistry*, Harper and Row, New York, 1964, pp. 98–111.
3. J.R. Partington, op. cit., pp. 713–714.
4. W.H. Brock, *The Norton History of Chemistry*, Norton, New York, 1993, pp. 133–147; 160–162.

"WE ARE HERE! WE ARE HERE! WE ARE *HERE*!"

Dr. Seuss's wonderful book *Horton Hears A Who*[1] tells of the Whos of Whoville, a town on a speck of dust. They are too small to be seen and only Horton the elephant can hear them. Before the dust speck is boiled in oil, Horton exhorts the entire town to make a loud unified noise to announce the Whos' existence and save their lives: "We are here! We are here! We are here!"

In many ways the invisible (and voiceless) atoms were calling attention to themselves early in the nineteenth century. Figure 232 is from Dalton's 1808 *A New System of Chemical Philosophy*. In illustration 1 we see Dalton's depiction of the structure of liquid water. Dalton postulated that when water freezes, the atoms in a layer move from the square to rhomboid arrangement of illustration 2. It is this hexagonal arrangement that Dalton perceived to be responsible for the well-known hexagonal symmetry of snow flakes and ice crystals (see illustration 5). He also tried to use these structures to explain the known fact that ice is less dense than water (ice floats). His arguments (using illustrations 3 and 4) were incorrect (this was noted with disapproval by Berzelius in 1812).[2] Also, liquid water is not an orderly array as depicted in diagram 1. Nevertheless, the core idea about ice structure was correct. We understand that water molecules are not per-

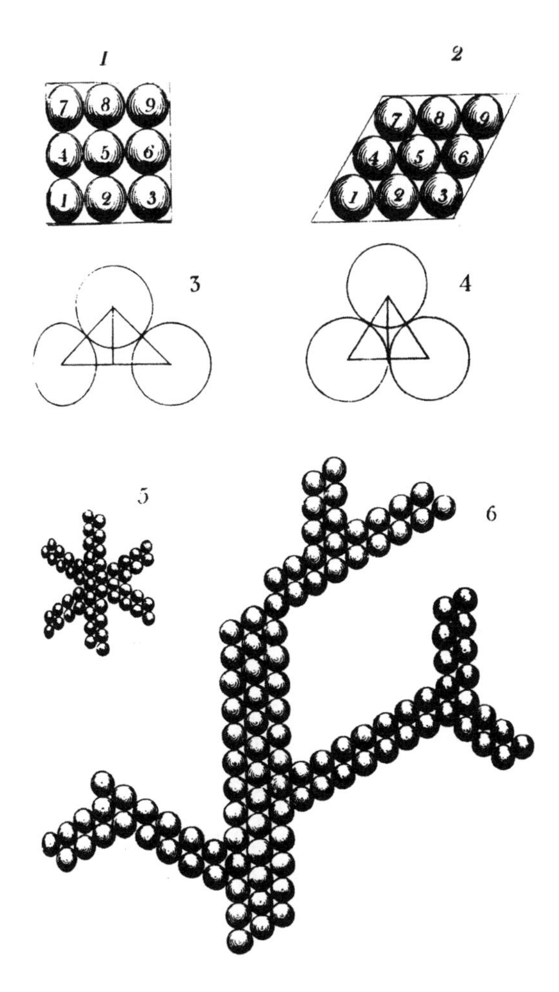

FIGURE 232. ■ A plate from Dalton's *A New System of Chemical Philosophy* explaining why ice is less dense than water using Atomic Theory. Although the details are not correct, Dalton's explanation of the sixfold symmetry of snowflakes and ice crystals originating at the molecular level was incredibly insightful. Berzelius and others noted the trigonometric error in his explanation of the decrease in density when water (*1* in this figure) becomes ice (*2*). Note the similarity to Hooke's explanation of ice structure (Figure 147).

fect spheres and that ice is less dense than water to allow full hydrogen bonding between water molecules. The overall molecular lattice of ice has sixfold symmetry and this is indeed reflected in the snowflake.

In 1808 Jean Louis Gay-Lussac[3] (1778–1850) summarized the results of experiments of others and a few of his own and discovered the law of combining volumes of gases. He realized that volumes of gases could only be compared if their pressures and temperatures were equal. (The first statement is Boyle's Law; the second is sometimes called Charles Law after the first discoverer or Gay-Lussac's Law after the person who first published it.) Thus, equal volumes of ammonia (NH_3) and muriatic acid (HC1) combine perfectly to form a solid salt; one volume of nitrogen and one volume of oxygen form two volumes of "nitrous gas" (NO). One volume of nitrogen and three volumes of hydrogen form two volumes of ammonia.

Dalton resisted Gay-Lussac's findings. They caused problems for his rule of greatest simplicity. Since an equal volume of hydrogen reacts with an equal volume of chlorine, it is reasonable that those volumes contain equal numbers of elementary particles. However, if two volumes of hydrogen react with one volume of oxygen, as observed by Gay-Lussac, this would not be consistent with Dalton's formulation of water as HO. (Note that hydrogen peroxide, H_2O_2, was not discovered until 1815.)

In August, 1804 Biot and Gay-Lussac, like Charles in 1783, ascended in a hydrogen-filled balloon to make measurements of the earth's magnetic field. In September, 1804, Gay-Lussac ascended to 23,000 ft above Paris, collected air samples and found them to have the same composition as air at sea level.[3] You have to admire Charles's and Gay-Lussac's confidence in the gas laws.

In 1811, Amedeo Avogadro[4] (in full, Lorenzo Romano Amedeo Carlo Avogadro di Quaregua e di Cerreto, 1776–1856) used Gay-Lussac's, Dalton's, and others' works to make his hypothesis: equal volumes of gases (at same temperature and pressure) have equal numbers of molecules. Interestingly, this contribution was largely forgotten until resurrected by Cannizaro in 1858.[5]

Another piece of macroscopic evidence favoring atoms was the concept of Isomorphism enunciated by Eilhardt Mitscherlich (1794–1863) around 1818–1819.[6,7] He related atomic composition to observable crystal structures. Thus, phosphates and arsenates (e.g., Na_2HPO_4 $12H_2O$ and $Na_2HAsO_4 \cdot 12H_2O$) as well as sulfates and selenates (e.g., Na_2SO_4 and Na_2SeO_4) had identical or very similar crystal structures because their atomic compositions were so similar. (Can you hear the pendulum of the future Periodic Table swinging here?) Berzelius used these relationships to help in his assignments of atomic weights.

Calorimetric studies of the type done by Lavoisier and Laplace (Figure 212) were continued by others including Pierre Louis Dulong (1785–1838) and Alexis Therese Petit (1791–1820). They discovered the law that bears their names: the product of the specific heat and the atomic weight of solid elements (e.g., lead, gold, tin, silver, and sulfur) is constant. This really implies that all atoms (independent of their identities) have the same capacity for heat. This result was later extended to solid compounds and ultimately cleared up confusions such as whether the binary oxides of copper were really CuO and CuO_2 or Cu_2O and CuO.

---

1. T.S. Geisel, *Horton Hears A Who* (By Dr. Suess), Random House, New York, 1954.
2. W.H. Brock, *The Norton History of Chemistry*, Norton, New York, 1993, pp. 158–162.

3. J.R. Partington, *A History of Chemistry*, MacMillan, London, 1964, Vol. 4, pp. 77–90.
4. J.R. Partington, op. cit., pp. 213–217.
5. J.R. Partington, op. cit., pp. 489–494.
6. J.R. Partington, op. cit., pp. 207–212.
7. A.J. Ihde, *The Development of Modern Chemistry*, Harper and Row, New York, 1964, pp. 147–149.

WAS AVOGADRO'S HYPOTHESIS A PREMATURE DISCOVERY?

Avogadro got it right in 1811 when, combining Dalton's Atomic Theory and Gay-Lussac's Law of Combining Volumes, he concluded that equal volumes of gas (same temperature and pressure) have equal numbers of ultimate particles. One critical aspect was Avogadro's *term half-molecules*, which were really atoms for diatomic molecules such as H_2, O_2, N_2, and Cl_2. Dalton had never accepted the combining volumes law. He was bothered, for example, by the "sesquioxide" of nitrogen (1 volume of nitrogen atoms; $1.^5$ volumes of oxygen atoms). In his Quaker style of speech he said:[1] "Thou knows . . . no man can split an atom." Avogadro's nomenclature was somewhat confusing. For example, the "integrant molecule" of water contains half a molecule of oxygen and one molecule (or two half molecules) of hydrogen:[2] $1H_2 + \frac{1}{2}O_2 \rightarrow 2H + 1O \rightarrow 1H_2O$. Avogadro's Hypothesis also vexed Dalton: how could nitrogen gas (N) be more dense than ammonia gas (NH) if the two gases had equal numbers of molecules in a unit volume? Of course the answer is that nitrogen gas is N_2 while gaseous ammonia is NH_3.

Andre Marie Ampere, Jean Baptiste Andre Dumas, and their student Marc Augustin Gaudin adopted Avogadro's Hypothesis in their work during the next three decades.[3] Until Dumas, gas densities could be measured only for permanent gases. Dumas developed a technique to measure densities for volatile liquids and solids, thus extending the range of molecular (and atomic) weights determinable by the Ideal Gas Law ($PV = nRT$). In this way, Gaudin discovered that elemental (white) phosphorus is actually P_4.[2] Still, it remained for Stanislao Cannizzaro to reintroduce Avogadro's hypothesis in 1858.

Why did it take almost 50 years to achieve widespread acceptance of Avogadro's hypothesis? Here it may be of some value to refer to Gunther S. Stent's concept of a *premature discovery*, defined as follows: "A discovery is premature if its implications cannot be connected by a series of simple logical steps to canonical or generally accepted knowledge."[3] Stent exemplifies a premature discovery using Oswald T. Avery's unambiguous experimental identification of DNA as the genetic material in 1944.[3] What was the conceptual problem? It was known that DNA was composed of only four different nucleotides. How could such a simple "alphabet" code for an unimaginably vast store of genetic information? Proteins, by contrast, had a 20-amino-acid "alphabet" and were obviously the better choice for information storage. Thus, although Avery's experimental conclusions were solid and unambiguous, scientists did not immediately accept the conceptual framework to understand them and they remained relatively unnoticed for about five or six years. Similarly, atoms were not universally accepted as real and Avogadro's nomenclature was somewhat confusing. Moreover, Avogadro, who practiced law and was Professor of Mathematics at Turin, although "a man of great learning and modesty," was said to be "little known in Italy."[2] Similarly, Avery was "a quiet, self-effacing, non-disputatious gentleman."[3] Had Avogadro access to a good Madison Avenue public relations firm, perhaps the Periodic Table would have been discovered a decade or two earlier.

---

1. J.R. Partington, *A History of Chemistry*, MacMillan, London, 1962, Vol. 3, p. 806.
2. J.R. Partington, *A History of Chemistry*, MacMillan, London, 1964, Vol. 4, p. 213–222.
3. G.S. Stent, *Scientific American*, Vol. 227, No. 6 84–93, 1972.

CHEMISTRY IS *NOT* PHYSICS

Dalton's atoms were derived from chemical experiments and explained chemical laws. Atoms were "adopted" by physicists only after many decades passed.

Chymical Lectures:
In which almoſt all the
OPERATIONS
OF
Chymiſtry
ARE
Reduced to their True PRINCIPLES,
and the LAWS of NATURE.
Read in the Muſeum *at* Oxford, 1704.

By JOHN FREIND, M.D. Student
of *Chriſt-Church*, and Profeſſor of *Chymiſtry*.

Engliſhed by *J. M.*

To which is added,
An APPENDIX, containing the Account
given of this Book in the *Lipſick Acts*, together with the Author's Remarks thereon.

LONDON:
Printed by *Philip Gwillim*, for *Jonah Bowyer* at the *Roſe* in *Ludgate-ſtreet*, 1712.

FIGURE 233. ■ Title page from Dr. John Freind's 1712 book in which he attempted to use Newtonian physics to explain physical and chemical properties of matter. Newton suspected that the forces holding matter together were electrical and magnetical.

Indeed, attempts 100 years earlier to apply the physics of the age—Newton's great work—to chemistry failed. Among the first to attempt these applications were mathematician John Keill (1671–1721) and physician John Freind (1675–1728).[1] Newton had expressed the force arising from gravitational attraction between two bodies with the formula:

$$F = \frac{Gm_1m_2}{d^2}$$

The distance (d) was calculated from the centers of mass (center of the earth, mass m_1; center of the apple, mass m_2), and the weakening of the force with the square of the distance ($1/d^2$) meant that if the distance doubled, the force was only one-quarter of the original.

Keill and Freind both recognized gravity as a weak force unless a planet was involved. In the very rare *Chymical Lectures* by Friend (Figures 233 and 234), he describes another similar attractive force, extremely strong at exceedingly minute distances and present on the surfaces of particles (points c and d) and having a higher-order relationship with distance ($1/d^{10}$, $1/d^{100}$, . . . , ?) and thus vanishing when the distance between c and d remains tiny.

Velocity they approach each other For the Attractive Force exerts it ſelf only in thoſe Particles which are very near one another ; as for inſtance, in d *and* c ; *The Force of ſuch as are remote is next to nothing. Therefore no greater Force is requir'd to move the Bodies* A *and* B, *than what would put into motion the Particles* d *and* c, *when diſengag'd from the reſt. But the Velocities of Bodies moving with the ſame Force are reciprocally, as the Bodies themſelves. Therefore the more the Body* A *exceeds the Particle* d *in Magnitude, the leſs is its Velocity ; and this Motion is ſo languid, that oftentimes 'tis overcome by the Circumambient Medium, and other Bodies. Hence it is that this Attractive Force does ſcarce exert it ſelf, unleſs in the ſmalleſt Particles, ſeparated from the reſt.*

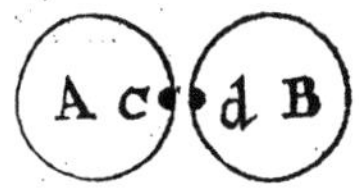

A c d B

FIGURE 234. ▪ A page from Freind's book depicting gravitational attraction between atoms.

Freind too recognized that a metal was lighter than its calx. He explained this observation by postulating the incorporation of igneous particles (particles of fire, see Boyle's "effluviums"), further separating the particles of metal; therefore weakening the forces between them. Thus, it is understandable that the melting point of silver metal is 962°C while its calx (Ag_2O) decomposes at only 230°C. Metal calxes, though not terribly water soluble, were more soluble than the metals themselves. However, lead melts at 327°C, while the white pigment litharge (PbO) melts at 886°C; mercury is a liquid, while HgO is a solid, albeit very slightly more water soluble than the metal. Adding to the confusion were calxes that were actually mixtures having components that readily decomposed.

During the twentieth century we have come to recognize that Newtonian physics explains the behavior of large, slow-moving objects like Nolan Ryan's fastball. The electrons that we know are responsible for holding atoms together need quantum mechanics to explain their behavior. They simply do not obey Newton's laws. Ironically, the forces that hold together salts, composed of ions such as Fe^{2+} and O^{2-} (ions were established by Arrhenius in the late nineteenth century), are almost entirely explained by the classical physics of Coulomb's law. However, negative electrons do not collapse into the positive nucleus.

---

1. J.R. Partington, *A History of Chemistry*, MacMillan, London, 1962, Vol. 2, pp. 478–482.

SECTION VI
A YOUNG DEMOCRACY AND A NEW CHEMISTRY

IF YOU *DO* FIND THE PHILOSOPHER'S STONE, "TAKE CARE TO LOSE IT AGAIN"—BENJAMIN FRANKLIN

Benjamin Franklin (1706–1790) was a brilliant and worldly polymath and a force behind both the American Declaration of Independence and the Constitution. The story of his arrival in Philadelphia at age 17 is well known—walking up Market Street on his first day, munching one "great Puffy Roll" with the other two under each arm and meeting his future wife Deborah Read.[1] According to Franklin, Miss Read ". . . thought I made as I certainly did a most awkward ridiculous Appearance."[1] Starting in the printing trade, he spent two years in England before setting up business in the American Colonies. Money-making ventures starting around 1730 included the printing of *Poor Richard's Almanacs* and the concessions for printing the currencies of Pennsylvania, New Jersey, Delaware, and Maryland. From this period through the 1740s Franklin accumulated wealth, became active in politics, and successfully promoted many ventures including the nascent University of Pennsylvania.[1] During the 1740s Franklin turned his attention increasingly to scientific pursuits.

Contemporary interest in electricity intrigued Franklin. He demonstrated that lightning and electricity were the same by flying a kite into an electrical storm and was lucky to have escaped electrocution. He believed electricity was a fluid that flowed from a body rich in it to a body poor in it. These considerations led to his invention of the lightning rod. The electrical terms *positive* and *negative*, *battery*, and *conductor* were coined by Franklin. His book *Experiments and Observations on Electricity* was first published in 1751 and went through four additional English editions as well as editions in French, German, and Italian (see Figure 235).

Franklin met Joseph Priestley in London around 1765 and encouraged him to write his book *The History and Present State of Electricity* (1767). One bit of correspondence between the two has Priestley writing to Franklin in 1777 that he "did not quite despair of the philosopher's stone"; Franklin's response was that if he (Priestley) found it, "to take care to lose it again."[2] Franklin spent parts of the late 1770s soliciting and receiving military support from the French. He became a popular figure and a virtual cult hero in France and spent the years immediately following the American Revolution as a diplomat and business agent in France. He was an intimate in Lavoisier's scientific and social circle and Madame Lavoisier painted his portrait—apparently one of his favorites.[3,4] Apparently, Madame Lavoisier painted the portrait, and a copy of it, following the portrait by Duplessis.[5] The painting (Figure 208) given to Franklin remains today (1999) in the possession of one of his descendents,[5,6] while Madame Lavoisier's personal copy appears to be unlocated.[5]

---

1. *Encyclopedia Brittanica*, Encyclopedia Brittanica, Chicago, 1986, Vol. 19, pp. 556–559.
2. J.R. Partington, *A History of Chemistry*, MacMillan, London, 1962, Vol. 3, pp. 241, 245–246.

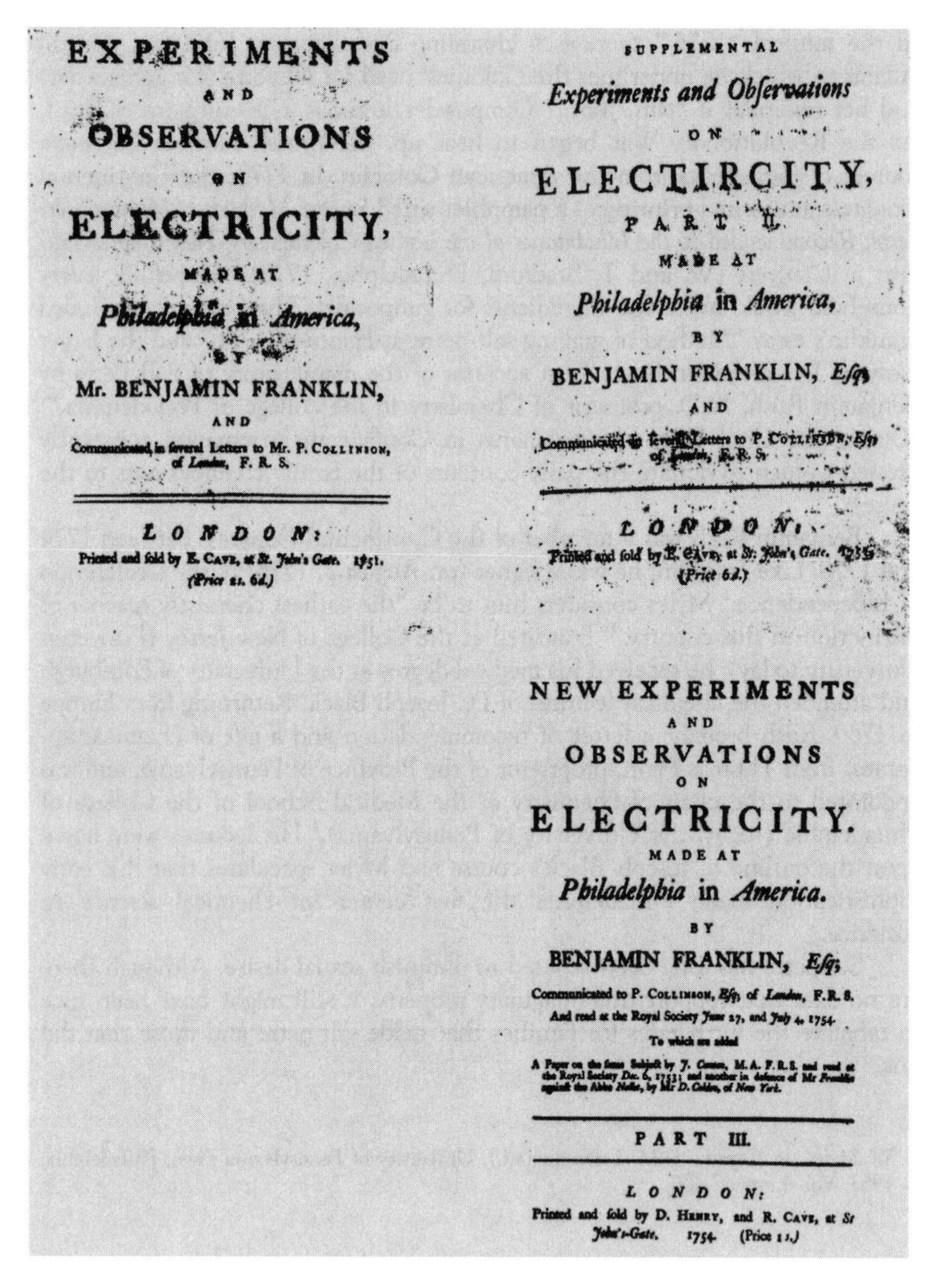

EXPERIMENTS
AND
OBSERVATIONS
ON
ELECTRICITY,
MADE AT
Philadelphia in *America*,
BY
Mr. BENJAMIN FRANKLIN,
AND
Communicated in several Letters to Mr. P. COLLINSON,
of *London*, F. R. S.

LONDON:
Printed and sold by E. CAVE, at St. *John's Gate*. 1751.
(Price 2s. 6d.)

SUPPLEMENTAL
Experiments and Observations
ON
ELECTRICITY,
PART II.
MADE AT
Philadelphia in *America*,
BY
BENJAMIN FRANKLIN, *Esq*;
AND
Communicated in several Letters to P. COLLINSON, *Esq*;
of *London*, F. R. S.

LONDON:
Printed and sold by E. CAVE, at St. *John's Gate*. 1753.
(Price 6d.)

NEW EXPERIMENTS
AND
OBSERVATIONS
ON
ELECTRICITY.
MADE AT
Philadelphia in *America*.
BY
BENJAMIN FRANKLIN, *Esq*;
Communicated to P. COLLINSON, *Esq*; of *London*, F.R.S.
And read at the Royal Society *June* 27, and *July* 4, 1754.
To which are added
A Paper on the same Subject by *J. Canton*, M.A. F.R.S. and read at the Royal Society *Dec.* 6, 1753; and another in defence of Mr *Franklin* against the Abbe *Nollet*, by Mr *D. Colden*, of *New York*.

PART III.

LONDON:
Printed and sold by D. HENRY, and R. CAVE, at *St John's-Gate*. 1754. (Price 1s.)

FIGURE 235. ■ Title pages of Benjamin Franklin's works on electricity. Franklin considered electricity to be a fluid and coined the terms "positive" and "negative" to denote electrical charges (courtesy of Jeremy Norman & Co., from Catalogue 5, 1978).

3. B.B. Fortune and D.J. Warner, *Franklin and His Friends—Portraying the Man of Science in Eighteenth-Century America*, Smithsonian Portrait Gallery and University of Pennsylvania Press, Washington, D.C. and Philadelphia, 1999.
4. D. McKie, *Lavoisier—Scientist, Economist, Social Reformer*, Collier, New York, 1962, p. 68.
5. C.C. Sellars, *Benjamin Franklin in Portraiture*, Yale University Press, New Haven, 1962, pp. 273–274.
6. Personal correspondence of the author with Franklin relative.

SALTPETRE, ABIGAIL. *PINS,* JOHN

In the musical *1776*, there is a charming duet between John and Abigail Adams in which he underlines the Colonies' need for saltpetre (for gunpowder) and her rejoinder is "*pins*, John." Gunpowder is about 75% saltpetre (KNO_3). As the Revolutionary War began to heat up, the British blocked European sources of ingredients from the American Colonies. In 1775, the Continental Congress authorized printing of a pamphlet titled *Several Methods of Making Saltpetre; Recommended to the Inhabitants of the United Colonies, by Their Representatives in Congress* (W. and T. Bradford, Philadelphia, 1775).[1] Hopefully, every household would make the ingredients for gunpowder. The pamphlet included Franklin's essay "Method of making salt-petre at Hanover, 1766" and the larger essay by Dr. Benjamin Rush: "An account of the manufactory of Salt-Petre by Benjamin Rush, M.D. professor of Chemistry in the college of Philadelphia."[1] (During the Civil War, advertisements in Confederate newspapers constantly pressed women to donate the daily contents of the family chamber pots to the cause).

Benjamin Rush was a member of the Continental Congress between 1774 and 1778. Like Franklin, he was a signer (on August 2, 1776) of the Declaration of Independence.[1] Myles considers him to be "the earliest chemistry teacher of distinction in this country."[1] Educated at the College of New Jersey (Princeton University today), he received his medical degree at the University of Edinburgh and attended the chemical lectures of Dr. Joseph Black. Returning from Europe in 1769. Rush brought a letter of recommendation and a gift of chemical apparatus from Thomas Penn, proprietor of the Province of Pennsylvania, and was appointed to the chair of chemistry at the Medical School of the College of Philadelphia (today, the University of Pennsylvania).[1] His lectures were based upon the outline of Joseph Black's course and Myles speculates that this early sophistication made Philadelphia the first center for chemical science in America.

Saltpetre has long been reputed to diminish sexual desire. Although there are no data that support this imaginary property, it still might have been nice to tabulate the birth rates for families that made salt-petre and those that did not.

1. W. Myles. in *Chymia*, H.M. Leicester (ed.), University of Pennsylvania Press, Philadelphia, 1953. Vol. 4, pp. 37–77.

"IT IS A PITY SO FEW CHEMISTS ARE DYERS, AND SO FEW DYERS CHEMISTS"

A daring entrepreneurial, yet practical, spirit imbued the citizens of the nascent United States of America, and it was exemplified by the precocious young physician John Penington. A student of Dr. Benjamin Rush at the University of Pennsylvania and a contemporary of Dr. Caspar Wistar, Penington completed his medical education at Edinburgh, where he wrote the following to his former teacher in 1790:[1]

> Alas, dear sir, I despair of meeting a Rush, or a Wistar, here, it is not the character of the professors at Edinburgh, to take the youthful inquirer by the hand & accompany him in the road of true knowledge.—Pride and reserve prevail among the professors, idleness & dissipation in the generality of the students: and for want of *proper company,* I have hitherto retreated to books and a solitary walk: in short I find nothing here likely to corrupt my patriotism.

In 1789, at the age of 20, Penington formed, in Philadelphia, the first chemical society in America[2–4] (possibly the world's first[3]) and authored the first American chemistry book—*Chemical and Economical Essays* (see Figure 236),[5,6] a work that had favorably impressed Thomas Jefferson (see p. 397). His chemical society lasted briefly and was succeeded by The Chemical Society of Philadelphia, which was founded in 1792 [Dr. James Woodhouse (M.D.)—first president].[2,3] In 1793, he was one of six co-signatories to an endorsement for the Hopkins process for making potash (KOH) and pearl ash (K_2CO_3), for which the first U.S. patent was granted.[1]

Chemical and Economical Essays is a spirited book, and its flavor may be sampled from his picturesque comparison between the "practical chemist" and the "mere theorist" that retains a bit of resonance even today:[7]

> Chemists themselves belong to two great and distinct classes, which, it is a pity are not connected; in the one class we may rank those who perform a great number of operations by heat and mixture, without ever knowing the secondary causes of the effects produced; these are called practical chemists, such as are dyers, who cannot account for, or conceive why, alum, for instance, should be of use in their art; or why galls and copperas should produce a black dye; such also are tanners, who cannot explain the action of oak bark upon the hides; such likewise are many apothecaries, who can make aqua fortis. &c. &c. but know nothing of the rationale of the process; the other class is the mere theorist, who is well acquainted with the "effects of heat and mixture" upon all bodies, and can account for them all, but never soils his fingers with a piece of charcoal, or has had occasion to break a crucible; such a chemist can inform us admirably how the changes of colour in dying are produced, but would be unable to produce them himself; he can account for the action of oak bark upon animal substances, without ever having smelt the odour of a tan-yard; he could explain the theory and process of making aqua fortis; and perhaps were he to attempt to make it, he would be two hours kindling a fire in his furnace, break his distillery apparatus, and suffocate himself with the fumes.

Figure 237, taken from Penington's book,[8] depicts an apparatus that is a rather striking hybrid of chemistry vessels (actually stoneware—30-gallon oil jars connected by leaden pipes) and bellows typical of a smith's shop. The purpose is the production of sulfuric acid. More than 10 years earlier, Lavoisier burned sulfur under oxygen in a closed vessel by using a powerful magnifying lens and collecting the resulting sulfuric acid. However, since small amounts of gas occupy huge volumes, very little sulfuric acid can be produced in practical vessels in this manner. Alternatively, saltpetre (KNO_3) can be used as a highly condensed source of oxygen, but it is very expensive. Penington's hybrid apparatus is a kind

CHEMICAL

AND

ECONOMICAL ESSAYS,

DESIGNED TO ILLUSTRATE

THE CONNECTION BETWEEN THE THEORY AND PRACTICE

OF CHEMISTRY, AND THE APPLICATION OF

THAT SCIENCE TO SOME OF THE ARTS

AND MANUFACTURES OF

THE

UNITED STATES OF AMERICA.

"IT IS A PITY SO FEW CHEMISTS ARE DYERS, AND SO FEW DYERS CHEMISTS."

BY JOHN PENINGTON.

PHILADELPHIA:

PRINTED BY JOSEPH JAMES.

M, DCC, XC.

FIGURE 236. ■ Title page of the first chemistry textbook, as opposed to a syllabus, pamphlet, translation, or reprinting of a foreign text, published in the United States. Its precocious author, John Penington of Philadelphia, published the book when he was 21 years old. The prior year he formed the first chemical society in the United States. Trained as a physician under Benjamin Rush at the University of Pennsylvania, Penington died at the age of 25, during the yellow-fever epidemic of 1793 as he struggled to save lives.

of flow reactor rather than batch reactor since it introduces fresh reagent oxygen continuously into the reaction. Penington notes the extreme exertions needed to push the bellows and suggests a modification in which a small iron still continuously passes steam into vessel *B* (he misstates it as C) in order to provide a continuous source of pressure to aid the bellows.[8]

Dr. Penington, who devised a method of heat-preserving milk (prior to pas-

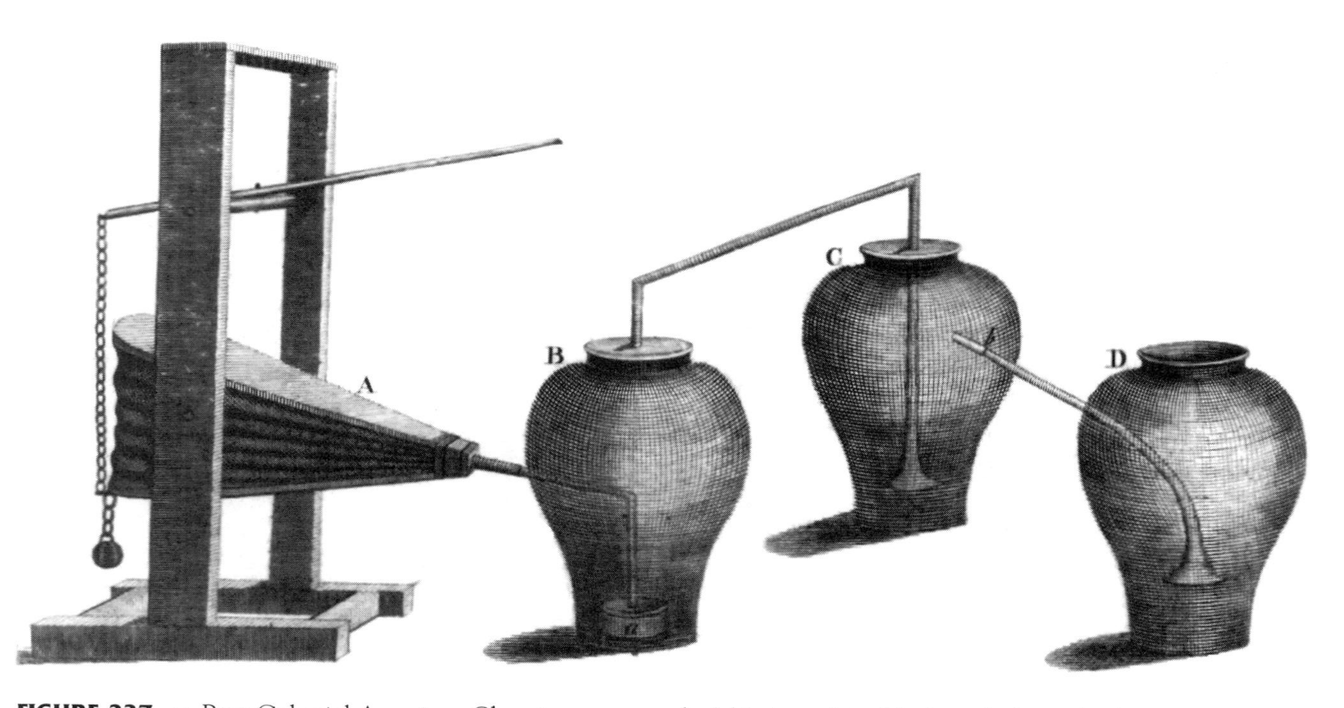

FIGURE 237. ■ Post-Colonial American Chemistry—a wonderful fusion of the blacksmith shop, the farm, and the laboratory. Early American (1790) apparatus for producing sulfuric acid—three 30-gallon crocks joined by lead pipe and connected to a bellows. (From Penington, *Chemical and Economical Essays*, see Figure 236.)

teurization), began his medical practice in Philadelphia in 1792. He was stricken in the yellow-fever epidemic of 1793 that claimed the lives of one-fifth of the city's population and "continued to attend patients until he also succumbed."[6] He was 25 years old.

---

1. W. Myles, in *Chymia,* No. 4 (H.M. Leicester, ed.), University of Pennsylvania Press, Philadelphia, 1953, pp. 76–77.
2. W. Myles, In *Chymia,* No. 3 (H.M. Leicester, ed.), University of Pennsylvania Press, Philadelphia, 1950, pp. 95–113.
3. E.F. Smith, *Chemistry in America,* D. Appleton & Co., New York and London, 1914, pp. 12–43.
4. The Chemical Society of Philadelphia disbanded between 1805 and 1810 and was succeeded in 1811 by the Columbian Chemical Society, also formed in Philadelphia (see notes 2 and 3 above).
5. J. Penington, *Chemical and Economical Essays,* Joseph James, Philadelphia, 1790.
6. W.D. Williams and W.D. Myles, *Bulletin of History of Chemistry,* Vol. 8, p. 18, 1990. Earlier American chemistry publications, by Rush and the blind lecturer Henry Moyes, were small specialized pamphlets, not books. I am grateful for correspondence concerning Penington and his book with Professor William D. Williams, Harding University, Searcy, Arkansas. Harding University houses the combined Americana chemical book collections of Dr. Williams and historian Dr. Wyndham D. Myles and has been officially designated an historic chemical site.
7. Penington, op. cit., pp. 4–5.
8. Penington, op. cit., pp. 146–150.

TWO EARLY VISIONS: OXIDATION WITHOUT OXYGEN AND WOMEN AS STRONG SCIENTISTS

The Chemical Society of Philadelphia began in 1792 and was succeeded by the Columbian Chemical Society in 1811.[1] There are few traces of the earlier society, but copies of the annual address delivered by Thomas P. Smith on April 11, 1798 (Figure 238)[2] are known.[3] Smith, only 21 or 22 years old, was a member of the Society's Nitre Committee.[1] The committee placed announcements in newspapers asking citizens to provide any information they had on niter, a component of gunpowder, by mail ("post paid", mind you) to Mr. Smith at No. 19 North Fifth Street or to the other four committee members, including Society President Dr. James Woodhouse (No. 13 Cherry Street).[1]

An amusing aspect of young Mr. Smith's oration was that he politely ignored the Society's expectation that "it shall contain all the discoveries made in the science of chemistry during the preceding year."[2] Instead, he gave a quite excellent brief history of chemical revolutions through the end of the eighteenth century. But we shall dwell briefly on two conjectures near the end of his presentation.

For the first one, Smith briefly summarizes Lavoisier's theory of combustion:

1. Combustion is never *known* to take place without the presence of oxigene.
2. In every *known* combustion there is an absorption of oxigene.
3. There is an augmentation of weight in the products of combustion equal to the weight of the oxigene absorbed.
4. In all combustion there is a disengagement of light and heat.

SKETCH

OF THE

REVOLUTIONS

IN

CHEMISTRY

By THOMAS P. SMITH.

PHIIADELPHIA:

PRINTED BY SAMUEL H. SMITH,
No. 118, Chefnut ftreet.

M,DCC,XCVIII.

FIGURE 238. ▪ Title page from Thomas P. Smith's lecture before The Chemical Society of Philadelphia on April 11, 1798. The precocious Mr. Smith dared to imagine a world of scientific contributions by female chemists and stretched the notion of oxidation to include (correctly, as it soon turned out) oxidation by gases other than oxygen.

He then poses a simple, but imaginative, question: "Should we conclude because those substances which burn the readiest in oxigene will not burn in any other gas, that no substances are to be found that will?"

Actually, another such gas was already known—but misunderstood. Scheele had isolated chlorine gas in 1773, by dissolving pyrolusite (MnO_2) with cold acid of salt (HCl or muriatic acid).[4] Scheele had learned that pyrolusite was "dephlogisticated" (a good oxidizing agent). He logically considered chlorine to be dephlogisticated acid of salt.[4] It was later found to support the glow of a taper better than air, explode with hydrogen when kindled by a glowing taper, and burn phosphorus, ammonia, bismuth, antimony, powdered zinc, and other active metals.[5] However, Lavoisier argued that all acids contain oxygen (oxygen means "acid former") and the French school called chlorine (Cl_2) *oxygenated muriatic acid*. Note the perfect mirror-image consistency between Scheele's and Lavoisier's nomenclature for chlorine: "oxygenated" = "dephlogisticated." Since sulfuric and similar acids released oxygen from pyrosulite, the idea that MnO_2 released its oxygen to muriatic acid was not far-fetched. Lavoisier's view remained dominant, if questioned, until Humphry Davy proved some 30 years later that chlorine contained no oxygen and was, therefore, a pure element—a view finally accepted by Berzelius in the 1820s.[6] So young Smith was correct—there is combustion without oxygen. And in another hundred years the "Tyrannosaurus Rex"[7] or "Tasmanian Devil" (see p. 493) of elements, fluorine gas, would be isolated and found to support vigorous, even explosive, spontaneous combustion. Indeed, its affinity for hydrogen is much stronger than oxygen's and it will liberate oxygen gas from water with much heat:

$$2\ H_2O\ (liq) + F_2\ (gas) \rightarrow 4\ HF\ (aq\ soln) + O_2\ (gas) + heat$$

Smith nears the end of his brief history of chemistry as follows:[2,8]

> I shall now present you with the last and most pleasing revolution that has occurred in chemistry. Hitherto we have beheld this science entirely in the hands of men; we are now about to behold women assert their just, though too long neglected claims, of being participators in the pleasures arising from a knowledge of chemistry. Already have Madame Dacier and Mrs. Macauly established their right to criticism and history. Mrs. Fulhame has now laid such bold claims to chemistry that we can no longer deny the sex the privilege of participating in this science also.[9] What may we not expect from such an accession of talents? How swiftly will the horizon of knowledge recede before our united labours! And what unbounded pleasure may we not anticipate in treading the paths of science with such companions?

(A vision of Marie and Pierre Curie 100 years into the future?)

Smith died, from an accidental gun mishap, in 1802 on an ocean voyage to Europe where he was to continue his chemical and mineralogical studies.[1,3] He was, like Dr. John Penington, only 25 when he perished.

1. W. Miles, in *Chymia*, Vol. 3, University of Pennsylvania Press, Philadelphia, 1950, pp. 95–113.
2. T.P. Smith, *Annual Oration Delivered before the Chemical Society of Philadelphia, April 11, 1798—A Sketch of the Revolutions in Chemistry*, Samuel H. Smith, Philadelphia, 1798.
3. E.F. Smith, *Chemistry in America*, D. Appleton and Co., New York and London, 1914, pp. 12–47 (reprints in full Thomas P. Smith's oration).
4. J.R. Partington, *A History of Chemistry*, Vol. 3, MacMillan & Co, Ltd., London, 1962, pp. 212–214.
5. Partington, op. cit., pp. 540–542, 572.
6. W.H. Brock, *The Norton History of Chemistry*, W.W. Norton & Co., New York and London, 1993, p. 154.
7. G. Rayner-Canham, *Descriptive Inorganic Chemistry*, Freeman, New York, 1996, pp. 349–352.
8. Smith is referring to Anne Dacier (1654–1720), a renowned European classicist, editor, and translator and Mary Ludwig Hays McCauley (1754–1832), who joined her husband Hays at the Battle of Monmouth (New Jersey) on June 28, 1778. Mrs. McCauley hauled pitchers of well water back and forth for the soldiers, and when Hays collapsed from the heat, she took his place at the cannon for the remainder of the battle (see *The New Encyclopedia Britannica*, Encyclopedia Britannica, Inc., Chicago, 1986, Vol. 3, pp. 841–842; Vol. 7, p. 611. "Molly Pitcher" has been honored by a battle monument and, profoundly, by the naming of the food court/gasoline station complex at Exit 7 of the New Jersey Turnpike.

9. Smith's footnote here is as follows: *Mrs. Fulhame has lately written an ingenious piece entitled "An Essay on Combustion, with a view to a new art of dyeing and painting, wherein the phlogistic and anti-phlogistic hypotheses are proved erroneous". Since the delivery of this oration she has been elected a corresponding member of this Society.* (see also K.J. Laidler, *The World of Physical Chemistry*, Oxford University Press, Oxford, 1993, pp. 250–252; 277–278. Laidler calls Mrs. Fulhame a "forgotten genius").

'TIS A BONNIE CHYMISTRIE WE BRRRING YE

Although Hessian crucibles and other sophisticated apparatus for practical metallurgy were employed in Jamestown, Virginia in the early seventeenth century, Philadelphia is the birthplace of modern chemistry in America. But Philadelphia's chemical roots extend back to the Scottish Enlightenment and the University of Edinburgh in particular. David Hume and Adam Smith spent a considerable part of their lives in Edinburgh, and historian Jan Golinski notes its "stimulating local environment" and quotes Tobias Smollett, who called it a "hotbed of genius."[1,2] The first Chair of Chemistry in America was awarded in 1769 to Dr. Benjamin Rush at the University of Pennsylvania (see Figure 239 for the title page to his 1770 syllabus of lectures).[3,4] Rush was a co-signer with Benjamin Franklin of the Declaration of Independence. He obtained his M.D. degree at the University of Edinburgh and studied chemistry with Dr. Joseph Black.[5] Black's most important discovery of course was the isolation and characterization of "fixed air" (carbon dioxide), published in 1756. However, his other great contribution was his influence on students who attended his lucid and up-to-date lectures. Black's lecture notes were published posthumously by his student John Robison in 1803.[6] Samuel Latham Mitchill, John Maclean, and Benjamin Silliman, Sr., the first professors of chemistry at Columbia (1792), Princeton (1795), and Yale (1802), all studied with Black at Edinburgh.[7,8] Prior to Rush's appointment, the first person to teach chemistry as part of the course at the medical college of the University of Pennsylvania was John Morgan, who, of course, learned his chemistry from Black at Edinburgh.[9]

It is fair to say that other colonies besides Pennsylvania showed an early interest in chemistry. Indeed, Jefferson of Virginia and Adams of Massachusetts weighed in with rather strongly held opinions (see p. 395). James Madison, the future author of the Constitution and fourth president of the United States, included chemistry in his natural philosophy lectures at the College of William and Mary. He even published a letter on his chemical experiments on the "Sweet Springs" in the *Transactions of the Philosophical Society*.[10]

Dr. Black studied chemistry at the University of Glasgow in the laboratory of William Cullen (1710–1790).[11] He moved to Edinburgh with Cullen and presented the dissertation for his M.D. degree in 1754.[5] Black succeeded Cullen as professor of chemistry at Glasgow in 1756 and at Edinburgh in 1766. Cullen had learned his chemistry by reading the works of Boerhaave in English translation. Cullen's original works included studies on bleaching and salt manufacture. His essay[12] "Of the Cold produced by evaporating Fluids, and of some other means of producing Cold" extended Boerhaave's studies of thermometry:

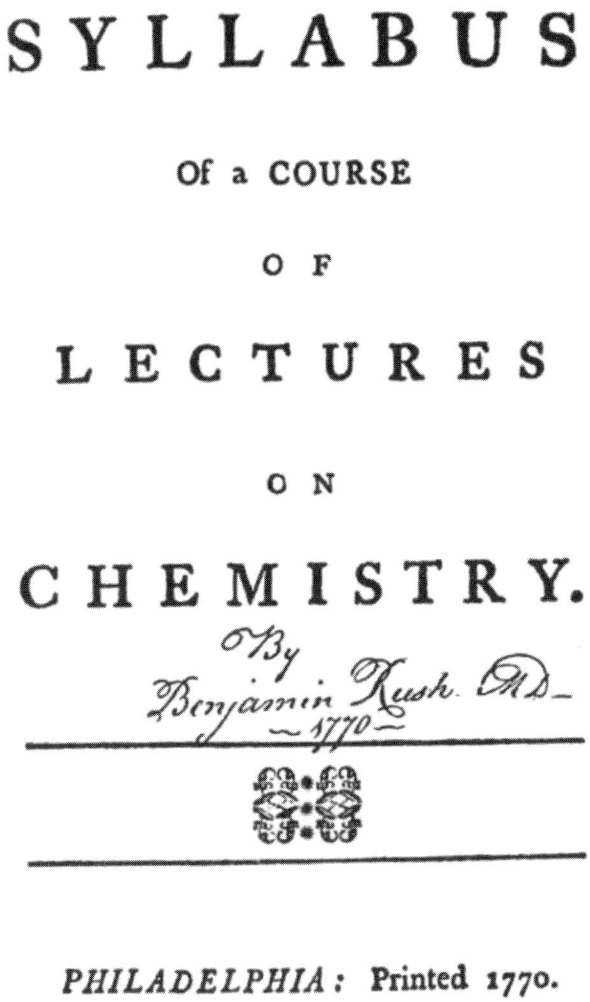

SYLLABUS

Of a COURSE

OF

LECTURES

ON

CHEMISTRY.

By
Benjamin Rush. M.D.
~1770~

PHILADELPHIA: Printed 1770.

FIGURE 239. ■ Dr. Benjamin Rush was educated by Dr. Joseph Black at the University of Edinburgh. In 1769 he was awarded the first Chair in Chemistry in America, at the University of Pennsylvania. Here is the title page for his course syllabus. (From Smith, *Chemistry in America*). Dr. Rush was a signer of the American Declaration of Independence and worked with co-signer Benjamin Franklin to perfect saltpetre.

> when a thermometer had been immersed in spirit of wine, tho' the spirit was exactly of the temperature of the surrounding air, or somewhat colder; yet upon taking the thermometer out of the spirit, and suspending it in the air, the mercury in the thermometer, which was of Fahrenheit's construction, always sunk two or three degrees.

Philadelphia was the home of the first chemical society, formed in 1789, by John Penington, who studied chemistry in Edinburgh.[9] Two years later, James Woodhouse, a student of Rush, and therefore “chemical grandson” to Black, founded the Chemical Society of Philadelphia.[9] In 1794, Rush tried to convince Joseph Priestley, persecuted for his political beliefs, to leave England and resettle in Philadelphia, home of Priestley's longtime friend Benjamin Franklin, since deceased.[13] Priestley eventually chose the more bucolic Northumberland. In summary, is it not appropriate that two of the most important resources for chemical historians in America, the Edgar Fahs Smith collection at the University of Pennsylvania and the Chemical Heritage Foundation, reside in the City of Brotherly Love?

---

1. J. Golinski, *Science as Public Culture—Chemistry and Enlightenment in Britain, 1760–1820*, Cambridge University Press, Cambridge, UK, pp. 13–25.
2. A. Herman, *How the Scots Invented the Modern World: The True Story of How Western Europe's Poorest Nation Created Our World and Everything in It*, Crown Publishers, New York, 2001.
3. E.F. Smith, *Old Chemistries*, McGraw-Hill, New York, 1927, pp. 11–14.
4. E.F. Smith, *Chemistry in America*, D. Appleton and Co., New York, 1914.
5. J.R. Partington, *A History of Chemistry*, MacMillan & Co. Ltd., London, 1962, Vol. 3, pp. 130–143.
6. J. Black, *Lectures on the Elements of Chemistry, delivered in the University of Edinburgh, by the late Joseph Black, M.D. . . . now published from his Manuscripts, by John Robison, LL.D.*, Edinburgh, Mundell & Sons for Longman & Rees, 1803. The American edition was published in 1807. My copy of this three-volume set was purchased at an auction of books de-accessed by the Franklin Institute in Philadelphia in 1986. The signature of its original owner, Adam Seybert, a student of Dr. Rush and thus the "chemical grandchild" of Dr. Black, dated 1807, is on the title page of Volume 1. In such a manner does a book collector enjoy direct links to history.
7. A.J. Ihde, *The Development of Modern Chemistry*, Harper & Row, New York, 1964, p. 268.
8. D.S. Tarbell and A.T. Tarbell, *Essays on the History of Organic Chemistry in the United States*, Folio Publishers, Nashville, 1986, pp. 17–23.
9. W. Myles, *Chymia*, Vol. 3, University of Pennsylvania Press, Philadelphia, 1950, pp. 95–113.
10. Smith (1914), op. cit., pp. 5–7.
11. Partington, op. cit., pp. 128–130.
12. J. Black, *Experiments on Magnesia Alba, Quick-Lime, and Other Alcaline Substances; by Joseph Black, M.D., to Which is Annexed, an Essay on the Cold produced by Evaporative Fluids, and of some other means of producing Cold, by William Cullen, M.D.*, William Creech, Edinburgh, 1782, pp. 117–118.
13. Smith (1914), op. cit., pp. 109–118.

♫♪ "FOR IT'S HOT AS HELL . . . IN PHILA-DEL'-PHI-A'" ♪♫

As already mentioned, the weather was almost unprecedentedly hot; and his laboratory was in sundry places perpetually glowing with blazing charcoal, and red-hot furnaces, crucibles and gun-barrels, and often bathed in every portion of it with the steam of boiling water. Rarely, during the day, was the temperature of its atmosphere lower than from 110° to 115° of Fahrenheit—at times, perhaps, even higher.

Almost daily did I visit the professor in that salamander's home, and uniformly found him in the same condition—stripped to his shirt and summer pantaloons, his collar unbuttoned, his sleeves rolled up above his elbows, the sweat streaming copiously down his face and person, and his whole vesture

> dripping wet with the same fluid. He, himself, moreover, being always engaged in either actually performing or closely watching and superintending his processes, was stationed for the most part in or near to one of the hottest spots in his laboratory.
>
> My salutation to him on entering his semi-Phlegethon[2] of heat not infrequently was: "Good God, doctor, how can you bear to remain so constantly in so hot a room! It is a perfect purgatory!" To this half interrogatory, half exclamation, the reply received was usually to the same purport. "Hot, sir—hot! Do you call this a hot room? Why, sir, it is one of the coolest rooms in Philadelphia. Exhalation, sir, is the most cooling process. And do you not see how the sweat exhales from my body, and carries off all the caloric? Do you not know, sir, that, by exhalation, ice can be produced under the sun of the hottest climates?"

So writes Benjamin Silliman, the elder during a period (1802–1803) spent attending chemistry lectures and visiting Dr. James Woodhouse, (M.D.), Professor of Chemistry of the University of Pennsylvania.[3] Silliman would become Professor of Chemistry at Yale University. The "salamander" reference evokes the mythical fire-resistant salamander of alchemical lore (see Figure 49). As noted previously, a decade earlier, Woodhouse had founded the Chemical Society of Philadelphia.[4] His lectures were enthusiastic if not riveting, and Silliman notes that "He appeared, when lecturing, as if not quite at his ease, as if a little fearful that he was not highly appreciated,—as indeed he was not very highly."[3] Still, his demonstrations were effective even on "humble" apparatus that had the virtue of affordability. Indeed, Figure 240 depicts "Dr. Woodhouse's Economical Apparatus"—a "Portable Laboratory; containing a Philosophical Apparatus, and a great number of Chemical Agents; by which any person may perform an endless variety of amusing and instructive experiments." The apparatus in Figure 240 appears in the Appendix to the first American edition (Philadelphia, 1802) of *The Chemical Pocket Book*[5] written by Dr. James Parkinson, (M.D.), who characterized the illness of the central nervous system that bears his name. After proclaiming the wide range of experiments available through his "economical apparatus," Dr. Woodhouse points out the inadequacies of the rival portable laboratory of Guyton de Morveau, for example:[6]

> It is less expensive. The lamp of Guyton, is one of the worst of the kind, for a Chemical Laboratory. There is no occasion for a number of screws, to elevate or depress the retort or lamp, for a great or low heat may be made, merely by raising or lowering the wick.

Advertising notwithstanding, Dr. Woodhouse was one of the most respected chemists of his day and was the first president of the Chemical Society of Philadelphia (1792). As an antiphlogistonist, he was critical of Dr. Joseph Priestley's phlogistic arguments, which were summarized in the latter's 1796 pamphlet, published in Philadelphia and reprinted in the London that he had fled a few short years earlier.[7] For example, Woodhouse noted that when he calcined metals in "pure air" (oxygen), the volume decreased but "I could not find that the air which remained behind was injured."[8] If one takes the phlogistic view, calcinations cause the metal to release its phlogiston to the air until the air is fully phlogisticated and no longer capable of supporting flame or life (the air is "mephitic"

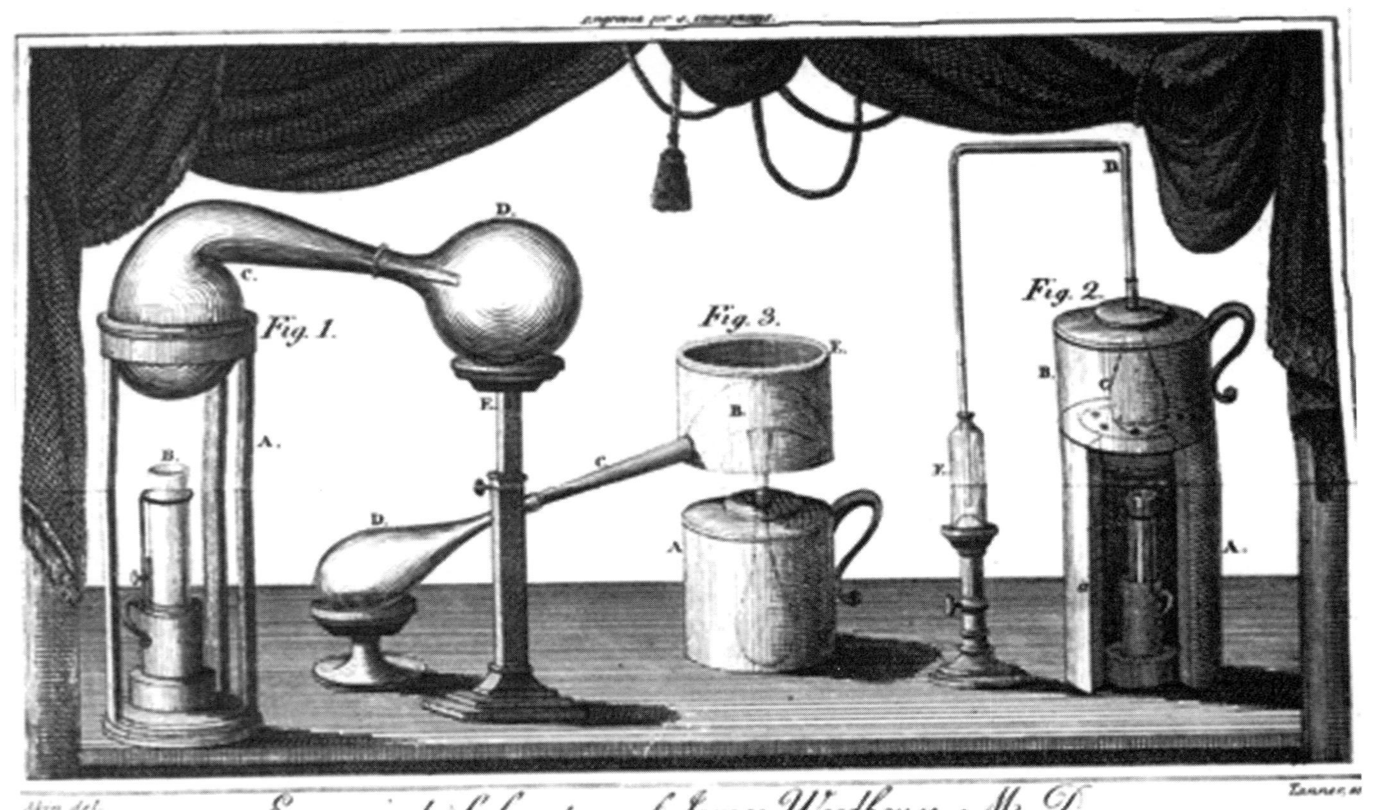

FIGURE 240. ■ Dr. James Woodhouse, Professor of Chemistry at the University of Pennsylvania, founded the Chemical Society of Philadelphia. He also had a profitable lecture series for which he sold his own chemical apparatus. He proclaimed it to be far superior and more economical than that of Guyton de Morveau in France. (From Parkinson, *The Chemical Pocket-Book*, 1802.)

or deadly). However, that is clearly not the case since, in an excess of oxygen, the metal takes up a specific quantity of this gas, leaving behind a smaller volume of equally pure oxygen.

However, Woodhouse was not doctrinaire, continued to harbor some uncertainties about the "new" chemistry" emanating from France, and was solicitous of the aging Priestley. In a letter answering an attack by Dr. John Maclean, professor at Princeton University, one paragraph reads[9]

> A judgement may be formed how well you have accomplished your purpose, and what right you have to condemn the experiments of Dr. Priestley in the authoritative manner you have done, having made none yourself, from the following particulars. You are not yet, Doctor, the conqueror of this veteran in philosophy.

In the 1802 *Chemical Pocket-Book*, Dr. Woodhouse[6] writes an Appendix titled: "An Account of the Principal Objections to the Antiphlogistic System of Chemistry." For example, he verifies experimentally a Priestley experiment in which red-hot "scales of iron" (iron oxide devoid of water and, thus, hydrogen) is mixed with red-hot charcoal (likewise devoid of water and, thus, hydrogen) and the result is an inflammable gas. (Hydrogen, i.e., phlogiston? Carbonated hydrogenous gas, i.e., methane?) So the gases produced must have removed some phlogiston (i.e., hydrogen) from the charcoal. (The confusion is that most of the gas produced is carbon monoxide, which *is* flammable.)

The rigors of his work undoubtedly contributed to the untimely death of Dr. Woodhouse, who Silliman noted never made "use of any of the facts revealed by chemistry, to illustrate the character of the Creator as revealed in his works" and Dr. Benjamin Rush, his former teacher, simply called "an open and rude infidel,"[3] in 1809 at the age of 39.

1. Homage to the Broadway production of *1776*.
2. In mythology, one of the five rivers of Hades.
3. E.F. Smith, *Chemistry in America*, D. Appleton and Co., New York and London, 1914, pp. 103–106.
4. Smith, op. cit., p. 12.
5. J. Parkinson, *Chemical Pocket-Book*, James Humphreys, Philadelphia, 1802.
6. Parkinson, op. cit., pp. 201–215.
7. J. Priestley, *Experiments and Observations Relating to the Analysis of Atmospherical Air*, Philadelphia, 1796.
8. Smith, op. cit., p. 83.
9. Smith, op. cit., p. 92.

ADAMS OPPOSES ATOMS

John Adams and Thomas Jefferson, the second and third Presidents of the United States, came from the two original "power colonies" Massachusetts and Virginia, respectively. They were allies and fundamental forces in the American

Revolution, became bitterly estranged later on, but attained a reconciliation in old age.[1] Amazingly, the two men died on July 4, 1826, precisely 50 years after the signing of the Declaration of Independence. Unaware of his friend's fate, Adams' last words were: "Thomas Jefferson survives."

Both of these great leaders were invested in the intellectual life of their young nation. Figure 241 is the dedication page from the book *Plain Discourses on*

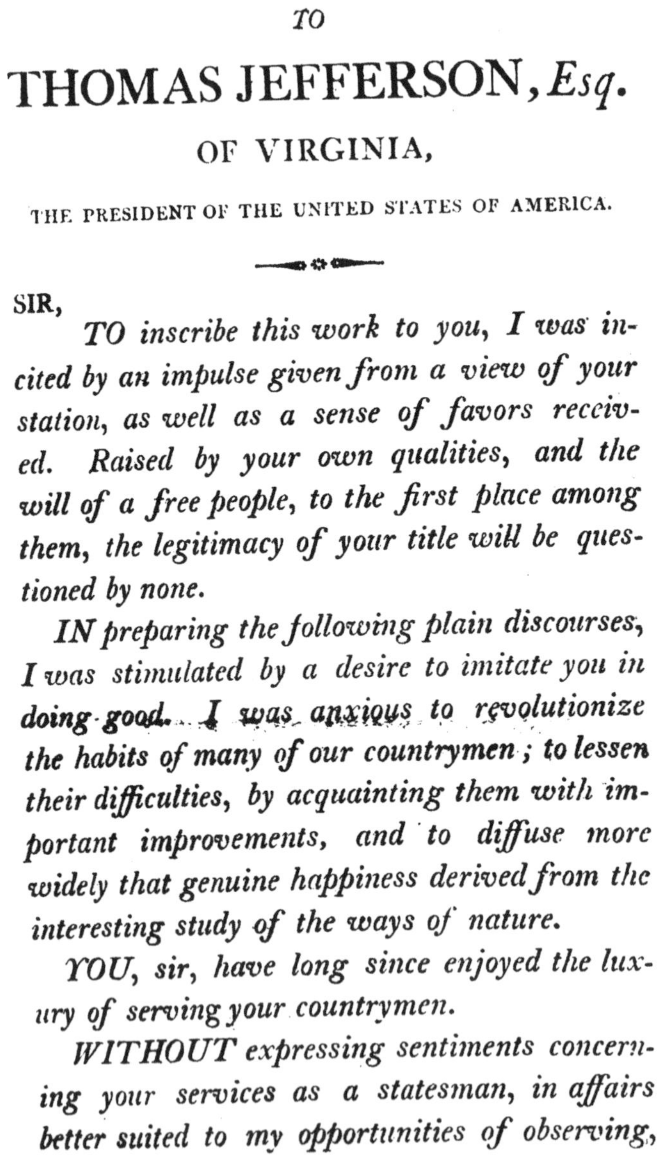

TO

THOMAS JEFFERSON, *Esq.*

OF VIRGINIA,

THE PRESIDENT OF THE UNITED STATES OF AMERICA.

SIR,

TO inscribe this work to you, I was incited by an impulse given from a view of your station, as well as a sense of favors received. Raised by your own qualities, and the will of a free people, to the first place among them, the legitimacy of your title will be questioned by none.

IN preparing the following plain discourses, I was stimulated by a desire to imitate you in doing good. I was anxious to revolutionize the habits of many of our countrymen; to lessen their difficulties, by acquainting them with important improvements, and to diffuse more widely that genuine happiness derived from the interesting study of the ways of nature.

YOU, sir, have long since enjoyed the luxury of serving your countrymen.

WITHOUT expressing sentiments concerning your services as a statesman, in affairs better suited to my opportunities of observing,

FIGURE 241. ■ The dedication page from Thomas Ewell's *Plain Discourses on the Laws or Properties of Matter* (New York, 1806). President Jefferson had complained that ". . . chemists have filled volumes on the composition of a thousand substances of no sort of importance to the purposes of life. . . ." He asked for a *useful* book and Dr. Ewell delivered.

the Laws and properties of Matter: containing the elements or principles of Modern Chemistry, &c, published in 1806 by Thomas Ewell, M.D. of Virginia, one of the surgeons of the United States Navy. In 1805, he had received from President Jefferson, the following letter:[2]

> Of the importance of turning a knowledge of chemistry to household purposes, I have been long satisfied. The common herd of philosophers seem to write only for one another. The chemists have filled volumes on the composition of a thousand substances of no sort of importance to the purposes of life; while the arts of making bread, butter, cheese, vinegar, soap, beer, cider, &c remain unexplained. Chaptal has lately given the chemistry of wine making; the late Dr. Penington did the same as to bread, and promised to pursue the line of rendering his knowledge useful to common life; but death deprived us of his labors. Good treatises on these subjects should receive general approbation.

When John Gorham assumed the Erving Chair of Chemistry at Harvard in 1817, he received a wonderful congratulatory letter[2] from John Adams. The retired President expressed the view that matter is "a mere metaphysical abstraction" and that he "could not comprehend" atoms and he "could not help laughing" at molecules. Near the end of his delightful letter he exhorts:

> Chymists! Pursue your experiments with indefatigable ardour and perserverance. Give us the best possible Bread, Butter, and Cheese, Wine, Beer and Cider, Houses, Ships and Steamboats, Gardens, Orchards, Fields, not to mention Clothiers or Cooks. If your investigations lead accidentally to any deep discovery, rejoice and cry "Eureka!" But never institute any experiment with a view or a hope of discovering the first and smallest particles of Matter.

1. J.J. Ellis, *American Sphinx*, Knopf, New York, 1997, pp. 12, 290, 292.
2. E.F. Smith, *Old Chemistries*, McGraw-Hill, New York, 1927, pp. 50–52; 60–64.

TWELVE CENTS FOR A CHEMISTRY LECTURE

Amos Eaton, A.M. was a busy man. In his 1826 book *Chemical Instructor*[1] he describes himself as "Attorney and Counsellor at Law; Professor of Chemistry and Natural Philosophy in Rensselaer School and in Vermont Academy of Medicine, &c. &c." As an entrepreneurial chemistry lecturer he followed a trail blazed by the likes of Henry Moyes, M.D., whose 1784 twenty-one-part lecture series "Heads of a Course of Lectures on the Philosophy of Chemistry . . ." was advertised for one guinea (or one shilling per lecture)[2] in the *Massachusetts Sentinel*. Here is Mr. Eaton's syllabus (I think we know whose text he used):[3]

COURSE OF LECTURES
TO BE DIVIDED INTO THIRTY-THREE

Lecture	To page	Lecture	To page
1st Lecture to page	22	18th Lecture to page	125
2nd	26	19th	130
3rd	32	20th	137
4th	38	21st	144
5th	42	22nd	152
6th	48	23rd	159
7th	54	24th	16S
8th	60	25th	17S
9th	66	26th	180
10th	70	27th	190
11th	76	28th	199
12th	82	29th	211
13th	88	30th	224
14th	94	31st	239
15th	102	32nd	246
16th	112	33rd	249
17th	119		

For this course charge $4. If it is condensed to 22 lectures, charge $3.

Our Mr. Eaton would never burden his students with esoteric theories about useless compounds:

> It is much to be regretted, that most of the celebrated treatises on chemistry, have so large a proportion of their pages devoted to useless compounds, which can never profit the scholar nor the practical man. Particularly those endless compounds with chlorine and iodine, which may be equally multiplied and extended with any other substance. This is surely trifling with the richest stores of human knowledge. Put such works into the hands of a student, and tell him to place full confidence in the authors, he would form strange views of the science. He would imagine, that the chloridic and iodic theory of Davy constituted the whole science of chemistry; and that all further knowledge of the subject should be pursued as a convenient , though not very important, appendage to chlorine and iodine. And even admitting all Davy's speculations to be well supported, are not those idle speculations as unimportant as any of the smallest mites of human knowledge? I would as soon set a student to commit to memory all the amulets of the dark ages, or the number of ways in which the letters of the alphabet can be arranged.

Well, who said that pandering to tuition-paying students is only a modern phenomenon? Furthermore, we have always been a pragmatic nation—listen to President Thomas Jefferson in 1805:[4]

> The chemists have filled volumes on the composition of a thousand substances of no sort of importance to the purposes of life; while the arts of making bread, butter, cheese, vinegar, soap, beer, cider &c remain unexplained.

The "chloridic and iodic theory of Davy" refers to his careful studies some 10 years earlier than the appearance of Eaton's book. These established that chlorine and iodine were pure elements, that hydrochloric (HCl) and hydriodic (HI) acids did not contain oxygen, and that Lavoisier's theory that all acids contain oxygen was incorrect—conclusions perhaps mightier than "the smallest mites of human knowledge." But Eaton was situated on the other side of "the big pond" from Davy and presumably could have defended himself ably if the ailing Davy had pursued a defamation suit. In fact, Davy finished his last days fishing and published *Salmonia: Or Days of Fly Fishing* in 1828, a year before he died, appropriately unaware of the criticisms emanating from the wilds of upstate New York and Vermont.

Actually, Amos Eaton's career took some fascinating twists and turns.[5] Born in Chatham, New York in 1776, Eaton studied law at Columbia College, was first attracted to chemistry by Samuel Latham Mitchill, and passed the bar exam in 1802. He practiced law, went into business, was convicted of forgery on evidence framed by some enemies, was imprisoned in 1811, and was pardoned by Governor Tompkins in 1815 with the pledge of never returning to New York State. He moved to New Haven, learned chemistry from Benjamin Silliman at Yale, became an itinerant chemistry lecturer throughout towns and villages in New England, and even occasionally dared to cross into New York State. His fame spread; he was invited by New York's new Governor, DeWitt Clinton, to give lectures to legislators; and he met the wealthy Stephen van Rensselaer, who founded the Rensselaer School largely to provide a home for the talented lecturer in chemistry and geology. Rensselaer Polytechnic Institute (RPI) is now a widely respected research university with an African-American female physicist, Dr. Shirley Jackson, at its helm as this book is going to press. RPI has just received a $300 million unrestricted donation. Not bad for a tiny upstate school with a convicted forger as its first chemistry professor.

The chemist-entrepreneur thrived during the early nineteenth century. In *Practical Facts in Chemistry*, Robert Best Ede has written a small instructional book bound as a virtual 193-page preface to his 48-page catalog *Robert Best Ede's Series of Chemical Laboratories and Chests . . .* , published in London in 1837 (see Figure 242).[6] The catalog is complete with four pages of testimonials including magazines, newspapers, and famous chemists such as Thomas Graham, Professor of Chemistry at Glasgow ("I have had occasion to look over the contents of Mr. Ede's Portable Laboratory and have formed a high opinion of it") and Thomas Clark, M.D. at Aberdeen ("Mr. Ede's Portable Laboratory, I consider a *cheap* and very *useful selection* for Students in experimental Chemistry")—high praise, indeed, from a Scotsman.

Note the simple functional apparatus in Figure 243. The simple globe retort and tube receiver (item *14*) is an apparatus made from two pieces of glass tubing. The retort is made by slightly bending a tube, sealing it at one end, and blowing a bubble in the sealed end while the tube is heated.[7] The tube receiver is slightly bent and nearly sealed, leaving a pinprick opening for the escape of gases and vapors. The two vessels are connected airtight by paste of linseed meal or a tube of sheet caoutchouc (natural rubber). The use of this apparatus in depicted in Figure 244.[7] The bend in the receiver is immersed in ice water. Lead nitrate [$Pb(NO_3)_2$] is placed in the retort, the apparatus joined, and the retort heated with an alcohol lamp to form mixed oxides, and a solution of nitrous acid is collected in the bend of the tube receiver. When a very small quantity of this liquid

Robert Best Ede,

In submitting this Catalogue to the Public in general, very respectfully begs to observe, that it is his intention from time to time, to make such an enlargement in his selection both of scientific and domestic articles, as the progressive improvement of the age, and other circumstances may point out, to be best calculated to meet the still advancing taste of the public, for the higher branches of knowledge and science, and secure that popular approbation which it has been, and ever will be his first object to attain.

Feb. 1837.

FOR CHEMICAL STUDENTS, AMATEURS & PROFESSORS.

R. B. EDE'S
SERIES OF
PORTABLE CHEMICAL
Laboratories, Cabinets
AND
BLOWPIPE APPARATUS,
CONTAINING
A CHOICE SELECTION OF THE MOST USEFUL
Tests, Re-Agents & Preparations,
AND AN ORGANIZED COLLECTION OF THE BEST CONTRIVED
MODERN APPARATUS,
Adapted for performing with facility, safety and success,
A Course of Instructive and Entertaining Experiments
AND
For the exhibition of those interesting phenomena in Chemistry, Mineralogy, &c.
WHICH RENDER THE STUDY OF THESE SCIENCES SO FASCINATING;
ALSO,
PRACTICAL FACTS IN CHEMISTRY,
an appropriate
COMPANION TO THE PORTABLE LABORATORIES AND CABINETS;
AND
WARD'S FOOTSTEPS TO CHEMISTRY,
EQUALLY SUITABLE AS A GUIDE TO THE YOUTH'S LABORATORY.

FIGURE 242. ■ Robert Best Ede prefaced his 48-page catalog with a 193-page textbook on chemistry. But selling is the main object—here is the start of his catalog. (From Ede, *Practical Facts in Chemistry*, London, 1839, 1837.)

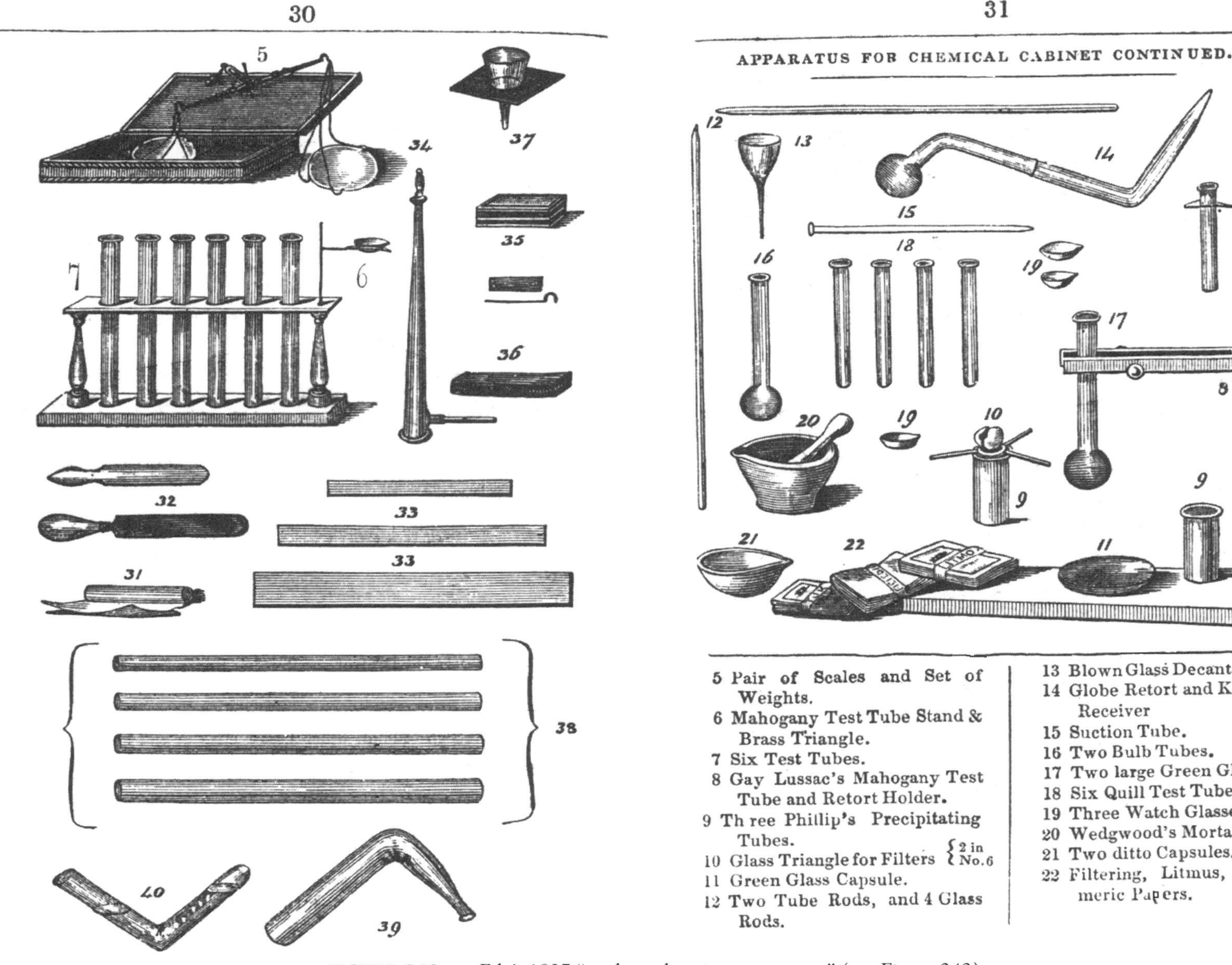

FIGURE 243. ■ Ede's 1837 "student chemistry apparatus" (see Figure 242).

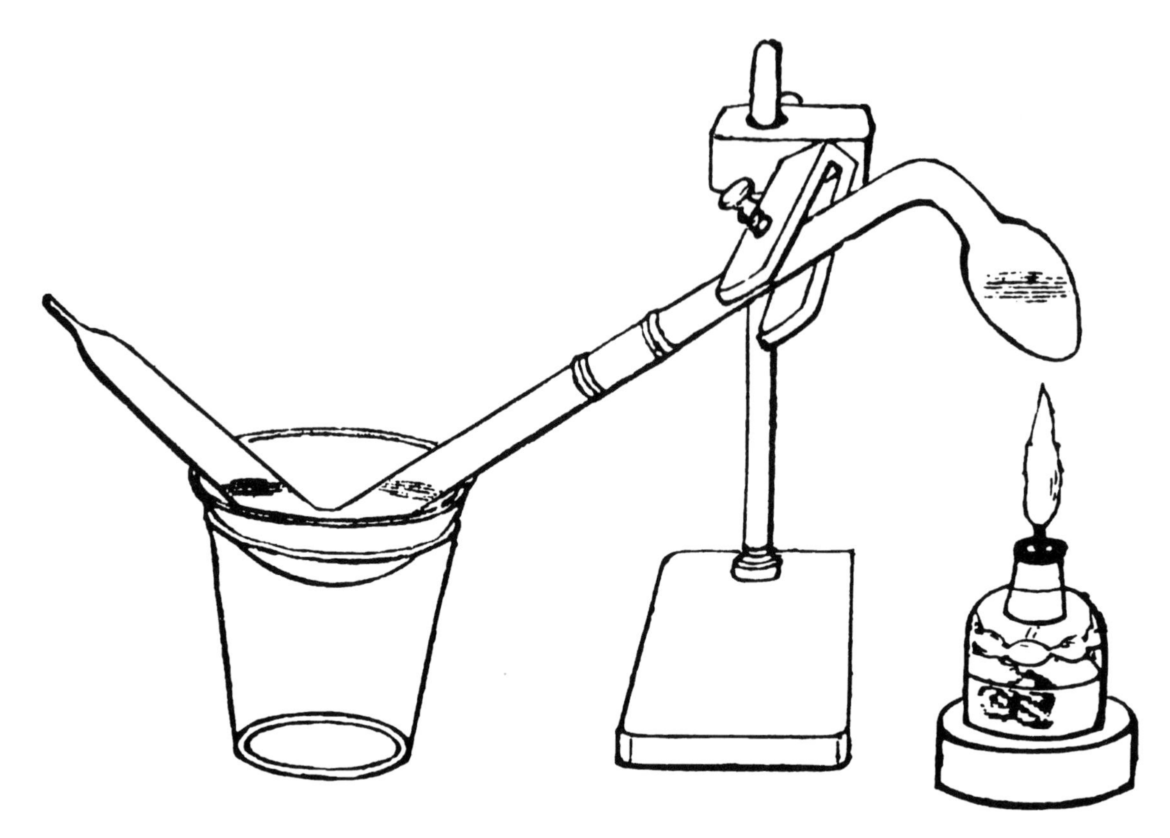

FIGURE 244. ■ Ede's very simple and elegant student experiment for obtaining and trapping at 0°C, the deep blue, highly unstable dinitrogen trioxide (N_2O_3), which decomposes at only 3°C. This would still be a nice experiment today, almost 170 years later. (See Figure 243.)

is added to a dry test tube and a single drop of distilled water added, a deep blue color is observed, probably due to the presence of dinitrogen trioxide (N_2O_3)—a rather uncommon and unstable substance that decomposes at only 3.5°C.[8] So here we see some pretty elegant chemistry performed in a cheap apparatus assembled using two glass tubes and linseed meal paste.

---

1. A. Eaton, *Chemical Instructor: Presenting a Familiar Method of Teaching the Chemical Principles and Operations of the Most Practical Utility to Farmers, Mechanics, Housekeepers and Physicians; and Most Interesting to Clergyman and Lawyers*, Websters and Skinners, Albany, 1826.
2. H. Moyes, *Heads of a Course of Lectures on the Philosophy of Chemistry*, Boston, 1784.
3. Eaton, op. cit., pp. 3–11.
4. E.F. Smith, *Old Chemistries*, McGraw-Hill, New York, 1927, pp. 50–52, 60–64.
5. H.S. Van Klooster, *Chymia*, Vol. 2, University of Pennsylvania Press, Philadelphia, 1949, pp. 1–15.
6. R.B. Ede, *Practical Facts In Chemistry, Exemplifying the Rudiments and Showing with What Facility the Principles of the Science May Be Experimentally Demonstrated at a Trifling Expense by Means of Simple Apparatus & Portable Laboratories, More Particularly in Reference to Those by Robert Best Ede*, Thomas Tegg, and Simkin, Marshall, and Co., London, 1839. Issued and bound with *Robert Best Ede's Series of Chemical Laboratories and Chests, with Appropriate Companions, Also, Mineralogical Boxes, Labels and Other Select and Approved Articles*, dated February 1837.
7. Ede, op. cit., pp. 144–159.
8. F.A. Cotton and G. Wilkinson, *Advanced Inorganic Chemistry*, fifth edition, John Wiley & Sons, New York, 1988, pp. 320–328.

SECTION VII
CHEMISTRY BEGINS TO SPECIALIZE, SYSTEMIZE, AND HELP THE FARM AND THE FACTORY

THE ELECTRIC SCALPEL

Count Rumford, whose efforts led to the chartering of the Royal Institution of Great Britain in 1799, took note of the accomplishments and verve of the 23-year-old Humphry Davy, and had him appointed lecturer in chemistry in 1801.[1,2] The fact that Davy had been critical of Lavoisier's caloric theory probably did not hurt his case.

The handsome, poetic Davy was an immediate hit at the Royal Institution, attracting women as well as men to his lectures. He also worked on practical problems including the chemistry of tanning and agriculture (*Elements of Agricultural Chemistry*, London, 1813). At the time, the scientific world and popular interest had been galvanized by Alessandro Volta's "artificial torpedo (*electric fish*)." It consisted of a pile of alternating circular disks of silver and zinc, each pair separated by a layer of cardboard soaked in brine. Volta (1745–1827) discovered methane in 1776 in Lake Como by stirring up the mud and collecting the bubbles in an inverted bottle filled with water. He described the voltaic pile for the first time in a letter to Sir Joseph Banks, President of the Royal Society, dated March 20, 1800.[3]

During the latter part of the seventeenth century, a variety of chemical experiments had been performed using electricity.[3] Only about a month after Volta's disclosure, Anthony Carlisle and William Nicholson constructed a voltaic pile of 36 pairs of silver and zinc plates (half-crown coins were sometimes used as the silver plates).[2,3] They attached a brass (copper–zinc alloy) wire to a silver plate (Volta's negative pole) and a brass wire to the zinc plate at the other end (positive pole) and dipped the wires into a test tube containing water. Bubbles of hydrogen were produced at the negative pole while the wire attached to the positive pole corroded. When both wires were made of platinum, hydrogen gas was formed at the negative pole and oxygen gas at the positive pole.[3] Davy started to apply himself to electrochemical studies including the electrolysis of water. Figure 245(a) is taken from Davy's *Elements of Chemical Philosophy* (London, 1812; Philadelphia, 1812). It depicts a voltaic pile consisting of 24 pairs of silver and zinc plates, each pair separated by cloth soaked in liquid. By 1806, Davy stated in public that the forces holding compounds together were electrical in nature.[1,2]

In 1807, Davy applied himself to a problem that had vexed Lavoisier, who was convinced that potash (KOH) was a compound even though it resisted "simplification."[1,2] He employed a huge, more powerful, voltaic pile (his "battery of the power of 250 of 6 and 4"—seemingly a pile of 150 pairs of 4-inch square plates connected to a pile of 100 pairs of 6-inch square plates).[4] His attempts to decompose aqueous solutions of potash merely electrolyzed water. However, when a piece of solid potash was placed on a disk of platinum (connected to the

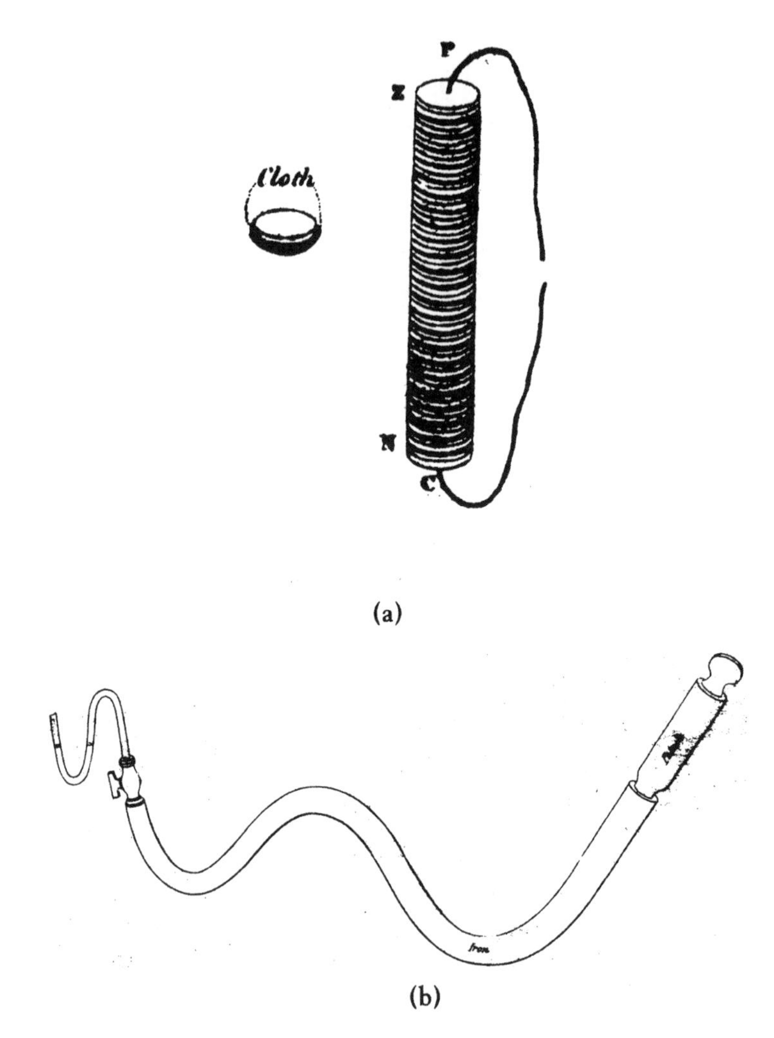

FIGURE 245. ■ From Humphry Davy, *Elements of Chemical Philosophy* (Philadelphia & New York, 1812; first English, London, 1812): (a) Davy's voltaic pile consisting of alternating zinc and silver plates separated by moistened cloth; (b) another gun-barrel experiment. In the white-hot gun barrel (free of air), potash yields metallic potassium; potassium was first generated using the voltaic pile. When he discovered potassium Davy ". . . actually bounded about the room in ecstatic joy."

negative pole) and a platinum wire (connected to the positive pole) was touched to the top of the potash, the solid fused at both points of contact. A violent effervescence at the upper surface (positive pole) was due to oxygen gas. At the lower part (platinum plate), beads of a silvery mercurylike liquid appeared, some of which exploded and burned with a bright flame. According to his cousin Edmund Davy, then working as an assistant (report by Humphry's brother John):[4]

> [When Humphry Davy] saw the globules of potassium burst through the crust of potash, and take fire as they entered the atmosphere, he could not contain his joy—he actually bounded about the room in ecstatic delight; and some little time was required for him to compose himself sufficiently to continue the experiment.

In a few days, Davy successfully isolated sodium. His voltaic pile also gave him the alkaline earths barium, strontium, calcium, and magnesium. (Lavoisier correctly identified their oxides as compounds but could not isolate the metals.) During this period Davy proved that chlorine gas (first isolated by Scheele in 1774) did not contain oxygen. Thus, hydrochloric acid did not contain oxygen, disproving Lavoisier's hypothesis that all acids contained oxygen.

In the year following Davy's electrochemical isolation of potassium, Gay-Lussac and Thenard obtained it chemically. Figure 245(b) (from Davy's *Chemical Philosophy*) shows an experiment performed (once again) in a gun barrel. Iron in an air-free environment is made white hot and the potash in the upper right tube is melted to produce potassium when the melt contacts the iron.

---

1. J.R. Partington, *A History of Chemistry*, MacMillan, New York, 1964, Vol. 4, pp. 29–75.
2. W.H. Brock, *The Norton History of Chemistry*, Norton, New York, 1993, pp. 147–153.
3. J.R. Partington, op. cit., pp. 4–5; 12–19.
4. J.R. Partington, op. cit., p. 46.

CHEMICAL SCALPELS THROUGH THE AGES

Until the middle of the twentieth century, Humphry Davy held the record for discovering the most chemical elements: six.[1] He succeeded because he was "the first kid on the block" to apply a new type of scalpel systematically (the voltaic pile or battery) to chemical problems. He very modestly attributed his discoveries to the instruments rather than to his own brilliance:[2]

> The active intellectual powers of man in different times are not so much the cause of the different successes of their labours, as the peculiar nature of the means and artificial resources in their possessions.

Fire is clearly the most ancient chemical scalpel. Indeed, Vulcan's release of Athena from the head of Zeus prior to her chemical marriage (Figure 83) can be taken as a metaphor for the role of fire in causing chemical change. Prior to 1600, the application of fire ultimately added four new elements (antimony, arsenic, bismuth, and zinc) to the nine elements known to the ancients (carbon, sulfur, and the seven metals: iron, tin, lead, copper, mercury, silver, and gold—one for each day of the week).[1] Flames were themselves dissected by blowpipes and the reducing and oxidizing parts of the flame used as scalpels in Sweden starting in the eighteenth century. Fire powered the stills that produced the new scalpels sulfuric acid (by distillation of green vitriol, $FeSO_4 \cdot 7H_2O$), nitric acid (distill the product produced by adding oil of vitriol to saltpetre), and *aqua regia* (nitric and hydrochloric acids). Oxygen and chlorine (isolated by Scheele) and fluorine (isolated over 100 years later by Moissan) were also potent scalpels.

Radiation, including α-particles and neutrons, eventually led to real transmutation. It is no coincidence that Glenn Seaborg and his associates at Chicago and Berkeley hold the record for discovery of elements since they used these particles to make brand new ones. Expanding the Periodic Table, like expanding the baseball season from 154 to 162 games, almost doesn't seem fair to Davy. Perhaps Seaborg's name should have an asterisk in the record books like the one for Roger Marts when he broke Babe Ruth's home-run record during the first extended season.[3] In the most recent two decades, the laser and the atomic force microscope have been successful in promoting reactions one atom at a time—seemingly the ultimate in chemical dissection.

---

1. A.J. Ihde, *The Development of Modern Chemistry*, Harper and Row, New York, 1964, pp. 747–749.
2. W.H. Brock, *The Norton History of Chemistry*, Norton, New York, 1993, pp. 187–188.
3. Absolutely no disrespect is meant here. My early adolescent interest in nuclear physics and chemistry made Seaborg the first living chemist whose name I knew. I never met him but I recall the thrill of being at an American Chemical Society lecture when Seaborg quietly and gracefully entered the room.

DAVY RESCUES THE INDUSTRIAL REVOLUTION

"Two great events amazed Britain in 1815: the victory of Wellington over Napolean and the victory of Davy over mine gases."[1] The Industrial Revolution was in danger of stalling in the early nineteenth century due to the dangers in mining with contemporary lamps that used flame and ignited explosions. A disaster near Newcastle in 1812 killed 101 miners, and more than two-thirds of the coal mines in England were considered too dangerous to work because of their levels of coal gas (primarily methane).[1]

In 1815, Humphry Davy was invited by the Chairman of a "Society for Preventing Accidents in Coal Mines" to invent a solution.[2] His elegant and simple invention is shown (Figure 246) in the frontispiece of his 1818 book *On The Safety Lamp For Coal Miners; With Some Researches On Flame*. Davy had earlier studied flames and their propagation and noted that flames could not propagate through small holes. Thus, his solution was merely to surround the lamp with a cylinder of wire mesh that still left the flame open to the atmosphere. The mesh conducted away the heat of the flame, thus cooling it so that the temperature methane would encounter at the lamp would be lower than its flash point. The flame itself could not penetrate the mesh.[2]

While on the topic of coal gas, we note that chemist Friedrich Accum (1769–1838) played a key role in support of the introduction of coal–gas lighting in England. It is hard to imagine the change in London nightlife upon its widespread use. "Full moon at night, lovers' delight," but what about the other 27 days? In a London fog on a moonless night, two lovers might hear each other, touch each other, but not see each other. Coal gas, obtained by destructive distillation of coal,[3] consists largely of hydrogen and methane, with smaller amounts of carbon monoxide, ethylene, and some acetylene as well carbon dioxide, hydrogen sulfide, and ammonia.

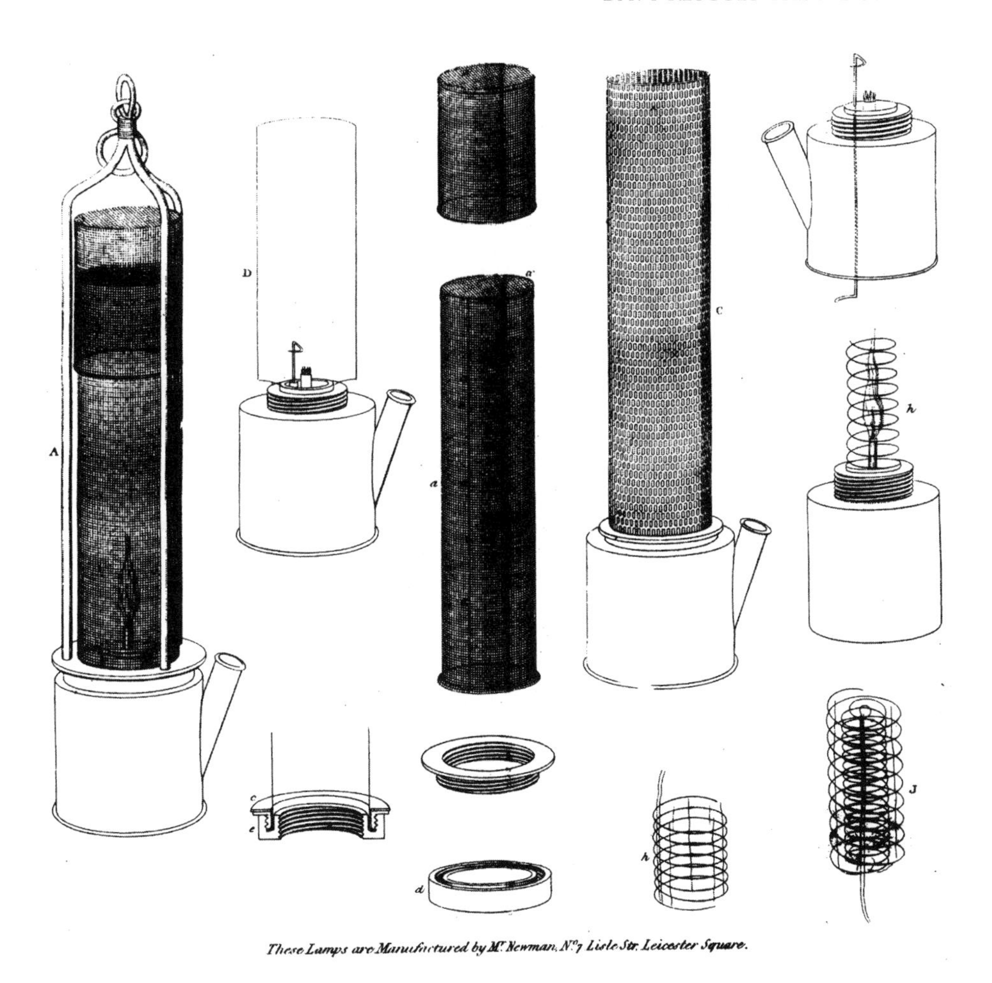

FIGURE 246. ■ Depictions of aspects of Humphry Davy's Safety Lamp for Coal Miners (London, 1818). His ingenious solution to lamps that would ignite coal gas with deadly results was incredibly simple. The fine metallic mesh would cool the coal gas below its flash point. Thus, although the flame and combustible gas were in open contact, there would be no explosion.

Figure 247 is from the third edition of Accum's book *A Practical Treatise on Gas-Light* (3rd ed., London, 1816; first and second eds., 1815). It shows a gas apparatus for exhibit and for testing different coals. At the right is a portable furnace with cast-iron retort for burning the coal; the center unit is a purifier having three chambers (one with water to trap ammonia; the second with aqueous potash (KOH) to trap carbon dioxide and hydrogen sulfide; the third receives other liquid products). The unit at the left of Figure 247 is the gasometer that stores coal gas over water at a moderate pressure. Note the elegant lamps or burners at the top. By 1815, there were already 26 miles of main gas pipes under London streets.

Accum wrote a number of interesting books on chemistry theory and practice and chemical amusements in addition to his *Practical Treatise on Gas-Light*. His book *Death in the Pot: A Treatise of Food, and Culinary Poisons* (London, 1820), made him many enemies. Some of these may have conspired in accusing

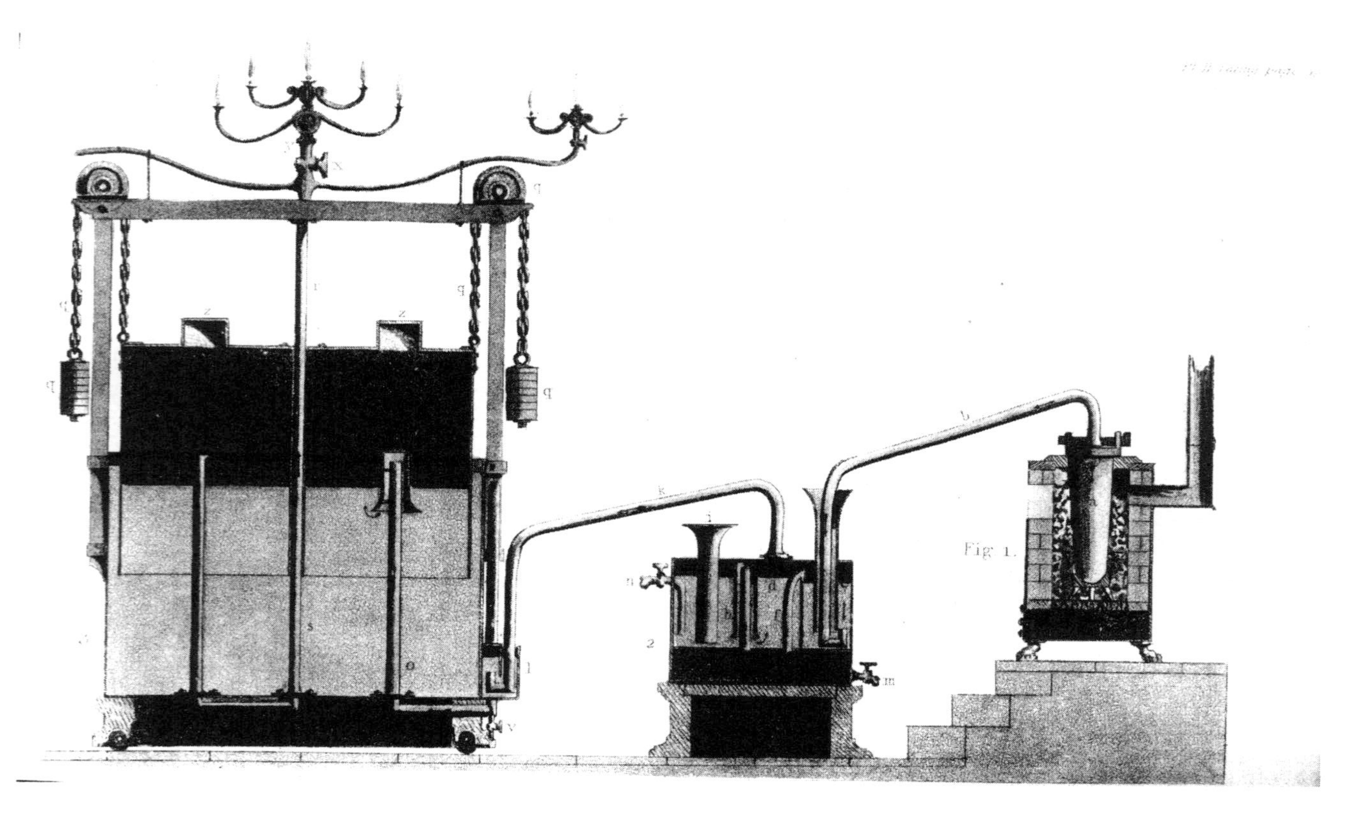

FIGURE 247. ■ A demonstration model of a coal-gas lighting system (Friedrich Accum, *A Practical Treatise on Gas-Light, Third edition, London*, 1816).

him of stealing and defacing books in the library of the Royal Institution. He was aquitted but left England in disgrace for a Professor's position in Germany.[4] His excellent 1824 work, *An Explanatory Dictionary of the Apparatus and Instruments Employed in the various Operations of Philosophical and Experimental Chemistry*, was published anonymously.

---

1. J. Stradins, *Chymia*, No. 9, 125–145, 1964.
2. J.R. Partington, *A Short History of Chemistry*, 3rd ed., Dover, New York, 1989, pp. 189–190.
3. J.R. Partington, *A History of Chemistry*, MacMillan, London, 1962, Vol. 3, pp. 826–827.
4. C.A. Browne, *Chymia*, No. 1, 1–9, 1948.

THE DUALISTIC THEORY OF CHEMISTRY

The early alchemists and natural philosophers believed in the duality of matter —sun and moon; male and female; sulfur (fixed) and mercury (volatile). When Davy electrolyzed pure potash (KOH) and produced a volatile (female) spirit (oxygen) at the positive pole and an explosive, fixed (male) matter (potassium) at the negative pole, this would have been intuitively obvious to them.

Jons Jacob Berzelius[1,2] (1779–1848) was born in Stockholm one year before his great countryman Scheele discovered lactic acid in rotting milk. The title page shown in Figure 248 is from Berzelius' first book. He found lactic acid in muscle ("flesh juice") and this appeared in the second volume of the two-volume set (1806; 1808). Lactic acid was to play a critical role in the development of stereochemistry some 75 years into the future. Like Scheele before him, Berzelius is omnipresent in the chemistry of his day and, indeed, in our modern textbooks. He developed the abbreviations we use for the elements (H, C, and Po, which was subsequently changed to K, Cl, etc.) and wrote versions of our modern-day formulas in which the numbers attached to the elements were superscripted. The modern subscripted formulas such as H_2O were introduced by Liebig and Poggendorff in 1834.[1,2] Berzelius was a great systematizer of chemistry and is credited with the discoveries of selenium and thorium, a share in the discovery of cerium, the first identification of silicon (actually generated earlier by Gay-Lussac and Thenard but not identified), and the first isolation of zirconium and titanium as metals—they had been earlier identified as new elements in their combined states.[1,3] Actually, titanium, currently the "sexiest element in the whole Periodic Table" was really obtained as the pure metal for the first time in 1910.[4] He contributed major work in chemical analysis, including the new and complex realm of organic analysis, and his extraordinarily careful studies (as many as 30 replications) verified Dalton's law of multiple proportions and strengthened atomic theory. He also demonstrated that Dalton's and Berthollet's findings were mutually compatible, differentiated what he termed "empirical formulas" (e.g., C_2H_6O) from what he termed *rational formulas* (e.g., C_2H_4 + H_2O), and defined the terms *isomers* and *allotropes*. In 1827, Berzelius "asserted that a peculiar vital force intervenes in the formation of or-

FÖRELÄSNINGAR

I

DJURKEMIEN,

AF

J. JACOB BERZELIUS.

FÖRRA DELEN.

STOCKHOLM,

Tryckte hos CARL DELÉN, 1806.

FIGURE 248. ■ Title page of Jons Jakob Berzelius' first book. He reports the isolation of lactic acid in "flesh juice" (muscle).

ganic compounds and their preparation in the laboratory can hardly be expected."[1]

The central tenet of Berzelius' world view was the dualistic theory that still pervades our understanding of chemistry—particularly for ionic compounds such as sodium chloride. Briefly, table salt is composed of a positive part (Na^+) and a negative part (Cl^-). Such dualism was already part of Lavoisier's thinking some 30 years earlier:[1]

Acid = radical + oxygen

Base = metal + oxygen

Salt = base + acid

The term radical, introduced by de Morveau and employed by Lavoisier, is defined as "of or from the root"; "foundation or source of something."[5] Berzelius divided ponderable bodies into an electronegative class and an electropositive class. Substances of the electronegative class are attracted to the positive pole (following Davy's convention) and substances of the electropositive class are attracted to the negative pole (Berzelius had initially defined the poles differently but bowed to the widespread acceptance of Davy's definitions).[2] Although organic salts such as sodium acetate fit the dualistic concept, the vast majority of organics did not.

1. J.R. Partington, *A History of Chemistry*, MacMillan, London, 1964, Vol. 4, pp. 142-177.
2. W.H. Brock, *The Norton History of Chemistry*, Norton, New York, 1993, pp. 150-159.
3. A.J. Ihde, *The Development of Modern Chemistry*, Harper and Row, New York, 1964.
4. D. Rabinovich, *Chemical Intelligencer*, October, 1999, pp. 60–62. Professor Rabinovich likes the "iridium" or "palladium" card as successor to the "titanium" card.
5. J.R. Partington, op. cit., pp. 258-262.

THE CHEMICAL POWER OF A CURRENT OF ELECTRICITY

The nineteenth century was a period of specialization in the sciences. Organic, inorganic, physical, and analytical chemistries emerged as disciplines. It is noteworthy that Michael Faraday (1791–1867) first saw Humphry Davy lecture at the Royal Institution in 1812, requested employment in his laboratory, and was appointed as laboratory assistant in 1813.[1] Davy was knighted in 1812 and married a wealthy widow during that year. Although he resigned his Professorship at the Royal Institution in 1813, he continued to visit and perform experiments and maintained his mentor relationship with Faraday. In 1813, Davy started a series of travels to the Continent, packing his chemical apparatus (he performed some experiments in his hotel rooms).[1] Although England and France were at war, there was always an eager audience for Davy. However, the complexities of the war situation had Davy appoint the young Faraday as his "temporary valet" —an appointment the regal and formal Lady Davy apparently took "too literally."[1] In 1815 Faraday received a higher position at the Royal Institution and started presenting public lectures. He started writing his first research papers in 1820, and, at his own request, they were edited by his respected mentor Davy. Although Faraday produced significant research in many areas of chemistry, his most important contributions were in electrochemistry. This was, of course, the field pioneered by Davy as well as Berzelius.

During the period 1831 through 1855 Faraday published a number of series of articles, "Experimental Researches in Electricity," in the *Philosophical Transactions of the Royal Society*. Partington notes that the major studies of electrolysis and the galvanic cell appeared between 1833 and 1840.[1] The most important discovery of these was the electrochemical equivalent:

> The chemical power, like the magnetic force, is in direct proportion to the absolute quantity of electricity which passes.

Figure 249 is from Faraday's Seventh Series of Lectures, presented to the Royal Society on January 9, 1834 and read on January 23, February 6 and 13, 1834. *Figs*. 64 to 66 (in Figure 249) are variants of the apparatus invented by Faraday to measure the quantity of gases generated by *electrolysis* (Faraday's term) of water. The gases were sometimes collected separately or together. He showed that the amount of water decomposed was directly proportional to the quantity of electricity employed, and he briefly defined "a degree of electricity" as that quantity that released 0.01 cubic inch of dry, mixed gas (corrected for temperature and pressure).[1] He realized that this apparatus was useful in determining quantity of

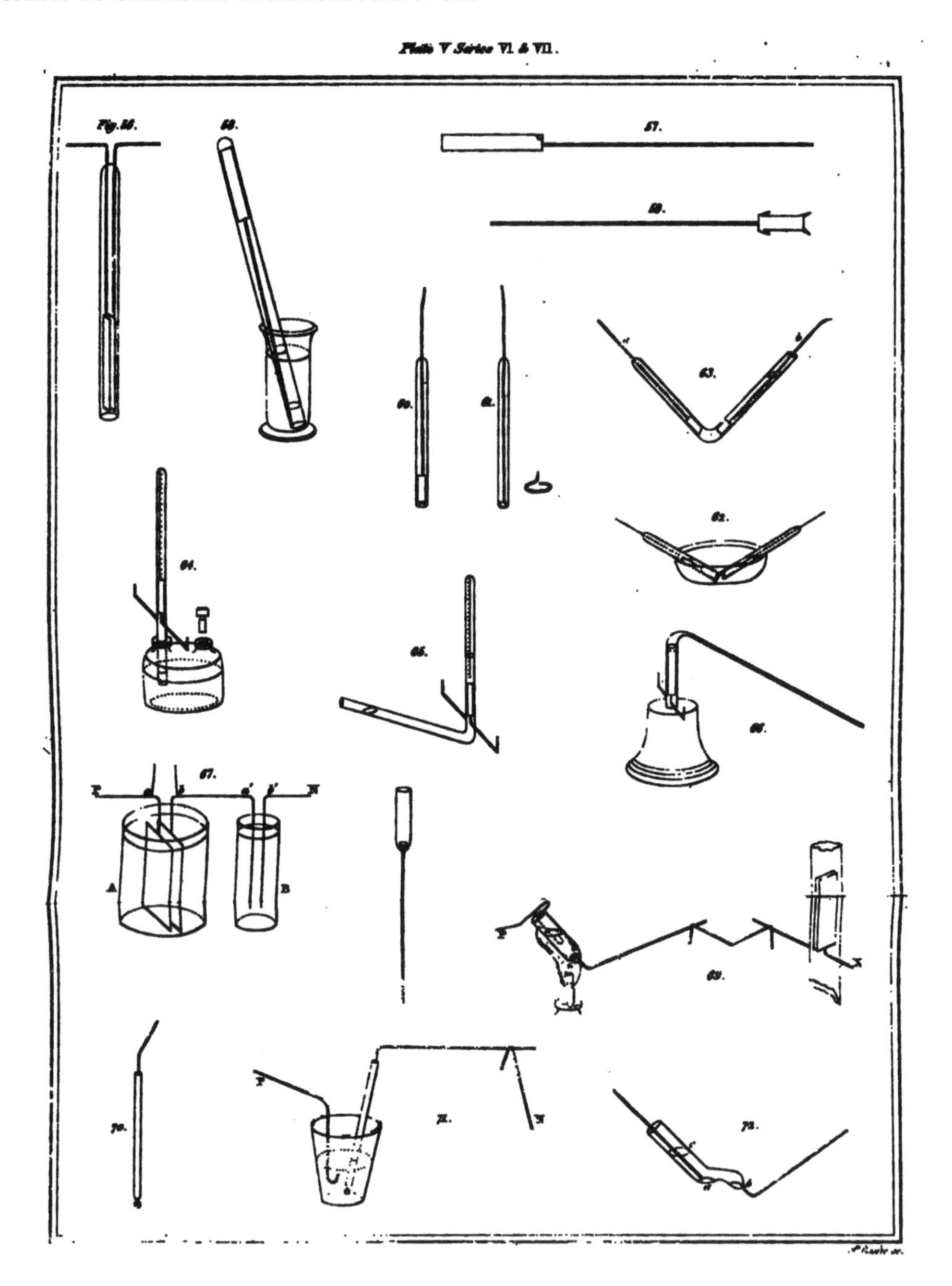

FIGURE 249. ■ This plate is from Michael Faraday's Seventh Series of Lectures to the Royal Society (January–February, 1834) and depict his Volta electrometer (coulometer since 1902) in which electrical current is measured by the volume of gas produced by electrolysis (Faraday's term) of water (see 64 to 66). In 69 to 72 we see apparatus for electrolysis of melts. Faraday discovered that the mass of matter produced by electrolysis was proportional to the current and demonstrated electrical equivalents of matter.

electricity and called it the "Volta-electrometer" (later termed by him the *voltameter*; since 1902 it has been termed the *coulometer*).[1]

Faraday discovered (accidentally) that ice is an insulator and also that, while salts are insulators, their melts are good conductors of electricity. *Figs*. *69*, *71*, and *72* depict three versions of Faraday's apparatus for electrolysis of molten salts. *Fig*. *69* includes a glass tube into which a platinum wire with a bulb at one end is fused. The other platinum wire *P* is dipped into the molten salt. The apparatus was connected to a battery through a voltameter. Starting with molten stannous chloride ($SnCl_2$), chlorine is released, combines with stannous chloride, and forms hot gaseous stannic chloride ($SnCl_4$, boiling point 114°C), which is collected. Metallic tin deposits on the preweighed platinum wire. Once the apparatus is allowed to cool, fused $SnCl_2$ is scraped off of the wire, and the increase in the wire's mass due to tin plating is determined. Faraday found that 3.2 grains of tin were collected and this coincided with collection of 3.85 cubic inches of gas. On a scale of $H = 1$, the equivalent mass of tin [Sn(II) in stannous chloride] was found to be 58.53 (four determinations).[1] This is quite close to the modern value of 118.7/2 = 59.35). Although, like his mentor Davy, Faraday was uncomfortable with the reality of atoms, he was forced to conclude that:[1]

> The equivalent weights of bodies are simply those quantities of them which contain equal quantities of electricity. Or, if we adopt the atomic theory . . . the atoms of bodies which are equivalent to each other in their ordinary chemical actions, have equal quantities of electricity associated with them. But I must confess I am jealous of the term "atom"; for though it is very easy to talk of atoms, it is very difficult to form a clear idea of their nature, especially when compound bodies are under consideration.

In addition to the term *electrolysis*, with the collaboration of William Whewell, a broadly trained scholar, Faraday developed the terms *electrode*, *anode*, *ion*, *cathode*, *anion*, *cation*, and *electrolyte*.[2] He is considered to be the inventor of the test tube.[3] He made early studies of the liquefaction of gases. For example, when a syringe was used to compress chlorine gas into a tube, a small amount of oily, green liquid was formed. He also used the newly discovered solid carbon dioxide in a bath of acetone (1835, by Thilourier; compression of CO_2 into a liquid and rapid expansion of the liquid rapidly cools the substance and forms dry ice!) to achieve a temperature of –78°C. This permitted Faraday to liquefy ethylene and other low-boiling-point gases using high pressure and cooling and led him to conclude that certain gases, such as hydrogen, were "permanent gases." For example, the critical temperature of methane (T_c) is equal to –82.6°C. At this temperature, the critical pressure (P_c) of 45.4 atm (4.60 MPa) will condense it to a liquid. However, at –78°C, no amount of pressure will condense methane, and hence it is a "permanent gas" at this and higher pressures.

1. J.R. Partington, *A History of Chemistry*, MacMillan, London, 1964, Vol. 4, pp. 99–128.
2. A.J. Ihde, *The Development of Modern Chemistry*, Harper and Row, New York, 1964, pp. 133–138.
3. W.H. Brock, *The Norton History of Chemistry*, Norton, New York, 1993, p. 191.

COLORFUL "NOTIONS OF CHEMISTRY"

Théophile Jules Pelouze (1807–1867)[1] and Edmond Fremy (1814–1894)[1] co authored one of the most beautifully illustrated chemistry books of the nineteenth century: *Notions Générale de Chimie*.[2] Pelouze was a student of Gay-Lussac at the École Polytechnique in Paris, and Partington describes his living conditions thus: "His lodging was so small that he humorously said he found it necessary to open the window to find space to put on a coat; he dined on bread and water, which he said tended to clear the mind."[1] Pelouze eventually succeeded Gay-Lussac at École Polytechnique, subsequently succeeded Thenard and Dumas at the Collège de France, and, in 1848, became president of the Commission of the Mint. In 1838, Pelouze was the first to react nitric acid with cotton and he produced spontaneously combustible nitrated cellulose. However, it was Christian Friedrich Schönbein who, eight years later, produced highly nitrated cellulose, an explosive commonly called "guncotton," using a mixture of nitric and sulfuric acids.[3]

Fremy began his chemistry career as first assistant to Pelouze at the École Polytechnique. He later was appointed Professor at this institution as well as in the Museum d'Histoire Naturelle. On the very day in 1850 that Pelouze's Chair at Collège de France was to be filled by election, Fremy read a paper attacking the widely favored successor, Auguste Laurent, and the Académie des Sciences did not elect him. Laurent died of poor health less than three years later at the age of 44.[4] Fremy may well have been the first to generate and catch a whiff of fluorine by electrolyzing calcium fluoride in 1854.[5] However, his student Henri Moissan benefitting from Fremy's experiences, isolated fluorine in 1886, and received the 1906 Nobel Prize in Chemistry.[5] Fremy was fascinated by the colors of cobalt salts for which he proposed an original (and long-dead) nomenclature.[6] Perhaps this fascination with colors led to the production of this lovely book.

Fig. 1 (Figure 250) depicts Lavoisier's experiment in which a matrass (a type of flask also known as a "bolt head") containing mercury is connected by its long curved neck into a graduated bell jar open over mercury in a basin. Mercury in the matrass is heated just to boiling for five days, until no further reduction of air volume is observed in the bell jar. Further heating occurs for a few days. Lavoisier's finding was that 27% of the gas volume in the bell jar diminished because of the loss of oxygen (the later accepted value was about 21%). The red crystalline substance found floating on the surface of the mercury in the matrass was mercuric oxide (HgO). In *Fig. 2* (Figure 250) an iron wire with a piece of lighted tinder at its end is placed into a jar of pure oxygen. The iron immediately inflames and throws sparks of iron oxide hot enough to melt and penetrate deeply into the glass. In *Fig. 3* (Figure 250) one pound of manganese dioxide (MnO_2) is heated very strongly in an earthen retort to produce oxygen. This is how Gahn first isolated manganese in 1774; less heat is required to produce oxygen from MnO_2 in the presence of sulfuric acid.

In *Fig. 4* (Figure 251) a small plaster cupel sits on a cork that floats, boatlike, on water in a trough. A small piece of phosphorus in the cupel is ignited and a bell jar inverted over the cupel. The gas remaining in the bell jar is nitrogen. *Fig. 5* (Figure 251) depicts a flask containing zinc into which water is first added

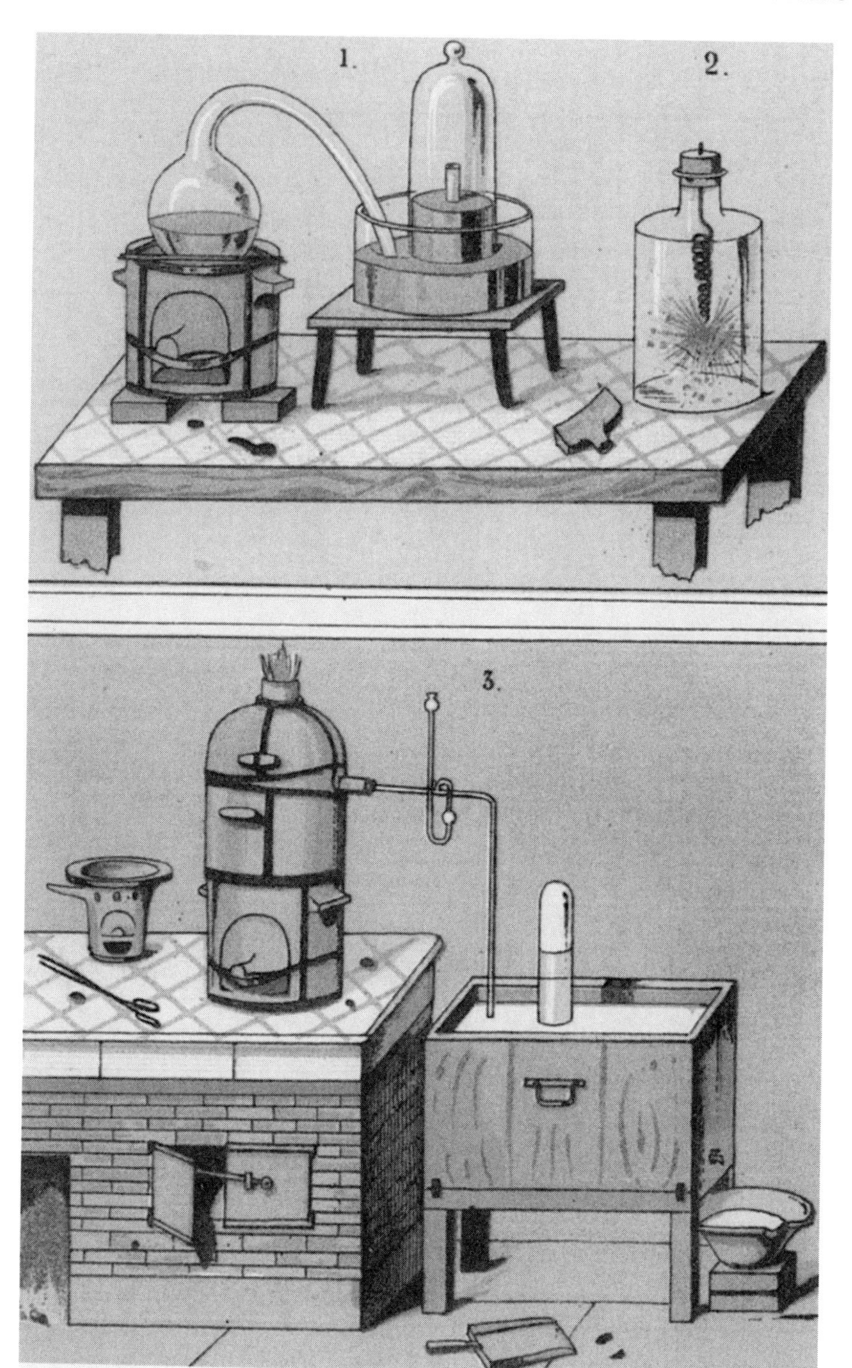

FIGURE 250. ■ Black and white image of a color plate from the American edition (1854) of *Notions Générale de Chimie* by Théophile Jules Pelouze and Edmond Fremy. See color plates. Depicted in *1* is Lavoisier's classic mercury oxidation experiment involving reflux of mercury in air; *2* illustrates the oxidation of an iron wire using a flame in pure oxygen; *3* depicts strong heating of manganese dioxide to release oxygen—an experiment first performed by Scheele.

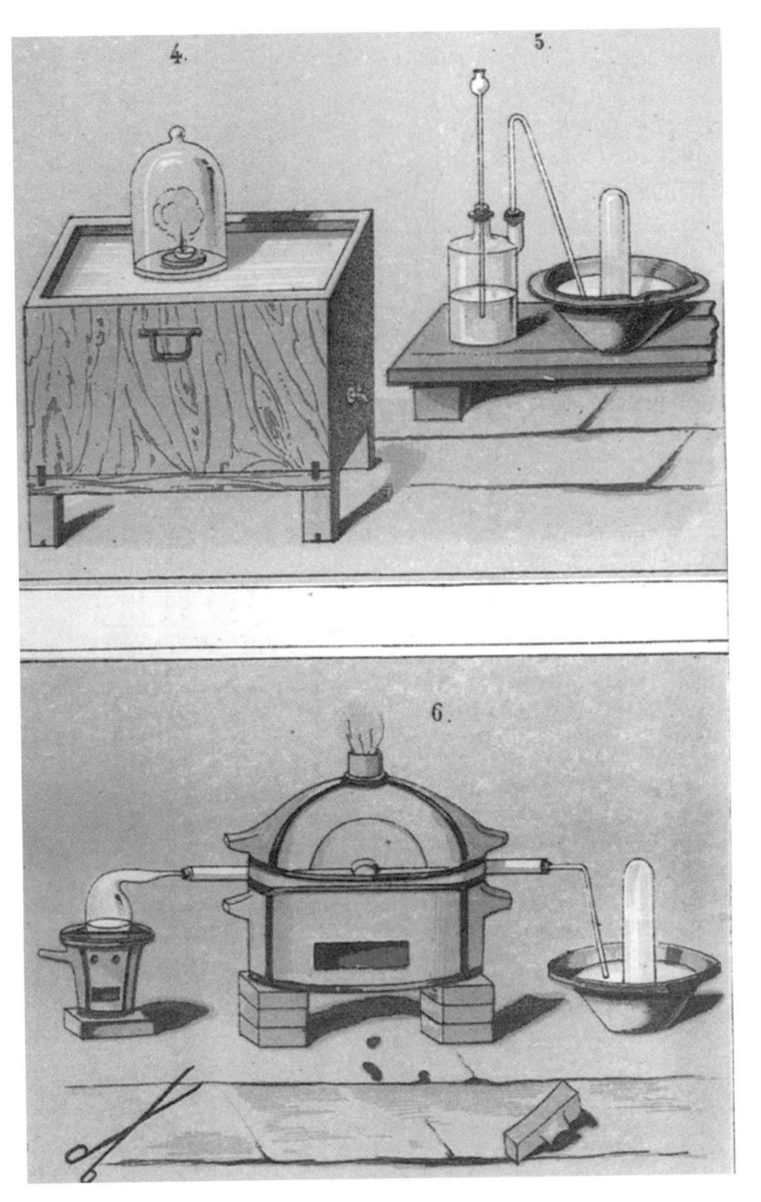

FIGURE 251. ■ Black and white image of a color plate from the 1854 American edition of *Notions Générale de Chimie* by Pelouze and Fremy. See color plates. Depicted in 4 is the combustion of phosphorus in air leaving unreactednitrogen; 5 shows the generation of hydrogen through reaction of zinc and sufuric acid—work first published by Henry Cavendish in 1766. In 6, we see Lavoisier's decomposition of water using an iron wire in a porcelain tube heated red hot in a furnace (see Figure 204).

through a funnel followed by sulfuric acid. Hydrogen gas is collected over water. *Fig.* 6 (Figure 251) shows Lavoisier's classic decomposition of water using pieces of iron wire in a porcelain tube heated red hot in a furnace. Oxide of iron forms in the tube and hydrogen gas is collected by a receiver inverted in water.

In *Fig.* 7 (Figure 252) hydrogen is generated (see *Fig.* 5, Figure 251), moves through a drying tube and is combusted with oxygen in common air, and the wa-

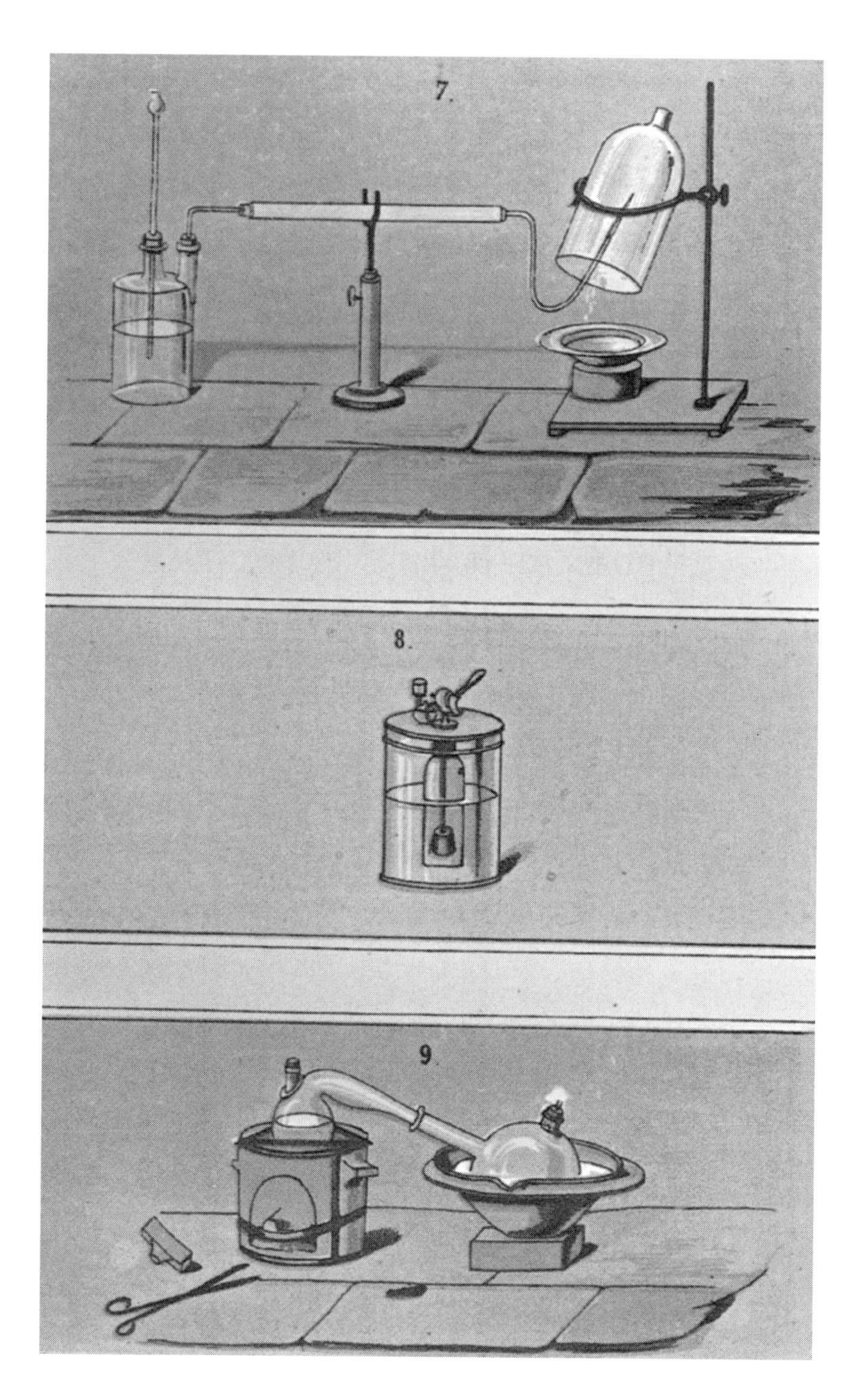

FIGURE 252. ■ Black and white image of a color plate from the 1854 American edition of *Notions Générale de Chimie* by Pelouze and Fremy. See color plates. Depicted in *7* is a synthesis of water involving generation of hydrogen and its combustion in air; 8 shows a clever self-controlling hydrogen gas generator; 9 illustrates a laboratory-scale distillation apparatus appropriate for students.

ter condensate drips into a collection bowl. *Fig. 8* (Figure 252) depicts a bell jar containing air and a cylinder of zinc suspended by a wire into acidified water. When the hydrogen formed pushes the acidified water from the bell jar, further reaction ceases. *Fig. 9* (Figure 252) shows a laboratory-scale distillation apparatus. In *Fig. 10* (Figure 253) we see a large-scale distillation apparatus using a copper boiler covered with a hood. The curved condenser tube is called a "worm." *Fig. 11* (Figure 253) shows an amazingly modern-looking laboratory-scale distillation apparatus developed by Gay-Lussac. Cooling water is added through the funnel at the lower right and leaves the condenser at the upper left.

Phosphorus in a cupel, suspended by a wire attached to a cork, burns blindingly in an atmosphere of pure oxygen (*Fig. 12*, Figure 254). Diamonds, known to be pure carbon, are depicted in *Fig. 13* (Figure 254). Heating saltpetre (KNO_3) and sulfuric acid in a glass retort and distillation produces "azotic" (nitric) acid (*Fig. 14*, Figure 254). Large-scale production of nitric acid (*Fig. 15*, Figure 254) employs niter ($NaNO_3$), which is cheaper to produce than saltpetre. Quantities of 100–150 kg (kilograms) of niter are heated in earthen cylinders to which sulfuric acid is added periodically. The distillate is received by a series of 12–15 three-necked flasks.

Fig. 27 (Figure 255) is a rather surrealistic rendition of stalactites (suspended from the ceiling) and stalagmites that rise to meet them in a cave that looks like the open maw of some hideous beast. Stalactites and stalagmites are composed of calcium carbonate ($CaCO_3$) arising from contact of carbonic acid dissolved in water with lime (CaO) in the earth's surface. The wide sky, strange clouds, and bizarre cave seem to anticipate the style of modernist René Magritte (1898–1967).

---

1. J.R. Partington, *A History of Chemistry*, Vol. 4, MacMillan & Co., Ltd., London, 1964, pp. 395–396.
2. T.J. Pelouze and E. Fremy, *Notions Generales de Chimie. Avec un Atlas de 24 Planches en Couleur, en 2 Volumes*, Victor Masson, Paris, 1853. The plates shown here are from the American edition: *General Notions of Chemistry*, Lippincott, Grambo & Co., Philadelphia, 1854.
3. A.J. Ihde, *The Development of Modern Chemistry*, Harper & Row, New York, 1964, p. 451.
4. Partington, op. cit., p. 376.
5. Ihde, op. cit., p. 367.
6. W.H. Brock, *The Norton History of Chemistry*, W.W. Norton & Co., New York, 1993, pp. 577–578.

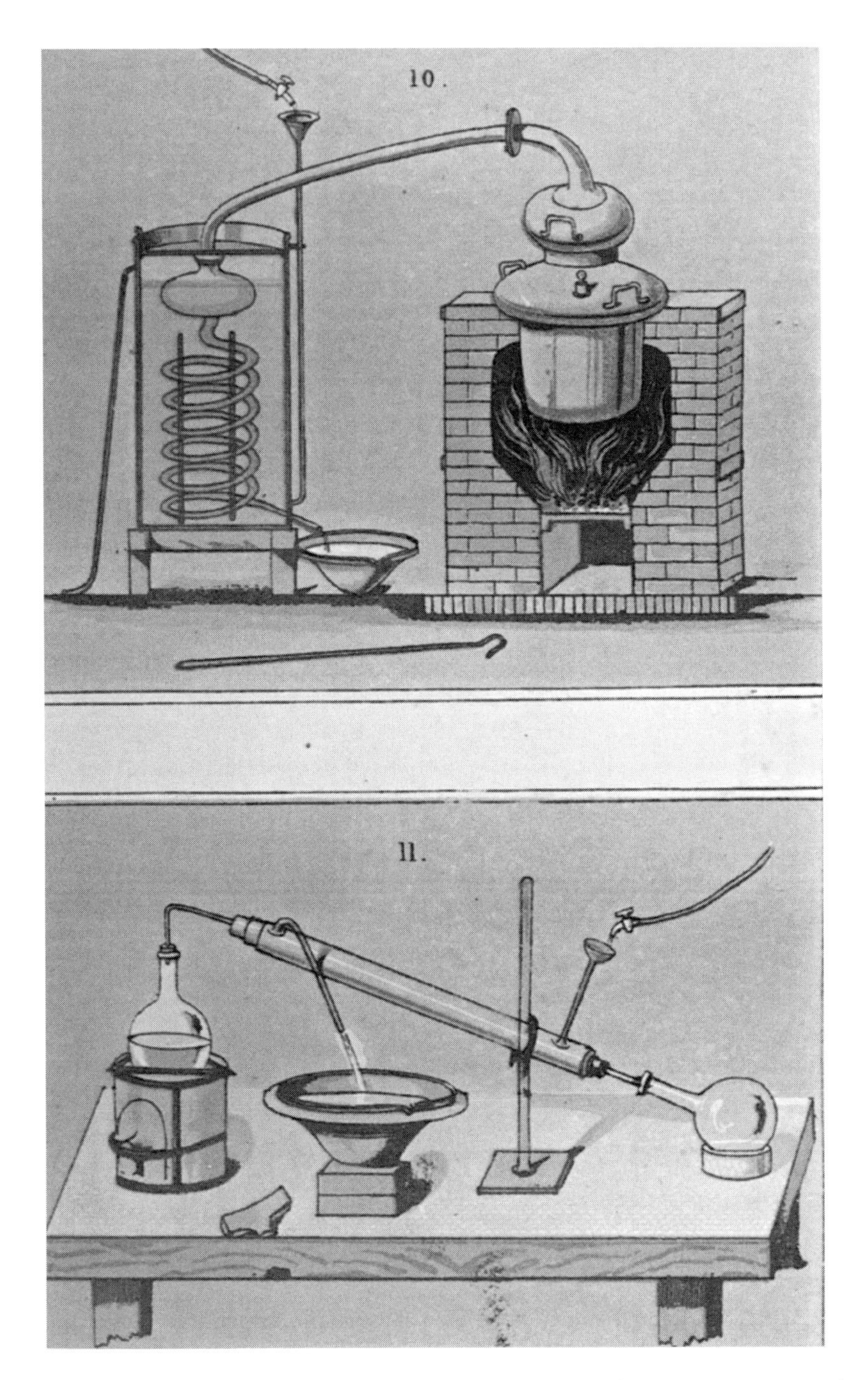

FIGURE 253. ■ Black and white image of a color plate from the 1854 American edition of *Notions Générale de Chimie* by Pelouze and Fremy. See color plates. Depicted in *10* is an industrial-scale still consisting of a copper boiler covered with a hood; *11* displays the very modern-looking water-cooled distillation apparatus designed by Gay-Lussac.

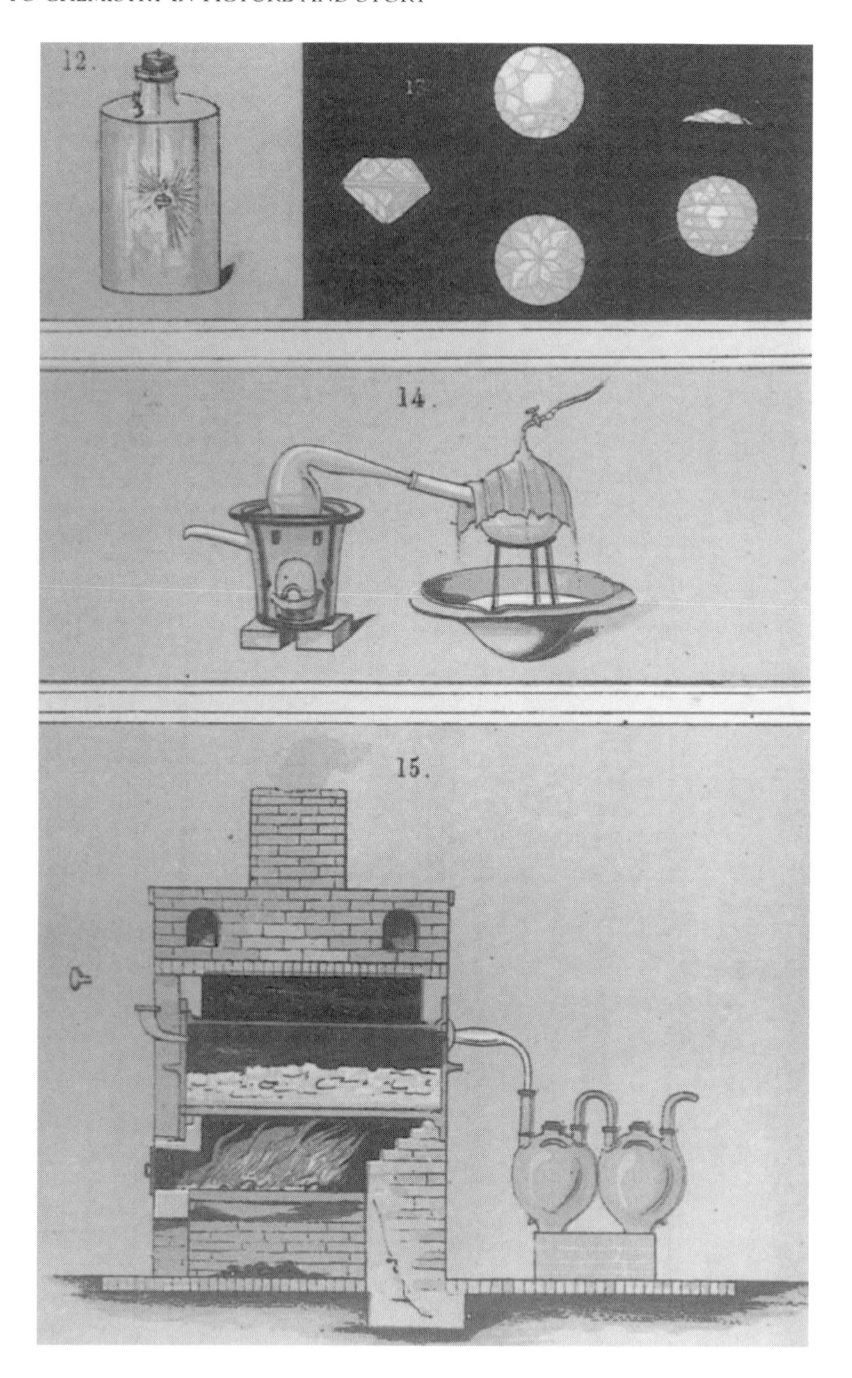

FIGURE 254. ■ Black and white image of a color plate from the 1854 American edition of *Notions Générale de Chimie* by Pelouze and Fremy. See color plates. In *12* phosphorus, suspended by a wire, burns with blinding brightness in pure oxygen; diamonds (*13*) are composed of pure carbon just like the humble mineral plumbago (graphite), another carbon allotrope, and totally different from ruby and other gemstones; heating saltpetre (KNO_3) and sulfuric acid produces nitric acid (*14*) but on an industrial scale (*15*) it is cheaper to use niter ($NaNO_3$).

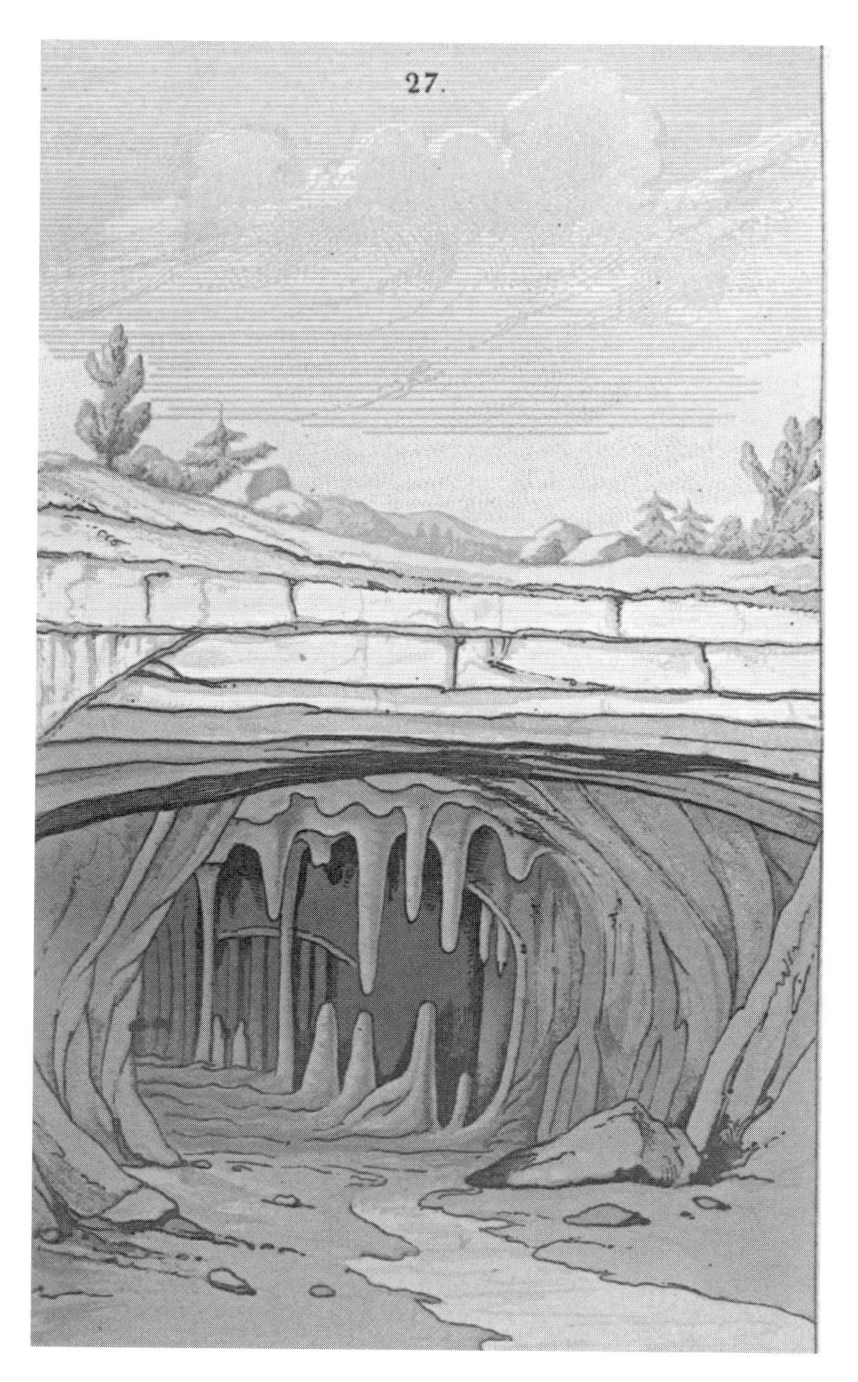

FIGURE 255. ■ A wonderfully surrealistic rendition of stalactites (suspended from the ceiling, in case you forgot) and stalagmites composed of limestone ($CaCO_3$) formed by contact between the CO_2 dissolved in groundwater and lime (CaO) in the soil. See color plates. This image seems to anticipate the artistic style of René Magritte some 45 years before his birth. (From the 1854 American edition of *Notions Générale de Chimie* by Pelouze and Fremy.)

A PRIMEVAL FOREST OF THE TROPICS

"Organic chemistry appears to me like a primeval forest of the tropics, full of the most remarkable things" wrote Friedrich Wöhler to Berzelius in 1835.[1] I remember receiving from my father his personal copy of Karrar's *Organic Chemistry* (3rd ed., 1947) on the eve of taking my first organic chemistry course along with the admonition that I needed to learn everything in it (almost 1000 large pages). Thirty-five years later I realize that he may have been pulling my leg a bit, but it took me almost half the semester to gain my footing in the course. I thus have a great deal of empathy with students in my own organic chemistry course.

Organic chemistry is the chemistry of carbon compounds. We note that minerals such as carbonates are not considered to be organic, nor are certain gases such as carbon dioxide (or carbon monoxide) that may be derived from them. Although Lavoisier did not make this differentiation, organic chemistry was regarded as different from the remainder of chemistry and, through the early nineteenth century, relegated to descriptive sections on "Animal Chemistry" and "Vegetable Chemistry" in chemistry texts. The sheer complexity of the mixtures, the complexities of the formulas, and the fact that organic compounds did not obey the dualism seen for inorganics such as water or sodium chloride added to these conceptual problems.

I visited the web page of Chemical Abstracts Service (CAS) (www.cas.org) on May 24, 1999. At precisely 11:17:11 A.M. EDT, there were 19,632,211 registered substances of which some 12 million were organic (68%). The remaining substances were biosequences (17%), coordination compounds (6%), polymers (4%), alloys (3%), and tabular inorganics (2%). Of these, about 160,000 substances are of sufficient practical importance to be on national or international chemical inventories and registry lists. As of 1997, 1.3 million new substances were being added to the list each year. The cause of these daunting numbers and incredible diversity is, primarily, the carbon atom, which can form bonds with almost all other elements including other carbons. It forms four bonds in combinations of single, double, and triple bonds, as well as chains, rings, and cages. Two minutes later (11:19:12 A.M. EDT), there were 19,632,221 substances (10 new ones!) in the CAS registry. A revisit of this site on February 17, 2006 at 7:33 A.M. EDT found a somewhat modified organization of substances. There was at that time and date a cumulative total of 27,345,897 "small molecules" and 57,213,566 biological sequences for a total cumulative substances count of 84,559,463. The "small molecules" are overwhelmingly organic and increasing by roughly 3 million per year. New techniques in combinatorial chemistry, computerized synthesis and sequencing, and the rapid growth of chemistry research worldwide are adding to the exponential growth of these numbers.

Figure 256, from the Youmans 1857 edition of the *Chemical Atlas* depicts the atmospheric part of the carbon cycle involving plants and animals. On the right we see animals that inhale (arrows down) oxygen (note it is written as monoatomic rather than O_2) with the food that nourishes them to produce carbon dioxide and water. In the Youmans text, confusion reigning for 50 years persists in the formula for water (HO) and atomic weights for oxygen (8),

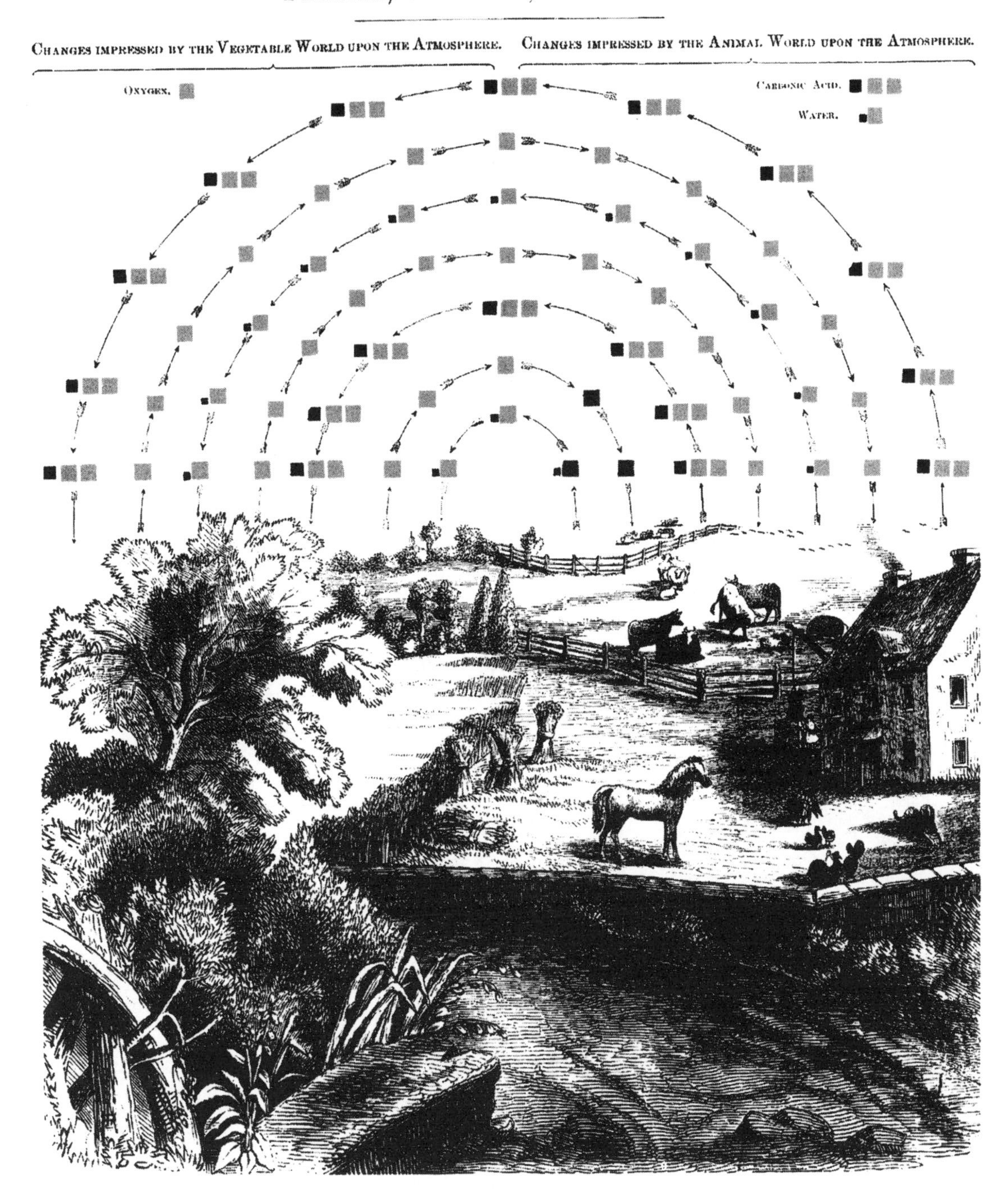

FIGURE 256. ■ This beautiful hand-colored figure (see color plates) is from the 1857 edition of Edward Youmans' *Chemical Atlas* (New York, first published in 1854).

carbon (6), sulfur (16), and others. The formula for ammonia (NH_3) is correct and the atomic weight of nitrogen is correctly 14. On the left are plants that incorporate carbon dioxide and water (arrows down) to produce oxygen (arrows up).

The confusion so evident in Youmans's book over formulas, atomic weights, isomers, and valence will all clear up within the following ten years or so.

1. J.R. Partington, *A History of Chemistry*, MacMillan, London, 1964, Vol. 4, p. 233.

TAMING THE PRIMEVAL FOREST

In nature, organic compounds are usually found in incredibly complex mixtures. Destructive distillation of a sample of coal produces hundreds of compounds in easily measurable quantities and thousands of compounds if one wishes to measure trace levels. If a chemist wishes to determine the formula of a compound, it must first be separated from other compounds and rigorously purified. Even today, absolute separation of compounds cannot always be achieved using *fin-de-millennium* techniques.

Lavoisier did not believe that organic compounds were outside the normal realm of chemistry and analyzed the amount of oxygen consumed and carbon dioxide formed in the combustion of charcoal using the usual apparatus (e.g., Figure 205, *Fig. 1*). He also burned alcohols, fats, and waxes.[1] However, his data on the composition of H_2O and CO_2 were inaccurate:[1]

	Lavoisier	Correct
CO_2	28% C; 72% O	27.2% C; 72.8% O
H_2O	13.1% H; 86% O	11.1% H; 88.9% O

These errors may appear to be negligible. While they would be for determining simple formulas such as CH_4, the errors would be significant for formulas such as $C_{18}H_{38}O$ and would interfere with the understanding of carbon's valence.

Gay-Lussac and Thenard made the first accurate determinations of carbon content of organic compounds by using potassium chlorate ($KClO_3$) as the oxidizing agent.[1] The sample for analysis and potassium chlorate were pressed together into a pellet, which was dropped carefully into a vessel heated by charcoal. The resulting CO_2 was absorbed by potash. They eventually replaced $KClO_3$ with cupric oxide (CuO), which was safer and did not oxidize organic nitrogen. Apparatus continued to evolve notably due to improvements by Berzelius.[1]

Justus Liebig (1803–1873) developed the method for C, H, and 0 analysis essentially in use today.[1–3] His first book on organic analysis was published in 1837 and is quite rare. Figure 257(a) is from the first English edition, *Hand-Book*

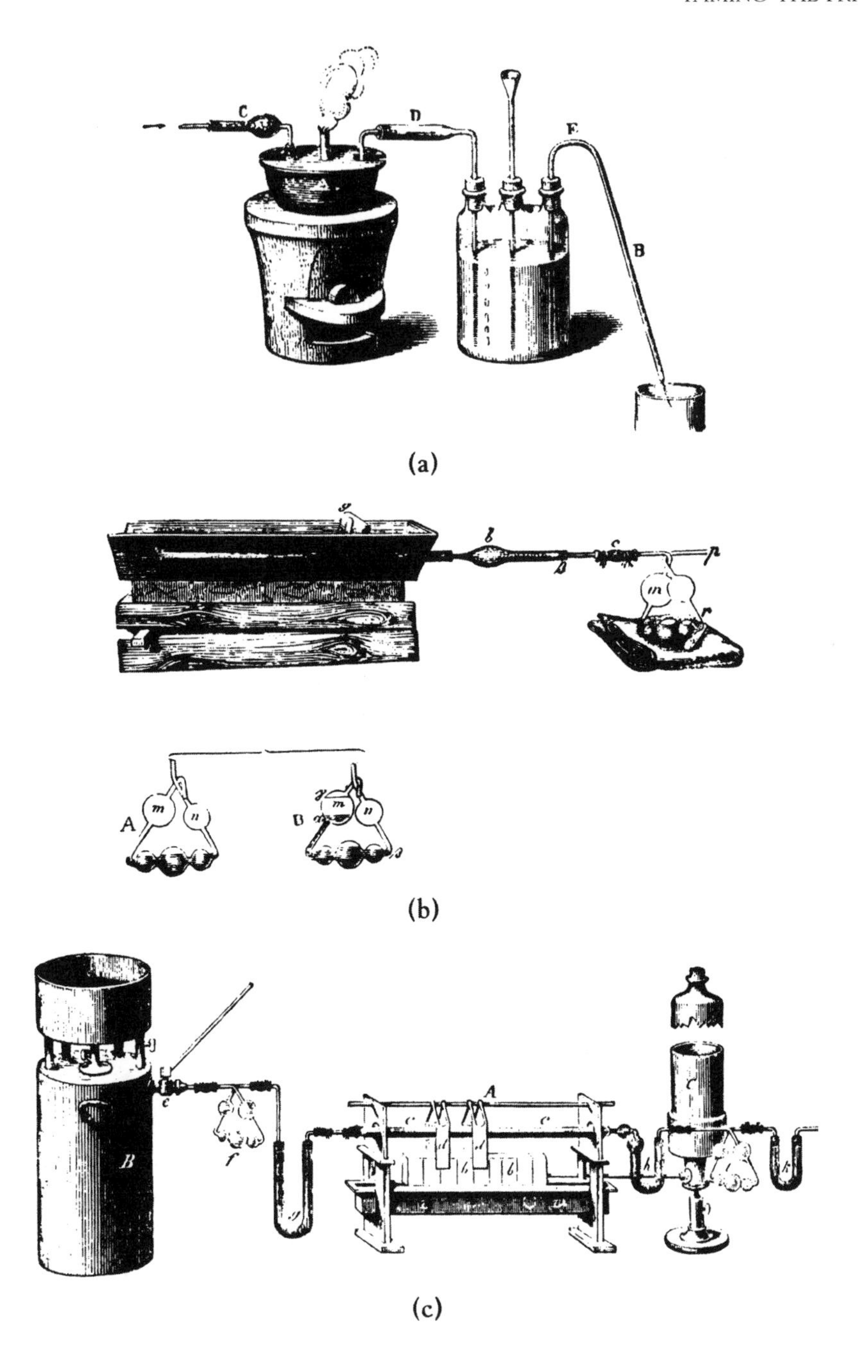

FIGURE 257. ■ Figures from Justus von Liebig's *Handbook of Organic Analysis* (first English Edition, London, 1853; Liebig's first book on organic analysis was published in 1837 in Braunschweig and is extremely rare): (a) Apparatus for quantitative drying of organic substance to be analyzed; (b) Apparatus for carbon and hydrogen determination; Liebig's ingenious *kaliapparat*, the five-bulbed glassware containing potash solution in the lower three bulbs for quantitation of CO_2 is shown as part of the apparatus and separately; (c) Apparatus capable of using pure oxygen as well as air for carbon/hydrogen analysis. The *kaliapparat* appears in the American Chemical Society logo (Figure 52).

of Organic Analysis (London, 1853). Liebig notes that organic substances often absorb water and that they must first be free of water prior to analysis. Figure 257(a) depicts a drying apparatus. The siphon attached to the three-necked flask on the right draws off water, creating a slight vacuum that pulls air through drying tube C (filled with calcium chloride) on the left. The sample itself is in a tube *A*, which is not seen here because it is in the hot bath above the furnace. It is connected by glass tubing to C as well as to tube *D*, which condenses any water released from the sample. Tube *A* is periodically removed from the heat and weighed until no further change occurs. Tube *D* can also be weighed if necessary to determine water content.

Figure 257(b) depicts Liebig's *kaliapparat* (*A*, *kalia* refers to potassium and the potash solutions that occupy the three lower bulbs of this five-bulbed piece of glassware). The sample for analysis is placed in a hard-glass tube sitting in an iron trough heated over a flame.[1,2] Water derived from combustion is trapped in the preweighed drying tube to the right of the iron trough. Carbon dioxide is trapped in the potash solutions present in the lower three bulbs of the *kaliapparat*. The upper two bulbs at both ends of the *kaliapparat* serve two functions: as depicted in Figure 257(b), prior to the start of the experiment, some air is drawn out by mouth from the apparatus and potash solution climbs into bulb *m*. If the vacuum is maintained and the liquid level in *m* does not drop, then there are no leaks in the apparatus. These bulbs also prevent loss of potash solution due to splashing. The preweighed *kaliapparat* is followed by a preweighed drying tube that traps any water vapors lost from the *apparat*.[1,2] The apparatus depicted in Figure 257(c) employs pure oxygen for combustion—the oxygen is generated in *B*, passed through the *kaliapparat f* containing concentrated sulfuric acid, and then tube *g* is filled with calcium chloride. The dry oxygen is introduced to combustion tube *cc* in a bed of magnesium oxide in an iron trough. The combustion tube has a thick plug of copper turnings at the left and is two-thirds filled with copper oxide. Dried atmospheric air, free of carbon dioxide can be introduced using the apparatus on the right.

Here are the results of a state-of-the-art analysis reported by Adolph Strecker in Liebig's laboratory at the University ofGiessen in 1848:[3] the formula for cholic acid was found to be $C_{48}H_{39}O_9$ (with the atomic weights C = 6; O = 8). The present day formula is $C_{24}H_{40}O_5$ (C = 12; O = 16).[3] It is obvious that the results are accurate but not enough to "hit the formula on the head." Yet that is precisely what is needed to make sense of carbon's valence.

Liebig was born and raised in rather poor circumstances. He was an intense, irascible man who, as a student, was arrested for his political activities. He was sponsored by Karl Wilhelm Kastner (1783–1857) at the University of Bonn and later Eriangen. Kastner persuaded the Eriangen faculty to award Liebig an honorary doctorate *in absentia* in 1822. As Brock states:[3] "It is one of the ironies of Liebig's teaching career that he himself never presented a thesis for his doctorate." He engaged in acrimonious debates throughout his career, was unkind in his later criticisms of his kind patron Kastner, and did not hesitate to attack Friedrich Wöhler when the two found the same formula for Liebig's silver fulminate and Wöhler's silver cyanate. The controversy was settled when the two performed their analyses together, discovered the first example of isomerism, and began one of the greatest friendships in the history of chemistry. The gentle and wise Wöhler counselled the "Type-A" Liebig in 1843 thusly:[1]

> To make war against Marchand, or indeed against anyone else, brings no contentment with it and is of little use to science. You merely consume yourself, get angry, and ruin your liver and your nerves—finally with Morrison's pills. Imagine yourself in the year 1900, when we are both dissolved into carbonic acid, water, and ammonia, and our ashes, it may be, are part of the bones of some dog that has despoiled our graves. Who cares then whether we have lived in peace or anger; who thinks then of the polemics, of the sacrifice of thy health and peace of mind for science? Nobody. But thy good ideas, the new facts which thou hast discovered—these, sifted from all that is immaterial, will be known and remembered, to all time. But how comes it that I should advise the lion to eat sugar?

Liebig had joined the Giessen faculty in 1824 and became Co-Editor of the *Magazin fur Pharmazie*. In 1832, he assumed sole editorship, changed the title to *Annalen der Chemie und Pharmazie* and the tough, caustic "lion" made it a vital chemical journal. He built a renowned research and teaching school at Giessen and by 1852 he had influenced about 700 students of chemistry and pharmacy.[3] In that year, he moved to the University of Munich but his health no longer permitted him to work in the laboratory. His intensity probably contributed to his poor health and he spent his final 20 years in bitter chemical controversies.[3] His friend Wöhler, whose sense of humor was reflected in an 1843 paper he published in the *Annalen* under the pseudonym "S.C.H. Windier,"[4] lived to reach the age of 82. He trained over 20 American chemists. Among these were Ira Remsen, who started at Johns Hopkins University the first American Ph.D. program in chemistry as well as Edgar Fahs Smith at the University of Pennsylvania.[5] It is delightful to realize that Wöhler synthesized urea in 1828 and E.F. Smith, his student, published the delightful book Old Chemistries in 1927.

---

1. A.J. Ihde, *The Development of Modern Chemistry*, Harper and Row, New York, 1964, pp. 173–183.
2. J.R. Partington, *A History of Chemistry*, MacMillan, London, 1964, Vol. 4, pp. 234–239.
3. W.H. Brock, *The Norton History of Chemistry*, Norton, New York, 1993, pp. 194–207.
4. W.H. Brock op. cit., p. 218.
5. A.J. Ihde, op. cit., p. 264.

THE ATOMIC WEIGHT OF CARBON AND RELATED CONFUSIONS

Confusion over molecular formulas and atomic weights was an unfortunate byproduct of the early atomic theory. Dalton's Rule of Greatest Simplicity provided incorrect formulas such as HO for water and NH for ammonia. Although Gay-Lussac's Law of Combining Volumes (1808), Avogadro's hypothesis (1811), the Law of Dulong and Petit (1819), and other studies began to clear up the confusion, it was not until Cannizzaro's 1858 paper and the 1860 Karlsruhe Conference that atomic formulas, equivalents, and atomic weights were really clarified.

Figure 258 is from Edmund Youmans' *Chemical Atlas* (New York, 1854; 1857 printing). It exemplifies the continuing confusion over the atomic weights of carbon and oxygen relative to hydrogen (assumed to be 1). Thus, Dulong determined that the ratio of densities of CO_2/O_2 = 1.38218.[1] Therefore, the same volume of oxygen gas containing 100 g would contain 138.218 g of carbon dioxide. If one accepts Avogadro's hypothesis, then the mass ratio of oxygen to carbon is 100.00/38.218. Using the assumption of Gay-Lussac and Dumas, the formula of "fixed air" being CO not CO_2, the atomic weight of carbon would be 6.12 if oxygen is 16.0. Berzelius determined that fixed air is CO_2 and assigned an

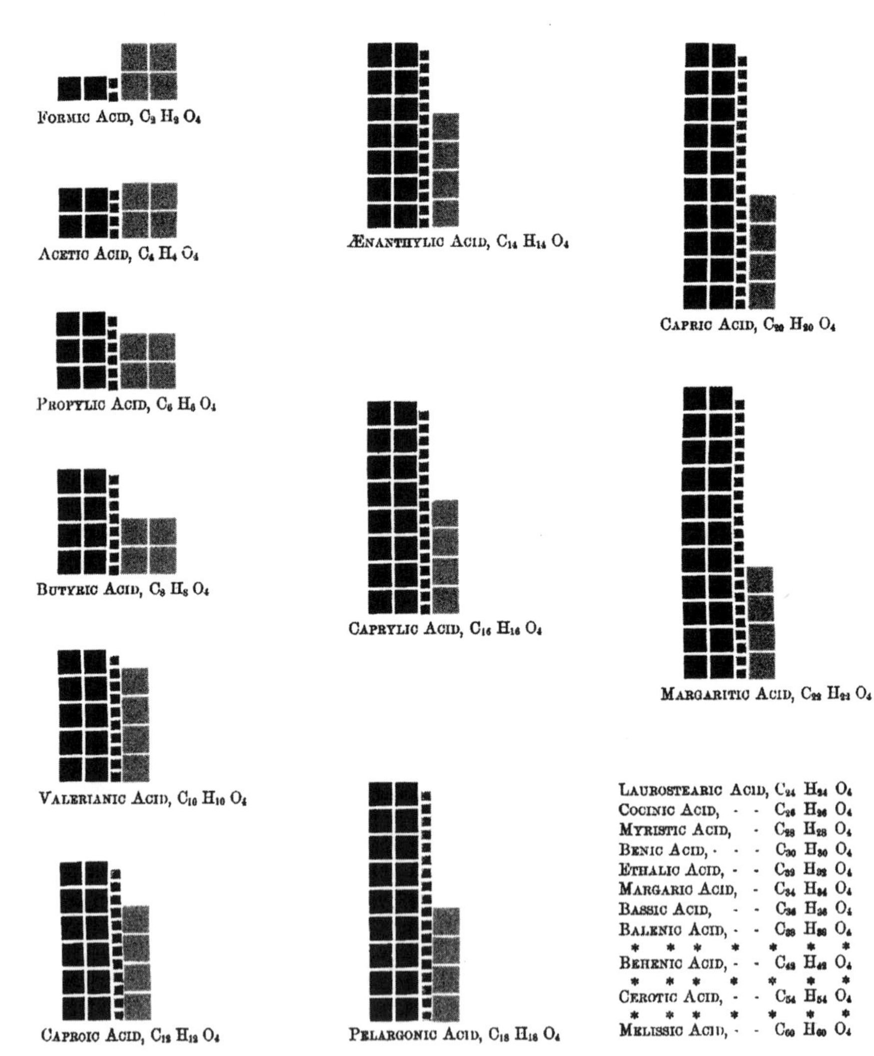

FIGURE 258. ■ Plate from Youmans' *Chemical Atlas* (Figure 256). See color plates. The organization of the "Primeval Forest" of organic chemistry by Laurent and Gerhardt included the concept homology. The units of homology are CH_2 rather than C_2H_2 as shown. The confusion was the result of discrepancies in the atomic weight of elements and assumed formulas. These would be cleared up very shortly in the Karlsruhe Congress of 1860.

atomic of 12.24 to carbon. However, in 1840 Dumas and Stas published their very precise studies of the combustion of purified graphite in a stream of pure oxygen. Weighing any unburned ash, they determined carbon's atomic weight at 12.0 (if oxygen = 16.0 and fixed air is CO_2). Nevertheless, it is clear from Figure 258 that confusion continued for about 20 more years. The problems of formulas and atomic weights would only be settled in Karlsruhe. In Youmans' *Chemical Atlas* the atomic weights of carbon (6) and oxygen (8) are half of the accepted (post-Karlsruhe) values while nitrogen is correct at 14. Thus, acetic acid, given as $C_4H_4O_4$ in Figure 258 is really $C_2H_4O_2$, butyric acid is really $C_4H_8O_2$, and the common difference in a homologous series is CH_2, not C_2H_2.

1. A.J. Ihde, *The Development of Modern Chemistry*, Harper and Row, New York, 1964, pp. 183–184.

WHY'S THE NITROGEN ATOM BLUE, MOMMY?

The beautiful hand-colored plates in Youmans' *Chemical Atlas* (e.g., Figure 258) have unique colors representing individual atoms. These are explained by Youmans: oxygen changes the color of blood to bright red in the lungs, hence it is represented as red; the sky, which is 80 nitrogen, is blue, thus nitrogen atoms are colored blue; carbon, sulfur, and chlorine are depicted in their natural elemental colors of black (sort of), yellow, and green. It is "cool" that most molecular models, made of wood, metal, or plastic, keep the same color scheme. Moreover, most modern computer programs that model molecules represent oxygen as blood red and retain the other traditional colors. I am further reminded of such traditions when I recall a professor's quip at a chemical meeting that "everybody knows that *p* orbitals are blue and green"—a reference to the influence of and colors used in the groundbreaking work on orbital symmetry by Robert Burns Woodward and Roald Hoffmann.[1]

1. R.B. Woodward and R. Hoffmann, *The Conservation of Orbital Symmetry*, Verlag Chemie, Weinheim, 1970.

I CANNOT HOLD MY CHEMICAL WATER—I CAN MAKE UREA!

Figure 259 is from Edward L. Youmans' *Chemical Atlas*[1] and depicts concepts of isomerism from the mid-nineteenth century. The field of organic chemistry is vast. As of the year 2006, there are over 20 million known organic compounds, not counting biological sequences. This enormous diversity is due, in large part, to the occurrence of isomers—molecules with the same formula but different arrangements of atoms.

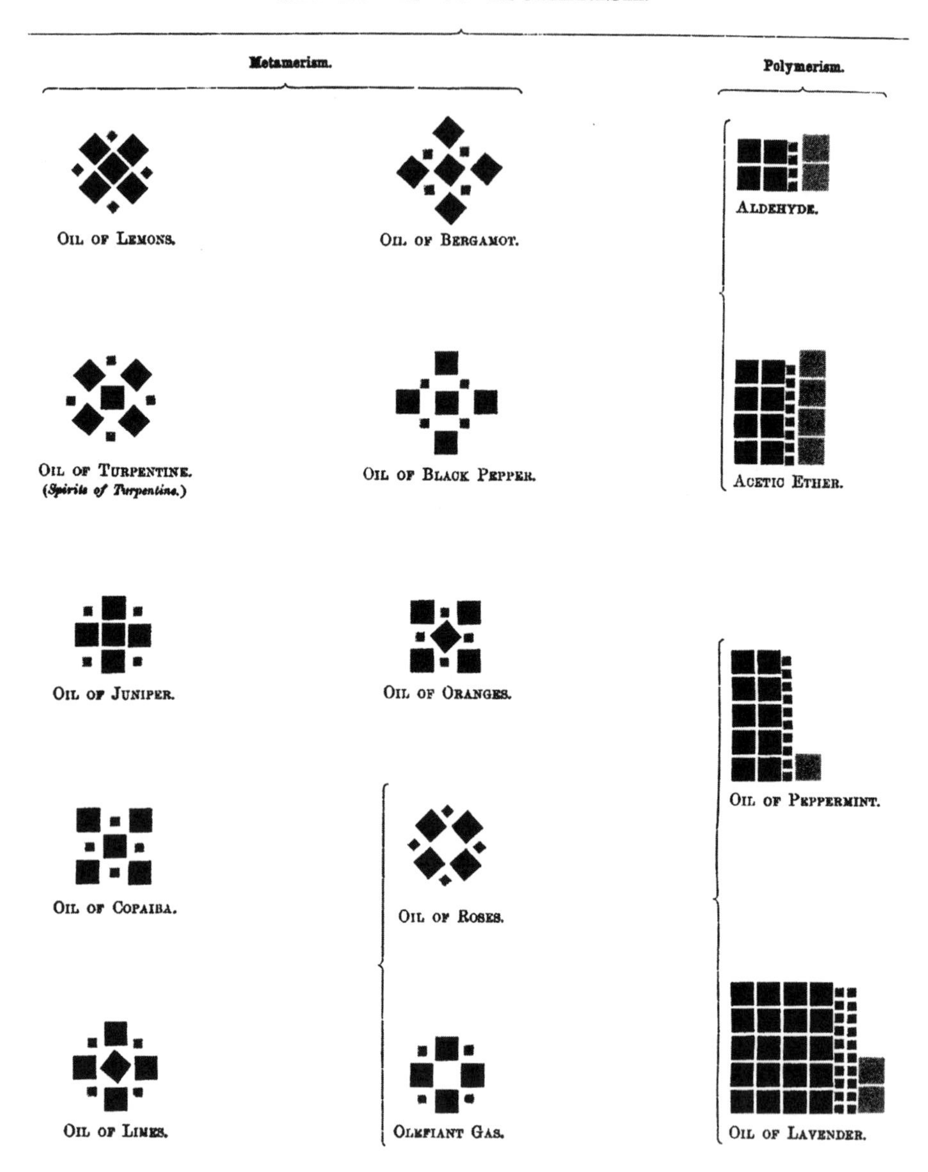

FIGURE 259. ■ Plate from Youmans' *Chemical Atlas*. See color plates. Although Wöhler and Liebig discovered that silver fulminate and silver cyanate were isomers (term coined by Berzelius in 1830) and it was suspected that the origin was the different arrangement of atoms, the concept of valence remained to be discovered. In this plate Youmans depicts isomers as different arrangements of atoms. But he postulates that their chemical history plays a role. For example, since the atomic arrangements in the carbon allotropes graphite and diamond are different (presumably different charcoals are also allotropes), then hydrocarbon isomers (butane and isobutane, for example) maintain the different carbon arrangements of the allotropes from which they were (presumably) derived.

In the 1820s two great organic chemists, Justus Liebig and Friedrich Wöhler, discovered that two very different substances, silver fulminate and silver cyanate, respectively, had the same composition. Liebig initially attacked (after fulminating?) Wöhler's results but after meeting and comparing results, they agreed that they were consistent—and thus very confusing.[2,3] The quandary was essentially resolved by Berzelius in 1830, who came up with the concept of *isomers*.[2] However, he differentiated isomers that were metamers, basically similar to our modern concept of isomers, and *polymers*, which had the same formula but different densities. Thus, the density of gaseous butylene (C_4H_8) is double that of gaseous ethylene (C_2H_4) even though the two have the same composition (85.7% C; 14.3% H or CH_2).[2] In Youmans' text, isomers are considered to arise from different allotropes of carbon.

It is interesting that one is more likely to find information on fulminates in a history of chemistry book than in a chemistry text. There are two readily accessible fulminates,[4] mercury fulminate, which has been used as a "primer" in percussion caps, and silver fulminate, which is considered to be too dangerous for use (kudos, therefore, to the famous Liebig!). Solid silver fulminate has the structure Ag-CNO (for some time in the twentieth century, it was *thought* to be CNO-Ag).[5] The structure of solid silver cyanate is Ag-NCO.[5,6] And here is some interesting irony. The cyanato ligand (NCO^-) could, in principle, combine with metals at N or O. Although the solid Ag-NCO should be named silver *isocyanate*, Brittin and Dunitz[6] bowed to history and indicated tolerance or at least resigned acceptance for retaining Liebig's nomenclature since the NCO-Ag isomer (the *real* silver isocyanate) is unknown.

In 1828, Wöhler attempted to synthesize ammonium cyanate (NH_4OCN) and found instead a substance having the same formula as the target compound but identical in all of its properties to urea (H_2NCONH_2). Urea is a component of mammalian urine and Wöhler wrote to his mentor Berzelius: ". . . I cannot, so to say, hold my chemical water, and must tell you that I can make urea, without thereby needing to have kidneys, or anyhow, an animal, be it human or dog. . . ."[7] This was the beginning of the end for the theory of Vitalism, which held that "organic" substances have a kind of vital force, since they had always been isolated from or at least related to living organisms. Thus, they could not be synthesized from nonorganic (really, elemental) substances.

Actually, Wöhler apparently may have first made ammonium cyanate as he had intended.[8] However, upon heating and evaporating off water, ammonium cyanate isomerizes *in solution* to urea.[8] Two years later, in 1830, Liebig and Wöhler actually synthesized solid ammonium cyanate by reaction of dry ammonia and cyanic acid.[9] In a sealed vessel under ammonia the crystals are now known to be stable, but if the vessel is opened, conversion to urea is complete in two days.[9] Dunitz and colleagues[9] also determined the structure of ammonium cyanate by x-ray crystallography. They note the difficulty in differentiating the N versus O end of cyanate by x-rays even using end-of- twentieth-century technology. The structure is found to be NH_4NCO. Just as in the silver case, in a formal sense, the solid could be called ammonium isocyanate but it is not.

Wöhler is said to have remained a believer in Vitalism.[8] It has been noted that just as Priestley, the phlogistonist, discovered oxygen, which was the downfall of phlogiston theory, Wöhler, the Vitalist, synthesized urea, which was the beginning of the downfall of Vitalism.[8] The true end for Vitalism came in the 1840s when the German chemist Hermann Kolbe effectively demonstrated the synthesis of acetic acid (the active component of vinegar, which is related to wine, which is related to sugar—therefore ORGANIC and ALL-NATURAL) from its constituent chemical elements through the following sequence:[10]

$$H_2 + O_2 \rightarrow H_2O$$
$$FeS_2 + C \rightarrow CS_2 + Fe$$
$$CS_2 + 2Cl_2 \rightarrow CCl_4 + 2S$$
$$2CCl_4 \rightarrow C_2Cl_4 + 2Cl_2$$
$$C_2Cl_4 + 2H_2O + Cl_2 \rightarrow CCl_3CO_2H + 3HCl$$
$$CCl_3CO_2H + 3H_2 \rightarrow CH_3CO_2H + 3HCl$$

where CH_3CO_2H is acetic acid.

There remained other issues to clarify in structural chemistry. In 1841, Berzelius developed the concept of allotropism—different arrangements of atoms in pure elements.[2] Modern examples include oxygen (O_2) and its allotrope ozone (O_3), sulfur, which commonly has octagons of sulfur atoms but can be heated to form "plastic sulfur" (long chains of sulfur atoms), and carbon, which is considered to have three allotropes (graphite, "infinite" sheets of carbon atoms; diamond, an "infinite" three-dimensional network of carbon atoms; and fullerenes such as C_{60}, buckminsterfullerene, a soccer-ball arrangement of carbon atoms). One could imagine each significant fullerene (C_{60}, C_{70}, C_{84}) as an allotrope. The confusing issue of different crystalline arrangements of the same substance was also solved by Berzelius who termed them *polymorphs*.[2]

We end this essay on an amusing note. John Darby, Professor of Chemistry and Natural Sciences in East Alabama College, describes isomeric bodies in his 1861 *Text Book of Chemistry*.[11] He correctly notes that ethyl formate and methyl acetate are "isomeric bodies" (their formulas are both $C_3H_6O_2$). He then goes on to say: "An explanation of these phenomena that attributes them to a different arrangement of atoms, is not satisfactory, as elementary bodies assume different states in inorganic chemistry, which is called allotropism, when such a cause is evidently impossible. We can only refer it, at present, to the will of the Creator."

1. E.L. Youmans, *Chemical Atlas*, Appleton, New York, 1857.
2. A.J. Ihde, *The Development of Modern Chemistry*, Harper and Row, New York, 1965, pp. 170–173.
3. W.H. Brock, *The Norton History of Chemistry*, Norton, New York, 1993, pp. 201–202; 214.
4. A.G. Sharpe, *Comprehensive Coordination Chemistry*, G. Wilkinson, R.D. Gillard, and J.A. McCleverty (eds.), Pergamon, Oxford, 1987, Vol. 2, pp. 12–14.
5. K. Vrieze and G. Van Koten, *Comprehensive Coordination Chemistry*, G. Wilkinson, R.D. Gillard, and J.A. McCleverty (eds.), Pergamon, Oxford, 1987, Vol. 2, pp. 227–236.
6. D. Brittin and J.D. Dunitz, *Acta Crystallographica*, 18: 424–428, 1965.
7. P.S. Cohen and S.M. Cohen, *Journal of Chemical Education*, 73: 883–886, 1996. I am grateful to Dr. Daniel Rabinovich for bringing this paper and the one in Reference 7 to my attention.
8. G.B. Kauffman and S.H. Chooljian, *Journal of Chemical Education*, 56: 197–200, 1979.
9. J.D. Dunitz, K.D.M. Harris, R.L. Johnston, B.M. Kariuki, E.J. MacLean, K. Psalidas, W.B. Schweitzer, and R.R. Tykwinski, *Journal of the American Chemical Society*, Vol. 120: 13 274–13 275, 1998.
10. W.H. Brock, op. cit., pp. 620–621.
11. J. Darby, *Text Book of Chemistry—Theoretical and Practical*, Cooper, Savannah and Barnes & Burr, New York, 1861, pp. 275–276.

TWO STREAMS IN THE PRIMEVAL FOREST

The darkness in Wöhler's "primeval forest" only deepened during the 1840s and early 1850s as the complexity of organic chemistry became ever more apparent.[1,2] Ironically, the vast majority of organic compounds are composed of only four elements: carbon, oxygen, hydrogen, and nitrogen.

For starters, there remained three sets of atomic masses for H, C, and O:[1,2] 1, 12, 16, Berzelius; 1, 6, 8, Liebig; I, 6, 16, Dumas. Berzelius' Dualistic Theory of chemistry remained an important organizing principle. Today we recognize that the vast majority of organic compounds, such as ethyl alcohol (C_2H_6O), are held together by covalent (electron-sharing) bonds; simple inorganic salts, such as sodium chloride, are composed of ions held together by electrostatic forces. However, it was not until 1884 that Svante August Arrhenius recognized ions as real entities independent of electrochemistry.[3]

The work of Davy and Berzelius clearly established the importance of electrical forces in holding compounds such as sodium chloride and water together. Electropositive elements substituted for other electropositive elements (e.g., HC1, KC1, $MgCl_2$); electronegative elements substituted for electronegative elements (e.g., Na_2O, NaCl, NaBr). In 1815, Gay-Lussac's studies of prussic acid (HCN) led him to discover cyanogen, $(CN)_2$, and a series of other compounds such as potassium cyanide (KCN) and silver cyanide (AgCN), which kept the CN radical intact as if it were an "atom." Indeed, cyanogen seemed to be as "elementary" as Cl_2.[1] Even more complex radicals were soon discovered: in studies of benzaldehyde and derivatives published in 1832, Liebig and Wöhler discovered the benzoyl radical (modern formula C_7H_5O)—exciting because it appeared to be an "intact" unit of three elements. But many vexing problems were coming to the fore, for example:

1. Gay-Lussac found that reaction of prussic acid (HCN) with chlorine gas produced cyanogen chloride (ClCN). How could an electronegative element replace an electropositive one?
2. Similarly, how is it that the hydrogen in chloroform (CHCL;) can be replaced by chlorine to produce CCl_4?
3. If benzoyl radical combines with chlorine to form benzoyl chloride, it should be electropositive. How can it include the electronegative element oxygen?
4. We understand that $SO_3 + H_2O \rightarrow H_2SO_4$. We know that $C_2H_4 + H_2O \rightarrow C_2H_6O$ (ethyl alcohol) and with HCl, C_2H_4 (the radical "etherin") forms C_2H_6Cl (ethyl chloride). But why does ethyl alcohol appear to release C_2H_3 radicals when reacted with sulfuric acid to form "sulfuric ether" ($C_4H_{10}O$)? Do the molecules break into different radicals when they please?

Figure 260 is from Youmans' *Chemical Atlas*. The year of this printing, 1857 (first printing 1854), occurs near the end of this chaotic period. The figure illustrates the prevailing confusion in theories as well as atomic weights.[1,2,4]

Brock's book has very nicely captured August Kekulé's metaphor of two streams of thought in the organization of organic chemistry.[2] The top illustration in Figure 260 shows the theory of compound radicals largely developed by Berzelius and favored by the German and English schools. In this stream of

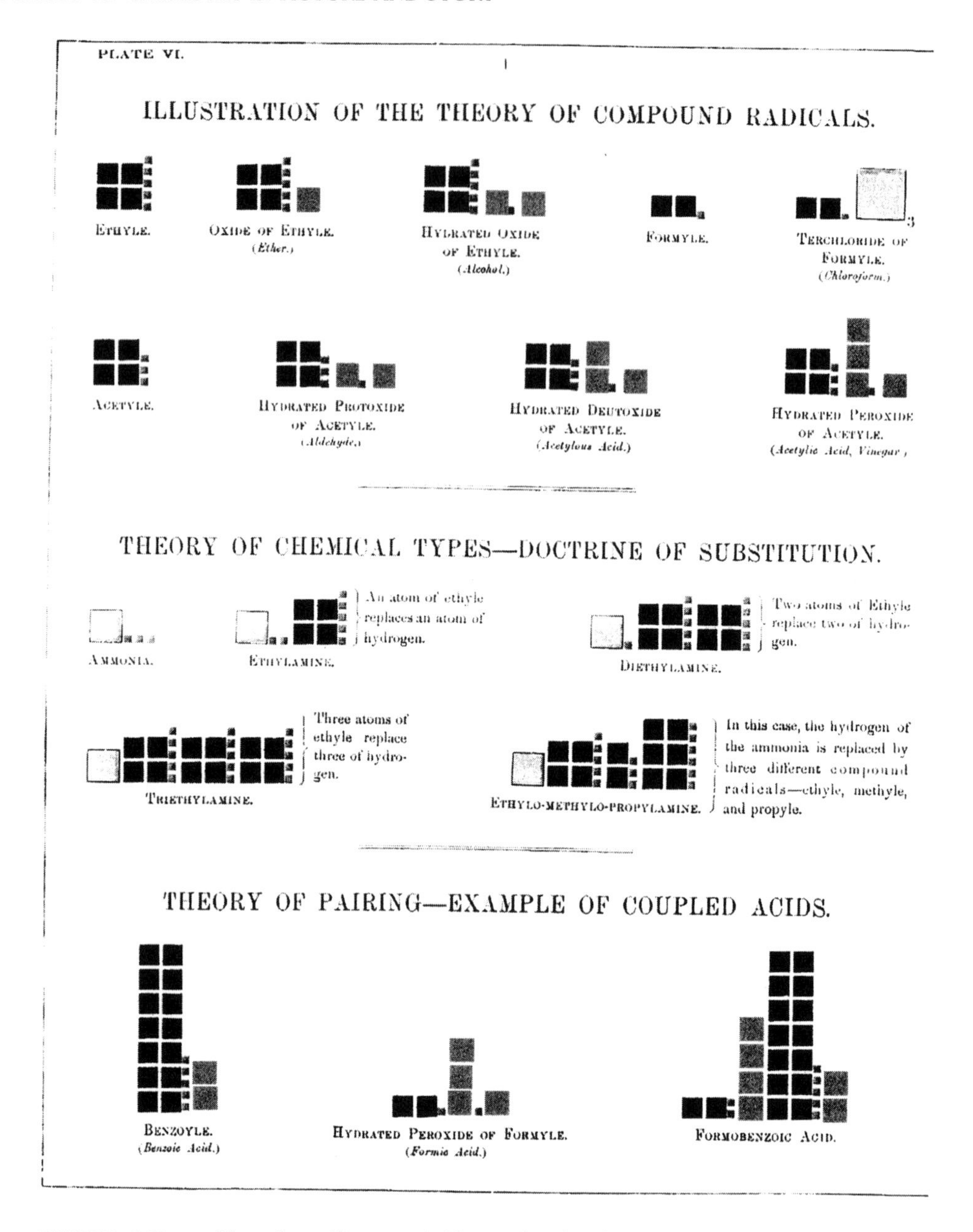

FIGURE 260. ■ Plate from Youmans' *Chemical Atlas* (original in color) depicting the three prevailing theories of organic chemistry structure in reactivity prior to Karlsruhe.

thought, organic radicals that exist independently join each other to form organic compounds. These radicals can be ethyl (C_2H_5, represented as C_4H_5 using atomic weight of C = 6 and the double carbon atom favored in Germany and used in Youmans' book) and hydroxyl (OH, represented as HO_2, using the atomic weight of oxygen as 8). The "formyl" radical, here described as C_2H—meaning CH—can combine with three chlorine radicals to form chloroform.

The middle illustration depicts the other stream of thought, the "theory of types," advanced initially by Dumas, in which compounds are related to a Chemical Type (or Class). Classes make sense. Acids, which share H in common, comprise a class—HCl, HBr, HCN. Similarly, salts such as NaCl, KCl, $MgCl_2$, NaBr, NaCN, and Na_2O also form a class. Here things get a bit "dicey" in light of being able to replace H by Cl (see above). Indeed, in an 1840 issue of *Annalen der Pharmacie und Chemie*, a certain S.C.H. Windier (also known as Friedrich Wöhler) published a satirical paper in which he reported replacing, in logical stages, all of the atoms in manganous acetate ("MnO + $C_4H_6O_3$") by chlorine, thus "demonstrating" that manganous acetate, a salt, and chlorine, a gas, were of the same "type."[1] Students saddled with the "Amines I" chapter of a typical modern-day organic text will have no trouble recognizing this figure: ethylamine, diethylamine, and triethylamine (and countless other amines) are in the "ammonia-type class" because they can be directly derived from ammonia. The "water-type class" was much more complicated. It included alcohols, ethers, carboxylic acids, esters, and acid anhydrides (at least five chapters in the modern Organic Chemistry text) and even extended to acids such as sulfuric and phosphoric. Gerhardt ultimately recognized four fundamental types: the nitrogen-type (amines and amides); water-type (see above); hydrogen-type (hydrocarbons, ketones, aldehydes); and hydrogen chloride-type (alkyi chlorides, acid chlorides, and related bromides).

The Theory of Pairing (Figure 260, bottom) was introduced by Berzelius to modify his radical theory. The idea was that one radical in the original compound retains its character in the new compound while the other changes its character (is "copulated") in the new compound, through substitution or rearrangement. This was used to explain, for example, Dumas' discovery in 1838 that chlorination of acetic acid produced trichloroacetic acid having essentially the same properties.[2] The acidic radical remained essentially constant, while the other associated radical changed.

Actually, the organization of organic chemistry by Auguste Laurent and Charles Frederic Gerhardt, *Les Enfants Terribles*,[2] ultimately more or less brought together aspects of the radical and type theories into the ten-pound organic texts lugged by the pre-med students of today and the multi-ounce CD ROMs of next year.

---

1. A.J. Ihde, *The Development of Modern Chemistry*, Harper and Row, New York, 1964, pp. 203–216.
2. W.H. Brock, *The Norton History of Chemistry*, Norton, New York, pp. 210–240.
3. A.J. Ihde, op. cit., pp. 413–415.
4. J.R. Partington, *A History of Chemistry*, MacMillan, London, 1964, Vol. 4, pp. 352–464.

NEVER SMILE AT A CACODYL

In 1835 Friedrich Wöhler called organic chemistry a "primeval forest of the tropics" (p. 422) and the metaphor, of an unimaginably complex living system, was seemingly an apt one. Organic compounds seemed to be isolable only from living creatures—plants and animals. Often, they had to be extracted from enormously complex matrices and were challenging to isolate pure. Even urine, a clear liquid, is extraordinarily complex. The simple organic compound urea (later understood to be CH_4N_2O) was reported in 1773 by Hilaire Martin Rouelle (it had earlier been described by Boerhaave).[1] The substance was impure, but its "soapy" texture and ease of decomposition upon "distillation" marked it as notably different from typical inorganic salts, which were crystalline and usually heat-stable. Although Wöhler quite accidentally synthesized urea from inorganic compounds in 1828, he retained, at the time, the prevalent opinion that organic compounds were imbued with a "vital force" and could thus never be made artificially (p. 429). Some two decades later Hermann Kolbe "killed vitalism" by quite effectively synthesizing acetic acid ("vinegar") from its chemical elements. There were three prevalent systems of relative atomic weights in common use by the midnineteenth century. These have been summarized by Aaron J. Ihde as follows:[2]

	H	C	O
Berzelius	1	12	16
Liebig	1	6	8
Dumas	1	6	16

Further confusion was caused by the difficulties in precise chemical analysis. Early nineteenth-century analyses involved measurements of the volume of CO_2 generated during combustion. This limited analytical samples to quite small quantities and had the effect of magnifying small errors. Even Liebig's kaliapparat, which captured carbon dioxide in condensed form for weighing, greatly increasing sample size and accuracy, did not completely solve these problems. This is well illustrated[3] by the disparity between the formula of cholic acid ($C_{48}H_{39}O_9$, Liebig atomic weights) reported by the famous chemist Adolph Strecker working in Liebig's Giessen laboratory in 1848, and the present-day formula ($C_{24}H_{40}O_5$). The disparities, albeit small, are highly significant since use of the Berzelius atomic weights for C, H, and O (very close to the modern values) would have given the formula $C_{24}H_{19.5}O_{4.5}$. This was, of course, incompatible with whole atoms as well as the rules of valence that were still about a decade into the future.

Organic compounds also posed numerous problems for early chemical theory. Davy's electrolytic studies, which produced electropositive potassium at one electrode and electronegative chlorine at the other, led Berzelius to postulate a theory of dualism. Hydrogen was clearly electropositive—it formed water and hydrogen chloride with the electronegative elements oxygen and chlorine. Electropositive carbon also formed compounds with oxygen and chlorine. However, organic compounds did not seem to fit this theory. How could one explain methane—a compound of carbon and hydrogen—both electropositive elements? How could an electronegative element such as chlorine fully replace the four electropositive methane hydrogen atoms to form CCl_4?

Among the early significant discoveries that helped clarify and systematize organic chemistry was the notion of a radical ("from the root") that had its earliest origins with Lavoisier: acid = radical + oxygen (where the radical could be the element sulfur whose combination with oxygen formed "sulfuric acid"—really SO_3).[4] This crude concept was followed by much more refined studies that disclosed the existence of the cyano radical (CN). It was Scheele who first treated the pigment Prussian Blue, which consists of iron compounds of ferrocyanide [today $Fe(CN)_6$], possibly in the presence of alkali metals or ammonia {e.g.,

$NH_4Fe[Fe(CN)_6]$—modern formula, of course}. Treating potassium ferrocyanide with sulfuric acid, he obtained "prussic acid" ("hydrocyanic acid" or hydrogen cyanide, HCN), and it was remarkable that he did not kill himself testing its odor. Scheele obtained potassium cyanide (KCN), mercury cyanide $[Hg(CN)_2]$, and silver cyanide (AgCN).[5] In 1787, Berthollet reacted "prussic acid" with chlorine and discovered cyanogen chloride (ClCN).[6] In 1815, Gay-Lussac discovered cyanogen $[(CN)_2]$ from his work on "prussic acid."[7] Thus, a body of evidence suggested that the CN radical acts virtually like an atom (i.e., "Cy") in passing unchanged from compound to compound.[8] Even more exciting was the 1832 publication by Liebig and Wöhler of the benzoyl radical ("$C_{14}H_{20}O_2$"—actually $C_7H_{10}O$), a stable unit containing three different types of atoms.[8]

Figure 260 is from the 1857 edition of Youmans' *Chemical Atlas*.[9] Radicals were initially considered to be stable "superatoms," which were joined, separated, and recombined to form molecules. The first radical in Figure 260 is ethyl, depicted as C_4H_5 (using the Liebig convention and thus C_2H_5 in modern terms). If we look at the third structure on the top, we see diethyl ether, which we recognize today as $C_4H_{10}O$ rather than C_4H_5O (or C_2H_5O using Liebig atomic weights). From a dualistic viewpoint, the C_4H_5O "molecule" is composed of the "ethyl" radical ("C_4H_5") as the electropositive part and O as the electronegative part. Seemingly, addition of water ("OH") to ethyl ether forms its hydrate, also known as "ethyl alcohol" (here as $C_4H_5 \cdot OH \cdot O$; really C_2H_6O using modern, rather than Liebig, atomic weights). In any case, the early theory of compound radicals (top third of Figure 260) depicted simple exchange of stable radicals to form different organic molecules.

This brings us to cacodyl, an early name for the awful smelling, spontaneously flammable, colorless liquid tetramethyldiarsine, obtained by heating arsenious oxide and potassium acetate:[10]

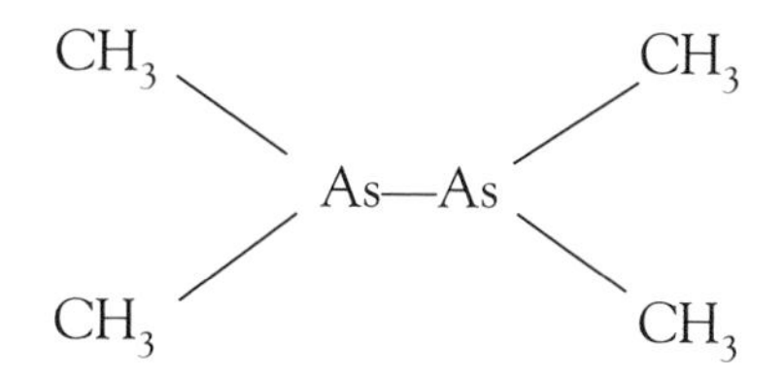

The cacodyl radical $[(CH_3)_2As]$[13] also appeared to Bunsen to be a stable "superatom" that could be exchanged amongst other radicals. Many cacodyl compounds are explosive as well as spontaneously flammable. One of these, cacodyl cyanide $[(CH_3)_2AsCN]$, exploded during Bunsen's exploratory studies, and he lost his right eye.[11]

In one of his studies, Bunsen synthesized cacodyl oxide from cacodyl and converted it to the chloride. Upon reaction with zinc, chlorine was lost and the pure arsenic, carbon, hydrogen compound remaining was thought by Bunsen to be the free cacodyl radical $(CH_3)_2As$. In fact it was liquid cacodyl $[(CH_3)_2As—As(CH_3)_2]$.[12] Similarly, reaction of ethyl iodide with zinc freed the organic molecule of iodine and was thought to produce the free radical "ethyl" but in fact yielded butane (C_2H_5—C_2H_5)—the dimer of ethyl. This work by Edward Frankland (1825–1899) actually produced some diethylzinc, $(C_2H_5)_2Zn$, a spontaneously flammable volatile substance, which heralded the beginning of organometallic chemistry.[12] Searches for these free radicals were fruitless, and they were thus assumed to be incapable of isolation until the unexpected observation of the triphenylmethyl radical by Moses Gomberg.[13] Gomberg reacted triphenylmethyl chloride with zinc dust expecting to obtain hexaphenylethane:[14]

$$2\ \Phi_3C\text{—}Cl + Zn \xrightarrow{??} \Phi_3C\text{—}C\Phi_3 + ZnCl_2 \qquad \text{(where } \Phi = \text{phenyl or } C_6H_5\text{)}$$

What occurred was a surprising reaction forming a colored solution. Addition of iodine, for example, produced triphenylmethyl iodide and a colorless solution. Gomberg had generated a stable, yet reactive, free radical—triphenylmethyl radical:

$$2\ \Phi_3C\text{—}Cl + Zn \rightarrow ZnCl_2 + 2\ \Phi_3C. \xrightarrow{I_2} 2\ \Phi_3C\text{—}I$$

Interestingly, although the dimer of triphenylmethyl radical was thought to be the expected hexaphenylethane for some 60 years, we now know the correct structure of the dimer that exists in equilibrium with triphenylmethyl radical—it is not an ethane derivative.[14] Indeed, although pentaphenylethane ($C\Phi_3$—$CH\Phi_2$) exists and has abnormally long C—C bonds,[15] almost a century after Gomberg's discovery, hexaphenylethane continues to elude clever modern chemists.[16]

It is a wonderful irony to note that there is indeed a truly stable (actually, "persistent") cacodyl free radical.[17,18] Some 20 years ago it was discovered that mild heating of the compound $[(CH_3)_3Si)_2CH]_2As$—$As[CH(Si\{CH_3\}_3)_2]_2$ produces two $As[CH(Si\{CH_3\}_3)_2]_2$ radicals that are stable and observable for indefinite periods in solution at 25°C.[17,18] The trick here is the group of four huge $[(\{CH_3\}_3Si)_2CH]$ groups that hinder recombination of the radicals and formation of the weak[19] As—As bond.

The middle part of Figure 260 depicts the type theory in which, for example, an amine type could be replaced in turn by alkyl radicals in a series clearly related to ammonia. This was a positive contribution since it recognized families of related compounds (functional groups). The bottom section in Figure 260 illustrates the "pairing" ("copula") theory advanced by Berzelius in one final attempt to rescue dualism. As Ihde illustrates,[20] acetic acid ($C_2H_4O_2$, modern formula) could be rationalized (using "double formulas") as a combination of an electropositive part (C_4H_6), an electronegative part (O_3), and water (H_2O). Trichloroacetic acid ($C_2HCl_3O_2$) was troubling to Berzelius. The substitution theory of the period visualized simple replacement of the hydrogens in the C_4H_6 radical by chlorines. The resulting formula, $C_4Cl_6 + O_3 + H_2O$, now has a serious "charge imbalance" since the carbon-containing radical is much less electropositive if not "downright electronegative" while the O_3 part is now fully electronegative. Berzelius felt that a dramatic rearrangement was needed such that the chlorinated carbon part (now C_2Cl_6) was "coupled" ("copulated") with "oxalic acid" (C_2O_3) with water a recognizable unit: $C_2Cl_6 + C_2O_3 + H_2O$. In this way, increased electronegativity in the C_2Cl_6 part was balanced by decreased electronegativity in the C_2O_3 part. The bottom section of Figure 260 depicts this kind of rearrangement of atoms in the "combination" of benzoyl radical with formic acid to produce formylbenzoic acid. However, a variety of chemical investigations clearly showed that acetic acid (CH_3COOH) and trichloroacetic acid (CCl_3COOH) were very closely related chemically and dualism was forced to disappear as a viable theory in organic chemistry. Resolution would begin to occur only with the development of the valence concept.

---

1. J.R. Partington, *A History of Chemistry*, MacMillan and Co. Ltd., London, 1962, Vol. 3, p. 78.
2. A.J. Ihde, *The Development of Modern Chemistry*, Harper & Row, New York, 1964, p. 191.
3. W.H. Brock, *The Norton History of Chemistry*, Norton, New York, 1993, pp. 194–207.
4. J.R. Partington, *A History of Chemistry*, MacMillan and Co. Ltd., London, 1964, Vol. 4, pp. 142–177.
5. Partington (1962), op. cit., pp. 233–234.
6. Partington (1962), op. cit., p. 511.
7. Partington (1964), op. cit., pp. 253–254.
8. Ihde, op. cit., pp. 184–189.
9. E. Youmans, *Chemical Atlas*, Appleton, New York, 1857.
10. D.H. Hey (ed.), *Kingzett's Chemical Encyclopedia*, ninth edition, Baillière, Tindall and Cassell, London, 1966, p. 149.
11. Partington (1964), op. cit., pp. 283–286.
12. J. Hudson, *The History of Chemistry*, Chapman & Hall, New York, 1992, pp. 114–116.
13. Ihde, op. cit., pp. 619–621.
14. F.A. Carroll, *Perspectives on Structure and Mechanism in Organic Chemistry*, Brooks/Cole Publishing Co., Pacific Grove, CA, 1998, pp. 257–258.
15. R. Destro, T. Pilati, and M. Simonetta, *Journal of the American Chemical Society*, Vol. 100, pp. 6507–6509, 1978.
16. C.R. Arkin, B. Cowans and B. Kahr, *Chemistry of Materials*, Vol. 8, pp. 1500–1503, 1996.
17. M.J.S. Gynane, A. Hudson, M.F. Lappert, P.P. Power and H. Goldwhite, *Journal of the Chemical Society Dalton Transactions*, pp. 2428–2433, 1980.
18. P.R. Hitchcock, M.F. Lappert, and S.J. Smith, *Journal of Organometallic Chemistry*, Vol. 320, pp. C27–C30, 1987.
19. Pauling provides a value of only 32.1 kcal/mol (kilocalories per mole) for the As—As bond energy, compared to 83.1 kcal/mol for a typical C—C bond (L. Pauling, *The Nature of the Chemical Bond*, third edition, Cornell University Press, Ithaca, 1960, p. 85.) Actually, there does not appear to be data allowing a good determination or estimate of the As—As bond in "cacodyl" according to J.F. Liebman, J.A. Martinho-Simões, and S.W. Slayden, in *The Chemistry of Organic Arsenic, Antimony and Bismuth Compounds*, S. Patai (ed.), John Wiley and Sons, Chichester, 1994, pp. 153–168. Interestingly, the carbon–carbon bond in cyanogen is quite strong (134.7 kcal/mol using National Institute of Standards data—see *http://nist.gov*). In this case, as strong as the C—C bond is, it is much weaker than the carbon–nitrogen triple bonds in cyanogen that maintain the integrity of the cyano "radical."
20. Ihde, op. cit., p. 198.

WANT A GREAT CHEMICAL THEORY? JUST LET KEKULÉ SLEEP ON IT

Toward the end of the 1850s, two major advances occurred to begin the taming of the "primeval forest." In 1858, Stanislao Cannizzaro emphasized the importance of Avogadro's Hypothesis, first published in 1811, that equal volumes of gas (same temperature and pressure) had equal numbers of ultimate units (molecules of Cl_2, O_2, P_4; atoms of Hg vapor). Using the ideal gas law ($PV = nRT$) and the Dumas technique, the atomic masses of atoms not volatile in the elemental state could also be measured in molecules if the other atoms' weights were known. Thus, at the end of the decade and at the Karlsruhe Conference in 1860, the atomic mass problem was largely solved. Not only did this set the table for the periodic law, but it also brought coherence to chemical formulas.

The other major advance was the realization that carbon is tetravalent, first enunciated narrowly by Friedrich August Kekulé (1829–1896) in 1857, and then for all carbon-containing compounds in 1858.[1] Kekulé started at the University of Giessen as an architecture student before he was drawn into chemistry by Liebig. The concept of valence, sometimes credited to Kekulé for work published in 1854, appears to be due to William Odling and others a bit earlier.[1] Equally important was his idea that carbons are directly linked—equally sharing their "affinities." This idea was in direct conflict with the electrochemical-dualism theory.[1] Kekulé recalled how he had integrated pictures with the extant data to arrive at the tetravalence theory in 1854 while riding a London omnibus:[2]

> I fell into a reverie. The atoms were gambolling before my eyes. I had always seen them in motion, these small beings, but I had never succeeded in discerning the nature of their motion. Now, however, I saw how, frequently, two smaller atoms united to form a pair; how a larger one embraced two smaller ones; how a still larger one kept hold of three or even four of the smaller; whilst the whole kept whirling in a giddy dance. I saw how the larger ones formed a chain and the smaller ones hung on only at the end of the chain.

History indicates that Archibald Scott Couper (1831–1892) discovered the tetravalence of carbon (and carbon–carbon bonding) simultaneously with and independently of Kekulé. His publication was delayed for technical reasons by his Director Adolph Wurtz. When he was "scooped" by Kekulé, he complained bitterly and was promptly fired by Wurtz. A physical breakdown before he was 30 effectively ended this promising scientist's career.[3]

The recognition that carbon is tetravalent established the foundation for structural organic chemistry. In 1861 Aleksandr Butlerov first stated that the particular arrangement of atoms in a molecule is responsible for the substance's physical and chemical properties:4

> Only one rational formula is possible for each compound, and when the general laws governing the dependence of chemical properties on chemical structure have been derived, this formula will express all of these properties.

This sentence still belongs in the first lecture of any modern course in organic chemistry. Figure 261 is taken from the 1868 Leipzig edition of Butlerov's *Lehrbuch Der Organischen Chemie* (original Russian edition, 1864). The formulas show the tetravalence of carbon and clearly express the differences in structures between isomers.

Benzene, first obtained from compressed oil gas in 1825 by Faraday, was an interesting enigma. With a formula of C_6H_6, it was highly "unsaturated" and would have been expected to undergo addition reactions like ethylene and other olefins. Its chemistry was remarkably different from that expected. Once again Kekulé claims to have dreamed up a solution:[5]

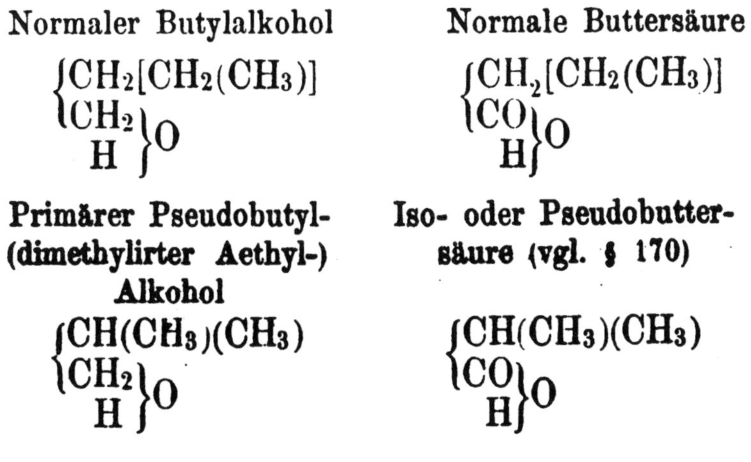

Normaler Butylalkohol

$$\left\{\begin{matrix} CH_2[CH_2(CH_3)] \\ \left.\begin{matrix} CH_2 \\ H \end{matrix}\right\} O \end{matrix}\right.$$

Normale Buttersäure

$$\left\{\begin{matrix} CH_2[CH_2(CH_3)] \\ \left.\begin{matrix} CO \\ H \end{matrix}\right\} O \end{matrix}\right.$$

Primärer Pseudobutyl- (dimethylirter Aethyl-) Alkohol

$$\left\{\begin{matrix} CH(CH_3)(CH_3) \\ \left.\begin{matrix} CH_2 \\ H \end{matrix}\right\} O \end{matrix}\right.$$

Iso- oder Pseudobuttersäure (vgl. § 170)

$$\left\{\begin{matrix} CH(CH_3)(CH_3) \\ \left.\begin{matrix} CO \\ H \end{matrix}\right\} O \end{matrix}\right.$$

FIGURE 261. ■ Figure from Butlerov's 1868 Leipzig edition of *Lehrbuch Der Organischen Chemie* (original Russian Edition, 1864). Butlerov first enunciated the modern structural basis for organic chemistry.

> I was sitting, writing at my textbook; but the work did not progress; my thoughts were elsewhere. I turned my chair to the fire and dozed. Again the atoms were gambolling before my eyes. This time the smaller groups kept modestly in the background. My mental eye, rendered more acute by repeated visions of the kind, could now distinguish larger structures, of manifold conformation: long rows, sometimes more closely fitted together; all twining and twisting in snake-like motion. But look! What was that? One of the snakes had seized hold of its own tail, and the form whirled mockingly before my eyes. As if by a flash of lightning I awoke; and this time also I spent the rest of the night working out the consequences of the hypothesis.

The sausagelike model of benzene is at the top right of the group of structures in Figure 262 from Kekulé's 1865 paper in the *Bulletin Société Chimiques*.[5] Benzene, chlorobenzene, and a dichlorobenzene are shown later. Later work by Ladenburg and Körner caused Kekulé to postulate two equivalent alternating structures of benzene in 1872. What was Kekulé reading before he dozed off? Perhaps Libavius' Alchymia (Figure 137) inspired his serpent dreams. Perhaps Porta's tortoise [Figure 71(b)] was the clue he needed for benzene's structure. I am convinced, however, that (1) Kekulé accomplished more asleep than I have awake; (2) I am going to seek a lighter course load to add napping to my yearly activity report.

Figure 263 is from a rare pamphlet distributed at an 1886 meeting of the German Chemical Society that celebrated Kekulé's structure work.[6] The monkeys adopt two rapidly alternating structures (tails entwined and not entwined). The modern representations for benzene [dotted or solid circle in the hexagon — see Figure 71(b)] were contributed by Johannes Thiele in 1899 and Sir Robert Robinson in 1925.[7]

The ability to explain the complex substitution chemistry of benzene and other related aromatic derivatives as the result of benzene's structure signaled the triumph of structural chemistry. When Hermann Kolbe, who hated Kekulé, died in 1884, the last real resistance to structural chemistry died with him.[8] In the eloquent words of Brock:[8]

108 BULLETIN DE LA SOCIÉTÉ CHIMIQUE.

ces principes permettra-t-elle de prévoir de nouvelles métamorphoses et de nouveaux cas d'isomérie.

Qu'il me soit permis, en terminant, de faire une observation sur les formules rationnelles par lesquelles on pourrait représenter la composition des substances aromatiques et sur la nomenclature qu'il conviendrait de leur appliquer.

Il est vrai que les substances aromatiques présentent sous plusieurs rapports une grande analogie avec les substances grasses, mais on ne peut pas manquer d'être frappé du fait que sous beaucoup d'autres rapports elles en diffèrent notablement. Jusqu'à présent, les chimistes ont insisté surtout sur ces analogies; ce sont elles qu'on s'est efforcé d'exprimer par les noms et par les formules rationnelles. La théorie que je viens d'exposer insiste plutôt sur les différences, sans toutefois négliger les analogies qu'elle fait découler, au contraire, là où elles existent réellement, du principe même.

Peut-être serait-il bon d'appliquer les mêmes principes à la notation des formules, et, quand on a de nouveaux noms à créer, aux principes de la nomenclature.

Dans les formules on pourrait écrire, comme substitution, toutes les métamorphoses qui se font dans la chaîne principale (noyau); on pourrait se servir du principe de la notation typique pour les métamorphoses qui se font dans la chaîne latérale, lorsque celle-ci contient du carbone. C'est ce que l'on a tenté dans ce Mémoire pour plusieurs formules, en supprimant toutefois des formules typiques la forme triangulaire que la plupart des chimistes ont acceptée, en suivant l'exemple de Gerhardt, et que l'on ferait bien, selon moi, d'abandonner complétement à cause des nombreux inconvénients qu'elle entraîne.

Je ne dirai rien sur les principes que l'on pourrait suivre en formant des noms. Il est toujours aisé de trouver des noms qui expriment une idée donnée, mais tant qu'on n'est pas d'accord sur les idées, il serait prématuré d'insister sur les noms.

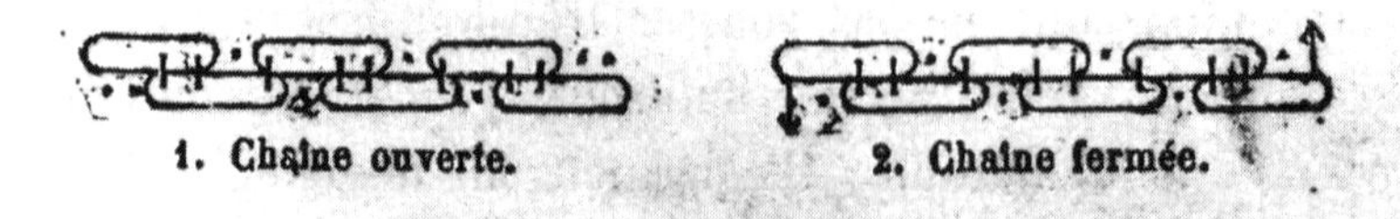

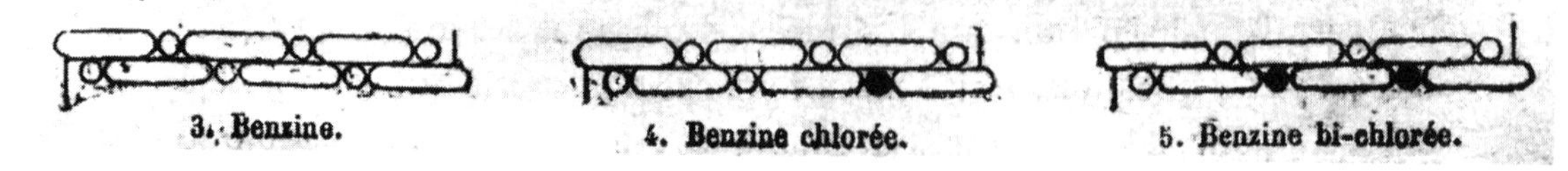

FIGURE 262. ■ Kekulé's "sausage formulas" for benzene and two benzene derivatives appearing in *Bulletin de la Société Chimique* (Paris), Vol. 3, p. 98, 1865) (courtesy Edgar Fahs Smith Collection, Rare Book & Manuscript Library, University of Pennsylvania).

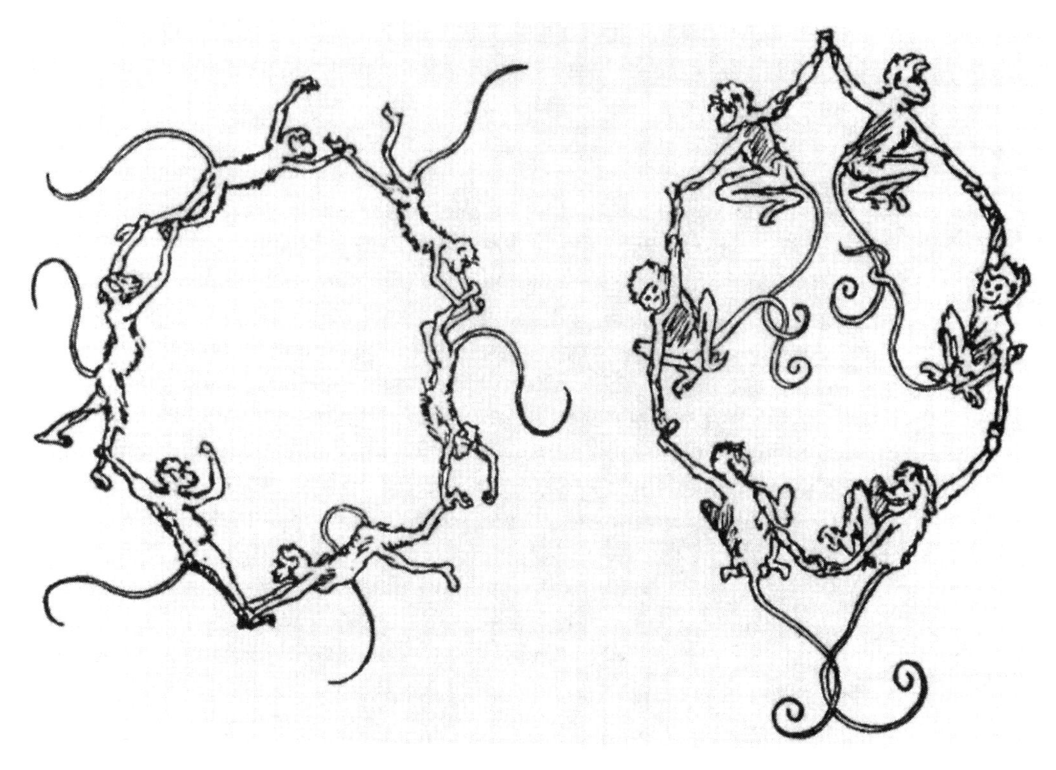

FIGURE 263. ▪ Satirical celebration of Kekulé's benzene structures by the German Chemical Society in 1886 (see E. Heilbronner and J.D. Dunitz, *Reflections on Symmetry*, VCH, Weinheim, 1993, p. 52; courtesy of John Wiley–VCH).

Just as Picasso had transformed art by allowing the viewer to see within and behind things, so Kekulé had transformed chemistry. Chemical properties arose from the internal structures of molecules, which could now be "seen" and "read" through the experienced optic of the analytical and synthetic chemist.

---

1. J.R. Partington, *A History of Chemistry*, MacMillan, London, 1964, Vol. 4, pp. 533–565.
2. J.R. Partington, op. cit., p. 537.
3. A.J. Ihde, *The Development of Modern Chemistry*, Harper and Row, New York, 1964, pp. 222–225.
4. W.H. Brock, *The Norton History of Chemistry*, Norton, New York, 1993, p. 256.
5. A.J. Ihde, op. cit., pp. 310–319.
6. E. Heilbronner and J.D. Dunitz, *Reflections On Symmetry*, VCH, Weinheim, 1993, p. 52.
7. W.H. Brock, op. cit., p. 555.
8. W.H. Brock, op. cit., pp. 263–269.

"MY PARENTS WENT TO KARLSRUHE AND ALL I GOT WAS THIS LOUSY TEE-SHIRT!"

I apologize, gentle reader, for this tacky take-off on the archetypal All-American souvenir tee-shirt. In all likelihood, the 140 chemists who came to the pleasant Rhineland town of Karlsruhe in 1860 did little boating and souvenir shopping. Human endeavors, such as science, are collective efforts. We need human interaction to jostle, needle, annoy, and inspire us—the sum 15 greater than the parts. Are we different from the termites that must first assemble in great number before they can develop a "collective idea" and construct a mound? Well, frankly yes—we're big, they're small; we get coffee breaks and TV, they don't. Still, we do touch "antenna," use the phone, send e-mail, send snail mail, give talks, attend seminars and symposia, read articles and books, chat at breakfast, and hide from assessment reports at professional meetings—the remoter, the better.[1]

Attempts to classify the chemical elements began in the early nineteenth century. Johann Wolfgang Döbereiner (1780–1849) noted during 1816 and 1817 that strontium, which is chemically similar to calcium and barium, had an atomic weight that was the arithmetic average of the other two.[2] By 1829, he had noted other such "triads" and claimed to have correctly predicted the atomic weight of the newly discovered bromine by averaging chlorine and iodine.[2] Johann Wolfgang von Goethe (1749–1832) spent extended periods in his teacher Döberiner's laboratory although Döbereiner does *not* appear to be a model for *Faust*.

On September 3, 1860 the Karlsruhe Conference convened in order to attempt to settle vexing issues pertaining to atoms, molecules, equivalents, nomenclature, and atomic weights.[3] The clarity on atomic weights provided by Cannizzaro's 1858 pamphlet and presentations at the conference moved Julius Lothar Meyer to comment:[2]

> The scales seemed to fall from my eyes. Doubts disappeared and a feeling of quiet certainty took their place. If some years later I was able myself to contribute something toward clearing the situation and calming heated spirits no small part of the credit is due to this pamphlet of Cannizzaro.

Stanislao Cannizzaro (1826–1910), born in Palermo, was the star of Karlsruhe. He recalled for all assembled the importance of Avogadro's hypothesis, combined it with the Law of Dulong and Petit and other findings and clarified the atomic weights that became the "*y* axis" for the future Periodic Table even as the chemical properties were to become the "*x* axis." In Figure 264, we see Cannizzaro's delineation of atoms, molecules, and atomic weights. The "half-molecule" concept derives from Avogadro's 1811 nomenclature. Figure 265 is a very straightforward exposition of the Law of Dulong and Petit. In this table, all triatomic solids have almost the same specific heat per atom, regardless of the identity of the atom. It is a powerful validation of atomic theory as well as the atomic weights employed.

1. I gratefully acknowledge Lewis Thomas' book, *Lives of a Cell*, Viking Press, New York, 1974, for its influence on this essay.
2. A.J. Ihde, *The Development of Modern Chemistry*, Harper and Row, New York, 1964, pp. 236–237.
3. A.J. Ihde, op. cit., pp. 228–229.

WHAT ARE ORGANIC CHEMISTS GOOD FOR?

Until the middle of the nineteenth century the dyes used in textiles and other commercial applications had their origins in plant and animal matter.[1] Indeed,

	Simboli delle molecole dei corpi semplici e formule dei loro composti fatte con questi simboli, ossia simb. e form. rappresentanti i pesi di volumi eguali allo stato gassoso		Simboli degli atomi de'corpi semplici, e formule dei composti fatte con questi simboli		Numeri esprimenti pesi corrispondent
Atomo dell'idrogeno . . .	$\mathfrak{H}^{1}/_{2}$	=	H	=	1
Molecola dell'idrogeno . .	$\mathfrak{H}$	=	H^2	=	2
Atomo del cloro	$\mathfrak{Cl}^{1}/_{2}$	=	Cl	=	35,5
Molecola del cloro. . .	$\mathfrak{Cl}$	=	Cl^2	=	71
Atomo del bromo	$\mathfrak{Br}^{1}/_{2}$	=	Ar	=	80
Molecola del bromo . . .	$\mathfrak{Br}$	=	Br^2	=	160
Atomo dell'iodo	$\mathfrak{I}^{1}/_{2}$	=	I	=	127
Molecola dell'iodo. . . .	$\mathfrak{I}$	=	I^2	=	254
Atomo del mercurio . . .	$\mathfrak{Hg}$	=	Hg	=	200
Molecola del mercurio . .	$\mathfrak{Hg}$	=	Hg	=	200
Molec. dell'acido cloridrico	$\mathfrak{H}^{1}/_{2}\mathfrak{Cl}^{1}/_{2}$	=	HCl	=	36,5
Mol. dell'acido bromidrico.	$\mathfrak{H}^{1}/_{2}\mathfrak{Br}^{1}/_{2}$	=	HBr	=	81
Mol. dell'acido iodidrico .	$\mathfrak{H}^{1}/_{2}\mathfrak{I}^{1}/_{2}$	=	HI	=	128
Mol. del protocl. di merc.	$\mathfrak{Hg}\mathfrak{Cl}^{1}/_{2}$	=	HgCl	=	235,5
Mol. del protobr. di merc.	$\mathfrak{Hg}\mathfrak{Br}^{1}/_{2}$	=	HgBr	=	280
Mol. del protoiod. di merc.	$\mathfrak{Hg}\mathfrak{I}^{1}/_{2}$	=	HgI	=	327
Mol. del deutoclor. di merc.	$\mathfrak{Hg}\mathfrak{Cl}$	=	$HgCl^2$	=	271
Mol. del deutobr. di merc.	$\mathfrak{Hg}\mathfrak{Br}$	=	$HgBr^2$	=	360
Mol. del deutoiod. di merc.	$\mathfrak{Hg}\mathfrak{I}$	=	HgI^2	=	454

FIGURE 264. ■ Cannizzaro's system of atomic weights based upon Avogadro's Hypothesis (of 1811) and recalled in his 1858 paper and presentation at the 1860 Karlsruhe Congress. This figure is from S. Cannizzaro, *Scritti Intorno Alla Teoria Molecolare Ed Atomica ed alla Notazione Chimica Di S. Cannizzaro* (Palermo, 1896).

Formule dei composti	Pesi delle loro molecole $=p$	Calorici specifici dell'unità di peso $=c$	Calorici specifici delle molecole $=p\times c$	Numeri di atomi nelle molecole $=n$	Calorici specifici di ciascun atomo $=\frac{p\times c}{n}$
$HgCl^2$	271	0,06889	18,66919	3	6,22306
$ZnCl^2$	134	0,13618	18,65666	3	6,21888
$SnCl^2$	188,6	0,10161	19,163646	3	6,387882
$MnCl^2$	126	0,14255	17,96130	3	5,98710
$PbCl^2$	278	0,06641	18,46198	3	6,15399
$MgCl^2$	95	0,1946	18,4870	3	6,1623
$CaCl^2$	111	0,1642	18,2262	3	6,0754
$BaCl^2$	208	0,08957	18,63056	3	6,21018
HgI^2	454	0,04197	19,05438	3	6,35146
PbI^2	461	0,04267	19,67087	3	6,55695

FIGURE 265. ■ Cannizzaro's use of the Law of DuLong and Petit to strengthen his system of atomic weights (see Figure 264).

indigo dyes from three species of snails were the basis of the ancient dye *tekhelet*, specified by Moses for coloring blue the fringes of the Hebrew prayer shawl or *tallit*.[2] This interesting history is related by Roald Hoffmann and Shira Leibowitz Schmidt, who note that *tekhelet* was likely a mixture of two closely related indigo dyes.[2] The Hebrews, according to their lore, lost the art of making *tekhelet* by the year 760. Since that time, the fringes have been white since no substitutes were allowable under religious law. Despite the subsequent discovery of plant sources for these dyes and modern chemical techniques that definitively validate their identities, the modern tradition of white fringes remains firm. And since, as the authors note, "there is no authentic Hebrew textile dyed in *tekhelet* that has survived," no attempt to re-create *tekhelet* is likely to be acceptable.[2]

In the middle of the nineteenth century, the precocious William Henry Perkin (1838–1907) entered the Royal College of Chemistry at the age of 15 and soon became an assistant to its Director, Professor August Wilhelm Hofmann.[3,4] By that time, coal tar had become an unwanted waste product and while commercial benzene and toluene had been obtained from coal tar by distillation, it was still considered a massive nuisance.[5] Working in his home laboratory in London in 1856, young Perkin tried unsuccessfully to synthesize the drug quinine but obtained instead dark tars. A modification, using the coal-tar component aniline, provided another dark substance that was found, again quite by accident, to be an excellent purple dye, that Perkin named *mauve*. Perkin left the university, much to Hoffmann's dismay, and built a factory to manufacture mauve financed by his father. Suddenly, a synthetic dye industry emerged and coal tar became a commodity rather than a waste product.[1,5]

Figure 266 is a family tree showing the development of synthetic dyes for about the first 75 years following the discovery of mauve by Perkin.[6] The limb branching to the left of mauve includes a series of chemically related dyes, some of which are named in Figure 267a. There is a fairly smooth transition in color from the purplish-red fuchsine through a series of three violet dyes to the two blue dyes whose structures are shown in this figure. In Figure 267(a), fuchsine is employed as the "core dye" and the five others differ slightly through substitutions of the highlighted groups for hydrogen atoms. The custom synthesis and fine tuning of the colors of these dyes illustrate one of the fundamental strengths of organic chemistry. Organic chemists are experts at "tweaking" the properties of complex molecules by substituting atoms or groups of atoms for each other. The difference between fuchsine and methyl violet B is a rather simple replacement of five hydrogen atoms by five methyl groups.

It is interesting to note that each of the six compounds in Figure 267(a) has one nitrogen atom forming five bonds with other atoms. Indeed, the beautiful book, published in 1935, containing the structures in Figure 267(a) also includes some rather "precious" cartoons of atoms with hands signifying valences.[6] Figure 267(b) depicts a water molecule in which one oxygen and two hydrogen atoms hold hands [rather like the "water fairies" depicted in Figure 295(b)]. In Figure 267(c) we note illustrations of the valence of each atom including a valence of 5 for nitrogen. The confusion may be illustrated for ammonium chloride (NH_4Cl). Since the valences of hydrogen and chlorine are commonly known to be 1, the only reasonable arrangement appears to have nitrogen forming single bonds with the four hydrogen atoms as well as with the fifth atom, chlorine. It had previous-

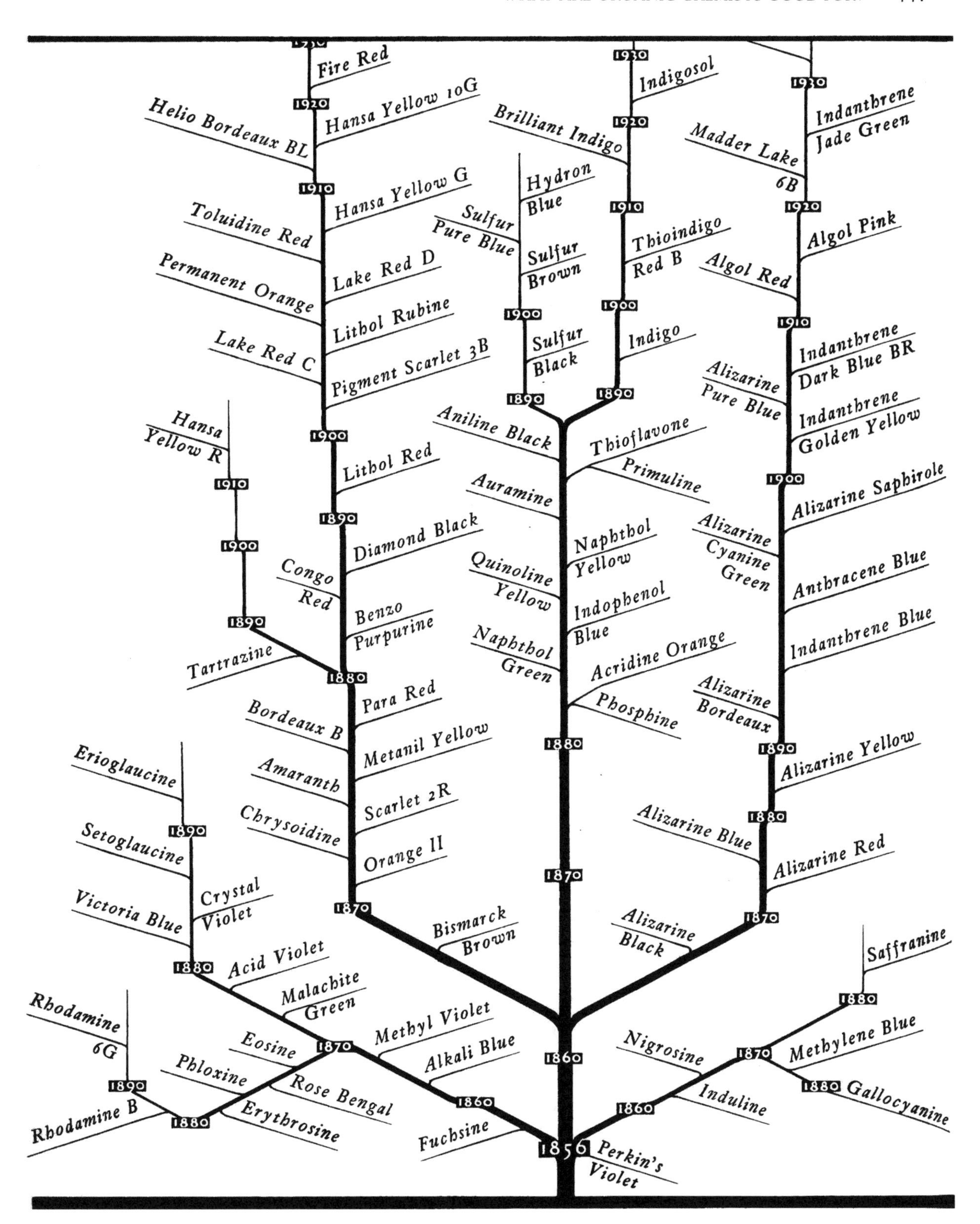

FIGURE 266. ■ The Tree of Dyes (?), showing the evolution of synthetic organic dyes through the early twentieth century following William Henry Perkin's discovery in 1856, at the age of 18, of "mauve," termed here "Perkin's Violet" (from *Color Chemistry*, No. 1, courtesy Ms. Lynne Crocker).

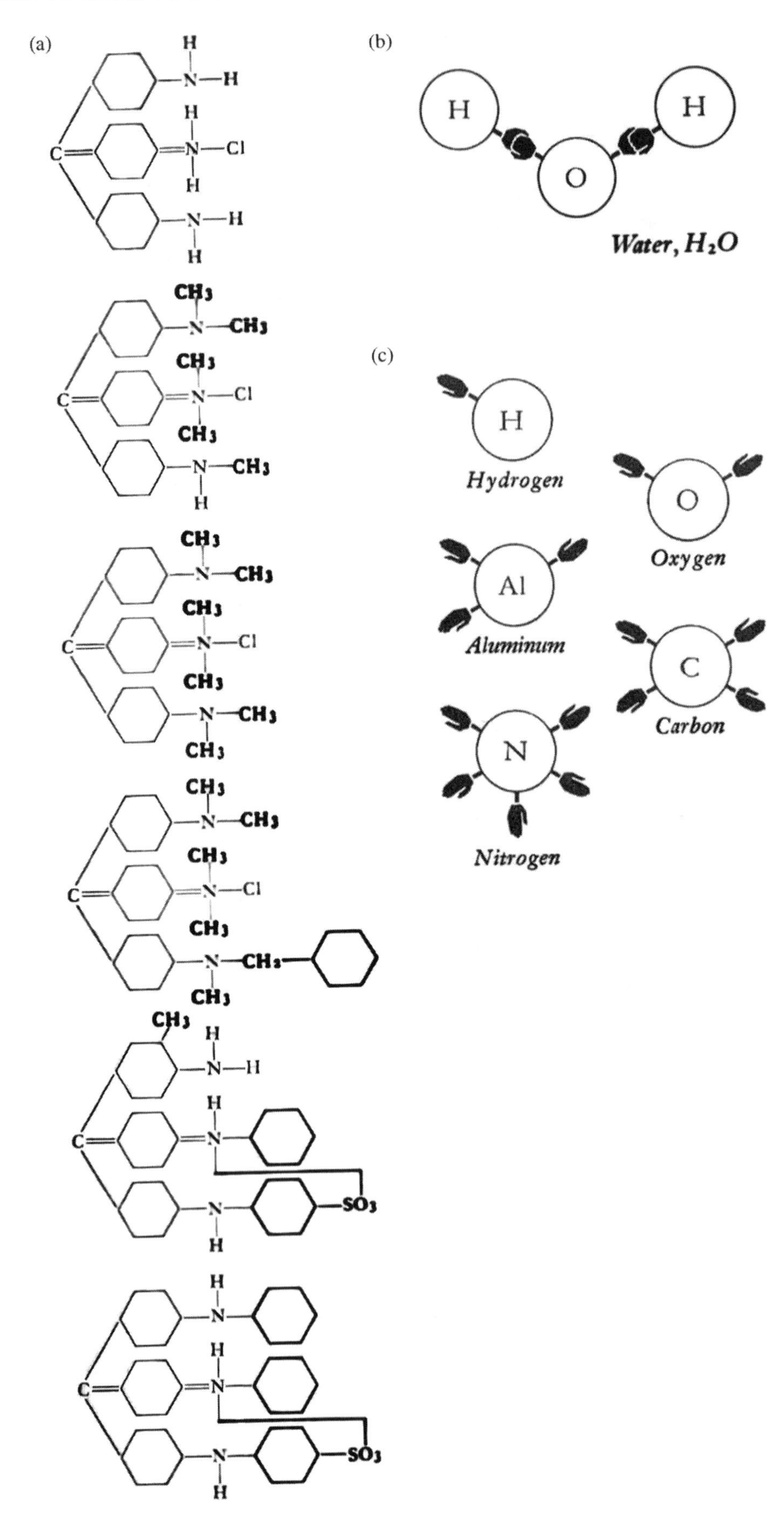
(a)
H
N—H
H
C
N—Cl
H
N—H
H
CH3
N—CH3
CH3
C
N—Cl
CH3
N—CH3
H
CH3
N—CH3
CH3
C
N—Cl
CH3
N—CH3
CH3
CH3
N—CH3
CH3
C
N—Cl
CH3
N—CH2
CH3
CH3
H
N—H
H
C
N
N
SO3
H
H
N
H
C
N
N
SO3
H
(b)
H
H
O
Water, H2O
(c)
H
Hydrogen
O
Oxygen
Al
Aluminum
C
Carbon
N
Nitrogen

FIGURE 267. ■ In (a) we see the synthetic organic chemist's skill in "tweaking" a molecular framework in order to fine-tune desired properties; substitution of the hydrogen atoms in the top structure (the purplish-red dye fuchsine) by methyl (CH_3) and other groups smoothly varies color through shades of violet, violet-blue, and then blue. This approach has long been used to modify penicillins and other drugs and now to synthesize laser dyes. Diagram (b) depicts two hydrogen atoms joining their friend oxygen atom to make water. Diagram (c) the hands (valences or combining "power") of five common atoms; the confusion about the valence of nitrogen [sometimes 3, sometimes 5—see the nitrogen atoms in (a)] was finally settled in the 1920s. (From *Color Chemistry*, No. 1, courtesy Ms. Lynne Crocker.)

ly been known that NH_4Cl separated into ammonia (NH_3) and hydrogen chloride (HCl) upon heating. The theory of ionic solutions developed by Svante Arrhenius[7] in the 1880s laid the basis for understanding the true nature of ammonium chloride, an ionic salt of formula $NH_4^+Cl^-$ that behaves as two distinct particles when the salt is dissolved in water, rather than one NH_4Cl molecule. The four bonds attached to nitrogen in the NH_4^+ ion were completely consistent with the octet rule and Lewis structures (1916). However, consolidation of ionic theory and the Lewis octets really occurred only in the 1920s,[8] barely a decade before Figure 267 was published.

The molecular fine tuning and "tweaking" by organic chemists found early application in the pharmaceutical industry. For example, the differences between morphine, codeine, and heroine are not very significant structurally but enormously significant pharmaceutically. The accidental discovery of penicillin in 1928 by Alexander Fleming stimulated a two-decade search for its chemical structure, ultimately obtained in the mid–1940s by the crystallographer Dorothy Crowfoot (later Hodgkin) (1910–1994),[9] who received the 1964 Nobel Prize in Medicine and Physiology for her determination of the structure of Vitamin B_{12}.[10] Once the penicillin structure was known, pharmaceutical chemists synthesized thousands of derivatives looking to increase efficacy, lower cost, and diminish undesired side effects, such as allergic responses.

1. W.H. Brock, *The Norton History of Chemistry*, W.W. Norton & Co., New York, 1993, pp. 297–301.
2. R. Hoffmann and S. Leibowitz Schmidt, *Old Wine New Flasks—Reflections on Science and Jewish Tradition*, W.H. Freeman and Co., New York, 1997, pp. 159–174.
3. J.R. Partington, *A History of Chemistry*, MacMillan and Co. Ltd., London, Vol. 4, 1964, pp. 772–774; 791–793.
4. S. Garfield, *Mauve: How One Man Invented a Color that Changed the World*, Faber and Faber Ltd., London, 2001.
5. A.J. Ihde, *The Development of Modern Chemistry*, Harper & Row, New York, 1964, pp. 452–458.
6. *Color Chemistry. Number One of a Series of Monographs on Color*, The Research Laboratories of the International Printing Ink Corporation and Subsidiary Companies, New York, 1935. I thank Ms. Lynne Crocker, Portsmouth, New Hampshire, for supplying this book.
7. Partington, op. cit., pp. 672–681.
8. G.B. Kauffman and I. Bernal, *Journal of Chemical Education*, Vol. 66, pp. 293–300, 1989.

9. D. Crowfoot, C.W. Bunn, B.W. Rogers-Low, and A. Turner Jones, in *The Chemistry of Penicillin*, H.T. Clarke, J.R. Johnson, and R. Robinson (eds.), Princeton University Press, Princeton, 1949, pp. 310–381.
10. G. Ferry, *Dorothy Hodgkin: A Life*, Granta Books, London, 1998.

MENDELEEV'S EARLY THOUGHTS ABOUT RELATIONSHIPS BETWEEN ELEMENTS

In 1945, 12-year-old Oliver Sacks was first entranced by the giant periodic table at the Science Museum in South Kensington as he examined individual samples of each element.[1] More than half a century later, a distinguished career as physician and author only seems to have enhanced his love of chemistry, metals in particular, and his ecstatic passion for the periodic table. The autobiographical *Uncle Tungsten* recounts young Sacks mentally spiraling the "sober, rectangular table" into "a sort of cosmic staircase or a Jacob's ladder, going up to, coming down from, a Pythagorean heaven."[2] The ancient Pythagoreans imagined a universe governed by pure mathematics, and Dr. Sacks perceived "Mendeleev's garden," the periodic table, as part of a cosmology vibrant with natural harmony.[1]

Dmitri Ivanovich Mendeleev (1834–1907)[3] was the youngest of 14 children. The heroism of his mother in obtaining a suitable education for him under tragic circumstances is well known.[3] Following her husband's physical collapse and the destruction in 1848 of the glass factory she had financially restored and managed, she took her gifted son to Moscow. Unsuccessful in enrolling him in the University because he was Siberian, she moved to St. Petersburg and succeeded in enrolling him in the Pedagogical School in 1850, the year she died. In a dedication published in 1887, Mendeleev wrote: "She instructed by example, corrected with love, and in order to devote him to science, left Siberia with him, spending her last resources and strength."[3] Mendeleev obtained his master's degree at St. Petersburg in 1856, and Figures 268–270 show the title page and four additional pages from his eight-sheet (15-page) dissertation.[4] Chemist/book collector Roy G. Neville has noted hints of the periodic law in the masters dissertation[5] and Mendeleev's interest in atomic masses and the relationships between elements is certainly apparent. Almost 40 years earlier, Eilhard Mitscherlich (1794–1863)[6] had originated the concept of isomorphism, noting the great similarities in form and measured angles between certain crystalline substances. For example, sodium hydrogen phosphate hydrate ($Na_2HPO_4 \cdot 12H_2O$) and the analogous arsenate ($Na_2HAsO_4 \cdot 12H_2O$) form isomorphic crystals. Crystalline ammonium sulfate [$(NH_4)_2SO_4$] and potassium sulfate (K_2SO_4) are also isomorphic (Figure 271). In the master's dissertation, the 22-year-old Mendeleev applied his interest in isomorphism to exploring relationships between the elements. This is demonstrated in statement 17 on page 6 (Figure 269), where he compared angles in crystals of related substances. Thirteen years later (in 1869), Mendeleev would place phosphorus and arsenic in the same chemical family and thus provide an understanding of the isomorphism of the aforementioned phosphate and arsenate. We now recognize that ammonium (NH_4^+) and potassium (K^+) are monovalent, positive ions of

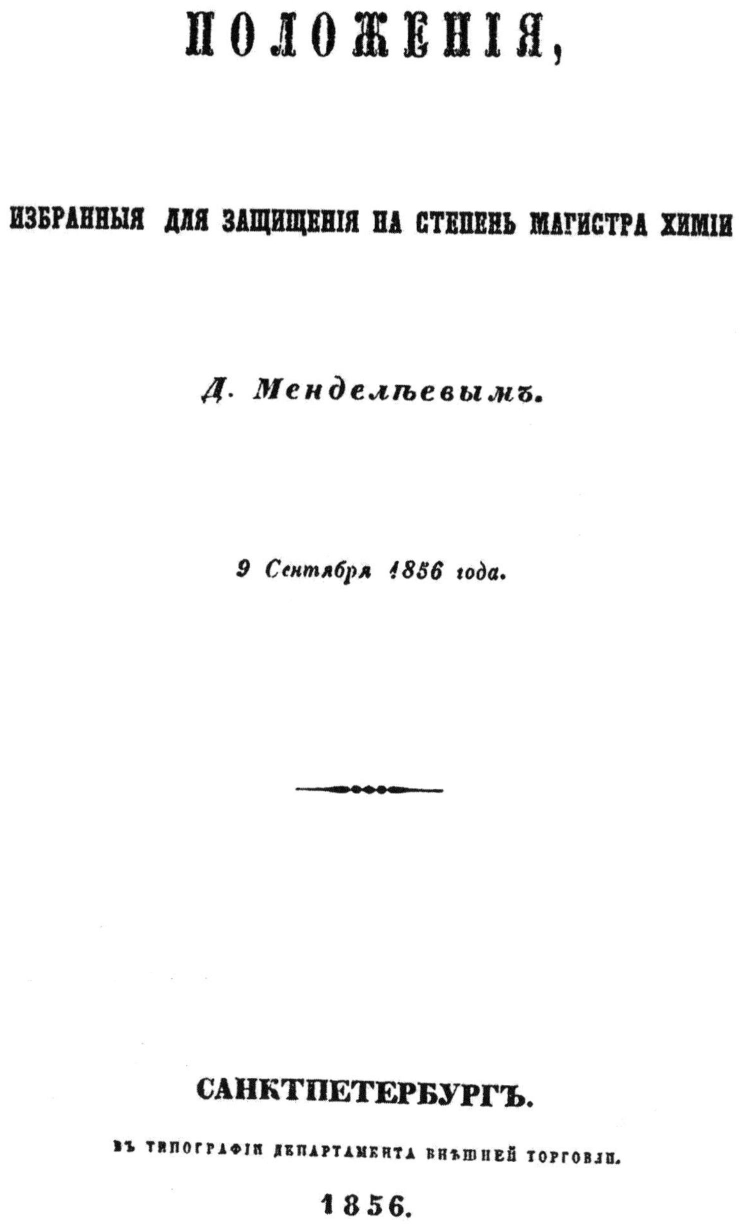

ПОЛОЖЕНІЯ,

ИЗБРАННЫЯ ДЛЯ ЗАЩИЩЕНІЯ НА СТЕПЕНЬ МАГИСТРА ХИМІИ

Д. Менделѣевымъ.

9 Сентября 1856 года.

САНКТПЕТЕРБУРГЪ.

ВЪ ТИПОГРАФІИ ДЕПАРТАМЕНТА ВНѢШНЕЙ ТОРГОВЛИ.

1856.

FIGURE 268. ▪ The title page of young Dmitri Mendeleev's 15-page master's dissertation (from The Roy G. Neville Historical Chemical Library, a collection in the Othmer Library, CHF).

similar size and this is the basis for the isomorphism of their sulfates (Figure 271). Statement 19 on page 7 (Figure 269) describes similarities in specific volume between chemically related substances.

In statement 20 on pages 7 and 8 (Figures 269 and 270), we observe atomic weight relationships, noted by Mendeleev, on a scale where the relative weight of oxygen is taken as 100. The numbers that follow "V" in statement 20 refer to relative volume, while the data following "Π" refer to relative mass. It was well known that in water, 100 grams of oxygen were combined with 12.5 grams of hy-

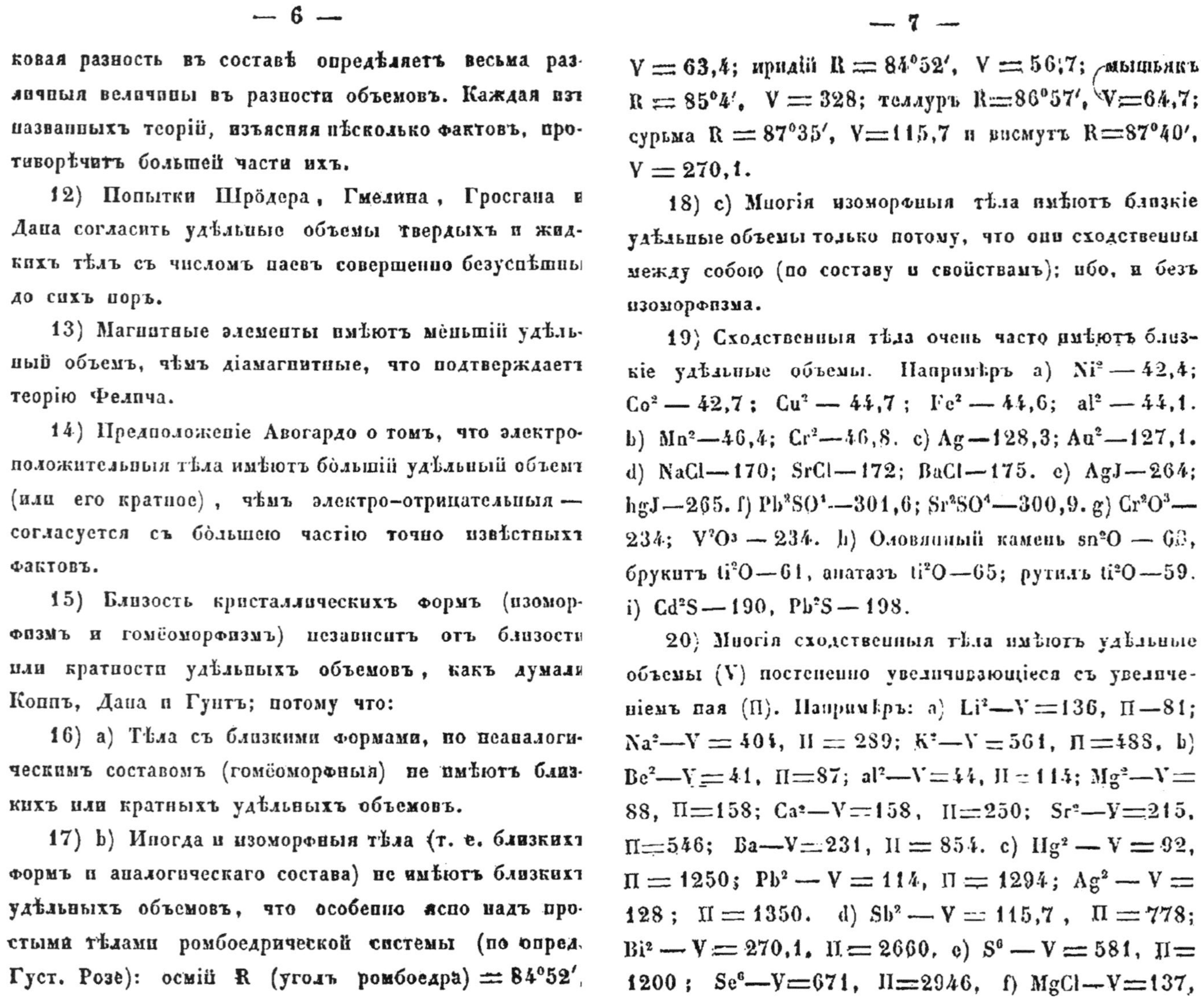

— 6 —

ковая разность въ составѣ опредѣляетъ весьма различныя величины въ разности объемовъ. Каждая изъ названныхъ теорій, изъясняя нѣсколько фактовъ, противорѣчитъ большей части ихъ.

12) Попытки Шрёдера, Гмелина, Гросгана и Дана согласить удѣльные объемы твердыхъ и жидкихъ тѣлъ съ числомъ паевъ совершенно безуспѣшны до сихъ поръ.

13) Магнитные элементы имѣютъ меньшій удѣльный объемъ, чѣмъ діамагнитные, что подтверждаетъ теорію Фелича.

14) Предположеніе Авогардо о томъ, что электроположительныя тѣла имѣютъ большій удѣльный объемъ (или его кратное), чѣмъ электро-отрицательныя — согласуется съ большею частію точно извѣстныхъ фактовъ.

15) Близость кристаллическихъ формъ (изоморфизмъ и гомёоморфизмъ) независитъ отъ близости или кратности удѣльныхъ объемовъ, какъ думали Коппъ, Дана и Гунтъ; потому что:

16) a) Тѣла съ близкими формами, но неаналогическимъ составомъ (гомёоморфныя) не имѣютъ близкихъ или кратныхъ удѣльныхъ объемовъ.

17) b) Иногда и изоморфныя тѣла (т. е. близкихъ формъ и аналогическаго состава) не имѣютъ близкихъ удѣльныхъ объемовъ, что особенно ясно надъ простыми тѣлами ромбоедрической системы (по опред. Густ. Розе): осмій R (уголъ ромбоедра) = 84°52′,

— 7 —

V = 63,4; иридій R = 84°52′, V = 56,7; мышьякъ R = 85°4′, V = 328; теллуръ R = 86°57′, V = 64,7; сурьма R = 87°35′, V = 115,7 и висмутъ R = 87°40′, V = 270,1.

18) c) Многія изоморфныя тѣла имѣютъ близкіе удѣльные объемы только потому, что они сходственны между собою (по составу и свойствамъ); ибо, и безъ изоморфизма.

19) Сходственныя тѣла очень часто имѣютъ близкіе удѣльные объемы. Напримѣръ a) Ni^2 — 42,4; Co^2 — 42,7; Cu^2 — 44,7; Fe^2 — 44,6; al^2 — 44,1. b) Mn^2—46,4; Cr^2—46,8. c) Ag—128,3; Au^2—127,1. d) NaCl—170; SrCl—172; BaCl—175. e) AgJ—264; hgJ—265. f) Pb^2SO^4—301,6; Sr^2SO^4—300,9. g) Cr^2O^3—234; V^2O^3 — 234. h) Оловянный камень sn^2O — 63, брукитъ ti^2O—61, анатазъ ti^2O—65; рутилъ ti^2O—59. i) Cd^2S—190, Pb^2S—198.

20) Многія сходственныя тѣла имѣютъ удѣльные объемы (V) постепенно увеличивающіеся съ увеличеніемъ пая (П). Напримѣръ: a) Li^2—V = 136, П—81; Na^2—V = 404, П = 289; K^2—V = 561, П = 488, b) Be^2—V = 41, П = 87; al^2—V = 44, П = 114; Mg^2—V = 88, П = 158; Ca^2—V = 158, П = 250; Sr^2—V = 215, П = 546; Ba—V = 231, П = 854. c) Hg^2 — V = 92, П = 1250; Pb^2 — V = 114, П = 1294; Ag^2 — V = 128; П = 1350. d) Sb^2 — V = 115,7, П = 778; Bi^2 — V = 270,1, П = 2660, e) S^6 — V = 581, П = 1200; Se^6—V = 671, П = 2946, f) MgCl—V = 137,

FIGURE 269. ■ Statement 17 on pages 6 and 7 of Mendeleev's master's dissertation discusses the angles measured in crystals of related substances. The Periodic Law, published by Mendeleev in 1869, would explain the similarities between isomorphic crystals, first noted by Eilhard Mitscherlich some 40 years before Mendeleev's master's dissertation was published. (From The Roy G. Neville Historical Chemical Library, a collection in the Othmer Library, CHF.)

— 8 —

П=292; CaCl—V = 156, П=347; NaCl—V = 170, П=366; KaCl—V=240; П—466. g) **Перекись водорода** V = 146, П = 212; **перекись барія** V—212, П—1054. h) Na^2O—V=139, П—389; K^2O—V=221, П=588. i) CaF V = 76, П = 243; PbCl—V=150, П=869; AgCl—V=165, П—897; hgCl—V=209, П = 1472. j) **Англійская соль** V=900, П=1540; **желѣзный купоросъ** V=934, П=1739. l) **Вода** H^2O—V=112,5, П=112,5; **древесный спиртъ** CH^4O—V=249, П=200; **алкоголь** C^2H^6O—V=359, П=287; **амиль-алкоголь** $C^5H^{12}O$—V=673, П=550; **гексиль-алкоголь** $C^6H^{14}O$—V=765, П=637; **октиль-алкоголь** $C^8H^{18}O$—V = 987, П = 817. m) **Муравьинокислый эфиль** $C^3H^6O^2$—V = 500, П = 462, **уксуснокислый эфиль** $C^4H^8O^2$—V=611, П=550; **пропіоновокислый эфиль** $C^5H^{10}O^2$—V = 712, П = 637; **маслянокислый эфиль** $C^6H^{12}O^2$ — V = 809, П = 725; **валеріановокислый эфиль** $C^7H^{14}O^2$ — V = 928, П = 812. n) **Трифанъ** $al^1Li^1Si^2O^3$—V=186, П=592; **пироксенъ** Ca Mg Si^2O^6 — V = 213, П = 685. o) **Виллемитъ** Zn^2SiO^4 — V = 167, П=700; **діоптазъ** Cu H Si O^4—V = 153, П = 496; **бериллъ** Si al be O^3 — V = 139, П = 378.

21) Эти два закона даютъ твердую опору естественной классификаціи. Ихъ указалъ Дюма въ 1821 и 1854 годахъ

22) Удѣльные объемы соединеній водорода больше удѣльныхъ объемовъ соотвѣтствующихъ соединеній

— 9 —

магнія и мѣди, но меньше чѣмъ калія и эфиля и близки къ уд. объемамъ соединеній барія и натрія. Напримѣръ (въ типѣ RCl) Cu Cl — 137, Mg Cl — 137, Na Cl — 170, Ba Cl — 175, HCl — 180 (?), Ka Cl — 240, C^2H^5Cl — 450; также (въ типѣ R^2O) Mg^2O — 69, Cu^2O — 80; H^2O — 112,5; Na^2O — 139; Ba^2O — 186; Ka^2O — 221; $(C^2H^5)^2O$ — 859, и также (въ типѣ R^2SO^4) Cu^2SO^4 — 280, Mg^2SO^4 — 283; Ba^2SO^4 — 326; H^2SO^2 — 331; Na^2SO^4 — 337, K^2SO^4 — 412; $(C^2H^5)^2SO^4$ — 859.

23) Когда вода вступаетъ въ двойное разложеніе съ окисломъ одноосновнаго радикала R^2O и образуетъ HRO, то при этомъ удѣльн. объемъ измѣняется очень мало.

$\frac{1}{2}(K^2O)$ и $\frac{1}{2}(H^2O)$ образуютъ $\left.\begin{matrix}K\\H\end{matrix}\right\}O$

$\frac{1}{2}(221) + \frac{1}{2}(112,5) = 167.$ 169

$\frac{1}{2}(Na^2O)$ и $\frac{1}{2}(H^2O)$ образуютъ $\left.\begin{matrix}Na\\H\end{matrix}\right\}O$

$\frac{1}{2}(139) + \frac{1}{2}(112,5) = 121.$ 121

$\frac{1}{2}(Ca^2O)$ и $\frac{1}{2}(H^2O)$ образуютъ $\left.\begin{matrix}Ca\\H\end{matrix}\right\}O$

$\frac{1}{2}(111) + \frac{1}{2}(112,5) = 112.$ 105

$\frac{1}{2}\left(\left.\begin{matrix}C^2H^3O\\C^2H^3O\end{matrix}\right\}O\right)$ и $\frac{1}{2}(H^2O)$ образуютъ $\left.\begin{matrix}C^2H^3O\\H\end{matrix}\right\}O$

окись ацетила — уксусная кислота

$\frac{1}{2}(592) + \frac{1}{2}(112,5) = 352,5.$ 352,8

$\frac{1}{2}\left(\left.\begin{matrix}C^4H^7O\\C^4H^7O\end{matrix}\right\}O\right)$ и $\frac{1}{2}(H^2O)$ образуютъ $\left.\begin{matrix}C^4H^7O\\H\end{matrix}\right\}O$

окись бутирила — масляная кислота

$\frac{1}{2}(1009) + \frac{1}{2}(112,5) = 561.$ 564

FIGURE 270. ■ Statement 20 on pages 7 (see Figure 269) and 8 deals with the relative atomic weights so vital to the development of the Periodic Law. The relative mass of water ("H^2O" on page 8), is 112.5, and this is based on the widely employed assumed relative mass = 100 for the oxygen atom. (From The Roy G. Neville Historical Chemical Library, a collection in the Othmer Library, CHF.)

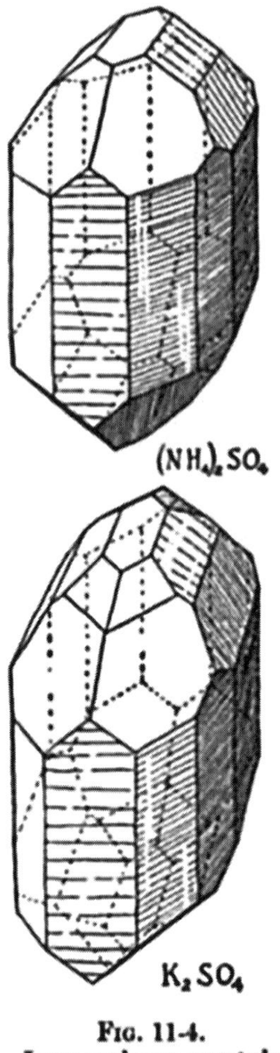

FIGURE 271. ▪ Isomorphous crystals of ammonium sulfate and potassium sulfate (illustration by artist Roger Hayward in the first edition of his friend Linus Pauling's text, *General Chemistry*, 1947, p. 202. © 1947 by Linus Pauling. Used with permission of W.H. Freeman and Company). Mitscherlich had discovered that some phosphate and arsenate crystals were isomorphic, and these observations were explained by Mendeleev as arising from arsenic and phosphorus being in the same chemical family. We now know that ammonium (NH_4^+) and potassium (K^+) ions are similarly congruent, thus accounting for the isomorphism of their sulfate crystals.

drogen and, thus, the relative mass for "H^2O" (see page 8 in Figure 270) on this scale is 112.5. If we employ the relative mass of "Na^2O," we see that a total mass of 289 combines with 100 of oxygen. Therefore, each sodium atom has a relative mass of 289/2 or 144.5. If we multiply 144.5/100 by 16.0 (the modern atomic mass of oxygen), we would obtain a value of 23.1 for sodium, in decent agreement with the modern value.

Despite the agreements of these particular relative atomic masses with modern data, there were considerable uncertainties during the 1850s in atomic masses and equivalent weights. At the international chemical congress in Karlshuhe, Germany, during 1860 some of the greatest chemists of the day gathered to debate these issues. Critical insights were provided by Stanislao Cannizzaro who enlightened the gathering by making the 50-year-old hypothesis of his fellow countryman Amedeo Avogadro accessible within the new chemical knowledge of the mid-nineteenth century. The youthful Mendeleev took a leisurely trip to Karlsruhe accompanied by another young chemist, Aleksandr Porfirevich Borodin (1833–1887).[7] Borodin became, of course, one of the world's great composers, and his opera *Prince Igor* is still immensely popular. But this precocious musician (self-taught on the cello; he composed musical pieces at the age of 14) was also a precocious chemist (a homebuilt laboratory and youthful efforts at making explosives).[7] It is fun to imagine a film depicting the 26-year-old Mendeleev and the 27-year old Borodin making their leisurely way to Karlsruhe, delighting in the music from the giant pipe organ at Freiburg.[7] Both had powerful Russian mothers[8] who lovingly guided their gifted sons' educations and even established lodging near their colleges. Although music would be his most powerful legacy, Borodin made some fundamental discoveries in organic chemistry. Today's overburdened college sophomores who must remember that the Aldol Condensation may be followed by dehydration have Borodin to blame for this additional fact.[9,10]

The Karlsruhe congress settled many controversies and placed atomic masses and equivalents on firm footing in the chemical community. Attempts to organize these data were not long in coming, and early periodic tables were proposed by, among others, John Newlands (1865), William Odling (1865), and Julius Lothar Meyer (1868).[11] "Eureka moments" are actually very rare in science, and an evolution of ideas is more typical. As noted earlier, attempts to organize the elements are apparent in Mendeleev's 1856 master's thesis. Figure 272 is from his 1863 text on organic chemistry.[12] It is thrilling to see an embryonic form of the periodic table here.

1. O. Sacks, *Uncle Tungsten—Memories of a Chemical Boyhood,* Alfred A. Knopf, New York, 2001.
2. Sacks, op. cit., pp. 187–211.
3. J.R. Partington, *A History of Chemistry,* MacMillan and Co. Ltd., London, Vol. 4, 1964, pp. 891–899. Partington notes here that the Russian chemist spelled his name "Mendeleeff" when he signed the register of the Royal Society and Ramsay recommended "Mendeléeff." We will employ the more commonly used transliteration "Mendeleev".
4. D.I. Mendeleev, *Polozhenija, izbrannya dlja zachschishchenija na stepen' magistra khimii,* St. Petersburg, 1856. I am grateful to The Roy G. Neville Historical Chemical Library (California) for supplying these images.
5. The Roy G. Neville Historical Chemical Library (California), catalog in preparation. I am grateful to Dr. Neville for making me aware of his views on Mendeleev's early thoughts on the relationships between elements manifested in his master's thesis.
6. Partington, op. cit., pp. 205–211.
7. S.A. Dianin, *Borodin,* Oxford University Press, London, 1963, pp. 12–13; 22–29.
8. The roots of my mother, Bella Greenberg, are similarly Russian. She first took me as a young child to see the Halls of Dinosaurs at the American Museum of Natural History, encouraged my early interest in reading, and gently helped remove hundreds of newborn praying mantids

Изъ этого видимъ, что радикалъ сѣрной кислоты есть SO^2. Точно также найдемъ, что радикалъ фосфорной кислоты есть PO. Этотъ способъ опредѣленія сложныхъ радикаловъ особенно ясно выводится при изученіи органичеекихъ соединеній.

Атомность сложныхъ радикаловъ опредѣляется тѣмъ же путемъ, какъ и атомность простыхъ радикаловъ. Къ одноатомнымъ сложнымъ радикаламъ относятся, напримѣръ, радикалъ амміачныхъ соединеній—аммоній NH^4, радикалъ азотной кислоты — NO^2, радикалъ ціановыхъ соединеній — CN, потому-что ихъ хлористыя соединенія содержатъ одинъ пай хлора.

Нашатырь .... NH^4Cl.
Такъ-называемая хлороазотная кислота .. NO^2Cl.
Газообр. хлор. ціанъ .... $NCCl$.

Радикалы SO^2 и CO сѣрной и углекислоты суть двуатомные, потому-что ихъ хлористыя соединенія:

такъ-называемая хлоросѣрная кислота или
второй хлорангидридъ сѣрной кислоты.. SO^2Cl^2 и
фосгенъ .... $COCl^2$

содержатъ 2 пая хлора въ одной частицѣ.

Радикалъ PO фосфорныхъ солей есть трех-атомный, потому-что хлорокись фосфора $POCl^3$, содержитъ въ одной частицѣ три пая хлора. Исчисленныя хлористыя соединенія посредствомъ реакцій замѣщенія, даютъ другія соединенія тѣхъ же радикаловъ. Такъ хлорокись фосфора съ водою дастъ фосфорную кислоту:

$$POCl^3 + 3H^2O = PH^3O^4 + 3HCl.$$

Зная атомность радикаловъ, легко предугадать ихъ обыкновеннѣйшія соединенія, наблюдая всегда чтобы сумма атомностей всѣхъ радикаловъ была четное количество.

Простѣйшіе виды соединеній будутъ:

$$R'R', \quad R'^2R'', \quad R'^3R'''.$$

Потому водородъ образуетъ слѣдующія типическія соединенія:

Главные типы: $\left.\begin{matrix}H\\H\end{matrix}\right\}$, $\left.\begin{matrix}H\\H\end{matrix}\right\}O$ и $N\left\{\begin{matrix}H\\H\\H\end{matrix}\right.$

Производные типы: $\left.\begin{matrix}H\\Cl\end{matrix}\right\}$ $\left.\begin{matrix}H\\H\end{matrix}\right\}S$ $P\left\{\begin{matrix}H\\H\\H\end{matrix}\right.$

$\left.\begin{matrix}H\\Br\end{matrix}\right\}$ $As\left\{\begin{matrix}H\\H\\H\end{matrix}\right.$.

Орг. химія, Менделѣева. 2

FIGURE 272. ▪ See the bottom of page XVII in Mendeleev's text on organic chemistry published in St. Petersburg in 1863. You will note this early organization anticipating the periodic law that Mendeleev will publish six years hence. (From Mendeleev, *Organischeskaja Khimia*, 1863.)

from the wall of my bedroom and placed them in the backyard (see the essay "A Natural Scientist" in the Epilogue of this book).
9. C.C. Gillispie (ed.), *Dictionary of Scientific Biography*, Charles Scribner's Sons, New York, Vol. II, 1970, pp. 316–317.
10. M.D. Gordin, *Journal of Chemical Education*, Vol. 83, pp. 561–565 (2006).
11. Partington, op. cit., pp. 883–891.
12. D.I. Mendeleev, *Organischeskaja Khimia*, St. Petersburg, 1863, p. xvii.

THE ICON ON THE WALL

In an observant Moslem household, a page of verses from the Koran handwritten in beautiful calligraphy may grace the wall. In a Catholic household, one might see a crucifix; in an observant Jewish household there will be a mezuzah affixed to the doorway; a Bodhisattva in a Buddhist household; an image of the family deity in a Hindhu household. And in every house of chemistry, every classroom, lecture hall, and laboratory, hangs our icon—the Periodic Table.

Shortly after the Karlsruhe Conference, John Alexander Newlands (1837–1898) published some papers on regularities in atomic weights.[1] In 1864 he published a version of a table of the elements and noted his law of octaves: ". . . the eighth element starting from a given one is a kind of repetition of the first, like the eighth note of an octave in music."[1] Newlands published a modified table in 1865 and further improved it in 1866. William Odling (1829–1921) published a table of the elements in order of atomic weights in 1865. Lothar Meyer made a table (unpublished) in 1868 that placed carbon, nitrogen, oxygen, fluorine, and lithium at the top of their respective groups. A modified version was first published in 1869.[1]

Credit for the Periodic Table is accorded to Dmitri Ivanovich Mendeleyev (Mendeleeff or Mendeleev, 1834–1907).[2] Subsequent to his training as a teacher in St. Petersburg, Mendeleev wrote a Masters thesis at the University of St. Petersburg and was given a position there. He later studied in Paris and Heidelberg, returned to St. Petersburg in 1861, became Professor at the Technological Institute and later Professor at the University. In 1868, while writing his *Principles of Chemistry*, Mendeleev is said to have started thinking about the periodic law, having been to Karlsruhe but was unaware of Newlands' work.[2] He completed his table on February 17, 1869 and it was presented to the Russian Academy of Science on March 6, 1869 by his colleague Nikolai A. Menschutkin (1842–1907), because Mendeleev was making a presentation elsewhere.

The first appearance of Mendeleev's periodic table in Russian occurred in 1869 (Figure 273),[3] and it also was printed in German (Figure 274) that same year.[2] Elements were arranged in order of atomic mass and grouped according to similarities in chemical properties as they recur periodically. We are accustomed to seeing a "horizontal" periodic table and this one was "vertical." Figure 275 shows both "vertical" and "horizontal" versions of the periodic table as they were published in 1872.[4] The brilliance and primacy of Mendeleev's Periodic Table rest upon his audacious act of leaving gaps in it, where he predicted that elements, as yet unknown, were missing.[5] In Figure 276, we see below aluminum

ОПЫТЪ СИСТЕМЫ ЭЛЕМЕНТОВЪ,

ОСНОВАННОЙ НА ИХЪ АТОМНОМЪ ВѢСѢ И ХИМИЧЕСКОМЪ СХОДСТВѢ.

			Ti=50	Zr=90	?=180.
			V=51	Nb=94	Ta=182.
			Cr=52	Mo=96	W=186.
			Mn=55	Rh=104,4	Pt=197,4
			Fe=56	Ru=104,4	Ir=198.
			Ni=Co=59	Pl=106,6	Os=199.
H=1			Cu=63,4	Ag=108	Hg=200.
	Be=9,4	Mg=24	Zn=65,2	Cd=112	
	B=11	Al=27,4	?=68	Ur=116	Au=197?
	C=12	Si=28	?=70	Sn=118	
	N=14	P=31	As=75	Sb=122	Bi=210?
	O=16	S=32	Se=79,4	Te=128?	
	F=19	Cl=35,5	Br=80	I=127	
Li=7	Na=23	K=39	Rb=85,4	Cs=133	Tl=204
		Ca=40	Sr=87,6	Ba=137	Pb=207.
		?=45	Ce=92		
		?Er=56	La=94		
		?Yt=60	Di=95		
		?In=75,6	Th=118?		

FIGURE 273. ■ Mendeleev's original 1869 periodic table.[3] (from the first edition of *Osnovy Khimii (Principles of Chemistry)* (from the collection of Gregory S. Girolami and Vera V. Mainz).

(Al) a blank space. Mendeleev predicted the existence of a new element he termed *eka*-aluminum as well as its atomic weight, density, melting point, and the formula of its oxide. (The term eka means "something added.") In 1875, the element gallium (Ga) was discovered by Paul Emile Lecoq de Boisbaudran and named after Gaul to soothe his countrymen's egos after their defeat in the Franco-Prussian War. In 1879, Lars Frederik Nilson of Sweden discovered *eka*-boron, well matching the properties predicted by Mendeleev, and named it scandium (Sc). In 1886, Clemens Winkler discovered *eka*-silicon, again matching Mendeleev's predictions, and named it germanium (Ge)—payback time for French chauvinism. Figure 277 displays an 1891 periodic table.

Mendeleev's predictions were not always correct. He courageously placed iodine *after* the heavier tellurium, incorrectly predicting that new experiments would correctly reverse their masses. He also predicted new elements that were never to be. Unbeknownst to Mendeleev, the source of order for the Periodic Table was not the atomic weight, but the atomic number, and this would be discovered by Henry Moseley just before the First World War.

In 1999, the penultimate year of the millenium, it was a sheer delight to discover a beautiful article by physician-writer (of *Awakenings* fame) Oliver Sacks who confesses his lifelong fascination with the Periodic Table[6,7]:

> My kitchen is papered with periodic tables of every size and sort—oblongs, spirals, pyramids, weather vanes—and on the kitchen table, a very favorite one, a round periodic table made of wood that I can spin like a prayer wheel.

Clearly, "chemistry is spoken" in the Sacks household and he even keeps two small periodic icons in his wallet. Perhaps at an appointed hour each day, Dr.

Ueber die Beziehungen der Eigenschaften zu den Atomgewichten der Elemente. Von D. Mendelejeff. — Ordnet man Elemente nach zunehmenden Atomgewichten in verticale Reihen so, dass die Horizontalreihen analoge Elemente enthalten, wieder nach zunehmendem Atomgewicht geordnet, so erhält man folgende Zusammenstellung, aus der sich einige allgemeinere Folgerungen ableiten lassen.

			Ti = 50	Zr = 90	? = 180
			V = 51	Nb = 94	Ta = 182
			Cr = 52	Mo = 96	W = 186
			Mn = 55	Rh = 104,4	Pt = 197,4
			Fe = 56	Ru = 104,4	Ir = 198
			Ni = Co = 59	Pd = 106,6	Os = 199
H = 1			Cu = 63,4	Ag = 108	Hg = 200
	Be = 9,4	Mg = 24	Zn = 65,2	Cd = 112	
	B = 11	Al = 27,4	? = 68	Ur = 116	Au = 197?
	C = 12	Si = 28	? = 70	Sn = 118	
	N = 14	P = 31	As = 75	Sb = 122	Bi = 210?
	O = 16	S = 32	Se = 79,4	Te = 128?	
	F = 19	Cl = 35,5	Br = 80	J = 127	
Li = 7	Na = 23	K = 39	Rb = 85,4	Cs = 133	Tl = 204
		Ca = 40	Sr = 87,6	Ba = 137	Pb = 207
		? = 45	Ce = 92		
		?Er = 56	La = 94		
		?Yt = 60	Di = 95		
		?In = 75,6	Th = 118?		

1. Die nach der Grösse des Atomgewichts geordneten Elemente zeigen eine stufenweise Abänderung in den Eigenschaften.
2. Chemisch-analoge Elemente haben entweder übereinstimmende Atomgewichte (Pt, Ir, Os), oder letztere nehmen gleichviel zu (K, Rb, Cs).
3. Das Anordnen nach den Atomgewichten entspricht der *Werthigkeit* der Elemente und bis zu einem gewissen Grade der Verschiedenheit im chemischen Verhalten, z. B. Li, Be, B, C, N, O, F.
4. Die in der Natur verbreitetsten Elemente haben *kleine* Atomgewichte

FIGURE 274. ▪ The first version of Mendeleev's 1869 periodic table (Figure 273) here translated into German; note the question marks in some places, notably following aluminum and silicon. The daring aspect of Mendeleev's table was the presence of these gaps. *Eka*-aluminum (i.e., gallium) and *eka*-silicon (i.e., germanium) were predicted by Mendeleev to exist and were discovered shortly thereafter. This is one of the most excellent illustrations of the power of the scientific method in human history. (From *Zeitschrift für Chemie;* with permission from The Edgar Fahs Smith Collection.)

(a)

Tabelle I.

Typische Elemente							
			K = 39	Rb = 85	Cs = 133	—	—
			Ca = 40	Sr = 87	Ba = 137	—	—
			—	?Yt = 88?	?Di = 138?	Er = 178?	—
			Ti = 48?	Zr = 90	Ce = 140?	?La = 180?	Th = 231
			V = 51	Nb = 94	—	Ta = 182	—
			Cr = 52	Mo = 96	—	W = 184	U = 240
			Mn = 55	—	—	—	—
			Fe = 56	Ru = 104	—	Os = 195?	—
			Co = 59	Rh = 104	—	Ir = 197	—
			Ni = 59	Pd = 106	—	Pt = 198?	—
H = 1	Li = 7	Na = 23	Cu = 63	Ag = 108	—	Au = 199?	—
	Be = 9,4	Mg = 24	Zn = 65	Cd = 112	—	Hg = 200	—
	B = 11	Al = 27,3	—	In = 113	—	Tl = 204	—
	C = 12	Si = 28	—	Sn = 118	—	Pb = 207	—
	N = 14	P = 31	As = 75	Sb = 122	—	Bi = 208	—
	O = 16	S = 32	Se = 78	Te = 125?	—	—	—
	F = 19	Cl = 35,5	Br = 80	J = 127	—	—	—

(b)

Tabelle II.

Reihen	Gruppe I. — R^2O	Gruppe II. — RO	Gruppe III. — R^2O^3	Gruppe IV. RH^4 RO^2	Gruppe V. RH^3 R^2O^5	Gruppe VI. RH^2 RO^3	Gruppe VII. RH R^2O^7	Gruppe VIII. — RO^4
1	H=1							
2	Li=7	Be=9,4	B=11	C=12	N=14	O=16	F=19	
3	Na=23	Mg=24	Al=27,3	Si=28	P=31	S=32	Cl=35,5	
4	K=39	Ca=40	—=44	Ti=48	V=51	Cr=52	Mn=55	Fe=56, Co=59, Ni=59, Cu=63.
5	(Cu=63)	Zn=65	—=68	—=72	As=75	Se=78	Br=80	
6	Rb=85	Sr=87	?Yt=88	Zr=90	Nb=94	Mo=96	—=100	Ru=104, Rh=104, Pd=106, Ag=108.
7	(Ag=108)	Cd=112	In=113	Sn=118	Sb=122	Te=125	J=127	
8	Cs=133	Ba=137	?Di=138	?Ce=140	—	—	—	— — — —
9	(—)	—	—	—	—	—	—	
10	—	—	?Er=178	?La=180	Ta=182	W=184	—	Os=195, Ir=197, Pt=198, Au=199.
11	(Au=199)	Hg=200	Tl=204	Pb=207	Bi=208	—	—	
12	—	—	—	Th=231	—	U=240	—	— — — —

11*

FIGURE 275. ■ Mendeleev's 1871 versions (actually published in 1872) of his periodic tables in (a) vertical form, and (b) horizontal form (from *Annalen der und Pharmazie*, 1872).

GRUPPE:	I.	II.	III.	IV.	V.	VI.	VII.	VIII.
Reihe: 1	. H			RH^4 .	RH^5	RH^2	RH	Wasserstoffverbindungen.
» 2	Li	Be .	B	C .	N .	O .	F .	
» 3	. Na	Mg	. Al	. Si	. P	. S	Cl	
» 4	K .	Ca .	Sc .	Ti	V .	Cr .	Mn .	Fe. Co. Ni. Cu.
» 5	. (Cu)	. Zn	. Ga	Ge	. As	. Se	. Br	
» 6	Rb. .	Sr .	Y .	Zr .	Nb	Mo .	— .	Ru. Rh. Pd. Ag.
» 7	. (Ag)	Cd	. In	. Sn	. Sb	. Te	. J	
» 8	Cs .	Ba .	La	Ce	Di? .	— .	— .	— — — —
» 9	. —	. —	. —	. —	. —	. —	. —	
» 10	— .	— .	Yb .	— .	Ta .	W .	— .	Os. Ir. Pt. Au.
» 11	. (Au)	. Hg	. Tl	. Pb	. Bi	. —	. —	
» 12	— .	— .	— .	Th .	— .	U .	— .	
	R^2O	R^2O^2 RO	R^2O^3	R^2O^4 RO^2	R^2O^5	R^2O^6 RO^3	R^2O^7	Höchste salzbildende Oxyde RO^4

FIGURE 276. ■ Mendeleev's first Periodic Table was published in 1869 (Figure 273). The one in this figure appeared in his 1891 book *Grundlagen der Chemie* (St. Petersburg, 1891).

Sacks faces St. Petersburg and meditates, contemplating the bearded, prophet-like Mendeleev.

1. J.R. Partington, *A History of Chemistry*, MacMillan, London, 1964, Vol. 4, pp. 886–891.
2. J.R. Partington, op. cit., pp. 891–899.
3. D.I. Mendeleev, *Osnovy Khimii* (Principles of Chemistry), Tovarishchestva 'Obshchestvennaia Pol'za' for the Author, St. Petersburg, 1869.[1] am grateful to Drs. Gregory S. Girolami and Vera V. Mainz for permission to use this image, and for helpful discussions.
4. D. Mendelejeff, *Annalen der Chemie Und Pharmazie*, VIII. Supplementband, Leipzig and Heidelberg, 1872, pp. 133–229. (This paper was received from St. Petersburg in August, 1871).
5. A.J. Ihde, *The Development of Modern Chemistry*, Harper & Row, New York, 1964, pp. 231–256.
6. O. Sacks, "Everything in its Place—One Man's Love Affair with the Periodic Table," *New York Times Magazine*, April 18, 1999, pp. 126–130.
7. O. Sacks, *Uncle Tungsten—Memories of a Chemical Boyhood*, Alfred A. Knopf, New York, 2001.

THE ELECTRIC OXYGEN

Renewal and rebirth—the bracing "electric" aroma of the seaside air following a thunderstorm; the evocative smell of an electric train set bringing back warm memories of childhood; even the New York City subway; traces of gaseous ozone, produced by electric arcs and sparks, remain in our memories. This "electric oxygen" has been used for over 100 years to purify drinking water and remove un-

FIGURE 277. ■ Renewal and rebirth, the bracing aroma of the sea, and purifier of water—the lore of ozone was quickly marketed. Here we see a century-old "Ozone Is Life" bottle for ozone-treated water produced in Canada. We have filled the bottle with life-giving milk in harmony with this theme (and also to improve photographic contrast). (Photograph courtesy of Ms. Susan J. Greenberg.)

pleasant odors. The century-old bottle in Figure 277 proclaims "Ozone is life" and probably contained treated water from which the ozone was long gone. And why not use this magical-sounding name to advertise a product having nothing whatsoever to do with ozone—say, soap (Figure 278)? Ozone Soap—rebirth and renewal—flowers; a happy, healthy baby.

It is truly remarkable that most of the significant facts about ozone, including its formula (O_3), were known by about 1872. Ozone is an allotrope of oxygen. Allotropes are different forms of an element in the same state. Diamond, graphite, and fullerenes (such as the "soccer ball" C_{60}) are carbon allotropes just as red phosphorus and white phosphorus are allotropes. However, ozone is typically found at only ppm (parts per million) levels in air [1 cubic meter of air contains about 1 mg (milligram) of ozone]. Moreover, it is much higher in energy than oxygen, is a much more reactive substance (stronger oxidizer), and readily

FIGURE 278. ■ Early advertising cards for Ozone Soap. See color plates. While we strongly doubt that even the minutest traces of ozone were ever to be found in Ozone Soap, it was still a wonderfully evocative trade name.

decomposes thermally or after absorbing ultraviolet (UV) light.[1] When concentrated in liquid form, it is explosive. So how was so much learned so early?

In 1785, a decade following the discovery of dephlogisticated air (oxygen) by Scheele and then Priestley, M. van Marum sparked a sample of this gas trapped in a tube above mercury.[2,3] He observed that the mercury in contact with this electrified oxygen tarnished.[4] In contrast, "plain oxygen" typically reacts with mercury at temperatures exceeding 300°C.[5] He further noted that the gas in the tube possessed the characteristic sulfurous odor that was already associated with electricity. However, it took over 50 years for Christian Friedrich Schönbein to postulate a distinct new substance and provide a name (derived from *ozon* = "I smell") and another 20 years to understand that it was composed solely of the element oxygen.[2,3] Indeed, in the latter part of the nineteenth century ozone–oxygen mixtures were liquefied[5] and concentrated by fractional distillation from oxygen. The compression of samples containing concentrated ozone had to be done with great care since heating upon compression could cause an explosion.[2] Pure ozone is a deep blue, explosive liquid boiling at –112°C.[1]

Figure 279 depicts a late-nineteenth-century ozone generator.[2] *BB* is an iron tube through which cold water is passed (tube CC). Glass cylinder *AA* has a slightly larger diameter than *BB*, and the small space between these two cylinders is filled with oxygen introduced through tube *DD*. Part of the outer (glass) cylinder is covered with tinfoil (GG). The outer tin jacket and the inner iron tube are connected at points *E* and *F* to an induction coil. This apparatus was designed to produce a "silent discharge" since sparks are also known to decompose ozone to oxygen. The early characteristic test for ozone was its ability to turn potassium iodide/starch-impregnated filter paper blue:[2,3]

$$O_3 + 2KI + H_2O = O_2 + I_2 + 2\ KOH \qquad \text{(iodine–starch complex = blue)} \qquad (1)$$

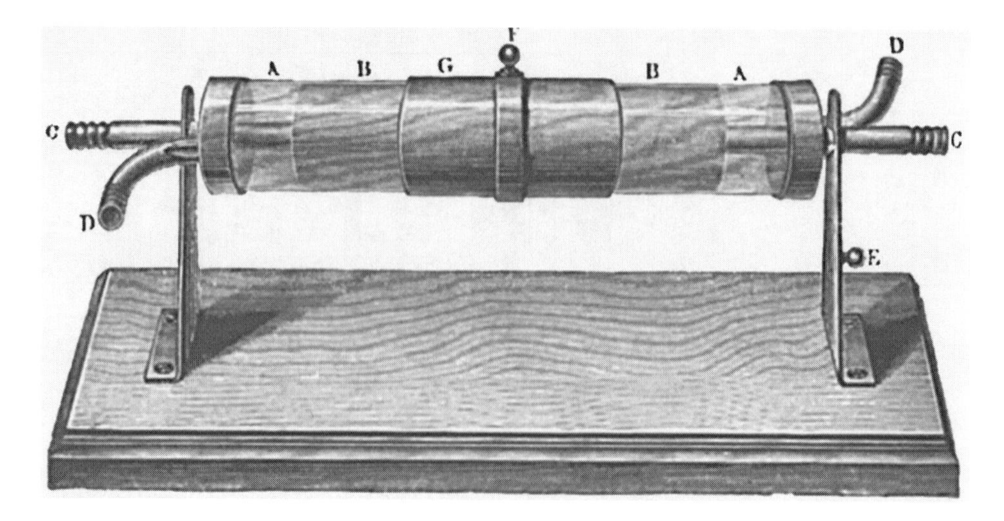

FIGURE 279. ■ Illustration of a late-nineteenth-century ozone generator. The apparatus was designed to expose flowing oxygen gas to a "silent discharge" of electricity since sparks decompose ozone (1 mol O_3) back to oxygen (1.5 mol O_2). (From Roscoe and Schorlemmer, *A Treatise on Chemistry*, 1894.)

It is noteworthy that this reaction and other many other reactions of ozone produced oxygen as a by-product (1 molecule of gas yielding 1 molecule of gas—i.e., no volume change). However, it was known that components in turpentine "absorbed" ozone completely:[3,6]

$$O_3 + \text{turpentine} \rightarrow (\text{turpentine–ozone "compound"}) \quad (2)$$

The ozone generator in Figure 279 does not produce complete conversion to oxygen, since decomposition to starting material is very significant. Ten percent ozone/oxygen mixtures are usually attained.[1] So the problem becomes one of establishing the formula of this highly reactive substance, present in only 10% quantity:

$$2O_3 \rightarrow 3O_2 \quad (3)$$

The problem was solved by J.L. Soret in 1872 using the apparatus schematically depicted in Figure 280.[2] The solution in the concentric vessel on the left-hand side of Soret's apparatus contains dilute sulfuric acid or copper sulfate with a wire

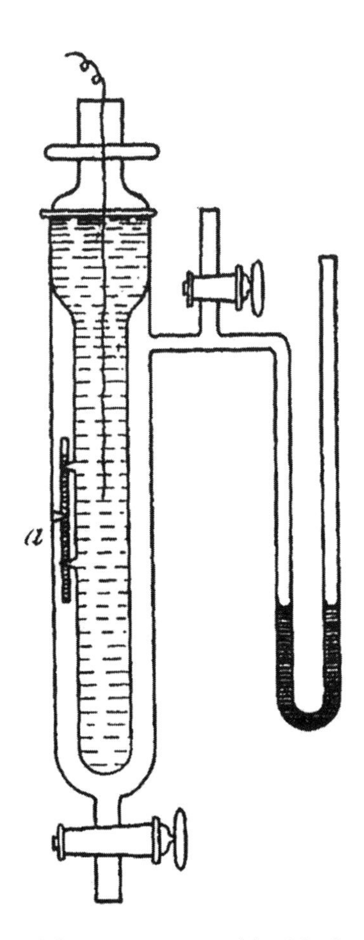

FIGURE 280. ■ Schematic of the apparatus used by J.L. Soret in 1872 to elucidate the formula (O_3) of ozone—see text for explanation (from Partington, *Everyday Chemistry*, 1929, MacMillan and Co., Ltd, London). It is quite amazing that the formula of this highly energetic, highly reactive trace component of the atmosphere was solved more than 130 years ago.

dipping into it. (Not shown is the bath vessel containing another wire dipped into ice water; the wires are connected to a source of electricity.) Oxygen is introduced into the concentric space, which also contains a sealed thin glass tube filled with turpentine. There is airtight communication of this glass tube with the outside so that it may be broken when desired. A manometer containing concentrated sulfuric acid with indigo dye is placed in series with the left-hand vessel.

Thus, if a 10% ozone mixture is produced, complete absorption by turpentine will reduce 100 cm^3 (cubic centimeters) of gas to 90 cm^3. On the other hand, if this 10% mixture is heated in order to decompose ozone to oxygen, the 10 cm^3 of ozone present yields 15 cm^3 of oxygen for a total gas volume of 105 cm^3. In other words, if reactions (2) and (3) hold (i.e., the formula of ozone is O_3), then the diminution of volume upon complete reaction with ozone must be twice the expansion of volume upon thermal decomposition of ozone. The reproducibility of this experiment was verified and allowed the assignment of the formula despite its relatively low abundance in the mixture.

Since the concept of valence was more than a decade old and oxygen was assigned the valence 2, the most reasonable structural formula in 1872 was **1** (remember, gentle reader, there was no octet rule or lone pairs of electrons and no appreciation of ring strain in 1872). This structure was favored at least through the late 1920s.[3] However, experimental structural chemistry of solids (via X-ray diffraction) and gases (via electron diffraction) began in the 1920s and 1930s, and ozone was found to be a bent molecule[1] (not an equilateral triangle) with an O—O—O angle of almost 117° and O—O bond lengths (1.28 Å) intermediate between single (1.49 Å) and double (1.21 Å). This structure was nicely rationalized by the resonance theory of Pauling in the 1930s as a hybrid of the two canonical Lewis structures, **2A** and **2B** (which obey the octet rule):[7]

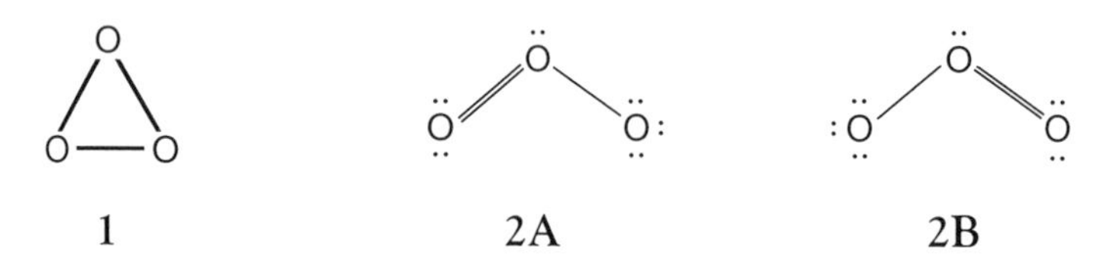

Cyclic O_3 (1) has never been observed. Although calculated to be about 30 kcal/mol higher in energy than ozone (2A ↔ 2B), it has a remarkably high barrier (ca. 22 kcal/mol) to the bond-stretching opening to ozone because the thermal reaction is symmetry-forbidden.[8]

Just as the ozone molecule is a resonance hybrid of two contributing (though identical) Lewis structures, so are ozone's properties a kind of hybrid of our conventional views of "good" and "bad." Ozone remains today an effective drinking water purification agent. However, its concentration in the lower atmosphere in human-made smog poses a significant health risk, particularly to asthmatics and the elderly. On the other hand, stratospheric ozone absorbs harmful ultraviolet light and lowers our risk of skin cancer. We have great cause to be concerned about the current decrease in stratospheric ozone caused by chlorofluorocarbons (CFCs) long used in aerosol cans. So to quote from a rock-and-roll era song—"Ozone," are you "Devil or Angel?"[9]

1. F.A. Cotton and G. Wilkinson, *Advanced Inorganic Chemistry*, fifth edition, John Wiley & Sons, New York, 1988, pp. 452–454.

2. H.E. Roscoe and C. Schorlemmer, *A Treatise On Chemistry*, Vol. 1, MacMillan and Co., London, 1894, pp. 235–243.
3. J.R. Partington, *Everyday Chemistry*, MacMillan and Co., Ltd, London, 1929, pp. 285–288.
4. This is mercury(I) oxide (Hg_2O). Heat will disproportionate it to Hg and HgO.
5. The liquefaction of "permanent gases" will be described in the essay on the discovery of neon.
6. Ozone reacts with alkenes (olefins) through addition to the double bond. The initially formed ozonides decompose further. Schōnbein appears to have performed the first such identified ozonolysis (on ethylene). Turpentine is a complex mixture of olefinic terpenes (see P.S. Bailey, *Ozonation in Organic Chemistry*, Academic Press, New York, 1978, pp. 1–4, for a brief historical perspective).
7. L. Pauling, *The Nature of the Chemical Bond*, Cornell University Press, Ithaca, 1939.
8. B. Flemming, P.T. Wolczanski, and R. Hoffmann, *Journal of the American Chemical Society*, Vol. 127: 1278, 2005. This article describes potential ways to stabilize O_3 in transition metal complexes. The author thanks Professor Roald Hoffmann for bringing this problem to his attention.
9. Or is ozone simply an oxymoron—"a human-made, naturally occurring, environment-protecting hazardous pollutant?" Actually, "oxygen" is derived from "acid maker"—Lavoisier's incorrect conclusion that all acids contain oxygen. Acidic properties are considered by early definition to be "sharp." "Oxymoron" means literally "sharp–dull"—an internal contradiction in a single word. But this would hardly surprise the ancients, who recognized contraries (male– female, good and evil) in all earthly things.

THE PEOPLE'S CHEMISTRY

A Muck Manual for Farmers (Figure 281) and 600 *Receipts Worth Their Weight in Gold* (Figure 282) are nineteenth-century American books that continue a tradition dating back to the early sixteenth century. Books of secrets,[1] such as Porta's *Natural Magick*, gave recipes for cosmetics, wines, and other concoctions. Books of recipes[2] and "household" books also provided practical home remedies, information on preserving food, and thousands of other bits of technical assistance for daily living.

A Muck Manual, by American chemist Samuel L. Dana, provides an intelligent, accessible and never-patronizing introduction to minerals, rocks, soils, manures, and composts. Dana (1795–1868) was an esteemed technical chemist and inventor of the "American System" of bleaching.[3] His book follows a tradition exemplified by works like *A Treatise Shewing the Intimate Connection that Subsists between Agriculture and Chemistry* (Archibald Cochrane, 9th Earl of Dundonald, London, 1795) and Humphrey Davy's *Elements of Agricultural Chemistry* (London, 1813). An interesting aspect of Dana's book is his introduction of a new term, *urets*, for minerals such as metal sulfides. He carefully accounts to his practical audience for his need to introduce a new term to the chemical lexicon. *Muck Manual* also discusses the chemical nature of geine—humus—whose complexity remains daunting. Today, state land-grant universities run Agricultural Extension Service programs to fulfill the practical teaching role for farmers assumed by Dana's useful manual.

I was at one time a transplanted Brooklyn Yankee[4] living in Charlotte, North Carolina. It is interesting to speak with people whose families have lived for long periods in this region. One friend[5] told me of the desperate importance of salt, for food preservation and refrigeration, as the Confederacy was losing the Civil War and in dire straits. Unsalted meat would often rot in transit. Heavily

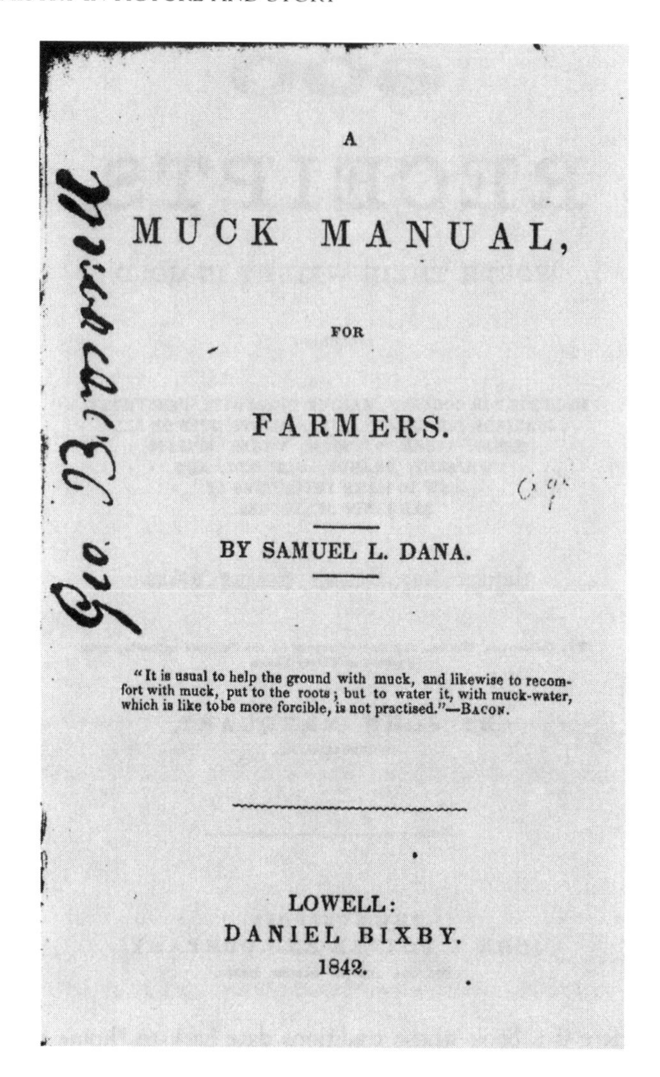
A
MUCK MANUAL,
FOR
FARMERS.
BY SAMUEL L. DANA.
"It is usual to help the ground with muck, and likewise to recomfort with muck, put to the roots; but to water it, with muck-water, which is like to be more forcible, is not practised."—BACON.
LOWELL:
DANIEL BIXBY.
1842.

FIGURE 281. ▪ Here is a nice title for a practical American book. Samuel L. Dana was a respected American chemist who authored a practical and never patronizing book for farmers on soils and compost (a kind of Agricultural Extension Service).

salted meat would arrive preserved but have to be repeatedly boiled in water in order to make it barely edible. Destroy the sources of salt and you have dealt the Confederate soldiers a crippling blow. Strategically, a major battle was fought in 1864 for control of Saltville, Virginia, a location of natural salt-licks.[6] Southern families were reduced to desperately digging up the soil from dirt floors under smokehouses and "boiling" it in water in order to recover the salt from previous seasons. This was a very sad life-or-death people's chemistry.

We end this essay on lighter, though not uplifting, notes. In Marquart's 1867 book, we find Recipe No. 83, "A remedy for Black Teeth": pulverized cream of tartar and salt; wash your teeth in the morning then rub them with this powder; Recipe No. 479, "To cure Hoven or Blown in Cattle [cattle over-eating rich food, bloating due to "overcharged" first stomach and incapable of expelling its contents—a life-threatening situation]—see Recipe No. 480": 1 pound of Glauber's salt ($Na_2SO_4 \cdot 10H_2O$—a cathartic); 2 ounces ginger powder; 4 ounces

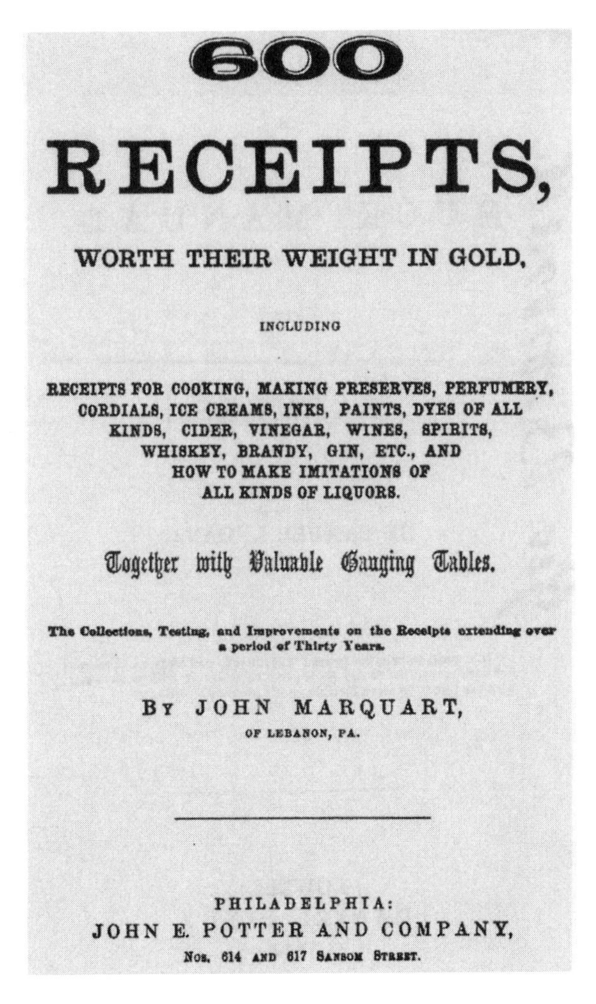

600

RECEIPTS,

WORTH THEIR WEIGHT IN GOLD,

INCLUDING

RECEIPTS FOR COOKING, MAKING PRESERVES, PERFUMERY,
CORDIALS, ICE CREAMS, INKS, PAINTS, DYES OF ALL
KINDS, CIDER, VINEGAR, WINES, SPIRITS,
WHISKEY, BRANDY, GIN, ETC., AND
HOW TO MAKE IMITATIONS OF
ALL KINDS OF LIQUORS.

Together with Valuable Gauging Tables.

The Collections, Testing, and Improvements on the Receipts extending over a period of Thirty Years.

BY JOHN MARQUART,
OF LEBANON, PA.

PHILADELPHIA:
JOHN E. POTTER AND COMPANY,
Nos. 614 AND 617 SANSOM STREET.

FIGURE 282. ■ Here is a book whose traditions date back to "home manuals" of the sixteenth century: how to prepare inks, drinks, and a catharsis for an overblown cow.

of molasses, mix, and then pour 3 pints of boiling water on the mass. When "new-milk warm" (i.e., fresh from the udder or "udderly" fresh), give the entire dose (cover your ears and hold your nose).

---

1. J.R. Partington, *A History of Chemistry*, MacMillan, London, 1961, Vol. 2, pp. 27–31.
2. J.R. Partington, op. cit., pp. 68–69.
3. E.F. Smith, *Chemistry In America, Appleton*, New York, 1914, p. 222.
4. A Brooklyn Yankee is an oxymoron—ask anybody who ever went to a baseball game at Ebbets Field. We hated "Yankees" as much as any Carolinian.
5. I learned of the history of boiling soil from floors under smokehouses from retired North Carolina state trooper, Harold Eaker, whose family has lived in the vicinity of King's Mountain, NC since the American Revolution. See also: Charles Frazier, *Cold Mountain*, Vintage, New York, 1997, p. 103.
6. G.G. Walker, *The War in Southwest Virginia: 1861–1865*, 6th ed., A & W Enterprise, Roanoke, 1985, pp. 10, 71–106.[1] thank Mr. R. Stewart Lillard for enlightening discussion and for making me aware of this book.

INK FROM PEANUTS AND THE FINEST SUGAR IN THE SOUTH

In 1947, Nobel laureate Roald Hoffmann was ten years old, in a displaced persons (DP) camp in Germany, when he was fascinated by the biographies (in German translation) of Marie Curie and George Washington Carver.[1] Carver (ca. 1860–1943) was born to slaves belonging to Moses Carver just prior to the start of the Civil War.[2] At the end of the War, Moses Carver discovered that his only living former slave was the five-year-old George who was seriously ill with whooping cough. Returned to his former Master's house, George remained almost ten years before traveling and developing his interests in plants and animals and talents in art and music. He obtained a high-school degree in his late 20s and became the first Afro-American graduate of what is now Iowa State University (1894), soon attaining the Masters degree there in 1896. His agricultural knowledge led him to act as a kind of extension service for African-American farmers. This dedication led Carver to Tuskegee Institute in Alabama, then under the Presidency of Booker T Washington.

As a university administrator running the newly-organized Agriculture Department at Tuskegee, Carver was not strong on bureaucratic practice or budget balancing. However, applied research was his real calling and Carver pioneered crop rotation and planting of legume products such as soybeans and peanuts to replenish the soil's nutrients. He and his collaborators at Tuskegee developed about 300 products derived from peanuts (e.g., inks, plastics, dyes, coffee) and over 100 others from sweet potatoes. Peanuts evolved from being a "noncrop" to the second leading cash crop (following "king cotton") in the South. In 1990, Carver and organic chemist Dr. Percy L. Julian, were the first African Americans to be inducted into the National Inventors Hall of Fame. Dr. Julian (1899–1975) pioneered the synthesis of physostigmine, used to treat glaucoma, perfected an economical route to the steroid cortisone, so effective for treating rheumatoid arthritis, and became the first African-American Director of Research for a major company (Glidden Company in Chicago).[3]

Norbert Rillieux (1806–1894) was the son of inventor Vincent Rillieux (the great uncle of artist Edgar Degas) and a free woman of color, Constance Vivant, with whom he had a long-standing relationship.[4] The younger Rillieux, a chemical engineer educated in Paris, developed the triple-effect-evaporator for sugar refining in the 1830s. In partnership with Jewish plantation owner Judah P. Benjamin (later Jefferson Davis's Secretary of War), the sugar produced by Rillieux's apparatus won awards and recognition and the apparatus was widely adopted. It is thought that Degas may have used the two men as models for one of his double portraits.[4]

---

1. R. Hoffmann and V. Torrence, *Chemistry Imagined—Reflections On Science*, Smithsonian Institution Press, Washington, D.C., 1993, pp. 30–32.
2. R. Holt, *George Washington Carver: An American Biography*, rev. ed., Doubleday, Doran and Co., Garden City, 1963.
3. E.J. McMurray (ed.), *Notable Twentieth-Century Scientists*, Gale Research Inc., Detroit, MI, 1995, Vol. 2, pp. 1045–1047.
4. *Chemical Heritage*, 16 (1):10, Summer, 1998.

SECTION VIII
TEACHING CHEMISTRY TO THE MASSES

GEODES[1]

The rhythms, rhymes, and imagery of poetry have preserved oral traditions since antiquity and they powerfully fix ideas in our minds. So why not apply this learning tool to contemporary teaching? And so, we have two small volumes of "Werneria" published in 1805 and 1806 by *Terrae Filius* (aka *Terrae Filius Philagricola*—"son of the earth lover Agricola").[2,3] "Werneria" refers to Abraham Werner, a German geologist who believed that all rocks originated in the oceans ("Neptunist school"), but was "a brilliant lecturer in geology."[4] Our mystery author is Reverend Stephen Weston (1747–1830),[5] a poet, a man of letters, having an incredible breadth of interests. Following the death of his young wife around 1790, he devoted the remainder of his life to art and literature.[5] His works included translations from Latin, French, Arabic, Persian, and Chinese languages and discussions of travels and religious thought. It is said that he "lived for some years among the dilettanti in London" and "had a numerous circle of lady admirers who fed his vanity."[5] In his preface to Werneria,[2] Weston quotes Aristotle: "Is it because men, before they discovered the art of writing, sang their laws, that they might not forget them?" and so he applies his own artistic talents to teaching mineralogy.

How about this poetic description of diamond that treats some physical properties and teaches that diamond is pure carbon—its combustion forms carbon dioxide with no other residue remaining?[2]

> In hardness, brilliance, and transparency,
> The diamond every mineral excels,
> Black, yellow, green, blue, brown, or grey, 'tis known,
> And colourless in quartzose sand is found
> In flat, or rounded grains, sometimes cube-shap'd,;
> But oft its form eight-sided is, or twelve:
> In texture laminous, but fibrous too,
> Irregularly so; to solar rays
> Expos'd the diamond is phosphoric;
> Rubb'd, it emits electric sparks: What gem
> But this from rich Golconda's shore, can e'er
> To carbone's acid be converted, and
> Leaves no wreck behind?

And here is the description of lime (CaO, calcium oxide or quicklime), obtained by heating calcium carbonate ($CaCO_3$, limestone)—it was the primary binder in concrete until the early 1800s. It can recombine with carbon dioxide and is strongly basic, but moisture converts it, with evolution of heat, to a milder calcium hydroxide powder:

This earth from carbonate of lime obtained
For various use, by application of
Incessant heat, in form is concrete, or
In powder; in colour white; in taste hot,
Pungent, and caustic; and when in water solved,
Will change the vegetable blue to green:-
When in the concrete state exposed to air,
Cohesion's force is lost; but when the gas
Of carbo from the atmosphere's absorbed,
Then all its pristine hardness is regain'd.
Add moisture, and to powder it returns,
Shines in the dark, its caloric evolves,
And doubly mild and temperate becomes.
Per se infusible, all others it can fuse;
With borax, and microsmic salt it melts,
And effervesces not.

Limestone is widely distributed in soils, and in Figure 283 we display the apparatus designed by Humphry Davy[6] to measure its abundance. Sulfuric acid in *B* is added dropwise to the soil in vessel *A*, the carbon dioxide released is carried through C and collected into the balloon in vessel *E*, which is filled with water, and the volume of expansion is measured in cylinder *D*.

1. The blame for this title rests on my brother Kenny Greenberg.
2. *Terrae Filius* [i.e., Stephen Weston], *Werneria; or, Short Characters of Earths: with Notes according to the Improvements of Klaproth, Vauquelin and Hauy, C. and R. Baldwin*, London, 1805. I am grateful to chemist and book collector Dr. Roy G. Neville for making me aware of *Werneria* and Stephen Weston.
3. *Terrae Filius Philagricola* [i.e., Stephen Weston], *Werneria, (Part the Second) or, Short Characters of Earths and Minerals According to Klaproth, Kirwan, Vauquelin, and Hauy*, C. and R. Baldwin, London, 1806.
4. *The New Encyclopedia Brittanica*, Vol. 12, Encyclopedia Brittanica, Inc., Chicago, 1986, pp. 582–583.
5. L. Stephen and S. Lee (eds.), *The Dictionary of National Biography*, Oxford University Press, Oxford, 1921 (reprint 1964/65), pp. 1283–1285.
6. H. Davy, *Elements of Agricultural Chemistry, in a Course of Lectures for The Board of Agriculture*, Longman, Hurst, Rees, Orme, and Brown, London, 1813, Figure 15 (facing p. 145).

MICHAEL FARADAY'S FIRST CHEMISTRY TEACHER

Mrs. Jane (Haldimand) Marcet (1769–1858) was bom in England and married a prominent Swiss physician and respected amateur chemist Alexander Marcet.[1,2] Influenced by Humphry Davy's public lectures she tried some experiments and decided to write a book to explain the science:

> In venturing to offer to the public, and more particularly to the female sex, an Introduction to Chemistry, the author, herself a woman, conceives that

FIGURE 283. ▪ Humphry Davy's apparatus for measuring the amount of limestone in soil. Sulfuric acid releases one equivalent of CO_2 from an equivalent of limestone ($CaCO_3$). The gas fills a balloon, displacing a volume of water that is measured to give the volume of gas generated. (From Davy, *Agricultural Chemistry*, London, 1813.)

> some explanation may be required: and she feels it the more necessary to apologize for the present undertaking, as her knowledge of the subject is but recent, and as she can have no real claims to the title of chemist.

(Compare this strategically diplomatic *Apologia* with the one cited earlier from Mrs. Fulhame's 1794 book (Figure 220). Mrs. Fulhame is openly contemptuous of narrow and ignorant people who would limit a woman's role). The first London edition of Conversations (Figure 284) is said to have appeared in 1805[1] (another opinion is 18062). Edgar Fahs Smith avers that about 160,000 copies of its numerous editions were sold before 1853.[1]

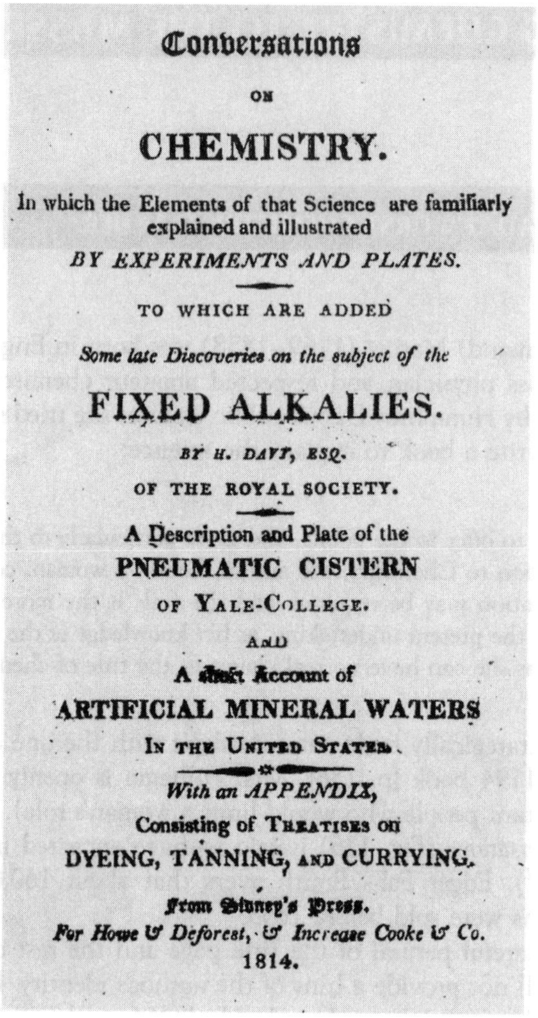

Conversations
ON
CHEMISTRY.
In which the Elements of that Science are familiarly explained and illustrated
BY EXPERIMENTS AND PLATES.
TO WHICH ARE ADDED
Some late Discoveries on the subject of the
FIXED ALKALIES.
BY H. DAVY, ESQ.
OF THE ROYAL SOCIETY.
A Description and Plate of the
PNEUMATIC CISTERN
OF YALE-COLLEGE.
AND
A Short Account of
ARTIFICIAL MINERAL WATERS
IN THE UNITED STATES.
With an APPENDIX,
Consisting of TREATISES ON
DYEING, TANNING, AND CURRYING.
From Sidney's Press.
For Howe & Deforest, & Increase Cooke & Co.
1814.

FIGURE 284. ■ *Conversations on Chemistry* was actually authored by Mrs. Jane Marcet. It is a beautiful teaching text that uses Socratic dialogue involving a Mrs. B. and two adolescent girls, Caroline and Emily. It inspired the young Michael Faraday's interest in chemistry and appeared in a number of editions over almost 50 years and sold over 160,000 copies.

The most careful perusal of the title page and the rest of the text of the early editions will not provide a hint of the author's identity. Part of the reason was Mrs. Marcet's own modesty about her lack of formal training. However, the etiquette of the day is also a likely cause. Most outrageously, later editions (e.g., 1822, 1826, 1829, and 1831. edited by Dr. J.L. Comstock) were published by men who, while crediting the "authoress," were quick to add their own criticisms. One defender of Mrs. Marcet wrote[1]:

> We are informed by one of the American editors of this work that his reason for not placing the name of Jane Marcet on the title-page, was because scientific men believed it fictitious!

Conversations on Chemistry is a delightful interplay between a Mrs. B. (sometimes referred to as Mrs. Bryan[2]) and Caroline and Emily[2] (ages 13 to 15). Its coverage of chemical principles, while accessible, is not at all superficial, and Mrs. Marcet updated her own editions by including the latest work of her correspondent Davy and other prominent chemists. Here is a selection found on pages 198–199 of the 1814 American edition.

Mrs. B.: From its own powerful properties, and from the various combinations into which it enters, sulphuric acid is of great importance in many of the arts. It is also used as a medicine in a state of great dilution; for were it taken internally, in a concentrated state, it would prove a most dangerous poison.

Caroline: I am sure it would burn the throat and stomach.

Mrs. B.: Can you think of any thing that would prove an antidote to this poison?

Caroline: A large draught of water to dilute it.

Mrs. B.: That would certainly weaken the power of the acid, but it would increase the heat to an intolerable degree. Do you recollect nothing that would destroy its deleterious properties more effectually?

Emily: An alkali might, by combining with it; but then, a pure alkali is itself a poison, on account of its causticity.

Mrs. B.: There is no necessity that the alkali should be caustic. Soap, in which it is combined with oil, or magnesia, either in a state of carbonat, or mixed with water, would prove the best antidotes.

Emily: In those cases, then, I suppose, the potash and the magnesia would quit their combinations to form salts with the sulphuric acid?

Mrs. B.: Precisely.

It appears that the novelist Maria Edgeworth read Mrs. Marcet's book and may have saved the life of her younger sister, who had swallowed acid, by administering milk of magnesia.[2] It is intriguing that in her 1998 novel of suspense,[3] historian Barbara Hambly provides a schoolteacher, a free woman of color, with a book titled *Conversations in Chemistry More Especially for the Female Sex* that is authored by a (presumably Mrs.) Mercer.

The great nineteenth-century scientist Michael Faraday came from a family of very modest means and worked as a bookbinder starting in 1804 at the age of 13. He was first introduced to chemistry by Mrs. Marcet's book[1]:

> So when I questioned Mrs. Marcet's book by such little experiments as I could find to perform, and found it true to the facts as I could understand them, I felt that I had got hold of an anchor in chemical knowledge, and clung fast to it. Hence my deep veneration for Mrs. Marcet: first, as one who had conferred great personal good and pleasure on me, and then as one able to convey the truth and principle of those boundless fields of knowledge which concern natural things, to the young, untaught, and inquiring mind.
>
> You may imagine my delight when I came to know Mrs. Marcet personally; how often I cast my thoughts backward, delighting to connect the past and the present; how often, when sending a paper to her as a thank-offering, I thought of my first instructress, and such like thoughts will remain with me.

Mrs. Marcet's influence on Faraday is probably doubly profound. In addition to his fundamental contributions to science, Michael Faraday was renowned for his public lectures to lay audiences and his book *A Course of Six Lectures on the Chemical History of a Candle* (1861) became a classic for popularizing chemistry.

---

1. E.F. Smith, *Old Chemistries*, McGraw-Hill, New York, 1927, pp. 64–71.
2. M. Rayner-Canham and G. Rayner-Canham, *Women in Chemistry: Their Changing Roles from Alchemical Times to the Mid-Twentieth Century*, American Chemical Society and the Chemical Heritage Foundation, Washington, D.C. and Philadelphia, 1998, pp. 32–35.
3. B. Hambly, *Fever Season*, Bantam, New York, 1998, p. 292. I thank Professor Susan Gardner for bringing this to my attention.

"CHEMISTRY NO MYSTERY"

At the very start of Marcet's *Conversations on Chemistry*, Caroline says: "To confess the truth, Mrs. B., I am not disposed to form a very favorable idea of chemistry, nor do I expect to derive much entertainment from it. I prefer those sciences that exhibit nature on a grand scale, to those which are confined to the minutiae of petty details." Four years after Dalton's Atomic Theory and already "I'm bored" from teen-age students!

John Scoffern, a surgeon and occasional chemical assistant at the London Hospital,[1] wrote a book titled *Chemistry No Mystery* (London, 1839) that offered excitement to young and old alike:

> If I were to present myself before you with an offer to teach you some new game:—if I were to tell you an improved Plan of throwing a ball, of flying a kite, or of playing leap-frog, oh, with what attention you would listen to me. Well, I am going to teach you many new games. I intend to instruct you in a science full of interest, wonder, and beauty; a science that will afford you amusement in your youth, and riches in your mature years. In short, I am going to teach you the science of chemistry.

How wonderfully fitting that the title page (Figure 285) depicts a scene outside a show-caravan wherein the imaginary narrator ("The Old Philosopher" or "O.P.") recalls a scene from his misspent youth. He enjoyed practical jokes and released hydrogen sulfide gas (rotten-egg odor) under the flooring of the stage driving out the show's giant and its dwarf. He apparently soon felt the giant's wrath and spent two days in the hospital afterward.[1]

Figure 286 depicts one of O.P.'s hypothetical lectures in which he makes and then foolishly distributes laughing gas to the students in his lecture hall.[1] The illustrator and caricaturist George Cruikshank (1792–1878), who produced these drawings, was probably the first to provide lively, humorous pictures for children's books, and he illustrated Charles Dickens' *Oliver Twist* (1838).[2]

---

1. J. Read, *Humour and Humanism in Chemistry*, Bell, London, 1947, pp. 208–214.
2. *Encyclopedia Brittanica*, Chicago, 1986, Vol. 3, p. 763.

FIGURE 285. ■ "Step right up, ladies and gentlemen, and get your nice hot tootsie-frootsie chemistry!" (homage to Marx—Chico, not Karl). This figure, drawn by George Cruikshank (who illustrated *Oliver Twist*), appears in Dr. John Scoffern's *Chemistry No Mystery* (London, 1839). A practical application of chemistry (a stinkbomb) has been released in a circus tent.

FIGURE 286. ▪ More chemical mischief in *Chemistry No Mystery:* the Old Philosopher ("O.P") has allowed his class to participate in the nitrous oxide experiment. I imagine the following dialogue afterward: "Lucky you are tenured," sayeth O.P.'s Department Chair; "Academic freedom," responds O.P; "Don't press your luck," responds the Chair who sees no excuse for laughter in a lecture hall.

THE CHEMICAL HISTORY OF A CANDLE

A *Course of Six Lectures on the Chemical History of a Candle* [London, 1861; New York, 1861—see Figure 287(a)], derived from notes at Faraday's public lectures, is the culmination of a wonderful 60-year heritage of popularizing chemistry involving three individuals: Humphry Davy, Jane Marcet, and Michael Faraday. We have already met Count Rumford, whose "boring experiment" (Figure 222) disproved Lavoisier's caloric theory. He married the widowed Madame Lavoisier in 1805 and they effectively separated within two months (a boring husband?). In 1799 Rumford's ideas for improving the education of the middle classes and improving arts and manufacturing led to the chartering of the Royal Institution of Great Britain. He brought in young Humphry Davy as Assistant Lecturer in Chemistry, Director of the Laboratory, and Editor of the Institution's chemical journal. Davy's public lectures were popular and well attended. One of those attending was Mrs. Jane Marcet. Davy's lectures inspired Mrs. Marcet's interest in chemistry and ultimately stimulated her to write *Conversations on Chemistry*, which went through numerous editions and sold over 160,000 copies. Mrs. Marcet included some of Davy's latest work in her editions and maintained scientific correspondence with him.

Michael Faraday (1791–1867) was born to the family of a poor blacksmith.[1] At the age of 13 he was apprenticed to a bookbinder. With the owner's permission, he read and was inspired by Mrs. Marcet's book. Faraday started to attend public chemical lectures and, in 1812, a customer rewarded him with a ticket to Davy's lecture at the Royal Institution. Shortly afterward, Faraday sent a copy of the lecture notes he wrote out to Davy and requested to be employed as his assistant. Davy hired the young man and by 1820 Faraday had published his first paper. Throughout his career, Faraday joyfully acknowledged his debt to Mrs. Marcet and remained her correspondent and friend. Faraday took a course in elocution in 1818 and was "a splendid lecturer."[1]

The *Chemical History of the Candle* was derived from Faraday's public lectures. The book was reprinted throughout the nineteenth century in many languages. In fact, the most recent reissue appears to be in 1993 (Cherokee Press, Atlanta). Here is Faraday's rationale presented in Lecture 1:

> I propose to bring before you, in the course of these lectures The Chemical History of a Candle. There is no better, there is no more open door by which you can enter the study of natural philosophy than by considering the physical phenomena of a candle. There is not a law under which any part of this universe is governed which does not come into play, and is not touched upon, in these phenomena. I trust, therefore, I shall not disappoint you in choosing this for my subject rather than any newer topic, which could not be better, were it even so good.

Figure 287(b) is from Lecture 2. The glass tube opens at one end into the dark middle part of a candle flame. At the other end, the invisible wax vapors from this part of the flame are seen to condense. Faraday then differentiates vapors from gases for his audience. He proceeds to heat some candle wax in another flask and pours the vapors into a basin and sets them on fire. In another

A

COURSE OF SIX LECTURES

ON THE

CHEMICAL HISTORY OF A CANDLE:

TO WHICH IS ADDED

A LECTURE ON PLATINUM.

BY

MICHAEL FARADAY, D.C.L., F.R.S.,

FULLERIAN PROFESSOR OF CHEMISTRY, ROYAL INSTITUTION; FOREIGN ASSOCIATE OF THE ACADEMY OF SCIENCES, ETC.

Delivered before a JUVENILE AUDITORY *at the* ROYAL INSTITUTION *of* GREAT BRITAIN *during the Christmas Holidays of* 1860–1.

EDITED BY WILLIAM CROOKES, F.C.S.

WITH NUMEROUS ILLUSTRATIONS.

NEW YORK:
HARPER & BROTHERS, PUBLISHERS,
FRANKLIN SQUARE.
1861.

(a)

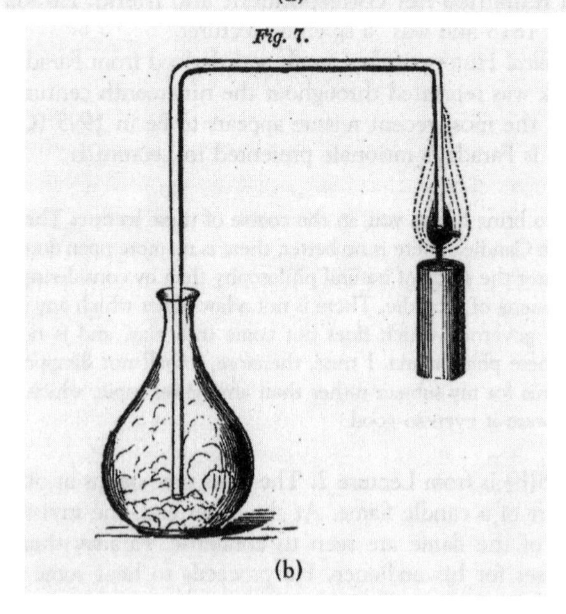

FIGURE 287. ■ (a) Title page from Michael Faraday's *Chemical History Of A Candle* (the London edition was also published in 1861). The book was not written by Faraday but derived using notes from his public lectures at the Royal Institution. Faraday's interest in teaching chemistry to the public follows a 60-year strand through Mrs. Marcet from Humphrey Davy. (b) Collecting the invisible vapors of a candle.

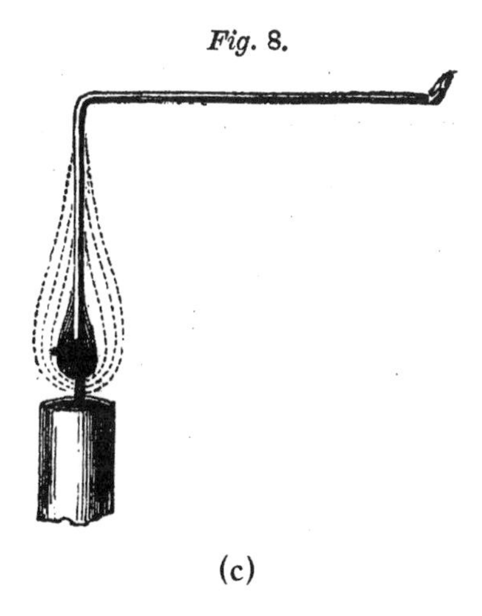

(c)

FIGURE 287. ▪ *Continued* (c) An articulated candle.

demonstration [Figure 287(c)] he uses a piece of glass tubing in communication with the middle of the flame and lights the other end of the glass tubing to form a kind of articulated candle. He notes further that if the glass tubing communicated with the top, rather than the middle, of the flame, there would be no vapor to carry through since it is burned in the upper region. He thus demonstrated the presence of invisible, flammable vapors present in the center of the flame but not at the top. Faraday quips: "Talk about laying on gas—why we can actually lay on a candle."

1. J.R. Partington, *A History of Chemistry*, MacMillan, London, 1964, Vol. 4, pp. 99–140.

INTO THE HEART OF THE FLAME

Figure 288 is the wild and wonderfully stylized illustration of a candle's flame in the 1857 edition of Edward Youmans' *Chemical Atlas*. The formula of carbon dioxide is shown correctly, but the author errs (see discussion of Figure 256) in describing water as HO, fuel as CH, and in depicting gaseous oxygen as atoms rather than as O_2 molecules. The lower interior region of the flame (which we see as blue) is shown as fuel rich and lacking oxygen. We now know that this part of the flame and the incandescent regions immediately above and around it are full of short-lived, exotic, and ultrareactive carbon-rich molecules, molecular fragments and particles.[1] These regions are reducing in nature since the carbon-rich species hungrily grab oxygen atoms from calxes, such as tin oxide, to produce the metals. (The carbon-oxygen bond in CO is the strongest bond in any neutral compound.[2]) In contrast, the outer blue edge of the flame is oxidizing—rich in the super-reactive hydroxyl radical (truly HO^-) as well as oxygen, carbon dioxide, and water.[1] In this region, tin would be immediately oxidized to its calx.

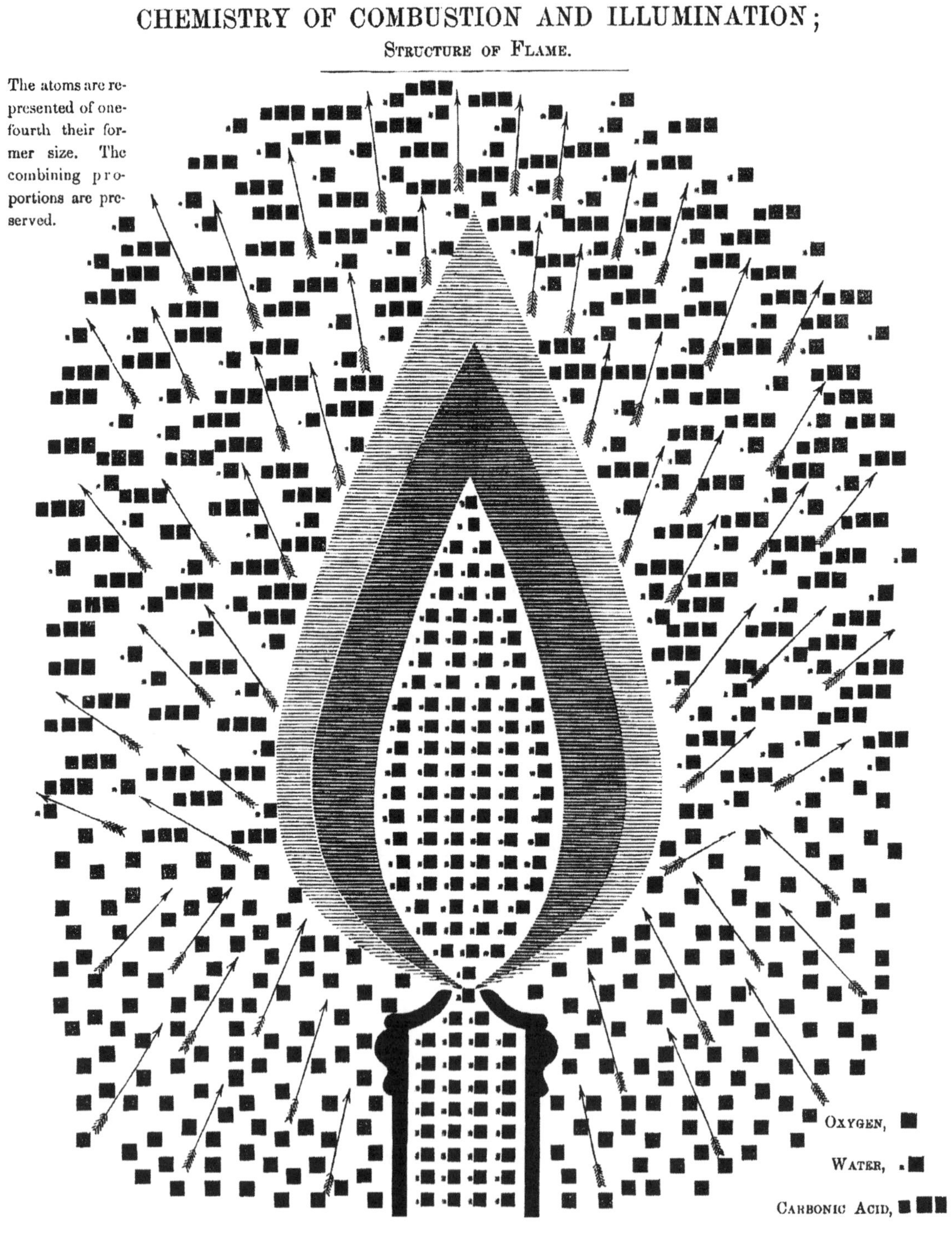

CHEMISTRY OF COMBUSTION AND ILLUMINATION;

STRUCTURE OF FLAME.

The atoms are represented of one-fourth their former size. The combining proportions are preserved.

OXYGEN, ■

WATER, ▪■

CARBONIC ACID, ■■■

FIGURE 288. ▪ An ebullient flame from the 1857 edition of Youmans' *Chemical Atlas* (see Figure 256; the errors in formulas such as HO for water are discussed in the text). See color plates.

These details have been known for almost 200 years through the application of a kind of "flame scalpel" called the blowpipe.

In his book *The Use of the Blowpipe in Chemical Analysis, and in the Examination of Minerals* (Stockholm, 1820; London, 1822), Jons Jacob Berzelius traces the history of the blowpipe to traditional applications by jewelers. He dates its earliest application to "dry" chemistry at about 1733. An ideal blowpipe is made of a brass tube with an ivory tip (having an opening of roughly ⅜-inch diameter) attached at one end, to facilitate the chemist's exhalation, with a fused platinum tip (about 1/16-inch opening diameter) following a 90° bend at the other end. The platinum tip is inserted into the flame and blowing is performed in a forceful but steady manner to excise the reducing or oxidizing parts for contact with the matter of interest. The author notes that inexperienced users seem to require exhausting bursts of lung power—"they might as well have proposed to play on a wind instrument with a bladder."[3] He details a technique in which the cheeks are filled with air and continuously replenished and used to generate a steady but forceful airstream. The blowpipe was a sensitive instrument for analysis of mineral samples and could provide evidence for metallic impurities at levels too low to weigh. For example, the ashes of a piece of paper, subjected to the reducing flame from a blowpipe, yielded microscopic particles of metallic copper.[3]

1. P.W. Atkins, *Atoms, Electrons and Change*, Scientific American Library, Freeman, New York, 1991, pp. 105–109.
2. I thank Professor Joel F. Liebman for calling this to my attention.
3. J.J. Berzelius, *The Use of the Blowpipe in Chemical Analysis, and in the Examination of Minerals* (translated by J.G. Children), Baldwin, Cradock and Joy, London, 1822, pp. 5, 8.

POOF! NOW YOU SMELL IT! NOW YOU DON'T!

Here is an imaginative way to teach chemistry from a highly imaginative person. In his 1823 book *Diagrammes Chimiques* Henri Decremps puts ideas into flow diagrams that dissect substances into parts and reassemble them following chemical reactions. Decremps was a lawyer and amateur magician.[1] In 1784, he published a book titled *La Magia Blanche Devoilee* ("White Magic Revealed"). A rival conjuror of great fame, Pinetti, who claimed to be a Knight, Professor of Mathematics and Natural Science, etc., borrowed liberally from *La Magia* without sharing credit. Decremps published many books attempting to debunk Pinetti but they only increased Pinetti's fame.[1] Finally, long in the tooth and gray in the beard, Decremps tried his hand at writing a chemistry text.

Figure 289 describes how "two odorless bodies placed in contact produce a very sharp odor and two other bodies form by their reunion a visible, palpable body." At the top left, we see sulfuric acid join with the components of limestone ($CaCO_3$), which are lime (CaO) and carbon dioxide (CO_2) liberated by addition of heat ("caloric"). Calcium sulfate ($CaSO_4$) and water remain in the top retort while carbon dioxide and "caloric" travel to the bottom-most flask. In the middle left, we add lime (CaO) and "caloric" to ammonium chloride (NH_4Cl), which

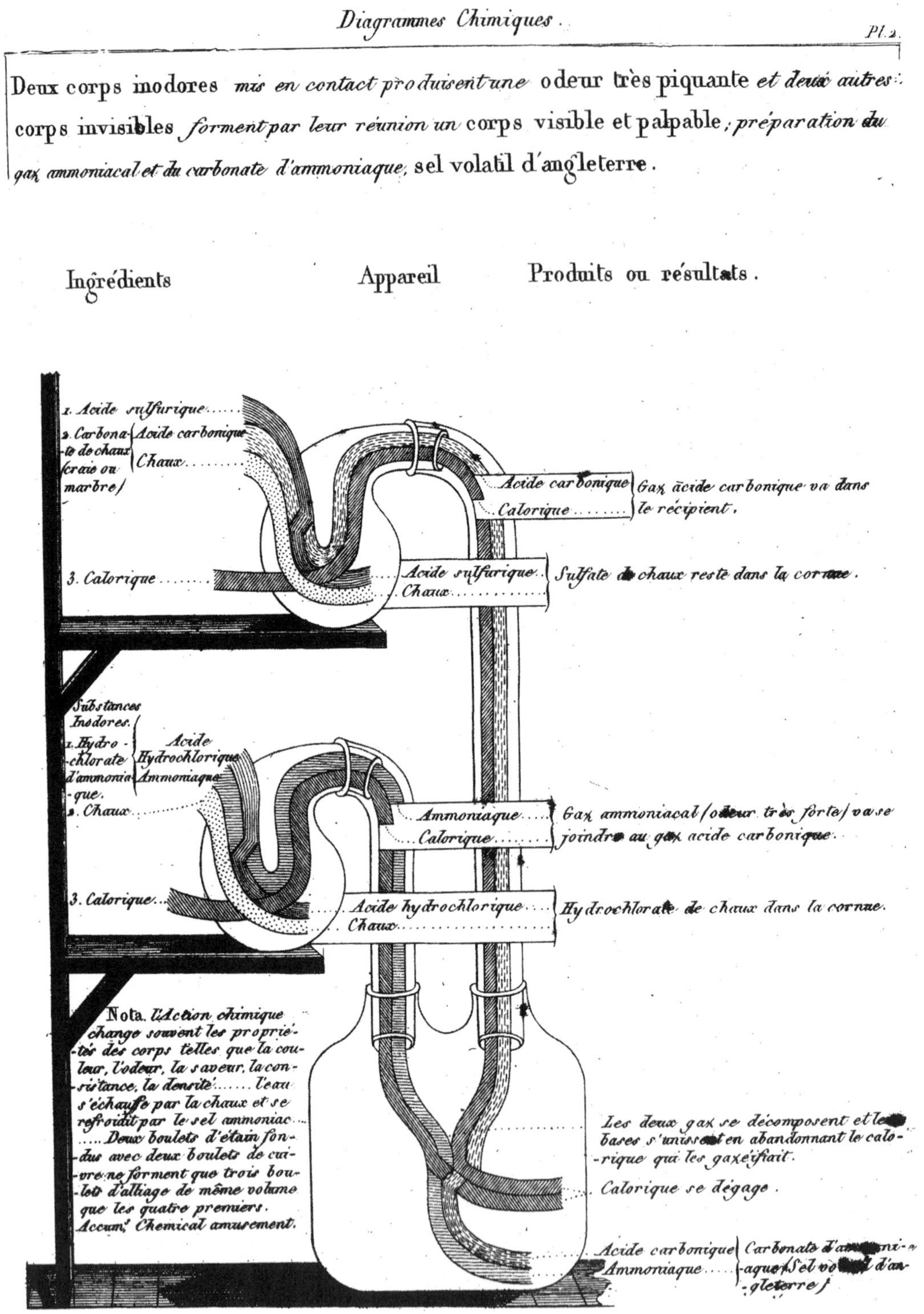

FIGURE 289. ■ Henri Decremps, the author of *Diagrammes Chimiques . . .* (Paris, 1823), was a famous magician for most of his life. The fascinating diagrams in this book seem to invent imaginary apparatus to conduct conceptual streams of chemicals, their "dissection," and subsequent reactions. In this figure, two odorless substances cause the formation of a "piquant" substance—ammonia.

will react under these conditions to release ammonia gas (NH_3)—also known as the "piquant odor," leaving calcium chloride ($CaCl_2$) and water in the retort in the middle left. Ammonia and "caloric" join carbon dioxide and "caloric" (undoubtedly in the presence of some water) to form ammonium carbonate $(NH_4)_2CO_3$—a visible, palpable body.

More magic in Figure 290: infusion of violets is actually an acid–base indicator (the first was discovered by Boyle in 1675). When vinegar, an acid, is added to the neutral blue infusion of violet, the solution turns red. When excess aqueous ammonia base is added, the solution goes from red to blue to green. The first human blow-hard neutralizes the solution back to blue by blowing in carbon dioxide, which forms carbonic acid in water. The second blow-hard returns the color to red by adding more carbonic acid.

Figure 291 shows the reader how to picture the molecular structure of cop-

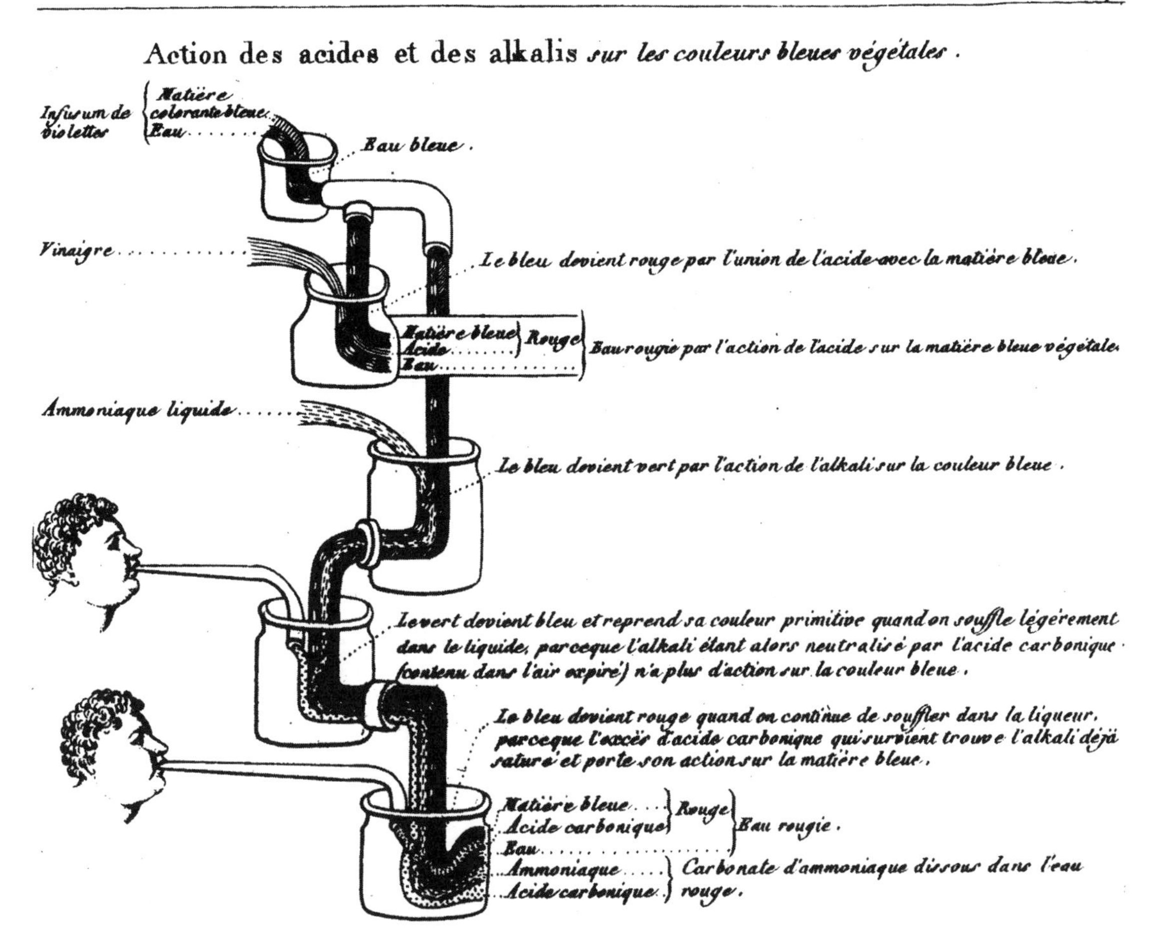

FIGURE 290. ■ Color changes in *Diagrammes Chimiques* brought about by adding vinegar to a neutral solution (colored blue by the indicator) and observing the solution turn red; ammonia is added and the solution goes back to blue then green (basic). Blowing carbon dioxide into the solution then neutralizes to blue. The second blowhard makes it acidic again (red).

per sulfate (one oxygen short). Ionic compounds were not understood until the work of Arrhenius late in the nineteenth century. The third figure in this drawing shows the addition of metallic iron (iron atoms) to copper sulfate. The iron atoms lose their electrons (are oxidized) to copper ions, which are reduced to atoms and precipitate out.

Figure 292 reminds the world that it was the French who defeated phlogiston. The top diagram shows metallic lead composed of "earth of lead" (lead calx or oxide) plus phlogiston. Heating of metallic lead causes loss of phlogiston, leav-

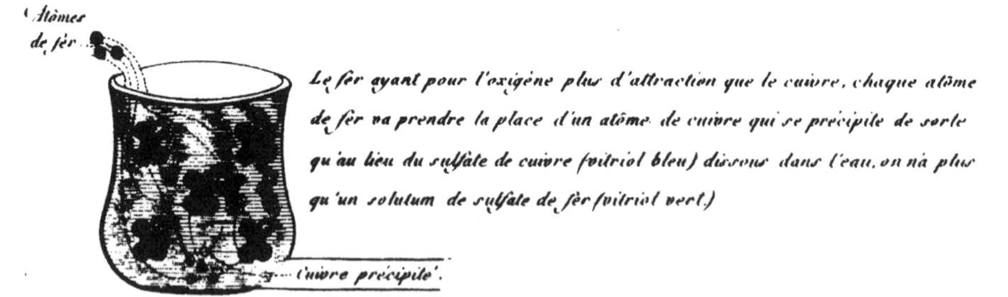

FIGURE 291. ■ Decremps did not know about ions. Also copper sulfate is $CuSO_4$. However, his diagram nicely shows that atomic iron will oxidize and reduce copper to the metal.

ing behind the calx. The diagram notes that the calx is heavier than the metal and this is impossible (unless phlogiston has negative mass). Thus "calorique," developed in France, receives kinder treatment than phlogiston, developed in Germany and championed in England.

1. C. Milbourne, *Panorama of Magic*, Dover, New York, 1962, pp. 27–31.

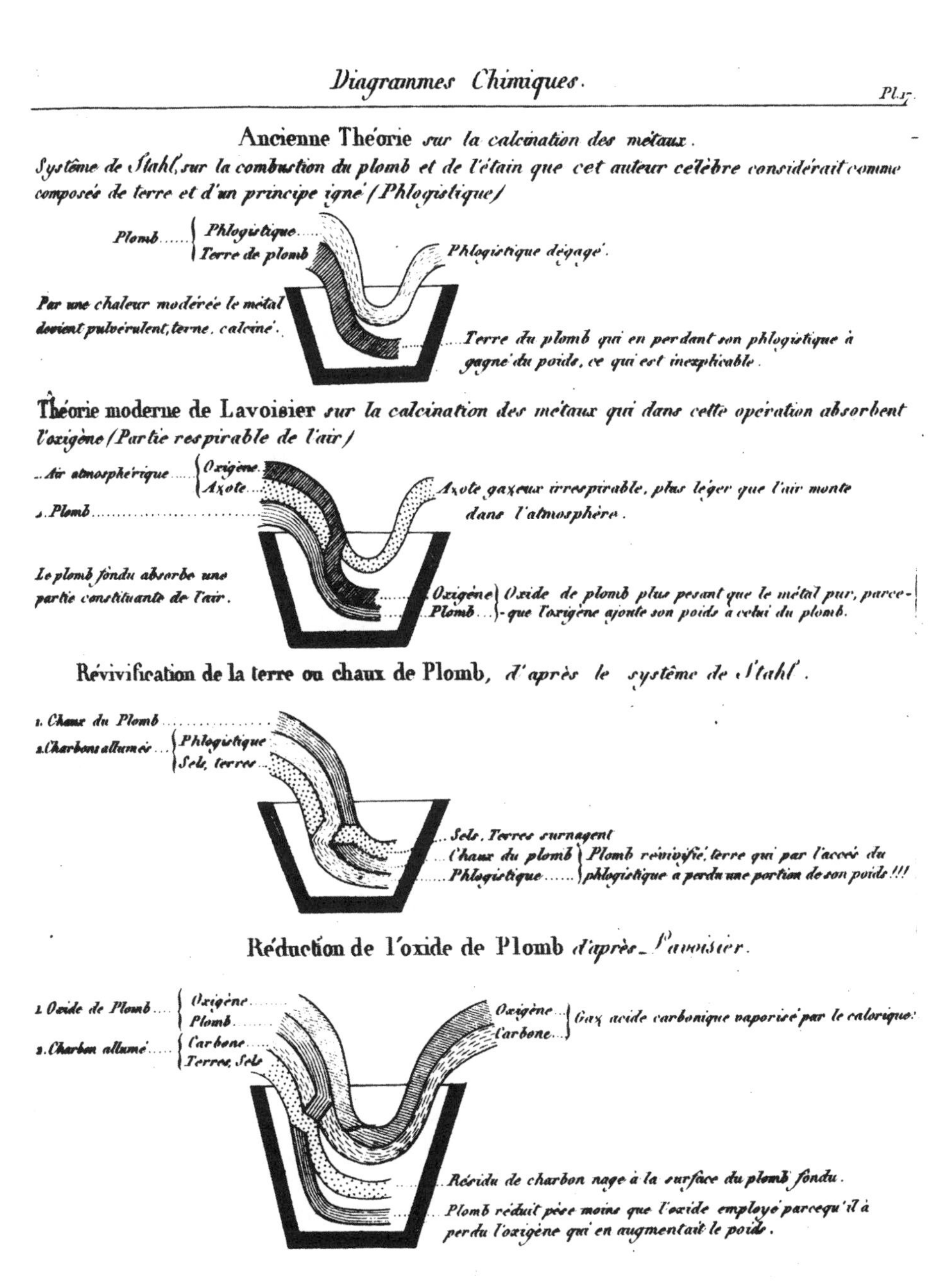

FIGURE 292. ■ Just in case anybody forgot, it was the French who defeated Phlogiston. This figure demonstrates that the gain in weight upon calcination of lead involves a gain in mass not a loss as would occur if the metal lost Phlogiston.

MY CHEM PROFESSOR TOOK THE FIRST PHOTOGRAPH OF THE MOON!

During the 1846–1847 academic year, a Silas H. Rathbone enrolled in the chemistry course offered by the University of New York (now New York University or NYU) Medical School. The admission/attendance card for the course (Figure 293) is signed by Mr. Rathbone and his professor, Dr. John William Draper, M.D. As a career academician, I find this quaint card heartwarming since it embodies the fundamental academic contract (to do one's best) between student and teacher. At a time when universities see students as "customers," employ marketing firms to create "brands," field deans of all manners and descriptions, Washington lobbyists and near-lobbyists, Associate Vice Presidents for "outreach scholarship" (whatever *that* is), collectors of frequent flyer miles who would hardly recognize an actual student, and teams of lawyers, it is a reassuring reminder of our fundamental mission.

Mister Rathbone was most fortunate to be a student of Professor Draper, a true "renaissance man." Dr. Draper was an experienced master teacher. His textbook,[1] first published in 1846 and undoubtedly used by Mr. Rathbone, was written to satisfy the need expressed in its Preface:

> The greatest service which can be rendered to our science, is for some person who has had the management of large classes for several years to sit down and write a book, setting forth what he said and what he did every day in his Lectures. That is the thing we want.

There is an all-too-familiar canard about teaching: "Those who can do, and those who can't teach." Let us test this statement by briefly reviewing some of Professor Draper's accomplishments during a teaching career spanning more than 40 years at the University of New York that ended in 1881, the year before he died.

Born in England in 1811, Draper obtained his MD degree at the Medical College of the University of Pennsylvania in 1836, yet another in the line of distinguished physician–chemists that began with Benjamin Rush (Figure 239). He received an initial appointment at the University of New York in 1837 and became Professor of Chemistry when the Medical College opened its doors in 1840. He became absorbed in the study of light and what we have come to call Photochemistry.[2] The Grotthus–Draper Law (only light that is absorbed can be effective in producing chemical change) is often called The First Law of Photochemistry.[2] His 1846 textbook describes the effects of the invisible light that is slightly less refracted by a glass prism than red light. Today we call this light "infrared." In his book, Draper described early experiments performed by scientists in which thermometers exposed to this invisible part of the spectrum registered higher temperatures than thermometers exposed to the visible region. But Draper's own contribution was the observation of the effects arising from the invisible light slightly *more* refracted than the violet. He discovered that these invisible rays were the actual source of exposure for the newly invented daguerreotype photography plate. Draper tentatively called this invisible light "tithonic rays" since he postulated that they were always associated with visible light (an allusion to the

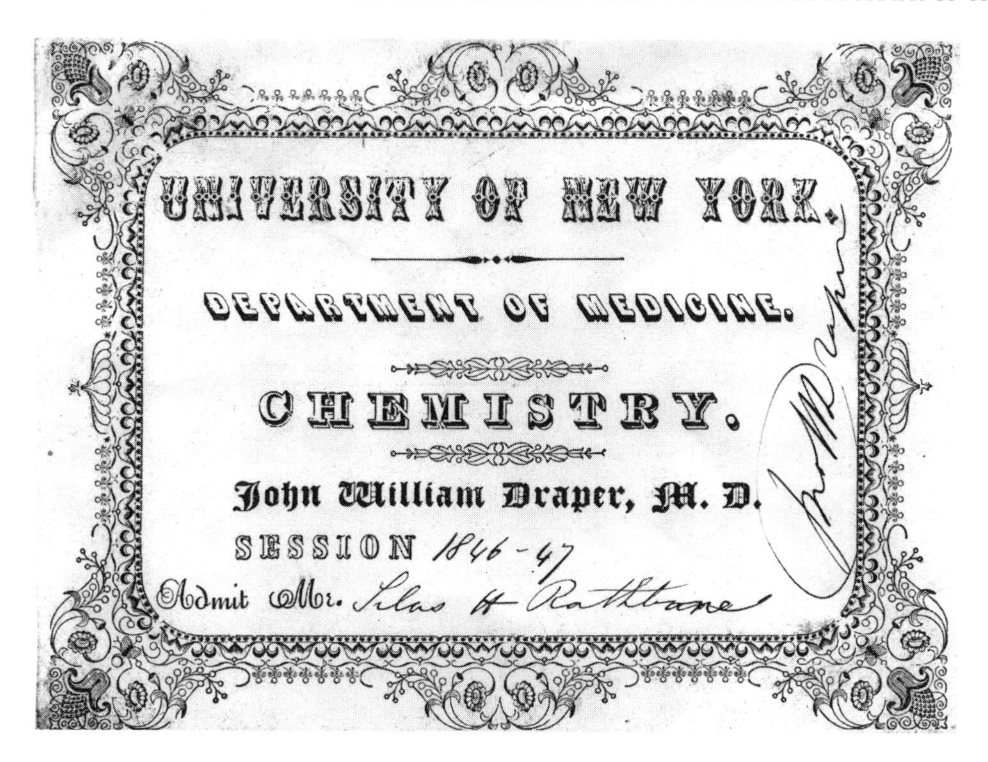

UNIVERSITY OF NEW YORK.

DEPARTMENT OF MEDICINE.

CHEMISTRY.

John William Draper, M. D.

SESSION 1846-47

Admit Mr. Silas H Rathbone

FIGURE 293. ■ The admission attendance card for a Mr. Silas H. Rathbone to attend the chemistry course presented during 1846–1847 by Dr. John W. Draper, M.D. at the University of New York (later New York University) Medical School. Dr. Draper was elected in 1876 as the first President of the American Chemical Society.

classical fable of Tithonus and Aurora). Today, we know this light as "ultraviolet." Draper invented the tithonometer, a precursor to the actinometer, a device for light measurement in photography.

Draper's greatest fame perhaps derives from his contributions to early photography. Although the first real photograph was reported in 1827, it was a very poor image. The first really useful photography was based upon a technique developed by Louis-Jacques-Mandé Daguerre (1789–1851) and publicly announced in 1839.

The principle of the daguerreotype is fairly simple. At its simplest, a copper plate is dipped into a silver nitrate solution, allowed to dry and then dipped into sodium chloride solution and dried to form a thin white film of insoluble silver chloride (AgCl). This plate can be placed at the rear of a camera obscura (a box totally dark inside except for light entering through a small hole covered by a lens on the outside). Exposure to light causes a chemical reaction:

$$AgCl + \text{Light} \rightarrow Ag + \tfrac{1}{2}\, Cl_2$$

The thin silver spots formed by exposure tarnish rapidly to form a black silver oxide; the brighter the exposure, the darker the image. Unexposed AgCl is, in fact, very sensitive to light so that installing and removing the plate from the camera would cause unwanted exposure of the plate. Images were initially "fixed" by solubilizing unreacted AgCl in ammonia. Early improvements involved the use of silver iodide (AgI) instead of AgCl to decrease unwanted background sensitivity, use of mercury vapor to obtain sharper images by forming amalgams with freshly deposited silver, use of sodium thiosulfate solution to remove unreacted AgI from the exposed plate, and a tinting process employing gold. Draper was unique in that his professional knowledge allowed him to understand the fundamental photographic process and improve it.

Draper's earliest partner in the development of photography was the inventor Samuel F.B. Morse (1791–1872). Morse had visited France in 1838 to promote his magnetic telegraph and met Daguerre who demonstrated in private his new process. There is some debate over who produced the first photographic portrait in the United States. One version has Morse using bright light and a long exposure time to produce, in 1839, the first portrait of a subject with eyes closed. Using improved techniques, it appears that Draper may have shot the first American portrait in which the eyes of the subject (his sister Dorothy) were actually open. However, it is generally accepted that Professor Draper obtained, in 1840, the first photograph showing details of the surface of the Moon. Incidentally, his son Henry stayed in "the family business," succeeding his father as Professor of Chemistry at NYU and obtaining, in 1871, the first photograph of the spectrum of a star.

Dr. John William Draper was appointed President of the University of New York Medical College in 1850. In 1864, he was elected President of the recently formed American Photographic Society. In 1876, the American Chemical Society was formed in the Washington Square building of the University of New York Chemistry Department and its Chair, Professor Draper (the elder), became the Society's first President. His wide-ranging interests produced books on physiology, botany, the history of the American Civil War, European intellectual thought, and the relationship between religion and science. He was truly a modern "renaissance man" and a pretty active "doer" for a teacher.

1. J.W. Draper, *A Textbook on Chemistry for the Use of Schools and Colleges*, Harper & Brothers Publishers, New York, 1846.
2. J.R. Partington, *A History of Chemistry*, Volume Four, MacMillan & Co. Ltd., London, 1964, pp. 707, 716–719, 722–724.

CHLORINE FAIRIES?

Real Fairy Folks or The Fairy Land of Chemistry (Lucy Rider Meyer, Boston, 1887) was a rather too precious take on Jane Marcet's marvelous *Conversations on Chemistry*, first published 80 years earlier. Twins (Joseph and Josephine or Joey and Jessie—sentimental descendants of Sol and Luna?) learn chemistry from their uncle Richard James, a chemist also known as "The Professor."

Chlorine fairies [Figure 294(a)] are the molecules in chlorine gas. The chlorine atoms each have one arm (monovalent); they wear green dresses; the fully spread wings signal volatility [remember the winged dragon in Basil Valentine's

FIGURE 294. ▪ From *Real Fairy Folks or The Fairy Land of Chemistry* (Lucy Rider Meyer, Boston, 1887): (a) Two chlorine fairies linked with one arm since chlorine's valence is one; (b) a chlorine fairy and a sodium fairy marry—their wings and legs are folded since salt is a solid; (c) hydrogen fairies each have one arm; (d) hydrochloric acid fairies form a sharp couple.

Third Key—Figure 38(c)?]. Bromine, a liquid, has one-armed fairies in red dresses with their wings folded; mild heating causes the bromine fairies to spread their wings and fly. The one-armed fairies in solid iodine wear purple dresses and have their wings folded and their legs tucked up. "My, my!" exclaimed Jessie "They must be just the *teentiest-weentiest* kind of people." Sodium and chlorine fairies wed to form salt [Figure 294(b)] and their dress is now white (what else?) and their wings folded and legs tucked up. Hydrogen fairies and hydrochloric acid (really gaseous HCl) fairies are shown in Figures 294(c) and 294(d). Figure 295(a) correctly depicts the atmosphere, which is 80% nitrogen fairies and 20 oxygen fairies. The oxygen fairies correctly have two arms [see water fairies in Figure 295(b)] but the nitrogen fairies should have three arms each rather than

FIGURE 295. ■ (a) Note that Ms. Meyer's fairies of the air are in the correct proportion of 4:1 N_2 to O_2; (b) wouldn't mermaids have been better for water than fairies?

one—perhaps a bit too monstrous? Come to think of it, how would Ms. Meyer know how many arms were linking the fairy atoms? The octet rule remained some thirty years into the future.

Uncle Richard has his niece and nephew and their neighborhood friends sniff chlorine, bromine, and hydrogen sulfide. He also keeps a bottle of strychnine in the house to show to the children. He composes poetry: "Hg, Mercuree, What a poet, I be!." I wouldn't want him near my children. Michael Faraday was inspired by Jane Marcet's book to become a chemist. Had he read *Fairy Land of Chemistry* he might have become a CPA.

"RASCALLY" FLUORINE: A FAIRY WITH FANGS?

Uncle Richard finishes his lesson about the halogens by talking about fluorine:

> Fluorine is the last of the cousins. Its fairies are very wilful [sic], harder to catch, and harder still to keep. it is supposed that they have very active feet and wings, and wear the invisible cloak, but they are such little rascals that no one is quite sure of ever having caught them, separate from everything else.

Did Ms. Meyer know that Henri Moissan isolated fluorine gas in 1886, the year before her book was published? Perhaps. Fluorine is the most reactive element—the bonds in F_2 are quite weak, those between carbon and fluorine and in HF are incredibly strong. The molecule will grab electrons from almost anybody. It does not react with argon but does react with xenon. XeF_2 is stable relative to Xe and F_2 while Kr and F; are stable relative to KrF_2.[1] The mineral fluorspar (CaF_2) had been employed for hundreds of years and the presence of a fourth halogen that could not be separated from its compounds was understood by 1830.[2] It was known by 1670 that addition of oil of vitriol (sulfuric acid) produced a gas (HF) that could etch glass.[2] At least two early nineteenth-century chemists died exploring the chemistry of gaseous fluorine compounds and many others became seriously ill. Although we might think of fluorine as a fairy with fangs, it has been called the *Tyrannosaurus rex* of the elements, although I prefer to call it the Tasmanian devil of the elements.[3,4] Finally, Moissan obtained fluorine gas from potassium acid fluoride (KF·HF or KHF_2) in liquid HF (–50°C),[4] using electrolysis with inert platinum–iridium alloy in an inert platinum vessel.[2,4] Moissan's efforts in fluorine chemistry took a toll on his health as well. He received the Nobel Prize in Chemistry in 1906, months before he died at the age of 55. He won the Prize by one vote over Mendeleev, who died the following year and thus would never win it.[4]

1. I thank Professor Joel F. Liebman for this insight.
2. A.J. Ihde, *The Development of Modern Chemistry*, Harper and Row, New York, 1964, pp. 366–369.
3. G. Rayner-Canham, *Descriptive Inorganic Chemistry*, Freeman, New York, 1996, pp. 349–352.
4. D. Rabinovich, *The Chemical Intelligencer*, 3: 64–65, October 1997.

A MID-SEMESTER NIGHT'S DREAM

The laboratory fairies at Haverford College recognized a talented artist in the 19-year-old Maxfield Parrish (1870–1966).[1] They did their best to whisk him through his chemistry course to his true calling as a painter, illustrator, and designer.[2] Indeed, he later placed these fairies on retainer and often used them as a leitmotif in his woodland scenes.[2]

Figure 296 is from Parrish's laboratory notebook, presently part of the Quaker Collection of the Haverford College Library.[3] Now, how should a professor respond to such a notebook? On the one hand, we ask for scrupulous accuracy in the description of an experiment. However, it is highly unlikely that more than one fairy at a time assisted an individual student at Haverford. Indeed, would the fairies have consented to their portraiture? On the other hand, his professor, Lyman Beecher Hall, duly noted that Parrish's "observations and experimental summaries are concise and carefully written."[1] Since Professor Hall made very few notations in the book (and these in light pencil) and since Parrish presented him with the book some 20 years later (in 1910), we can safely assume that the course ended amiably for the young artist.

1. J. Chesick, *Chemical Heritage*, Vol. 17, No. 2, p. 42 (1999). The original figure (and drawing) is in color.
2. J. Turner (ed), *The Dictionary of Art*, Vol. 24, 1996, p. 210, New York: MacMillan.
3. Although Parrish's family were Quakers, he married a non-Quaker and, although a declaration of sincere interest would have allowed him to remain a Quaker, it may be supposed that he elected not to rejoin the Quakers. I am grateful for discussions with Diana F. Peterson, Haverford College Library, and Barbara Katus, Pennsylvania Academy of Fine Arts, which ran the first-ever critical retrospective of Maxfield Parrish in 1999.

AND NOW TURN TO PAGE 3 OF OUR CHEMICAL PSALM BOOK

The illuminated title page of the gentle 1873 English Christian psalm book *Chemistianity* is depicted in Figure 297.

> This work may prove a memory burnisher.
> To teen-youth or octagenarian,
> And act as match or chemistian torch
> For needed light to order Ignorance.

Its clarion call to study chemistry is a bit "forced":

> Chemistry lore should be
> Well known on land and sea
> To sow the seed of *Chemistry*, so heigh, so ho, so hee

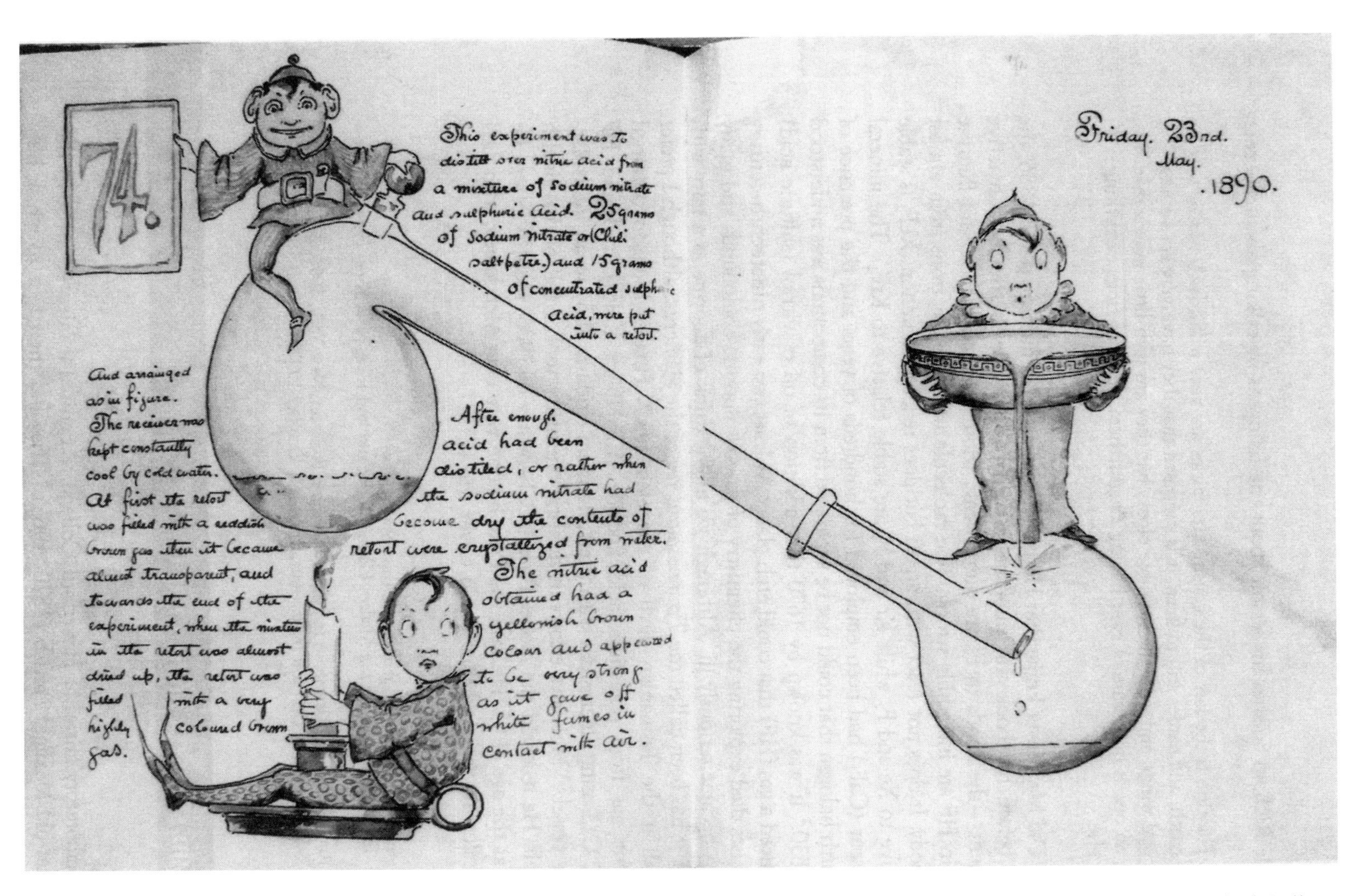

FIGURE 296. ■ Pages from Maxfield Parrish's beginning notebook. See color plates. Courtesy of the Quaker Collection, Haverford College Library.

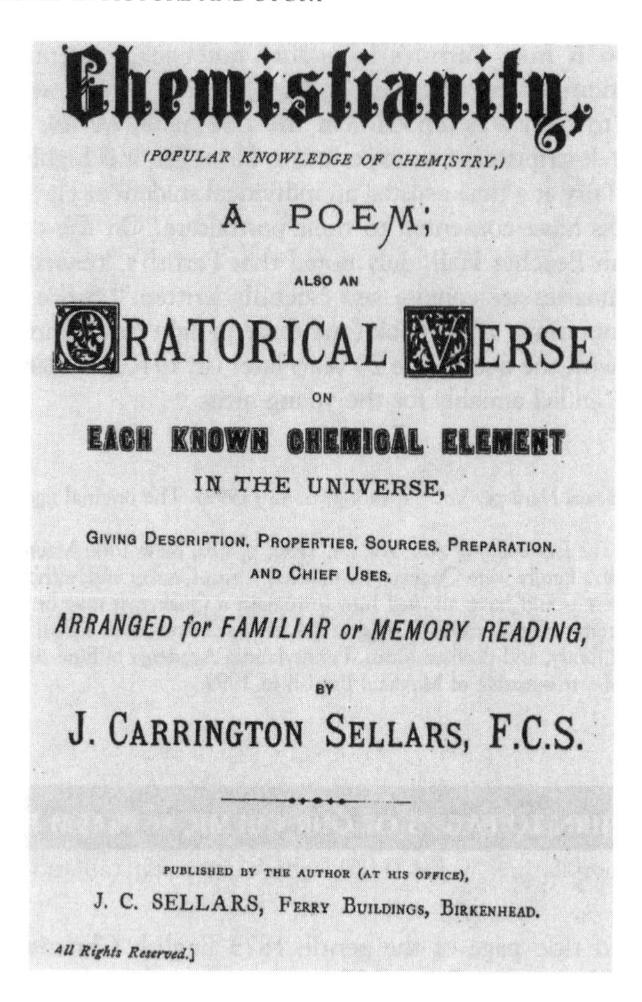
Chemistianity,
(POPULAR KNOWLEDGE OF CHEMISTRY,)
A POEM;
ALSO AN
ORATORICAL VERSE
ON
EACH KNOWN CHEMICAL ELEMENT
IN THE UNIVERSE,
GIVING DESCRIPTION, PROPERTIES, SOURCES, PREPARATION, AND CHIEF USES.
ARRANGED for FAMILIAR or MEMORY READING,
BY
J. CARRINGTON SELLARS, F.C.S.

PUBLISHED BY THE AUTHOR (AT HIS OFFICE),
J. C. SELLARS, FERRY BUILDINGS, BIRKENHEAD.
All Rights Reserved.]

FIGURE 297. ■ The idea of this chemical Psalm book is to teach teens and octagenarians, who both supposedly have short memories, chemistry by reciting psalms. The poetry in this book is among the worst published and if you prefer calling glass "die-bee-day," then this is the book for you!

Our Service begins on page 3 (ALL RISE):

> MATTER, is the body of the universe,
> That, by the aid of Chemical Science,
> With the best of all known appliances,
> Has been resolved into *Sixty-three bodies*
> (Or conditions of free, real essence)
> Term'd ELEMENTS, or Simple Substances;
> These, we have been unable to split up.
> Or subdivide, into more Primal being.
> Named in order of their combining weights,
> And forty-three known, proved, real Metals,
> Arranged under Chemist Roscoe's system,
> By classing in ten *families* or *Klans*;
> The bodies appertaining to each Klan
> Are writ in order of their *combining weight*
> Or type of their Chemical energy.

Please turn now to page 61:

> OXYGEN, the Queen of Body Affection;
> The supporter of man's Earthual life;
> The needed Air-puff for all common forms
> Of combustion in term'd live Animals,
> In ordinary burning Wood or Coal;
> And the prime mover in most heat-felt goceptions,
> Is a colorless gaseous metalloid,
> Tasteless and devoid of odour.

PLEASE BE SEATED

(The author has coined the term *goception* for chemical action and God is called *The Great Goceptor.*)

Sellars writes that: "In reading the names of chemical compounds, many persons are disappointed at their length and unmeaningness to them." (This remains a common complaint among students in Freshman Chemistry courses.) He, thus, develops a simpler alphabetical nomenclature which will be very briefly illustrated. For the five lightest elements known to the author we have:

Alphabetical Name	Composition Name in Brief		Pronounced Present Name
ABGEN	Ab	Abb	Hydrogen
AMYAN	Am	Amm	Boron
ATYAN	At	Att	Carbon
BAGEN	Ba	Bay	Nitrogen
BEGEN	Be	Bee	Oxygen

Using this nomenclature, water H_2O, which we *could* call today dihydrogen oxide but don't) would be pronounced "die-abb-bee." Common glass (silicon dioxide) would have the pleasing sound "die-bee-day" and P_2O_3 the jolly "try-bee-die-dee." However, nitrous oxide or laughing gas (N_2O) is "die-bay-bee," not likely to encourage a dental patient, but fortunately it is not N_3O, pronounced "try-bay-bee."

This gentle and heartfelt effort, doomed by its doggerel and nomenclature, is a compelling argument for separation of Church and Oxidation State.

WHAT ELSE COULD A WOMAN WRITE ABOUT?

Don't be fooled by the quaint title of Ellen Henrietta Swallow Richards' *The Chemistry of Cooking and Cleaning* (Figure 298),[1] published in 1882. Richards was the first female student at Massachusetts Institute of Technology (B.S., 1873), became an instructor at MIT and founded its women's laboratory. She bridged pure and applied chemistry with social science and founded the field of scientific

FIGURE 298. ■ Title page from Ellen Henrietta Swallow Richards' ("Ellen Richards") 1882 book *The Chemistry of Cooking and Cleaning*. Ms. Richards was in the first graduating class at Vassar, inaugurated the women's chemistry laboratory at MIT, and was the founder of the field of sanitary chemistry, a co-founder of the American Association of University Women, and an environmentalist many decades ahead of her time. She would not have been amused by the deceptive advertising for Ozone Is Life bottled water (Figure 277) or Ozone Soap (Figure 278).

home economics. She was a co-founder in 1882 of what would eventually become the American Association of University Women.[2–6]

Born to teachers in rural Massachusetts in 1842 (d. 1911), the precocious Ellen Swallow received a rural education, taught locally, and saved sufficient money to enter an experimental school for women's higher education in Poughkeepsie, New York. Her interest in analytical chemistry was stimulated by Professor A.C. Farrar. Notes Swallow after her first laboratory exercise: "Prof. Farrar encourages us to be very thorough there, as the profession of an analytical chemist is very profitable and means very nice and delicate work, fitted for ladies' hands."[5] She was a member of the first graduating class of Vassar College in 1870 and is honored by a plaque in Blodgett Hall.[5] The new graduates promised "cheerful submission to authority, compliance, diligence, and lady-like deportment."[5] In 1871 she entered MIT, excelled in her studies, and met Professor Robert Hallowell Richards, whom she married in 1875 after she had become a member of the faculty. A children's book dramatizes her student days at MIT with a quaint scene in which she wins the acceptance of the males in her class by baking cookies for them.[6] Richards' early work in the analytical chemistry of minerals and water earned her wide respect, but the work in bringing sanitary chemistry into the home eventually won her worldwide renown.

The Chemistry of Cooking and Cleaning is slim, economical, and very effective in its straightforward presentation to its female audience. Here is Mrs. Richards on the perfidy of manufacturers whose claims about "secret ingredients" she debunks:

> There is, lingering in the air, a great awe of chemistry and chemical terms, an inheritance from the age of alchemy. Every chemist can recall instances by the score in which manufacturers have asked for recipes for making some substitute for a well-known article, and have expected the most absurd results to follow the simple mixing of two substances. Chemicals are supposed by the multitude to be all-powerful, and great advantage is taken of this credulity by unscrupulous manufacturers.

Well, even Ivory Soap is only "ninety-nine and forty-four-one-hundredths percent pure." And what would she have said about Ozone Soap (Figure 278)?

Discussion of the chemical composition of foods is accompanied by analyses of their energy contents and the dietary habits of different cultures. It is duly noted that rice, a carbohydrate, is much lower in energy content than fat, explaining why the former is the dietary staple of tropical cultures while the latter is an important staple in arctic climates. Indeed, an astute woman, noting that her children (or husband) might be accumulating too much "residue" from their diet, can chemically "titrate" them with oxygen to burn off the excess as CO_2 and H_2O through outdoor activity:

> Cooking has thus become an art worthy the attention of intelligent and learned women. The laws of chemical action are founded upon the law of definite proportions, and whatever is added more than enough, is in the way. The head of every household should study the condition of her family, and tempt them with dainty dishes if that is what they need. If the ashes have accumulated in the grate, she will call a servant to shake them out so that the fire may burn. If she sees that the ashes of the food previously taken are clog-

> ging the vital energies of her child, she will send him out into the air, with oxygen and exercise to make him happy, but she will not give him more food.

The 1910 MIT convocation address written by Ellen Swallow Richards was cited over 60 years later by Yale University scientist Bill Hutchinson as an early clarion call to conservation and respect for the environment:[5,7]

> The quality of life depends on the ability of society to teach its members how to live in harmony with their environment—defined first as the family, then with the community, then with the world and its resources.

Ellen Swallow Richards was an early pioneer in the education of women in chemistry. Figure 299 shows a photograph, from the Frank Lloyd Wright Foundation, taken between 1900 and 1910.[8] The young women are from the Hillside Home School, Spring Green, Wisconsin. The famous architect's aunts were supporters of the school, and when it was closed, it eventually became part of the Frank Lloyd Wright Estate. The picture was probably taken routinely in the school and eventually became part of the estate's holdings. Another early pioneer was Dr. Edgar Fahs Smith, Professor of Chemistry, University of Pennsylvania. Ten women completed their Ph.D.s with Professor Smith between 1894 and 1908, a number of whom became college faculty members.[9] Smith's collection of

FIGURE 299. ■ Photograph of a chemistry class for women at the Hillside Home School in Spring Green, Wisconsin near the beginning of the twentieth century (courtesy The Frank Lloyd Wright Archives, Scottsdale, Arizona).

chemical books and artwork presently forms the core of the University of Pennsylvania History of Chemistry collection. His sublime book *Old Chemistries*[10] is a work of warmth and erudition that helped inspire the present book. It includes a fine discussion of Jane Marcet and her "conversations in chemistry."

---

1. E.H. Richards, *The Chemistry of Cooking and Cleaning—a Manual for Housekeepers*, Estes & Lauriat, Boston, 1882.
2. C.L. Hunt, *The Life of Ellen H. Richards*, Whitcomb and Barrows, Boston, 1912.
3. R. Clarke, *Ellen Swallow: The Woman Who Founded Ecology*, Follett Publishing Co., Chicago, 1973.
4. *The New Encyclopedia Britannica*, Vol. 10, Encyclopedia Britannica, Inc., Chicago, 1986, p. 45.
5. *http://departments.vassar.edu/~anthro/bianco/hidden/ellen.html* (6/14/01).
6. E.M. Douty, *America's First Woman Chemist—Ellen Richards*, Julian Messner, Inc., New York, 1961.
7. B. Hutchinson, "Swallow Warned Us All Years Ago," *Miami Herald*, January 17, 1974.
8. This photograph was provided courtesy The Frank Lloyd Wright Archives, Scottsdale, AZ. I am also grateful for conversations with Ms. Margo Stipe, The Frank Lloyd Wright Archives.
9. J.J. Bohning, *Chemical Heritage*, Vol. 19, No. 1, pp. 10–11, 38–44, Spring 2001.
10. E.F. Smith, *Old Chemistries*, McGraw-Hill Book Co., Inc., New York, 1927.

SECTION IX
CHEMISTRY ENTERS THE MODERN AGE

RIDING PEGASUS TO VISIT CHEMISTRY IN SPACE

Optical activity was a fundamental mystery of matter during most of the nineteenth century. Jean Baptist Biot discovered that certain minerals were optically active—they rotated the plane of polarized light. In 1815 he found that certain liquids, oil of turpentine and camphor in alcohol solution for example, were also optically active.[1] However, it was Louis Pasteur's genius that perceived the molecular connection in 1848 even though rational structural chemistry remained some fifteen years or so in the future.

Pasteur first stated the oft-quoted: "Chance favors only the prepared mind."[2] Indeed, serendipity was working in his favor in a (fortunately) cold laboratory in Dijon when he crystallized sodium ammonium tartrate. A close look at the large hemihedral crystals indicated that they were "right-handed" and "left-handed" in the sense of being mirror images (like our hands or feet) that cannot be superimposed point-for-point on each other. (Structures VIII and IX in Figure 300 are flat pictures of right-handed and left-handed hemihedral crystals of ammonium bimalate—the three-dimensional structures are not superimposable.) Meticulously separating the two sets of crystals by hand and dissolving each set in separate solutions, Pasteur discovered that each solution was optically active—but in an equal, yet opposite sense. One solution rotated the plane of polarized light clockwise (called *dextrorotatory*); the other solution was *levorotatory*. Pasteur had affected the first resolution of an equal mixture of enantiomers termed the racemate.

Pasteur's observations began to connect with others.[1] For example, in 1770 Scheele had isolated lactic acid [$CH_3CH(OH)COOH$] from fermented milk. In 1807, Berzelius isolated lactic acid from muscles. Subsequently, lactic acid from fermented milk was found to be optically inactive while that from muscle was found to be optically active. What was the origin of this dichotomy?

The solution to the problem was discovered in 1874 by Jacobus Henricus van't Hoff, 22 years old, and Joseph Achille Le Bel, age 27. Although they both worked in the laboratory of Adolph Wurtz in Paris in 1874, their discoveries were completely independent.[1,2] Van't Hoff would continue to make major contributions to physical chemistry and won its first Nobel Prize (1901) for his discovery of laws of osmotic pressure of solutions.

In Figure 300 we see the plate printed in the first English edition[3] of van't Hoff's work, translated from the second French edition. The two young chemists postulated that a carbon atom at the center of a tetrahedron with four different atoms or groups attached to it (at the corners of the tetrahedron) would be asymmetric, existing as nonsuperimposable mirror images. These were the enantiomers earlier described. Structures I and II in Figure 300 show flat formulas of generalized enantiomers with four different groups (R_1 to R_4) attached to the asymmetric carbon. Structures III and IV are the corresponding three-dimen-

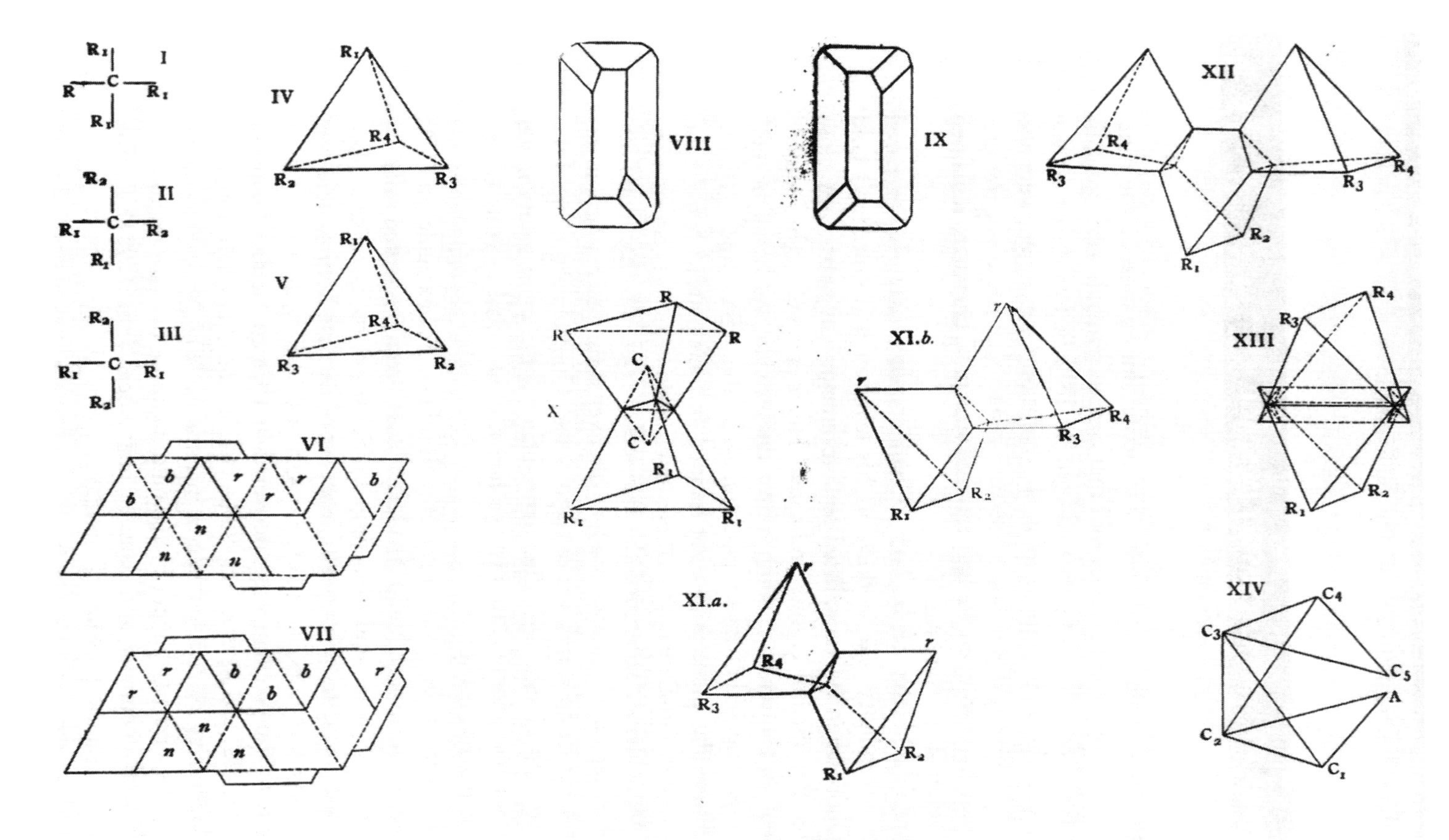

FIGURE 300. ■ Figures from the first English edition of J.H. van't Hoff's *Chemistry in Space* (Oxford, 1891). His discovery simultaneously with (independently of) LeBel may have played some role in his receipt of the first Nobel Prize in Chemistry (1901). However, the prize was awarded specifically for his discovery of laws of osmotic pressure of solutions.

sional tetrahedral representations that are not superimposable. (Structure V depicts a nonasymmetric carbon since it is attached to two identical groups—no enantiomers are possible.) Structures VI and VII are cut-outs for making models of structures IV and V (*n* is black; *r* is red; *b* is blue; unmarked parts are white; see the beautiful book by Heilbronner and Dunitz[4] for a photo of van't Hoffs personal set of handmade models).

The solution to the lactic acid dichotomy was now clear. Lactic acid has an asymmetric carbon atom. The four different groups (R_1 to R_4 in structures I and II or III and IV of Figure 300) are H, CH_3, OH, and COOH. Scheele's lactic acid from fermented milk had both enantiomers in equal quantity (the racemate) and was optically inactive, while Berzelius' lactic acid from muscle was optically active because only one enantiomer was present.

In 1876, van't Hoff was appointed to a junior-faculty position at the Veterinary College of the University of Utrecht in Holland. In 1877, the 1875 French translation of his work was translated into German. He received very strong support from Johann Wislicenus at Wurzberg but a far different reception from Herr Professor Doktor Hermann Kolbe at Leipzig:

> A Dr. J.H. Van't Hoff, of the Veterinary College, Utrecht, appears to have no taste for exact chemical research. He finds it a less arduous task to mount his Pegasus (evidently borrowed from the Veterinary College) and to soar to his Chemical Parnassus, there to reveal in his "La Chimie dans L'Espace" how he finds the atoms situate in the world's space.[5]

Nasty stuff! Sadly, this may be Kolbe's most quoted passage although he was an accomplished scientist and over 30 years earlier hammered the nails into Vitalism's coffin (see pp. 431–432). Ironically, it was Wislicenus who succeeded Kolbe in the Chair at Leipzig in 1885.

Structure X in Figure 300 shows interpenetrating tetrahedra with carbon centers and a single bond between these carbons. Van't Hoff correctly postulated that there is free rotation about such single bonds. Structures XIa, XIb, XII, and XIII rationalize *cis* and Irons isomerism (e.g., the difference between maleic acid and fumaric acid). Structure XIV explains the widespread occurrence of six-membered rings in chemistry and aspects of Baeyer's strain theory.[6]

We conclude this tour of the molecular third dimension with a bit of verse by occasional poet, full-time theoretician and long-time friend Joel F. Liebman[7]:

Owed, to van't Hoff and Le Bel

Lacking magnifying scopes and magic wands
We cannot see molecules and their bonds.
No need though, for it's plain to see
That tetracoordination means planar C*
(What else can it be?)

Enter van't Hoff, Le Bel and their dissensions:
Molecules, invisible, but in three dimensions.
How so? It's really plain to see;
Four-bonded carbon links tetrahedrally**
(What else can it be?)

*Now we clearly don't mean as square,
**Now clearly we don't mean T_d

For that, of course, would be unfair.
Should not the groups with bigger heft
Get more room, small ones get what's left.
Four groups form a quadrilateral;
Any disputation is caterwaul.

Bigger groups remain greedy.
Should not the groups with bigger heft
Get more room, small ones get what's left.
All angles not arccosine minus one third.
That's it, no need for another word.

Joel F. Liebman

---

1. J.R. Partington, *A History of Chemistry,* MacMillan, London, 1964, Vol. 4, pp. 749–759.
2. W.H. Brock, *The Norton History of Chemistry*, Norton, New York, 1993, pp. 257–264.
3. J.E. Marsh, *Chemistry In Space, from Professor J.H. van't Hoff's "Dix Annees Dans L'Histoire D'Une Theorie,"* Clarendon, Oxford, 1891.
4. E. Heilbronner and J.D. Dunitz, *Reflections on Symmetry,* VCH, Weinheim, 1993.
5. J.E. Marsh, op. cit., p. 16.
6. For a discussion of strain theory see A. Greenberg and J.F. Liebman, *Strained Organic Molecules*, Academic, New York, 1978. For a masterful treatment of stereochemistry, see E.L. Eliel and S. Wilen, *Stereochemistry*, Wiley, New York, 1996.
7. *Journal of Molecular Structure (Theochem)*, Vol. 338 frontis matter (1995). Courtesy Professor Joel F. Liebman.

LÆVO-MAN WOULD ENJOY THE "BUZZ" BUT NOT THE TASTE OF HIS BEER

Louis Pasteur's brilliant insight that the chirality ("handedness") of crystals has its origins in the underlying molecular structure stood unexploited for a quarter of a century. However, following the totally independent postulations by Joseph Achille LeBel and Jacobus Henricus van't Hoff of tetrahedral carbon in 1874, chemical investigator's moved rapidly into the third dimension. Pasteur had first discovered in 1848 that some crystalline substances were "chiral" or "handed" (like a left hand and a right hand in that some crystals were mirror images that could not be superimposed). When he dissolved "right-handed" and "left-handed" crystals of potassium ammonium tartrate, derived from winemaking, in separate vessels of water, the resulting transparent solutions were optically active in equal but opposite senses. Pasteur concluded that this "handedness" or dissymmetry was present in the molecules that made up the crystal but were free in solution. This abstract idea was formulated some ten or so years before the concept of valence, so Pasteur could not have had a clue about the rules governing how atoms were connected.

The next 25 years witnessed one of those rapid and dramatic revolutions so often seen in science—the concept of valence explained formulas and isomers; the realization that chemical behavior is often governed by molecular structure and finally the expansion of chemical theory into the third dimension. Figure 301 is derived from the first German edition (1877)[1] of van't Hoff's work on stereochemistry and shows some of the cutouts for his "fold your own" cardboard molecular models—a kind of "molecular origami" consistent with the artistic aspirations of the book you are *now* reading. The top two forms in Figure 301 (*Fig. 39* and *Fig. 40*) are cutouts for two tetrahedra in which the corners are painted different colors (red, blue, white, and yellow), each color representing a different type of atom or group of atoms bonded to a central carbon (*Cwrsb*). *Figs. 41* and *42* are cutouts for two tetrahedra in which the four triangular faces, rather than the corners, are painted four different colors. Assemble these two pairs of figures, and you will discover that they form two pairs of three-dimensional mirror-image figures that are nonsuperimposable and therefore chiral or "handed." (Color photographs of van't Hoff's original cardboard models may be found in the attractive book co-authored by Edgar Heilbronner and Jack D. Dunitz.[2]) Actually, the two pairs of figures shown in two dimensions at the top of Figure 301 themselves are obviously mirror images that cannot be superimposed by any movement or rotation in *two* dimensions. In "Flatland,"[3] they would be non-superimposable mirror images [as would be two-dimensional (2D) tracings on paper of your left and right hands]. If the colored areas were the same on each side of these four 2-D figures, then we three-dimensional (3D) people could cut one out, rotate it by 180° and superimpose it on the related figure. Similarly, if each of your hands were identical top and bottom (10 knuckles per hand—a "benefit" allowing delivery of both forehanded and backhanded punches), the 2D mirror images would be superimposable in three dimensions. The same principle indicates that we spatial people cannot superimpose two (3D) tetrahedra of opposite chirality or, for that matter, our left and right hands. However, a person living in four-dimensional space could clearly have lots of fun with us. For starters, it might be fun to imagine a right-handed barber setting his scissors down for a brief moment, having them projected into the fourth dimension, and returned to him in a flash as left-handed scissors. The results would probably not amuse the barber or the client.

Figure 302 is from a series of four articles published in 1901 for the purpose of enlightening those working in the arts and manufacturing about the breakthroughs in stereochemistry.[4] The author, William Jackson Pope, was Professor of Chemistry at Cambridge, and a remarkable contributor to the field of stereochemistry.[4] The three pictures in Figure 302(a) depicts the three-dimensional, tetrahedral structure of methane,[4] and the tetrahedral structures of the two enantiomers (nonsuperimposable mirror images) of lactic acid (the central carbon sits in the center of the tetrahedron).[4] The central carbon in lactic acid is bonded to four different substituents (atoms or groups of atoms). This asymmetric carbon center is a sufficient although not necessary condition for chirality. A helix (e.g., a spring or a screw) is also "handed."

Lactic acid played a central role in the development of stereochemistry. It was first isolated by Scheele from fermented milk in 1770. Berzelius isolated lactic acid from muscles in 1807. Following the development of polarimetry in the early nineteenth century, Scheele's lactic acid was found to be optically inactive while Berzelius' lactic acid, identical with Scheele's in all other respects, was optically active. Van't Hoff and LeBel both explained these phenomena by postulating that Berzelius' lactic acid, derived from muscles, contained only one enan-

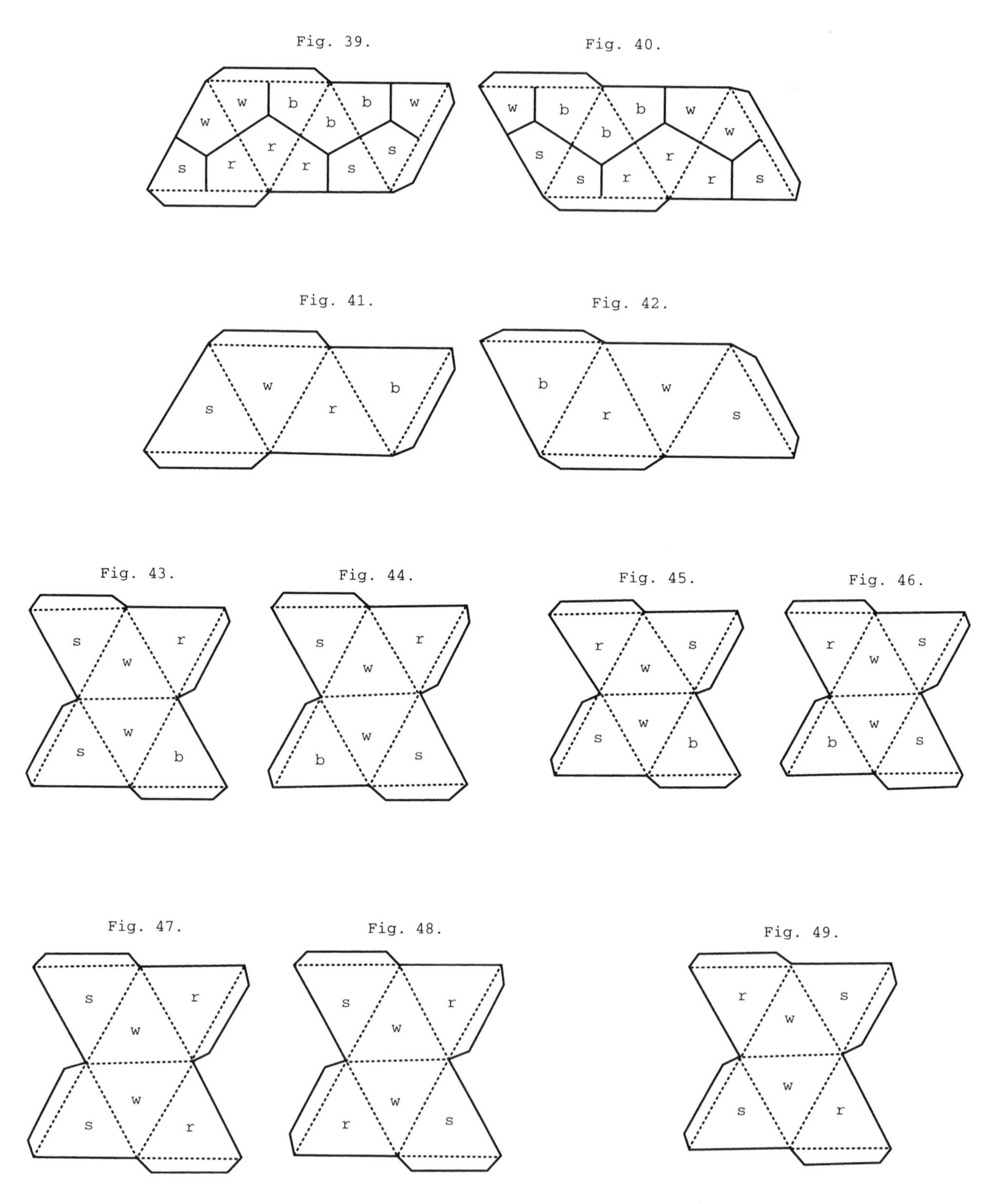

FIGURE 301. ■ Cutouts for paper or cardboard molecular models from the first German edition of van't Hoff's *The Arrangement of Atoms in Space* (*Die Lagerung Der Atome Im Raume*, 1877). Color photographs of the assembled models are found in Heilbonner and Dunitz, *Reflections on Symmetry*, 1993).

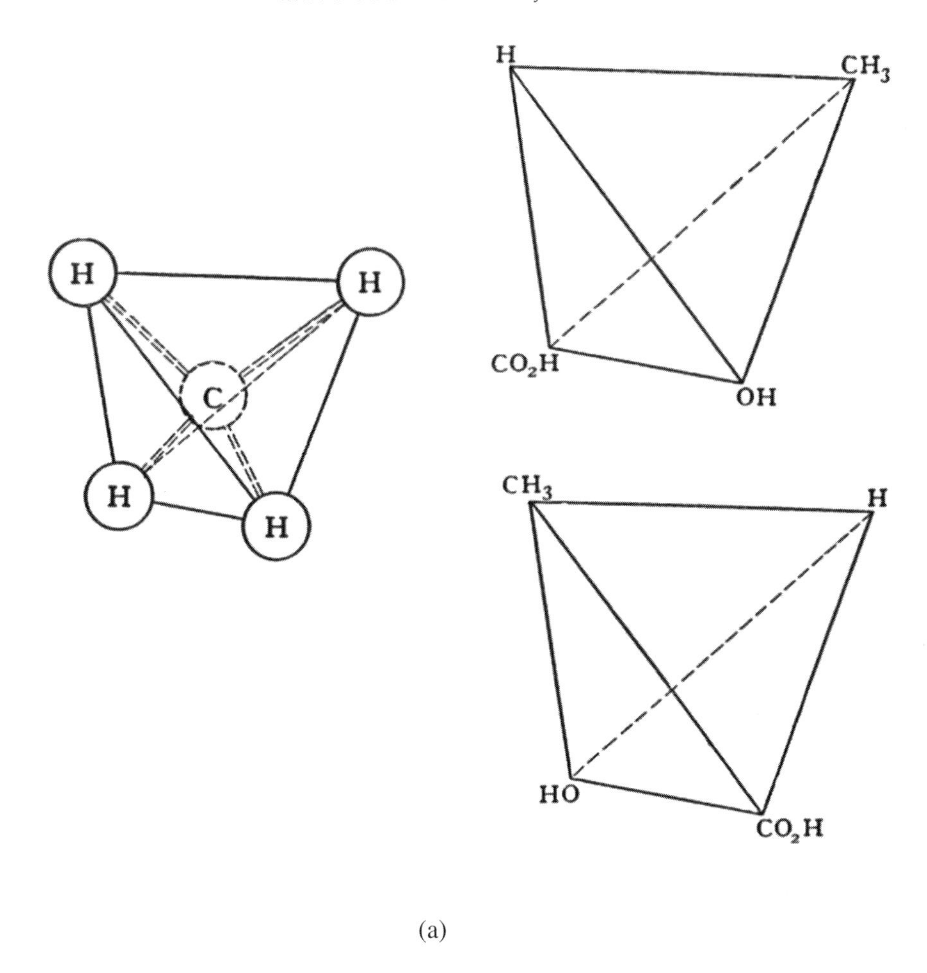

(a)

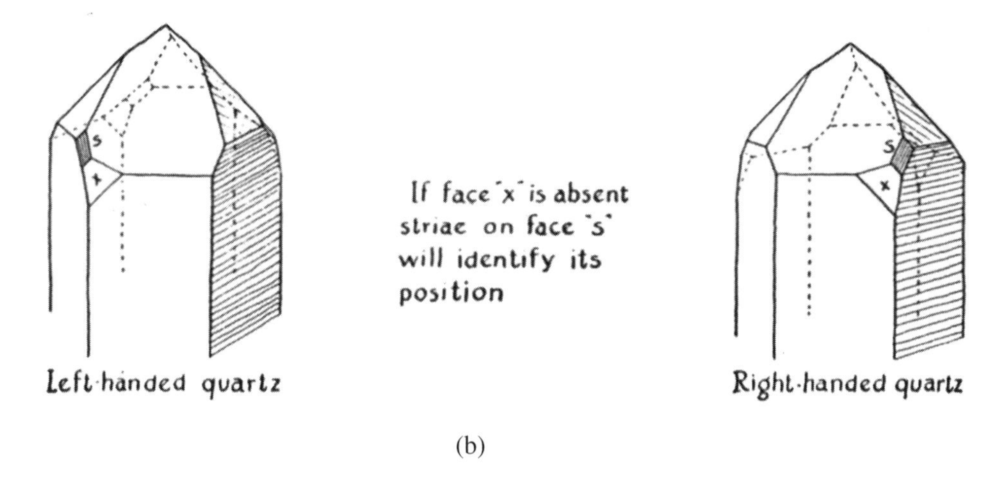

(b)

FIGURE 302. ■ (a) Depictions of tetrahedral methane as well as the nonsuperimposable mirror images (enantiomers) of lactic acid (from Pope, *Journal of the Society of the Arts*, 1901); (b) enantiomeric crystals of quartz, arising from its hidden helical structure. (From *General Chemistry* by Linus Pauling © 1947 by Linus Pauling. Used with the permission of W.H. Freeman and Company.)

tiomer while Scheele's lactic acid was the racemic mixture—both enantiomers in exactly equal quantities.

Pasteur had first noted the chirality of crystalline potassium ammonium tartrate, an organic substance, which he laboriously separated by hand into "left-handed" and "right-handed" crystals. Many naturally occurring minerals also exhibit macroscopic "handedness," and this was duly noted by Pasteur. The two structures in Figure 302 are drawings of "left-handed" and "right-handed" crystals of quartz.[5] Pasteur also discovered that while the physical and chemical properties of enantiomers appeared to be identical in virtually all ways, yeasts, molds, and bacteria could differentiate them. Thus, he incubated racemic tartaric acid with *Penicillum glaucum* and found that the mold "resolved" the mixture by metabolizing one enantiomer while passing the other.[4] Pasteur thus realized that the chemistry of life was chiral even though he could understand it only in the most abstract manner. Tartaric acid contains two attached asymmetric carbons. In principle, such an arrangement could allow 2×2 or four stereoisomers. *Figs. 43–46* in Figure 301 provide molecular models that can be assembled to provide all four stereoisomers of a molecule of generalized formula srwC-Cwbs. However, only three stereoisomers (*Figs. 47–49*) are possible for srwC-Cwrs. Assemble the figures and try them out. Pasteur's tartrates correspond to two of these three possible structures. Figure out which ones.

As Pope noted in 1901,[4] the observed optical activity of quartz crystals allowed late nineteenth-century scientists to reason that its underlying molecular structure is helical. They did understand at that time that the bonding connections in quartz involved no chiral centers, yet the crystal was chiral. The crystalline nature of quartz demanded a regular, periodic structure. The helix is the only regular, chiral structure that can meet these demands. While we now know that quartz is indeed composed of long helical structures[6] and clearly understand the logic of the chemists who discerned its molecular structure, I remain awed by these remarkably prescient predictions that predated X-ray crystallography by decades. The origin of optical activity on earth remains to this day a mystery. One view is that crystals such as quartz or calcite (calcium carbonate) formed chiral templates that, by pure chance, formed an excess of "handed" molecules of only one type on the planet.[7]

Pope informs his readers that, by the end of the nineteenth century, it was known that the sugars and amino acids that compose our bodies are specifically "right-handed" and "left-handed" respectively. He then has some fun by imagining the sudden appearance of a "lævo-man" ("left-handed" person) based upon "left-handed" sugars and "right-handed" amino acids.[4] Pope concludes his essay series rather grimly:

> if a human being enantiomorphously related to ourselves—the lævo-man of whom we have spoken—made his appearance on our planet, he would in all probability quickly starve to death, owing to his inability to assimilate the foodstuffs we were able to place before him.

To be a bit less grim about it, we might imagine lævo-man's first day on the planet starting nicely enough. Happily, he "gets up on the wrong side of the bed," has a refreshing glass of water, and takes a bracing shower.[8] Brewing coffee, he notes an off-odor and samples his first cup—black as usual. It tastes awful.[8] One teaspoon of sugar is followed by another, another, and then two more.[8] There is no improvement in the taste of the coffee. His toast tastes like chocolate-flavored mashed potatoes.[8] Agitated now, he pours a cold beer and gulps it down. Utterly awful.[8] Next, a stiff vodka, and this begins to taste familiar and starts to sooth his nerves. By late morning a headache begins to develop, but two aspirins and a glass of water quickly soothe it.[8] Back to bed for a short nap, and then he suddenly awakens, abruptly "gets up on the right side of the bed," and immediately realizes that the rest of the day will be a disaster.

---

1. J.H. van't Hoff, *Die Lagerung Der Atome Im Raume*, Friedrich Vieweg Und Sohn, Braunshweig, 1877, p. 48.
2. E. Heilbronner and J.D. Dunitz, *Reflections on Symmetry*, VCH-Wiley, New York, 1993, pp. 72–73.
3. E.A. Abbott, *Flatland: A Romance of Many Dimensions* (with Foreward by Isaac Asimov), Barnes & Noble, New York, 1983.
4. W.J. Pope, *Journal of the Society of Arts (London)*, Vol. 49, pp. 677–683, 690–697, 701–708, 713–718.
5. L. Pauling, *General Chemistry*, W.H. Freeman and Co., San Francisco, 1947, p. 521.
6. A.F. Wells, *Structural Inorganic Chemistry*, fifth edition, Clarendon Press, Oxford, 1984, pp. 1004–1006.
7. R.M. Hazen, *Scientific American*, Vol. 284, No. 4, pp. 76–85, April 2001.
8. Obviously we are joking when we talk about "left-hand and right-hand sides of the bed." Right-handed people and left-handed people all contain the same "right-handed" sugars and "left-handed" amino acids. However, the chirality or nonchirality of substances should affect, sometimes dramatically, how they interact with living beings. Fortunately for lævo-man, the oxygen and nitrogen in air as well as water are achiral (not "handed"). Ethyl alcohol and aspirin (acetylsalicylic acid) are also achiral. However, table sugar is chiral, as are the flavor and aroma ingredients in coffee and beer. Chocolate mashed potatoes is a secret family recipe best kept secret for the sake of humanity. It appears that Oliver Sacks' family had an old secret family recipe for Passover "matzoh balls of an incredible tellurian density, which would sink like little planetismals below the surface of the soup" (see O. Sacks, *Uncle Tungsten—Memories of a Chemical Boyhood*, Alfred A. Knopf, New York, 2001, p. 97).

IS THE *ARCHEUS* A SOUTHPAW?

Pasteur's brilliant work with enantiomers and racemates included his realization that a single enantiomer of one substance (e.g., an optically active base) could be used to separate the enantiomers of another (e.g., a racemic acid). The analogy with hands is simple—we refer to molecules that have nonsuperimposable mirror images as chiral (handed). A right glove can differentiate ("separate") a right hand from a left hand. Pasteur and others soon realized that all of their optically active compounds had come from living organisms. Lactic acid from muscle was optically active; synthetic lactic acid was not. Moreover, living organisms readily resolved racemates by selectively metabolizing one enantiomer. Thus, Pasteur incubated 3 g of optically inactive secondary amyl alcohol (the four groups attached to the asymmetric carbon are H, OH, CH_3, and C_3H_7) with a suspension containing yeast mold. After one month, the alcohol distilled from the mixture was found to be dextrorotatory.[1] The yeast had selectively metabolized the levorotatory enantiomer.

Was there a Vital Force in living organisms that allowed them to be the only source of optically active substances? Was this Vital Force ultimately the only means for resolution of racemates? Do you remember our earlier discussion of the *Archeus*, the Spiritual Alchemist, a Vital Spirit, thought by Paracelsus to reside near our stomachs (see Figure 97)? The *Archeus* was thought to have a head and two hands only and to function by separating the nutritional from the poisonous parts of food and air. Now, if the *Archeus* were left-handed, for example, we might have understood earlier how the body separates left from right. Happily, there was no return to Vitalism by serious scientists and today billions of dollars are earned by companies that have learned to make the pure enantiomer of a drug without contamination from the other enantiomer and without paying the salaries of any *Archei*.

1. J.E. Marsh, *Chemistry In Space From Professor J.H. van't Hoff's "Dix Annees Dans L'Histoire D'Une Theorie,"* Clarendon, Oxford, 1891, p. 41.

JOHN READ: STEREOCHEMIST

I have expressed my admiration at other points in this book for the prodigious intelligence, scholarship, and wit of Dr. John Read who wrote the wonderful trilogy on alchemy and chemistry. Read was an early stereochemist "present at the creation" of at least two very important discoveries in the field.

Interestingly enough, while both van't Hoff and Le Bel postulated a single asymmetric carbon as necessary for optical activity, by the end of the nineteenth century no optically active compounds had been isolated having fewer than three carbon atoms in a chain.[1] This prompted Norwegian chemist F. Peckel Möller to postulate his "Screw-Theory" in a section of his book *Cod-Liver Oil and Chemistry*[2] titled "Position of Atoms In Space."[3] A believer in the Universal Ether, disproven by the Michelson–Morley experiment published in 1887 but still adhered to by famous scientists including Mendeleev, Möller postulated that a three-carbon chain is the minimum requirement for optical activity. The idea was that three carbons in a zigzag chain were the minimum for a chiral corkscrew capable of creating right-handed or left-handed vortices through the ether thus accounting for dextro- or levorotatory properties. Ironically, the second English edition of van't Hoffs work,[1] published in 1898, three years after Möller's work, adds the new axiom of the three-carbon requirement. I would have liked to have seen van't Hoffs face when he learned of this piece of editing.

The first optically active compound containing only one carbon atom [$CHClI(SO_3H)$] was reported in 1914.[4] It was synthesized and optically resolved by the stereochemist William Jackson Pope and our own John Read at Cambridge University.[3] Pope and his co-workers extended stereochemical concepts from carbon only to nitrogen, phosphorus, sulfur, selenium, silicon, and tin.[3]

Read had earlier obtained his Ph.D. with Alfred Werner at the University of Zurich. Werner received the 1913 Nobel Prize in Chemistry for his extension of chirality to metallic compounds. Strictly speaking, there were six carbons in the cobalt compound whose resolution was reported in 1911.[5] However, the molecule's chirality was due to the spatial relationship about the hexacoordinate cobalt not to the carbons. Werner's collaborator in this revolutionary resolution—John Read.[6]

However, the history of this discovery is not so simple. It appears that Edith Humphrey, one of the very few women engaged in doctoral research 100 years ago, is likely to have actually made the original resolution 10 years before John Read, although it was not realized at the time.[6] Dr. Humphrey died in 1977 at the age of 102. At her 100th birthday she is quoted as saying "There were very few women students in Zurich, but fairly soon I was made assistant to the professor. I think being English helped—and also I knew more physical chemistry than most people there."[6]

1. J.H. van't Hoff, *The Arrangement of Atoms in Space*, 2nd ed., A. Eiloart (translator), Longmans, Green, London, 1898.
2. F.P. Möller, *Cod-Liver Oil and Chemistry*, Peter Möller, London, 1895.
3. A. Greenberg, *Journal of Chemical Education*, **70:** 284–286, 1993.
4. W.J. Pope and J. Read, *Journal of the Chemical Society (London)*, **105-1:** 811, 1914.
5. A. Werner, *Berichte*, **44:** 2447, 1911.
6. I. Bernal, *The Chemical Intelligencer*, **5,** (1):28–31, January, 1999.

FINDING AN INVISIBLE NEEDLE IN AN INVISIBLE HAYSTACK

Atmospheric air should be colorless and odorless although over certain segments of the New Jersey Turnpike it can be seen and even tasted. During the 1770s Scheele and Priestley demonstrated that the atmosphere is roughly 80% phlogisticated air (nitrogen) and 20% dephlogisticated air (oxygen). [Figure 295(a) depicts four nitrogen fairy couples and one oxygen fairy couple.]

During the 1890s, Lord Rayleigh (John William Strutt, Third Baron), a physicist, and chemist William Ramsay noted inconsistencies between the densities of "chemical nitrogen" and "atmospheric nitrogen." The density, at 0°C and 760 mm of "atmospheric nitrogen" (1.2572 g per liter) was apparently about six-tenths of 1% greater than that of "chemical nitrogen" (1.2505 g per liter). "Chemical nitrogen" had been synthesized through reaction of nitric oxide (NO) or nitrous oxide (N_2O, laughing gas) with hydrogen gas, heating of ammonium nitrite (NH_4NO_2), or reaction of urea (NH_2CONH_2) with sodium hypochlorite (NaOCl, pool disinfectant). The source of the discrepancy could be the presence of a light impurity, such as traces of residual hydrogen, in "chemical nitrogen" or, more likely, a heavy impurity in "atmospheric nitrogen." Rayleigh and Ramsay roughly estimated that this impurity might be present at a level of around 1%. It was hard to imagine that the very air they breathed could contain 1% of a hitherto-unknown substance[1]:

> The simplest explanation in many respects was to admit the existence of a second ingredient in air from which oxygen, moisture, and carbonic anhydride had already been removed. The proportional amount was not great. . . . But in accepting this explanation, even provisionally, we had to face the improbability that a gas surrounding us on all sides and present in enormous quantities could have remained so long unsuspected.

The meticulous work of Rayleigh and Ramsay led them to discover the gaseous element argon in 1894. They withheld announcement while they submitted their paper for the Smithsonian Institution's Hodgkin's Prize for the most important discovery related to atmospheric air.[2] They published their work in the *Philosophical Transactions of the Royal Society* during 1895 and the prize-winning paper was published by the Smithsonian in 1896.[1] Among their numerous careful experiments was the generation of "chemical nitrogen" from "atmospheric nitrogen" by removal of carbon dioxide and water from air using soda-lime and phosphoric anhydride and removal of oxygen through exposure to red-hot copper. The remaining "atmospheric nitrogen" was then ignited over magnesium at a "bright-red" heat to form powdery magnesium nitride (Mg_3N_2). Addition of water to the nitride produced ammonia (NH_3), which oxidized with calcium hypochlorite [$Ca(OCl)_2$] to produce "chemical nitrogen." Oxygen reacts rapidly with copper to form a salt. Nitrogen, being much less chemically reactive than oxygen, escapes red-hot copper unscathed. Magnesium is a much more reactive metal than copper. Indeed, it was unknown as a free metal until freed by Davy from its compounds in 1808 using a voltaic pile.

The apparatus in Figure 303(a) (see the Smithsonian report[1]) includes combustion tube *A* filled with magnesium turnings and heated over a wide-flame burner and combustion tube *B* filled with copper oxide (to remove residual hydrogen gas generated by reaction of magnesium in tube A with residual water vapor) and also heated with a wide-flame burner. Tube *CD* contains soda-lime and phosphoric anhydride, *E* is a gas volume measuring vessel, *F* is connected with the "atmospheric nitrogen" gas holder, and G stores unabsorbed gas after each cycle. Figure 303(b) shows a larger-scale apparatus in which gas can be introduced via C into gas holder A. Tube *D* is filled with soda-lime [in Figure 303(a)] and phosphoric anhydride [in Figure 303(b)]; combustion tube *E*, heated with a wide flame, is half-filled with porous copper and half with granular copper oxide; tube *F* contained granular soda-lime and G contains magnesium turnings heated to bright red over a wide-flame burner; *H* contains phosphoric anhydride and *I* soda-lime. Nitrogen prepared by passing atmospheric air through red-hot copper is introduced via C into vessel A. Over the course of 10 days this nitrogen is passed slowly back and forth between *A* and *B*. Magnesium is replenished as needed. The remaining small residue of gas was transferred to the apparatus in Figure 303(c), which was designed to exclude atmospheric air in the remaining operations.

It was difficult to accurately determine the density of argon since there were impurities, mainly nitrogen, associated with it. Values were typically in the range of 1.75 to 1.82 g per liter, approximately 20 times that of hydrogen (H_2) gas. Since the molecular weight of hydrogen is 2.0 amu, then the "molecular" weight of argon should be about 40 amu.

Rayleigh and Ramsay characterized the new gas by observing its light spectrum: "The spectrum seen in this tube has nothing in common with that of

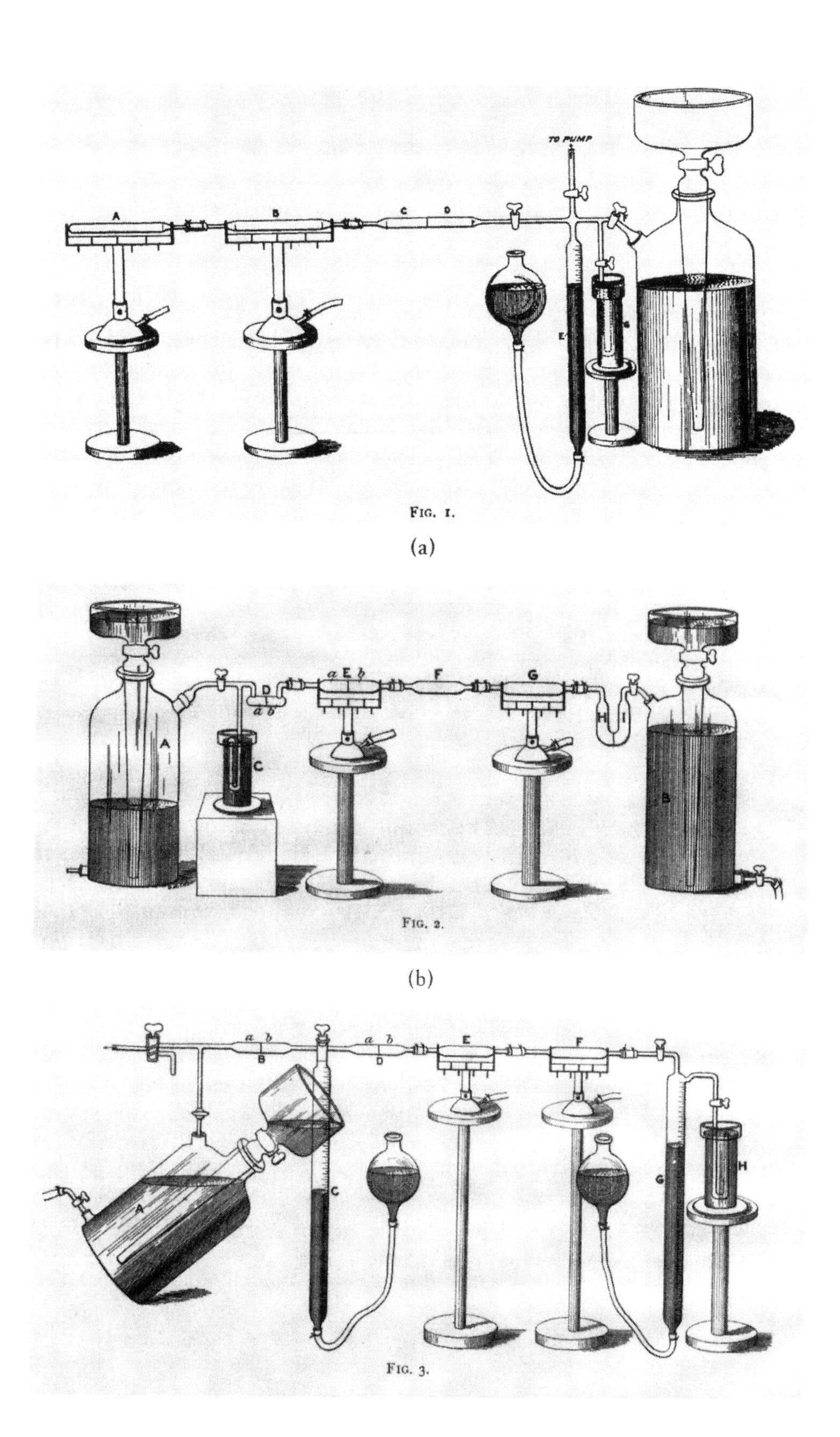

FIGURE 303. ■ (a) to (c) are described in the text. They are from the prize-winning essay published by Lord Rayleigh and William Ramsay (*Argon, a New Constituent of the Atmosphere*, Washington, D.C., 1896). Rayleigh had noted that atmospheric nitrogen is very slightly more dense than "chemical" nitrogen. After removing water and carbon dioxide from air, oxygen was removed with red-hot copper and then magnesium burned in the remaining nitrogen. The unreacted gas, comprising less than 1%, was mostly argon.

nitrogen, nor indeed, so far as we know, with that of any known substance."[1] They tested the reactivity of this new element with about every nasty chemical they could and found it totally unreactive. They gave this new element the name argon derived from the Latin *a* (without) and *ergon* (work), meaning "idle."

And in an eloquent salute to Henry Cavendish, who first reported in 1785 the isolation of an unreactive gas comprising $^1/_{120}$ of the phlogisticated air, Rayleigh and Ramsay write[1]:

> Attempts to repeat Cavendish's experiment in Cavendish's manner have only increased the admiration with which we regard this wonderful investigation. Working on almost microscopical quantities of material and by operations extending over days and weeks, he thus established one of the most important facts in chemistry. And what is still more to the purpose, he raises as distinctly as we could do, and to a certain extent resolves, the question above suggested.

1. Lord Rayleigh and Professor William Ramsay, *Argon, A new Constituent of the Atmosphere*, Smithsonian Institution, Washington, D.C., 1896.
2. W.H. Brock, *The Norton History of Chemistry*, Norton, New York, 1993, pp. 331–340. This is an especially enjoyable and accessible discussion.

BUT ARGON IS A MONATOMIC GAS—AND THERE ARE OTHERS!

There is another amazing aspect in the discovery of argon beyond its total chemical inertness. Rayleigh and Ramsay reported measurements of the speed of sound in argon that indicated that the ratio of its heat capacity at constant pressure to that at constant volume (C_P/C_V) was too high for a diatomic molecule. The only other similar observation was for monatomic mercury (vapor) whose atomic weight was known since it forms compounds. At constant volume, heat added to a diatomic molecule such as N_2 goes into both movement of the molecule (translation) as well as vibration of the bond. In a monatomic substance there is no bond vibration and, thus, less capacity to absorb heat.

The finding that argon is a monatomic gas and has an atomic weight of 40 dealt a severe jolt to the established order.[2] First, if it was a diatomic molecule, its atomic weight would be about 20 (see above), thus fitting it confusingly well between fluorine (19) and sodium (23). However, a new monatomic substance with an atomic weight of 40 would not only require a new and totally unanticipated family in the Periodic Table, it coincided with the atomic weight of calcium and messed up the order that Mendeleev first employed to organize his table. These findings did indeed upset Mendeleev and his students.[2] Rayleigh and Ramsay themselves noted: "If argon be a single element then there is reason to doubt whether the periodic classification of elements is complete."[1] Their report[1] concluded: "We would suggest for this gas, assuming provisionally that it is not a mixture, the symbol A" (later changed to Ar).

At the end of the nineteenth century techniques were developed to liquefy air by cooling and expansion. The front page of the Sunday, December 30, 1900 issue of *The Brooklyn Daily Eagle* gives the following page-wide headline: "LIQUID AIR WILL OPEN UP A NEW WORLD OF WONDERS" and under it a subheadline: "Pictet, Foremost of Savants, Calls the Liquid the Elixir of Life, and Declares It Will Banish Poverty From the Earth."[3,4] Using similar techniques to condense air, in 1898 Ramsay discovered the related inert or "noble" gases neon (Ne), krypton (Kr), and xenon (Xe). Helium (He), as its name bears witness, was discovered on the sun in 1868 through its light spectrum measured during a solar eclipse. It was isolated by Ramsay in 1895 through heating uranium ores. For their studies, Rayleigh received the Nobel Prize in Physics in 1904 and Ramsay the 1904 Nobel Prize in Chemistry. In 1908, Ramsay isolated the last of the inert gases, radioactive radon (Rn) from radium-containing minerals.

In his enjoyable book *The Periodic Kingdom,*[5] P.W Atkins describes the Periodic Table as a land of mountains, valleys, lakes, and shores. The noble gases are termed a strip of land on the eastern shore and Atkins notes that ". . . no other complete strip of land of the kingdom owes so much to a single person"—Ramsay.

1. Lord Rayleigh and Professor William Ramsay, *Argon, A New Constituent of the Atmosphere,* Smithsonian Institution, Washington, D.C., 1896.
2. W.H. Brock, *The Norton History of Chemistry,* Norton, New York, 1993, pp. 331–340. This is an especially enjoyable and accessible discussion.
3. Special *Newsday* reproduction of *The Brooklyn Daily Eagle,* Vol. 60, No. 360, Sunday, December 30, 1900.
4. Not to be too curmudgeonly about it, but note that in Brooklyn at least, the end of the century was properly celebrated and not snuck in at the end of 1899—mathematical authority still held sway over Madison Avenue if the latter indeed existed.
5. P.W. Atkins, *The Periodic Kingdom,* Basic Books, New York, 1995, pp. 53–54.

SEARCHING FOR SIGNS OF NEON

The discoveries of argon and the rest of the rare gases are conveyed on a very personal level by Morris W. Travers, who, three decades earlier, was a young graduate student of Ramsay's at Bristol.[1] I see an ironical aspect noted early in Travers' book. He quotes Van't Hoff, the first chemistry Nobel laureate, from a contemporary Dutch review as follows:[2]

> How then was this discovery made? Year after year Lord Rayleigh! Poor Rayleigh! had been weighing nitrogen: nitrogen from urea, nitrogen from ammonium nitrate, nitrogen from the air, ever did he find the latter heavier: 1.2572 against 1.2505 gram per litre of nitrogen. Nitrogen from the air was thus somewhat different, it contained something different from chemical ni-

> trogen; and starting from the latter supposition Ramsay removed all possible substances from the air, and there remained the celebrated little bubble of gas of Cavendish, colourless, without taste or smell.

Travers finds this article to be "in a bitterly sarcastic vein throughout, and makes light of the work of both Lord Rayleigh and Ramsay."[2] Indeed, Travers notes that Rayleigh did not "make a career" of weighing nitrogen, and his study of the relative densities of gases was part of a broad investigation of the relative atomic weights of the elements with, I might now add over a century later, unseen yet profound future implications for the understanding of the nuclear structures of atoms.[3] But wasn't this the same van't Hoff who, as a 26-year-old instructor at the Veterinary College at Utrecht some 22 years earlier, postulated tetrahedral carbon and chemistry in three-dimensional space? The same Van't Hoff who was famously "trashed" by the nearly 60-year-old doyen of German organic chemistry, Professor Doctor Adolph Wilhelm Hermann Kolbe?[4] Ah, the foibles of gifted humans. Happily, Travers describes his erstwhile mentor Ramsay as a kind, fatherly mentor, solicitous of his students and moderate and fair in argumentation.

A new series of spectroscopic lines were observed by Norman Lockyer during the solar eclipse of 1868. He recognized a new element and named it "helium" (from the Greek *helios*, the sun). In 1888, an unreactive gas obtained from the mineral clevite, which contains uranium, was isolated by Dr. W.F. Hildebrande of the U.S. Geological Survey, who erroneously characterized it as nitrogen. In 1895, a year after the discovery of argon, Ramsay, following a suggestion from Mr. Henry Miers of the British Museum, examined clevite as well as some other minerals containing uranium, collected the gaseous components as described above, and found an inert, monoatomic gas of mass 4 as the major component.[5] William Crookes, who had obtained the first spectrum of argon in collaboration with Rayleigh and Ramsay, discovered that the spectrum of the new light gas was identical with that of Lockyer's helium.[5] As noted above, Ramsay postulated that helium and argon formed a new periodic family. At this point, it is vital to remark that Ramsay's work on helium occurred about one year before Henri Becquerel's discovery of the phenomenon of radioactivity using a uranium salt; three years before the Curies coined the term *radioactivité* and isolated polonium and radium; and some eight years before alpha particles started to be generally recognized as helium lacking two electrons.[6] So helium was apparently to be found in certain exotic minerals containing uranium and thorium. There was no association with radioactivity since it was unknown.

And now arose a powerful challenge for Ramsay—the new family of noble gases had a member in the first row of the periodic table (helium) and in the third row (argon). Missing was the second-row noble gas calculated to have an atomic weight of 20. Missing, too, were possible heavier inert gases, but the primary goal was to fill the Mendeleevian gap. Attempts to find new noble gases in the atmosphere were initially unsuccessful. Mysterious and exotic minerals such as Norwegian clevite, meteorites, and gases from the bowels of the earth bubbling out of hot springs in Iceland and elsewhere were investigated to no avail. [5,7]

And now, very briefly, we need to learn how to liquefy (even solidify) gases at extremely low (cryogenic) temperatures. Hints of very low levels of at least another noble gas in argon motivated Ramsay and his co-workers to investigate the possibility of liquefying air (or selected fractions) followed by fractional distillation. Again, it is remarkable to note that temperatures very close to absolute zero [zero degrees Kelvin (0 K) or –273.16°C or about –460°F] were achieved before the end of the nineteenth century. The important principle here is called the Joule–Thomson effect,[8] discovered in the mid-nineteenth century. If a gas expands from a vessel into a vacuum, the remaining gas (and other matter in the flask) will be cooled. If a gas is first compressed, the gas will heat up, but a cooling jacket can remove this heat, leaving cold compressed gas, which can cool down further when exposed to reduced pressure. Indeed, Michael Faraday accidentally condensed chlorine gas into a green, nasty liquid by injecting it via syringe into a closed tube.[9] Carbon dioxide can be compressed into liquid form at pressures above 5.11 atmospheres (atm), but no degree of coldness will condense it to a liquid under atmospheric pressure. A. Thilourier discovered in 1835 that when pressurized, liquefied carbon dioxide is quickly exposed to atmospheric pressure, the cooling expansion (which also "steals" the heat of vaporization from the remaining material and surroundings) forms solid CO_2 or dry ice.[9] Now gas samples could be condensed in a "thermodynamic sink" of dry ice in diethyl ether maintained at –78°C. It wasn't long before oxygen [boiling point (bp) –183°C] and nitrogen (bp –196°C) and atmospheric air could be liquefied and even solidified. And boiling liquid oxygen could condense helium and even hydrogen under pressure. The masters of this technology were two Polish scientists, Olszewski and Wroblewski.[10,11] James Dewar, working in England, developed the vacuum vessels that today bear his name, and while he kept this apparatus secret for years, by 1892 it was widely disclosed.[11] The path was now clear for W. Hampson, who developed a process for liquefying air in collaboration with Mr. K.S. Murray, the managing director of the British Oxygen Company,[10] to supply Ramsay with liquid air sufficient to provide many liters (!) of gaseous argon.

At the end of May, 1898, Hampson brought a 750 cm^3 sample of liquid air, and at Ramsay's suggestion, over the course of about a week Travers allowed it to boil away until only about 25 cm^3 of gas (i.e., perhaps 0.025 cm^3 of the original liquid) remained. Travers describes his interplay with a young friend and colleague who gently teased him:[12] "'It will be the new gas this time, Travers.' 'Of course it will be,' I replied, and passed on upstairs to Ramsay's room. I was beginning to think that the discovery of the new gas would correspond with the Greek *Kalends*;[13] but Ramsay still had faith in the periodic law, and perhaps I had even stronger faith in Ramsay. However, we had to put up with a good deal of kindly chaff both within and without the department."

But later that day, following the usual residual cleanup of the remaining gas, a small quantity was introduced into a Plucker tube, the electricity turned on, and the light viewed using direct-vision spectroscopes. Lo! A distinctly new yellow band appeared—a new element—in a *less volatile* fraction derived from argon. The gas was not, of course, the missing atomic weight 20 species, but was a monoatomic gas of weight close to 80. The new element was named krypton (from the Greek *kryptos* = "hidden"). The dry spell was broken. Another large sample of liquid air was fractionally distilled and chemically treated to give high-boiling and low-boiling distillation cuts containing argon. The low-boiling cut was carefully fractionated and provided a gas that did not require the subtlety of spectroscopic glasses to disclose its secret, but let's allow Travers to tell it:[14]

> we each picked up one of the little direct-vision spectroscopes which lay on the bench. But this time we had no need to use the prism to decide whether or not we were dealing with a new gas. The blaze of crimson light from the tube told its own story, and it was a sight to dwell upon and never to forget.

Holy Krypton, Batman! A Neon Sign for Neon![15] Neon is derived from the Greek *neos* ("new"). And in a very short time xenon (Greek xenos = "strange") was similarly discovered [and five years later in a collaboration between Ramsay and Frederick Soddy—radon (originally called "niton"), a by-product of the decay of radium was identified].[16] In summary, the percent by volume of the noble gases in the atmosphere are now known: argon, 0.93%; neon, 0.0018%; krypton, 0.0011%; helium, 0.00052%; xenon, 0.0000087%.[17] Small wonder that neon, krypton, and xenon required very large quantities of liquid air in order to detect their presence. But it is fascinating to note, too, that every cubic meter of air contains nearly 10 grams of argon; each adult inhales almost 200 grams of argon per day. Yet we were unaware of argon's existence until 1894.

Ramsay was awarded the 1904 Nobel Prize in Chemistry, and Rayleigh was awarded the 1904 Nobel Prize in Physics. In Figure 304, we see an early twentieth-century caricature of the fatherly Ramsay blissfully posing with his own periodic family.

---

1. M.W. Travers, *The Discovery of the Rare Gases*, Edward Arnold & Co., London, 1928.
2. Travers, op. cit., pp. 1–4.
3. In 1815 William Prout postulated, on the basis of the small number of atomic weights known, that all were simple whole-number multiples of the lightest element—hydrogen. At one level, this seemed to be an incredibly prescient hypothesis. We have known for about 100 years that hydrogen has one proton [mass based on carbon–12 is 1.0073 atomic mass unit (amu)] and one electron of relatively negligible mass (0.0005486 amu). However, careful determinations by Berzelius, Dumas, and Stas throughout the midnineteenth century indicated significant discrepancies. And our modern understandings include the occurrence of isotopes, the fact that the neutron is slightly heavier than the proton (1.0087 amu), departures from ideal-gas behavior at higher pressures, nuclear binding energies, and numerous other flaws in the hypothesis. But it was a useful construct and remains conceptually helpful today in a very simplistic way.
4. J.E. Marsh, *Chemistry in Space, from Professor J.H. van't Hoff's "Dix Années Dans l'Histoire d'Une Théorie,"* Clarendon, Oxford, 1891, p. 16.
5. Travers, op. cit., pp. 56–57.
6. J.R. Partington, *A History of Chemistry*, Vol. 4, MacMillan & Co. Ltd., London, 1964, pp. 936–947.
7. Travers, op. cit., pp. 82–84.
8. W. Kauzmann, *Thermodynamics and Statistics: With Applications to Gases*, W.A. Benjamin, Inc., 1967, pp. 53–58.
9. Partington, op. cit., pp. 105–108.
10. Travers, op. cit., p. 87.
11. Partington, op. cit., pp. 904–906.
12. Travers, op. cit., p. 90.
13. *Kalends* = the first day of the month in the ancient Roman calendar.
14. Travers, op. cit., pp. 95–96.
15. This essay is dedicated to my brother Kenny, who immersed himself in *Superman* and *Batman* comics folklore as a boy, became a neon artist, and is the proprietor of *Krypton Neon* in New

FIGURE 304. ■ An early-twentieth-century caricature in *Vanity Fair* of William Ramsay pointing with fatherly pride to his chemical family—the rare gases. See color plates.

York City. I have tried, unsuccessfully, to convince him to post a sign "Krypton Neon Argon" when he goes to lunch.

16. Travers, op. cit., pp. 105, 110, 126.
17. F.A. Cotton and G. Wilkinson, *Advanced Inorganic Chemistry*, fifth edition, John Wiley & Sons, New York, 1988, p. 588.

JUST HOW MANY DIFFERENT SUBSTANCES ARE IN ATMOSPHERIC AIR?

How many substances there are in air depends upon how low you will go (in measuring concentrations). At the percent (part-per-hundred or pph) level, there is only nitrogen (78.08%) and oxygen (20.95%).[1,2] If we stretch a bit and add argon (0.93%), over 99.9% of the dry atmosphere is accounted for by just three substances. Water concentrations can vary over five orders of magnitude and actually reach percent levels in tropical rain forests.[1] These percentages are volume/volume (v/v), and since equal numbers of gas molecules occupy equal volumes under the same pressure and temperature, that means that one thousand molecules of dry air will have on average 780 N_2 molecules, 210 O_2 molecules, and 9 argon atoms. Carbon dioxide is present at about 350 parts-per-million (ppm). Other gases at or near the low-ppm levels include Ne, He, methane (CH_4), and Kr giving a total of nine substances including water. At the parts-per-billion (ppb) level, we start adding hydrogen, carbon monoxide, sulfur dioxide, ammonia, and ozone. Below that, in the ppb to ppt (parts-per-trillion) range we encounter oxides of nitrogen and hundreds of organic vapors such as benzene, toluene, and tetrachloroethylene.[3] Indeed the number of expected organic air pollutants at the trace level numbers in the thousands.[4]

What is a part-per-billion? Imagine adding a drop of alcohol to a pool of water 6 ft deep × 12 ft wide × 18 ft long and stirring thoroughly. Alternatively, imagine a golf foursome compared to the world's total population.[5]

---

1. T.E. Graedel and P.J. Crutzen, *Atmospheric Change: An Earth System Perspective*, Freeman, New York, 1993, p. 8.
2. J.H. Seinfeld, *Atmospheric Chemistry and Physics of Air Pollution*, Wiley, New York, 1986, p. 8.
3. B.B. Kebbekus and S. Mitra, *Environmental Chemical Analysis*, Blackie Academic and Professional, London, 1998, pp. 229–230.
4. T.E. Graedel, D.T. Hawkins, and L.D. Claxton, *Atmospheric Chemical Compounds: Sources, Occurrence, and Bioassay*, Academic, Orlando, 1986.
5. Thanks to Professor Joel F. Liebman for this suggestion.

ATOMS OF THE CELESTIAL ETHER

Early hints of the wave nature of light included the seventeenth-century discovery of diffraction by Hooke and other manifestations of interference. It was obvious that dropping a rock into a pond created waves, and Boyle showed that air was necessary for the transmission of sound waves. Thus, it appeared that there had to be a medium for transmitting light waves and it was thought to be a kind of "universal ether"—present everywhere, yet imperceptable. During the 1880s, the physicists Michaelson and Morley disproved, experimentally, the existence of the ether. Nevertheless, the concept continued to influence many outstanding scientists for perhaps two more decades. In a book published in 1895 titled *Cod-Liver Oil and Chemistry,* the author Friedrich Möller explains the rotation of plane-polarized light, clockwise or counterclockwise, by invoking clockwise or counterclockwise rotation of a bond in the molecule producing clockwise or counterclockwise "wakes" in the ether.

Mendeleev was clearly a believer in the ether. His explanation was straightforwardly chemical and constructed from his Periodic Table and the newly discovered inert gases.[1] The 1904 English edition of Mendeleev's book *An Attempt Toward a Chemical Conception of the Ether* appeared when the Russian master was 70. He postulates that the ether is composed of atoms of an as-yet-unknown superlight inert gas. Clearly, the gas must be inert in order to penetrate all matter without being reacted or absorbed and clearly it must be superlight not to be perceived.

He fits the "ether element" into his Periodic Table in the manner shown in Figure 305. Mendeleev placed the inert gases in Group 0, to the left of hydrogen and the alkali metals. This places helium in Period 2 and leaves a gap to the left of hydrogen in Period 1. Our modern Periodic Tables place the inert gases in Group 18 (8A in some versions) and thus helium now sits in Period 1 for reasons

Series	Zero Group	Group I	Group II	Group III	Group IV	Group V	Group VI	Group VII	Group VIII
0	*x*								
1	*y*	Hydrogen H=1·008							
2	Helium He=4·0	Lithium Li=7·03	Beryllium Be=9·1	Boron B=11·0	Carbon C=12·0	Nitrogen N=14·04	Oxygen O=16·00	Fluorine F=19·0	
3	Neon Ne=19·9	Sodium Na=23·05	Magnesium Mg=24·1	Aluminium Al=27·0	Silicon Si=28·4	Phosphorus P=31·0	Sulphur S=32·06	Chlorine Cl=35·45	
4	Argon Ar=38	Potassium K=39·1	Calcium Ca=40·1	Scandium Sc=44·1	Titanium Ti=48·1	Vanadium V=51·4	Chromium Cr=52·1	Manganese Mn=55·0	Iron Fe=55·9 Cobalt Co=59 Nickel Ni=59 (Cu)
5		Copper Cu=63·6	Zinc Zn=65·4	Gallium Ga=70·0	Germanium Ge=72·3	Arsenic As=75·0	Selenium Se=79	Bromine Br=79·95	
6	Krypton Kr=81·8	Rubidium Rb=85·4	Strontium Sr=87·6	Yttrium Y=89·0	Zirconium Zr=90·6	Niobium Nb=94·0	Molybdenum Mo=96·0	—	Ruthenium Ru=101·7 Rhodium Rh=103·0 Palladium Pd=106·5 (Ag)
7		Silver Ag=107·9	Cadmium Cd=112·4	Indium In=114·0	Tin Sn=119·0	Antimony Sb=120·0	Tellurium Te=127	Iodine I=127	
8	Xenon Xe=128	Cæsium Cs=132·9	Barium Ba=137·4	Lanthanum La=139	Cerium Ce=140	—	—		— — — (—)
9		—	—		—	—	—	—	
10	—	—	—	Ytterbium Yb=173	—	Tantalum Ta=183	Tungsten W=184	—	Osmium Os=191 Iridium Ir=193 Platinum Pt=194·9 (Au)
11		Gold Au=197·2	Mercury Hg=200·0	Thallium Tl=204·1	Lead Pb=206·9	Bismuth Bi=208	—	—	
12	—	—	Radium Rd=224	—	Thorium Th=232	—	Uranium U=239		

FIGURE 305. ■ In *An Attempt Towards A Chemical Conception of the Ether* (London, 1904), the aged Mendeleev postulates that the "universal ether" is composed of unimaginably light inert gas atoms (x) in series zero—group zero of his Periodic Table. Below x, there would have to be another new inert gas (y) with an atomic mass of 0.4 (H = 1.0).

theoretical as well as practical. Mendeleev postulated a new Group 0–Period 1 element, element *y* in the accompanying figure, which he calculated to have a relative atomic weight of 0.4 (hydrogen = 1.0) and notes that while this is clearly far too massive for atoms of the ether, it may correspond to unassigned lines in the solar spectrum (remember, helium was already known). He then postulates another new element *x* (see Figure 305) in the Group 0–Period 0 space, which he reasons has a relative mass in the range 0.00000096–0.000000000055, the atom comprising the celestial ether.

A MASSIVE ANTEDILUVIAN ANIMAL—THE MEGALOSAURUS.

IMMENSE PRE-HISTORIC ANIMALS—THE IGUANODON AND MEGALOSAURUS.

FIGURE 306. ■ The all-too-human attempt by Mendeleev to "cram" his periodic law into an explanation of the defunct ether theory is similar to the attempts by nineteenth-century paleontologists to "cram" the bones of dinosaurs into the shapes of bears and other known land animals.

This all-too-human attempt by Mendeleev to cram the ether concept into his Periodic Table illustrates our very human limitations in trying to fit our own world views to facts. Figure 306 depicts mid-nineteenth-century illustrations of dinosaurs. The bones were "crammed" into the shapes of bear-like or ox-like creatures because these were the largest land carnivores and herbivores then known. Indeed, the planetary model of the atom, developed by Bohr in 1913 and later completely eclipsed, was probably based upon his desire for a unity in the universe and an analogy with the solar system.

1. A. Greenberg, *The Chemical Intelligencer*, April, 1995, pp. 31–36.

NON-ATOMUS

Nonindivisible! The Greek philosophers conceived of the smallest unit of matter as atomos (Latin *atomus*): indivisible. John Dalton had said: "Thou knows . . . no man can split the atom" (see earlier discussion of Dalton). However, toward the end of the nineteenth century, this view had to be completely modified.[1,2] In 1859, Julius Plucker discovered that the visible discharges in vacuum tubes could be deflected by a magnetic field. The term *cathode ray* was coined around 1883 and William Crookes established that they were negatively charged. Joseph John (J.J.) Thomson established the particulate nature of these emissions and he determined a charge-to-mass ratio $e/m = 1.2 \times 10^7$ emu/g; present value, 1.7×10^7 emu/g = 5.1×10^{17} esu/g, for his "corpuscles." The term electron was introduced by G.J. Stoney over Thomson's objections. It was also known at this time that the e/m value for the electron was about 1300 times that of the hydrogen ion (modern ratio ca. 2000).

In 1908 Robert Millikan (1923 Nobel Prize in Physics) first performed his famous oil droplet experiment in which he determined a unit charge of 4.77×10^{10} esu (later 4.80×10^{10}) esu. With the modern e/m value (1.7×10^7 esu/g), the mass of the electron was found to be only $^1/_{1837}$ that of the lightest atom, hydrogen.

The cathode-ray tubes were also found to eject positive ions in the opposite direction from the electrons. These *canal rays* were comprised of much more massive particles. J.J. Thomson (1906 Nobel Prize in Physics) used a magnetic field to bend the paths of these ions and record their collisions on film. He discovered that pure neon gas produced two masses, 20 and 22, due to isotopes. The term was coined by Frederick Soddy (1921 Nobel Prize in Chemistry) during his studies of radioactive elements having the same chemical but different radioactive properties.[3] The separation of positive ions using a magnetic field followed by recording them on a photographic plate is the basis of mass spectrometry, developed by Francis W. Aston (1922 Nobel Prize in Chemistry).[1]

1. J.R. Partington, *A History of Chemistry*, MacMillan, London, 1964, Vol. 4, pp. 929–934.

2. A.J. Ihde, *The Development of Modern Chemistry,* Harper & Row, New York, 1964, pp. 478–483; 486.
3. J.R. Partington, op. cit., pp. 941–947.

A "GROUCH" OR A "CRANK"?

Did Mendeleev and Priestley Become Scientific Grouches in Old Age?

The discovery of the noble gases between 1894 and 1898 presented Dmitri Mendeleev the opportunity to apply his periodic law to develop a fully chemical explanation of the universal ether, the all-penetrating imponderable medium surrounding and imbuing all matter (see Figure 305).[1] Although the Michaelson–Morley experiment in 1887 disproved the existence of the ether, many important scientists, including a number of prominent physicists, still accepted its existence at the start of the twentieth century. Part of the problem was that the experimental result—that the velocity of light is equal in all directions, although accepted, could not be explained by current theory. In that sense, it could be considered to be an "anomaly."[2,3] Einstein's relativity theory would eventually furnish the explanation. Mendeleev stretched his periodic law to postulate the existence of an undiscovered inert gas of atomic mass (relative to $H = 1.0$) on the order of 0.00000096 to 0.000000000055, that would form the substance of the ether.[1] However, in order to postulate this ethereal element, he also needed to postulate one additional inert gas of atomic mass 0.4. There was no serious evidence for this element, which would represent an *extrapolation* rather than the interpolations that had worked so brilliantly in predicting new elements during the 1870s and 1880s. Perhaps it is unfair to say this, "hindsight always being 20 : 20," but we may view the aging Mendeleev as becoming a bit "grouchy," scientifically, with age.

There are other precedents for great scientists who "grouchily" persisted in retaining theories beyond their useful lives. A prominent example is Dr. Joseph Priestley,[4] whose discoveries of new gases including oxygen were so critical to the development of chemistry. Priestley was an early adherent of phlogiston theory, and his final chemical publication, *The Doctrine of Phlogiston Established and that of the Composition of Water Refuted,* was published in 1800 (Figure 307),[4] and the second edition was published in 1803, a year before he died and two decades after discovery of the true composition of water sank phlogiston theory. (Partington terms the prominent Swedish chemist Anders Retzius, who died in 1821—"probably the last phlogistonist."[5]) Such conservatism is not necessarily an unhealthy thing for science. It protects scientific theories from rapidly shifting with the prevailing winds and demands stronger proof and even generational change before what philosopher of science Thomas Kuhn calls a "paradigm shift"[6] is widely accepted.

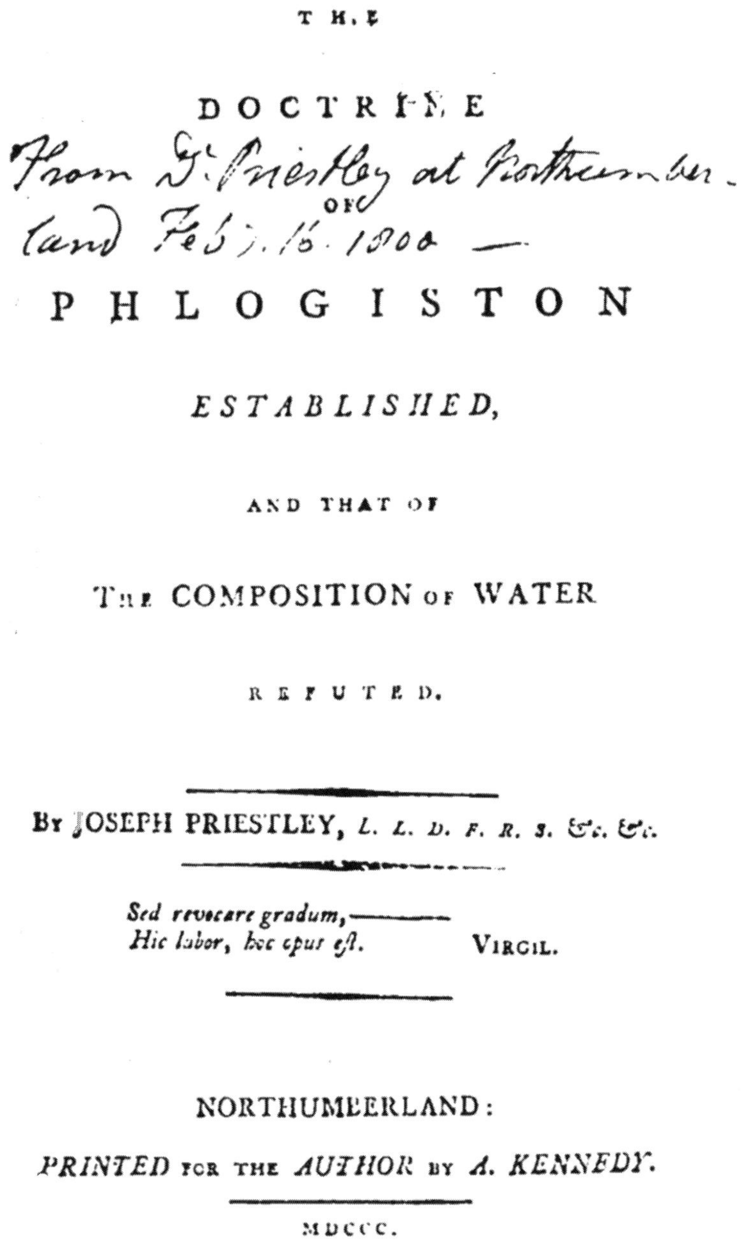

THE

DOCTRINE

From Dr. Priestley at Northumberland Feb. 16. 1800 —

OF

PHLOGISTON

ESTABLISHED,

AND THAT OF

THE COMPOSITION OF WATER

REFUTED.

BY JOSEPH PRIESTLEY, *L. L. D. F. R. S. &c. &c.*

Sed revocare gradum,——
Hic labor, hoc opus est. VIRGIL.

NORTHUMBERLAND:

PRINTED FOR THE *AUTHOR* BY *A. KENNEDY.*

MDCCC.

FIGURE 307. ■ Joseph Priestley "grouchily" retained his belief in phlogiston theory through the end of his life. Here is a copy of his spirited defense that he signed and presented to an acquaintance. (From The Roy G. Neville Historical Chemical Library, a collection in the Othmer Library, CHF.)

Rejecting Atomic Theory and Dismissing Continental Drift

In this light, it is important to note that Dalton's atomic theory, which we blithely inform our students was introduced and accepted at the start of the nineteenth century, was resisted by some prominent chemists (and many physicists) until the first decade of the twentieth century when Einstein and later Jean Perrin explained the molecular basis of Brownian motion. Thomas Sterry Hunt,[7] Professor of Geology at the Massachusetts Institute of Technology, member of the National Academy of Sciences (1873), President of the American Association for the Advancement of Science (1871), and two-time President of the American Chemical Society (1879, 1888), wrote a book in 1887 that totally rejected atomic theory. He retained this belief until he died five years later. The great German chemist Friedrich Wilhelm Ostwald (1853–1932) firmly resisted atomic theory throughout the first four decades of his scientific career. However, in 1909, the year he was awarded the Nobel Prize in Chemistry, Ostwald finally admitted that the work of Perrin as well as J.J. Thomson "justify the most cautious scientist in now speaking of the experimental proof of the atomic nature of matter."[8]

The boundary line between a "scientific grouch" and a "scientific crank" is sometimes not a very clear one. Simply imagining a person stubbornly resisting the prevailing views of the scientific establishment and "howling alone in the desert" is not sufficient reason to label the "infidel" a "crank." For example, countless people, including children, who have viewed maps of the world have undoubtedly noted the apparent jigsaw puzzle fit between the South American and African continents. It remained a curious, but not very informative, observation since no mechanism for explaining this "anomaly" existed until fairly recently. However, in 1912 the German geophysicist Alfred Wegener (1880–1930) noted similarities in fossils collected in the two continents, combined this with geophysical data, and postulated the theory of "continental drift."[2] He was isolated, and his views were considered "highly controversial," a description sometimes applied diplomatically to the work of "cranks," until he was vindicated by the theory of plate tectonics developed in the late 1960s.[2]

A Crank Who "Trashed" Phlogistonists and Antiphlogistonists with Great Gusto

I believe that a "crank" adheres to an ideology and it is this ideology rather than an inherent scientific conservatism or radicalism that defines a "crank" (i.e., unless conservatism or radicalism for its own sake is the ideology). Partington[9] clearly labels Robert Harrington,[10] an English surgeon, a "crank." Although a believer in phlogiston theory, Harrington's "ideology" seemed based on a scientific enterprise in which he was correct and all others wrong. He believed in "equal opportunity" and, in a pamphlet published in 1786, joyfully "trashed" phlogistonists of the English school (Priestley, Cavendish, and Richard Kirwan) and antiphlogistonists of the French school (Lavoisier):[10]

> Letter . . . to Dr. Priestley, Messrs. Cavendish, Lavoisier, and Kerwan . . . to prove that their . . . opinions of Inflammable and Dephlogisticated Airs forming Water, and the Acids being compounded of different Airs are fallacious, London, 1786.

And one must appreciate the titles of two of his subsequent works:[10]

> The Death-warrant of the French Theory of Chemistry . . . with a Theory fully . . . accounting for all the Phenomena. Also a full . . . Investigation of . . .

> Galvanism, and Strictures upon the Chemical Opinions of Messrs. Weiglet, Cruickshanks, Davy, Leslie, Count Rumford, and Dr. Thompson; likewise Remarks upon Mr. Dalton's late Theory and other Observations, 1804.

Or perhaps, "How To Make Friends and Influence Chemists" and finally, and most modestly, Harrington on Harrington:[10]

> An Elucidation and Extension of the Harringtonian System of Chemistry, explaining all the Phenomena without one single Anomaly, London, 1819.

Let us briefly examine Harrington's "hash" with his fellow phlogistonists. In his 1785 book on different kinds of air, he notes that it appears to be appropriate to "conclude that phlogiston is fire and light, or a certain subtile elastic fluid, upon the modifications of which the phænomena of heat and light immediately depend."[11] But he then observes that fresh air is required to receive the phlogiston from the combustible body until the air is "injured" by the fire and incapable of further supporting combustion. Harrington sees a "timing" issue here. If fresh air must first attract phlogiston and then fire and light follow, how could phlogiston actually be fire and light? Indeed, he ridicules other phlogistonists by indicating the logical extension of loss of phlogiston followed by heat and light: "For if a body be saturated with water it will not burn. But as the air attracts the water from the burning body it will burn, and agreeable to its quick or slow attraction of this moisture."[11] Thus, rather than being firelike, phlogiston would seem to be, idiotically enough, waterlike.

Here is another problem for phlogistonists less savvy than Dr. Harrington. Inflammable air (actually H_2) and nitrous air (actually NO) are known to contain equal quantities of phlogiston. Let us trace this correct phlogistonist logic for a moment using modern chemical equations. Aqueous sulfuric acid will "dissolve" metals because the acids hydrogen ions (H^+) are readily reduced by all metals except the most inert:

$$2H^+ + SO_4^{2-} + \text{Cu (metal)} \rightarrow H_2 \text{ (gas)} + Cu^{2+} + SO_4^{2-}$$

However, in dilute nitric acid, less active metals such as copper and iron will reduce the nitrate group rather than the hydrogen ion to produce nitric oxide ("nitrous air"):

$$3\text{Cu (metal)} + 2NO_3^- + 8H^+ \rightarrow 3Cu^{2+} + 2\text{NO (gas)} + 4H_2O$$

In both cases, calxes of copper remain, so it is "clear" that the phlogiston has been carried off in "flammable air" in the first case and in "nitrous air" in the second case. Now, in comparing "flammable air" and "nitrous air," it is "clear" using eudiometry that each contains the same quantity of phlogiston since they each react with the same quantity of "dephlogisticated air" (oxygen):

Inflammable air: $H_2 + \frac{1}{2}O_2 \rightarrow H_2O$

Nitrous air: $NO + \frac{1}{2}O_2 \rightarrow NO_2$ (reddish gas)

The question Harrington poses is why fire and light are generated with "inflammable air" but not with "nitrous air" even as the two release the same quantity of phlogiston.

And here is another inconsistency—we "know" that exhaled air is somewhat "injured" since it is has absorbed phlogiston and is not as breathable as the air freshly inhaled moments earlier. "Clearly," phlogiston is being captured in the lungs. A snake eats phlogiston-rich animals (i.e., containing fats).[11] Its body is cool, and so the lungs should be huge in order to rapidly expel large quantities of phlogiston extracted from the meat. In fact, the lungs in a snake are small—why does the snake not incinerate from the trapped heat? In contrast, herbivores like cows eat phlogiston-poor (fat-free) food, are warm, and have large lungs to rapidly expel phlogiston.[11] Why do cows not starve? Frankly, Harrington does not have any good answers himself, but that is not his mission.

Views of a Libertarian Chemist: The Nefarious Smithsonian Institution and Other Plots

Toward the end of the nineteenth century Gustavus Detlef Hinrichs, M.D., LL.D., Professor of Chemistry, St. Louis College of Pharmacy, had embarked on a desperate crusade to save America's chemical industry and its educational establishment from the incorrect atomic weights forced on our unsuspecting nation by its nefarious government.[12] The source of all evil was one Frank Wigglesworth Clarke, Chief Chemist of the Geological Survey, whose

> recalculations have been formally endorsed by the Secretary of the Smithsonian Institution and published officially at the expense of the Smithson Fund; it has, finally, been sent out under the official frank as registered matter. The deficiency of the postal service—partly so resulting—is made up by Congressional appropriations.
>
> The same author Clarke is also habitually sent by authority of the National Government and at public expense, as delegate to the Congresses of Chemists, and put in charge of National Exhibits at home and abroad. This highest possible official consideration has enabled him to exercise a ruling influence in the American Chemical Society.[13]

Sounds pretty alarming—"the new world order of atomic weights." So what's the "big hoo-hah" here? Seems like Professor Hinrichs had adopted Prout's hypothesis[14–16] as a strict fundamentalist.

In 1815 and 1816 William Prout (1785–1850)[14] published two papers in which he asserted that the densities of gases were simple whole-number multiples of hydrogen.[14–16] This was only a decade after Dalton first postulated atomic weights, and there were considerable uncertainties in experiments and formulas. Nonetheless, the concept was an attractive one since it implied the possibility of a simplest "primary material" out of which all other atoms were composed. Prout postulated the existence of this "protyle" from which all other atoms would be composed. It had an almost religious simplicity. In 1819, Berzelius published a complete series of relative masses of elements and compounds, which included many fractional atomic weights (relative to $H = 1$) that would not agree with Prout's hypothesis. However, there were considerable uncertainties in formulas and atomic weights. In 1825, Thomas Thomson published his table of atomic weights: all, including chlorine (36), for example, were whole-number multiples of hydrogen (1). Berzelius said of Thomson's atomic weights:[14]

> Much of the experimental part, even of the fundamental experiments, appears to have been made at the writing desk; and the greatest civility which his contemporaries can show its author, is to forget that it was ever published.

Berzelius soon regretted his accusation, and Partington notes that Thomson was scrupulously honest.[14] Attempts to salvage Prout's hypothesis included suggestions that hydrogen might contain exactly two or perhaps four "protyles."[17] Other attempts included suggestions that, statistically speaking, while most atoms of chlorine, for example, might have an atomic weight of 36, a few might be 35 and 37; even fewer, 34 and 38; and so on. Partington even notes ideas reminiscent of Newton's "worn atoms."[18]

However, the coffin of Prout's hypothesis was nailed shut around 1865 by the careful analytical studies of Jean Servais Stas. Figure 308 depicts a magnificent apparatus for total analysis of silver iodate ($AgIO_3$).[19] How could one possibly argue with a "Rube Goldberg–looking" apparatus like that? The gasometer (*A*) on the left delivered a steady stream of nitrogen gas purified by a *kaliapparat* (*B*) filled with concentrated sulfuric acid, followed by drying tubes (C) containing anhydrous calcium chloride, and then a gas furnace *D* containing a fused-glass tube filled with finely divided copper. Nitrogen, for flushing the system, could be made via combustion of ammonia. Its impurities would include unreacted ammonia (trapped in *B*), water (removed by C) and residual oxides of nitrogen (reduced to nitrogen in *D*). The balloon flask (*H*), containing silver iodate, is heated slowly over a bunsen burner and the oxygen generated (leaving molten AgI) is captured in gas furnace I containing finely divided copper.

Unimpressed by Stas and his apparatus, Hinrichs was nothing if not straightforward:[20]

> Ever since I understood the conditions of the chemical elements in reference to a single, primitive substance, (that is, since 1855), I have most faithfully labored in this field.

His idea was almost a half-century old when his book *Absolute Atomic Weights*[12] was published in 1901. His opinion of Jean Stas was summarized in a paragraph titled:[21] "The Greatest False Scientist."

Hinrichs diagnosed his contemporary John William Mallet, whose atomic weights of gold, lithium, and aluminum were not to his liking, as suffering from *Morbus Stasii* complicated by "the incipient stages of '*Furor Clarkii*'."[21] The work of Edgar Fahs Smith's student W.L. Hardin "represent nothing but his own imagination."[21] Under Henri Moissan "good French laboratory work is spoilt or falsified, by reducing it by German atomic weights."[21] Berzelius' problems are summarized as: "Great Chemist, Poor Balance."[21] William Ramsay's studies of atomic weights were infected by his use of Clarke's tables. And Hin-

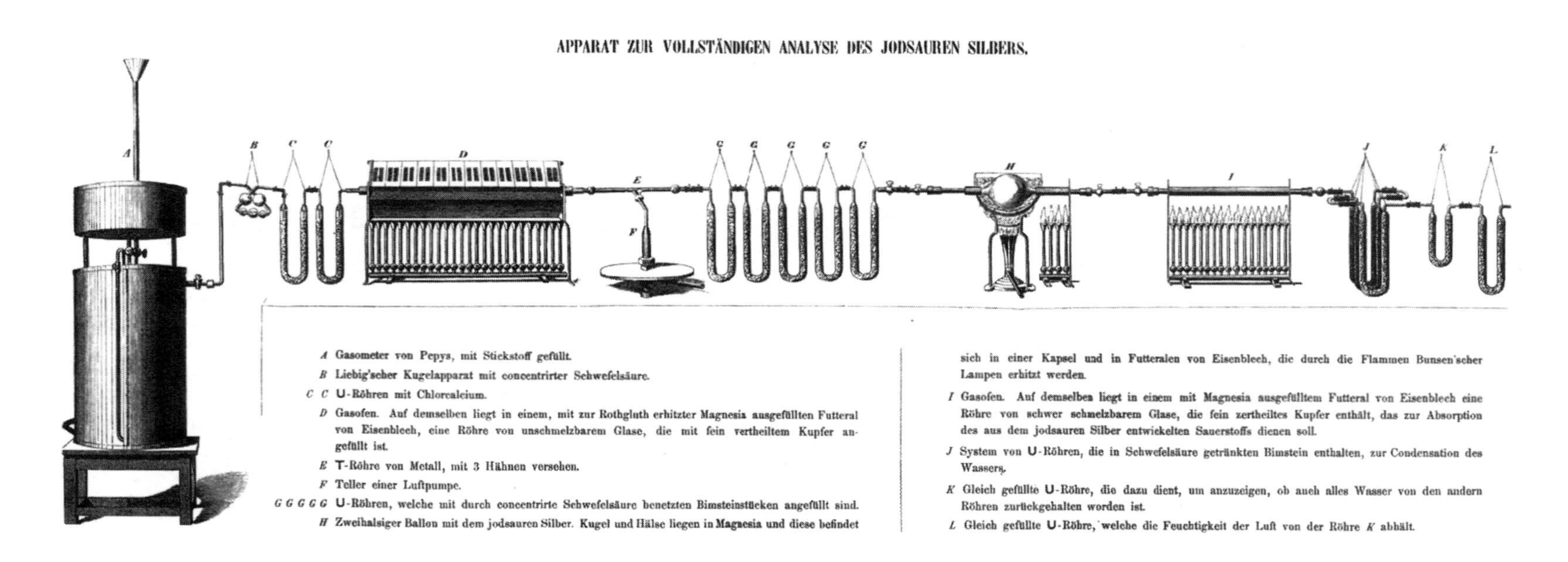

FIGURE 308. ■ Apparatus used by Jean Servais Stas for total analysis of silver iodate. Stas' exacting and precise work on atomic weights laid a firm basis for development of the periodic table. However, Gustavus Detlef Hinrichs referred to Stas as "The Greatest False Scientist." Late in his career, Hinrichs became a certifiable scientific crank who perceived a vast governmental conspiracy to hide the true atomic weights of the elements. (From Stas, *Untersuchungen Über Die-Gesetze Der Chemischen Proportionen*, 1867).

richs calculates using his particular statistical techniques that[21] "*Berzelius* was in 1826, a 10,000,000,000 times better chemist as Ramsey in 1893.

Thirty years before the aforementioned atomic weights book was published, Hinrichs was professor of physical science at the State University of Iowa. Although he did not accept the concept of periodicity, Hinrichs' "chart of the elements" has been credited as the first spiral classification of the elements.[22] In his 1871[23] text he listed atomic weights (pre-Stas) that included fractional weights such as aluminum (27.4), chlorine (35.5), copper (63.4), platinum (197.4), selenium (79.5), strontium (87.6), and zinc (65.2). The remaining 37 elements in his table (a total of 63 were known by this time) were pleasing whole-number multiples of hydrogen (=1). There is no explicit hint that Hinrichs was concerned about these departures from Prout's hypothesis in 1871. Perhaps he did not want to confuse his students. However, it is certainly clear that the Clarke-adopted 35.45 (rather than 35.5) for chlorine, for example, would have "bugged" Hinrichs, as did Darwin's theory of evolution.[20]

Hinrichs may have also been a bit "grouchy" in addition to being "certifiably cranky." Here is the first of his five "laboratory rules:"[24]

1. "BE QUIET—Talk not to your fellow students, and only in low whispers to your teacher. Walk to and from the balance so that your steps are not heard. Early thus learn to show reverence for truth and its investigation; the laboratory should be a temple of science."

Or, perhaps a monastery. Still, quiet whispers and silent footsteps in the teaching lab seem more attractive to me the older and grouchier I get.

---

1. A. Greenberg, *A Chemical History Tour*, John Wiley and Sons, New York, 2000, pp. 257–259.
2. A. Lightman and O. Gingerich, *Science*, Vol. 255, pp. 690–695, 1991. I am grateful to Dr. Joel F. Liebman and Dr. Larry Dingman for helpful discussions on this topic.
3. Scientific "anomalies" appear to me to be very closely related to Stent's "premature discoveries"; see G. Stent, *Scientific American*, Vol. 227, No. 6, pp. 84–93, 1972.
4. J.R. Partington, *A History of Chemistry*, MacMillan and Co. Ltd., London, 1962, Vol. 3, pp. 237–271.
5. Partington, op. cit., p. 200.
6. T.S. Kuhn, *The Structure of Scientific Revolutions*, second edition, University of Chicago Press, Chicago, 1970.
7. E.R. Atkinson, *Journal of Chemical Education*, Vol. 20, pp. 244–245, 1943.
8. Partington, op. cit., 1964, Vol. 4, p. 597.
9. Partington (1962), op. cit., p. 490.
10. L. Stephen and S. Lee, *The Dictionary of National Biography*, Vol. VIII, Oxford University Press, London, 1921–1922 (reprinted 1963–1964), pp. 1320–1321.
11. (R. Harrington), *Thoughts on the Properties and Formation of the Different Kinds of Air; with Remarks on Vegetation, Phosphori, Heat, Caustic Salts, Mercury, and on the Different Theories upon Air*, R. Faulder, J. Murray, and R. Cust, London, 1785, pp. 278–285.
12. G.D. Hinrichs, *The Absolute Atomic Weights of the Chemical Elements*, C.G. Hinrichs, Publisher, St. Louis, 1901.
13. Hinrichs, op. cit., p. iv.
14. Partington (1964), op. cit., pp. 222–226.
15. A.J. Ihde, *The Development of Modern Chemistry*, Harper & Row, New York, 1964, pp. 154–155.
16. Partington (1962), op. cit., pp. 713–714.

17. Partington (1964), op. cit., pp. 875; p. 886.
18. Partington (1964), op. cit., p. 882.
19. J.S. Stas, *Untersuchungen Über Die-Gesetze Der Chemischen Proportionen Über Die Atomgewichte Und Ihre Gegenseitigen Verhältnisse*, Verlag Von Quandt & Händel, Leipzig, 1867, pp. 187–200; see folding plate at the end of the book. This is the first German edition. The first French edition was published in 1865.
20. Hinrichs, op. cit., pp. 292–295.
21. Hinrichs, op. cit., pp. 20–35.
22. G.N. Quam and M.B. Quam, *Journal of Chemical Education*, Vol. 11, p. 288, 1934.
23. G. Hinrichs, *The Elements of Chemistry and Mineralogy*, Griggs, Watson, & Day, Davenport, 1871, p. 101.
24. Hinrichs (1871), op. cit., p. (161)—second leaf following p. 158.

WHY IS PROUT'S HYPOTHESIS STILL IN MODERN TEXTBOOKS?

Prout observed in 1815 that gas densities were whole-number multiples of the density of hydrogen gas. This led to his idea that all atomic weights are whole-number multiples of the atomic weight of hydrogen and that hydrogen might well be the "primary substance" from which all other elements are made. However, subsequent observations such as the atomic weight of chlorine (ca. 35.5) led to "protyles" having half the weight of the hydrogen atom. Chemical analyses and formulas at the time of Prout and for the next 50 or so years contained sufficient errors to cast doubts on published decimal-place accuracies of atomic weights. However, as analytical chemistry improved, it became abundantly clear toward the end of the nineteenth century that the deviations from unity, $\frac{1}{2}$ or even $\frac{1}{4}$, were real and significant and our "cranky" friend Hinrichs (previous essay) should have accepted this. As so often happens in science, these tiny discrepancies were hinting at things much more profound—the subatomic structure of matter.[1–3]

When chemistry is first taught to a student, the first exam may have "fill-in-the blanks questions" such as

The atomic number is the—number of protons

The atomic mass number is the—number of protons plus neutrons

The number of protons equals the number of—electrons

These simplified concepts tend to exemplify the apparent utility of Prout's hypothesis as an organizing principle. The atomic mass number, often incorrectly truncated to "atomic mass," treats protons and neutrons as equals. Taken too literally, the mass of uranium–238 would appear to be roughly equal to that of 238 hydrogens (protium or hydrogen–1 atoms) with some tiny discrepancy understood as arising from the 0.1% difference in mass between protons and neutrons. In fact, if we take the masses of 92 protons, 146 neutrons, and 92 electrons, the total mass is 240.0 amu.

The subatomic structure of the atom began to emerge in the late nineteenth and early twentieth centuries, thanks to the development of the Crookes [after Sir William Crookes (1832–1919)] tube and the work of J.J. Thomson. Isotopes, atoms of the same element (atomic number) having different atomic masses, emerged independently from two lines of investigation.[1] The newly discovered radioactive elements often included species that were chemically identical but had distinct radioactive decay properties.[1] Thus, two distinct species of uranium were discovered, each having a unique decay pattern. B.B. Boltwood at Yale discovered a new element, "ionium," which was an intermediate in the decay of uranium-II, but not uranium-I.[1] However, he learned that "ionium" was chemically identical with thorium. Frederick Soddy coined the term "isotope" ("same place"—i.e., in the periodic table) in 1913.[1] In 1919, F.W. Aston modified a technique of Thomson's and discovered that ionized neon atoms produced two different ions, one of mass 20, the other of mass 22.[1] It was the difference in atomic mass that accounted for isotopes. Not long afterward, chlorine was found to be a mixture of two isotopes (mass numbers 35 and 37) with a statistical average atomic mass of 35.45 amu.

Following the discovery of the long-suspected neutron in 1932, almost 25 years after Einstein's theory of relativity, here is where things stood with regard to Prout's hypothesis:

1. The simplest hydrogen isotope is protium. It has a proton (relative charge +1) in its nucleus and electron (relative charge –1) outside the nucleus.

2. The masses and (relative) charges of the three major subatomic particles are as follows (with the mass of the carbon-12 atom for reference):[3]

Particle	Charge	Mass (amu Relative to Carbon 12)
Proton	+1	1.0073
Neutron	0	1.0087
Electron	–1	0.0005486
Carbon-12 atom	0	12.0000000 (assumed)

The fact that the proton is 0.1% lighter than the neutron is not consistent with the literal interpretation of Prout's hypothesis.

3. Since we now know that a neutron free of the nucleus has a half-life of 17 minutes as it decays to a proton, an electron and an antineutrino of negligible mass, it might be tempting to think of the source of Prout's hypothesis as effectively the total mass of the proton and one electron. The total for the two isolated particles, 1.0078, is still less than that of the neutron. Nonetheless, it might also be tempting to consider the "protyle" to have the mass of the neutron.

4. Clearly, isotopes are the major source of the discrepancy with Prout's hypothesis. Hydrogen is 99.986% protium, the lightest isotope (1.0078 amu = mass of proton plus electron) and only 0.014% deuterium (2.0141 amu).[2] The amount of tritium is ultratrace. This coincidence is why hydrogen so often "works" as the apparent "protyle" in Prout's hypothesis. If, for example, protium were 80% and deuterium 20% in naturally occurring hydrogen, Prout's hypothesis would never have existed. For chlorine the two natural isotopes occur in significant amounts: chlorine–35, 75.53%; chlorine–37, 24.47%. The observed mass in naturally occurring chlorine is the weighted average: 35.45 amu—impossible to rationalize using Prout's hypothesis.

5. Another major discrepancy is the packing effect of nuclear particles. Thus, if we sum up the masses of all four nuclear particles in helium–4 (two protons + two neutrons = 4.03190 amu) and compare the sum to the observed mass (i.e., minus the two electrons) of the helium nucleus (4.00150 amu), the discrepancy (0.03040 amu) furnishes the energy (strong force, $\Delta E = \Delta mc^2$) that binds the nucleus.[3] This energy is about a million times more powerful than the chemical forces released by explosion of dynamite or TNT. The "mass defect" in uranium-238 (92 protons, 146 neutrons) is an amazing 1.9353 amu. Luckily, the calculated "excess" (1.9356 amu) from the sum of the masses of the nuclear particles (239.9356 amu), reduced by the binding energy equivalent mass (1.9353 amu), leaves us blissfully happy that the nuclear mass is 238.0003 amu (virtually identical to the atomic mass number). Now if we add those 92 electrons, another 0.050 amu is added, still too little to shake our blissful, sloppy complacency in using the atomic mass number to specify atomic weight.

As stated earlier, one might try imagining the neutron to be the "protyle" of matter or even the "primary material." However, a physicist would respond today that quarks are the primary material. The mass of the neutron is said to comprise these quarks plus their energy. It is a scary thing for a chemist to learn, however, that physicists admit that they do not yet really understand the fundamental nature of mass.

1. J.R. Partington, *A History of Chemistry*, MacMillan and Co. Ltd., London, 1964, Vol. 4, pp. 929–947.
2. *The New Encyclopedia Britannica*, Encyclopedia Britannica, Inc., Chicago, 1986, Vol. 14, pp. 343–348.
3. T.L. Brown, H.E. LeMay, Jr., and B.E. Bursten, *Chemistry The Central Science*, seventh edition, Prentice-Hall, Upper Saddle River, NJ, 1997, pp. 43–46; 771–791.

CRYSTALS CAN DIFFRACT X-RAYS

X-rays were discovered accidentally by William Röntgen in 1895.[1] He had a cathode-ray tube inside a cardboard box and nearby there was, by chance, a sheet of paper coated with phosphorescent material. When the tube was on, the phosphorescent material glowed in the dark. Röntgen found that the same penetrating radiation fogged photographic plates. He called the radiation x-rays and even took images of his own hand using them.[1] Röntgen won the first Nobel Prize in Physics (1901).

Light diffraction was a well-known and well-understood phenomenon by the end of the nineteenth century. It was known that if a transparent film is scored with lines separated by a distance close to the wavelength of light, interference (diffraction) occurs. For example, sodium light (wavelength = 0.0000589 cm or 589 nm) is diffracted by a grating having 7000 lines per centimeter (0.000143 cm spacing).[2] However, x-rays are not diffracted by such gratings despite the fact that they are electromagnetic radiation just like light. In 1912, Max

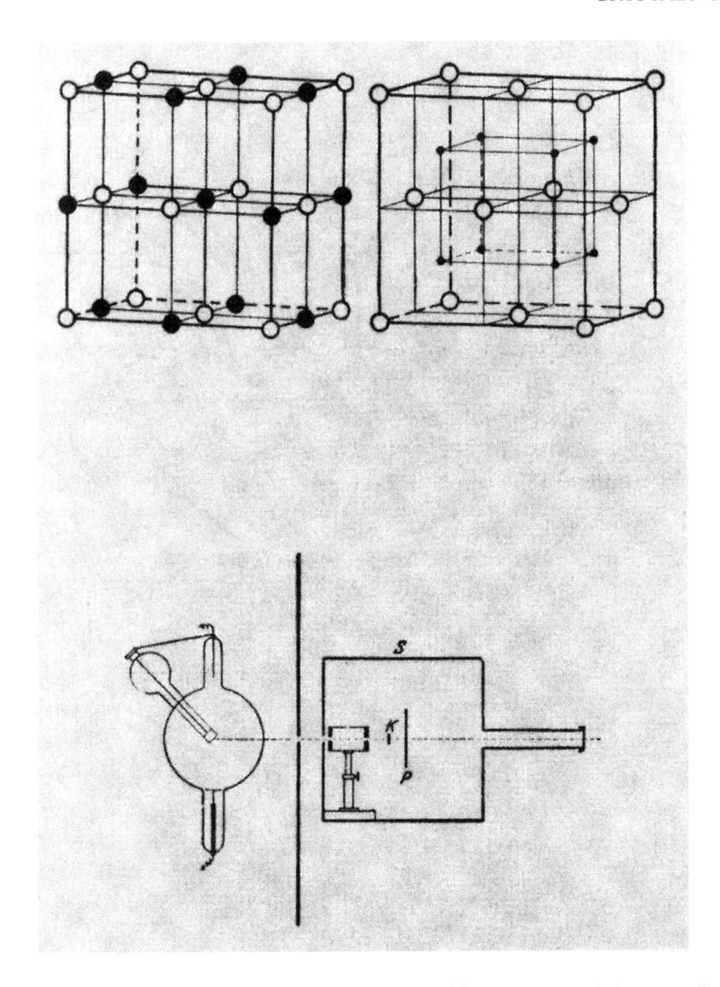

FIGURE 309. ▪ Shortly after x-rays were discovered by Röntgen, Max von Laue postulated that their wavelengths were similar to the separations between atoms in ionic crystals such as rock salt and fluorspar (top). His x-ray unit is pictured at bottom (from Max Born, *The Constitution of Matter* (London, 1923).

von Laue (1879–1950) correctly hypothesized that the wavelengths of x-rays, thought to be about 10^{-8} or 10^{-9} cm (1×10^{-8} cm = 1 angstrom), might be comparable to the distances between atoms (and ions) in crystals. He discovered that these crystalline lattices were capable of diffracting x-rays. In the upper part of Figure 309 we see depictions of the crystalline lattices of sodium chloride (rock salt) and calcium fluoride (fluorspar).[4] The lower half of Figure 309 depicts von Laue's x-ray apparatus: focused x-rays meet crystal C and then impinge on photographic plate P. The diffraction of the x-rays (theoretical construct, top of Figure 310),[4] produces a pattern on the photographic plate (bottom of Figure 310) that provides immediate clues to the crystal's symmetry. Von Laue won the 1914 Nobel Prize in Physics.

1. J.R. Partington, *A History of Chemistry*, MacMillan, London, 1964, Vol. 4, pp. 934–935.
2. W.H. Bragg and W.L. Bragg, *X-Rays and Crystal Structure*, 4th ed., Bell, London, 1924, pp. 1–5.

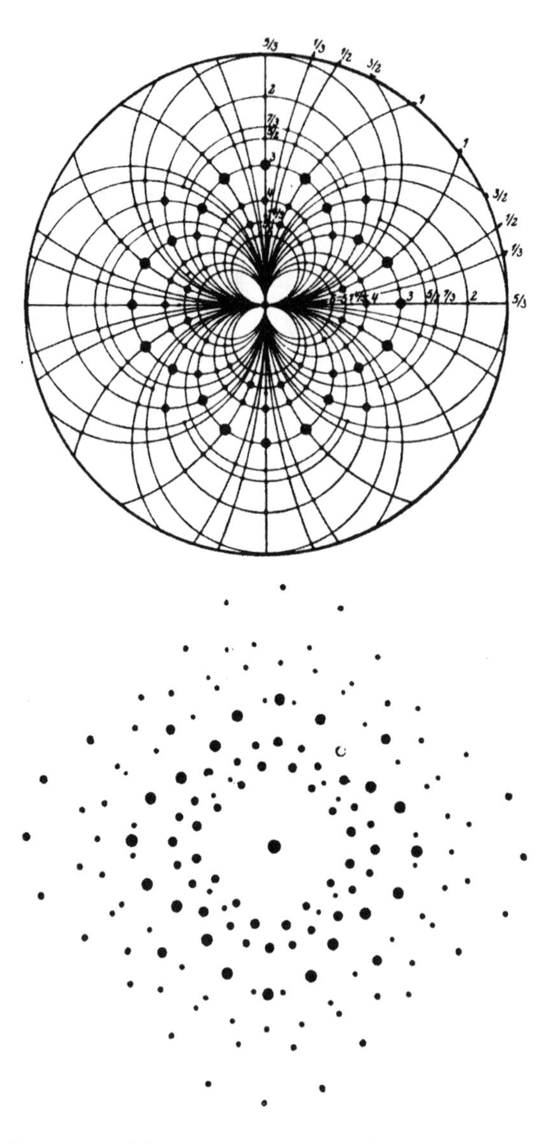

FIGURE 310. ■ Schematics of the x-ray pattern produced by von Laue's diffraction experiment (from Born, see Figure 309).

3. A.J. Ihde, *The Development of Modern Chemistry,* Harper & Row, New York, 1964, pp. 483–486.
4. M. Born, *The Constitution of Matter,* Methuen, London, 1923, pp. 12–19.

TWO NOBEL PRIZES? NOT GOOD ENOUGH FOR THE ACADEMIE DES SCIENCES!

Stimulated by Röntgen's discovery of x-rays, Henri Becquerel (1852–1908) postulated a relationship between x-rays and fluorescence. He placed a variety of fluorescent crystalline samples in contact with photographic plates that were wrapped and well protected from sunlight. Upon exposing the samples to sunlight, he discovered that potassium uranyl sulfate caused fogging of the photographic plates. Seemingly, sunlight stimulated these compounds to release x-rays just as high-energy electrons kicked x-rays from anti-cathodes. However, Becquerel also made a surprising discovery. When the zinc uranyl sulfate–photographic film combination was kept in the dark, the film was also fogged. Becquerel had discovered radioactivity.[1,2]

The term *radioactive* was apparently! introduced by Marie and Pierre Curie in their paper in *Comptes Rendus*[3] in which they reported the discovery of the element polonium (see Figure 311). Marya Sklodowska (1867–1934) came to Paris from Poland in 1891 to study mathematics and physics. Despite considerable privation she completed the equivalent of a Masters degree in physics at the Sorbonne in 1893 (top of her class) and a similar degree in mathematics in 1894. In that year she met Pierre Curie (1859–1906), a professor at the Municipal School of Industrial Physics and Chemistry.[4] She had plans to return to her beloved Poland and teach, and she rejected Pierre's proposals of marriage. She accepted his proposal when he offered to give up his research career and move with her to Poland.[4] Following their marriage in 1895, the couple decided to remain in Paris. Pierre finished his doctorate and his wife, Marie Sklodowska Curie, completed a license in teaching. They were given the opportunity to jointly pursue research at the Municipal School.

While Pierre performed research on piezoelectricity, Marie began her studies in the newly discovered field of radioactivity using her husband's electrometer as a detector. Madame Curie soon discovered that thorium (discovered by Berzeiius in 1829) was radioactive like uranium, a finding made independently by Gerhardt Carl Schmidt. In 1898, she found that the ore pitchblende was much more radioactive than its uranium content (80% U_3O_8) would predict. She suspected the presence of an unknown and intensely radioactive element. At this point, Pierre joined Marie in her studies. Pitchblende was very expensive and the Curies were forced to use the insoluble waste material they received from a pitchblende mine in Bohemia.[4] In order to perform chemical separations on tons of material poor in pitchblende, they worked in an abandoned dissection shed of the Municipal School. Pierre's work centered on studies of radioactivity, while Marie's work concentrated on chemical separation and analysis. She comments that "Sometimes I had to spend a whole day mixing a boiling mass with a heavy iron rod nearly as large as myself. I would be broken with fatigue at the end of the day."[4] In one chemical fraction laboriously derived from the impure pitch-blende, Marie Curie discovered in July, 1898 a new element, polonium, named after her native land (Figure 311 shows the title page of this article).

However, another chemical fraction that contained barium and other alkaline earth salts exhibited intense radioactivity. When Madame Curie had purified this fraction to a point where the specific radioactivity was 60 times that of uranium, a new spectral line was detected in the fraction. As sensitive as the spectroscope (developed[5] by Robert Wilhelm Bunsen and Gustav Robert Kirchoff around 1860) was in its detection of emitted light, the electrometer was even more sensitive to the detection of radioactivity. Further fractionation to a level of 900 times the specific radioactivity was accompanied by a correspon-

(175)

tube à potentiel très élevé, elle se charge négativement, ce qu'il est aisé de vérifier. De même, en touchant du doigt un tube de Crookes loin de la cathode, la paroi touchée devient cathode et il y a répulsion.

» Soit maintenant un tube à cathode plane centrée, de même diamètre que le tube. Les surfaces équipotentielles sont sensiblement planes et le faisceau est cylindrique. Vient-on à réduire le diamètre de la cathode, les surfaces de niveau se courbent et le faisceau est divergent. Si la cathode présente la forme d'un rectangle allongé, les rayons cathodiques doivent s'étaler en éventail dans un plan perpendiculaire à la plus grande dimension du rectangle, et c'est en effet ce qui a lieu.

» Supposons, au contraire, une cathode sphérique concave : à un vide peu avancé, les rayons émis forment un cône creux; menons un plan tangent à ce cône, le rayon contenu dans ce plan est repoussé d'une manière prépondérante par la partie de la cathode située du même côté de ce plan que le centre. De cette dissymétrie résulte une déviation du rayon qui tend à devenir parallèle à l'axe du cône. On peut également dire que les projectiles cathodiques, rencontrant obliquement les surfaces de niveau, se comportent comme des corps pesants lancés obliquement de haut en bas. De là cet allongement bien connu du foyer cathodique, d'autant plus marqué que le vide est plus avancé et le champ, par suite plus intense, près de la cathode. Plaçant au-devant de celle-ci un diaphragme à deux trous, on a deux faisceaux concourants, rectilignes à partir du diaphragme, se coupant cependant au delà du centre de courbure de la cathode; c'est donc surtout au voisinage de celle-ci que se produit l'inflexion des trajectoires, là précisément où le champ a son maximum d'intensité. »

PHYSICO-CHIMIE. — *Sur une substance nouvelle radio-active, contenue dans la pechblende* (1). Note de M. **P. Curie** et de Mme **S. Curie**, présentée par M. Becquerel.

« Certains minéraux contenant de l'uranium et du thorium (pechblende, chalcolite, uranite) sont très actifs au point de vue de l'émission des rayons de Becquerel. Dans un travail antérieur, l'un de nous a montré que

(1) Ce travail a été fait à l'École municipale de Physique et Chimie industrielles. Nous remercions tout particulièrement M. Bémont, chef des travaux de Chimie, pour les conseils et l'aide qu'il a bien voulu nous donner.

C. R., 1898, 2e *Semestre*. (T. CXXVII, N° 3.) 24

FIGURE 311. ■ First page of Pierre and Marie Curie's paper announcing the discovery of polonium in pitchblende and inventing the word *radioactive* (*Comptes Rendus*, 127: 175, 1898).

ding increase in the intensity of the new spectral line. This gave the Curies the assurance to report the new chemical element, radium, in the *Comptes Rendus*, in December, 1898.[1,2,4] It was only in July, 1902 that further separation provided pure radium. Several tons of pitchblende waste had been employed to yield 0.1 g of pure radium chloride.[1] Using the chemical analogy with its alkaline earth contaminant barium, very much in the manner of Mendeleev, the Curies assumed that the chloride was $RaCl_2$ and assigned its atomic weight at 225, thus leaving yawning gaps in the Periodic Table. Marie Curie presented her doctoral thesis in 1902 and it was published in 1903 (*Recherches sur les Substances Radioactives*).[1]

The Curies and Becquerel shared the Nobel Prize for Physics in 1903. The French Academy of Sciences had nominated Pierre Curie and Henri Becquerel for the Prize but Swedish scientist Magnus Costa Mittag-Leffler was able to add Marie to the nomination.[4] Pierre was appointed to the faculty at the University of Paris in 1904 while Marie was promoted to Professor at the women teacher's college in Sevres.[4] Already suffering from the effects of radiation poisoning, Pierre died in a street accident in 1906. Marie was then appointed to the faculty of the University of Paris—the first woman on its faculty in its 650-year history.[4] Incredibly, in 1911 she failed to be elected to the French Academy of Sciences, but later in the year she received the Nobel Prize in Chemistry—the only person to win two Nobels until Linus Pauling did so in 1963. Although she had received only two nominations, one was by the Swedish chemist and 1903 Nobel Laureate Svante Arrhenius, who was an enlightened advocate for women in science.[4]

Marie Curie's story is very dramatic and the discussion of her by the Rayner-Canhams[4] is succinct, sensitive and balanced. During World War I, Marie Curie stopped her research and she and daughter Irene (born in 1897; Eve was born in 1904) served as x-ray technicians with mobile units in the battlefield. Marie began investigations of the medical applications of radiation including cancer therapy at about this time. Irene Joliot-Curie[6] and her husband Frederic Joliot-Curie would eventually share the 1935 Nobel Prize in Chemistry for their discovery of artificial radioactivity. Irene's intense left-wing political activities furnished at least one excuse for rejection of her nomination to the French Academy of Sciences. The Rayner-Canhams note that although the evidence of radiation poisoning and cancers among her co-workers was clear, Marie Curie resisted the obvious conclusions about the health hazards. Daughter Irene died at 59 of leukemia.[6] Marie died of leukemia at age 67.[4] The Rayner-Canhams note the profound influence of Marie Curie in attracting a kind of "critical mass" of intellectually gifted women into nuclear chemistry and physics. One of these, Marguerite Perey discovered element 87 (francium) and became, in 1962, the first woman to be elected to the French Academy of Sciences.[6] She died of cancer at the age of 65.[6] The Rayner-Canhams further note the development of "critical masses" (my term) of women scientists in crystallography[7] as well as biochemistry.[8] The impact of these newly established and formidable "old-girl" networks in chemistry will be an interesting topic for future sociologists of science. For the record we note that the National Academy of Sciences (U.S.) was formed in 1863 and had an initial membership of 50. The first woman was admitted in 1925—Dr. Florence R. Sabin, Professor of Histology, Johns Hopkins University. As of April 27, 1999, there were 2,222 members of whom 132 are women.[9]

1. J.R. Partington, *A History of Chemistry*, MacMillan, London, 1964, Vol. 4, pp. 936–939.
2. A.J. Ihde, *The Development of Modern Chemistry*, Harper & Row, New York, 1964, pp. 487–490.
3. M. Curie and P. Curie, *Comptes Rendus*, **127**:175, 1898.
4. M. Rayner-Canham and G. Rayner-Canham, *Women In Chemistry: Their Changing Roles From Alchemical Times To The Mid-Twentieth Century*, American Chemical Society and Chemical Heritage Foundation, Washington, D.C. and Philadelphia, 1998, pp. 97–107.
5. A.J. Ihde, op. cit., pp. 233–235.
6. M. Rayner-Canham, op. cit., pp. 112–116.
7. M. Rayner-Canham, op. cit., pp. 67–91.
8. M. Rayner-Canham, op. cit., pp. 135–164.
9. Public Information Office, National Academy of Sciences.

IT'S THE ATOMIC *NUMBER*, DMITRI!

The first explicit use of atomic number is attributed to John Newlands, who arranged his 1864 table of elements by "the number of the element" in the order of their "equivalents" using Cannizzaro's system.[1] At the time, Professor George Carey Foster "humorously enquired of Mr. Newlands whether he had ever examined the elements according to the order of their initial letters."[1]

The fact that atoms have their identities locked inside their nuclei was only discovered at the beginning of the twentieth century. The Curies first postulated that radiation emitted from uranium and other radioactive substances was particulace in nature.[2] In 1906, Rutherford and Geiger determined a value for the charge-to-mass ratio of the α particle that was one-half that for the hydrogen ion (H^+). Therefore, the α particle could have been either H_2^+ or He^{2+}.[2] The latter was confirmed in 1911 and was, of course, consistent with the emission of helium gas from radioactive nuclei.[2] In 1909, Geiger and Marsden, working in Rutherford's laboratory found that many a particles pass through 0.01-mm-thick gold leaf with little deflection while only a few suffer major deflections or rebounds. Similar results had been obtained by Rutherford and Geiger a year earlier.[2] These and related studies using Wilson's new cloud chamber led Rutherford to conclude in 1911 that atoms were mostly empty space with a tiny, positively charged nucleus (term he introduced in 1912) at the center.[2,3]

The measurement of deflection angles using the cloud chamber led Geiger and Marsden to the conclusion that the positive charge in the nucleus (in whole-number multiples of the charge on an electron) tended to be about half the atomic weight.[2,3] A. Van den Broek, in 1913, suggested that the nuclear charge, in electron-charge units, is equal to the ordinal number (1, 2, 3, . . .) of the element in the Periodic Table.[2]

It was Henry G.J. Moseley (1887–1915) who, in 1913, used the term *atomic number* and established its significance.[2,3] Moseley made a study of the vibrational frequencies of certain x-rays (the *K* series) emitted from different metallic anticathodes. In Figure 312 (right), we see a decent-looking correlation between the square root of the frequencies of the *K* radiations with the atomic weights of the corresponding elements. However, the correlation with the atomic number (Figure 312, left) was virtually perfect. Clearly, the atomic number was more than a counting device. Ultimately, it explained certain troubling anomalies—

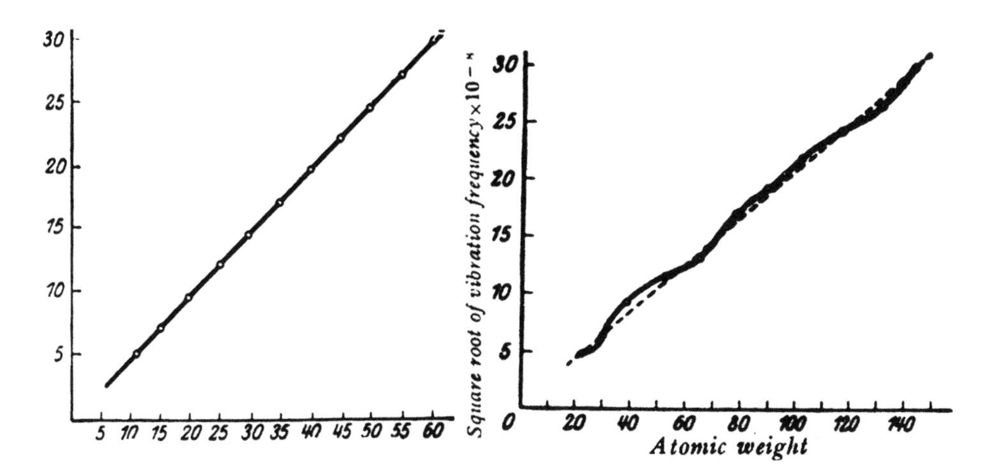

FIGURE 312. ■ The fundamental basis of the Periodic Table is the Atomic Number and not the Atomic Weight. The square root of the frequency of emitted x-rays from different metallic cathodes is imperfectly related to Atomic Mass but directly proportional to Atomic Number. This immediately explained certain anomalies in the Periodic Table. Henry G.J. Moseley, who made this critical discovery, was drafted in World War I and died at Gallipoli at the age of 28 (figure from Born; see Figure 309).

the reversal in placement between tellurium and iodine that worried Mendeleev and the apparent anomaly that the recently discovered argon (which almost equaled calcium in atomic weight) had to be placed before the lighter potassium. It verified the placement of cobalt before nickel on the basis of chemical properties despite the inversion of their atomic weights and confirmed gaps in the Periodic Table for as-yet-undiscovered metals.[3] Moseley was drafted during World War I and was killed at the age of 28 in the battle of Gallipoli.[2,3]

Starting around 1920, it was assumed that the difference between the atomic mass and the atomic number was due to protons *combined in the nucleus* with electrons. Thus, chlorine-35 would have 17 protons in the nucleus, 18 protons combined with 18 nuclear electrons with 17 electrons outside the nucleus.[4] This picture changed when Chadwick discovered the neutron in 1932. But remember, a free neutron decomposes to a proton and an electron (plus an antineutrino).

1. J.R. Partington, *A History of Chemistry*, MacMillan, London, 1964, Vol. 4, pp. 887–888.
2. J.R. Partington, op. cit., pp. 942–953.
3. A.J. Ihde, *The Development of Modern Chemistry*, Harper & Row, New York, 1964, pp. 485–486.
4. J.R. Partington, *Everyday Chemistry*, MacMillan, London, 1929, pp. 245–249.

THE PERIODIC HELIX OF THE ELEMENTS

In 1869, Mendeleev ordered the 63 then-known elements according to increasing atomic mass and placed them in rows having related chemical properties. This original vertical periodic table (Figures 273 and 274) was soon replaced by the horizontal form familiar to us today. Other representations were also feasible, and these included spirals and helices[1]. In 1916 W.D. Harkins and R.E. Hall published a wondrous periodic helix of the elements (Figure 313).[2] A month after Harkins and Hall submitted their paper to the *Journal of the American Chemical Society*, Gilbert N. Lewis submitted his paper "The Atom and the Molecule" to the same journal. His simple electron-dot formulas allowed researchers to merely glance at the periodic table and predict the nature of bonding (single, double, or triple bonds) between atoms of the main-group elements.

Three years earlier, Henry G.J. Moseley (1887–1915) discovered that the square root of the frequencies of X rays emitted from different metallic cathodes was directly proportional to simple cardinal numbers that he termed "atomic numbers." The atomic number—the integer number of positive charges in an atom's nucleus, and not the atomic weight—is the true determinant of an element's identity. It is the atomic number that provides the continuous one-by-one "roll call"[3] of elements that underlies periodicity. The eighth element after lithium (#3) is sodium (#11), and eight elements later comes potassium (#19). All three share very similar chemical properties. If the first 19 elements had been placed in strict order of atomic weights, #19 would have been argon (atomic weight = 39.95) and potassium would have been #18 (atomic weight = 39.10). Chemically, of course, this would have been nonsense. It was never a problem for Mendeleev back in 1869 because, happily for him, the inert gases had not yet been discovered.

Moseley's stepwise "walk" through the periodic table clearly indicated missing members of a larger and more complex family of 85 then-known elements. Descend Harkins' "staircase" (Figure 313) to the very bottom, and you finally arrive at the heaviest element known to Moseley—uranium (atomic # 92). (Uranium had first been isolated as an oxide in 1789; the pure metal was reported in 1841.[4]) This is the source of the "magical number" 92, part of our "Chemical Kabbala"—the number of "naturally occurring" elements fixed in our minds by Moseley. The reality is much more complex. Note in the foreground of Figure 172 a vacancy, corresponding to element #87 just below cesium (Cs), two vacancies below manganese (Mn) for #43 and #75, and one below iodine (I) corresponding to element #85. If we descend into the dark, dingy, and dangerous "basement," we discover, upon passing thallium (Tl), clutter and confusion. Radiation, first discovered in 1898 by Henri Becquerel, was a by-product of naturally occurring transmutations of elements exchanging identities before our very eyes. There are six different lead (Pb) isotopes in Figure 313. Below xenon (Xe) we see mysterious #86, an emission from thorium (Th Em) and also from Marie Curie's radium (Ra Em), the latter briefly named "Nitonium" (Nt).

Let us escape the radioactive basement and ascend the staircase. Just above tantalum (Ta, #73) there is a break and we must "scuttle up" a "rope ladder" of 15 elements. The topmost of these, lanthanum (La, #57), is connected by a strange loop to #58 (cerium, Ce). We have encountered the "rare earth" elements that are today recognized to include lanthanum, the 14 "lanthanides" (#58–71), as well as the lighter elements yttrium (Y, #39) and scandium (Sc, #21). When Harkins and Hall first published their helical representation, it was assumed that elements #57 through #72 were all rare earths.

The marked differences in chemical reactivities between adjacent elements (e.g., sulfur versus chlorine) that guided Mendeleev were largely absent in the 17 rare earths. Their chemistry was so similar (all commonly formed valence 3,

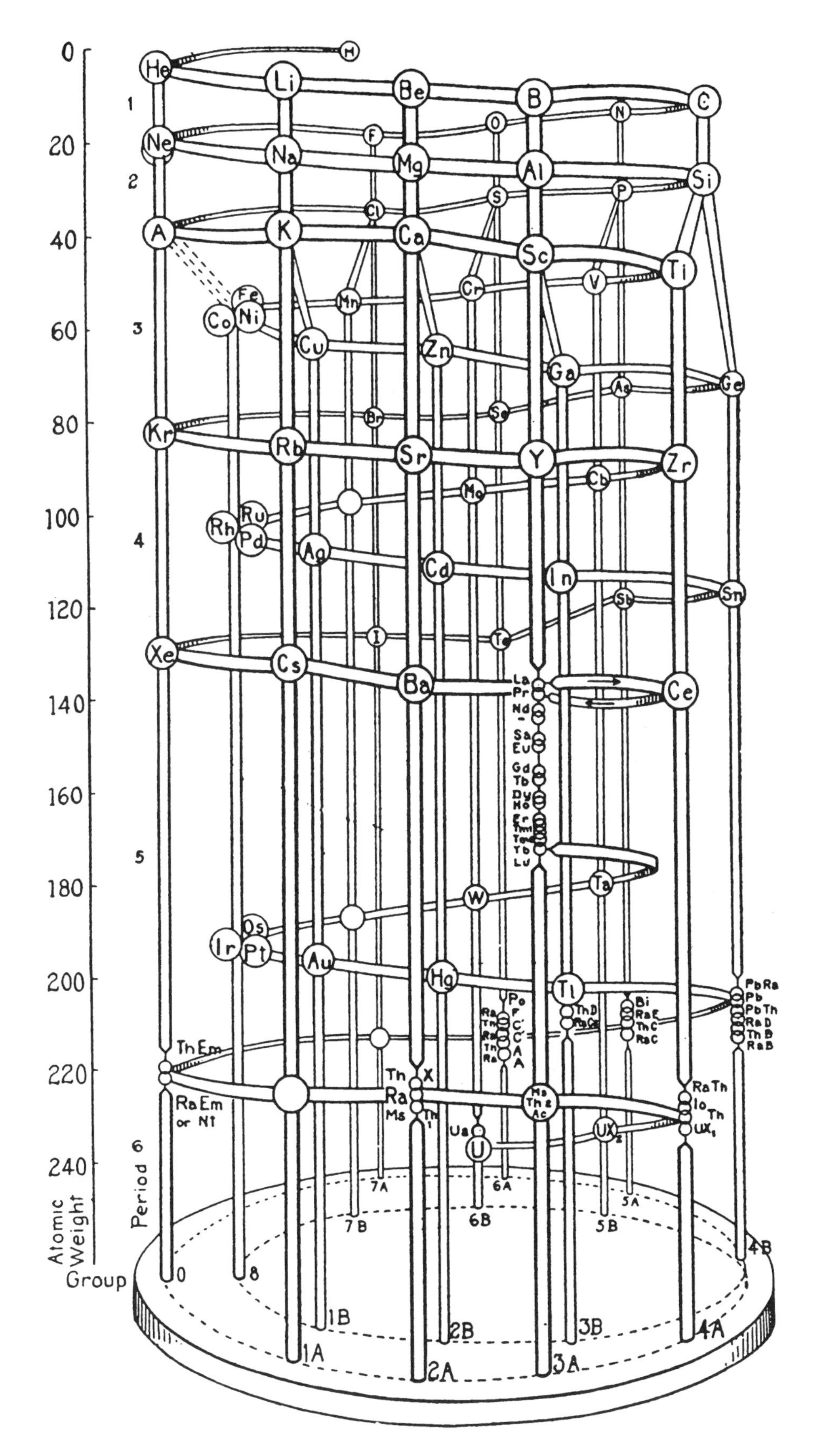

FIGURE 313. ■ The 1916 Harkins–Hall periodic helix of the elements (from *Journal of the American Chemical Society*, 1916, with permission of The American Chemical Society).

MX3-type compounds) that they were exceedingly difficult to separate. This is the source of a 150-year-long saga in the history of chemistry.[5,6] In 1794 John Gadolin obtained an unknown "earth" (a now-extinct term for oxides) from a black ore called *ytterbite* in the Swedish village of Ytterby and discovered the metal yttrium. In 1803, Jons Jacob Berzelius and Wilhelm Hisinger (Sweden) and Martin Klaproth (Germany) isolated and announced almost simultaneously another new element, cerium, from the mineral cerite. Credit for the discovery is given to Klaproth because he was the first to *seemingly* purify it. In fact, improvements in separation techniques, the development of new techniques, and the use of the spectroscope developed by Gustav Kirchhoff and Robert Wilhelm Bunsen guided the isolation of all 17 rare earths (Figure 314) from the original "two." In 1907, Georges Urbain (France), Carl Auer von Welsbach (Austria), and Charles James (United States) almost simultaneously announced separation of the final rare earth. This, too, was not without adventure and controversy, but ultimately Urbain was credited with the discovery of lutetium (Lu, #71).[5,6] Moseley confirmed the placement of these rare earths in the family of elements and noted one missing element (atomic number 61) and predicted its existence. In Figure 313, the mysterious element 61 is a blank space between neodymium (Nd) and samarium ("Sa"). Another missing element, #72, assumed by Urbain and Moseley and others to be a rare earth, was sought in vain from ore samples that had yielded the tight-knit family of 17.

Figure 315 is an updated Harkins' helix published in 1934. The truly distinctive aspect of science is its ability to make predictions and test them—the

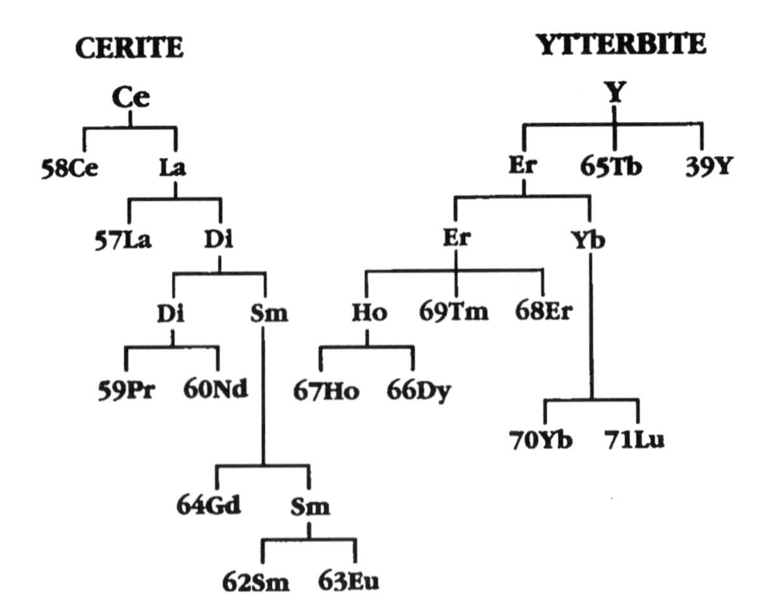

FIGURE 314. ■ Reprinted with permission from "A Natural Historical Chemical Landmark: Separation of Rare Earth Elements, University of New Hampshire, October 29, 1999." Copyright 1999 by the American Chemical Society.

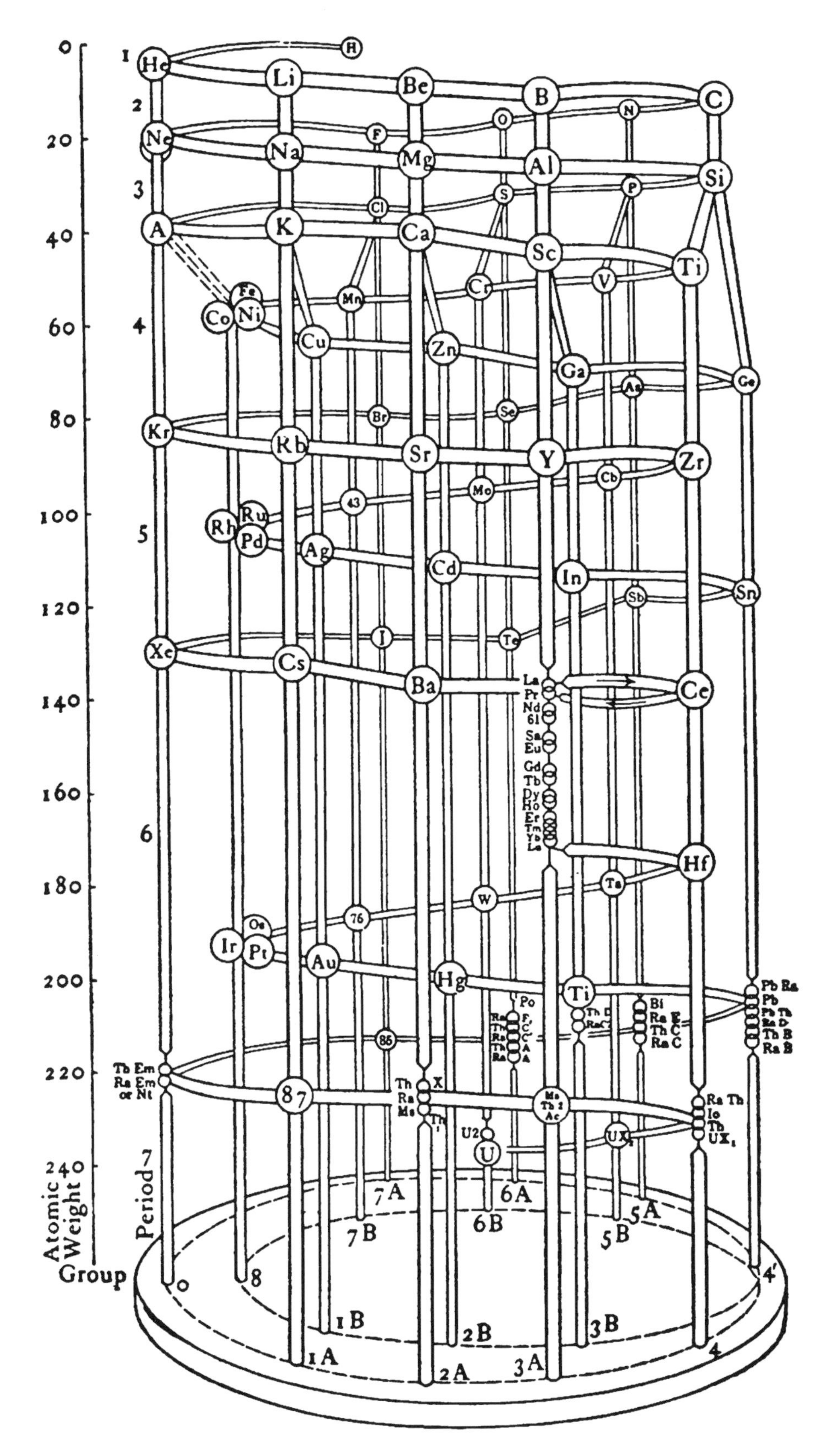

FIGURE 315. ■ Later version of the Harkins–Hall periodic helix including the 1923 discovery of halfnium (Hf, #72—*not* a rare earth as some originally thought) and the isolation of rhenium (Re, #75) in 1925, although it is not explicitly named here. (Used with permission from *Journal of Chemical Education*, 1934.)

more daring, the better. The blank spaces in Mendeleev's original vertical periodic table (Figures 273 and 274) were audacious predictions of new elements such as gallium (*eka*-aluminum) and germanium (*eka*-silicon). All of the new elements predicted by Moseley were found within the next 24 years. The mystery of hafnium (Hf, #72) was solved when Niels Bohr applied his version of quantum numbers and realized that #72 was not similar to #57–#71 but should be quite similar in chemical behavior to zirconium (Zr), which is commonly tetravalent. Once this was realized, the new element was found in tiny quantities in zirconium ores, laboriously separated and reported in 1923.[6] Similar considerations led to the isolation of rhenium (Re) in 1925 (although it is still listed as #75 in Figure 313).[6] The era of nuclear chemistry began in earnest in the late 1930s. Technetium (Tc, #43), francium (Fr, #87), astatine (As, #85), and the final rare earth, promethium (Pm, #61—see Figure 315) were isolated over the course of a decade. Moseley would have been 60 years old had he lived to witness the discovery of promethium in 1947. Sadly, he was drafted during World War I and was shot and killed in the battle of Gallipoli.

Transuranium elements (with atomic numbers greater than 92) were first discovered in 1940. The first two, neptunium (Np, #93) and plutonium (Pu, #94) actually do occur naturally in ultratrace amounts. However, during the 1940s significant quantities were made using nuclear bombardment. A series of increasingly heavy, highly unstable synthetic new elements, reported over the course of the next five decades, provided a very extended periodic table. If we continue to use the spiral staircase analogy, then nuclear chemists have been digging below the radioactive basement into an even stranger subbasement. However, there is no reason to assume that the helix must start lightest at the top and descend with ascending atomic number. As an organic chemist, the arrangement in Figure 315 is just fine—my elements, hydrogen, carbon, nitrogen, and oxygen are closest to heaven. However, one person's heaven is another person's hell. I suspect that nuclear chemists would better appreciate an ascent with increasing atomic number. Their goal has long been an "island of stability" (perhaps a "cloud of stability"?) of superheavy elements.[7] An organic chemist such as I might view their goal to be "Atlantis."[7] In 1999, it appeared that the promised land (or cloud) might have been reached with the nuclear synthesis of element #118.[7] However, the claim was subsequently withdrawn, and the goal remains elusive.[7] However, element #114 discovered by Yuri Oganessian and his co-workers in Dubna has an isotope with a half-life of 27 seconds,[7] amazingly long for one of the "super-heavies."

1. G.N. Quam and M.B. Quam, *Journal of Chemical Education*, Vol. 11, pp. 27–32, 217–223, 288–297 (1934).
2. W.D. Harkins and R.E. Hall, *Journal of the American Chemical Society*, Vol. 38, pp. 169–221 (1916).
3. J.R. Partington, *A History of Chemistry*, MacMillan and Co. Ltd., London, 1964, pp. 950–951.
4. A.J. Ihde, *The Development of Modern Chemistry*, Harper & Row, New York, 1964, pp. 747–749.
5. W.H. Brock, *The Norton History of Chemistry*, W.W. Norton & Co., New York, 1993, pp. 327–330.

6. American Chemical Society, *A National Historic Chemical Landmark—Separation of Rare Earth Elements, University of New Hampshire, Durham, New Hampshire, October 29, 1999*, Division of the History of Chemistry, American Chemical Society, Washington, DC, 1999.
7. M. Schädel, *Angewandte Chemie Internation Edition*, Vol. 45, 368–401 (2006).

X-RAYS MEASURE THE DISTANCES BETWEEN ATOMS OR IONS

While Max von Laue used crystals to perform an experiment with x-rays, William H. Bragg (1862–1942) and his son William Lawrence Bragg (1890–1971) used x-rays to determine the structures of crystals. In 1912 and 1913 the Braggs developed and applied the diffraction equation that bears their name:

$$nL = 2d \sin \theta$$

where $n = 1, 2, 3, \ldots$ L is the wavelength of the x-rays; d is the distance between layers of atoms (ions), and θ is the angle of incidence to the surface.

Figure 316(a) from the Braggs' text[1] demonstrates the reinforcement of x-ray waves that obey Braggs' Law. In Figure 316(b),[1] we see a schematic of their x-ray apparatus in which single crystals (or powders) were placed on a rotating table so that reflections could be collected from all angles.

Not only did the Braggs' x-ray diffraction apparatus allow these critical measurements of distances, they helped confirm the reality of ions since, as determined by J.J. Thomson, the intensity of scattering was proportional to the number of electrons (furnishing, in effect, an additional confirmation of atomic numbers).[2,3]

X-ray crystallography soon became the most important "optic" for structural chemistry in the solid state. It laid the basis for Linus Pauling's crystallographic studies that led to the synthesis of the principles expounded in his 1939 text, *The Nature of the Chemical Bond.* These principles were applied by him to solve the α-helical structure of proteins through simple use of homegrown molecular models. J.D. Watson and Francis Crick used Pauling's model-building approach, combined with x-ray data to beat him at his own game and arrive at the structure of DNA. When I was a graduate student in the late 1960s, the complete solution of a crystalline structure by x-ray data was a relatively rare event. It was then used primarily for structural chemistry studies in which researchers desired accurate bond lengths, bond angles, and other related data. Improved instrumentation and especially the incredibly increased power of computers have now made x-ray crystallography a fairly routine tool for structure confirmation of fairly large molecules that form good crystals. Large globular molecules are still, however, immense challenges.

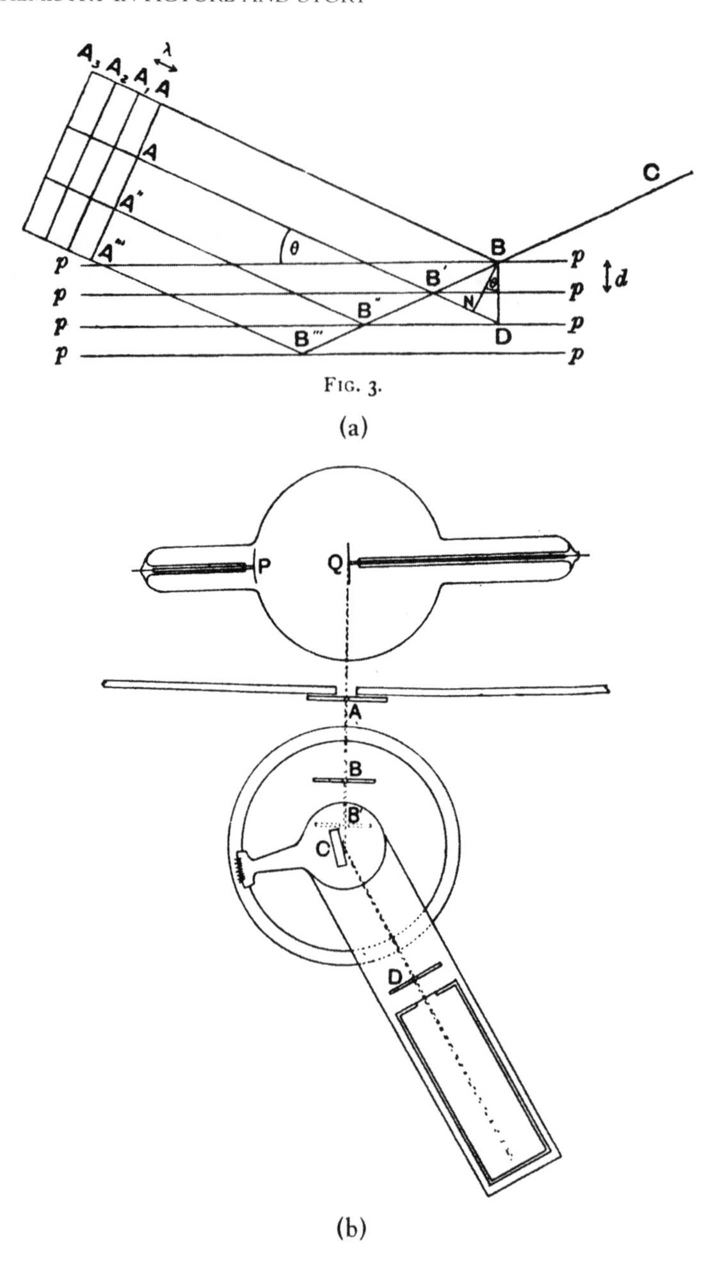

FIGURE 316. ■ William H. Bragg and his son William L. Bragg reversed von Laue's experiment and used x-rays to measure the distances between ions or atoms in crystals (see text), (a) Depicts the conditions for constructive interference of x-rays termed *Braggs' Law;* (b) schematic of Braggs's x-ray apparatus (W.H. Bragg and W.L. Bragg, *X-Rays And Crystal Structure*, 4th ed., London, 1924).

1. W.H. Bragg and W.L. Bragg, *X-Rays And Crystal Structure*, 4th ed., Bell, London, 1924.
2. J.R. Partington, *A History of Chemistry*, MacMillan, London, 1964, Vol. 4, pp. 934–936.
3. W.H. Brock, *The Norton History of Chemistry*, Norton, New York, 1993, pp. 393; 492–494.

WHERE DID WE DIG UP THE MOLE?

The modern concept of valence explains the early concept of chemical "equivalents" first introduced by Cavendish around 1767.[1] Thus, 36.5 g of acid of salt (HC1) gas neutralizes 56.0 g of potash (KOH); 49.0 g of acid of vitriol (H^) or 32.7 g of acid of phosphorus (H_2SO_4) equivalently neutralizes the same 56.0 g of potash. The ratio of equivalent weights for $HCl/H_2SO_4/H_5PO_4$ is always 1.00/1.34/0.90. Similarly, 53.0 g of soda (Na_2CO_3) neutralizes 36.5 g of HC1 gas: the ratio of equivalent weights for KOH/Na_2CO_3 is always 1.06. The same 36.5 g of HCl neutralizes 29.1 g of "milk of magnesia" [$Mg(OH)_2$; note that $Ca(OH)_2$ was once called "milk of lime"] to produce 47.6 g of $MgCl_2$ and 18.0 g of water. If we place platinum electrodes into 47.6 g of molten $MgCl_2$, 12.1 g of magnesium will electroplate, and 35.5 g of chlorine gas [12.2 liters at standard temperature and pressure (25°C and 1.00 atm)] will be released. Thus, the ratio of equivalent weights of Cl_2/Mg is always 2.93.

Equivalent masses (and the related concept of "normality") have gradually disappeared from modern chemistry texts in favor of a definition based directly on numbers of "particles" (atoms, molecules, ions, electrons), and this is truly ironic.

The term *mole* was first introduced by Wilhelm Ostwald in 1901.[1] It is derived from the Latin for "mass, hump, or pile"[1] (the term *molecule*, introduced by Pierre Gassendi[2] in the early seventeenth century has the same root; presumably it means a mass of atoms). Specifically, Ostwald used the term to represent the formula weight of a substance in grams: 36.5 g of HCl is one mole. The formal definition of the mole adopted by the Fourteenth Conference Generale des Poids et Mesures in 1971 is: "the amount of a substance of a system that contains as many elementary entities as there are atoms in 0.012 kilograms of carbon-12."[1] The rich irony is that Ostwald fiercely resisted the atomic concept at the time Boltzmann committed suicide in 1906 but his mole is now defined explicitly in terms of atoms.

The number of atoms in 0.012 kg of carbon-12 is Avogadro's Number ($6.02213670 \times 10^{23}$). Perrin's 1908 experiments on dust particles and particles of gamboge and mastic in water gave a value of about 6×10^{23}. Once Millikan had determined the charge of an electron (modern physical value, $q = 1.6021773 \times 10^{-19}$ coulombs or C) and this was combined with the modern value for the faraday (1 F = 96,485.31 C, the total charge in one mole of electrons), another completely independent determination was available for Avogadro's Number. Here's another: 1 g of radium yields 11.6×10^{17} α particles in one year and these produce 0.043 liters of helium gas at standard temperature and pressure (STP).[1] Indeed, the Rayleigh scattering that causes our sky to be blue allows calculation of Avogadro's Number. The current accepted value is based upon density, atomic mass, and x-ray diffraction measurements of pure crystalline silicon. Its uncertainty[1] is only 3.5×10^{17}! Let's remember Avogadro's Number as "six-point-oh-two-and-twenty-three-oh-oh-oh's" (like "Pennsylvania-6-5-oh-oh-oh" for aging Glen Miller buffs including lapsed hippies whose memories of Glen Miller are only prenatal). Although I suggested this in 2000, six years later nobody uses it!

1. J.J. Lagowski (ed.), *MacMillan Encyclopedia of Chemistry*, Simon & Schuster, MacMillan, New York, 1997, Vol. 1, pp. 198–199; Vol. 3, pp. 951–955.
2. J.R. Partington, *A History of Chemistry*, MacMillan, London, 1961, Vol. 2, p. 462.

XENON IS SLIGHTLY IGNOBLE AND KRYPTON IS NOT INVINCIBLE

The inertness of the noble gases as well as Richard Abegg's law of valence and counter valence[1] were important leads to understanding of valence and bonding. In 1916 Walther Kossel proposed that atoms used their valence-shell electrons to form bonds and that they tried to attain the electronic structure of the rare gas preceding them (electropositive elements) or immediately following them (electronegative elements). Kossel's theory is depicted in Figure 317.[2] Gilbert N. Lewis formulated the octet rule in his article "The Atom And The Molecule."[3] In Figure 318(a) we see his representations of the valence electrons of the first complete row of elements in families IA to VIIA.[3] The noble gas neon occurs at the end of this period and has each corner of the cube "occupied" by an electron. Thus, inertness corresponds to a completed octet and this "filling" of the valence

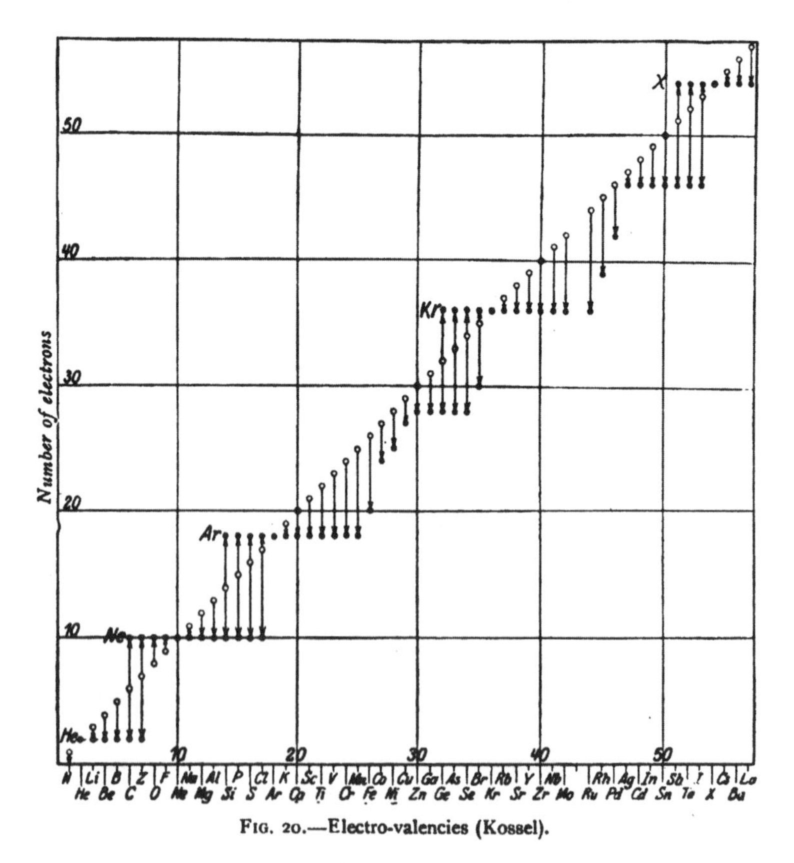

FIG. 20.—Electro-valencies (Kossel).

FIGURE 317. ■ Walther Kossel's theory in which atoms adopt the valence shell of the nearest inert gas by loss or gain of electrons (from Born, see Figure 309).

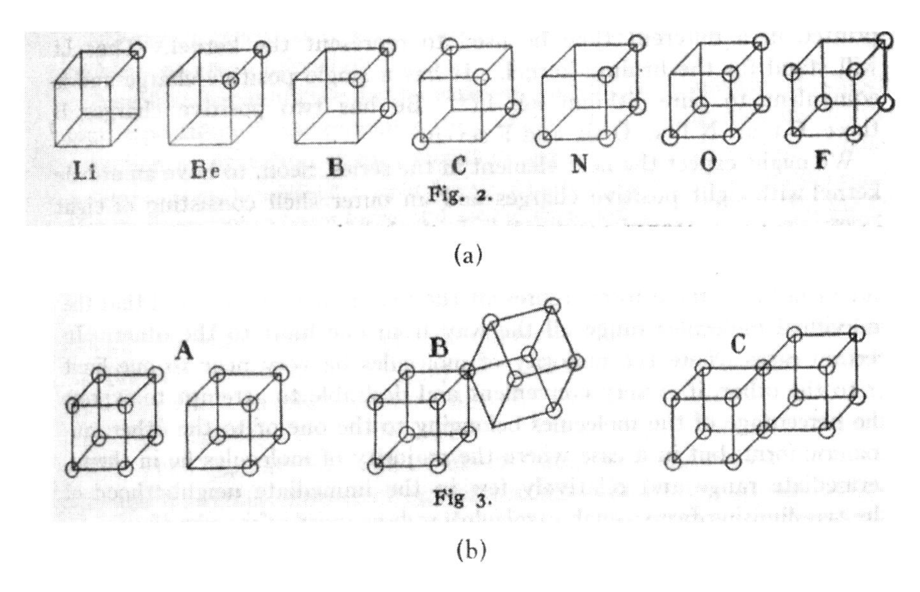

FIGURE 318. ▪ The original Gilbert N. Lewis dot structures (*Journal of the American Chemical Society*, **38**:762, 1916).

shell explains an atom's valence. The unreactivity of noble gases had become an article of faith, verified forcefully when Henri Moissan turned his newly found "Tasmanian devil" fluorine loose on a sample of argon sent by Ramsay in 1894. The result: Nothing![4]

However, in 1962 Neil Bartlett discovered that molecular oxygen (O_2) is oxidized by (loses an electron to) hexafluoroplatinum (PtF_6) to give the new compound $O_2^+PtF_6^-$.[5] He realized that the oxygen molecule holds its electron about as tightly as a xenon atom does. Therefore, Xe just might also lose an electron to PtF_6. The resulting red crystalline solid was originally thought to be $Xe^+PtF_6^-$,[4] but it is now thought to be $[XeF^+][Pt_2F_{11}]$.[6]

In any case, the conceptual threshold had been crossed and a family of fluorine-containing xenon compounds is now known. For example, a 1:5 mixture of Xe and F_2, heated in a nickel vessel, produces XeF_4 (melting point 117°C).[6] XeF_6, formed by Xe and F_2 at high temperature and pressure, attacks quartz and reacts violently with water to produce XeO_3, itself a high explosive.[6] Obviously, this is not "The Friendly World of Chemistry Neighborhood."

Krypton reacts with F_2 under an electric discharge at –183°C to form a solid (KrF_2) that decomposes slowly at room temperature.[6] There are also salts of KrF^+, such as $KrF^+SbF_6^-$.[6] Although radon loses electrons much more easily than xenon, its most stable isotope has a half-life of 3.8 days, and not much chemistry is done although compounds thought to be RnF_2, $RnF^+TaF_6^-$, and possibly RnO_3 are known.[6] As to the rumors concerning a reputed green ore of krypton—doubtful. In 2000, researchers at the University of Helsinki synthesized argon fluorohydride (HArF). The molecule was formed in a solid matrix of HF in argon and decomposed above 27 K (–411°F). Still, HArF is a real molecule with a vibrational spectrum.[7]

Figure 318(b) shows "steps" in the sharing of an edge (two electrons) to form a single bond between two cubic iodine atoms. If two atoms share a face

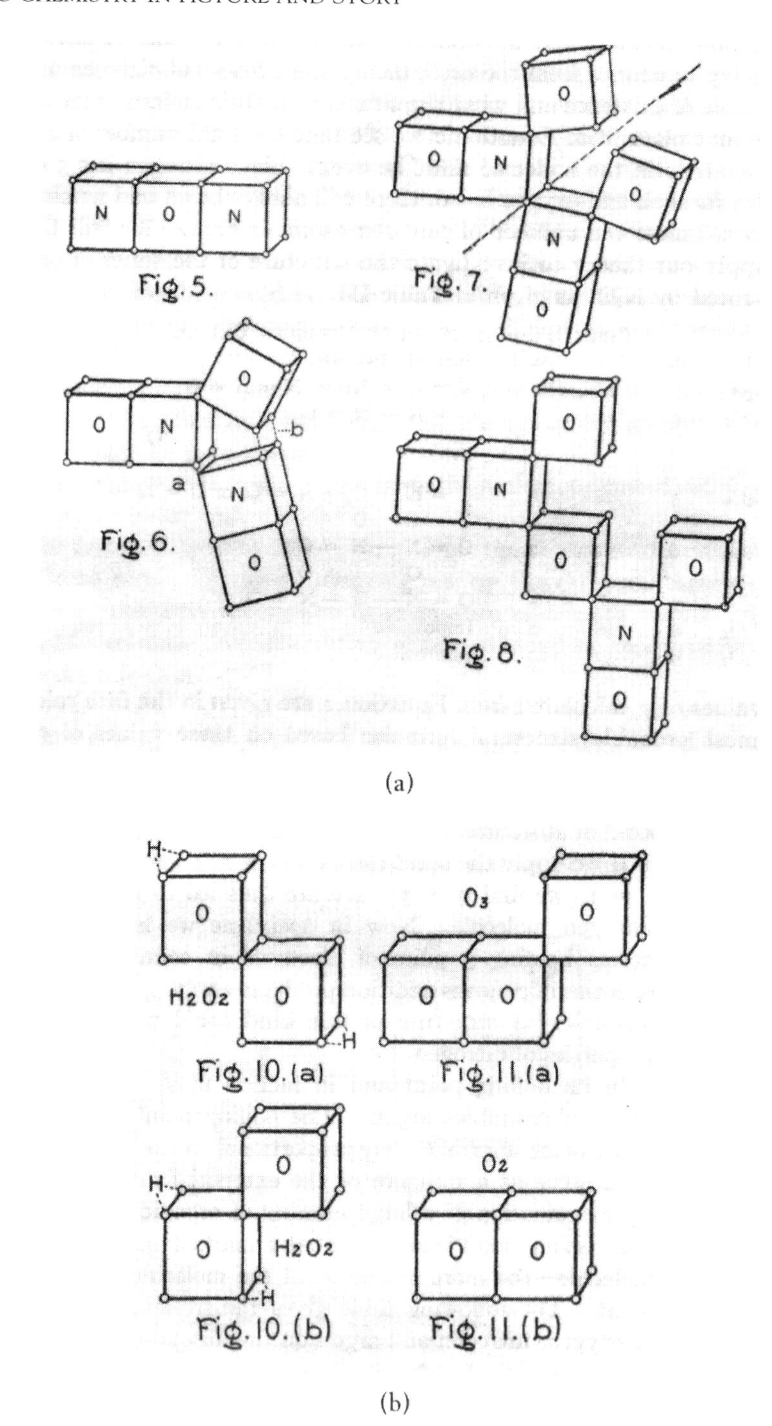

FIGURE 319. ■ Irving Langmuir's modification of the Lewis structures [*Journal of the American Chemical Society*, **41**, No. 6 and No. 10 (1919) and **42** (No. 2) (1920)].

formed by fusing the two cubes, they share four electrons and form a double bond. Tetrahedral bonding in methane (CH_4) was explained by sharing of two opposite edges on the top of cubic carbon with two cubic hydrogens and, similarly, sharing the two alternate edges with hydrogens on the bottom of the cube. While van't Hoff explained triple bonds through sharing faces of two tetrahedra, Lewis's picture is a bit more strained. In 1919 Irving Langmuir slightly modified Lewis's picture and extended coverage to the transition metals. Figure 319 depicts oxides of nitrogen, the two allotropes of oxygen as well as possible isomers of hydrogen peroxide (never found), and oxides of phosphorus. It is interesting to note that Langmuir handles hydrogen's completed valence shell "duet" by having it bridge along the edge of a cube.

In the 1960s *quadruple* bonds between metallic atoms were discovered in species such as $Re_2C1_8^{2-}$.[8] There would be no obvious way to explain quadruple bonds with cubic atoms or tetrahedra. In 2005, evidence was presented for a quintuple bond between the two chromium atoms in a crowded dichromium complex (RCrCrR).[9]

Lewis's paper proposed the bookkeeping dot structures (e.g., H:H) that bear his name. How does one convey to an introductory chemistry student just how ridiculously simple and powerful Lewis structures are for prediction?

---

1. Lewis (Ref. 3) states Abegg's Law as: "the total difference between the maximum negative and positive values or polar numbers of an element is frequently eight and is in no case more than eight."
2. M. Born, *The Constitution of Matter,* Methuen, London, 1923, pp. 21–23.
3. G.N. Lewis, *Journal of the American Chemical Society,* **38:**762–785, 1916.
4. W.H. Brock, *The Norton History of Chemistry,* Norton, New York, 1993, p. 337.
5. See N. Bartlett, *American Scientist,* **51:**114, 1963.
6. A. Cotton, G. Wilkinson, C.A. Murillo and M. Bochmann, *Advanced Inorganic Chemistry,* 6th ed., Wiley. New York, 1999, pp. 588–597.
7. K.O. Christie, *Angewandte Chemie International Edition,* **40:**1419, 2001.
8. F.A. Cotton, *Accounts of Chemical Research,* **2:**242, 1969.
9. S. Ritter, *Chemical & Engineering News,* **83,** No. 39: 9, 2005.

THE ATOM AS A SOLAR SYSTEM

The line spectra obtained by heating elements and refracting the light through a prism was employed by Bunsen to identify salts. His gas-powered burner was first used to obtain colorless flame to study light emissions of these salts—not for heating flasks.[1] The spectroscope designed by Bunsen and Kirchoff immediately led to the discovery of cesium in 1860 and rubidium in 1861.[1] In 1868, the emission spectrum of another new element, helium, was discovered in the spectrum of the solar chromosphere.[2] But what was the origin of line spectra— light having very precise frequencies (or wavelengths) unique to each element? What was the origin of the photoelectric effect: A small quantity of high-energy (high-frequency) light waves could kick an electron off of a metal surface, but a huge quantity of light of a lower frequency could not? Apparently the *quality* of the energy, not its quantity was the issue. These phenomena were addressed by Max Planck (1858–1947), who developed quantum theory around the year 1900.[3] The simple equation he advanced, $E = h\nu$, indicated that the frequency of light emitted by an excited atom, for example, was directly proportional to the energy decrease of the emitting atom (h is Planck's constant).

Shortly after the Rutherford model of the atom was established, it was obvious to wonder where the electrons were. In 1913, Niels Bohr[3] (1885–1962) used Planck's quantum theory, combined with the line spectra (visible, ultraviolet, infrared) of hydrogen, to postulate the circular planetary model of the atom. If negative electrons were orbiting the positively charged nucleus, classical physics required them to spiral into the nucleus. Bohr postulated that electrons could only have certain discrete energies (occupy only certain circular orbits) and never "in-between" values. These orbits corresponded to quantum numbers $n = 1, 2, 3, \ldots$. The model was revolutionary and even subversive. Where were the electrons when they moved between orbits? They could never be found in the "in-between." The model beautifully explained the spectrum of the hydrogen atom and the helium ion (He^+), and failed for all other atoms. Arnold Sommerfeld modified Bohr's orbits to allow both circular and elliptical orbits.[3] He explained the fine structure of the spectrum of hydrogen by adding a second quantum number for angular momentum. Now there was occasional reference to *orbitals*—really *suborbits*. Sommerfeld's theory enjoyed success in explaining H and He^+ and other atomic spectra. Figure 320, from a 1923 book by Born,[4] depicts the Bohr model for H, He, and He^+ and its extension to He_2. Figures 321 and 322, from Smith's 1924 book,[5] depict the "Rococo era" of the "old" quantum theory a few "hours before the dawn" of quantum mechanics in 1926. The pretty image of an electron spiraling on the surface of a 4*s* orbital would be seen to violate Heisen-

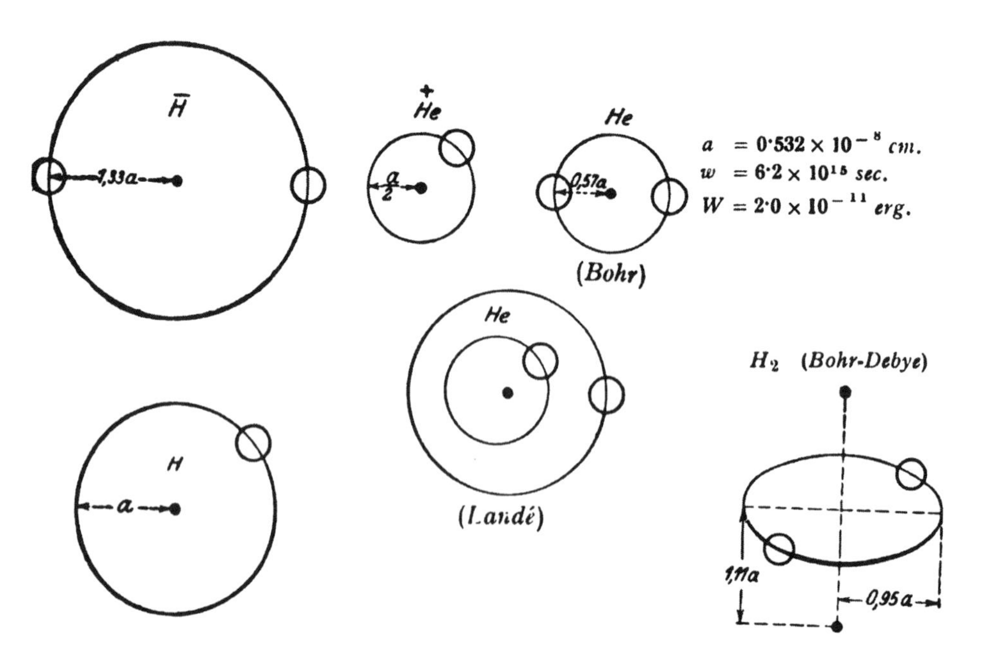

FIGURE 320. ■ Variations of the Bohr model of the atom that was really much more "subversive" than it looks. If an electron is forbidden to exist between orbits, how does it pass from one orbit to the next? (from Born; see Figure 309).

The Dynamic Atom 171

DIAGRAM XV

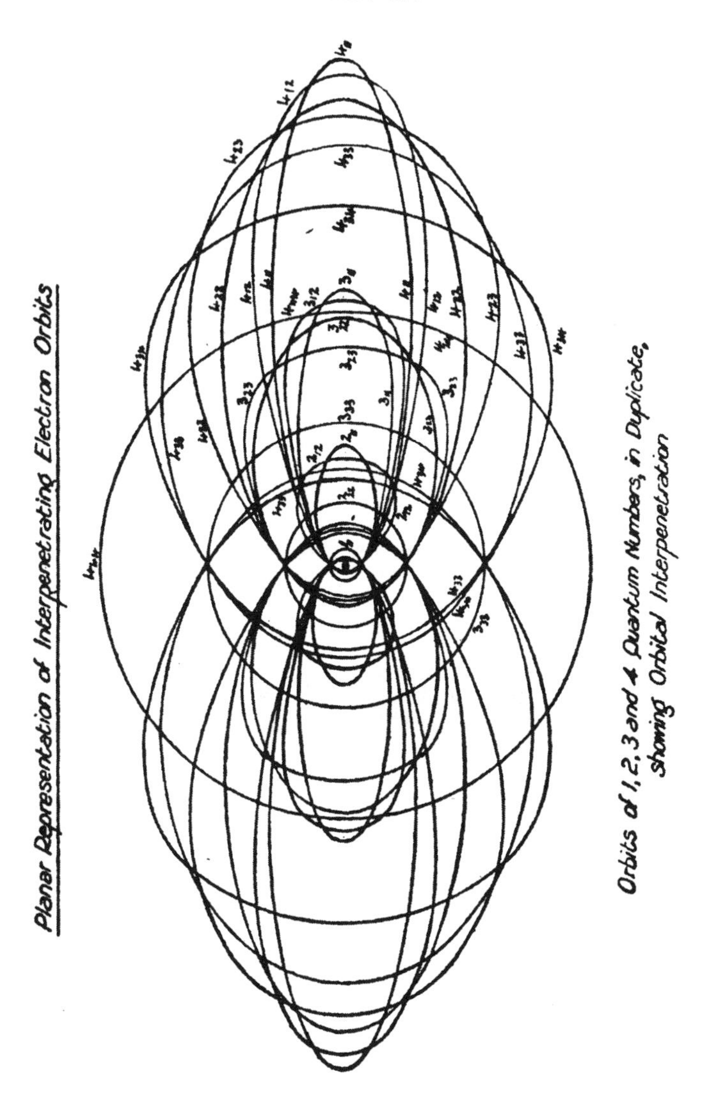

FIGURE 321. ■ The "Rococo" era of the "old" quantum theory comes to an end (J.D. Main Smith, *Chemistry and Atomic Structure*, New York, 1924).

berg's Uncertainty Principle. The picture of atoms that emerged in 1926 would almost seem to be more suited to abstract art than "hard" science.

---

1. A.J. Ihde, *The Development of Modern Chemistry*, Harper & Row, New York, 1964, pp. 231–235.
2. A.J. Ihde, op. cit., p. 373.
3. A.J. Ihde, op. cit., pp. 499–507.
4. M. Bom, *The Constitution of Matter*, Methuen, London, 1923, pp. 24–32.
5. J.D. Main Smith, *Chemistry and Atomic Structure*, D. Van Nostrand, New York, 1924, pp. 160–176.

DIAGRAM XVI

Spatial Representation of Electron Orbital Domain

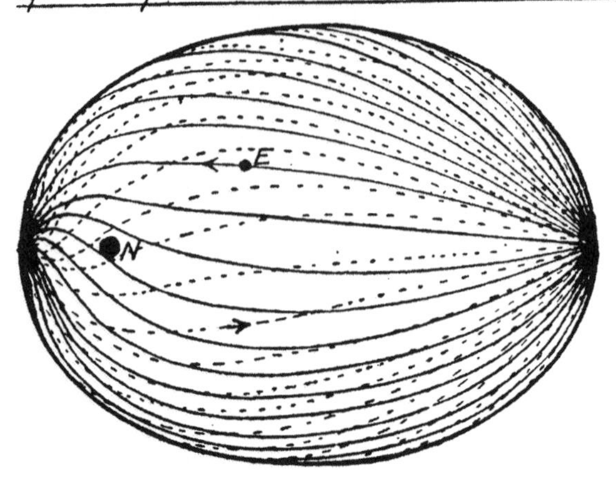

Precession Paths of a 4s Electron moving on the Surface of an Imaginary Solid Ellipsoid of Revolution with Nucleus at the Focus

FIGURE 322. ■ An electron spiraling down a 4*s* orbital before the Heisenberg Uncertainty Principle smudged the picture (from Main Smith, see Figure 321).

'TIS A GIFT TO BE SIMPLE

Simple "counting rules" are useful, powerful, and hint at underlying structure. When Gregor Mendel reported the laws of heredity in 1865, the "two-ness" of his results were incredibly simple and powerful, yet made little initial impact. The source was the as-yet-unknown genes and the ultimate origin: the double helix of DNA. What emerged from quantum mechanics in 1926 were four quantum numbers (n, l, m_l and m_s). Allowed values for these quantum numbers specified the energy, orbital ("domain"), and spin for every electron in an atom. The periodicities of the properties of the elements were manifestations of the quantum numbers. The transition metals corresponded to filling 3*d*, 4*d*, and 5*d* orbitals; the lanthanides to filling 4*f* orbitals; the actinides to filling 5*f* orbitals. The octet rule, which explained why H_2 and F_2 have single bonds, why N_2 has a triple bond, and why sodium chloride has Na^+ and Cl^- ions, while magnesium oxide has Mg^{2+} and O^{2-} ions, is consistent with quantum mechanics. There are lots of other "counting rules" in chemistry with quantum mechanics at the core. Nyholm and Gillespie's valence-shell electron pair repulsion (VSEPR) theory is incredibly good at predicting molecular geometries (CO_2 is linear, H_2O is bent). All one does is count electron pairs, obtained from straightforward Lewis octets or "expanded octets" (e.g., PF_5, SF_6) around the central atom. The stability of benzene is understood by Hückel's $4n + 2$ rule. The Woodward-Hoffmann rules follow similar $4n + 2$ and $4n$ alternation with the ability to predict thermal chemistry and photochemistry as further alternatives. I do not mean to imply that quantum mechanics (or chemistry) is easy. But the occurrence and power of simple counting rules in chemistry continues to amaze and delight me.

TRANSMUTING QUANTUM MECHANICS INTO CHEMISTRY

The contributions of Linus Pauling (1901–1994) to twentieth-century chemistry are arguably as fundamental as those of Lavoisier to the late eighteenth and early nineteenth centuries. Pauling received his Ph.D. under the guidance of Roscoe Dickinson, California Institute of Technology's first Ph.D. in chemistry, who literally took him by the hand and taught him crystallography.[1] Arthur Amos Noyes had been recruited from MIT to direct the Gates Chemical Laboratory just three years before Pauling's arrival as a student at the newly energized Caltech. Following the completion of Pauling's Ph.D., Noyes wished to retain the brilliant young man on his faculty and fretted over his growing friendship with Gilbert N. Lewis, the Department Chair at Berkeley.[2] The solution was to send him to Europe (immediately if not sooner). He arranged a dinner for the 24-year-old Pauling with the Director of the Guggenheim Foundation (a friend of Noyes). A fellowship for work at the great quantum theory centers of Europe followed. Still, there remained a waiting period and Pauling's planned visit to Berkeley. Noyes encouraged an early departure: "If the Paulings left early," he proposed, "then they would have time for stopovers in Madeira, Algiers, and Gibralter before docking in Naples, then a few weeks for touring Italy." Italy! Noyes spoke glowingly of the glories of Rome, the fabulous ruins at Paestum. "I'll give you enough money to pay the fare in Europe," he said, "and support you from the end of March until the beginning of the Guggenheim Fellowship."[2] Noyes retained his "franchise player."

Arriving in Munich during the Spring of 1926, Pauling immediately contacted Arnold Sommerfeld. He would later spend time with Niels Bohr in Copenhagen. However, the "old" quantum theory underlying the Bohr–Sommerfeld atom was just starting to crumble in late 1925 and Pauling bore witness to the work of physicists Louis De Broglie, Erwin Schrodinger, Wolfgang Pauli, Paul Dirac, Max Born, Walther Heitler, and Fritz London. At one point, Pauling excitedly presented his ideas on the power of the Bohr–Sommerfeld model to Pauli. "Not interesting" was the terse response.[3] But Pauling learned the new quantum mechanics and the application of the Schrodinger equation and made them accessible to chemists.

The Bible of mid-twentieth century chemistry was Pauling's *The Nature of the Chemical Bond* (Ithaca, 1939; 2nd ed., 1940; 3rd ed., 1960). In *The Double Helix*, J.D. Watson writes: "The book I poked open the most was Francis' copy of *The Nature of the Chemical Bond*. Increasingly often, when Francis needed to look up a crucial bond length, it would turn up on the quarter bench of lab space that John had given to me for experimental work. Somewhere in Pauling's masterpiece I hoped the real secret would lie. . . ."[4] Pauling's book was based on a series of articles, "The Nature of the Chemical Bond," that were published starting in 1931. Figure 323 shows the title page of the first article in

April, 1931 THE NATURE OF THE CHEMICAL BOND 1367

[Contribution from Gates Chemical Laboratory, California Institute Technology, No. 280]

THE NATURE OF THE CHEMICAL BOND. APPLICATION OF RESULTS OBTAINED FROM THE QUANTUM MECHANICS AND FROM A THEORY OF PARAMAGNETIC SUSCEPTIBILITY TO THE STRUCTURE OF MOLECULES

By Linus Pauling

Received February 17, 1931 Published April 6, 1931

During the last four years the problem of the nature of the chemical bond has been attacked by theoretical physicists, especially Heitler and London, by the application of the quantum mechanics. This work has led to an approximate theoretical calculation of the energy of formation and of other properties of very simple molecules, such as H_2, and has also provided a formal justification of the rules set up in 1916 by G. N. Lewis for his electron-pair bond. In the following paper it will be shown that many more results of chemical significance can be obtained from the quantum mechanical equations, permitting the formulation of an extensive and powerful set of rules for the electron-pair bond supplementing those of Lewis. These rules provide information regarding the relative strengths of bonds formed by different atoms, the angles between bonds, free rotation or lack of free rotation about bond axes, the relation between the quantum numbers of bonding electrons and the number and spatial arrangement of the bonds, etc. A complete theory of the magnetic moments of molecules and complex ions is also developed, and it is shown that for many compounds involving elements of the transition groups this theory together with the rules for electron-pair bonds leads to a unique assignment of electron structures as well as a definite determination of the type of bonds involved.[1]

I. The Electron-Pair Bond

The Interaction of Simple Atoms.—The discussion of the wave equation for the hydrogen molecule by Heitler and London,[2] Sugiura,[3] and Wang[4] showed that two normal hydrogen atoms can interact in either of two ways, one of which gives rise to repulsion with no molecule formation, the other

[1] A preliminary announcement of some of these results was made three years ago [Linus Pauling, *Proc. Nat. Acad. Sci.*, 14, 359 (1928)]. Two of the results (90° bond angles for *p* eigenfunctions, and the existence, but not the stability, of tetrahedral eigenfunctions) have been independently discovered by Professor J. C. Slater and announced at meetings of the National Academy of Sciences (Washington, April, 1930) and the American Physical Society (Cleveland, December, 1930).

[2] W. Heitler and F. London, *Z. Physik*, 44, 455 (1927).

[3] Y. Sugiura, *ibid.*, 45, 484 (1927).

[4] S. C. Wang, *Phys. Rev.*, 31, 579 (1928).

FIGURE 323. ■ The first article in the series *The Nature of the Chemical Bond* by Linus Pauling (*Journal of the American Chemical Society*, 53:1367, 1931). These articles formed the foundation of his book of the same title (Ithaca, 1939), in turn the core of twentieth-century structural chemistry.

the series. So much of what we teach in the first year of chemistry is presented in these works.

Although the title has an almost magical sound to it, the nature of the chemical bond was truly the domain Pauling began to explore. He formulated the concept of *hybridization* to explain how localized atomic orbitals best overlap to form two-electron bonds. The Kossel–Lewis–Langmuir picture explained ionic and covalent bonding in terms of the octet rule. An interesting question was whether the transition from covalent to ionic bonding (the nature of the chemical bond) was smooth and continuous or abrupt. In his work, Pauling examined the abrupt change in melting points of the second-row series of fluorides:[5] NaF (995°C); MgF_2 (1263°C); AlF_3 (1257°C); SiF_4 (–90°C); PF_5 (–94°C); SF_6 (–51°C). The seemingly obvious conclusion is that the first three are ionic (electrons cleanly transferred, not shared) and the next three are covalent (electrons shared). However, Pauling noted that structure is the key here and that while AlF_3 is polymeric, SiF_4 exists as individual molecules. He concluded that the Al–F and Si–F bonds were both polar covalent and not that dissimilar in nature. Pauling further explored what were termed one-electron bonds (thought to be present in diborane B_2H_6, the "nonclassical" structure reported in 1951 by Hedberg and Schomaker, using electron diffraction; its three-center bonding explained by Lipscomb) and three-electron bonds in species such as nitric oxide (NO). He developed the concept of *electronegativity* to quantitate the transition in nature from pure covalent to pure ionic bonding. His concept of *resonance* rationalized the transition from the highly polar covalent bond in hydrogen fluoride (comparable contributions from the H^+I^- and H–F resonance contributors) to the less polar HI (less H^+I^- contribution). It also furnished the explanation for the 70-year-old quandary of the relationship of benzene's structure to its reactivity. Much as two tuning forks embedded in the same wooden block exchange vibrations—one vibrates and then transfers its vibration to the other and the exchange reverses—so too can benzene be represented as two equivalent structures "in resonance." This is only an analogy and benzene is thought of as a resonance "hybrid" of the two limiting classical Lewis-type ("canonical") structures. It is worth mentioning here that the rival molecular orbital approached championed by Robert Mulliken has, with the aid of computers, probably become the more powerful technique in present-day research.

Pauling's audacious scientific career included the use of first principles and molecular models to intuit the structure of the protein α-keratin. He also was the first to characterize the basis for a disease at the molecular level—sickle-cell anemia—the result of a substitution of one amino acid for another in hemoglobin. Pauling was awarded the Nobel Prize in Chemistry in 1954.

Pauling's political activities, characterized as left-wing during the McCarthy era of the 1950s, caused him difficulties at Caltech as well as with the State Department. Its denial of a passport caused him to miss a 1952 meeting of the Royal Society in which critical information about DNA was exchanged.[6] Ultimately, his political activities were critical in obtaining agreement on a ban in atmospheric testing of nuclear weapons and he was awarded the 1962 Nobel Peace Prize on October 1, 1963—the date of the test-ban treaty. It is not unreasonable to consider Pauling to be one of the parents, along with Rachel Carson, of the environmental movement. Pauling's resonance theory was considered "revisionist" in Stalin's Soviet Union and he was vilified by staunch Communists. Anybody who can simultaneously upset Communists and McCarthyites must be doing something right!

---

1. T. Hager, *Force of Nature: The Life of Linus Pauling*, Simon & Schuster, New York, 1995, pp. 82–85, 88–91.
2. T. Hager. op. cit., pp. 107–109.

3. T. Hager, op. cit., pp. 116–117.
4. J.D. Watson, *The Double Helix,* A Norton Critical Edition, edited by G.S. Stent, W.W. Norton & Co., New York, 1980.
5. L. Pauling, *The Nature of the Chemical Bond,* 3rd ed., Cornell University Press, Ithaca, 1960, pp. 71–73.
6. T. Hager, op. cit., pp. 400–407.

PAULING'S CARTOON CARNIVAL

> When Linus did a traditional demonstration, dropping bits of sodium into a bowl of water and igniting the hydrogen formed, he added an instructive twist. He became extremely excited, in imitation of a stereotypical mad chemist. He would shout and jump about, run to the other end of the lecture table pour gasoline into a bowl, leap back, and throw in chunks of sodium. Amazed at the lack of an explosion, or any reaction, his frightened students had an unforgettable lesson. Pauling also often posed questions to the class, and the first student to answer was rewarded by receiving a candy bar tossed forth by Linus with gusto.[1,2]

Linus Pauling (1901–1994) was the first to effectively bring the modern quantum mechanics of Schrödinger, Heisenberg, and Pauli in the 1920s to the bench chemists of the 1930s and the university students of the 1940s.[3] The mathematical arguments of the quantum mechanics were beyond the reach of the vast majority of the contemporary chemists of the day. Furthermore, the "elementary" problems tackled by the physicists (H atom, He atom, H_2 molecule, He_2+ ion–molecule) lacked practical chemical utility and interest. One aspect of Pauling's genius lay in his ability to develop simple conceptual methods for learning—models useful to chemists who lacked the full theoretical foundation. So many of these heuristic concepts and models still grace our twenty-first-century textbooks almost unchanged over three score years. Among these concepts are

1. Electronegativity
2. Hybridization
3. Resonance

Pauling's classic text, *The Nature of the Chemical Bond,*[4] first edition in 1939, third edition in 1960, has a title that is equally magisterial and mysterious. It evokes Lucretius' ancient classic De Rerum Natura (On the Nature of Things). Indeed, his goal was perhaps no less than leading a diving expedition into the depths of chemical bonds to sample their electronic contours, waves, and currents.

Pauling's 1947 college textbook, *General Chemistry,*[5] was arguably even more influential than *The Nature of the Chemical Bond.* Professors will often employ a draft of a forthcoming textbook in the course they are teaching to test its effectiveness. In a 1944 draft of his future text[6] we see two rather basic, if pedestrian, schematic diagrams likely drafted by the author—one illustrating Avogadro's law of combining volumes of gases [Figure 324(a)] and the other,

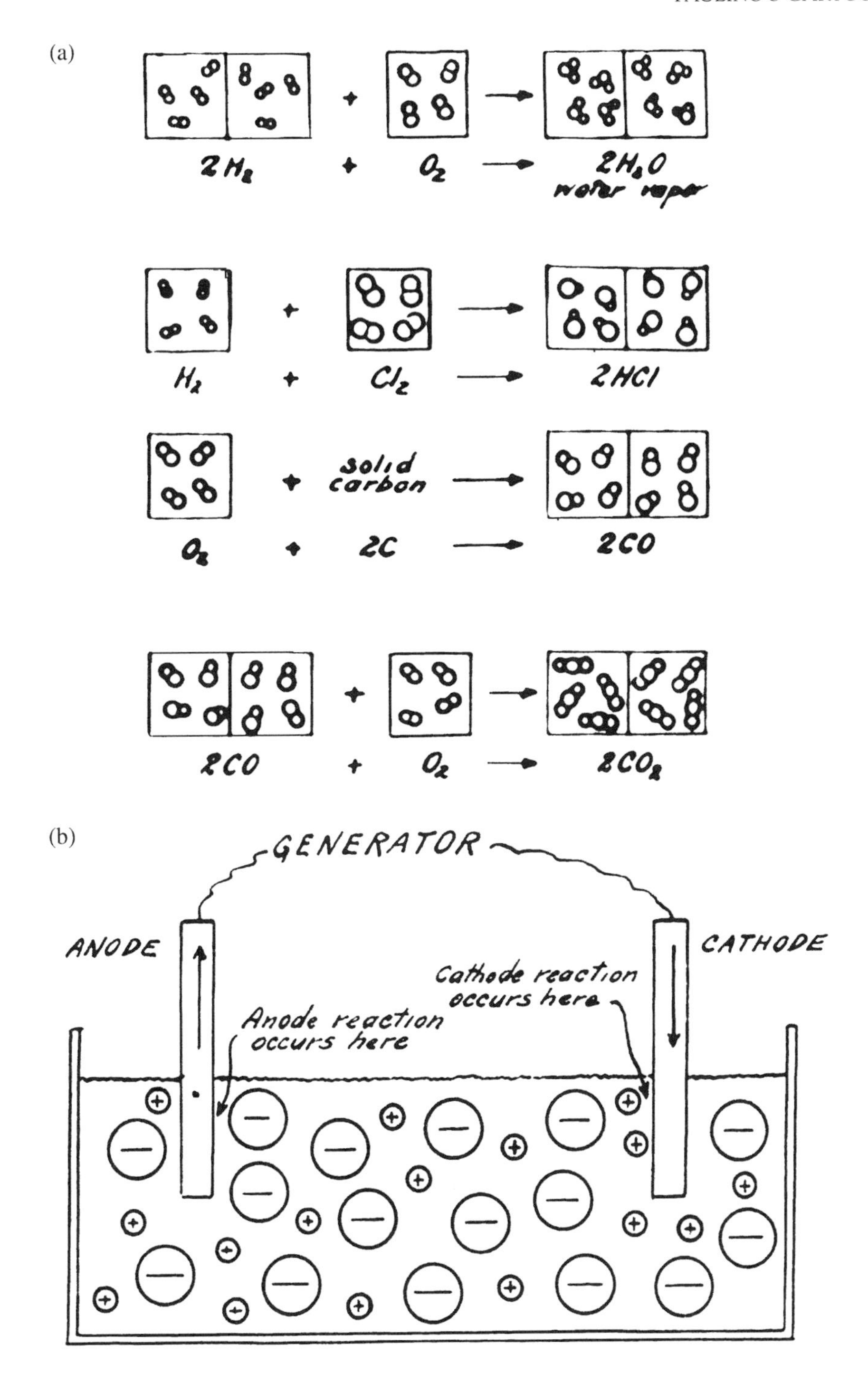

FIGURE 324. ▪ Linus Pauling's *General Chemistry*, first edition published in 1947, was a landmark in the teaching of chemistry. The two figures (a and b) shown here are from his self-published draft, printed in 1944, of his famous future textbook [Pauling, *General Chemistry* (privately printed, 1944, courtesy of the family of Linus Pauling)].

(a)

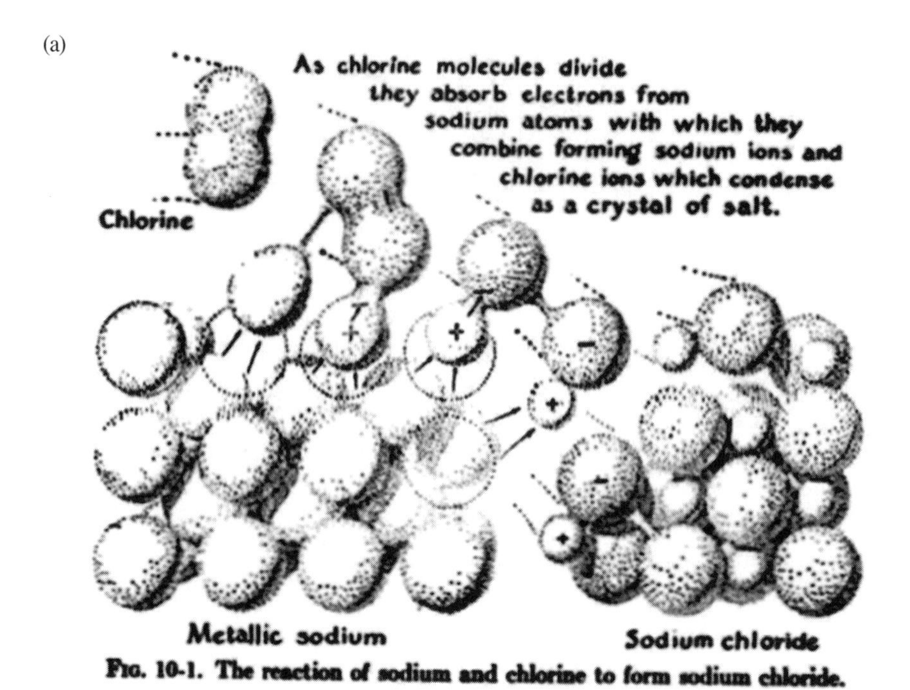

(b)

FIGURE 325. ■ The first edition of Pauling's General Chemistry united Pauling with artist Roger Hayward. *General Chemistry* by Linus Pauling © 1947 by Linus Pauling. Used with permission of W.H. Freeman and Company. The simple elegance and the dynamics of Hayward's illustrations depicted here certainly contrast with the static drawings in Figure 324. Pauling, as usual, was reaching beyond conventional representations to try to imagine the very rearrangements of nuclei and electron clouds that we understand today occur in femtoseconds (quadrillionths of a second).

an electrochemical cell [Figure 324(b)]. And here we might try to imagine Pauling, as always, impatiently reaching beyond the possible. Crystallography and electron diffraction had provided data on the spatial arrangements of bonded atoms, allowing the modeling of nearly static molecules. Pauling's heuristic models allowed one to plumb these molecules' electronic depths and contours.

So why not go beyond the static models of molecules and picture their birth and death stories—their dynamics—the collisions and rearrangements occurring at dimensions 10,000 times smaller than microscopic, on a timescale of femtoseconds (0.000000000000001 second) at air speeds approaching a mile per second? Impossible during the 1940s (and for decades afterward)! But why not imagine such events—say, the change of electron-density contours during the slow-motion collision of chlorine molecules with the surface atoms of metallic sodium en route to forming crystalline table salt (Figure 325a). In this endeavor Pauling was assisted by the artist-architect-engineer Roger Hayward. And note Hayward's rendition of the electrolysis of water (Figure 325b) and compare it with Pauling's earlier skeletal schematic (Figure 324b).

Figure 326 is from the 1964 book *The Architecture of Molecules*,[7] co-authored by Pauling and Hayward, a coffee-table art book for nonscientists and scientists alike. At the time this book was printed, computer-generated molecular graphics was nothing but a gleam in the eyes of computer scientists. But the paintings by Hayward often included visions of the molecular boundaries and contours derived from the best theoretical studies of the day. Figure 326, fittingly enough, depicts the heme molecule. There are four such molecules embedded within four protein chains in the huge hemoglobin supermolecule: two a chains of 141 amino acid residues each and two β chains of 146 amino acid residues each. Pauling and his co-workers discovered that sickle-cell anemia is an inherited disease characterized by replacement of a single polar amino acid in the β chain, glutamic acid, by a nonpolar amino acid, valine. Pauling's discovery of the α helix structure of proteins, based upon the principles of structural chemistry he helped to pioneer, and the discovery of the absolute molecular basis of sickle-cell anemia were astounding capstones to a career that included winning the 1954 Nobel Prize in Chemistry and the 1962 Nobel Peace Prize (received in 1963).[3]

---

1. This is a description of Pauling's teaching style furnished second hand by Professor Dudley Herschbach, a pioneer in molecular dynamics and 1986 Nobel Laureate in Chemistry; see Z.B. Maksić and W.J. Orville-Thomas (eds.), *Pauling's Legacy—Modern Modeling of the Chemical Bond*, Elsevier Press, Zurich, 1999, p. 750.
2. For more jubilant dancing, see Edmund Davy's description of brother Humphry's discovery of potassium metal earlier in this book (p. 404).
3. T. Hager, *Force of Nature*, Simon & Schuster, New York, 1995.
4. L. Pauling, *The Nature of the Chemical Bond*, Cornell University Press, Ithaca, 1939.
5. L. Pauling, *General Chemistry*, W.H. Freeman and Co., San Francisco, 1947.
6. L. Pauling, *General Chemistry* (privately printed in Pasadena, CA, lithoprinted by Edwards Brothers, Inc., Ann Arbor), 1944.
7. L. Pauling and R. Hayward, *The Architecture of Molecules*, W.H. Freeman and Co., San Francisco and London, 1964.

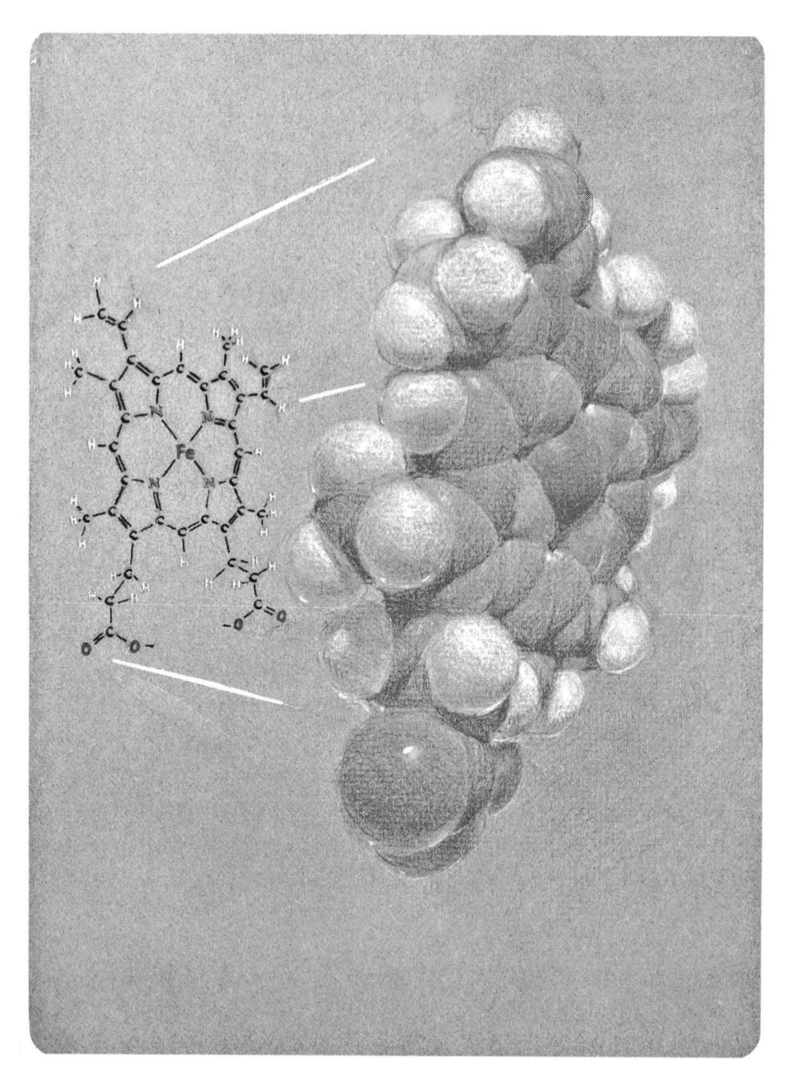

FIGURE 326. ■ Linus Pauling continued to be "way ahead of the curve" as he co-authored with long-time artist friend Roger Hayward *The Architecture of Molecules* in 1964. Before the era of molecular graphics, the depiction of the heme molecule, one of 57 color drawings in this lovely book, exemplified their enlightened attempt to convey the beauty of chemistry to the public. See color plates. (From *The Architecture of Molecules* by Linus Pauling and Roger Hayward © 1964 W.H. Freeman and Company. Used with permission.)

HERE'S TO LONG LIFE (L'CHAIM)!

Linus Pauling (1901–1994) wrote extensively about maintaining good health, and he lived his life accordingly.[1,2] I recall attending his lecture commemorating the Seeley Mudd Chemistry Building at Vassar College in 1987. At the age of 86 he presented an energetic, stimulating hour-long lecture that included a number of excursions, all of which neatly reconnected with the main theme of his talk about diet and health.

As admirable as Pauling's life and health were, at least two famous chemists lived to be centenarians: Michel Eugène Chevreul (1786–1889) and Joel H. Hildebrand (1881–1983). The two overlapped for eight years, and thus the lives of these two chemists combined to span the entire period of 1786–1983. In his 1964 monograph, Partington ponders and fantasizes a bit about the Frenchman Chevreul: "He died in my lifetime and he could have spoken to Lavoisier."[3] Chevreul began his chemical studies under the eye of Nicolas Vauquelin in 1803 at the *Muséum d'Histoire Naturelle* in Paris and retained his association with the museum for almost 90 years.[3,4] Chevreul's first publication appeared in 1806 when he was 20 and treated the analysis of bones. This was two years before Dalton published his atomic theory. He worked at the dawn of organic chemistry and was a pioneer in the daunting world of animal chemistry. In a decade starting around 1813, Chevreul discovered the true nature of the biblical art of soap-making. Saponification, the reaction of lard with lye, yielded fatty acids as well as glycerol. By examining numerous animal fats, he collected a "library" (in modern combinatorial terms) of fatty acids. Chevreul established the test of an unchanging melting point as the measure of purity for his new substances. As early as 1818, Chevreul effectively anticipated Berzelius' definition of isomers, 12 years hence, when he "defined a 'chemical species' as formed from the same elements in the same proportions and in the same arrangements."[3]

Chevreul's most profound impact derived from his studies of colors and dyes. The French established dominance in the textile dyeing industry during the reign of Louis XIV in the late seventeenth century. In 1691 the factory works of the Gobelins family became the official location of the government's dye industry and dominated the European industry for well over a century. In 1824 Chevreul succeeded Claude Berthollet as Director of dyeing at the *Manufactures Royales des Gobelins*. He developed a color wheel in which a third dimension was introduced where white forms the base and black the apex. The wheel was divided into 72 sectors, and the arc connecting the periphery of the circle to the apex was divided into 10 sectors. (Figure 327 shows Chevreul's color circle, in black and white, along with the "vertical arc."[6]) Taken together, the two form a hemisphere with all possible color and shade combinations. His work on the juxtaposition of colors had a profound impact on neo-Impressionists such as George Serat.[4] Chevreul's final paper (concerning vision) was published in 1883, at the age of 97, and his last scientific communication with his beloved *Muséum d'Histoire Naturelle* was presented at the age of 102.[3,4]

Joel H. Hildebrand[7] published his first paper, derived from his doctoral dissertation at the University of Pennsylvania (1906), in the *Journal of the American Chemical Society* in 1907.[8] It was abstracted in Volume 1 of *Chemical Abstracts*.[8] Following a period as an instructor at "Penn," Hildebrand was recruited to Berkeley by Gilbert N. Lewis in 1913—the beginning of a 70-year association with that campus. His lifetime of research work focused in part on electrolytes, their ionic nature only first disclosed by Arrhenius in 1884. However, his most profound impact was as a chemical educator. In the first edition of his influential *Principles of Chemistry*, published in 1918, Hildebrand was the first to include the more recently published (1916) Lewis dot structures in a textbook. In Figures 328(a)–328(c) we see drawings from G.N. Lewis' 1923 text,[9] including a page from Lewis' 1902 notebook. Hildebrand's lucid work was published in seven editions, with the final one appearing in 1964—a 46-year run! Hildebrand himself

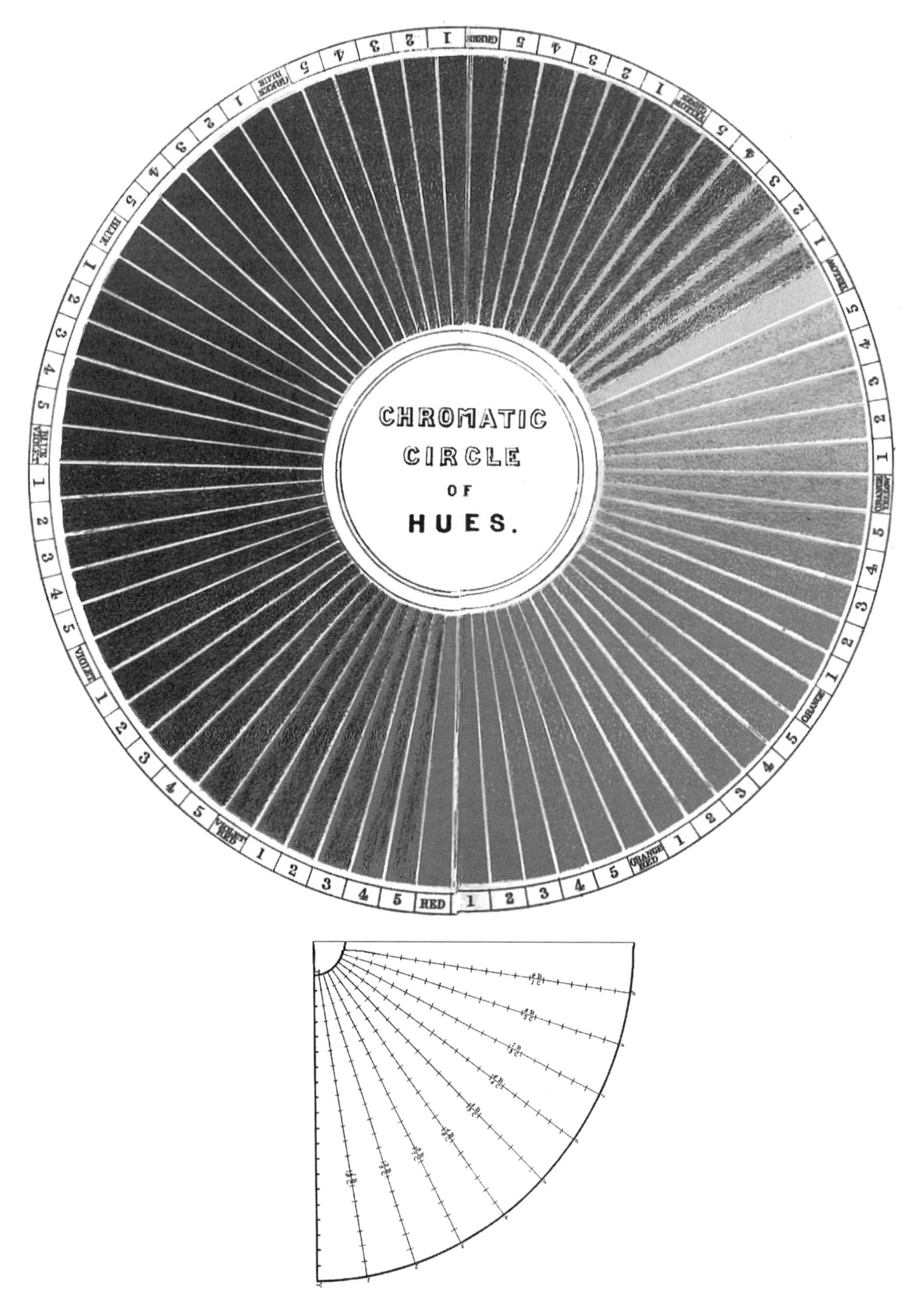

FIGURE 327. ■ This is the color wheel, in black and white, pioneered during the nineteenth century by the famous early organic chemist Michel Eugène Chevreul (1786-1889), who first published in 1806, and later published his final paper at the age of 97 and sent his last scientific communication to his beloved *Muséum d'Histoire Naturelle* at the age of 102.

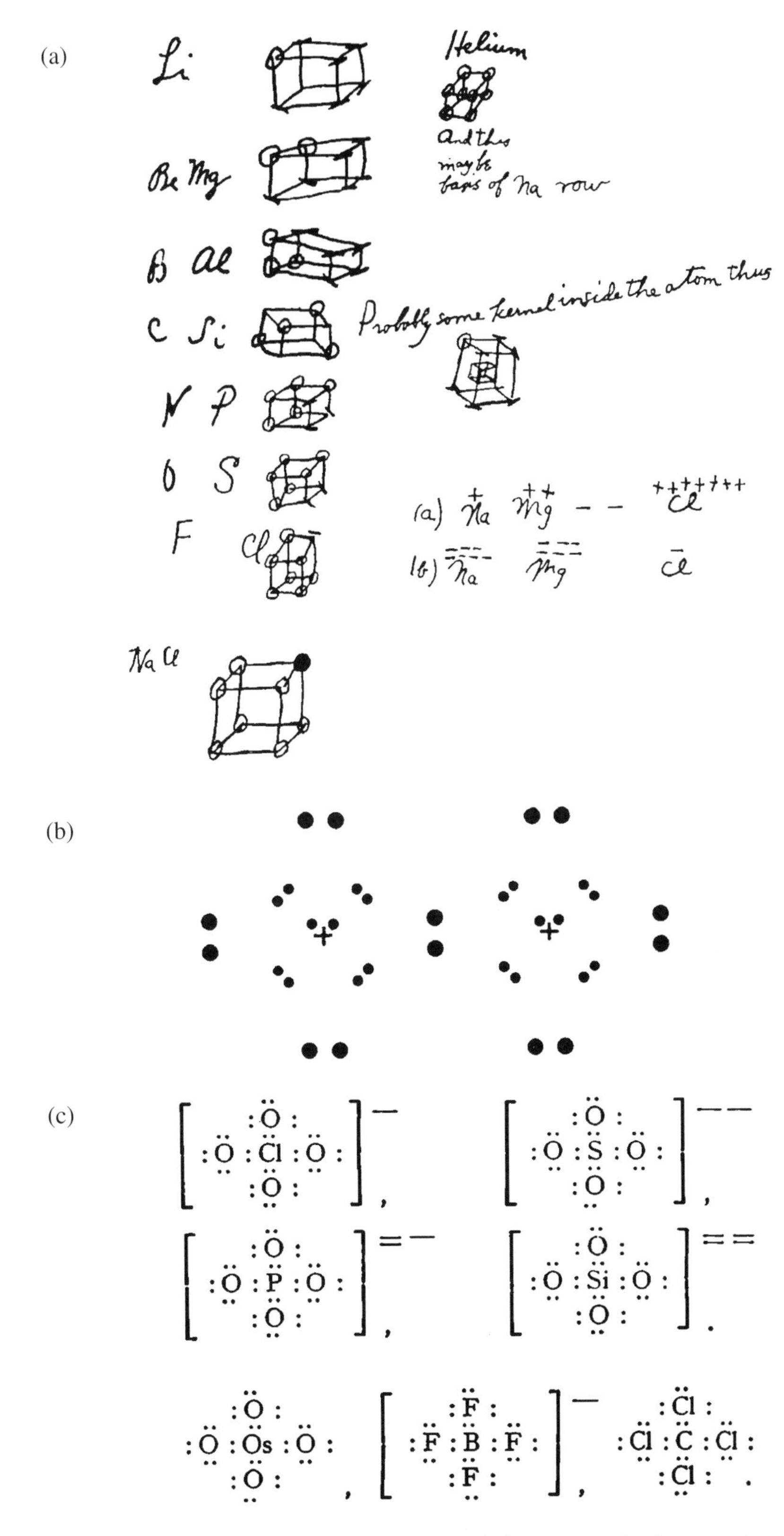

FIGURE 328. ■ Joel H. Hildebrand (1881–1983) became a faculty member under Gilbert N. Lewis at the University of California, Berkeley in 1913. In his influential *Principles of Chemistry*, first published in 1918, Hildebrand was the first to use the 1916 Lewis structures in a textbook (figures from Lewis' 1923 book are shown here). Hildebrand's seventh and final edition was published in 1964, and his last chemistry paper was published during the year of his hundredth birthday.

loved running and skiing. He coached the U.S. Ski Team at the 1936 Olympics in Berlin. He celebrated his 77th birthday with a rapid half-mile swim. Hildebrand was president of the American Chemical Society in 1955. His final published work was a history of electrolytes published in 1981,[10] the year of his 100th birthday. In 1982, the Berkeley Chemistry building housing his office was named Hildebrand Hall and the 101-year-old Professor Emeritus commented "The regents got tired of waiting for me to die before naming it."[7]

1. L. Pauling, *Vitamin C and the Common Cold*, W.H. Freeman and Co., San Francisco, 1970.
2. L. Pauling, *How to Live Longer and Feel Better*, W.H. Freeman and Co., New York, 1986.
3. J.R. Partington, *A History of Chemistry*, MacMillan and Co. Ltd., London, Vol. 4, 1964, pp. 246–249.
4. C.C. Gillispie, *Dictionary of Scientific Biography*, Charles Scribners Sons, New York, 1971, Vol. III, pp. 240–244.
5. C. Singer, *The Earliest Chemical Industry*, The Folio Society, London, 1948, pp. 255–260.
6. M.E. Chevreul, *The Principles of Harmony and Contrast of Colours and Their Applications to the Arts* (transl. C. Mantel), third edition, George Bell and Sons, London, 1899, pp. 56–57.
7. W.D. Miles and R.F. Gould (eds.), *American Chemists and Chemical Engineers*, Vol. 2, Gould Books, Guilford, 1994, pp. 128–130.
8. J.H. Hildebrand, *Journal of the American Chemical Society*, Vol. 29, pp. 447–455, 1907 [Chemical Abstracts, Vol. 1, No. 1832 (1907)].
9. G.N. Lewis, *Valence and the Structure of Atoms and Molecules*, The Chemical Catalog Co., Inc., New York, 1923, pp. 29, 82, 86.
10. J.H. Hildebrand, *Annual Reviews of Physical Chemistry*, Vol. 32, pp. 1–23, 1981.

MERCURY *CAN* BE TRANSMUTED TO GOLD

Transmutation happens! And mercury can be transmuted to gold—but not by chemistry or alchemy. The *Strong Force* that binds an atomic nucleus is on the order of millions of electron volts (MeV) per nuclear particle (proton or neutron). A neutron isolated from a nucleus has a half-life of a mere 17 minutes before disintegrating into a proton and an electron. Since the mass of the neutron equals the mass of a proton plus 2.5 electrons, the lost mass of 1.5 electrons is equivalent (Einstein's $E = mc^2$) to 0.78 MeV.[1] Now, chemistry involves the gain or loss of electrons only and thus chemistry happens with energies on the order of a few eV at most—roughly a millionth of the nuclear binding force. The nuclei in the carbon, hydrogen, nitrogen, and oxygen atoms doze peacefully when TNT explodes and all chemical hell breaks loose.

Radioactivity is emitted from atomic nuclei that are unstable and spontaneously change their structure. In 1896, Henri Becquerel first discovered radioactivity when he placed a piece of zinc uranyl sulfate wrapped in paper on a photographic plate. Two years later, Marie and Pierre Curie discovered two highly radioactive elements, polonium and radium, in pitchblende.[2] α particles, the nuclei of helium atoms, were among the radiations emitted by these substances which were spontaneously transmuting. Indeed, since the earth had billions of years ago lost its original complement of light, inert helium, all helium in our present environment comes from radioactive decay. The sun makes its own helium fresh every day (and night) by fusing hydrogen atoms. The first man-made transmutation was achieved by Rutherford in 1919 when he bombarded nitrogen (^{14}N) with α particles and made oxygen (^{17}O).[3] In 1932, Chadwick observed the neutron, thus explaining the existence of isotopes and largely unifying knowledge about the nucleus.

The neutron plays a pivotal role in manmade transmutations. In the words of Bronowski:[4] "At twilight on the sixth day of Creation, so say the Hebrew commentators to the Old Testament, God made for man a number of tools that gave him also the gift of creation. If the commentators were alive today, they would write "God made the neutron." Is it far-fetched to consider the neutron to be the Stone of the Philosophers (and atom smashers to be athanors—the furnaces of the Philosophic Egg)? Frankly, yes. But, in 1941, fast neutrons were used to transmute mercury into a tiny quantity of gold.[5] Was the age-old dream realized? Would a modern day version of the Roman Emperor Diocletian have to burn all of the notebooks and journal articles and destroy the atom smashers in order to protect the world's currency? Well, probably not. It is likely that an ounce of such gold would cost more than the net worth of the planet. Also, the gold so obtained is radioactive[6] and lives for only a few days at most.[5] But, we are not always logical when it comes to gold. In the words of Black Elk, a holy man of the Oglala Lakota-Sioux on the Pine Ridge Reservation in South Dakota:[7]

> Afterward I learned that it was Pahuska[8] who had led his soldiers into the Black Hills that summer to see what he could find. He had no right to go there, because all that country was ours. Also the Wasichus[8] had made a treaty with Red Cloud (1868) that said it would be ours as long as grass should grow and water flow. Later I learned too that Pahuska had found there much of the yellow metal that makes the Wasichus crazy; and that is what made the bad trouble, just as it did before, when the hundred were rubbed out.
>
> Our people knew there was yellow metal in little chunks up there; but they did not bother with it, because it was not good for anything.

---

1. *Encyclopedia Brittanica,* 15th ed., Chicago, 1986, Vol. 14, p. 332.
2. J.R. Partington, *A History of Chemistry,* MacMillan, London, 1964, Vol. 4. pp. 936–947, 953–955.
3. P.W. Atkins and L.L. Jones, *Chemistry: Molecules, Matter and Change,* 3rd ed., Freeman, New York, 1997, pp. 858–860.
4. J. Bronowski, *The Ascent of Man,* Little, Brown, Boston, 1973, p. 341.
5. R. Sherr, K.T. Bainbridge, and H.H. Anderson, *The Physical Review,* **60**:473–479, 1941.
6. In the 1964 James Bond movie *Goldfinger,* the arch-villain Auric Goldfinger tries to detonate a nuclear weapon inside Fort Knox to make the U.S. gold supply radioactive in order to increase the value of his own gold horde. Apparently, novelist Ian Fleming knew there is only one stable isotope of gold.
7. P. Riley (ed.), *Growing Up Native American,* William Morrow, New York, 1993, p. 99. Thanks to Professor Susan Gardner for this suggestion.
8. *Pahuska* is "Long Hair"—General George Armstrong Custer; *Wasichus,* the term for white settlers and soldiers, translates as "greedy ones." (Thanks to Professor Susan Gardner for the suggestion and background for this topic.)

MODERN ALCHEMISTS APPROACH ATLANTIS

In Figure 329 we see a 1944 formulation of the Periodic Table by Glenn T. Seaborg.[1] We are commonly told that there are 92 naturally occurring elements. Logically, this would seem to end with uranium (atomic number 92), and it is true that uranium is the highest atomic number element found naturally in any significant amount and that only ultratrace quantities of neptunium and plutonium occur in uranium ores. When Henry Moseley discovered the atomic number in 1913, there were gaps at numbers 43, 61, 72, 75, 85, 87, and 91. The final two stable (non-radioactive) elements, hafnium (#72) and rhenium (#75) were discovered in 1923 and 1925 respectively. However, there remained gaps: for example, at element 43. That element, technetium (Tc) was the first synthetic element, although it was later discovered that exceedingly minute (trace) quantities occur naturally due to uranium decay.[2,3] Perrier and Ségré succeeded in 1937 by bombarding molybdenum (Mo) with deuterons (nuclei of deuterium). The half-life of ^{97}Tc is 2.6 million years. Today, the Tc-99m (m = metastable) ($t_{1/2} \sim 6$ hr) isotope is used for heart imaging.[3] Element 87 is francium, synthesized from actinium by Marguerite Perey in 1939.[2] Its most stable isotope, ^{223}Fr, has a half-life of 21.8 minutes. It is also found in ultratrace (2×10^{-18} ppm) quantities in uranium ores since new francium is made as the "old" decays leaving a minute steady-state concentration.[3] Its properties, though little studied, resemble those of rubidium and cesium. Element 85, astatine, was produced by bombarding bismuth with accelerated a particles.[2] Its most stable isotope, ^{210}As, has a half-life of 8.3 hours and is also found in ultratrace quantities in uranium ores.[3] The largest quantity made of astatine (50 billionths of a gram) allowed a limited amount of study: It is concentrated in the thyroid like iodine and AtI_2^- is even more stable than the commonplace I_3^-.[3] Element 61, promethium, was discovered in 1945 and conclusively reported in 1947 as a trace by-product of uranium fission ($<1 \times 10^{-11}$ ppm).[2,3] The isotope reported (^{147}Pm) has a half-life of only 2.6 years. Subsequently, ^{145}Pm was found to have a half-life of 17.7 years.[3]

So, it seems that of these 92 "natural" elements, only 88 can be considered naturally occurring since the above four are transient species, newly formed by radioactive decay. Neptunium and plutonium can similarly be found in ultratrace quantities due to de novo synthesis coupled to rapid decay. We could stretch a point by noting that all helium on our planet is also formed *de novo*. However, although these fresh helium atoms are lost into space, the nuclei are totally stable.

The true stars of Seaborg's 1944 Periodic Table are the transuranium elements neptunium (Np) and plutonium (Pu) as well as elements 89 to 92 (actinium, thorium, protactinium, and uranium). Neptunium was synthesized by McMillan and Abelson at Berkeley in 1940.[1] In late 1940 and early 1941 McMillan, Kennedy, Wahl, and Seaborg made ^{238}Pu through bombardment of uranium with deuterons in early 1941, and ^{239}Pu was obtained by bombarding uranium with neutrons.[1] It was Seaborg who, in 1944, proposed a new series of compounds for the Periodic Table—the actinides—analogous to the rare earths or lanthanides. In his book *The Periodic Kingdom*, Atkins describes the lanthanides

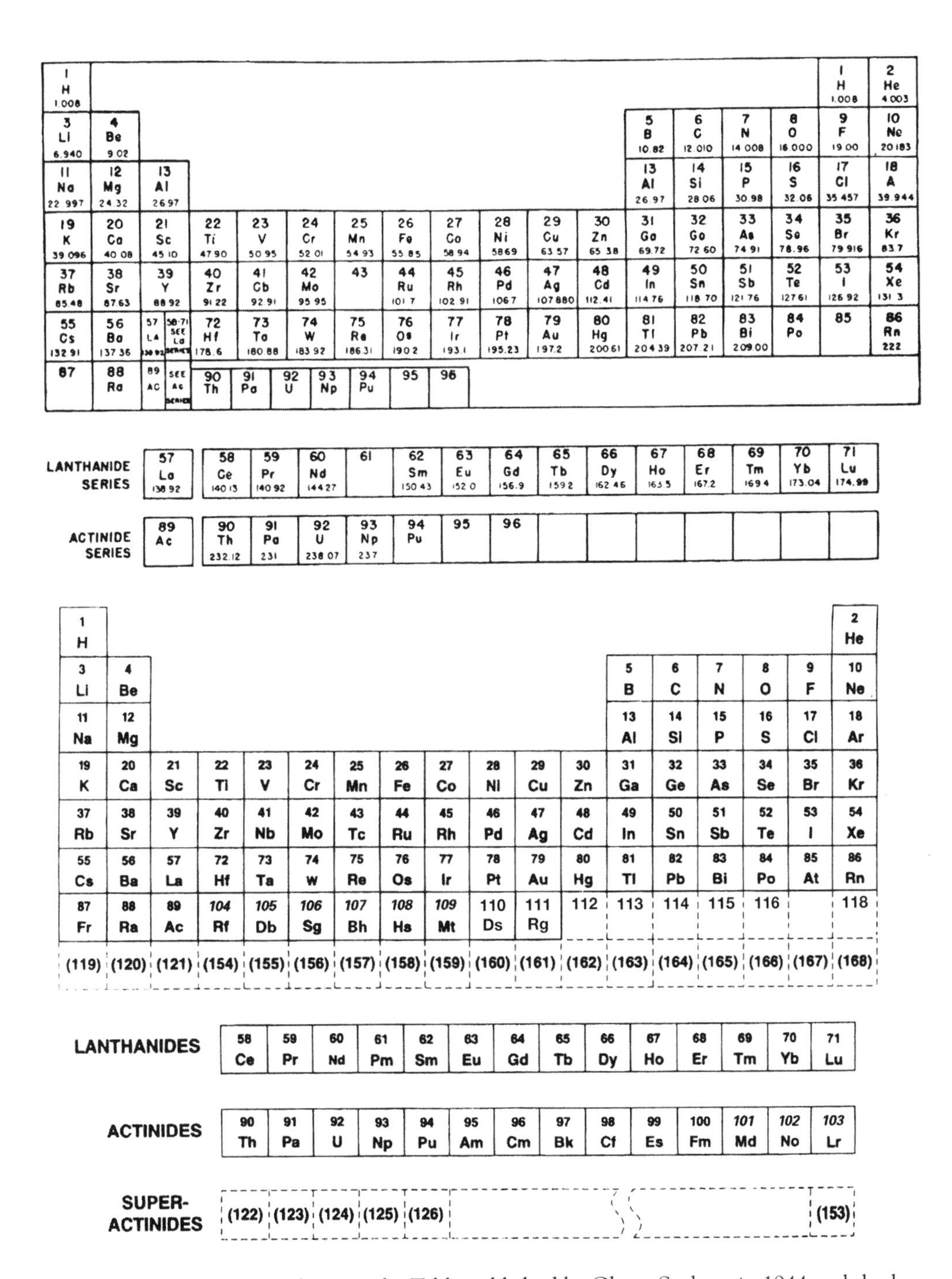

FIGURE 329. ■ The top figure is the Periodic Table published by Glenn Seaborg in 1944 and the bottom is a futuristic table (slightly modified from his diagram in *Accounts of Chemical Research*, **28**:257, 1995). Seaborg died in March, 1999 following a stroke some six months earlier. Sadly, he was unaware of a Russian group nearing the anticipated "Isle of Stability" with element 114 in January 1999 and the subsequent reports of elements 113, 115, 116, and 118 (courtesy American Chemical Society).

as the northern shore of an island off the south coast of the Periodic Kingdom.[4] Seaborg, thus, discovered the southern shore of this island.

The transmutations described here are nuclear physics and not chemistry (or alchemy). But, if we play with our earlier metaphor and liken the neutron to the Philosopher's Stone, then ^{239}Pu could be likened to gold as both a blessing and a curse. It was the fuel in the atomic bomb used on Nagasaki that ended World War II while killing and maiming a city's population. The incredible stockpile assembled during the Cold War now leaves the earth with hundreds of tons of this scary yet useful substance: the curse of King Midas on the one hand, a source of energy on the other.

In Figure 329 we also see a modified version of a futuristic Periodic Table published by Seaborg in 1995.[5] The southern shore of the coastal island is completed through lawrencium (Lr). Seaborg used Mendeleevian logic[5] to predict the properties of element 101 as "*eka*-thulium," just as Mendeleev predicted an *eka*-silicon (germanium) below silicon. Appropriately, it is named Mendelevium (Md). Seaborg's actinide series explicitly assumed chemical similarities with lanthanides. Indeed, the heavier actinides ending with lawrencium also favor the +3 oxidation state. As predicted, elements 104 and 105 (rutherfordium and dubnium) resemble their d-block relatives hafnium and tantalum in their chemistries. There was, unfortunately, a controversy about naming elements with atomic numbers over 100, which was finally settled in 1997:[6] 101, mendelevium (Md); 102, nobelium (No); 103, lawrencium (Lw); 104, rutherfordium (Rf); 105, dubnium (Db); 106, seaborgium (Sg); 107, bohrium (Bh); 108, hassium (Hs); 109, meitnerium (Mt). Naming 106 for Glenn Seaborg was a particularly significant gesture honoring his massive contributions. Transuranium elements up to 106 were made by colliding light nuclei with increasingly rare and unstable actinide nuclei. The Dubna laboratory pioneered "cold fusion" in which medium nuclei ($Z = 40 - 70$) collide with minimum energy into very stable ^{208}Pb or ^{209}Bi. This technique was successful for $Z = 107$ through 112. In order to make nuclei with more neutrons "hot fusion" techniques were employed starting with $Z = 114$. Elements 107 to 109 have half-lives of milliseconds.[5] Fittingly, element 106 was the last of the series to have a lifetime (one to ten seconds) to permit on-line (following creation) chemical study. Perhaps there will be seaborgic sulfate or calcium seaborgate.[5] Ten years after the discovery of hassium in 1984, elements 110, 111, and 112 were identified in that order. Elements 110 to 112 have half-lives on the order of microseconds and milliseconds.[5] These findings are consistent with the nuclear shell-structure theory of Maria Goeppart-Mayer and Hans Jensen, who shared the Nobel Prize in Physics in 1963.[2]

Göppart-Mayer and Jensen's theory predicted the existence of "Islands of Stability" among the superheavy elements. Atkins calls these Atlantis.[4] An audacious experiment by the Joint Institute for Nuclear Research in Dubna, Russia electrified the scientific world in early 1999 with a cautious announcement of the synthesis of element 114 (atomic mass 289) with a half-life (α-decay) of 27 seconds by "hot fusion," bombarding ^{244}Pu with "doubly-magic" ^{48}Ca ions.[7,8] Albert Ghiorso said, "This is the most exciting event in our lives."[7] The neutron number ($N = 175$) approached the "magic" closed neutron shell ($N = 184$) and $Z = 114$ is a "magic number." The Lawrence Berkeley Laboratory claimed synthesis of elements 116 and 118 in 1999 but had to withdraw them in 2001. The Dubna group reported element 116 in 2001 and elements 113 and 115 in 2004. The latter study even included a very tentative possible observation of 118.[9] In 2003 the Joint Working Party (JWP), comprised of the International Union of Pure and Applied Chemistry (IUPAC) and the International Union of Pure and Applied Physics (IUPAP), officially recognized and named element 110, darmstadtium (Ds), and element 111 (roentgenium, Rg) and 112 (ununbium or Uub) should be accepted soon.[9] Sadly, Glenn T. Seaborg suffered a crippling stroke in August, 1998 and died in early 1999, unaware of the near approach to Atlantis and chemical studies on bohrium and hassium.[9]

1. G.T. Seaborg, *Chemical and Engineering News*, **23** (23), December 10, 1945.
2. J.R. Partington, *A History of Chemistry*, MacMillan, London, 1964, Vol. 4, pp. 953–955.
3. H. Rossotti, *Diverse Atoms: Profiles of the Chemical Elements*, Oxford University Press, Oxford, 1998.
4. P.W. Atkins, *The Periodic Kingdom*, Basic Books, New York, 1995, pp. 25–26, 56.
5. G.T. Seaborg, *Accounts of Chemical Research*, Vol. **28**:257–264, 1995.
6. *Chemical and Engineering News*, Vol. **75**, September 8:10, 1997.
7. R. Stone, *Science*, **283**:474, 1999.
8. Yu.Ts. Oganesian, A.V. Yeremin, A.G. Popeko, S.L. Bogomolov, G.L. Buklanov, M.L. Chelnokov, V.I. Chepigin, B.N. Gikal, V.A. Gorshkov, G.G. Gulbekian, M.G. Itkis, A.P. Kabachenko, A. Yu Laurentev, O.N. Malyshev, J. Rohac, R.N. Sagaidak, S. Hofmann, S. Saro, G. Giardina and K. Morita, Nature, Vol. 400, p. 242 (1999).
9. M. Schädel, *Angewandte Chemie International Edition in English*, Vol. 45: 368–401, 2006.

THE CHEMISTRY OF GOLD IS NOBLE BUT NOT SIMPLE[1]

Now here is an interesting point: it was obvious to the ancients that gold was "noble" in the same sense that we think of the Group 8A gases (helium, neon, etc.) as "noble"—namely, unreactive. It did not tarnish like its valuable cousin silver. Silver tarnish is due to atmospheric hydrogen sulfide that forms a coating of black silver sulfide (Ag_2S) and it could be heated repeatedly with no change. It did not dissolve in hydrochloric acid or nitric acid. It did "dissolve" in *aqua regia,* a mixture of 3:1 HCl/HNO_3, but evaporation of the solution and intense heating recovered the gold unchanged, unlike the baser metals where a calx remained. However, gold chemistry is not obvious, even today.

We know that gold does exhibit reactivity (but so does xenon).[2,3] For example, the "dissolution" in *aqua regia* is really a chemical reaction in which HNO_3 and HC1 act synergistically (as a team). The oxidation of elemental gold to Au(III) can only happen because of the stability of the $AuCl_4^-$ ion:

$$Au + HC1 + HNO_3 \rightarrow AuCl_4^- + NO_2 + H_3O^+$$

$$Au + HC1 + Cl_2 + H_2O \rightarrow H_2O^+[AuCl_4^-]\,(H_2O)_3$$

where the final product is chloroauric acid. Similarly, in the presence of dilute cyanide solutions, it will oxidize in air at room temperature to form stable $Au(CN)_2^-$ ions. This reaction is used to extract gold from ores. Although ancient Andean metallurgists (see pages 46–47) apparently lacked aqua regia, they succeeded in dissolving gold, possibly by using warm aqueous solutions of salt

(NaCl), saltpeter (KNO_3), and alum [$KAl(SO_4)_2 \cdot 12H_2O$] that reproduced the action of aqua regia.[4]

The explanation of gold's nobility is not obvious.[2] In modern terms, we note that the outermost electrons in the atoms of the coinage metals copper, silver and gold are $4s^1$, $5s^1$, and $6s^1$. Thus, they appear at first to be close cousins of the decidedly *ignoble* alkali metals potassium, rubidium, and cesium (also $4s^1$, $5s^1$, and $6s^1$, repectively)—cesium would "date" almost anybody. The alkali metals described "underlie" their outermost electrons with a completed subshell structure—the octets of the preceding noble gas. These octets shield the outermost electrons (ns^1) from nuclear attraction and make them easy to ionize (lose) and, thus, render the alkali metals reactive. Cesium has its outermost electron furthest from the nucleus and is most reactive. In contrast, the coinage metals "underlie" their ns^1 electrons with a completed 18-electron shell that would also imply great stability. However, the *d* electrons are not particularly effective in shielding the ns^1 electrons, which are thus strongly attracted to the nucleus and hard to ionize.[2] In further contrast to the alkali metals, the order of reactivity is smallest to largest. Copper is most reactive, and gold the least reactive. Apparently, relativistic physics is required to explain the behavior of the $6s^1$ electron in gold.[3] So it appears that, in this instance at least, chemistry has been rescued by "the triumphal chariot of physics."

---

1. I thank my wife, Susan Greenberg, for suggesting this essay.
2. B.E. Douglas, D.H. McDaniel, and J.J. Alexander, *Concepts and Models of Inorganic Chemistry*, 3rd ed., Wiley, New York. 1994, pp. 724–725.
3. F.A. Cotton and G. Wilkinson, *Advanced Inorganic Chemistry*, 5th ed., Wiley, New York, 1988, pp. 937–939.
4. H. Lechtman, "Pre-Columbian Surface Metallurgy," *Scientific American* Vol. 250 No. 6 (June 1984): pp. 56–63. The author thanks Professor Roald Hoffmann for awareness of this work.

THE "PERFECT BIOLOGICAL PRINCIPLE"

> We wish to suggest a structure for the salt of deoxyribose nucleic acid (D.N.A.). This structure has novel features which are of considerable biological interest.

So reads the first paragraph of the ground-breaking communication by James D. Watson and Francis H.C. Crick in *Nature* reporting their double-helical structure for DNA.[1] The third paragraph from the end says:

> It has not escaped our notice that the specific pairing we have postulated immediately suggests a possible copying mechanism for the genetic material.

In his now-classic personal narrative, *The Double Helix*,[2] J.D. Watson imagines this understated eloquence of the paper-to-be. The book is a wonderfully idiosyncratic history, from Watson's perspective, of the race to discover the structure of DNA. It shows lay readers that scientists are human, for better or worse.

The thesis in Watson's narrative is that understanding the function of DNA may hint at its structure. Hopefully, the structure of DNA will be "beautiful" and make its function self-evident. In the book, Watson recalls Francis Crick's postulation, after sharing a few beers, of a "perfect biological principle"—the perfect self-replication of the gene.[3] It is this kind of overarching interest in function and a willingness to "play" with molecular models that give Watson and Crick an advantage in the search. Of continuing historical debate is their use of the x-ray crystallographic data obtained by Rosalind Franklin without her permission or knowledge. There remain to this day troubling questions[4,5] in spite of the acknowledgement of her data in the Nature paper and the fact that the next

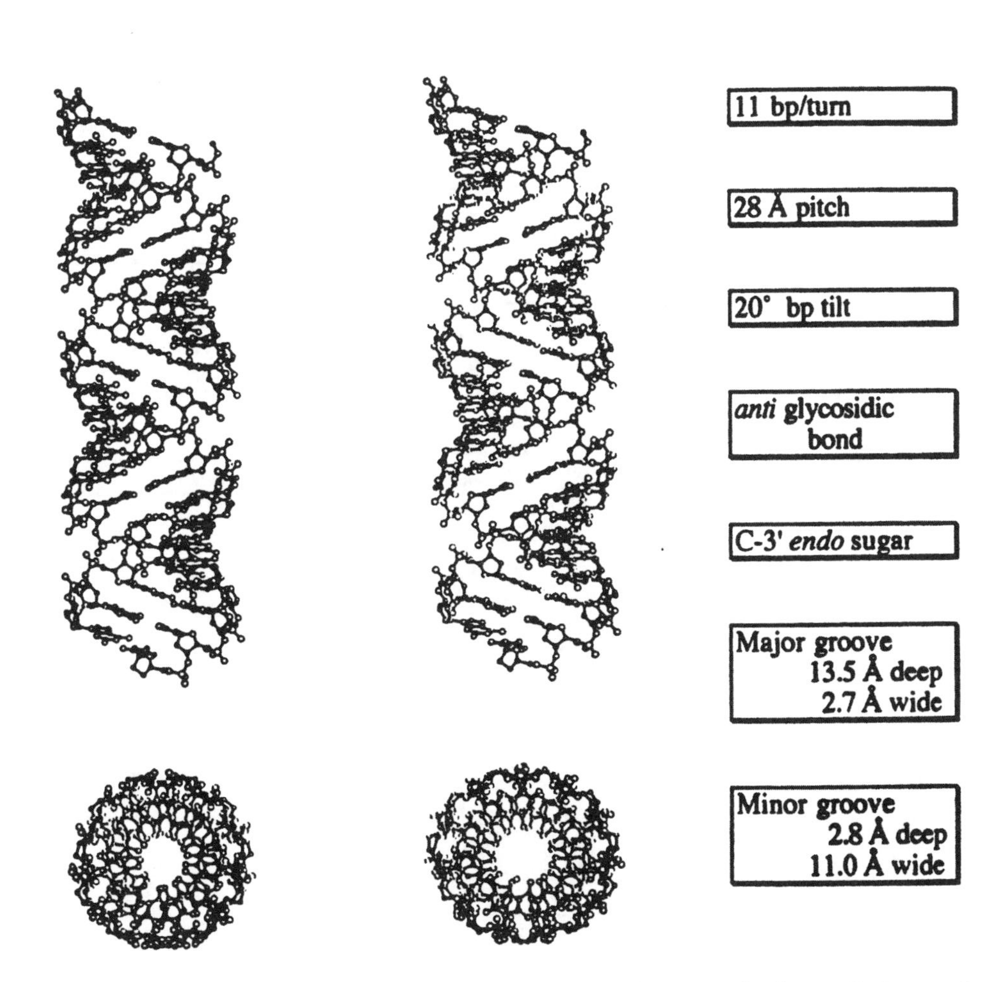

FIGURE 330. ▪ The compact structure of A-DNA (courtesy Professor Catherine J. Murphy based on structures in Arnott and Chandrasekaran, *Proceedings of the Second SUNYA Conversation in the Discipline Biomolecular Stereodynamics*, R. Sarma (ed.), Vol. 1, Adenine Press, 1981, pp. 99–122; courtesy Adenine Press).

two papers in the issue were authored by Wilkins, A.R. Stokes, and H.R. Wilson followed by Franklin and R.G. Gosling. It is clear that Franklin correctly concluded that the phosphates were on the outside of the helix. Furthermore, she understood that the data indicated helicity.[4,5] Her approach was a rigorous solution of the structure based upon straightforward data analysis although she had used molecular models in the past.[4,5]

Linus Pauling had solved the structure of α-keratin using the principles of bonding in his *Nature of the Chemical Bond* to construct models and take a shortcut to the laborious and incredibly complex interpretation of the x-ray data.

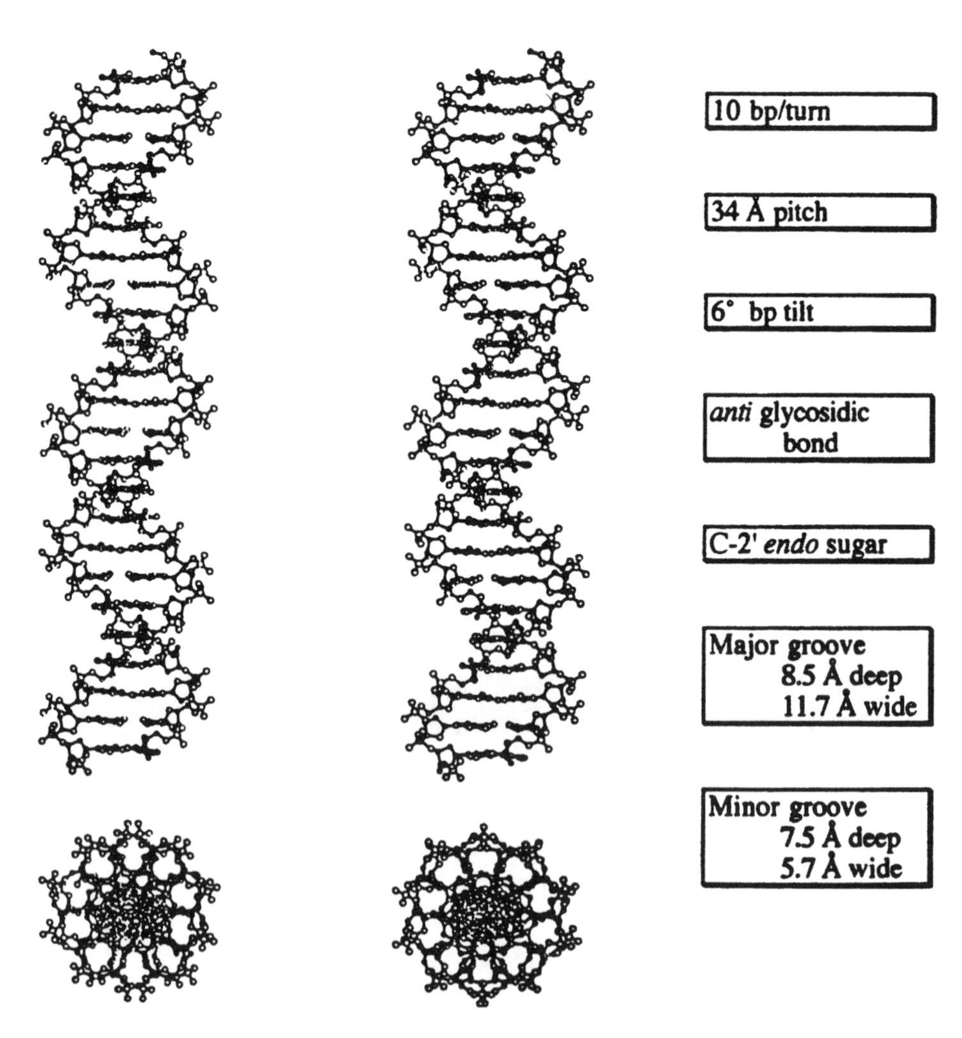

FIGURE 331. ■ The elongated B structure of DNA which provided Rosalind Franklin with a clear x-ray diffraction diagram indicating helical structure (courtesy Professor Catherine J. Murphy based on structures in Arnott and Chandrasekaran, *Proceedings of the Second SUNYA Conversation in the Discipline Biomolecular Stereodynamics*, R. Sarma (ed.). Vol. 1, Adenine Press, 1981, pp. 99–122; courtesy Adenine Press).

Watson and Crick succeeded at beating him at his own game. There is a delicious moment in Watson's book when Peter Pauling, Linus' son who is also at Cambridge, informs Watson and Crick that his father has solved the structure and that it is a triple helix.[6] Watson grabs the manuscript from the younger Pauling's coat pocket and he and Crick read it with trepidation. To their surprise and delight, Linus has goofed. The phosphates are protonated as if phosphoric acid were not an acid. Their next concern is that he would be chagrined by his mistake, redouble his determination, solve the problem, and win their Nobel! (The year is 1953. Pauling would win the Nobel Prize in Chemistry in 1954.)

Figure 330 shows the compact structure of A-DNA. Franklin discovered the technique of moistening A-DNA, which forms the more hydrated and elongated B-DNA (Figure 331) whose x-ray pattern spoke so eloquently of its helical structure. Watson, Crick, and Wilkins shared the Nobel Prize in Medicine in 1962. After Cambridge, Rosalind Franklin joined the efforts of J.D. Bernal at Birkbeck College in London, where she was given charge of her own research group. She was an effective group leader and became a world-renowned expert in the crystallography of viruses. Her work established that viruses are hollow-cored. She was diagnosed with ovarian cancer in 1956 and, during her final months, performed studies on the incredibly dangerous polio virus. She died in 1958 at the age of 37.[4,5]

In 1976, BBC released a film titled *The Race For The Double Helix.*[7] It was a very intelligent film that was essentially a dramatization of Watson's book although the interpretations were not identical. Jeff ("Jurassic Park") Goldblum played James D. Watson beautifully.

---

1. J.D. Watson and F.H.C. Crick, *Nature,* **171**:737–738, 1953.
2. J.D. Watson, *The Double Helix,* Simon & Schuster, New York, 1968.
3. J.D. Watson, op. cit., p. 126.
4. A. Sayre, *Rosalind Franklin and DNA,* Norton, New York, 1975.
5. M. Rayner-Canham and G. Rayner-Canham, *Women In Chemistry: Their Changing Roles from Alchemical Times to The Mid-Twentieth Century,* American Chemical Society and Chemical Heritage Foundation, Washington, D.C. and Philadelphia, 1998, pp. 82–90.
6. J.D. Watson, op. cit., pp. 157–163.
7. Now available: Films for the Humanities and Sciences, Box 2053, Princeton, NJ 08543-2053.

SO YOU *WEREN'T* JOKING, MR. FEYNMAN![1]

Way back in the middle of the last century (1959, to be more specific), Nobel Laureate physicist Richard P. Feynman gave an after-dinner talk ("There's Plenty of Room at the Bottom") that challenged scientists to explore the uncharted realms of nanotechnology.[2,3] He observed that the limits to miniaturization were truly reached only at the level of molecules and atoms. Among his wonderfully prescient, and typically bold, speculations was the following thought:[3]

> But I am not afraid to consider the final question as to whether, ultimately . . . in the great future . . . we can arrange the *atoms* the way we want, the very atoms, all the way down! What would happen if we could arrange the atoms

> one by one the way we want them (within reason, of course; you can't put them so that they are chemically unstable, for example).

At this atomic as well as slightly larger, "mesoscale,"[4] levels, the physical laws would be a tantalizing and unpredictable mixture of classical physics and quantum mechanics.[3]

Two years before Professor Feynman gave his talk, Russia had successfully launched the 184-pound space satellite Sputnik into earth orbit. This small instrument package, launched atop a huge multistage rocket, presaged the incredible advances in miniaturization during the following decade that would culminate in Americans walking on the moon's surface. Miniaturization produced, almost as a by-product, advances in technology that have placed high-powered computers in most modern homes. Indeed, it is now the computer industry that is perhaps the primary driver for nanotechnology, although biomedical science will surely drive this revolution during the twenty-first century. Profound questions on the limits of storage capacity and speed of communication have, in turn, raised the most fundamental questions about matter. For example, the hydrogen atom nucleus has the property of magnetic spin—hydrogens may be spin $\frac{1}{2}$ or $-\frac{1}{2}$, a binary choice of virtually equal probability. Could individual hydrogen atoms attached to molecules be the basis for molecular computers? What about DNA?[5]

Nanotechnology has, until recently, been dominated by a "top–down approach."[2,6] For example, a bulk material such as silicon may be etched to form a complex microchip, by using ultraviolet photolithography, to a level of 100 nanometers (nm).[5] Dimensions smaller than that are extremely expensive to achieve.[6] Feynman's dream of moving atoms one by one was achieved some two decades later by Heinrich Rohrer and Gerd K. Binnig of IBM Zurich who subsequently shared the 1986 Nobel Prize in Physics. The atomic force microscope (AFM), a modification of the scanning tunneling microscope (STM), was successfully employed to physically rearrange matter by pushing atoms one by one. The STM image of the "quantum corral," formed by using an AFM to place 48 iron atoms one at a time into a circle, has become an icon of modern science (see Figure 340 later in this book). Indeed, visionaries have imagined nanoassemblers ("nanobots") capable of assembling nanostructures from atoms or molecules. Here, of course, Avogadro's number is not a friend. As Nobel Laureate Richard E. Smalley—a co-discoverer of C_{60} ("buckyball")—has noted, assemblage of one mole of chemical bonds [the number in just 9 mL (milliliters) of water] would take a harried "nanobot," working at a billion new bonds per second, about 19 million years,[8] although I expect that productivity incentives and overtime pay could reduce this by 10%. Smalley notes that if "nanobots" could be designed to self-replicate prodigiously, then this problem might be overcome. However, this could introduce some potentially dangerous problems—it brings to my cartoonish mind an image of *The Sorcerer's Apprentice in Fantasia:* Mickey Mouse threatened by, say, a trillion dumb, frenetic, and possibly even malevolent brooms. There are other interesting problems as well. Smalley refers to the "fat fingers" and "sticky fingers" problems.[7] He observes that, typically, a new chemical bond will be influenced by about 5–15 atoms near the reaction site. A "nanobot's arms," themselves made of atoms, would have to get close to the bond-making site, thus forcing aside the neighboring atoms needed to determine the bond's fate. Also, one would expect that the atoms to be moved are likely to "stick" to the arms—surface effects are much more significant at nanoscale than at typical macroscopic levels. And finally, Smalley[8] notes the "love" problem. One might try to force atoms together but, as Feynman also remarked, in essence, that the rules of bonding and thermodynamics will ultimately determine whether the atoms "marry" or "cohabitate" as "significant others."

The answer to these limitations might well lie in using the "bottom–up approach"—the self-organization of molecules. A self-help program might paraphrase this approach thus: "Make Avogadro's mumber work for *you*!" And indeed countless trillions of molecules lining up repetitively in harmony with Nature's rules of chemical bonding may be an effective strategy for mass assembly of nanoscopic motors in chemical beakers. Let us examine one elementary example of the bottom–up approach—the molecular switch[8,9] depicted in Figure 332.[8] This molecule is an example of a simple catenane—in this case, two cyclic molecules interlooped much as two links in a chain. Each loop is a separate molecule fully capable of an independent existence. Now this particular molecule is designed so that there may be attraction (or repulsion) between an inner segment of one loop and an inner segment of the other loop. When the molecule in conformation [A°] ("switch open") is deliberately oxidized to [A^+], the center segment of one loop loses an electron and assumes a positive charge, is repelled by the four sets of positive charges in the interior of the other ring and circumrotates to conformation [B^+] in which electrostatic repulsion is reduced. Perhaps a bit surprisingly, when [B^+] gains its electron, by reduction to near-zero bias, to form [B°] ("switched closed"), it does not immediately circumrotate to the original [A°].[8] Both "switch positions" are thus stable and further, controlled, reduction of [B°] returns it to [A°].

There is an interesting point to be made about catenanes. The first catenane

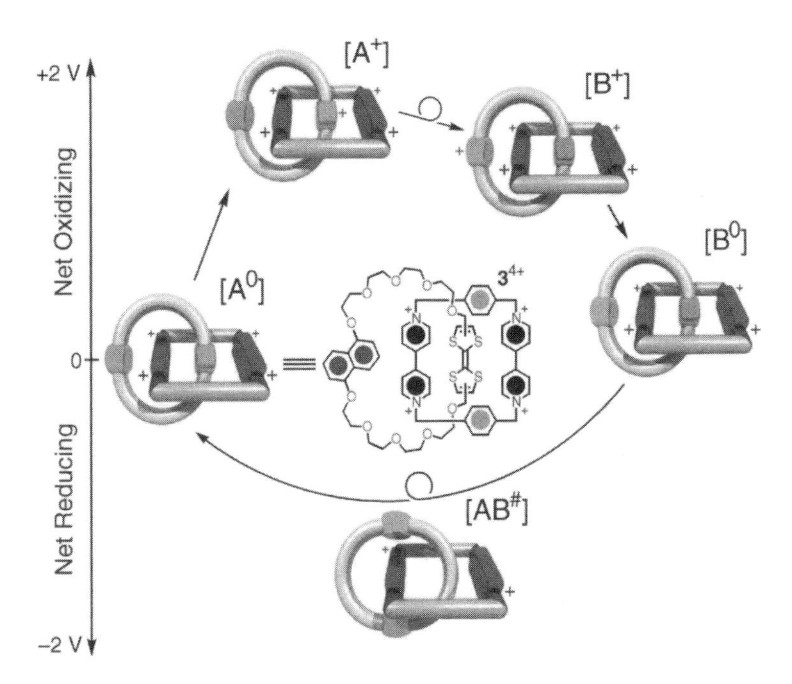

FIGURE 332. ■ A catenane molecule developed by Professor Stoddart and colleagues to function as a controllable (bistable) nanoswitch (see text). [Reprinted with permission from *Accounts of Chemical Research 1* (2001), copyright (2001) American Chemical Society.]

was made in 1960, as deliberately as then possible, by Edel Wasserman.[10] He cleverly closed a 34-carbon open-chain molecule, with reactive ester linkages at the two termini, in the presence of an equimolar amount of the corresponding 34-membered cyclic alkane. It is hard to form large rings and the overall yield of the freshly closed 34-membered ring was typically under 20%. But of this 20%, about 1% of the new rings closed while threaded in the "partner" cycloalkane, thus forming a loop about the other ring resulting in a catenane. This extremely low yield reflected the very low probability of catching a long chain through the center of a ring and then closing the threaded chain into its own cycle. Such catenanes[11] were barely more than interesting curiosities for years. They raised questions that seemed to be mere chemical semantics—are two neutral unreactive catenated cyclo-$C_{34}H_{68}$ really two molecules physically concatenated, or does this truly represent a new species? Wasserman noted different chromatographic behavior for the catenane relative to its "topological isomer" (the two separated rings taken together). It certainly was a definition for "isomer" unimaginable to Berzelius 130 years earlier. The application of esoteric species such as catenanes to serious problems in technology is a wonderful example of the benefits of pure research not immediately imaginable to practical "administrative types." One of my favorite illustrations of this principle derives from the search for the exotic, short-lived, "administratively-uninteresting" 1,4-benzenediyl by Jones and Bergman in 1972.[12] Fifteen years later, a new class of natural and synthetic anticancer agents were discovered with the reactive 1,4-benzenediyl nucleus at their core.[13]

The modern catenane in Figure 332 was designed so that the loops in the separated rings interact strongly. Furthermore, an aspect of this strong interaction "demands" that the appropriate long-chain precursor "recognize" and move into the interior of the complementary ring compound (see Figure 333).[9] This self-organization precedes the final synthetic step that closes the long chain into a ring and forms catenanes in an incredible 70% yield. Indeed, this is much the same way nature "thwarts" entropy by preorganization via molecular recognition prior to a catalyzed reaction. Figure 333 also illustrates a similar self-organization pathway for forming another exotic topological isomer—a rotaxane in which a long chain threading a cyclic molecule (in high yield due to molecular recognition) is then chemically capped with large terminal groups.

Nanotechnology raises some very interesting questions. For starters, is one gold atom truly gold? That is, is it metallic? The answer must be no since "metallicity" requires total delocalization of electrons over many (hundreds?) of metal atoms. So, when does a cluster of gold atoms begin to resemble gold? Here is another thought–when we think of a machine that stamps out a gear from a metal plate, we would not think of the stamping machine as a "catalyst." Clearly, the machine causes change and is itself unchanged after the operation except for slight wear. However, in real life enzymes eventually "wear down" and lose their potency. But of course the stamping machine is not a "catalyst" because it is causing mechanical, not chemical, change. However, suppose that a nanomachine or "nanobot" somehow facilitates very rapid assembly (i.e., chemical ring closure following self-organization of the two components) of the catenane in Figure 332 from its precursors. This would clearly be a chemical change and the nanomachine, nanoassembler, or "nanobot" would be both chemical catalyst and machine.[14]

But why should this seem so strange? It has been done for billions of years.[14] Molecular recognition drives the organization of the double helix, its replication

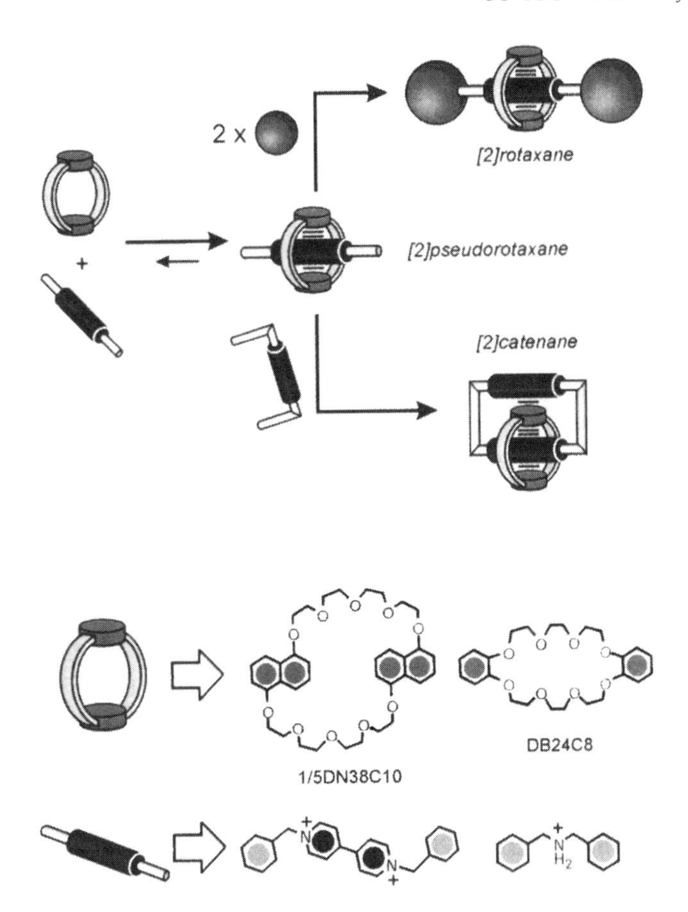

FIGURE 333. ▪ The use of molecular recognition and self-organization to form catenanes and rotaxanes capable of performing as controllable nanoswitches (see text). [Reprinted with permission from *Accounts of Chemical Research* (2001), Copyright (2001) American Chemical Society].

as well as its transcription to produce specific proteins. Indeed, it is molecular recognition that forms the complex of protein and RNA in the ribosome. It is molecular recognition that attracts transfer RNA (t-RNA) and their specific passenger proteins to the surface of the ribosome. The result is a supermolecular complex that links single amino acids into chains to form specific proteins. Clearly, the ribosome may simultaneously be regarded as a machine and a catalyst.[14] And what about assembling "nanobots" by these same self-organizational principles? Where will the resulting "gray goo" (novelist K. Eric Drexler's term[15]) stop? Will they continue to serve us or mutate into supercolonies having their own agendas and TV programs?

---

1. This is, of course, derived from the title of the autobiographical book by the late Professor Richard P. Feynman: R.P. Feynman, *Surely You're Joking, Mr. Feynman!—Adventures of a Curious Character,* W.W. Norton & Co., New York, 1985.
2. G. Stix, *Scientific American,* Vol. 285, No. 3, pp. 32–37, Sept. 2001.
3. Feynman's 1959 talk "There's Plenty of Room at the Bottom" can be found at *www.its.caltech.edu/~feynman.*
4. The definition of "nanotechnology" offered by Mihail C. Roco of the National Science Foundation indicates, among other things, that materials and systems must "have at least one di-

mension of about one to 100 nanometers" (see Stix, op. cit.). One nanometer (nm) is, of course, one-billionth of a meter (1×10^{-9} m). The sizes of atoms are typically presented in most textbooks in Ångstroms (Å). An ångstrom (1×10^{-10} m or 1×10^{-8} cm) is one-tenth the size of a nanometer. The diameter of an iron atom (in the metal) is roughly 2.5 Å or 0.25 nm. Thus, 100 nm would correspond to about 400 iron atoms in a line. A white blood cell is about 10 micrometers (10 μm) in diameter (see P. Morrison, P. Morrison, and the Office of Charles and Ray Eames, *Powers of Ten,* Scientific American Books, Inc., New York, 1982. This also corresponds to 10,000 nm or 40,000 iron atoms in a straight line. Objects on the order of 1–100 nm could be termed "mesoscale" if we consider subatomic particles to be at the lower end of the scale.

5. C.M. Lieber, *Scientific American*, Vol. 285, No. 3, pp. 59–64, Sept. 2001.
6. G.M. Whitesides and J.C. Love, *Scientific American*, Vol. 285, No. 3, pp. 39–47 (Sept. 2001).
7. R.E. Smalley, *Scientific American*, Vol. 285, No. 3, pp. 76–77, Sept. 2001.
8. A.R. Pease, J.O. Jeppesen, J. Fraser Stoddart, Y. Luo, C.P. Collier, and J.R. Heath, *Accounts of Chemical Research*, Vol. 34, pp. 433–444, 2001.
9. R. Ballardini, V. Balzani, A. Credi, M.T. Gandolfi, and M. Venturi, *Accounts of Chemical Research*, Vol. 34, pp. 445–455, 2001.
10. E. Wasserman, *Journal of the American Chemical Society*, Vol. 82, pp. 4433–4434, 1982.
11. G. Schill, Catenanes, *Rotaxanes and Knots*, Academic Press, New York, 1971.
12. R.R. Jones and R.G. Bergman, *Journal of the American Chemical Society*, Vol. 94, p. 660, 1972.
13. M.D. Lee, T.S. Dunne, M.M. Siegel, C.C. Chang, G.O. Morton, and D.B. Borders, *Journal of the American Chemical Society*, Vol. 109, pp. 3464–3465, 1987.
14. G.M. Whitesides, *Scientific American*, Vol. 285, No. 3, pp. 78–83, Sept. 2001.
15. K.E. Drexler, *Scientific American*, Vol. 285, No. 3, pp. 74–75, Sept. 2001.

NANOSCOPIC "HEAVENS"

In the movie "Fantastic Voyage" actress Raquel Welch is among a group of scientists and doctors tasked to remove an inoperable brain tumor from a VIP. The team enters a submarinelike vessel which is then reduced to microscopic dimensions and injected into the patient's bloodstream. A moment of elevated drama occurs when Ms. Welch, outside of the vessel, is attacked by blobby antibodies and the men vie for the honor of removing them from her bodysuit. Needless to say, after some tense moments, the tumor operation eventually succeeds and the movie ends happily.

Microscopic refers to objects that can be detected in common optical microscopes—they are microns (micrometers = 10^{-6} m) in dimension. Individual atoms are angstroms (10^{-10} m or 10^{-8} cm) in size. Large enough clusters of atoms form molecules or aggregates of molecules (such as viruses) that are tens of angstroms or nanometers (1 nm = 10^{-9} m) in scale. What if we could make computers, machines and even robots out of nanoscale parts? Clearly, Nature has already mastered nanotechnology, why can't we?

Figure 334 depicts two molecules that were each synthesized by merely mixing equal quantities of a *linear* bifunctional molecule (the "edges" of the squares end in nitrogen atoms) and an *angular* bifunctional molecule that makes a 90° bend [the "corners" of the square are centered on metallic (M) atoms].[1] This synthesis is depicted in equation (b) of Figure 335. Figures 335 and 336 show other possibilities [equations (a) to (h)] for joining bifunctional linear (1) and/or bifunctional angular (a) molecules to form regular polygons.[1] If one of the two

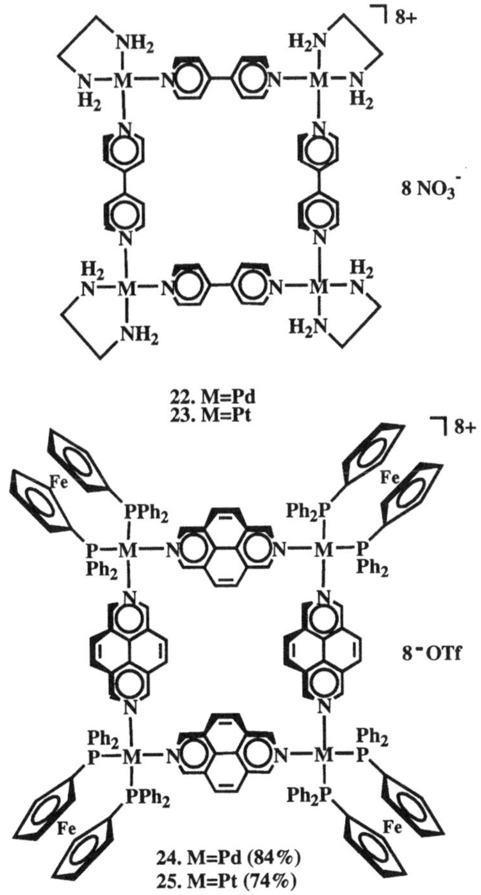

Chart 2.

FIGURE 334. ■ Chemical squares joined by bifunctional linear molecules and bifunctional angular (90°) molecules joined by dative bonds about 20% as strong as covalent bonds. Nature allows the four molecules that join to form each of these structures to form–break–reform until they "get it right" (P.J. Stang and B. Olenyuk, *Accounts of Chemical Research*, Vol. 30:502, 1997) (courtesy American Chemical Society).

molecules is trifunctional and it is combined with a bifunctional molecule, the result is a regular polygon [equations (j) to (m) in Figure 336].

Using this approach (see Figure 337), *planar* trifunctional molecule **1** was merely mixed with *angular* bifunctional molecule **3** in a 2:3 molar ratio in methyiene chloride solution. In 10 minutes a virtually perfect reaction yielding pure cuboctahedron **5** was complete.[2] The dimension of the huge molecule is about 5 nm (or 50 ångstroms). Furthermore, this remarkable approach was successful in making a nanoscopic dodecahedron of 5880 atoms (see Figure 338) by merely mixing a 2:3 molar ratio of *nonplanar* trifunctional molecules and linear bifunctional molecules;[3] its formula: $C_{2900}H_{2300}N_{60}P_{120}S_{60}O_{200}F_{180}Pt_{60}$.

We have now come full circle over the course of 2500 years. The ancient Pythagoreans envisioned a mathematical basis to matter and the four earthly elements and the fifth, heavenly element (the "ether") were represented by the five

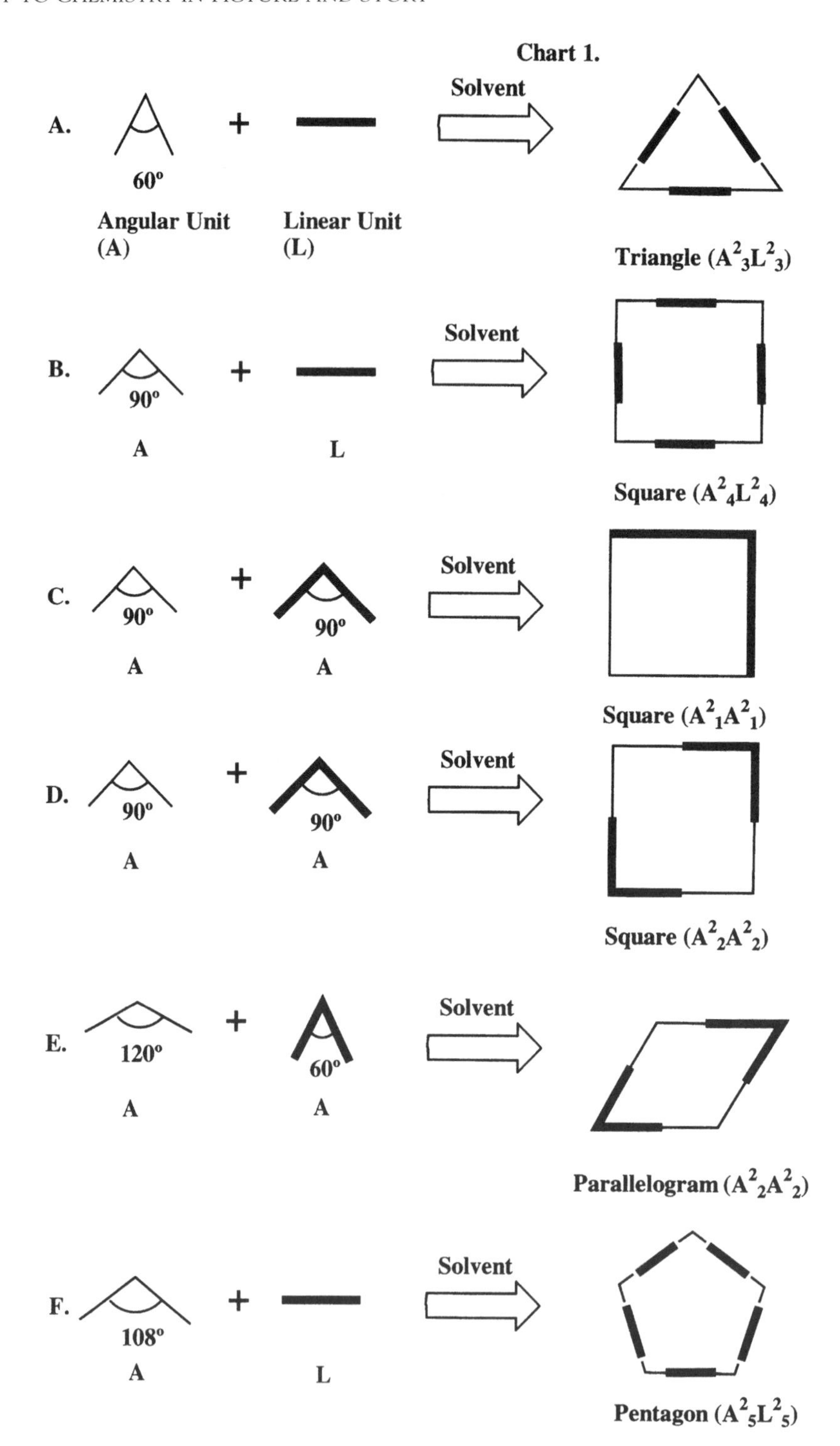

FIGURE 335. ■ Molecules that join to form polygons (see Stang and Olenyuk, Figure 334; courtesy American Chemical Society).

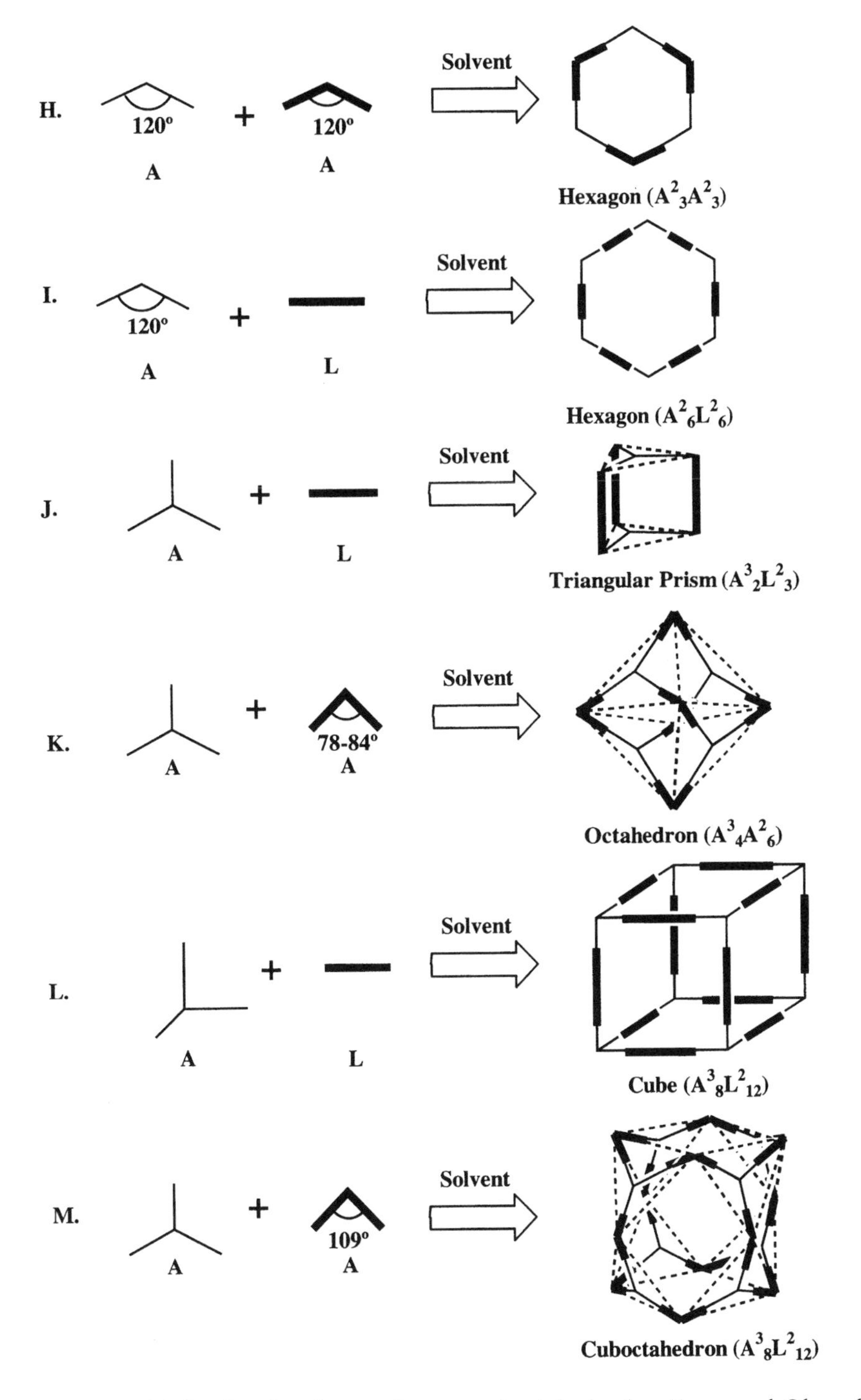

FIGURE 336. ■ Molecules that form polygons and polyhedra (see Stang and Olenyuk, Figure 334; courtesy American Chemical Society).

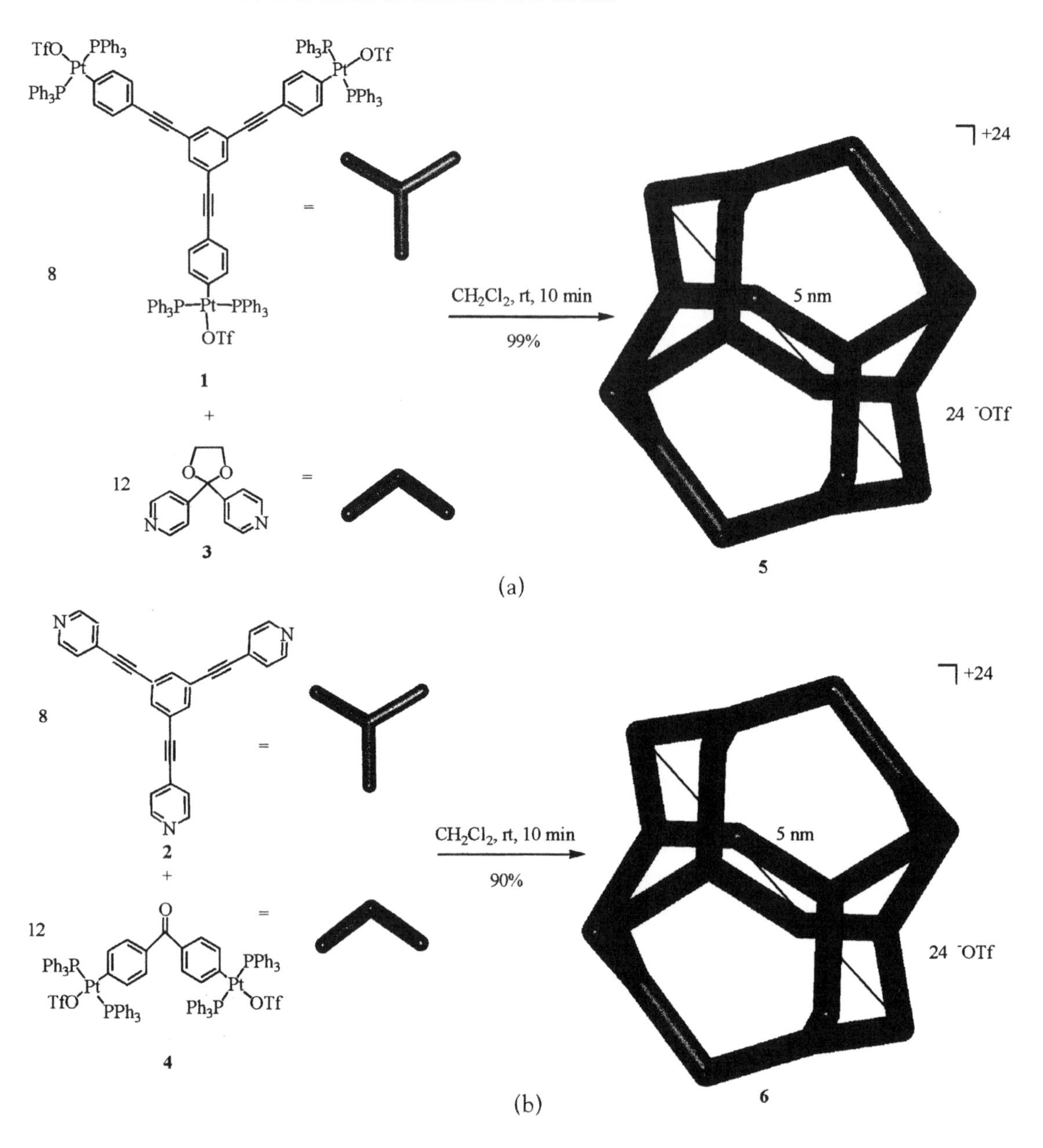

FIGURE 337. ■ The formation of a nanoscopic cuboctahedron (5 nanometers across) in 99% yield in 10 minutes using the scheme outlined in Figure 336 (B. Olenyuk, J.A. Whiteford, A. Fechtenkotter, and P.J. Stang, *Nature*, Vol. 398:794, 1999; courtesy *Nature;* the author thanks Peter J. Stang for this figure).

Platonic solids (see Kepler's *Harmonices Mundi*, Fig. 4). Platonic solids held together by strong covalent bonds have been known for some time. In white phosphorus (P_4), the atoms occupy the corners of a tetrahedron.[4] How *did* the ancients know that the tetrahedron was appropriate for fire and also for fiery white phosphorus? Starting almost 40 years ago, clever organic chemists laboriously "tricked" nature, seemingly "thwarted" entropy, and assembled cubes, dodecahedra, and tetrahedra of covalently attached carbons.[5] Nature, however, had its

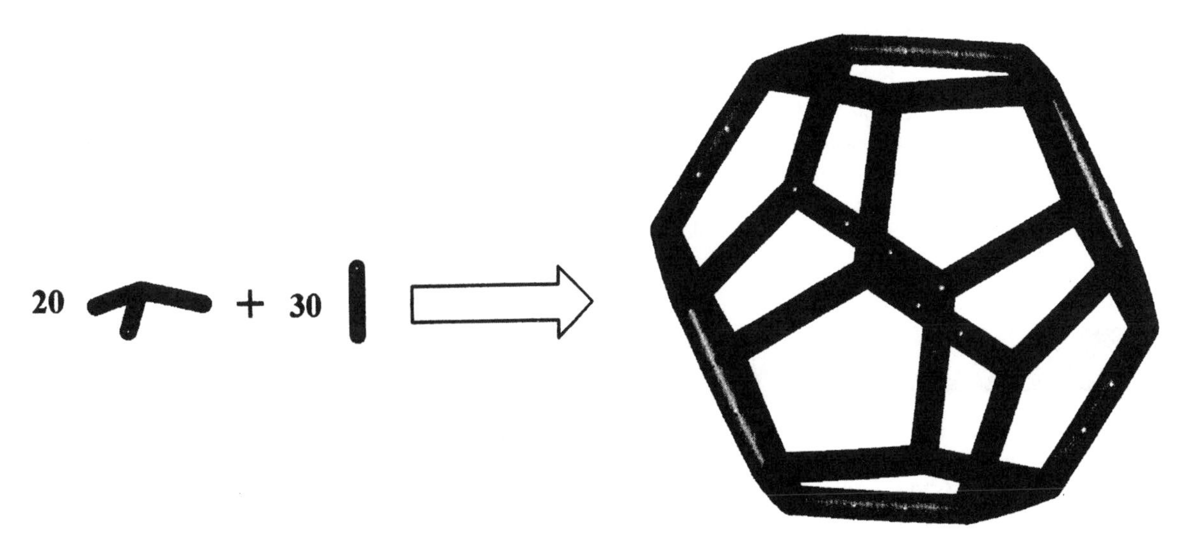

FIGURE 338. ■ The formation of a nanoscopic dodecahedron ("heaven," see Figure 4) using the conceptual approach shown of Stang and Olenyuk.[3] It is the largest abiological system made by self-assembly.[3]

own tricks in mind and in the late 1980s, the soccer ball of carbon atoms (C_{60}), "buckminsterfullerene" or "buckyball," a truncated icosahedron, was discovered in soot.[6] The truncated icosahedron ("soccer ball") and the cubocta-hedron are two examples of the 13 Archimedean semiregular solids. The five Platonic solids all have one type of polygonal face—triangles for an icosahedron, for example. The truncated icosahedron, in contrast, has both pentagonal and hexagonal faces. Although octahedra and icosahedra are not stable structures for carbon, the former are well represented by molecules containing transition metals such as rhenium and cobalt and the latter by elemental boron and a variety of boron-containing molecules and ions.[4]

How does nature (and we chemists are part of the natural world) assemble large orderly nanostructures like viruses and nanoscopic dodecahedra? First, it prefabricates complex units such as proteins using the genetic code. Exacting chemical synthesis is required for the synthetic structural units shown in Figures 334 to 338. These units then self-assemble using weak forces such as van der Waals interactions, dipole–dipole forces, and hydrogen bonds to organize spontaneously and self-order into an optimal structure. In the case of the nanostructures described in Figures 334 to 338, ligand–metal "dative" bonds that are perhaps only 20% as strong as covalent bonds are employed. If strong covalent bonds are formed in a chemical reaction, the final product may well depend upon the initial reaction conditions (e.g., temperature, pressure) since sometimes the product formed fastest will prevail, sometimes the most stable product will prevail and sometimes mixtures of the two (if they are indeed different) will be found. This is precisely the issue that confounded Berthollet (Figure 225) on the eve of the Atomic Theory. By contrast, structures held together by weak bonds will associate–dissociate–reassociate (build–repair, anneal) until the best structure is formed and the entire process may well be complete in minutes. In short, Nature will find a way.

1. P.J. Stang and B. Olenyuk. *Accounts of Chemical Research*, **30**:502–518, 1997.
2. B. Olenyuk, J.A. Whiteford, A. Fechtenkotter, and P.J. Stang, *Nature*, **398**:794–796, 1999.
3. B. Olenyuk and P.J. Stang, *Journal of the American Chemical Society*, **121**, 10, 434, 1999. See also: *Chemical & Engineering News*, November 15, 1999, p. 11.
4. F.A. Cotton and G. Wilkinson, *Inorganic Chemistry*, 5th ed., Wiley, New York, pp. 18–21.
5. A. Greenberg and J.F. Liebman. *Strained Organic Molecules*, Academic, New York, 1978.
6. R.F. Curl and R.E. Smalley. *Scientific American*, October. 1992. p. 54.

MOVING MATTER ATOM-BY-ATOM

Chemistry textbooks inform us that John Dalton formulated Atomic Theory in 1803 and imply that atoms were accepted from then on. Actually, such acceptance was far from universal and late-nineteenth-century books such as Brodie's *The Calculus of Chemical Operations* (London, 1866, 1877) and Hunt's *A New Basis for Chemistry: A Chemical Philosophy* (Boston, 1887), although antiatomic in nature, were not written by cranks or "nutters." The eminent physicist Ernst Mach and the famous chemist Wilhelm Ostwald resisted the reality of atoms into the beginning of the twentieth century. Jacob Bronowski strongly implies that the suicide in 1906 of Ludwig Boltzmann, who successfully explained heat as atomic and molecular motion, stemmed in part from his failure to totally convince the scientific community that atoms are real.[1]

However, at just about the same time, Albert Einstein developed a mathematical theory of the movement of microscopic particles in liquids (Brownian movement first analyzed by R. Brown in 1828) that modeled them as gas molecules.[2] In 1908, Jean Perrin explained the Brownian motion of microscopic particles in liquids and tobacco smoke, and used his data to make an excellent estimate of Avogadro's Number.[2] His book *Les Atomes* (Paris, 1913; London, 1916) laid out the case for the absolute reality of atoms and brought together a number of different ways of determining Avogadro's Number. These studies gained him the 1926 Nobel Prize in Physics.

Roughly 80 years after Boltzmann died by his own hand, we are imaging atoms, picking them up, moving them, and depositing them one at a time. Ernst Ruska, Gerd Binnig, and Heinrich Rohrer shared the 1986 Nobel Prize in Physics for their invention of the scanning tunneling microscope (STM). The STM "skates" a metallic tip of atomic dimensions to near atomic distances from surfaces of atoms or molecules. At these close distances, there is "crosstalk" between the electrons that "tunnel" between the two populations of atoms. The STM senses the miniscule changes in pressure required to keep a constant current and thus traces images of the atoms. Under certain conditions, an "energy trap" can be created under the STM probe tip that will allow the capture of an individual atom and its transfer across a surface. Figure 339 is a computer-generated model of an STM tip moving a xenon atom.[3] The STM had become a vital instrument in nanotechnology research.

Is the image in Figure 340 an extraterrestrial landscape, a fluted pie crust, the eye of a chameleon or the work of an abstract artist? Incredibly, it is an STM image of a "quantum corral" formed by moving 48 iron atoms one by one into a

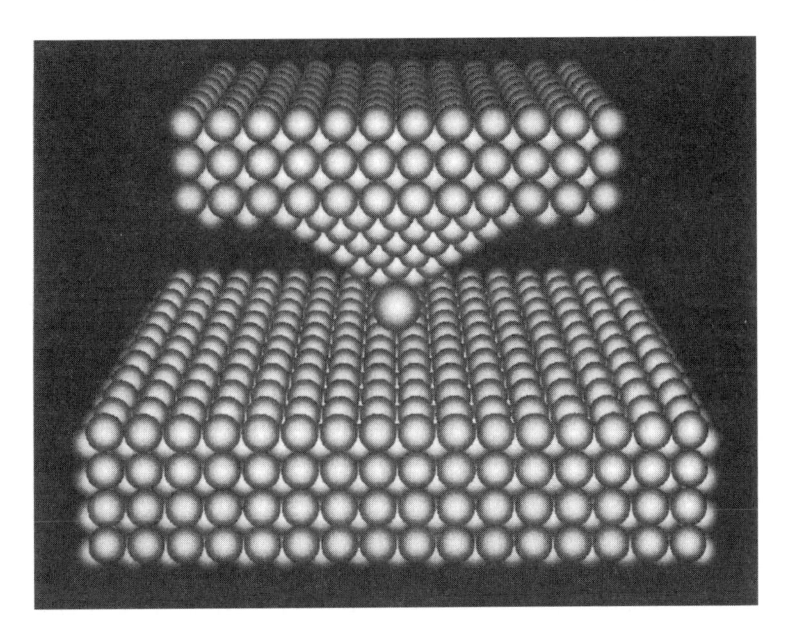

FIGURE 339. ▪ Schematic of the scanning tunneling microscope (STM) the tip of which is of atomic dimension (P. Avouris, *Accounts of Chemical Research*, **28:**95, 1995; courtesy American Chemical Society; the author thanks Dr. Phaedon Avouris, IBM Research Division, for this figure).

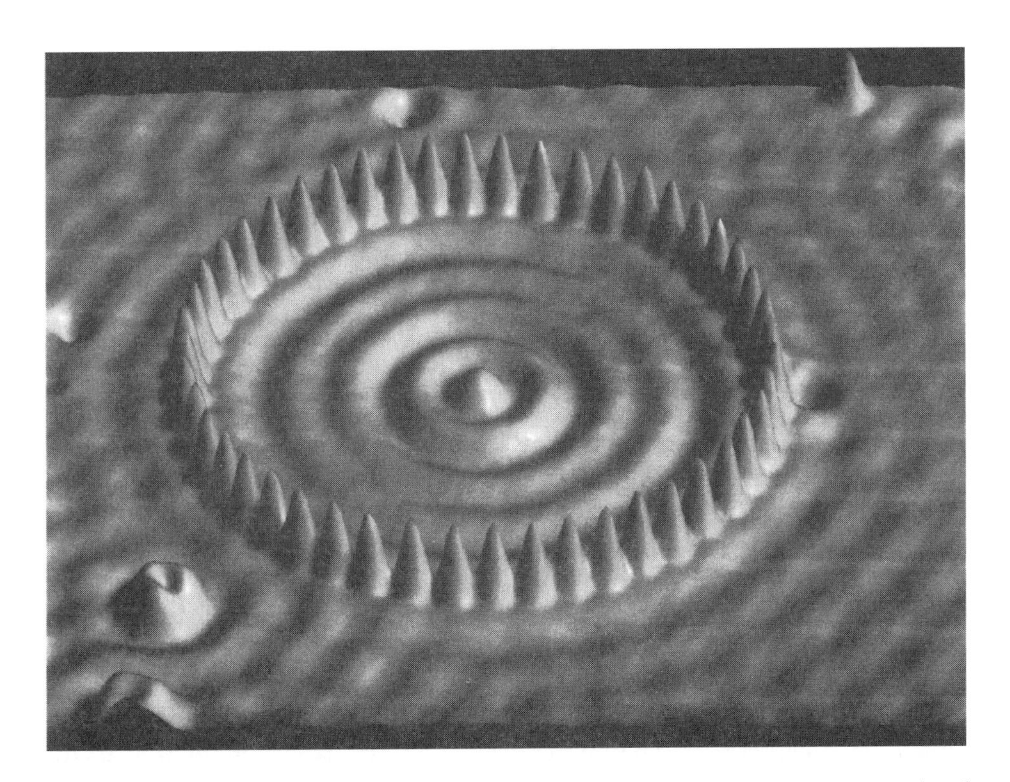

FIGURE 340. ▪ STM image of the "quantum corral" consisting of 48 iron atoms placed one at a time. The image shows the particle–wave nature of electrons (P. Avouris, *Accounts of Chemical Research*, **28:**95, 1995; courtesy American Chemical Society; the author thanks Dr. Phaedon Avouris, IBM Research Division, for this figure).

circle.[3] The ripples in the center reflect a standing wave produced by surface electrons confined by the circle of atoms and "provide a striking demonstration of the wave-particle nature of the electron."[3]

And what does that mean? In the 1920s Louis DeBroglie described electrons as both particles and waves because they have precise mass, go "splat-splat-splat" (or "click-click-click") into Geiger counters yet show interference like radio and light waves. It is one thing to say "particle-waves" and quite another to really picture them. Try it. Our problem is that electrons are outside of both our direct senses and experiences. As Bronowski notes, twentieth-century physics introduced abstraction and uncertainty and the need for what he describes as "tolerance" in modeling nature.[4] The nineteenth-century satire Flatland by Shakespearean scholar Edwin A. Abbott illustrates our limitations.[5]

A sphere, resident of the three-dimensional world of "Spaceland," visits the two-dimensional world of "Flatland" where he meets a square Flatlander. The square perceives the sphere in limited ways but only starts to truly understand his own limits in perception when the two visit one-dimensional "Line-land." Ironically, the square quite innocently turns the tables on the seemingly omniscient sphere as follows:[6]

Square: But my Lord has shewn me the intestines of all my countrymen in the Land of Two Dimensions by taking me with him into the Land of Three. What therefore more easy now than to take his servant on a second journey into the blessed Land of the Fourth Dimension . . . ?
Sphere: But where is this Land of Four Dimensions?
Square: I know not: but doubtless my Teacher knows.
Sphere: Not I. There is no such land. The very idea of it is inconceivable.

Incidentally, the sphere and the square finally visit zero-dimensional "Pointland" where they hear the sole resident singing hymns of self-praise:[6] "It fills all Space and what It fills, It is. What It thinks, that It utters; and what It utters, that It hears; and It itself is Thinker, Utterer, Hearer, Thought, Word, Audition; It is the One and yet the All in All. Ah, the happiness, ah, the happiness of Being." Have you ever met this type of person? Such self-satisfaction and isolation are inimical to all human endeavors including science.

Our mental images of matter continue to evolve. In late 1999, a group of scientists coupled x-ray and neutron diffraction techniques with quantum mechanical calculations to physically "see" the shape of an electron orbital.[7,8] The technique involved comparing an experimental electron density distribution with a calculated electron density distribution and plotting a difference density map. "We were just amazed when it first came up on the screen," exclaimed one of the scientists.[9] As we continue to probe the innermost secrets of chemical bonding, first explained in part by Bohr's solar system atom almost 90 years ago, I am reminded of the closing movements of Gustav Hoist's symphonic opus *The Planets*. The mysterious outermost planets are evoked in the music which gradually disappears into the void leaving an open-ended sense of wonder—a metaphor for the very human curiosity that urges scientific exploration.

1. J. Bronowski, *The Ascent of Man*, Little, Brown, Boston, 1973, pp. 347–351.
2. J.R. Partington, *A History of Chemistry*, MacMillan, London, 1964, Vol. 4, pp. 744–746.
3. P. Avouris, *Accounts of Chemical Research*, **28:** 95–102, 1995. I am grateful to Dr. Phaedon Avouris, T.J. Watson Research Center, IBM Research Division, for kindly supplying original of Figures 339 and 340.
4. J. Bronowski, op. cit., Chap. 11.
5. E.A. Abbott, *Flatland: A Romance of Many Dimensions* (with Foreward by Isaac Asimov), Barnes & Nobles, New York, 1983.
6. E.A. Abbott, op. cit., pp. 102–103, 109–110.
7. J.M. Zuo, M. Kirn, M. O'Keefe, and J.H. Spence, *Science*, **401**:49–52, 1999
8. C.J. Humphreys, *Science*, **401**:21–22, 1999.
9. M. Jacoby, *Chemical and Engineering News*, September 6, 1999, p. 8. See also M.W. Browne, *New York Times*, September 7, 1999, pp. D1–D2.

A NANOCAR ROLLING ON A GOLD-PAVED ROAD

In 1985, Richard E. Smalley (1943–2005), Robert F. Curl, Jr. (1933–), and Harry W. Kroto (1939–) made the startling discovery of the soccer-ball-like C_{60} (named informally "buckminsterfullerene," after the architect–engineer Buckminster Fuller, and often called "buckyball" for short). By the end of the decade, "buckyball" was available in quantity and a related class of "buckytubes" started to play a major role in the rapidly developing field of nanotechnology. Smalley, Curl, and Kroto would share the 1996 Nobel Prize in chemistry.

Although one can design on paper motor-like devices constructed of a large single molecule or associated molecules, reality is more complex. On a molecular scale, quantum effects, irrelevant for even micron-sized (10^{-6} m) assemblies, can be significant. Moreover, the energies available in the "thermal bath" that surrounds such assemblies may be comparable to their binding and/or kinetic energies. Thus, in these earliest decades of exploration in nanotechnology much effort is devoted toward developing the art of constructing simple devices and demonstrating what they can and cannot do.[1]

At Rice University, where C_{60} was discovered, James M. Tour and Kevin F. Kelly and their co-workers published in 2005 the synthesis and properties of a "single-molecule nanocar."[2] The molecule ["Nanocar 1" in Figure 341(a)] was synthesized by fairly conventional organic chemistry given the availability of C_{60}. The final step involves "snapping on" the four C_{60} wheels chemically. "Nanocar 1" has the chemical formula $C_{430}H_{274}O_{12}$ and while it is tempting to call it a "Formula One" car, we will not do so since the pun is simply too painful. Just like any other large organic molecule, it has been fully characterized by FTIR, ^{1}H-, as well as ^{13}C-NMR, and MALDI-TOF mass spectrometry.[2] Although the large molecule is quite stable thermally, at 300°C its four C_{60} wheels fall off. "Nanocar 1" decomposes above 350°C. The distance between the centers of the spherical front C_{60} wheels is about 3.3 nm (33 Å), whereas the distance between the centers of a front and a rear wheel is about 2.1 nm. The area of a small, visible pencil dot should accomodate about 10 million tightly parked nanocars. Even as an open "go-kart" this nanocar is not a

$OC_{10}H_{21}$ $OC_{10}H_{21}$
H $C_{10}H_{21}O$ $C_{10}H_{21}O$ H
$C_{10}H_{21}O$ $OC_{10}H_{21}$
$C_{10}H_{21}O$ $OC_{10}H_{21}$
$OC_{10}H_{21}$ $OC_{10}H_{21}$
H $C_{10}H_{21}O$ **Nanocar 1** $C_{10}H_{21}O$ H

FIGURE 341. ■ Synthesis and motion on a gold surface of a molecular nanocar. See color plates. (Courtesy of James Tour.)

very comfortable ride for one of the smallest known viruses (the human rhinovirus, diameter ca. 30 nm). It is, however, certainly the right size to shuttle various molecules to and fro.

The most interesting test provided by "Nanocar 1" is its physical behavior on a thin gold film coated onto mica. Monitored by a scanning tunneling microscope (STM), the C_{60} wheels of each "Nanocar 1" adhere to the gold surface, keeping it stationary up to 170°C. At 200°C, "Nanocar 1" rolls (not slides) slowly in a direction perpendicular to its axles [Figure 341(b)]. Above 225°C, the motion becomes rapid and erratic."[2] However, an STM tip was used successfully to pull the nanocar forward in one direction as the C_{60} wheels turned.

It is fair to say that assembly line manufacture of larger versions of "Nanocar 1" by a General Nanomotors will not occur in the near future, nor will some *E. coli* bacterium soon be kicking each C_{60} prior to a purchase. However, the knowledge gained in observing physical interactions of structures like "Nanocar 1" with surfaces will certainly be of great value.

---

1. *Scientific American Special Issue: Nanotech—The Science of Small Gets Down to Business* (September, 2001).
2. Y. Shirai, A.J. Osgood, Y. Zhao, K.F. Kelly, and J.M. Tour, *Nano Letters*, **5**: 2330–2334, 2005. Figure 341(b) was supplied by Professor Tour for publication in *Chemical & Engineering News*, December 19, 2005, p. 20.

FEMTOCHEMISTRY: THE BRIEFEST FLEETING MOMENTS IN CHEMISTRY

The time scales on which chemists could monitor chemical reactions shortened from milliseconds (10^{-3} s) in the 1920s and 1930s, using chemical kinetics, to microseconds (10^{-6} s) during the 1950s and early 1960s, thanks to flash photolysis and kinetic relaxation techniques. The invention of the ruby laser in 1960 provided chemists and physicists with a new tool that shortened the "clock" to nanoseconds (10^{-9} s) during the 1960s and picoseconds (10^{-12} s) during the 1970s.[1,2] Picoseconds are approximately the lifetimes of molecular vibrations and are on the order of the lifetimes of the transition states that connect reactants to products. In 1987, the Egyptian-born chemist Ahmed H. Zewail (1946–), of Caltech, would exploit newly developed techniques employing ultrashort laser pulses and a process called mode locking. He was able to generate laser pulses with lifetimes on the order of femtoseconds (10^{-15} s) that allowed the observation of the relative motions of individual atoms in a molecule. He noted that atoms typically move at speeds around 1 km/s (10^5 cm/s) and monitoring motion at the atomic scale or less (1 Å or 10^{-8} cm) requires a time scale on the order of 100×10^{-15} s or about 100 femtoseconds (fs). While seemingly nudging the limits imposed by the Heisenberg Uncertainty Principle, Zewail demonstrated the ability to spectroscopically observe transition states, a feat previously thought by many to be impossible. Professor Zewail received the 1999 Nobel Prize in chemistry for pioneering femtochemistry.

In his experiment, Zewail used a common laser source to both initiate a chemical change and to monitor it. When a femtosecond *pump* pulse passes through the sample, chemical change begins at time = zero. Femtosecond *probe* pulses are sent on a slightly different path whose length may be slightly varied. The difference in the path lengths of the pump and probe pulses divided by the speed of light (299,792 km/s) provides the time scale for the response (e.g., fluorescence) produced by the chemical species being monitored. Zewail called the technique "laser stroboscopy" and likened it to the late-nineteenth-century studies by Eadweard Muybridge at the University of Pennsylvania, who employed light flashes with a regular time interval of 0.052 seconds to demonstrate that a galloping horse does pass through moments in which all four hooves are off the ground. During the middle of the twentieth century Harold Edgerton at MIT shortened the time interval and produced stop-motion pictures of microsecond duration.

Figure 342 illustrates key aspects of one of Zewail's first successful picosecond experiments. Figure 342(a) is a very simplified energy surface for the ground state and an excited state of the molecule I-CN. The vertical scale is energy and the horizontal scale essentially represents the distance R separating I and CN. The bottom curve is the Morse potential curve for vibration of the I-CN covalent bond in the stable (ground-state) molecule and R at the minimum of the curve is the I-CN bond length. The pump pulse at t = 0, places I-CN into its un-

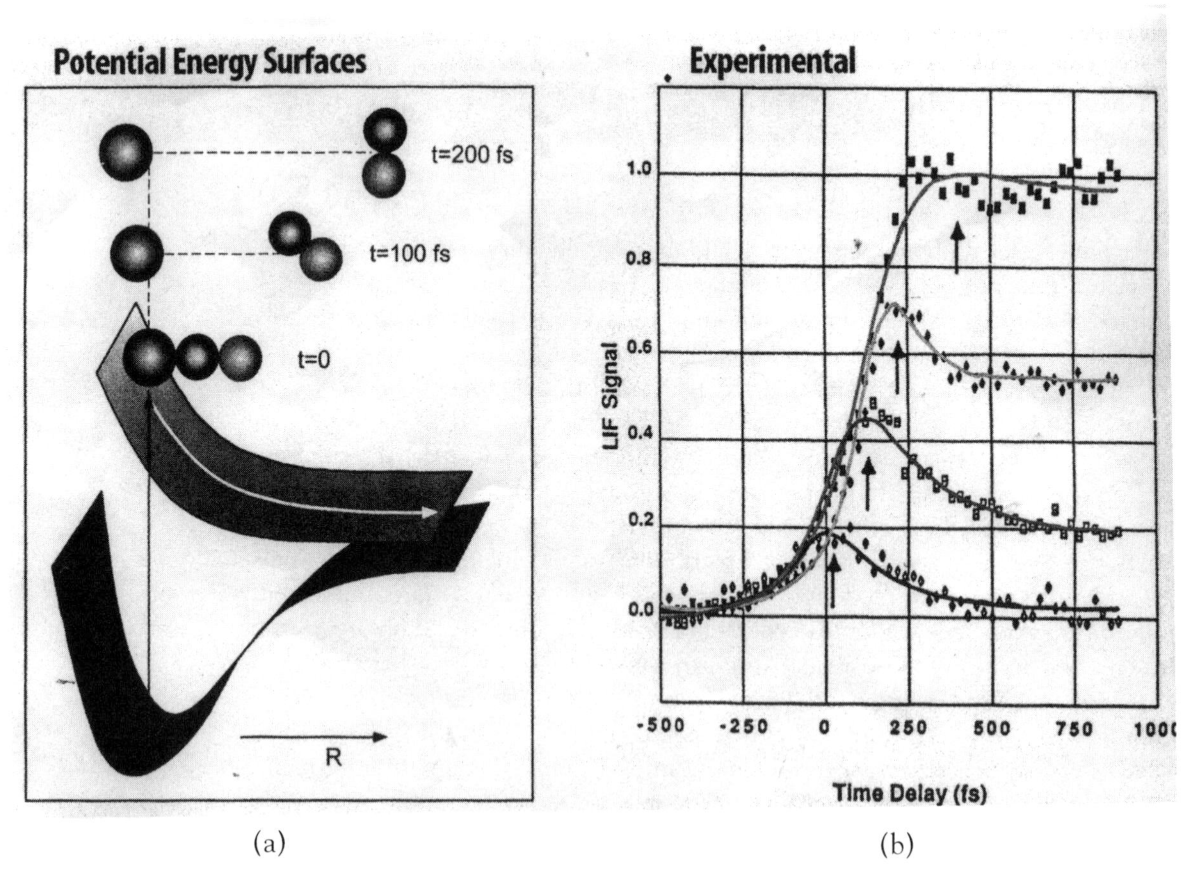

(a) (b)

FIGURE 342. ■ Representations of Zewail's 1987 experiment using femtosecond laser pulses to study the decomposition of I-CN (see text).[1] (Courtesy American Chemical Society and Professor Ahmed H. Zewail.)

stable excited state (upper curve) and the probe pulse causes fluorescence of this species as time passes. Figure 342b shows the disappearance of I-CN and the appearance of free CN. The two species are essentially free at 200 fs, with a transition state near 100 fs. Another early experiment involved monitoring the continuous change in "the nature of the chemical bond" in gas-phase NaI as it changed from covalent to ionic on the reaction coordinate. It is certainly most fitting that Zewail's title at Caltech is now Linus Pauling Professor of Chemistry and Professor of Physics. Much as Pauling clarified the nature of the chemical bond, Zewail is clarifying the "dynamics of the chemical bond."[3]

---

1. A.H. Zewail, *Journal of Physical Chemistry A*, **104:** 5660–5694, 2000.
2. J.S. Baskin and A.H. Zewail, *Journal of Chemical Education*, **78:** 737–751, 2001.
3. The author acknowledges helpful discussions with Ahmed H. Zewail.

SECTION X
SOME BRIEF CHEMICAL AMUSEMENTS

CLAIRVOYANT PICTURES OF ATOMS—A STRANGE CHYMICAL NARRATIVE

> I remember the occasion vividly. Mr. Leadbeater was then staying at my house, and his clairvoyant faculties were frequently exercised for the benefit of myself, my wife and the theosophical friends around us. I had discovered that these faculties, exercised in the appropriate direction, were ultra-microscopic in their power. It occurred to me once to ask Mr. Leadbeater if he thought he could actually *see* a molecule of physical matter. He was quite willing to try, and I suggested a molecule of gold as one which he might try to observe. He made the appropriate effort, and emerged from it saying the molecule in question was far too elaborate a structure to be described. It evidently consisted of an enormous number of some smaller atoms, quite too many to count; quite too complicated in their arrangement to be comprehended. It struck me at once that this might be due to the fact that gold was a heavy metal of high atomic weight, and that observation might be more successful if directed to a body of low atomic weight, so I suggested an atom of hydrogen as possibly more manageable. Mr. Leadbeater accepted the suggestion and tried again. This time he found the atom of hydrogen to be far simpler than the other, so that the minor atoms constituting the hydrogen atom were countable. They were arranged on a definite plan, which will be rendered intelligible by diagrams later on, and were eighteen in number.[1]

This narrative appears early in the second edition (published in 1919) of the strange but fascinating book *Occult Chemistry—Clairvoyant Observations on the Chemical Elements*, authored by Annie Besant and Charles W. Leadbeater and first published in 1908.[2] The fact that a deluxe edition of the book was published as late as 1951[3] is evidence of the enduring allure of this richly illustrated text. Well, what are we to make of its contents? Figure 343 depicts the structure of a "chemical atom" of sodium. *The method of examination employed was that of clairvoyance.*[4] This chemical atom consists of upper and lower parts (each composed of a globe and 12 funnels) and a connecting rod. The parts inside the funnels, globes, and rod and counted below are the "smaller atoms" referred to above:

> We counted the number in the upper part: globe–10; the number in two or three of the funnels—each 16; the number of funnels—12; the same for the lower part; in the connecting rod—14; Mr. Jinarajadasa reckoned: 10 + (16 × 12) = 202; hence 202 + 202 + 14 = 418: divided by 18 = 23.22 recurring. By this method we guarded our counting from any pre-possession, as it was impossible for us to know how the various numbers would result on addition, multiplication and division, and the exciting moment came when we waited to see if our results endorsed or approached any accepted weight.[5]

Et, voilá! The accepted atomic weight of sodium is 23.0—in pretty darned good agreement with 23.2, it seems. In total, 57 of the 78 recognized elements were ex-

PLATE I.

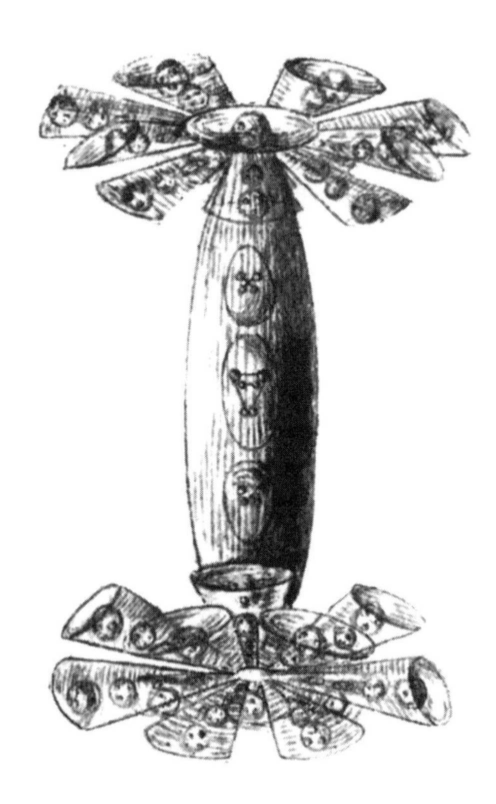

SODIUM.

FIGURE 343. ■ The structure of the sodium atom obtained through clairvoyance and published in the classic book (*Occult Chemistry*, 1908). There are a total of 418 smaller "physical atoms" in the sodium "chemical atom." Clairvoyance also showed that a hydrogen atom consisted of 18 smaller physical atoms. So . . . 418 divided by 18 = 23.22, in pretty darned good agreement with 22.99! (Q.E.D.)

amined as well as one previously unknown, "occultum," a "chemical waif" tucked between hydrogen and helium. Also, six new "varieties" of known elements were reported—not a bad day's haul. The agreement between accepted atomic weights and the clairvoyant count of "smaller atoms" is impressive, and this is one major component of the scientific validation of the atoms derived through clairvoyance.

The clairvoyant uncertainty in obtaining an observation is noteworthy. Note that only two or three funnels are sampled for counting, and uncertainties in counting of 1 or 2 "smaller atoms" are admitted by the investigators. Pretty difficult to keep this vision from flickering in and out without getting a severe headache. So the limitations here might illustrate a kind of "clairvoyance uncertainty principle."[6]

But what of the constituent "smaller atoms?" They are shown in Figure 344—these fundamental ("smaller" or "ultimate physical") atoms are found to be male and female. For the male atoms, force whorls in from fourth-dimensional space (the astral plane) and *out* into the physical world. The female atoms take force *in* from the physical world and whorl it, in the opposite screw sense, back into the astral plane. *Hmm.* The relationship to the male (sulfur) and female (mercury) imagery of early alchemy is pretty apparent—Sol and Luna, the "atoms family."

The other powerful scientific validation of the chemical atoms derived by clairvoyance is their seeming consistency in explaining the chemical and physical properties so neatly organized by the periodic law. Figures 345 and 346 depict family types of chemical atoms, not protozoans as they might appear to the uncritical eye (or even *d* and *f* orbitals to a wishful-thinking chemist!). The point can be illustrated succinctly—the structural type I in Figure 345 is classified as the "dumbbell" class and includes copper, silver, and gold, three coinage metals found in group 11 of the periodic table. However, Besant and Leadbeater also

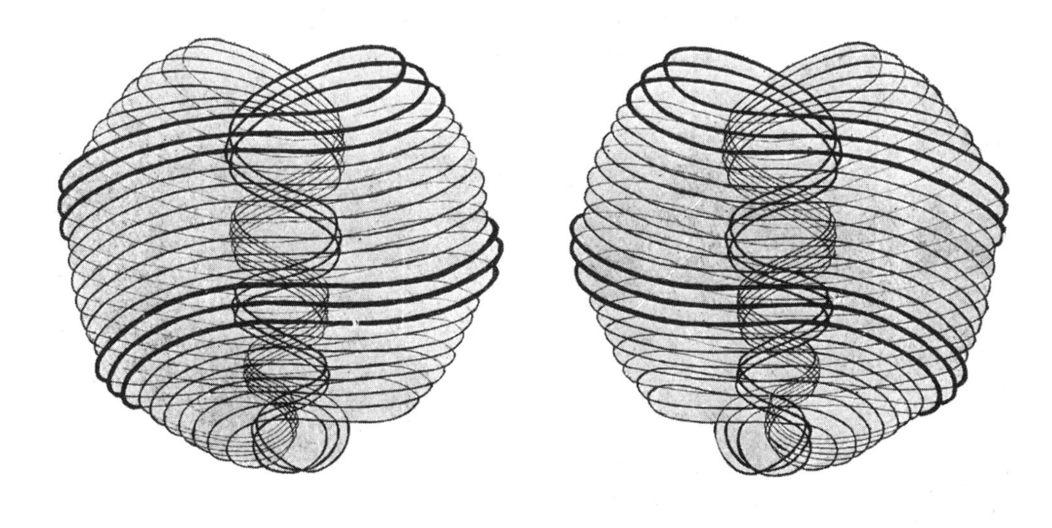

FIGURE 344. ■ The smaller physical atoms are, by the way, male and female. We have returned to Sol and Luna, sulfur and mercury. Thus, the hydrogen atom has nine female smaller physical atoms and nine male physical atoms. (From *Occult Chemistry*, 1908.)

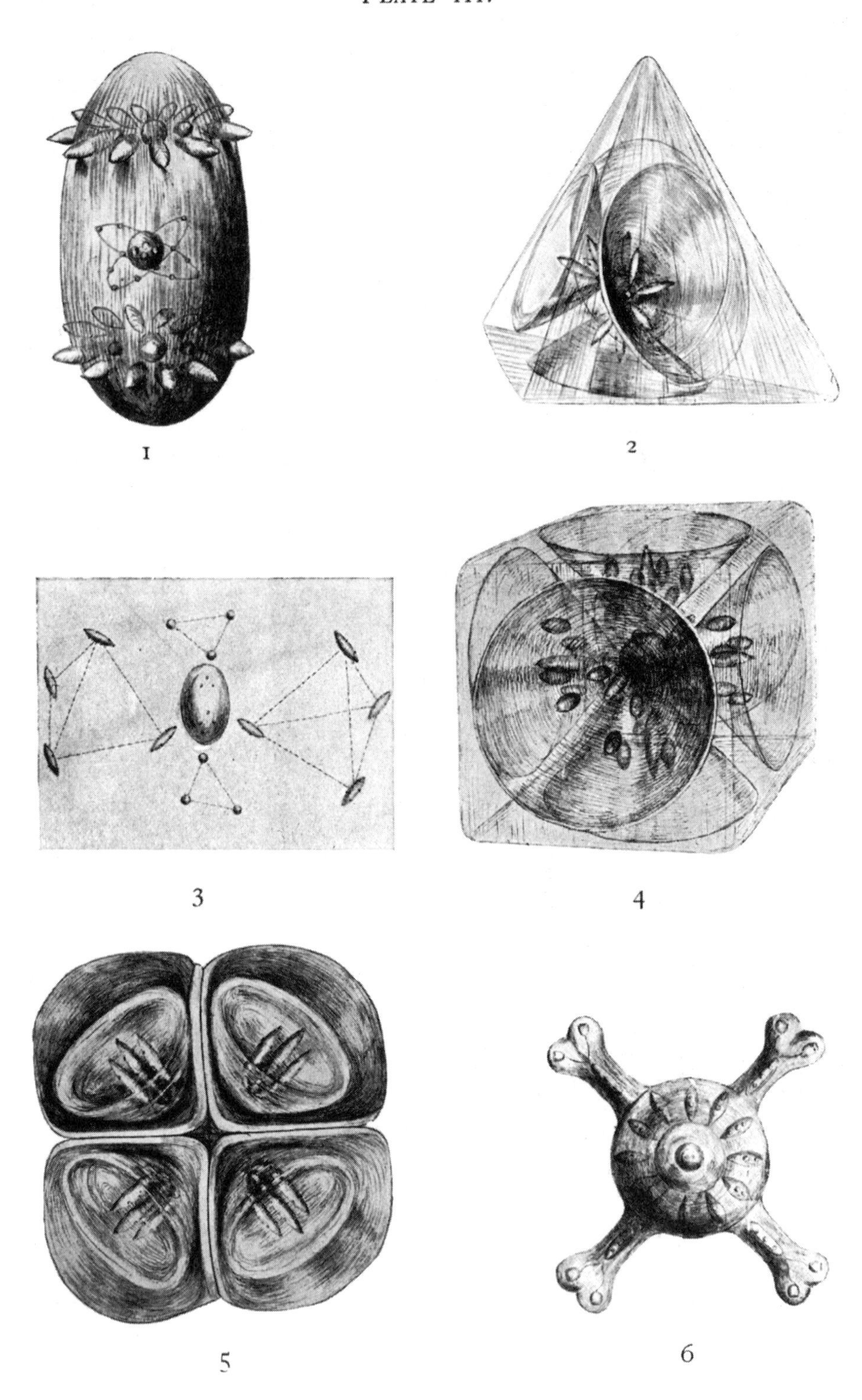

FIGURE 345. ■ Here is why the Periodic Law works according to Besant and Leadbeater—atoms in the same families share similar shapes. For example, look at structure 1 in this figure—representing the structure of copper, silver and gold atoms. Since some early classifications actually grouped sodium with copper, silver, and gold (see Figures 275 and 347), it would appear perfectly sensible that these four atoms would bear some familial resemblance (see Figure 343). Fortunately, we do not make coins of sodium, so our change does not burn holes in our pockets. (Figures from *Occult Chemistry*, 1908.)

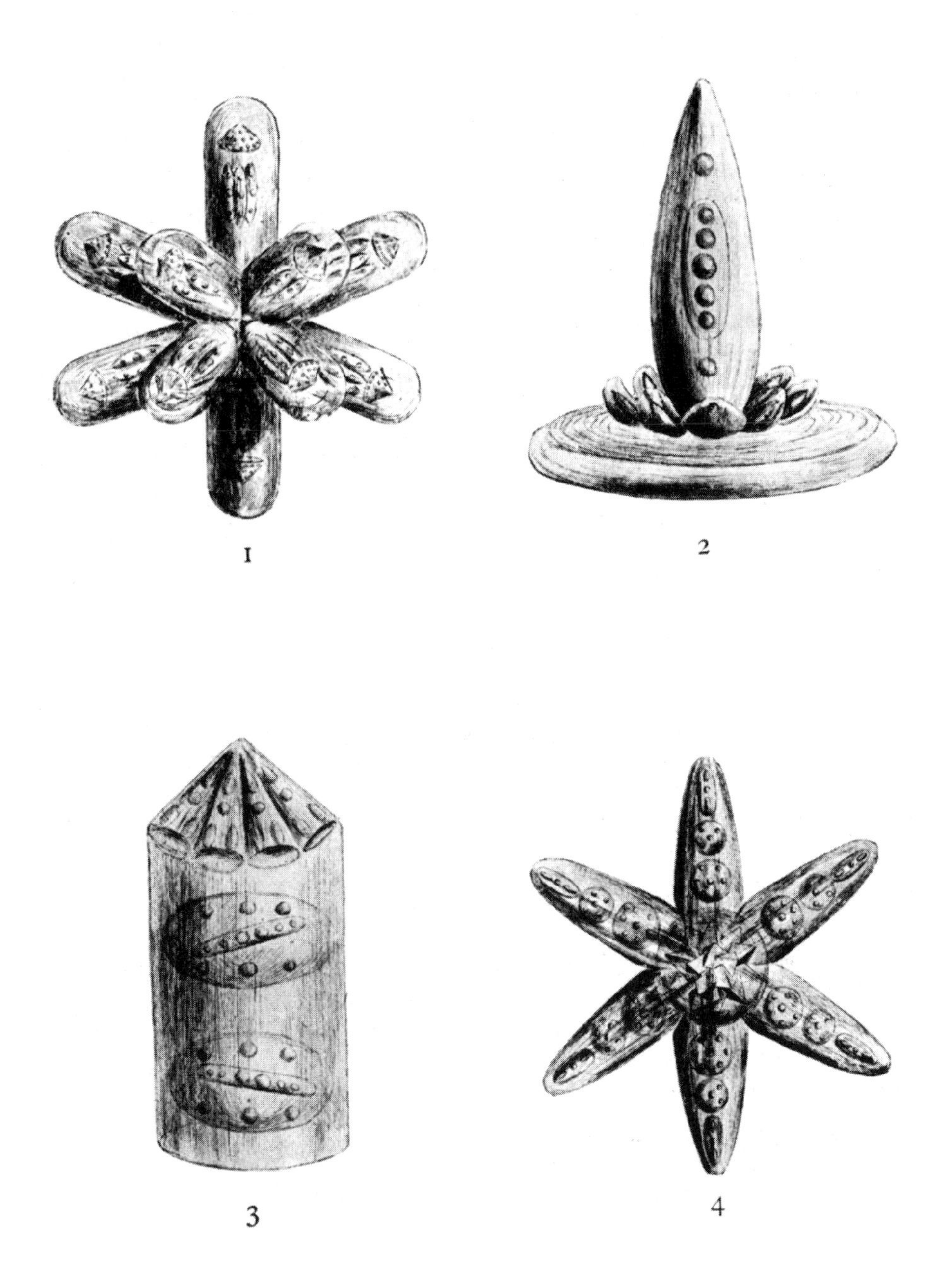

FIGURE 346. ■ More chemical atom structural types (from *Occult Chemistry*, 1908). Professor Pierre Laszlo commented on their uncanny resemblance to atomic orbitals.

place sodium in this class—the structure of the sodium atom (see Figure 343 again) makes it one of the dumbbell class. However, I suggest, gentle reader, that you avoid accepting a sodium penny unless you love to burn money (and your hand as well). The confusion is clarified by the errors incurred by the authors' use of the periodic roller coaster supplied to them by the truly eminent scientist Sir William Crookes (1832–1919), with whom they had maintained some correspondence (see Figure 347). (In a later book[7] by the aforementioned Jinarajadasa another version of Crookes' periodic table appears, but this one specifically includes "elements discovered first by clairvoyant investigation"—talk about chutzpah!) Crookes, the inventor of the vacuum tube that led to the discovery of electrons as well as X rays, developed an interest in spiritual and highly speculative ideas and what William Brock terms "metachemistry"[8]—clearly an appropriate scientific correspondent for the authors of *Occult Chemistry*.

In summary, it appears that even in the early twenty-first century, the theory may require some further study and some modifications. Although the Bohr atom and the Lewis–Kossel–Langmuir explanation of the octet rule came after the first edition and before the second edition of *Occult Chemistry*, the latter was little modified. There is also no evidence that Besant and Leadbeater saw any need to include the concept of atomic numbers developed by Moseley during this same period. Models that can survive the assault of more modern theories, including quantum mechanics and their supporting experimental data, must be powerful indeed. I am thus recommending grant support and imagining the nature of a research budget:

National Séance Foundation
Proposal Title: *Atoms and Astral Plane Interactions*
Budget

I. Personnel:
Clairvoyant (100% academic load)
Artist/Recorder (50% academic load)
Physical atom counter/mathematician (50% academic load)
II. Equipment
Clairvoyaniscope (one)
Astral plane direction detector (clairvoyaniscope option kit Y2K) (one)
Calculator (one)
III. Supplies
Aspirins (10 gross)
Herbal teas (10 gross)
IV. Facilities and administration (85% of direct costs)

I confess that I am having difficulty in mechanically interpreting the manner in which clairvoyant pictures of atoms are obtained. But I have always had similar problems trying to fathom how the Philosopher's Stone changes lead to the "purer" metal gold through a mysterious process termed "projection." Although Robert Boyle was credulous about alchemy and wrote "a strange chymical narrative" describing a reverse transmutation (Figure 153), he probably would not have been enthused about clairvoyance as an experimental technique.[9] Well, methinks Mr. Leadbeater has a wonderfully appropriate name, and it would not surprise me to learn that *he*, at least, understood both clairvoyance and projection.

Actually, the more interesting co-author is Mrs. Annie Besant (1847–1933), who Emsley describes as "a fiery social reformer with socialist tendencies and boundless energy."[10] Mrs. Besant, originally married to a vicar but eventually separated, secretly published a pamphlet questioning the divinity of Jesus Christ

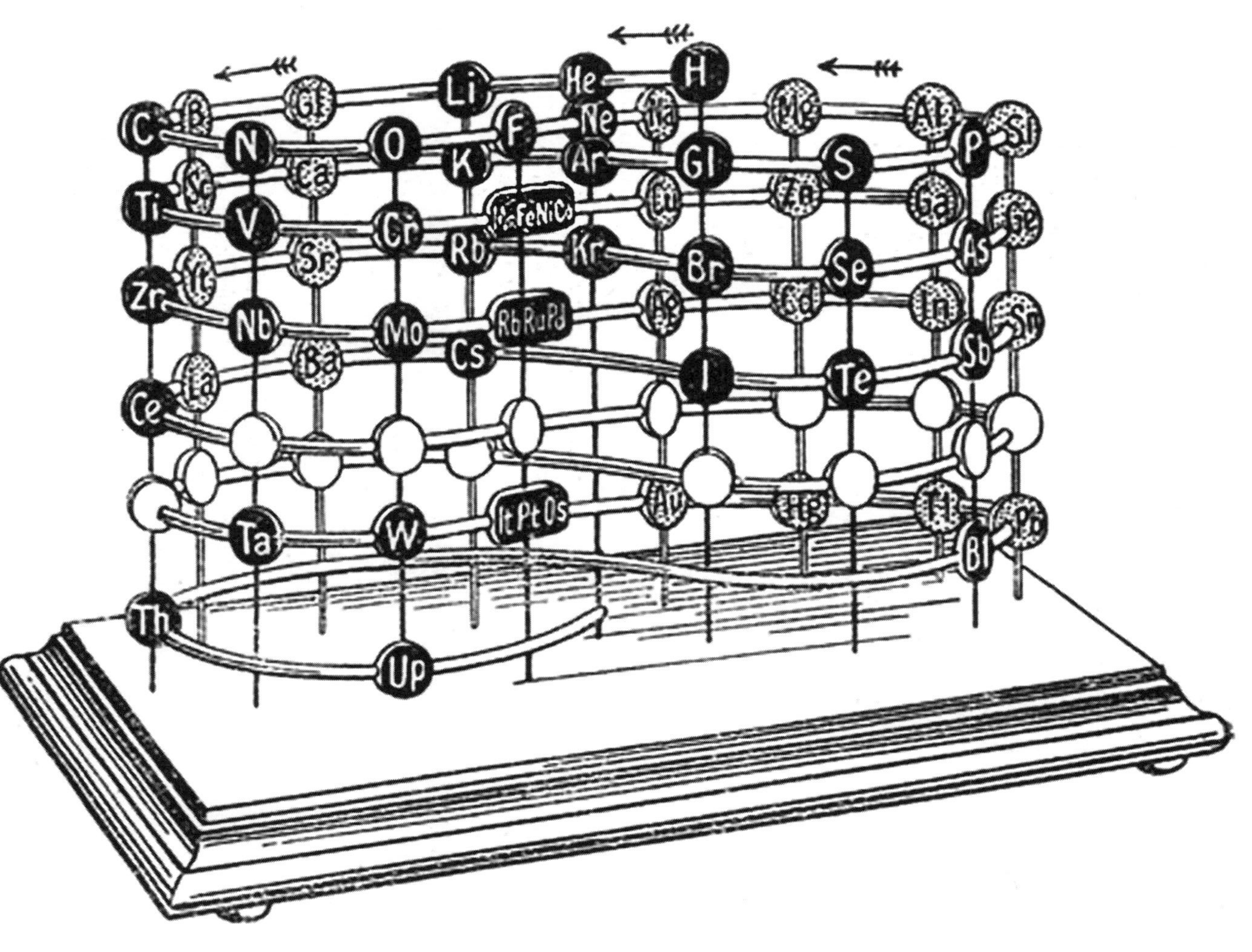

FIGURE 347. ■ The figure-eight periodic roller coaster of Sir William Crookes. A truly eminent physicist whose vacuum tube helped to establish the properties of the electron in the late nineteenth century, Crookes was also sympathetic to occult beliefs. (From *Occult Chemistry*, 1908.)

and later an article on birth control. She helped organize a strike by the "poorest of the poor and the lowest of the low" (mostly women and children) against a London manufacturer of matches in 1888 and scored a smashing victory for the rights of the workers.[10] In 1889, she converted to the doctrines of the Theosophical Society, which emphasized human service and spiritualism and served as its president from 1907 to 1933, living in its home city in Madras, India.[10] Mrs. Besant was an early advocate of India's independence and formed the Indian Home Rule League in 1916.[11] In sum: a totally "difficult" and wonderful woman.

---

1. A. Besant and C.W. Leadbeater, *Occult Chemistry—Clairvoyant Observations on the Chemical Elements*, revised edition, Theosophical Publishing House, London, 1919, pp. 1–2.
2. A. Besant and C.W. Leadbeater, *Occult Chemistry—a Series of Clairvoyant Observations on the Chemical Elements*, Theosophist Office, Adyas (Madras), 1908.
3. A. Besant and C.W. Leadbeater, *Occult Chemistry—Clairvoyant Observations on the Chemical Elements*, third edition, Theosophical Publishing House, Adyas (Madras), 1951.
4. Besant and Leadbeater (1908), op. cit., p. 2.
5. Besant and Leadbeater (1908), op. cit., p. 3.
6. Unpublished discussions with Professor Joel F. Liebman, who places occult chemistry in the realm of "arts and séances" and further suggests the National Séance Foundation as a potential research funding source.
7. C. Jinarajadasa, *First Principles of Theosophy*, third edition, Theosophical Publishing House, Adyar (Madras), 1923, pp. 156–181.
8. W.H. Brock, *The Norton History of Chemistry*, W.W. Norton & Co., New York and London, 1993, pp. 454–459.
9. R. Boyle, *An Historical Account of a Degradation of Gold, Made by an Anti-Elixir: A Strange Chymical Narrative*, R. Montagu, London, 1739 (the original 1678 edition was anonymous).
10. J. Emsley, *The Thirteenth Element; The Sordid Tale of Murder, Fire and Phosphorus*, John Wiley & Sons, Inc., New York, 2000, pp. 89–96.
11. *Encyclopedia Britannica*, Vol. 2, Encyclopedia Britannica, Inc., Chicago, 1986, p. 165.

WHITE LIGHTNING IN AN ATOM, A KISS, OR A STAR

Chemist Primo Levi's powerful book, *The Periodic Table* (see pp. 12, 14),[1,2] employed 21 elements as chapter titles, to explore symbolically his experiences, memories, and dreams as an Italian-born Jew working in World War II Turin. For example, in the opening chapter, Argon, Levi likens his Italian renaissance ancestors and their heirs to the inert gases:[3]

> The little that I know about my ancestors presents many similarities to these gases. Not all of them were materially inert, for that was not granted them. On the contrary, they were—or had to be—quite active, in order to earn a living and because of a reigning morality that held that "he who does not work shall not eat." But there is no doubt that they were inert in their inner spirits, inclined to disinterested speculation, witty discourses, elegant, sophisticated, and gratuitous discussion. . . . Noble, inert, and rare: their history is quite poor when compared to that of other illustrious Jewish communities in Italy and Europe.

(Incidentally, Levi mistakenly assumed that Professor Neil Bartlett was awarded a Nobel Prize in Chemistry for discovering in 1962 that the inert gas xenon reacts to form chemical compounds.[3] Although Levi's history is wrong, I think his judgment is sound.)

Fifty years before publication of *The Periodic Table*, Edwin Herbert Lewis, "an eccentric English Professor at Chicago's Lewis Institute,"[4] authored *White Lightning*[5] (Figure 348), a 354-page novel divided into 92 chapters each named for elements in order of (the recently discovered) atomic number. It is a Mendeleevianly confident book—Chapters 43, 61, 75, 85, and 87 are unnamed and are particularly mysterious.[6] The novel relates the coming of age of a Marvin Mahan, his tempering through the bombs and gas of World War I, and his emergence in the 1920s as a brilliant young radiochemist. "White lightning" is employed as a metaphor throughout the book for the energy hidden in matter—Marvin is "this imp of bottled lightning";[7] earth viewed from Venus is "a steady point of white lightning."[8] To let the metaphor explode from the bottle, Marvin ponders:[9]

> the cheek of a girl, which feels so smooth to the lips, is really a starry sky full of electric suns and moons. The tension between each sun and its moons is all that keeps the cheek from exploding when you kiss it. And here he had been calling them all "darlin'"! Well, he might have known that girls were composed of electricity. He had often felt it thrilling up his arm.

More ominously, Lewis predicts the use of nuclear weapons:[10] "Nothing but subatomic lightning will teach the Germans anything." Ironically, the author also predicts inevitable war with Japan over natural resources and colonization in Asia. Marvin's left hand is also a powerful symbol since it contains, receives, and releases lightning (Figure 348). A bomb blows it off in the war, and, thus maimed, he has lost youthful perfection and innocence.

While I am not a licensed "lit-crit" (literary critic), I think that a 1916 reviewer of another Lewis' book had him properly pegged as a novelist: "The plot moves swiftly with the help of incredible coincidences and improbable romances."[11] Nevertheless, let us give the author very considerable credit as a knowledgeable and sophisticated observer of chemistry. He was amazingly well informed and current about the complex revolution in the understanding of the structure of the atomic nucleus that was very much in motion as he wrote *White Lightning*. Marvin reads of Henry G.J. Moseley's discovery of atomic numbers in 1914:[12] "This unknown Moseley had found it—a sure way to determine the amount of electricity concealed in the heart of an atom . . . Think of it—an atom of lead is a small universe of compressed lightning carrying eighty-two electric moons in its sk. . . . If a gram of radium emits enough energy to lift five hundred tons a mile high, a gram of disintegrated lead ought to turn every wheel in a great factory!" (And when Moseley is shot and killed at the age of twenty-seven at the battle of Gallipoli—"Lead driven through the one brain that really understood lead"[13]). Marvin attends Yale and works under the supervision of (the very real) Professor Bertram Borden Boltwood, discoverer of "ionium" (soon identified as a thorium isotope from the radioactive decay of uranium).[14] Indeed, Lewis cites the work of Soddy and Aston and their discoveries of isotopes, excitedly relates Rutherford's nuclear transmutation of nitrogen and provides the contemporary understanding of isotopes that rationalizes extra unit masses as due to nuclear

FIGURE 348. ■ The front cover and spine section for the 1923 novel *White Lightning* by the mildly "eccentric" Edwin Herbert Lewis. The author was quite knowledgeable and current about chemistry (particularly nuclear chemistry), and his 354-page novel consisted of 92 chapters starting with hydrogen and ending with uranium. Chapters 43, 61, 75, 85, and 87 were, of course, untitled, but Chapter 72 was titled "Hafnium" since it was reported in 1923, the year of publication. Lewis' two most lasting gifts to posterity were the University of Chicago Alma Mater and his daughter, the distinguished poet and novelist Janet Lewis, who died in 1998 at the age of 99.

protons neutralized by nuclear electrons. The discovery of the neutron by Chadwick occurred in 1932, almost a decade after *White Lightning* was published.

Occasionally, Lewis *does* "hit the mark." Marvin's first two "Darlin's" are Cynthia and Gratia. Jean, the woman who becomes his wife in Chapter 92, witnesses her mother's early death from a sudden stroke on the same day she learns that her brother was killed in the war. Although that chapter (10) is highly contrived, it likens neon to a cold, indifferent and amoral universe—"But all the time the noble gas called neon remained unmoved. Like some quiet-eyed chemist looking down the future, it heard no explosion."[15] And Jean, emotionally numbed, vows chastity in Chapter 18 (Argon) and will allow no man to woo her—"She would be ready for them, as inert as a nun."[16]

But our author cannot resist a periodic stretch and so in Chapter 31 (Gallium, an element predicted by Mendeleev to fill a gap in his periodic table), we have Marvin lightning-struck by his future wife:[17]

> Just as Mendeleyeff had prophesied an element like boron and an element like aluminum, so had he unconsciously known that there must be a girl as impassioned as Cynthia and as exquisitely self-contained as Gratia.

And it gets much worse:

Chapter 25	"Jimmy's face grew much pinker than manganese salts." (Ouch!)
Chapter 27	Somehow I knew just as I started to read Chapter 27 that the Laurentians *had* to be cobalt blue. (What else?)
Chapter 31	"She might not exactly melt in his hand as the metal gallium melts, but she would yield." (Help!)
Chapter 38	Begins: "The happy youth rowed off to his own hired island and for a time sat watching the port lights coming up the river, red as a nitrate of the thirty-eighth element." (I'd prefer—"It was a dark and stormy night . . .")
Chapter 50	(Tin, if you are still paying attention): "I like canned milk first rate" (Of course you would).
Chapter 59	Praseodymium. How *do* you capture the reader's interest with this one?

Well, it all ends happily. Argon *can* form a compound.[18] And (I nearly forgot) Jean initially spurns Marvin's proposal and invites him to return and visit her three years to the day with his wife. During this three-year purgatory, Marvin begins to make his mark, Jean develops an interest in chemistry and sets up a simple laboratory, and her gifted intellect leads her to admire Marie Curie and discover, on her own, some of the fundamental chemical questions of the day. Marvin returns, learns that he will occupy an endowed chair in chemistry in Palo Alto (i.e., Stanford) and is finally accepted by Jean. Although Pierre and Marie Curie are perhaps suggested here, Antoine and Marie Anne Pierrette Lavoisier are probably more apropos.

And what of our idiosyncratic author Mr. Lewis? His most important literary contribution was arguably his daughter—Janet Lewis (1899–1998).[4,19] She was a renowned poet, playwright, and novelist whose most famous work remains *The Wife of Martin Guerre*. She wrote the libretto for William Bergsma's opera of the same title, and her work might be reasonably counted as one of the sources for the French Film *The Return of Martin Guerre*. It is a wonderful thing to imagine a father–daughter dialog that included a mutual interest in Native Americans—they are ubiquitous in *White Lightning* and are the subject of Ms Lewis' first book of poetry (*The Indians In The Woods*, 1922). And part of their loving conversation might have included a weaving together of science and poetry as she has done so beautifully in this brief work:[20]

Early Morning

The path
The spider makes through the air,
Invisible,
Until the light touches it.
The path
The light makes through the air,
Invisible,
Until it finds the spider's web.

---

1. P. Levi, *The Periodic Table*, Schocken Books, Inc., New York, 1984 (original Italian edition published in 1975).
2. A. Greenberg, *A Chemical History Tour*, John Wiley & Sons, New York, 2000, pp. 10,12.
3. Levi, op. cit., pp. 3–4. For a discussion of Neil Bartlett and the discovery of xenon compounds, see P. Laszlo and G.J. Schrobilger, *Angewandte Chemie*, *International Edition in English*, Vol. 27, pp. 479–489, 1988.
4. *Los Altos Town Crier*, Dec. 9, 1998. See also The University of Chicago Library Catalog Webpage for Edwin Herbert Lewis (1866–1938), writer and rhetorician, University of Chicago alumnus, and faculty member from 1896 through 1934 at the Lewis Institute in Chicago, now part of Illinois Institute of Technology. His most lasting work is the words to the University of Chicago "Alma Mater." See *http://webpac.lib.uchicago.edu/webpac-bin*.
5. E.H. Lewis, White Lightning, Covici-McGee, Chicago, 1923. Herein also lies a brief story. Pascal Covici, was a relative of my wife Susan (née Covici). He owned a bookstore in Chicago and started to publish books in 1922 (Covici-McGee; Pascal Covici; Covici-Friede). Although *White Lightning* was obscure, Covici became widely respected for the quality of books and their artwork. He is quite fairly said to be the discoverer of John Steinbeck, whose first successful novels were published by Covici-Friede in the 1930s (see T. Fensch, *Steinbeck and Covici: The Story of a Friendship*, Paul S. Eriksson, Burlington, 1979). I, too, have a famous relative—my father's cousin whose biography is also in print: T. Carpenter, *Mob Girl—a Woman's Life in the Underworld*, Simon & Schuster, New York, 1992. But the less said about that, the better.
6. Element 43: technetium (Tc, discovered 1939); 61, promethium (Pm, 1945); 75, rhenium (Re, 1925); 85, astatine (At, 1940); 87, francium (Fr, 1939). Lewis' book was quite up-to-date—Hafnium (Hf) was discovered in 1923, the year *White Lightning* was published, and one can imagine the author happily updating the title of Chapter 72 in the galley proofs. Chapter 86 is titled "Niton" (now Radon); Chapter 91 is titled "Brevium" (now Protactinium). For a brief table on the discovery of the chemical elements, see A.J. Ihde, *The Development of Modern Chemistry*, Harper & Row, New York, 1964, pp. 747–749.
7. Lewis, op. cit., p. 4.
8. Lewis, op. cit., p. 32.
9. Lewis, op. cit., p. 9.

10. Lewis, op. cit., p. 79.
11. *The Book Digest*, 1916, p. 337.
12. Lewis, op. cit., p. 16.
13. Lewis, op. cit., p. 38.
14. J.R. Partington, *A History of Chemistry*, Vol. 4, MacMillan & Co., Ltd., London, 1964, pp. 944, 946.
15. Lewis, op. cit., p. 54.
16. Lewis, op. cit., p. 75.
17. Lewis, op. cit., p. 132.
18. The HArF molecule, is a ground-state molecule observable only at very low temperatures in a solid matrix. Its decomposition to HF and Ar is hugely favored thermodynamically but a tiny (8 kcal/mol) activation barrier allows its covalently held atoms to "shake, rattle and roll" (i.e., vibrate) under these unearthly conditions (see K.O. Christie for a brief discussion of "A Renaissance in Noble Gas Chemistry" in *Angewandte Chemie, International Edition*, Vol. 40, pp. 1419–1421 (2001). One can only hope that Marvin and Jean have greater affinity.
19. See Stanford University American Literary Studies homepage: *http://www-sul.stanford.edu/depts/hasrg/ablit/amerlit/lewis.html*. Note that Ms Lewis taught creative writing and literature at Stanford and co-founded with her husband, author Yvor Winters, a professor at Stanford, a literary journal Gyroscope. All of this seems to have occurred three or four years after the fictional Marvin Mahan accepted the endowed chair at Stanford. 'Tis a mystery.
20. J. Lewis, From *The Selected Poems of Janet Lewis*, edited by R.L. Barth, p. 91. Reprinted with the permission of Swallow Press/Ohio University Press, Athens, Ohio. It is most interesting that Janet Lewis was a writer-in-residence at the Djerassi Resident Artists Program. (See C. Djerassi, *This Man's Pill: Reflections on the 50th Birthday of the Pill*, Oxford University Press, Oxford, 2001, p. 239). Professor Djerassi co-authored with Professor Roald Hoffmann the play Oxygen cited elsewhere in the present book. Djerassi coined the term "science-in-fiction" (see pp. 151–167 in *This Man's Pill*) and Edwin Herbert Lewis' book *White Lightning* was perhaps, something of an early contribution to this genre.

THE SECRET LIFE OF WANDA WITTY[1]

This essay is dedicated to the countless chemistry students who occasionally (very occasionally, mind you) allow their minds to drift off during class as they doodle and dream—for example, the young lady who owned Cooley's *Chemistry* (Figure 349)[2] over a century ago and speaks to us, on a sultry southern day in late May, perhaps, through her drawings.

> So class, we have already learned how to obtain saltpetre (potassium nitrate, you know) from barns and charcoal by burning wood under oxygen-poor conditions. Now we discuss the final component of gunpowder—sulfur. Sulfur is an amorphous yellow solid that is commonly obtained from pyrites by heating in a closed environment and watching the vapors drift lazily to the upper sides of the vessel and

"She is so smart yet so severe in her manner," thought Wanda. "I wonder if she ever had more choices than teaching in this tiny old high school. We call her 'Professor' but I think she could have been an army general and I her adjutant."

"Colonel Witty, we are low on ammunition, short on food and bandages, there are no explosives and a Yankee regiment is advancing in the valley below!" The steam engine that powered the regiment's locomotive could be heard very

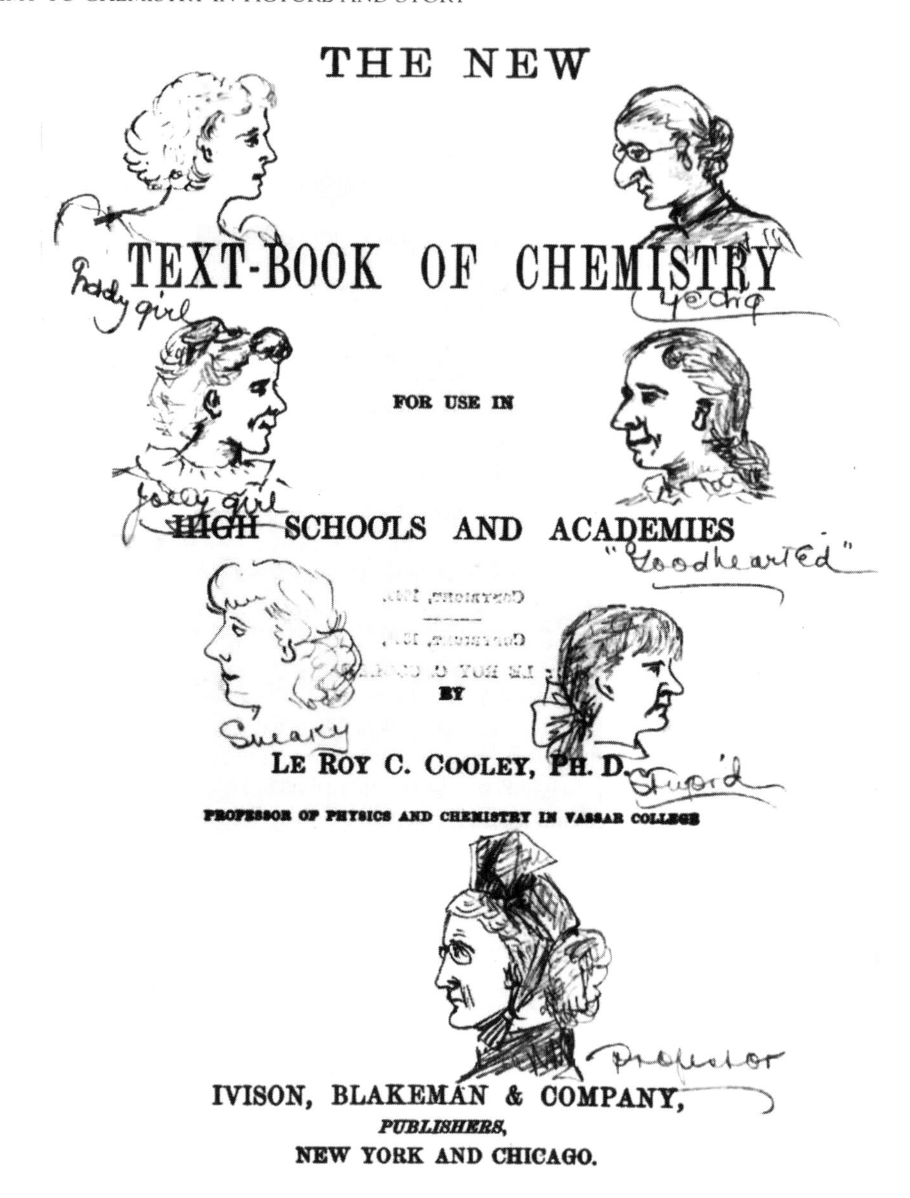

THE NEW

TEXT-BOOK OF CHEMISTRY

FOR USE IN

HIGH SCHOOLS AND ACADEMIES

BY

LE ROY C. COOLEY, PH. D.

PROFESSOR OF PHYSICS AND CHEMISTRY IN VASSAR COLLEGE

IVISON, BLAKEMAN & COMPANY,

PUBLISHERS,

NEW YORK AND CHICAGO.

FIGURE 349. ■ What became of the young lady who doodled so imaginatively on her high school chemistry textbook well over a century ago? What did she daydream about in class? Read the accompanying essay and discover her "secret life."

softly in the distance (*pocketa-pocketa-pocketa*) as advance troops used mortars and sniper fire to clear out Rebel resistance near the dairy farm. "We must stop that ammo train," said the General with grim determination. "I know, m'am, but we are under fire and . . . My God! Goodheart Ed has been wounded! The men are so brave, but they are squeamish." And with that Colonel Wanda Witty ripped the sleeve off of her shirt, grabbed a bottle of whiskey from one of the sergeants, ran with her rifle in the other hand to Ed, drank a swig, poured the remaining contents over the wound, and dressed it with a patch of her uniform. It would still be another two hours before the train would pass by and then Salisbury would be lost and all of those soldiers released from the prison to rejoin their compatriots. *No explosives—a hopeless situation*. And then Colonel Witty was remembering the pyrite formation she had casually observed near the barn a few days earlier. Suddenly, she gave orders to one of the men to collect a few pounds of pyrite, place it into a copper still, and heat it over the fire. Another soldier was ordered to burn a few pounds of wood completely in a vessel having only a small opening for air. She then commandeered three more men to collect the oldest dung from shady moist parts of the barn, expose it to the air for 15 minutes, and then place the mass in boiling water. She then collected campfire ashes (rich in pearl ash or potassium carbonate) and added them to the cooling pot. A mass of white solid appeared, and the solution cooled and was poured through mosquito mesh. *Pocketa-pocketa-pocketa*—but louder now. Every man in the outfit was ordered by Wanda to pour some of the solution into his mess kit and boil off the water until dryness. In this way, saltpetre magically appeared as white crystals in every mess kit were promptly scraped out and collected. Powdery yellow sulfur was scraped from the top of the pyrite heating kettle and the charcoal remains from the wood collected in a vat. *Pocketa-pocketa-pocketa*. "What was that formula that you taught us last week General?!" screamed Wanda. Armed with the formula she mixed the gunpowder, ran for the tracks, felt the wind of a sniper's bullet as it barely grazed her, and placed the gunpowder in a can on the tracks. *Pocketa-pocketa-pocketa*!! The wires were speedily connected to the plunger and that was immediately pushed just in the nick of time—*POCKETA-POCKETA*. . . . Bang! . . . "Bang!?" . . . Not . . . *KABOOOM*!!?

Bang! "Professor" again rapped the ruler on Wanda's desk as the class giggled. "And what was my last sentence?" she asked. "But General . . ." the giggles became gales of laughter. Wanda then noticed that her shirt was fully intact and no cow dung was in sight. "And how many times must I tell you 'Never let your mind wanda' . . . witty, eh?" and with that, "the General" victoriously dismissed the class for the day. And so, on that warm spring day over a century ago, Wanda sat pondering the fate of a S/hero born 100 years too soon and started to dream. "The General's been kidnapped by Prussians!" shouted "Giddy Girl" as she ran into the classroom toward Wanda. Standing up, touching the hilt of her sword, and flattening an errant hair curl with her fingers, she again prepared for the rescue—Wanda Witty—the indomitable, undefeated to the last.

---

1. This is written in homage to humorist James Thurber, author of *The Secret Life of Walter Mitty*, and with happy recollections of reading Thurber's works with my daughter Rachel.
2. L.C. Cooley, *The New Text-Book of Chemistry for Use in High Schools and Academies*, Ivison, Blakeman & Co., New York & Chicago, 1881. The drawings on the title page were drawn by a student, who I think is female by the nature of these drawings and written names of friends; the year 1891 is also written in the same hand.

"TRADE YA BABE RUTH FOR ANTOINE LAVOISIER!"

Babe Ruth is the "Father of Modern Baseball" because his home-run hitting revolutionized the game. In 1918, none of the 16 Major League *teams* hit more than 27 home runs.[1] That year, the 23-year-old Ruth pitched the Boston Red Sox to the World Championship by winning two of their four World Series victories *and* tied for the League lead in home runs by an individual (11).[2] In 1919, his final year with the Red Sox, Ruth hit 29 of his entire team's season total of 33 home runs. The cash-poor Red Sox promptly sold Ruth to the wealthy New York Yankees and "wandered in the desert" for 86 years without winning the World Series (the infamous "Curse of the Bambino"). As a Yankee in 1920, Ruth hit 54 home runs—greater than the season totals of 14 of the 15 other Major League teams. He hit 59 home runs in 1921 and established, in 1927, the modern record of 60 that held for 35 years. In 1930 Ruth signed for a salary of $80,000 per year and, when told that he was making more than the president of the United States, was said to have responded: "Well, I had a better year than he did." History suggests that Ruth was probably right. Antoine Laurent Lavoisier, the "Father of Modern Chemistry," totally redefined the field of chemistry and revolutionized it. One of the 40 "Farmers" in the Ferme Générale, Lavoisier might conceivably have had a higher salary in 1789 than did Louis XVI, although I doubt it, but he certainly would not have dared to brag about it. It is also fair to say that Lavoisier had a better year in 1789 than the king did. It is thus eminently fair to call Babe Ruth "The Antoine Lavoisier of Baseball." One might even consider calling Lavoisier "The Babe Ruth of Chemistry."

Trading a Babe Ruth baseball card for a Lavoisier card would seem like "a steal" to *me*. However, the grim reality is that there is today a very active investor's market in baseball trading cards, while chemistry trading cards, "hot" in Belgium and Holland over sixty years ago, are not exactly "selling for a premium." And since I know of no chemists signing 10-year $252 million contracts[3] with no-cut clauses, there will probably be no renaissance in chemistry cards in the near future.[4]

The handsome portraits in Figure 350 are from tobacco trading cards issued by *La Cigarette Oriental de Belgique* in 1929 or 1930 (the narrations on the backs of the cards are in French and Flemish). Figure 350(a) depicts Carl Wilhelm Scheele, a brilliant Swedish apothecary of very modest means who made the original discovery of oxygen but failed to understand its role in combustion. Figure 350(b) is a portrait of the aristocratic Antoine Laurent Lavoisier, the "Father of Modern Chemistry." Figure 350(c) shows the boyishly handsome Humphry Davy whose chemical demonstrations entranced women as well as men at Royal Institution "chemistry nights" starting in 1801. Figure 350(d) depicts Claude Berthollet. His discovery of the "mass law effect" raised questions that would vex Dalton's atomic theory until the inconsistencies were fully understood. In contrast, the law of combining gas volumes of Gay-Lussac [Figure 350(e)] strongly supported the atomic theory of the modestly attired Quaker, John Dalton [Figure 350(f)]. Justus Liebig [Figure 350(g)] was one of the founders of "animal chemistry" (i.e., biochemistry). His work in analytical chemistry helped tame the "primeval forest" of organic chemistry. Robert Bunsen [Figure 350(h)], German chemist and physician, set about making a spectrosope for analysis of trace metals, but the light source is now the Bunsen burner known to everybody who has taken high school chemistry. Figure 350(i) is a portrait of Alfred Nobel, whose wealth was derived from the manufacture of explosives and who willed his fortune to establish the world-renowned prizes bearing his name that include the Nobel Peace Prize.

Figure 35(1) displays six laboratory scenes. The Chocolat Poulain card de-

FIGURE 350. ■ Collectors' cards portraying famous chemists issued in 1938 for *La Cigarette Oriental de Belgique*. See color plates. Although Topps issued bubblegum trading cards in the early 1950s that included Marie Curie and Louis Pasteur, there seems to be no current market for a "Stars of Chemistry" bubblegum trading card series. *Quel domage!* (I am grateful to Jamie and Steve Berman for this information.)

PURO ESTRATTO DI CARNE LIEBIG
CHIMICI CELEBRI.
1) Geber col suo maestro Giafar el Ssâdik. (VIII° secolo).
Riproduzione vietata.
Spiegazione a tergo.
CHOCOLAT POULAIN
GOUTEZ ET COMPAREZ !
QUALITÉ SANS RIVALE
180
GAY-LUSSAC
SÀPIS
SÀPIS, ESTRATTO CARNE AROMATIZZATO
CHIMICI CELEBRI.
3) Visita di Berthollet a Lavoisier alla Sorbona a Parigi.
Riproduzione vietata.
Spiegazione a tergo.
DADI LIEBIG
DADI PER MINESTRA LIEBIG
CHIMICI CELEBRI.
5) Il laboratorio di J. v. Liebig a Giessen. (1840).
Riproduzione vietata.
Spiegazione a tergo.

picting Gay-Lussac on the top right of Figure 351 raised the art of the trading card to a new level. The middle right card, dating from the 1930s, is a German rendition of perhaps a turn-of-the-century laboratory, and the card at the bottom right shows Louis Pasteur in his laboratory. The three colorful cards at the left of Figure 351 are from a 1930s series published by an Italian company advertising Liebig's Meat Extract. The top card paints an imagined scene in the laboratory of the legendary eighth-century Arab physician and alchemist Geber (Jabir ibn Hayyan).

The second card at the left of Figure 351 depicts Lavoisier[5] and Berthollet at the Sorbonne in Paris, although neither held an appointment there. Berthollet defended phlogiston theory between 1780 and 1783. During 1783 the true nature of water was discovered by Cavendish and the compound carefully synthesized from the elements and decomposed back into the elements by Lavoisier. Water was found to be a compound consisting of precisely eight parts by weight of oxygen and one part hydrogen. It was not, after all, "dephlogisticated phlogiston." In April 1785, Berthollet became the first French chemist of prominence to support Lavoisier's new theory of oxidation.[6] He remained a friend of Lavoisier, survived the French Revolution while maintaining his integrity, and accompanied Napoleon on his military and scientific expedition to Egypt in 1798. It was during this expedition that Berthollet made the curious discovery of deposits of soda (sodium carbonate) on the shores of salt lakes that led to his formulation of the mass law theory (p. 362). With his characteristic integrity as a senator, Berthollet voted to depose his friend in 1814 in order to end the disastrous war led by Napoleon.[6]

The scene depicted at the bottom left of Figure 351 is that of Justus Liebig's laboratory in Giessen, and the original is in the Museum of the University of Giessen. Most of the figures have been identified.[7] The figure seated in the front center, dreamily applying the mortar and pestle, is Liebig's student Adolph Friedrich Ludwig Strecker, my great-great-great-great-great-grandfather, chemically speaking (yes, I am also a chemical descendant of Liebig—see the Epilogue in this book). Just to the left of Strecker is Heinrich Will, who will soon succeed Liebig as Director at Giessen. At the rightmost part of this picture is August Wilhelm Hofmann, Liebig's greatest student, who accepted the position of Professor at the Royal College of Chemistry after it was declined by his mentor.[8]

Now what do we make of the six trading cards in Figure 352 that advertise Justus Liebig food products with a facsimile of his autograph (just like those on baseball cards)? Liebig had been writing scientific books about food chemistry since the 1840s. He held a theory about the vital importance of "meat juice" for diet and health.[9] He prepared a "chicken tea" by allowing minced chicken to sit in cold water for hours with a few drops of hydrochloric acid added to soften the meat. Frequent drinks were shown to cure all manner of illness. However, his most popular preparation was his meat extract. In 1865, a German railway engi-

FIGURE 351. ■ Collectors' cards issued during the 1920s and 1930s depicting chemistry laboratories and famous chemists. See color plates. The figure of Gay-Lussac published for *Chocolat Poulain* ("Taste and Compare! Quality without Rival") is particularly well done. The card at the lower left is a version of the famed drawing of Justus Liebig's laboratory housed at the University of Giessen. The chemist in the front-center, dreamily applying the mortar and pestle, is my *chemical* great-great-great-great-great-grandfather, Adolph Strecker.

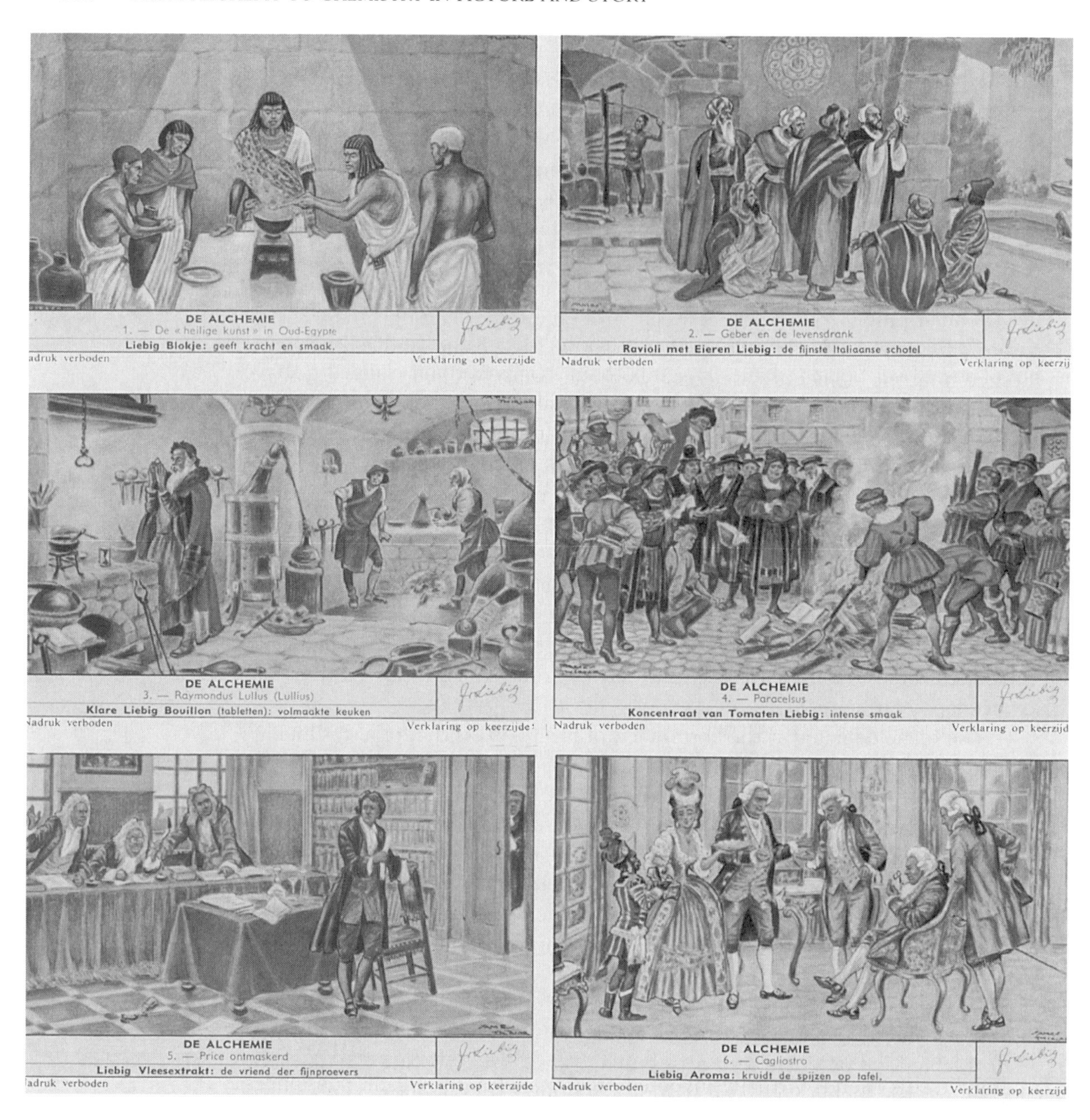

FIGURE 352. ■ Justus Liebig was one of the fundamental pioneers in biochemistry (animal, plant, and food chemistry). He held strong views about the value of "meat juice" in the diet and lent his name to commercial endeavors. (Today there is a company that sells Linus Pauling vitamin C tablets.) The Justus Liebig Company sponsored these Belgian cards printed during the 1930s that tout their line of food items. See color plates. I am grateful to Jamie and Steve Berman for this information.

neer named Georg Christian Giebert hired Liebig as a director of the newly created Liebig Extract of Meat Company. Shares were sold on the London Stock Exchange.[9] To this day, Liebig Extract of Meat is still sold in Germany. The collectors cards in Figure 352, narrated in Dutch, were issued in the 1930s. They are attractive and on the backs have quite informative histories of chemistry.[10] Card 1 ("Liebig Blocks, Give Strength and Taste") tells of "The Sacred Art in Ancient

Egypt" in a succinct but quite informative manner. Card 2 ("Liebig's Ravioli with Egg: The Finest Italian Dish") describes the history of panaceas beginning in Arab lands, including the work of the eighth-century alchemist Geber. Card 3 ("Clear Liebig Bouillon (Cubes): Perfect Chicken") describes the thirteenth-century Catalan mystic Ramond Lully (Ramon Llull) to whom numerous alchemical books have been (fictionally) attributed. Card 4 ("Liebig Tomato Concentrate: Intense Taste") provides an informative glance at the life of Paracelsus, who made and prescribed synthetic medicines consisting of metallic compounds rather than the traditional extracts and distillates derived from plants and animals. Card 5 ("Liebig Meat Extract: The Friend of the Connoiseur") provides an interesting discussion of a late-eighteenth-century English physician and member of the Royal Society, James Higginbotham. Higginbotham claimed to have found the "Philosopher's Stone." When the Royal Society took him to court to substantiate his claims, he poisoned himself in front of his colleagues. Card 6 ("Liebig Aroma: Seasons Food") tells of one Joseph Balsamo, also known as Count Alexander of Cagliostro. He created a stir at the Court of Louis XVI as a miracle worker and "gold-maker." One could only imagine Lavoisier "making minced meat out of him," but the opportunity apparently never came. Marked by numerous scandals, Balsamo moved to Rome, where he was seized during the Inquisition and died in captivity in 1795.

The backs of the portrait cards in Figure 350 describe details of the various chemists' lives, but we admire the statistical line format on the backs of baseball cards that also usually include some catchy little career summary lines. Let us try our hand for an "improved" presentation for the back of the Justus Liebig card (Figure 350g):

No. 57

BARON JUSTUS "THE GATEKEEPER" VON LIEBIG

b. 1803 in Darmstadt, Germany   Height: 5' 9"*   Weight: 145 lb*

Writes blackboard: Right-handed*   Erases blackboard: Left-handed*

Justus, who prefers to be called "Herr Professor Doktor" or sometimes just plain "Baron" to his friends virtually invented precise analysis of organic substances, laid the foundation for understanding organic chemistry, and is one of the fathers of biochemistry. His exacting standards as Editor of the *Annalen der Chemie und Pharmazie* earned him the nickname of "The Gatekeeper." He was the first Major League chemist to sign a lucrative product endorsement contract. His hobby is "xtreme" chemical debate in which he proudly notes "I take no prisoners."

CAREER STATISTICS*

Lectures Started	Lectures Completed	Students Influenced	Analyses Completed	Debating Penalties (in minutes)	Journal Started
3251	3251	705	2348	3655	1

*These are fictional except for the number of students influenced and journal started.

Chemical historian William H. Brock points out that Liebig was a very public scientist who took strong positions on issues of great interest to the public such as farming, nutrition, and public health.[11] Brock compares him,[11] in this regard, to Linus Pauling, who dominated twentieth-century chemistry and also played a public role in the debate on atmospheric testing of nuclear weapons, the environment, vitamin C, peace, and public health.[12] Pauling was also no stranger to controversy, and the back of his (hypothetical) collectors' card would include students influenced (tens of thousands), Nobel Prizes (2), and orthodox Communists and McCarthyites offended (all).

---

1. D.S. Neft and R.M. Cohen (eds.), *The Sports Encyclopedia: Baseball*, St. Martin's Press, New York, 1989.
2. For those amiable readers who are not baseball *cogniscenti*, pitchers are notoriously weak hitters, and to have one lead the league in home runs borders on the outrageous. Had Ruth been a weak hitter, he probably would have been elected to the Hall of Fame on the basis of his pitching alone.
3. This is an actual contract, signed in December 2000 by baseball player Alex Rodriguez and the Texas Rangers. Incidentally, the 2001 Texas Rangers finished last in their division. Before the 2004 season began, Texas unloaded this huge contract and the Yankees once again outbid the Red Sox for the services of a great star. However, the 2004 season ended with the Sox first world championship in 86 years. Go figure.
4. A tobacco company in the Canary Islands, *Obsequio De La Fabrica De Cigarillos*, published a series of collectors' cards of Nobel Prize winners, including chemists, in 1952. Closer to home, the Topps Company printed collectors' cards in 1952 of famous people, including Marie Curie that closely resembled the company's wonderful baseball cards.
5. What did Lavoisier really look like? There are, of course, no photographs, but see M. Beretta, *Imaging a Career in Science—the Iconography of Antoine Laurent Lavoisier*, Science History Publications/USA, Canton, 2001.
6. J.R. Partington, *A History of Chemistry*, MacMillan and Co. Ltd., London, 1962, Vol. 3, pp. 496–516.
7. A.J. Ihde, *The Development of Modern Chemistry*, Harper & Row, New York, 1964, p. 263.
8. W.H. Brock, *Justus Von Liebig—the Chemical Gatekeeper*, Cambridge University Press, Cambridge, UK, 1997, pp. 112–114.
9. Brock, op. cit., pp. 216–233.
10. I am grateful to my former chemistry professor Dr. Arno Liberles for obtaining this translation.
11. Brock, op. cit., pp. viii-ix.
12. L. Pauling, *How to Live Longer and Feel Better*, W.H. Freeman and Co., New York, 1986.

JIVE MOLECULES DOIN' THE JITTERBUG

Maurice Sendak was born in Brooklyn, New York in 1928 to poor Polish immigrants, the youngest of three children. His brother and sister described to him his birth date thusly:[1] "Papa gave us two plums apiece and sent us out to play—then you were born." From these modest beginnings emerged one of the world's most beloved illustrators (and writers) of children's books. While his books are now classics, one of them, *Where The Wild Things Are*, has become a part of our cultural canon and won for him the Cadecott Medal.[2] It was later developed into a Broadway musical for which Mr. Sendak wrote the libretto.

Now, esteemed reader, a question for you: What is the title of the very first book illustrated by Mr. Sendak? The answer (and I'll bet that very few chemists and physicists know it) is *Atomics for the Millions*, published in 1947.[3] The book was co-authored by Dr. Maxwell Leigh Eidinoff, Professor of Chemistry, Queens College of the City University of New York (CUNY), and Mr. Hyman Ruchlis, young Mr. Sendak's physics teacher at Lafayette High School. The book's Introduction was written by Harold C. Urey, the 1934 Nobel Laureate in Chemistry. Figure 353 is Sendak's delightful depiction of "jive" girl and boy atoms lined up as they might at a "sock-hop" and then jitterbugging together as molecules—certainly a "cool" (oops . . . "hep") way of teaching chemistry.[4]

And how did the gifted young Maurice Sendak get his first big break? (His early talents were already employed in an after-school job drawing background for *Mutt and Jeff* comics.) Let us read the artist's narrative:

> Getting out of high school was my entire goal in life. And the only way I got out of high school, because I was failing everything, was by illustrating the first book on the atomic bomb and called *Atomics for the Millions*. The bomb had just been dropped. I graduated in 1946 and my physics teacher, Dr. Heiman Ruckless [*sic*], wrote the first book explaining it to the layman, chose me as his illustrator—the dumbest kid he ever had in his class. I could draw, but he had to explain each picture. The deal was a hundred dollars and a passing grade so I could graduate.[5]

FIGURE 353. ■ Nineteen forties atoms and molecules as portrayed by author/illustrator Maurice Sendak in 1947.

1. From *More Junior Authors,* H.W. Wilson Co., New York, 1963. See Educational Paperback Association site: www.edupaperback.org/showauth.cfm?authid=42.
2. I enjoyed reading *Where The Wild Things Are* and *In The Midnight Kitchen* to my young children, David and Rachel. Mr. Sendak has described how much he enjoyed the stories his father would make up and tell him. Happily, I share a bit of this cultural experience: my father, Murray, would compose stories and tell them to me at bed time, my favorite being "Arthur and the Ants."
3. M.L. Eidinoff and H. Ruchlis, *Atomics for the Millions,* McGraw-Hill, New York, 1947.
4. Eidinoff and Ruchlis, *op. cit.*, p. 13. Note the commentary on this specific drawing by Selma G. Lanes on p. 38 of her sumptuous book *The Art of Maurice Sendak* (Harry N. Abrams, New York, 1980).
5. From "Mozart, Shakespeare, and the Art of Maurice Sendak," Occasional Papers, The Dureen B. Townsend Center for the Humanities, University of California at Berkeley, 1995. See http://ls.berkeley.edu/dept/townsend/pubs/OP05_Changelings3_top.html. I am grateful to Ms Joyce Hanrahan, author of Works of *Maurice Sendak—Revised and Expanded to 2001* (Custom Communications, Saco, Maine, 2001) for making me aware of Mr. Sendak's history with *Atomics for the Millions*.

EPILOGUE

This book concludes with two essays that are somewhat personal in nature. Although appropriate to themes developed earlier, they do not fit smoothly into the historical flow of the book. Their placement at the end might at first appear to be exercises in self-indulgence and self-aggrandizement. In fact, although the second essay describes my own chemical genealogy, I am not a significant player in the history of our field. The real points of the genealogy essay are the flow of chemical history, the evolution of education, the fact that at some level these connections matter, and, finally, the sheer delight of discovery. The first essay describes my memories of a young genius, Robert E. Silberglied, during our early and middle teenage years. It is relevant to this book at two levels. Chemistry describes hidden reality. Robert's studies of the communications between butterflies uncovered the hidden reality of sexual selection in the ultraviolet, a light range invisible to us, but apparently like neon signs to them. The chemical pigments and material structures of the wings govern these behaviors. But the true *raison d'être* of this essay is the fun in imagining the youths of some of the geniuses visited so briefly here.

A NATURAL SCIENTIST

Do we recognize truly creative scientific talent when we witness it at an early age? Should we nurture it or just get out of the way and let it develop on its own?

Robert E. Silberglied died on January 13, 1982 in the crash of Air Florida flight 90 in Washington, DC at the age of 36—ironically on the day he proposed marriage (a proposal joyfully accepted).[1] I last saw him when he was 16, yet he remains for me a vital force—the combination of quirkiness and creativity so typical of a natural-born scientist.[2]

We first met as 12-year-olds at a Brooklyn junior high school. I had then a fairly high opinion of my own scientific abilities: I read natural history, insect, and dinosaur books as well as books by Isaac Asimov, and walked around with complex books on nuclear physics, very prominently displayed, that I hoped would enlighten me if I carried them long enough. At least, they might impress girls! I drew designs of impossible-to-build rockets. My "specialty" was liquid fuels, and I assumed that some day I'd get a hold of hydrazine and liquid oxygen or red fuming nitric acid. My valve designs consisted of multiple layers of cardboard—load the liquids (how?), then run like hell. Fortunately, I was a rocket-design "theoretician" in contrast to other enthusiasts who used available solid fuels and sometimes injured themselves performing real experiments.[3] Around this time, I first heard rumors of a kid who, according to our crowd, was a "scientific brain," and I had to "check him out."

Robert was short, wore plain glasses, and was hopeless in gym class. His best

From Alchemy to Chemistry in Picture and Story. By Arthur Greenberg.

defenses in the Brooklyn schoolyard were his wit and the fact that there was no glory in beating him up. Early in our friendship he took me up to his room in an aged and very modest apartment house and showed me his insect collection. Unlike my own random walks through insect collecting, Robert had a systematic laboratory with homemade nets, insect-killing jars, killing fluid (actually lighter fluid—more on that later), relaxing fluid, and mounting boards and pins. His insects were scrupulously mounted with proper pins sticking through labels bearing their scientific names and places of capture, written in a very tiny but neat hand (more on that, too). Clearly, Robert was doing science on a much higher plane than I was. We hunted insects at the Brooklyn Botanical Gardens, and he taught me how to make a sweep net (metal clothing hanger plus a fine-mesh curtain). He would sweep the opening of the net back and forth along the side of a bush and then place it opening-down on the grass. The results were thrilling: an entomological grab bag—literally hundreds of beetles, bugs, aphids, leaf hoppers, flies, wasps, and ants to harvest at our leisure.

Robert was also known among our junior high group for his famous "scroll wristwatch." He had removed the works from an old wristwatch and had scrolled "gyp sheets," written in a tiny yet neat entomologist's hand, onto the watch's rollers. Why didn't we report him? I guess we were taken with his ingenuity; we probably received a vicarious thrill from this bold, if secret, challenge to the authorities who ran the school, and, in any case, his grades were barely Bs. Frankly, I suspect he never used the watch "in battle" but kept it for security liked a nuclear-tipped ICBM.

Here are some other highlights of Robert's activities; for instance, he would carry (insect-killing) lighter fluid around on rainy days and spray some onto puddles hoping to see an old, religious lady discard a lit cigarette butt into the puddle, and witness "a miracle." He obtained a catalog from a Florida reptile supply house and fantasized about releasing 750 chameleons (only $15.00!) into the junior high. In fact, he did "collect" (I suspect purchase) a load of praying mantis cocoons and place them in hidden spots throughout the school and was generous enough to give me a dozen. They never did hatch in the malevolent junior high school environment. However, during an abnormally warm February they did hatch in my bedroom, covering the walls with hundreds of mini-mantids, forever traumatizing my then-baby sister Roberta.

During our senior year at Erasmus Hall High School I saw little of Robert. I was totally "into" sports, and he had come under the sway of a gifted zoology teacher. Through the grapevine I learned that he had retired the infamous wristwatch and was making As and applying to Cornell's "Ag School." I moved to Englewood, New Jersey during the last half of my senior year and, having read that fossils could be found along the Navesink River, invited him tocross the Hudson and take a bus to Red Bank. Our fossil hunt was unsuccessful. However, with his ever-present gear, Robert caught a bumblebee, meticulously removed the stinger, tied one end of a thread to the abdomen of the disarmed bee and the other to his shirt button, and rode victoriously home, "bee-buzzing" all the way. This occurred in Spring, 1963 and it was the last time I saw Robert.

Some eighteen years later my two children, David and Rachel, who were eight and six years old, respectively, gave me the excuse to reexperience insect collecting. I purchased for them the insect collecting paraphernalia that *I* had always wanted. I thought of Robert and, on a hunch, looked him up in *American*

Men and Women in Science. There he was—B.S., Cornell; Ph.D., Harvard; he was now a faculty member at Harvard and a curator of the university's butterfly collection. I wrote to him and reminded him of our earlier experiences together in minute (if painful) detail. He wrote back, congratulated me on my memory, and invited us all to visit him in Cambridge. He signed off now as "Bob." My family was too busy at the time to accept his offer. About five years later I again looked him up and found the word "deceased." Only years later did I learn that he died in the icy waters of the Potomac.

As a teenager, Robert had won a science fair with a study of variations in butterfly markings in different parts of New York City. I suspect that the recognition was incidental to his scientific interests, and I am certain that he had no assistance from his parents or an established scientific laboratory.[1] A Smithsonian Institution Website[1] informed me that Robert rose to the rank of Associate Professor and Associate Curator of Lepidoptera, Museum of Comparative Zoology at Harvard. He first obtained an appointment at the Smithsonian Tropical Research Institute in 1976 and finished his career there. He was also an environmental activist who devoted himself to the protection and management of Lignum vitae Key in Florida. Among many scientific accomplishments, Robert was particularly recognized for his studies of the importance of ultraviolet light in the mating habits of butterflies.

His last paper, "Visual Communication and Sexual Selection among Butterflies," was completed a few days before he died and published posthumously.[4,5] Silberglied noted that butterflies have the widest spectral range of vision among animals—the full human visible region as well as ultraviolet down to 300 nm. The male *Colias eurytheme* was found to reflect ultraviolet light (invisible to humans) from the dorsal (back or top) surface of its wings, and this reflection, not its color, attracted females. A control male of *Colias eurytheme* was dyed yellow by magic marker on the underside of its wing. Although its color and appearance, even on the dorsal surface, have been altered, it still reflects UV light from the dorsal surface and mates successfully. The other five males were dyed various colors on the dorsal surfaces of the wings, both changing visible colors and suppressing UV reflection. These five males were rejected by females. Males of *Colias philodice*, are known to absorb UV light on the dorsal surfaces of their wings rather than reflect it. Yellow dye was applied to the surface of the control male and the same group of colored dyes as earlier applied dorsally. Since all six dorsal surfaces absorbed UV light, all six males mated successfully.

In his autobiographical book *Naturalist*, the renowned Professor Edward O. Wilson, Robert's Ph.D. mentor, called him "a gifted naturalist and a polymath taxonomist."[6] His undergraduate teacher and friend at Cornell, Thomas Eisner, described him as "A gentle, extremely funny and considerate person. Bob was a naturalist through and through, at once observant and inquisitive."[7] In a posthumous dedication to the symposium book containing his final written paper, the editors note in part:[8] "Bob Silberglied captivated all who met him with his infectious enthusiasm and boundless energy. This was never more true than at the Symposium meeting, when he was in great form, buzzing with ideas, information and humour. His terrible death, in the Washington air disaster of 13th January 1982, not only robbed biology of a considerable talent but also took from us a delightful friend." Amen to that!

I was most fortunate to be touched by this embryonic genius at such an early stage of my own life, and I have been trying throughout my adult teaching life to find him again.

---

1. The Smithsonian Institution Archives retain Robert's papers (1960–1982) and related materials (to 1984) in a collection (Record Unit 7316) described by Rebecca V. Schoemaker (see *http://www.si.edu/archives/faru7316.htm*). An article on the aftermath of Air Florida Flight 90 appeared (*The New York Times Magazine*, August 4, 2002, pp. 36–41). I am grateful for the details about Robert's life that it provides.
2. Here I happily acknowledge my friendship with Joel F. Liebman dating back to Fall 1967, when we met as first-year graduate students in chemistry at Princeton University. Although Joel and I were 20 when we met, we were "young adults," scientifically speaking, and thus mostly "formed." It would have been fun to have known Joel at, what entomologist Silberglied might have referred to as, his pupal stage. Joel, too, won science fairs as an early teenager without help of parents or any scientific establishment. Finally, Joel's continued scientific creativity, his "manic" punning, his appreciation for the absurd, and his innate kindness perhaps allow me to imagine Robert as an adult.
3. In 1997 I had the good fortune to meet Dr. Slayton A. Evans, Jr., Professor of Chemistry at the University of North Carolina at Chapel Hill. Slayton, an African-American, grew up in rural Alabama, and like so many other boys in the "Sputnik generation," also developed an interest in rockets. He related to me how one of his (solid-fuel) rockets blew up near the house, leaving a small crater in the yard. Alarmed by the explosion, his mother called out to her precocious son "What happened?" To which he replied "Nothing, Mom." To which she replied "Well, make sure that 'nothing' never happens again." Professor Evans died on March 24, 2001 following a prolonged illness. On the university Webpage the following words appear: "His grace and dignity touched everyone he met in a unique way." Girls were not particularly encouraged into the sciences and engineering during those days. But many were not to be denied and I think of my contemporary Dr. Marye Anne Fox, Chancellor of University of California, San Diego, and a member of the National Academy of Sciences, as one formidable example. Dr. Joan Valentine, the first female graduate student in chemistry at Princeton, Professor of Chemistry at UCLA, and journal editor, is another. Indeed, my wife Sue would, as a young girl, unfashionably watch and assist her dad, Wilbert Covici, in his building and repair activities. As a result, her mechanical abilities put mine to shame. Similarly, my sister Ilene ("Dee Dee") Franklin was not encouraged to imagine a career as a scientist, but she became a gifted chemistry teacher and is now stimulating her second generation of high school students. My other sister, Roberta, a talented dancer, might have become a business executive.
4. R.E. Silberglied, in *The Biology of Butterflies. Symposium of the Royal Entomological Society of London*, No. 11, R.I. Vane-Wright and P.R. Ackery, Academic Press, London, 1984, pp. 207–223.
5. See Plate 4 and its description in the book cited in Siberglied, op. cit.
6. E.O. Wilson, *Naturalist*, Island Press/Shearwater Books, Washington, DC, 1994, pp. 276–279.
7. Eisner, T., *For Love of Insects*, Belknap Press of Harvard University Press, Cambridge and London, 2003, pp. 155–164.
8. R.I. Vane-Wright and P.R. Ackery, op. cit., p. xxi.

DESCENDED FROM FALLOPIAN TEST TUBES?

A Long Overdue Question

I am embarrassed to admit that I was 55 years old when I finally researched my chemical genealogy. It is not, however, completely my fault. My "chemical father," Pierre Laszlo, who directed and signed my Ph.D. thesis at Princeton University, was educated in France, had a couple of slightly unconventional twists in his postgraduate education and never really talked with our research group about his chemical lineage, nor were we curious enough to ask. But Pierre's visit to New Hampshire in October 2001, provided some relaxing moments, and I finally popped the question: "Who was my 'chemical grandfather'?" He informed me that it was Edgar Lederer, at the Sorbonne in Paris, an organic chemist and pioneer in chromatography. Pierre also supplied the identity of Lederer's advisor—Richard Kuhn at Heidelberg.

That night, I leaped into the World Wide Web and within minutes discovered that Kuhn, who was a Nobel Laureate, completed his doctorate under Richard Willstätter, who was awarded the Nobel Prize in 1915, that Willstätter studied with Adolph von Baeyer, the 1905 Nobel Laureate, who studied with August Kekulé. I was overjoyed to find this distinguished "family" history and proud to discover familial traits in myself. Kekulé was one of the fathers of structural organic chemistry. He first realized that carbon forms four bonds and that benzene, the fundamental aromatic molecule, is composed of hexagonal rings. I have long considered myself a structural organic chemist and published articles about aromatic compounds and "aromaticity." I have even had the occasional "snake dream," although never so useful as Kekulé's. Among Baeyer's numerous accomplishments was the development of the first theory explaining the high reactivity of angle-strained organic molecules such as cyclopropane. I co-authored, with long-time friend Joel F. Liebman, the book *Strained Organic Molecules* in 1978 and published other papers in this research area. So, this rapid discovery of distinguished chemical lineage back into the mid-nineteenth century (with more to come), was a particular thrill to one whose real family heritage prior to the twentieth century was quite literally demolished, buried, and paved over in some shtetels in eastern and central Europe.

From Germany to France to America—a Twentieth-Century Odyssey

The twentieth century history of my chemical forefathers is fascinating and was dramatically impacted by events in Germany between 1920 and 1950. Following his 1915 Nobel Prize, for purifying chlorophyll and laying the basis for determining its structure, my "great-great grandfather" Richard Willstätter, a Jew, was appointed Professor of Chemistry at the University of Munich.[1,2] Willstätter was a close friend of Fritz Haber, another Jew, who performed the chemical miracle of "fixing" atmospheric nitrogen to form fertilizers (and explosives) for which he won the 1918 Nobel Prize.[2] Haber was a devoted patriot and developed poison gas as a weapon to save the Fatherland during World War I. During the 1920s, he tried to develop a method to extract gold from seawater to allow Germany to pay its war reparations.[3,4] When the Nazis assumed power in 1933, "Jew Haber"[4] was forced to quit his Directorship of the Kaiser Wilhelm Institute for Physical Chemistry in Berlin. He accepted a position in Cambridge, England but died of a heart attack the following year.

In 1924, at the age of 53, Willstätter quit his professor's chair to protest rising anti-Semitism that he felt was the reason for denying an appointment at Munich to Dr. V.M. Goldschmidt.[1,2] His career, abruptly truncated, became increasingly difficult. His life endangered after the Nazis' anti-Semitic campaign of

1938, Willstätter was forced to flee, settling in Switzerland in 1939, but only after considerable difficulty. The principled and gentle Willstätter described his eventful life in an excellent work of scientific autobiography, *Aus meinem Leben* (*From my Life*),[5] published in 1948, some six years after his death.

Richard Kuhn completed his Ph.D. under Willstätter's direction at Munich in 1922, performing early exploratory work on enzymes and carotene.[6,7] Following a period in Munich, then Zurich, Kuhn returned to Germany as a Professor and a Director in the Kaiser Wilhelm Institute (later the Max Planck Institute) for Medical Research in Heidelberg. His work on carotene and vitamin A earned him the 1938 Nobel Prize. However, he had to refuse the award. Hitler had banned acceptance of Nobel Prizes by Germans after the 1935 Nobel Peace Prize was awarded posthumously to a concentration camp prisoner, German pacifist Carl von Ossietzky, who had died of tuberculosis.[8] In 1949, Kuhn finally accepted his medal and certificate in Stockholm. Kuhn's student Lederer developed chromatography in order to separate different isomers of carotene. Lederer was Jewish and fled Germany in March 1933, just four days ahead of the Gestapo's visit to the Kaiser Wilhelm Institute.[9] He was aided by Kuhn's assistant André Lwoff[9] and emigrated to France, and that is how my genealogy became a *généalogie*. Lwoff shared the 1965 Nobel Prize in Physiology.

Confusions and Conceits

And now came the enjoyable task of tracing my distinguished "family" history from Baeyer to Kekulé to . . . but wait! Closer reading of historical sources disclosed that when Willstätter tried to join the inspirational Baeyer's group at Munich, the great master steered him to his department colleague Alfred Einhorn with whom he then completed his doctoral dissertation.[3–5] Einhorn had deduced a structure for cocaine. Willstätter suspected that it was incorrect and asked his director if he could work on the problem. Einhorn refused and gave him an unrelated research problem. Willstätter consulted his true inspirational advisor Baeyer, who finessed this problem in academic politics. Baeyer suspected that tropine was very similar in structure to cocaine and suggested its investigation to Willstätter. After hours, so to speak, Willstätter solved the tropine structure that eventually led to the correct structure of cocaine. Furious, Einhorn refused to speak with Willstätter for years, that is, until his former student became his department director, at which time his views moderated.[3,4] Einhorn gets only the briefest possible mention in Partington's comprehensive history of chemistry.[10] However, he did invent novocaine,[5] and this legacy at least reduces some of the pain from having the steady Einhorn rather than the brilliant Baeyer in my lineage.

In the Willstätter–Einhorn relationship we have an example of a problem that arises occasionally in these genealogical searches—who "gets credit" for the famous scientist—the dissertation advisor or the intellect of true influence? This gets even trickier as we delve deeper into the past. And what about postdoctoral supervisors? Why are they not considered "parents"? Perhaps it is fair to say that it is the dissertation advisor who first spots the accidental puddles in the chemical toddler's lab and provides nurturing to chemical adulthood.

Three Hundred Years of German Chemical Heritage

The academicization of science and awarding of the Doctor of Philosophy (Ph.D.) degree was pioneered in nineteenth-century Germany.[11] The Doctor of Medicine degree is many centuries older. My search, at this point, led to a useful chemical genealogy page on the University of Illinois Website (*www.scs.uiuc.edu/ ~mainzv/Web_Genealogy*). Alfred Einhorn received his Ph.D. under Wilhelm Staedel at Tòbingen, and Staedel completed his doctoral dissertation with Adolph Friedrich Ludwig Strecker, also at Tòbingen. Strecker completed his Ph.D. at Giessen with Justus Liebig, arguably the most important organic chemist of the nineteenth century.

Justus Liebig developed the *kaliapparat* that revolutionized organic analysis. Only after precise analysis could complex formulas of compounds be obtained, leading ultimately to conclusions about valence and structure derived decades later.[12,13] In Figure 351 (bottom left), we see a rendering of the Liebig lab in Giessen. Do I spot great-great-great-great-great-grandfather Strecker? Yes, he's seated center front, mortar and pestle in hand.

Justus Liebig—the father of animal, vegetable, and food chemistry! "O frabjous day! Callooh, Callay!" I chortled in my joy.[14] Liebig was an academic failure in his early schooling and an overly passionate and emotional adult (pp. 426–427)[15]—I recognized clear genetic similarities to myself. In Liebig's history, we encounter another "paternal" ambiguity, closely resembling that of Willstätter. Liebig himself recognized Joseph-Louis Gay-Lussac in Paris as his primary mentor. However, history informs us that Liebig completed his doctoral dissertation with Karl Friedrich Wilhelm Gottlob Kastner, with whom he was dissatisfied, at Erlangen.[12,13] And so we continue back in time:

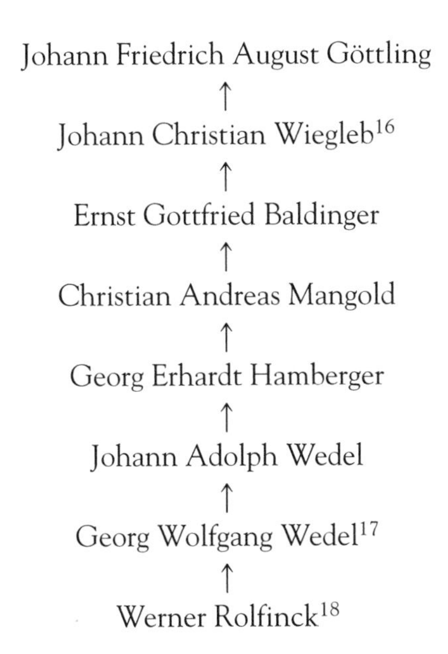

Werner (or Guerner) Rolfinck! Rolfinck was born in 1599 in Hamburg, was educated in Wittenberg, Leiden, Oxford, Paris, and Padua, receiving his M.D. in 1625.[18] In 1638 he established the chemical laboratory in Jena and, in 1641, be-

GVERNERI ROLFINCII,
PHIL. AC MED. DOCTO-
RIS ET PROFESSO-
RIS PUBLICI

CHIMIA
IN ARTIS FOR-
MAM REDACTA,
Sex Libris
comprehenſa.

JENÆ,
SAMUEL KREBS CURABAT.
ANNO M DC LXII

FIGURE 354. ■ Title page from the chemistry text published in 1662 by Dr. Guerner Rolfinck, a chemical forefather of the author. Rolfinck represented a critical transition from physicians educated in Padua to chemists teaching in Germany. He achieved some notoriety for his public dissections in Jena of executed criminals, for the purpose of medical instruction. For some time, human dissections were referred to as "rolfincking."

came the first chemistry professor at this university. Rolfinck's *Chemia in Artis Forman Redacta* (Figure 354), first published in 1661, opposed alchemy and presented medicinal treatments. Rolfinck achieved notoriety for his public dissections, in the anatomical theater he constructed at Jena, of executed criminals. For a period, such dissections were referred to as "rolfincking."[18]

Sixteenth-Century Venetian Anatomists

Rolfinck was the transitional figure in my "family" history. Born in Germany and enjoying a distinguished medical and scientific career in Germany, he had completed his M.D. in Padua under the supervision of Adriaan van den Spiegel (also Spieghel).[19] While Rolfinck could equally be thought of as iatrochemist and surgeon, van den Spiegel clearly belonged to the fields of anatomy and surgery. Born in Brussels in 1578, he was trained at Padua by Girolamo Fabrici and Giulio Casseri, and completed sometime between 1601 and 1604.

The University of Illinois genealogy Website continues to transport me back in time. Having reentered late-sixteenth-century Italy, I start to wonder where this will all end (or really, begin). Geographically closer to biblical lands, I begin to wonder whether Moses, a worker of gold, or Tubal-Cain,[20] the earliest metallurgist described in the Old Testament, were part of my chemical lineage, too.

In fact, my chemical genealogy runs through almost the entire sixteenth century in Padua and then fades. Why the transition from the northern Republic of Venice to Germany in the southern part of the Holy Roman Empire? The City of Venice and its Republic formed a thriving mercantile and cultural region during the fifteenth century. When the French invaded through the northwestern Republic of Milan at the end of that century and the Spanish were enlisted as allies, the two powers began an occupation of Italy that weakened the region. While it remained independent throughout the sixteenth century, the Venetian state was weakened economically and culturally and lost its place as Europe's intellectual center. Casseri[21] received his M.D. at Padua in 1580, and his teacher, Fabrici,[22] took his degree at Padua around 1569. As a faculty member at Padua, Fabrici shared all the traits of a modern university prima donna. In 1588 his students accused him publicly of neglecting his teaching in favor of his research. In 1611 he became notably embroiled with a colleague over the scheduling of courses. Nonetheless, his academic reputation and consulting practice thrived and he took Galileo on as a patient starting in 1606.[22]

Fabrici studied under Gabrielle Fallopio (1523?–1562)[23] at Padua. Fallopio became such a famous figure that many books were falsely attributed to him following his death. Only the *Observationes anatomicae* (1561) can be attributed to him with any certainty.[23] His pioneering studies in anatomy extended the work of an earlier Chair of Anatomy at Padua, Andreas Vesalius. Of course, we know Fallopio best as the discoverer of the fallopian tubes that deliver eggs into the uterus. Fallopio's contemporary, Realdo Colombo[24] (often incorrectly called Matteo Realdo Colombo), was born around 1510. Both he and Fallopio studied under Giovanni Antonio Lonigo, although there are many uncertainties here. Colombo succeeded Vesalius as Chair at Padua. Shortly thereafter, responding to rumors of his criticism by Colombo, Vesalius "denounced him as an ignoramus and a scoundrel."[24] In 1548, Colombo visited Rome and studied anatomy with Michelangelo.[24] and later described the role of the heart in the pulmonary system, thus predating William Harvey.[24] Fallopio succeeded Colombo as the Chair of Anatomy at Padua and also enjoyed angry relations with him just as Vesalius had.[24]

A modern novel, *The Anatomist,*[25] by Frederico Andahazi, recreates the pioneering surgeon Colombo amid the religious inhibitions and hypocrisies of sixteenth-century Padua. In the novel, he has discovered the true function of the *Amor Veneris*, further discussion of which is outside of the realm of my very proper and decent chemistry book. The fictional Colombo enrages the Church and meets a horrific end. In historical fact, Fallopio studied the *Amor Veneris*, and one wonders whether Andahazi invented a composite "Matteo" from Colombo and Fallopio.

Yet further back, before Lonigo, we find Antonio Musa Brasavola (1500–1555), a major contributor to Renaissance pharmacy and physician to Pope Paul III.[26] And while I still have another 2800 years or so to account for in my future efforts to connect Brasavola and Moses, it has, to date, been a thrilling journey.

1. C.C. Gillespie (ed.), *Dictionary of Scientific Biography*, Charles Scribner & Sons, New York, Vol. XIV, 1976, pp. 411–412.
2. B. Narins (ed.), *Notable Scientists From 1900 to the Present*, The Gale Group, Farmington Hills, Vol. 5, 2001, pp. 2438–2441.
3. J.R. Partington, *A History of Chemistry*, MacMillan & Co. Ltd., London, Vol. 4, 1964, pp. 636.
4. *The New Encyclopedia Britannica*, Encyclopedia Britannica, Inc., Chicago, 1986, Vol. 5, pp. 601–602.
5. R. Willstätter, *Aus mein Leben*, A. Stoll (ed.), Verlag Chemie, Weinheim, 1948; transl. L.S. Hornig as From My Life, W.A. Benjamin, New York, 1965.
6. Gillespie, op. cit., Vol. VII (1973), pp. 517–518.
7. Narins, op. cit., Vol. 3, pp. 1281–1282.
8. *The New Encyclopedia Britannica*, op. cit., Vol. 8, p. 1031.
9. See: *http://sun0.mpimf-heidelberg.mpg.de/History/Kuhn1.html* (obtained December 7, 2001).
10. Partington, op. cit., p. 860.
11. J. Ziman, *The Force of Knowledge—the Scientific Dimension of Society*, Cambridge University Press, Cambridge, 1976, pp. 57–62.
12. Partington, op. cit., pp. 294–334.
13. Gillespie, op. cit., Vol. VII, 1973, pp. 329–350.
14. With sincerest apologies to Lewis Carroll for bawdlerizing *Jabberwocky*.
15. A. Greenberg, *A Chemical History Tour*, John Wiley & Sons, New York, 2000, pp. 196–199.
16. Gillespie, op. cit., Vol. XIV, 1976, pp. 332–333.
17. Gillespie, op. cit., Vol. XIV, 1976, pp. 212–213.
18. Gillespie, op. cit., Vol. XI, 1975, p. 511.
19. Gillespie, op. cit., Vol. XII, 1975, p. 577.
20. J. Read, *Humour and Humanism in Chemistry*, G. Bell & Sons Ltd, London, 1947, p. 3.
21. Gillespie, op. cit., Vol. III, 1971, pp. 98–100.
22. Gillespie, op. cit., Vol. IV, 1971, pp. 507–512.
23. Gillespie, op. cit., Vol. IV, 1971, pp. 519–521.
24. Gillespie, op. cit., Vol. III, 1971, pp. 354–357.
25. F. Andahazi, *The Anatomist* (transl. by A. Manguel), Doubleday, New York, 1998.
26. J.R. Partington, *A History of Chemistry*, MacMillan & Co. Ltd., London, Vol. 2, 1961, p. 96.

INDEX